AF323921

SPECIAL TOPICS IN
ACCELERATOR PHYSICS

ALEXANDER WU CHAO
SLAC National Accelerator Laboratory, USA

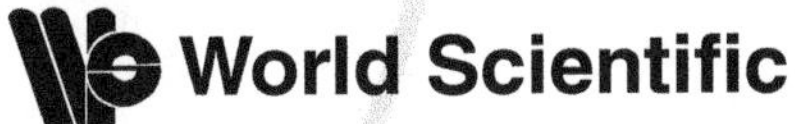

World Scientific

NEW JERSEY · LONDON · SINGAPORE · BEIJING · SHANGHAI · HONG KONG · TAIPEI · CHENNAI · TOKYO

Published by

World Scientific Publishing Co. Pte. Ltd.

5 Toh Tuck Link, Singapore 596224

USA office: 27 Warren Street, Suite 401-402, Hackensack, NJ 07601

UK office: 57 Shelton Street, Covent Garden, London WC2H 9HE

Library of Congress Cataloging-in-Publication Data
Names: Chao, Alexander Wu, author.
Title: Special topics in accelerator physics / Alexander Wu Chao,
    SLAC National Accelerator Laboratory, USA.
Description: New Jersey : World Scientific Publishing Co., [2022] |
    Includes bibliographical references and index.
Identifiers: LCCN 2021056666 | ISBN 9789811253492 (hardcover) |
    ISBN 9789811253508 (ebook for institutions) | ISBN 9789811253515 (ebook for individuals)
Subjects: LCSH: Particle accelerators.
Classification: LCC QC787.P3 C476 2022 | DDC 539.7/3--dc23/eng/20220120
LC record available at https://lccn.loc.gov/2021056666

British Library Cataloguing-in-Publication Data
A catalogue record for this book is available from the British Library.

For any available supplementary material, please visit
https://www.worldscientific.com/worldscibooks/10.1142/12757#t=suppl

Preface

Accelerators as research and industrial tools are increasingly becoming a key driver of the advances of a modern society. As accelerator science evolves to meet the ever-demanding needs of the society, the field of accelerator physics has grown and deepened over the past few decades, and many of its branches have developed into special topics of research by their own rights. As researches in these topics mature, it is appropriate to start documenting this hard-earned expertise by the accelerator community for readers inspired to venture into this rich field of science. With this view, a selection of special topics is presented in this volume, Special Topics in Accelerator Physics.

This volume is dedicated to a relatively small selection of these special topics. The choice of these topics is somewhat arbitrary, no doubt influenced by the interest and experience of the author, but they are believed to present the accelerator physics as a diversified and exciting field. Eight special topics are included in this volume with no intention to pretend the list to be exhaustive. Many more special topics can readily be added to the ones happen to be chosen.

The topics are not arranged in any particular order. The material on each topic has been intended to be self-contained. Individual chapters have varying lengths and assume different levels of basic knowledge of the readers. However, the reader is assumed to be knowledgeable of basic accelerator physics to put the material in some context. The book is presented as a textbook, with homework at the end of each section. The contents were derived from lectures delivered at various schools and universities. The style will be emphasizing the physical principles of the various subjects.

I would like to thank the students for their interest and their many inquisitive questions, from which I learned. Warm and heart-felt thanks go to many of my colleagues, with whom the learning has been such a joy over the years. Special thanks also go to Karl Bane, Gennady Stupakov, Xiujie Deng, Linda Spentzouris, Eberhard Keil, Kohji Hirata, Yukihide Kamiya, Miguel Furman, and Rudiger Schmidt, who generously allowed me to use their illustrations.

Alexander Chao
Stanford, January 2022

Contents

5 SPIN DYNAMICS OF PROTON — 328

6 SPIN DYNAMICS OF ELECTRON — 395

7 ECHO — 479

8 BEAM-BEAM INTERACTION — 541

Chapter 1

The Fokker–Planck Equation

The description of the motion of charged-particle beams in an accelerator proceeds in steps of increasing complexity. The first step is to consider a single-particle picture in which the beam is represented as a collection of noninteracting single-particles moving in an environment of prescribed external electromagnetic fields. The electromagnetic fields generated and carried by the particles are ignored. Knowing the external fields, it is then possible to calculate the beam motion to a high accuracy by repeating the calculation for each and every one of the single-particles in the beam.

The actual beam consists of a large number, perhaps 10^{11}, of particles. Although possible in principle, it is clearly not practical to treat the actual beam behavior using this single-particle approach. This is even more the case if one includes the effects of the electromagnetic fields the beam generates, as would be necessary if one wants to deal with the collective instability effects of an intense beam.

One way to approach this problem is to supplement the single-particle picture by another qualitatively different picture. The commonly used tools in accelerator physics for this purpose are the Vlasov equation[1] and the Fokker–Planck equation.[2] These equations assume a continuum of beam distribution and are strictly valid in the limit of infinite number of microparticles, each carrying an infinitesimal charge. The hope is that by studying the two extremes, the single-particle picture and the continuum picture, we gain a handle of the behavior of our 10^{11}-particle beam, at least approximately.

As mentioned, the most notable use of the continuum picture is the study of collective beam effects. However, the purpose of this chapter is not to address

---

[1] A.A. Vlasov, J. Phys. USSR 9, 25 (1945).

[2] A.D. Fokker, Ann. Phys. 348, 810 (1914); M. Planck, Sitzungsberichte der Preussischen Akademie der Wissenschaften zu Berlin, 24, 324 (1917); S. Chandrasekhar, Rev. Mod. Phys. 15, 1 (1943).

this vast subject. Rather, it has the more limited goal to introduce the analytical tools, namely the Vlasov and the Fokker–Planck equations. It is further assumed that the reader has been somewhat familiar with the Vlasov equation, so that more of our emphasis shall be on the Fokker–Planck equation, although both are discussed. We will first derive these equations and then illustrate their applications by several examples which allow analytical solutions. The actual treatment of collective instabilities is not included, but the examples selected hopefully still represent useful applications to several other commonly encountered accelerator problems.

1.1   Vlasov equation

1.1.1   Conservative system

Both the Vlasov and the Fokker–Planck equations describe the beam motion in its phase space. To construct these equations, one invariably starts with defining the phase space and establishing the Hamilton equations for a single-particle motion in the phase space. This is also what we will do.

Consider the 1-D case with phase space (q, p). Let the single-particle equations of motion be

$$\dot{q} = f(q, p, t), \qquad \dot{p} = g(q, p, t).$$

In the Vlasov framework, we consider the system to be conservative, i.e. there is a Hamiltonian $H(q, p, t)$ underlying the system with

$$\dot{q} = \frac{\partial H(q, p, t)}{\partial p}, \qquad \dot{p} = -\frac{\partial H(q, p, t)}{\partial q}.$$

It follows that, for a conservative system, we have the condition

$$\frac{\partial f}{\partial q} + \frac{\partial g}{\partial p} = 0. \tag{1.1}$$

An immediate trivial consequence of Eq. (1.1) is that if f is independent of q, then g must be independent of p. Also, if f depends on q, then g must depend on p. Note that for a Hamiltonian system, condition (1.1) holds at all phase space point (q, p) at all time t.

Equation (1.1) describes the condition of phase space conservation of the Hamiltonian system. The quantity $\frac{\partial f}{\partial q} + \frac{\partial g}{\partial p}$ specifies the phase space expanding or shrinking rate. If $\frac{\partial f}{\partial q} + \frac{\partial g}{\partial p} = 0$, phase space area is conserved.

If we consider an area element $A = \Delta q \Delta p$ in the phase space near a position (q, p), for example, its time evolution is given by

$$\dot{A} = \left[\frac{\partial f(q, p)}{\partial q} + \frac{\partial g(q, p)}{\partial p} \right] A.$$

We will return to this point when discussing the Fokker–Planck equation, but for now, in the Vlasov framework, we stay with the conservative system obeying condition (1.1). Phase space area is conserved.

1.1.2   Derivation

As mentioned, we only briefly describe the Vlasov equation and its derivation here.[3] The derivation later on the Fokker–Planck equation can also serve the purpose if one switches off the noise and damping effects.

Consider a continuum beam distribution in phase space. Let the number of particles in the phase space element $\Delta q \Delta p$ near the position (q, p) be $\psi(q, p, t) \Delta q \Delta p$, where ψ is the distribution density depending on q, p and t, and is normalized by

$$\int dq \int dp \, \psi(q, p, t) \; = \; N \, , \tag{1.2}$$

with N the total number of particles the continuum distribution ψ is designated to describe.

Because phase space trajectories cannot cross each other in a conservative system, the particles enclosed within this phase space element $\Delta q \Delta p$ are not going to leak out, and neither can particles outside the element sneak in. The number of particles within the phase space element is conserved. Since the phase space area itself is also conserved, it follows that the phase space density ψ must be conserved if one moves along with the flow of the particle in the phase space. When observed at time $t + dt$, for example, it follows that

$$\psi(q, p, t) \; = \; \psi(q + f\,dt, p + g\,dt, t + dt) \, . \tag{1.3}$$

Expanding to 1st order in dt, we obtain

$$\frac{\partial \psi}{dt} + f \frac{\partial \psi}{dq} + g \frac{\partial \psi}{dp} \; = \; 0 \, . \tag{1.4}$$

Equation (1.4) is referred to as the Vlasov equation — especially when the forces are electromagnetic in origin. It can also be cast in the form

$$\frac{d\psi}{dt} \; = \; 0, \qquad \text{or} \qquad \psi \; = \; \text{constant in time} \, , \tag{1.5}$$

which states that the local particle density does not change with time if the observer moves with the flow of the phase space element.

Note that the right-hand-side of Eq. (1.3) does not read $\psi(q, p, t + dt)$. It is important that the invariance of the distribution density is applied only when the observer moves with the phase space element, and that is reflected by the inclusion of $q + f\,dt$ and $p + g\,dt$ in the arguments.

Strictly speaking, f and g here are given by the external fields; collisions among discrete particles in the system are excluded from Eqs. (1.4) and (1.5). On the other hand, if a particle interacts more strongly with the collective long-range fields of the other particles than its nearest neighbors, Vlasov equation can still be applied by treating the collective fields on the same basis as the

---

[3]For a more detailed derivation and discussion, including the price to pay to consider a continuum model, see, for example, A.W. Chao, Lectures on Accelerator Physics, World Scientific (2020), Chapter 8.

external fields. This in fact forms the basis of treating collective instabilities using the Vlasov technique.[4] Short-range forces among neighboring particles, however, must be excluded.

When written in the form (1.5), it is sometimes loosely referred to as the Liouville theorem — as we will also do. However, the Liouville theorem applies to an ensemble of many systems, each containing many particles. It describes the density conservation of the ensemble system in a $2N$-dimensional space (the Γ-space), where N can be as large as 10^{11}. It applies to situations much more general than that of our Eq. (1.5), such as when collisions among discrete particles are included.

Homework 1.1 Consider a beam whose distribution ψ evolves according to the Vlasov equation (1.4). Show that if ψ is normalized by the condition (1.2) initially, then it will be normalized at all times.

Solution This should be expected because the phase space density does not change according to the Liouville theorem, so the total integral of ψ over the phase space should not change. This homework, however, asks for a mathematical proof.

1.1.3   Solution of Vlasov equation

One special case when the Vlasov equation can be solved exactly is when the system Hamiltonian $H(q,p)$ does not have an explicit time dependence. In this case, using the properties $f = \frac{\partial H}{\partial p}, g = -\frac{\partial H}{\partial q}$, a solution of stationary beam distribution to the Vlasov equation (1.4) in the phase space is found to be

$$\psi(q,p) = \text{any function of } H(q,p). \tag{1.6}$$

In this Hamiltonian system, particles stream rapidly along constant Hamiltonian contours. Distribution (1.6) is one that does not change in time — it is a stationary solution. Even as each individual particle rapidly streams along its respective contour in the phase space, the overall distribution of the beam looks stationary to a nonsuspecting observer.

Simple harmonic oscillator   For a simple harmonic system with Hamiltonian

$$H(q,p,t) = \frac{\omega}{2}(q^2 + p^2),$$

the Vlasov equation reads

$$\frac{\partial \psi}{dt} + \omega p \frac{\partial \psi}{dq} - \omega q \frac{\partial \psi}{dp} = 0. \tag{1.7}$$

Making a transformation from (q,p) to polar coordinates (ϕ,a) by

$$q = a\sin\phi, \qquad p = a\cos\phi, \tag{1.8}$$

---

[4]The fact that the long-range collective field is considered on the same footing as the external fields is not to be taken for granted.

Equation (1.7) can be written as

$$\frac{\partial \psi}{\partial t} + \omega \frac{\partial \psi}{\partial \phi} = 0 \,,$$

which has the general solution

$$\psi(\phi, a, t) = \text{any function of } (\phi - \omega t, a) \,. \tag{1.9}$$

Once the initial distribution of the beam is given at $t = 0$, Eq. (1.9) means that the distribution at time t in the simple harmonic case is given by rigidly rotating the initial distribution in phase space angle ϕ at a constant angular speed of ω. The stationary distribution, a special subset of the general solution (1.9) that does not change in time, is given by any function of H, or equivalently any function of amplitude a, without any ϕ dependence.

Damped simple harmonic oscillator As a second example, consider a damped simple harmonic motion with

$$f = \omega p \,, \qquad g = -\omega q - 2\alpha p \,, \tag{1.10}$$

where α is the damping rate, and we assume $|\omega| > \alpha > 0$. The single-particle motion is easily solved and is found to be described by

$$q(t) = \frac{\omega A}{\sqrt{\omega^2 - \alpha^2}} e^{-\alpha t} \sin(\sqrt{\omega^2 - \alpha^2}\, t + \Phi) \,,$$

$$p(t) = -\frac{\alpha}{\omega} q(t) + A e^{-\alpha t} \cos(\sqrt{\omega^2 - \alpha^2}\, t + \Phi) \,, \tag{1.11}$$

where A and Φ are constants specified by the initial conditions.

Equation (1.11) gives the general solution to the single-particle equations of motion (1.10). But here we want to solve for the beam distribution $\psi(q, p, t)$ and we need to solve the Vlasov equation. A slight modification of the Vlasov equation is required here because the damping means the conservation of phase space area is violated. In this case,

$$\frac{\partial f}{\partial q} + \frac{\partial g}{\partial p} = -2\alpha \,,$$

meaning the phase space area shrinks a little from time t to time $t + dt$. Equation (1.3) then reads

$$\psi(q, p, t) = \psi(q + f dt, p + g dt, t + dt) \times \left[1 + \left(\frac{\partial f}{\partial q} + \frac{\partial g}{\partial p} \right) dt \right]$$

$$= \psi(q + f dt, p + g dt, t + dt) \times (1 - 2\alpha dt) \,,$$

or, keeping to 1st order in dt,

$$\frac{\partial \psi}{dt} + \omega p \frac{\partial \psi}{dq} - (\omega q + 2\alpha p) \frac{\partial \psi}{dp} = 2\alpha \psi \,. \tag{1.12}$$

Note that a straightforward substitution of f and g into Eq. (1.4) would miss the term on the right-hand-side. Note also that $\frac{\partial f}{\partial q} + \frac{\partial g}{\partial p} = -2\alpha$ is negative, meaning the area of phase space shrinks with time.

The general solution of Eq. (1.12) is related to the initial conditions in the single-particle solution (1.11), and is given by

$$\psi(q,p,t) \;=\; e^{2\alpha t} \times [\text{any function of } A \text{ and } \Phi]\,, \tag{1.13}$$

where

$$A \;=\; e^{\alpha t}\left[q^2 + p^2 + \frac{2\alpha qp}{\omega}\right]^{1/2},$$

$$\Phi \;=\; \tan^{-1}\left[\frac{q(\omega^2 - \alpha^2)^{1/2}}{\omega p + \alpha q}\right] - (\omega^2 - \alpha^2)^{1/2}t\,.$$

The overall factor $e^{2\alpha t}$ in Eq. (1.13) means the distribution is becoming denser as a result of damping. Quantities A and Φ reduce to a and $\phi - \omega t$ when $\alpha = 0$. The general solution is that of a distribution spiraling inward with time as one would expect. The only stationary solution is when the entire distribution collapses into a δ-function at the phase space origin.

Spin statistics is ignored   One effect ignored so far for the multiparticle system is the spin of the particles. To make sure the system is described by classical statistics rather than Fermi statistics, we need to show that the number of quantum states far exceeds the number of particles for all practical accelerators of interest.

The beam occupies a volume in the 6-D phase space of the order of

$$V \;=\; \sigma_x \sigma_{x'} \sigma_y \sigma_{y'} \sigma_z \sigma_\delta P_0^3\,,$$

where P_0 is the beam momentum in the laboratory frame. The number of quantum states available to the beam particles (assuming the particles have spin-$\frac{1}{2}$) is

$$N_q \;=\; \frac{2V}{\hbar^3}\,,$$

where a factor of 2 is for the two spin states (up and down).

Consider for example an envisioned 20-TeV proton beam for the Superconducting Super Collider with $\gamma = 2 \times 10^4$, and let the beam be radiation damped to reach the equilibrium parameters $\sigma_\delta = 2 \times 10^{-6}$, $\sigma_x = 5\,\mu\text{m}$, $\sigma_y = 0.01\,\mu\text{m}$, $\sigma_z = 3$ mm, $\sigma_{x'} = 0.04\,\mu\text{rad}$, $\sigma_{y'} = 0.1$ nrad. This gives $N_q = 2 \times 10^{21}$. Since the number of protons per bunch is about 10^{11}, the number of quantum states exceeds the number of protons by about 10 orders of magnitude; there is no problem ignoring Fermi statistics.

Homework 1.2 Derive Eq. (1.13), including the overall factor $e^{2\alpha t}$.

Solution The homework asks to derive Eq. (1.13) by solving Eq. (1.12). A lesser solution is to verify that (1.13) satisfies (1.12) by back substitution.

1.2   Potential well distortion

As a first application of the Vlasov technique, we will study the effect of longitudinal potential well on the equilibrium shape of a beam bunch in a storage ring. To set up the problem, as mentioned, we first need to define the phase space, and then to specify how single particles move in the phase space.

Before launching the subject of potential well distortion, we emphasize at the beginning that the picture of potential well distortion due to wakefields has an underlying assumption that synchrotron motion is slow enough that the longitudinal coordinates of the particles do not change much from revolution to revolution. This is called the adiabatic approximation. In particular, it requires the synchrotron oscillation frequency be much smaller than the revolution frequency, i.e. the longitudinal dynamics must be weak focusing. In case the longitudinal dynamics is also strong focusing like in the transverse dimensions, a potential well cannot be defined. When the adiabatic approximation breaks down, the picture of potential well distortion also breaks down. In what follows in this section, we assume adiabatic approximation holds.

One consequence of the adiabatic approximation is that the single particle equations of motion will assume the form of differential equations instead of discrete difference equations, and the coefficients in the differential equations of motion will not depend on the space coordinate s around the circular accelerator, as we shall see explicitly below.

1.2.1   RF bucket

Consider a particle executing longitudinal synchrotron oscillation in a beam bunch. The physical quantities of interest are the longitudinal displacement z and the momentum error $\delta = \frac{\Delta P}{P}$ of the particle, both evaluated relative to the ideal synchronous reference particle in the laboratory frame. The phase space coordinates q and p of the previous section are related to these quantities by

$$q = z \quad \text{and} \quad p = -\frac{\eta c}{\omega_s}\delta,$$

where η is the phase slippage factor and ω_s is the synchrotron oscillation angular frequency. The equations of motion are

$$z' = -\eta\delta \quad \text{and} \quad \delta' = K(z), \tag{1.14}$$

where a prime means taking derivative with respect to the distance s along the accelerator circumference. We have left the δ'-equation open for the time being by designating it to be a function K, except that we do know the function K cannot depend on δ, because the system is conservative; and we have already assumed that it does not depend on s per the adiabatic approximation.

We aim to apply to high energy accelerators and accordingly have assumed an ultrarelativistic case $v \approx c$, or $\gamma \gg 1$.

The Vlasov equation reads

$$\frac{\partial \psi}{\partial s} - \eta \delta \frac{\partial \psi}{\partial z} + K(z) \frac{\partial \psi}{\partial \delta} \;=\; 0\,,$$

where we will set $\frac{\partial \psi}{\partial s} = 0$ since we are looking for a stationary distribution. The general stationary solution can be written as[5]

$$\psi(z,\delta) \;=\; \text{any function of the Hamiltonian } H(z,\delta)\,,$$

$$H(z,\delta) \;=\; \frac{\eta^2 c^2}{\omega_s} \left[\frac{\delta^2}{2} + \frac{1}{\eta} \int_0^z K(z')\,dz' \right].$$

The second integral term in the Hamiltonian is the potential well term. A simple harmonic system would have a parabolic potential well.

The fact that this Hamiltonian is independent of s (and therefore is a constant of the motion) follows from the adiabatic approximation. All changes of δ have been considered smoothly distributed over revolution around the circular accelerator.

If the potential well is provided by an external RF voltage $V_{\rm rf}(z)$, we have

$$K(z) \;=\; \frac{eV_{\rm rf}(z)}{CE} \;=\; \frac{\omega_s^2}{c^2 \eta V'_{\rm rf}(0)} V_{\rm rf}(z)\,,$$

where E is the particle energy, C is the accelerator circumference, $V'_{\rm rf}$ is the derivative of $V_{\rm rf}$ with respect to z, and we have used the expression

$$\omega_s^2 \;=\; \frac{\eta c^2\, eV'_{\rm rf}(0)}{CE}\,.$$

If $V_{\rm rf}(z) = V'(0)z$, then we have the case of a simple harmonic oscillator.

A practical case is given by

$$V_{\rm rf}(z) \;=\; \hat{V} \left[\sin\left(\phi_s - \frac{\omega_{\rm rf}}{c} z \right) - \sin \phi_s \right]\,, \tag{1.15}$$

where ϕ_s is the acceleration synchronous phase; $\hat{V}, \omega_{\rm rf}$ are the amplitude and angular frequency of the RF system. Deviation of $V_{\rm rf}(z)$ from a linear dependence on z is a cause of potential well distortion.

The beam is being accelerated if $\phi_s \neq 0$. When $\phi_s = 0$, the beam receives an RF voltage[6]

$$V_{\rm rf}(z) \;=\; -\hat{V} \sin\left(\frac{\omega_{\rm rf}}{c} z \right).$$

---

[5]Strictly speaking, we should write this expression of the Hamiltonian as $H(q,p,t)$ to specifically identify (q,p) as the canonical coordinates and t as the independent time variable. But here, we loosen up on this formality and simply call it $H(z,\delta)$. It should be kept in mind, however, that the time variable of our Hamiltonian here is t, not s.

[6]If you are curious about the overall minus sign, it is because $z > 0$ refers to the front of the bunch, and a particle having $z > 0$ arrives the RF with a negative phase lag relative to the synchronous particle.

Figure 1.1: Equilibrium beam distribution in phase space conforms to constant-Hamiltonian contours for the case of an RF bucket. Cases shown are for $\phi_s = 0$ (upper, nonaccelerating) and $\phi_s = -\frac{\pi}{4}$ (lower, accelerating). Plotted are the contours of $\frac{\omega_{\mathrm{rf}}^2}{\omega_s c^2} H$ in the $(\frac{\omega_{\mathrm{rf}}}{c} z, \frac{\eta \omega_{\mathrm{rf}}}{\omega_s} \delta)$ space. The horizontal axis spans from $-\pi$ to π.

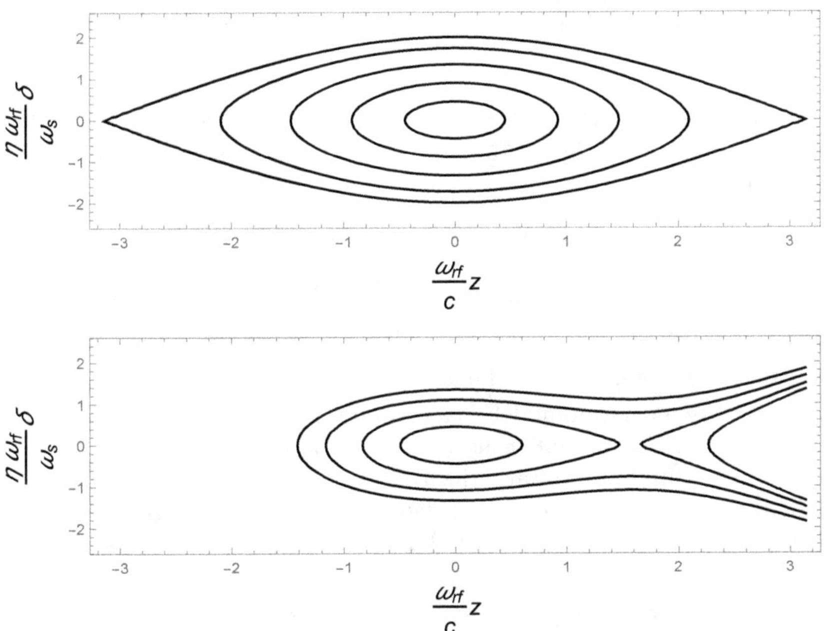

The general stationary distribution is given by any function of the Hamiltonian,

$$H(z, \delta) \;=\; \frac{\eta^2 c^2}{2\omega_s}\delta^2 + \frac{\omega_s c^2}{\omega_{\mathrm{rf}}^2}\left[\cos\phi_s - \cos\left(\phi_s - \frac{\omega_{\mathrm{rf}} z}{c}\right) + \frac{\omega_{\mathrm{rf}} z}{c}\sin\phi_s\right]. \quad (1.16)$$

This Hamiltonian also describes the form of the RF bucket. A stationary distribution must conform to the contours of constant Hamiltonian (1.16) inside the bucket, as illustrated in Fig. 1.1.

For small oscillation amplitudes, we have

$$K(z) \;\approx\; \frac{\omega_s^2}{\eta c^2} z\,, \qquad\qquad (1.17)$$

the case of simple harmonic motion. For those small-amplitude particles, their constant-Hamiltonian contours are elliptical, as also is seen in Fig. 1.1.

One noteworthy special case of the stationary beam distribution is that given by

$$\psi(q,p) \;\propto\; \exp[-\text{const} \times H(q,p)] \,. \tag{1.18}$$

This distribution is always Gaussian in δ. In case the bunch length is much shorter than the RF wavelength ($|z| \ll \frac{c}{\omega_{\rm rf}}$), the familiar quadratic form of the Hamiltonian is reestablished, and the distribution would also be Gaussian in z. As the bunch length increases, the bunch shape deviates from Gaussian; the potential well is distorted by the RF bucket, although the distribution remains Gaussian in δ in this case.

Homework 1.3 The RF bucket treated in the text is for the case when the RF voltage is given by Eq. (1.15). Extend the analysis to the case when (see footnote 6)

$$V_{\rm rf}(z) \;=\; -\hat{V}_1 \sin\left(\frac{\omega_{\rm rf}}{c} z\right) - \hat{V}_3 \sin\left(\frac{3\omega_{\rm rf}}{c} z\right) \,.$$

The case applies when a third harmonic is added to the fundamental RF. By adjusting the ratio $\frac{\hat{V}_3}{\hat{V}_1}$, the bunch length can be varied. In particular, the bunch length can be substantially lengthened by choosing $\frac{\hat{V}_3}{\hat{V}_1} = -\frac{1}{3}$ so that $V_{\rm rf}'(0) = 0$.

Adding a higher harmonic RF is a common technique to control the equilibrium bunch length in a circular accelerator. A longer bunch helps to deal with collective instabilities and to raise the Touschek lifetime. A shorter bunch length could help reducing synchrobetatron beam-beam effects. Figure 1.2 shows the effect of a third harmonic RF on the bunch distribution.

1.2.2   Nonlinear phase slippage factor

The RF bucket of the previous section is considered when the phase slippage factor η is a constant parameter. In practice, it could be a function of δ, i.e. there could be nonlinearity in the storage ring dynamics in such a way that the phase slippage factor contains a nonlinear term,

$$\eta \;=\; \eta_0 + \eta_1 \delta \,.$$

This nonlinearity of phase slippage in δ, not unlike a potential well distortion in z, also distorts the beam distribution in phase space.

Take the case of an RF bucket with $\phi_s = 0$, and include the term of η_1, the Hamiltonian (1.16) becomes

$$H(z,\delta) \;=\; \frac{\eta_0^2 c^2}{2\omega_s}\left(\delta^2 + \frac{2\eta_1}{\eta_0^2}\delta^3\right) + \frac{\omega_s c^2}{\omega_{\rm rf}^2}\left(1 - \cos\frac{\omega_{\rm rf} z}{c}\right).$$

Figure 1.3 shows the effect of η_1. The middle panel reproduces the upper panel of Fig. 1.1. From up to down, the five panels show the phase space distortion due to $\frac{\eta_1 \omega_s}{\eta_0^2 \omega_{\rm rf}} = 1, 0.3, 0, -0.3, -1$, respectively.

It should perhaps be commented here that in the context of Vlasov equation, we are mainly interested in the beam distribution in the phase space. As a result,

Figure 1.2: Equilibrium beam distribution in phase space when a third harmonic RF is added. Cases shown are for $\phi_s = 0$. Plotted are the contours of $\frac{\omega_{\mathrm{rf}}^2}{\omega_s c^2} H$ in the ($\frac{\omega_{\mathrm{rf}}}{c} z$, $\frac{\eta \omega_{\mathrm{rf}}}{\omega_s} \delta$) space, where ω_s refers to the synchrotron frequency without the third harmonic. The upper and the lower panels show bunch shortening and lengthening when $\frac{\hat{V}_3}{\hat{V}_1} = \frac{1}{3}$ and $-\frac{1}{3}$, respectively. The middle panel is when $\hat{V}_3 = 0$, reproducing the upper panel of Fig. 1.1.

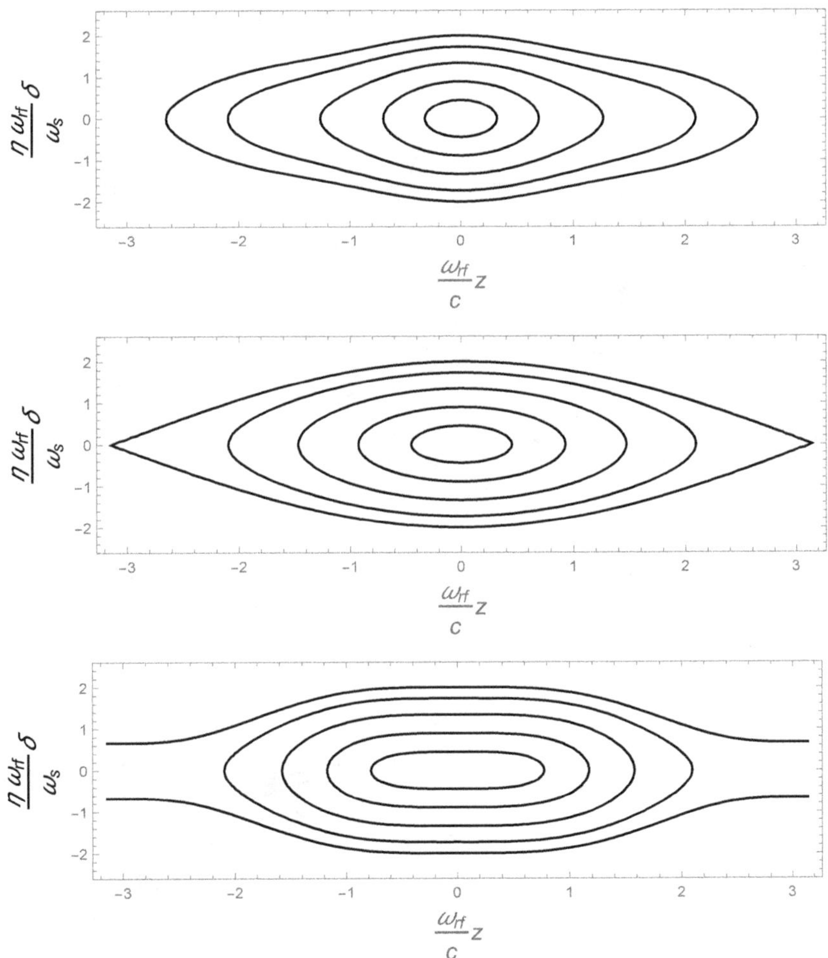

the focus is paid to the Hamiltonian contours. Detailed single-particle dynamics under the influence of linear and nonlinear phase slippage factors, a rich subject by itself, is not focused here.[7]

---

[7]See, for example, K.Y. Ng, Sec. 2.3.11, Handbook Accel. Phys. & Eng. 2nd ed., World Scientific (2013); S.Y. Lee, Accelerator Physics, 3rd ed., World Scientific (2011).

Figure 1.3: Equilibrium beam distribution in phase space when a nonlinear phase slippage factor η_1 is added to a nominal RF bucket. Cases shown are for $\phi_s = 0$. Plotted are the contours of $\frac{\omega_{\rm rf}^2}{\omega_s c^2} H$ in the $(\frac{\omega_{\rm rf}}{c} z, \frac{\eta \omega_{\rm rf}}{\omega_s} \delta)$ space, where ω_s refers to the synchrotron frequency without the η_1 consideration. The five panels show the phase space distortions due to $\frac{\eta_1 \omega_s}{\eta_0^2 \omega_{\rm rf}} = 1, 0.3, 0, -0.3, -1$, respectively. The middle panel for $\eta_1 = 0$, reproduces the upper panel of Fig. 1.1. To show the phase space clearly, the horizontal range is expanded to cover from -2π to 2π. Note the vertical scales of different panels are different.

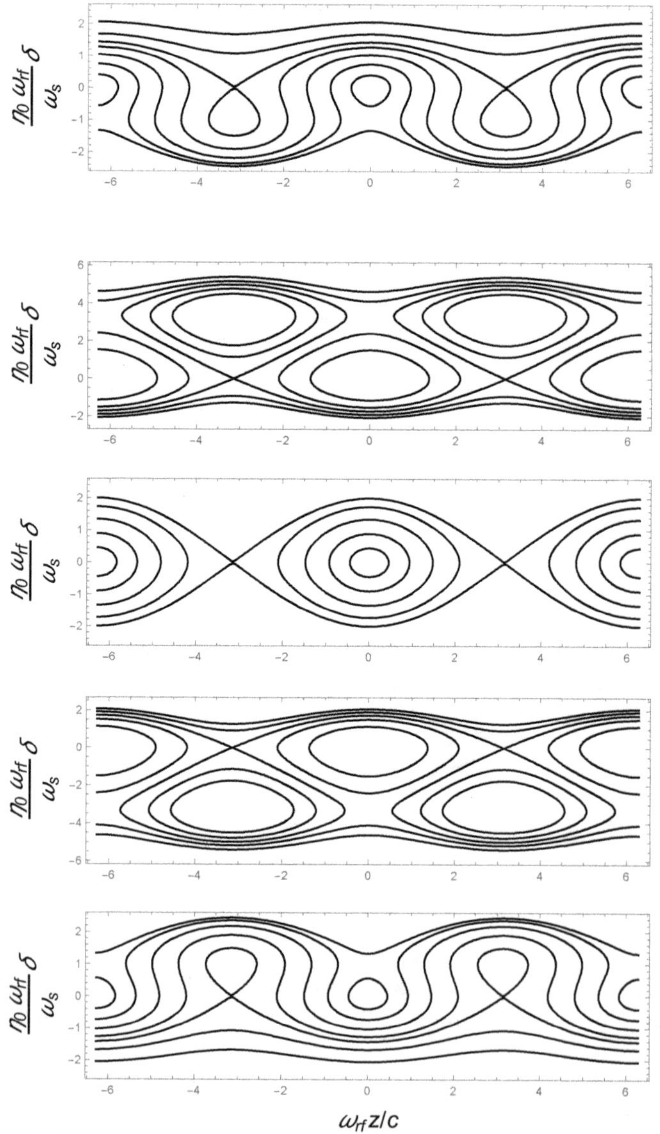

Homework 1.4 The text applied the potential well distortion to the case when the linear behavior in z is broken by an RF bucket. It also applied it to the case when the linear behavior in δ is broken by the nonlinear phase slippage. In case both z and δ linearities are broken, e.g. when both the RF bucket and the phase slippage are nonlinear, can you find a potential well distorted beam distribution? (See also Homework 1.16.)

1.2.3 Wakefield

Other than the nonlinear RF voltage and nonlinear phase slippage factor, there is another reason for the Hamiltonian to deviate from the quadratic form, and thus to cause potential well distortion, namely, the wakefields induced by high-intensity beams.[8]

To illustrate this, consider a storage ring with a beam bunch that is short compared with the RF wavelength so that the nonlinear RF bucket is not an issue. Let the wakefields be characterized by a wake function $W_0'(z)$ (integrated over the accelerator circumference — an averaging procedure as required by the adiabatic approximation), and assume that the wake has dissipated before the beam completes one revolution. The bunch shape will be distorted by the wakefields. The mechanism is a static one; no part of the beam bunch is executing collective oscillation. The extent of the bunch shape distortion depends on the beam intensity; higher beam intensities cause larger distortions.

Fourier transform of the wake function $W_0'(z)$ is the impedance $Z_0^{\parallel}(\omega)$,

$$Z_0^{\parallel}(\omega) \; = \; \int_{-\infty}^{\infty} \frac{dz}{c}\, e^{-i\omega z/c}\, W_0'(z)\,.$$

We assume that both $W_0'(z)$ and $Z_0^{\parallel}(\omega)$ are known.

When the bunch length is much shorter than the RF wavelength, the unperturbed $K(z)$ is given by Eq. (1.17). Adding the wakefield contribution, the single-particle motion is described by Eq. (1.14) with

$$K(z) \; = \; \frac{\omega_s^2}{\eta c^2} z - \frac{4\pi\epsilon_0 r_0}{\gamma C} \int_{z}^{\infty} dz'\, \rho(z') W_0'(z - z')\,,$$

where $r_0 = \frac{e^2}{4\pi\epsilon_0 m c^2}$ is the classical radius of the beam particle; the second term is given by the voltage seen by a particle at longitudinal location z due to the retarding wake force produced by all particles in front of it; $\rho(z)$ is the particle density at location z and is normalized by

$$\int_{-\infty}^{\infty} dz\, \rho(z) \; = \; N\,. \tag{1.19}$$

[8]C. Pellegrini and A.M. Sessler, Nuovo Cimento 3A, 116 (1971); B. Zotter, Proc. 4th Advanced ICFA Beam Dynamics Workshop on Collective Effects in Short Bunches, Tsukuba, 1990, KEK Report 90-21; A.W. Chao, Physics of Collective Beam Instabilities in High Energy Accelerators, John Wiley (1993).

Given $K(z)$, the corresponding Hamiltonian is

$$H(z,\delta) \;=\; \frac{\eta^2 c^2}{2\omega_s}\delta^2 + \frac{\omega_s}{2}z^2 - \frac{4\pi\epsilon_0\eta c^2 r_0}{\omega_s\gamma C}\int_0^z dz'' \int_{z''}^\infty dz'\, \rho(z')W_0'(z''-z')\,. \quad (1.20)$$

Again, the fact that this Hamiltonian is independent of s follows from the adiabatic approximation.

The stationary solution to the Vlasov equation must be a function of $H(z,\delta)$. The complication here, compared with the case of RF bucket distortion, is that the complicated z-dependence of H now involves the beam density ρ, which in turn is determined by the stationary distribution itself. Clearly some self-consistency requirement is involved in solving the problem. Below we will give a few explicit examples of this procedure.

Haissinski distribution   An inspection of the Gaussian example mentioned in Eq. (1.18) shows that even though the z-dependence is complicated, the stationary distribution maintains its Gaussian distribution in δ,

$$\psi(z,\delta) \;=\; \frac{1}{\sqrt{2\pi}\sigma_\delta}\exp\left(-\frac{\delta^2}{2\sigma_\delta^2}\right)\rho(z)\,, \quad (1.21)$$

where σ_δ is the rms beam energy spread.

The Gaussian form and the value of σ_δ in Eq. (1.21) can be arbitrarily chosen as long as the collective beam behavior is governed by the Vlasov equation, as in the case of a proton beam. However, if the beam behavior is governed, as for an electron beam, by the Fokker–Planck equation, then Eq. (1.21) with a specific value for σ_δ will be the unique solution of the stationary beam distribution. We will demonstrate the electron case in Sec. 1.3.3.

Equation (1.21) matches the stationary solution

$$\psi(z,\delta) \;\propto\; \exp\left(-\frac{\omega_s}{\eta^2 c^2 \sigma_\delta^2}H\right)\,. \quad (1.22)$$

Substituting Eq. (1.20) into Eq. (1.22), we obtain a transcendental Haissinski equation for the line density $\rho(z)$,[9]

$$\rho(z) \;=\; \rho(0)\exp\left[-\frac{1}{2}\left(\frac{\omega_s z}{\eta c\sigma_\delta}\right)^2 + \frac{4\pi\epsilon_0 r_0}{\eta\sigma_\delta^2\gamma C}\int_0^z dz'' \int_{z''}^\infty dz'\, \rho(z')W_0'(z''-z')\right]\,.$$
$$(1.23)$$

In the limit of zero beam intensity, the wakefield term vanishes, the solution reduces to the bi-Gaussian form, where the rms bunch length is related to σ_δ by

$$\sigma_z \;=\; \frac{|\eta|c\sigma_\delta}{\omega_s}\,.$$

For high beam intensities, $\rho(z)$ deforms from Gaussian shape. Together with Eq. (1.19), Eq. (1.23) can in principle be solved numerically for $\rho(z)$ once the

---

[9]J. Haissinski, Nuovo Cimento 18B, 72 (1973).

Figure 1.4: Potential well distortion of bunch shape for various beam intensities for the SLC damping ring. The horizontal axis is $x = -\frac{z}{\sigma_{z0}}$, where σ_{z0} is the unperturbed rms bunch length. The vertical scale gives $y = \frac{4\pi e \rho(z)}{V'_{\rm rf}(0)\sigma_{z0}}$. [Courtesy Karl Bane (1992).]

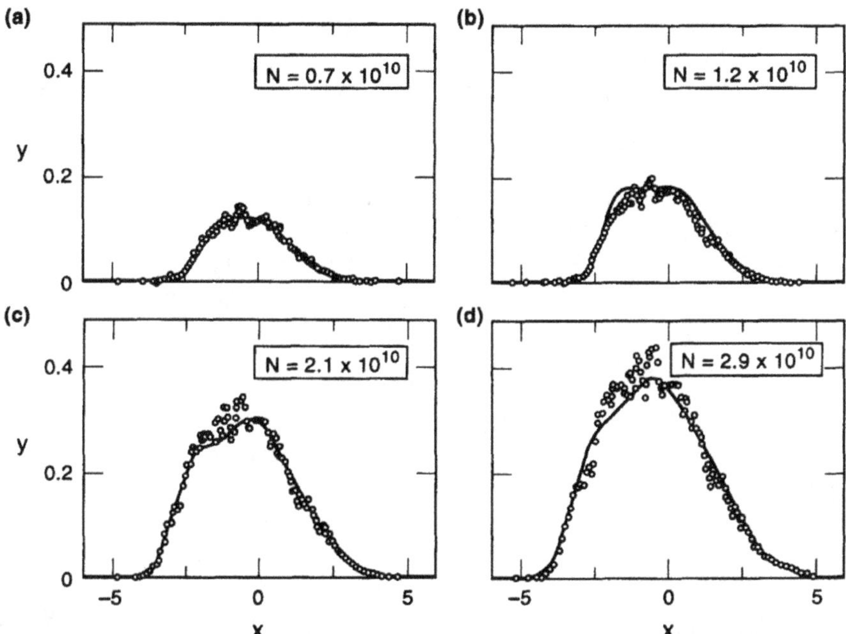

wake function $W'_0(z)$ is known and σ_δ specified. To give a practical realistic example, Fig. 1.4 shows the result for the actual case of the electron damping ring of the SLAC Linear Collider.[10] Bunch shape is Gaussian at low beam intensities, and it distorts as the beam intensity is increased. The calculated bunch shapes (solid curves) agree well with the measured results shown as open circles in Fig. 1.4.

One feature of Fig. 1.4 is that the distribution leans forward ($z > 0$) as the beam intensity increases. This effect comes from the parasitic energy loss of the beam bunch, which in turn comes from the real part of the impedance. Since the SLC damping ring is operated above transition, the bunch moves forward so that the parasitic energy loss can be compensated by the RF voltage. Another feature of Fig. 1.4 is that the bunch length increases as the beam intensity increases. The bunch shape distortion comes from the imaginary part of the impedance. That the bunch lengthens in Fig. 1.4 is a consequence of the fact that the imaginary part of the impedance seen by the beam is mostly inductive.

[10]K.L.F. Bane and R.D. Ruth, Proc. IEEE Conf. Part. Accel., Chicago, 1989, p. 789; L. Rivkin, et al., Proc. Euro. Part. Accel. Conf., Rome, 1988, p. 634.

Impedance	Effect	Below transition	Above transition
Real part $\operatorname{Re} Z_0^{\parallel}$	centroid shift	shift backward	shift forward
Imaginary part $\operatorname{Im} Z_0^{\parallel} > 0$ (capacitive)	bunch length	lengthen	shorten
Imaginary part $\operatorname{Im} Z_0^{\parallel} < 0$ (inductive)	bunch length	shorten	lengthen

The Haissinski equation (1.23) applies to a general wake function $W_0'(z)$. Its solution is obtained numerically. Closed-form solutions, however, exist for specific wake functions. We give a few special analytically soluble examples below.

Resistance wake — Haissinski beam   Consider the wake

$$W_0'(z) \; = \; S\delta(z) \,,$$

or equivalently a pure resistance impedance,

$$Z_0^{\parallel}(\omega) \; = \; \frac{S}{c} \,.$$

It follows from the Hassinski solution Eq. (1.23) that

$$\rho(z) \; = \; \rho(0) \exp\left[-\frac{z^2}{2\sigma_z^2} + \alpha \int_0^z dz' \rho(z') \right]$$

$$\implies \quad \frac{\rho'(z)}{\rho(z)} \; = \; -\frac{z}{\sigma_z^2} + \alpha\rho(z) \,, \tag{1.24}$$

where

$$\alpha \; = \; \frac{4\pi\epsilon_0 r_0 S}{\eta\sigma_\delta^2 \gamma C} \,,$$

and $\sigma_z = \frac{|\eta| c \sigma_\delta}{\omega_s}$ is the unperturbed rms bunch length.

A closed form solution for the beam distribution is given by[11]

$$\rho(z) \; = \; \frac{\sqrt{\frac{2}{\pi}}\, e^{-z^2/2\sigma_z^2}}{\alpha\sigma_z \left[\coth\left(\frac{\alpha N}{2} \right) - \operatorname{erf}\left(\frac{z}{\sqrt{2}\sigma_z} \right) \right]} \,, \tag{1.25}$$

where $\operatorname{erf}(x) = \frac{2}{\sqrt{\pi}} \int_0^x e^{-t^2} dt$ is the error function.

---

[11] A.G. Ruggiero, IEEE Trans. Nucl. Sci. NS-24, 1205 (1977).

Figure 1.5: Potential well distortion for various beam intensities for a resistance wake. The horizontal axis is $x = \frac{z}{\sqrt{2}\sigma_z}$. The vertical scale gives $f(x) = \frac{\sigma_z \rho(z)}{N}$. All five curves are normalized so that they cover the same area $\frac{1}{\sqrt{2}}$.

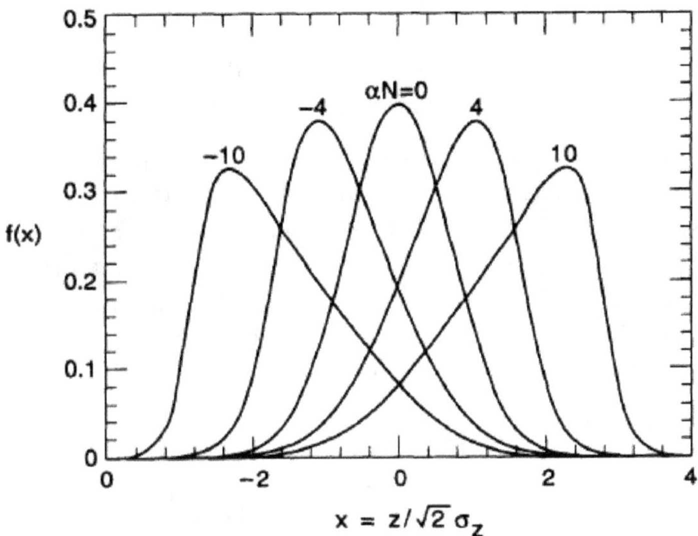

Figure 1.5 shows the function $f(x) = \frac{\sigma_z \rho(z)}{N}$ as a function of $x = \frac{z}{\sqrt{2}\sigma_z}$. The five curves are for the cases $\alpha N = -10, -4, 0, 4, 10$, respectively. The case $\alpha N = 0$ is the unperturbed Gaussian distribution. The peak location of the beam distribution moves forward above transition $\alpha N > 0$ and backward below transition $\alpha N < 0$ as the beam intensity increases. The head of the bunch is toward the right.

Inductance wake — Elliptical beam for proton   We do not have to limit ourselves to the Gaussian (Haissinski) form (1.18) if we consider protons. Consider the wake

$$W_0'(z) = S\delta'(z), \tag{1.26}$$

where $\delta'(z)$ is the derivative of the δ-function. The wakefield (1.26) can be produced by a purely imaginary impedance,

$$Z_0^{\parallel}(\omega) = i\frac{S\omega}{c^2}. \tag{1.27}$$

The quantity S can be related to the familiar parameter $\frac{Z_0^{\parallel}}{n}$ by

$$S = \frac{cC}{2\pi}\frac{Z_0^{\parallel}}{n}.$$

Strictly speaking, the frequency dependence of the impedance (1.27) says it is inductive with an inductance $L = -\frac{S}{c^2}$, and one has $S < 0$; but we will extend the meaning of this impedance — admittedly somewhat awkwardly — to capacitive impedances by allowing $S > 0$.

One physical effect that produces such a wake function (1.26) is that due to the space charge Coulomb repulsion,

$$S = \frac{Z_0 c C}{4\pi\gamma^2}\left(1 + 2\ln\frac{b}{a}\right),$$

where b is the vacuum chamber pipe radius, a is the transverse beam radius (assuming a uniform circular cylindrical beam in a perfectly conducting cylindrical pipe), $Z_0 = \sqrt{\frac{\mu_0}{\epsilon_0}} = 377\,\Omega$ is the vacuum impedance, C is the storage ring circumference, and γ is the Lorentz energy factor. Indeed for the example of space charge impedance, $S > 0$, i.e. it has a capacitive nature.

With the inductance wakefield (1.26), the retarding wake at location z can be related to the local derivative of the line density,

$$\int_z^\infty dz'\,\rho(z')\,W_0'(z - z') = S\rho'(z).$$

In the present example, we will not assume a Gaussian (Haissinski) solution, but will write the stationary distribution of the beam bunch in the ansatz form

$$\psi(z,\delta) = \begin{cases} \frac{3|\eta|cN}{2\pi\omega_s \hat{z}_0^3}\sqrt{\hat{z}_0^2 - \frac{1}{\kappa}\left(\frac{\eta c}{\omega_s}\delta\right)^2 - \kappa z^2}, & \text{if } \frac{1}{\kappa}\left(\frac{\eta c}{\omega_s}\delta\right)^2 + \kappa z^2 < \hat{z}_0^2, \\ 0, & \text{otherwise,} \end{cases} \tag{1.28}$$

where the dimensionless parameter $\kappa > 0$, yet to be found as a function of beam intensity, specifies the degree of distortion of the beam distribution due to the wakefields. The unperturbed beam would have $\kappa = 1$.

The form (1.28) is such that the unperturbed beam has an elliptical distribution in the phase space; and when perturbed by the wakefield, the beam distribution remains elliptical but distorted in such a way that its phase space area (the emittance) is independent of κ, i.e., independent of the beam intensity. The fact that the bunch centroid is located at $z = 0$ for all beam intensities assumes there is no net parasitic energy loss of the bunch, and this is a consequence of the wake being considered has a purely imaginary impedance (1.27).

Integrating over δ, the distribution (1.28) has a parabolic line density,

$$\rho(z) = \frac{3N\sqrt{\kappa}}{4\hat{z}_0^3}(\hat{z}_0^2 - \kappa z^2), \quad \text{if} \quad z < \frac{\hat{z}_0}{\sqrt{\kappa}}. \tag{1.29}$$

The unperturbed beam has a total length $2\hat{z}_0$; the perturbed beam has a total length $2\hat{z} = \frac{2\hat{z}_0}{\sqrt{\kappa}}$. The perturbed total spread in δ is equal to $2\hat{\delta} = \frac{2\hat{z}_0\omega_s\sqrt{\kappa}}{|\eta|c}$. The product $\hat{z}\hat{\delta}$ is independent of κ.

To be self-consistent, the distribution has to be a function of the Hamiltonian (1.20). In the present case, we have

$$H \; = \; \frac{\eta^2 c^2}{2\omega_s}\delta^2 + \frac{\omega_s}{2}\left(1 + D\kappa^{3/2}\right)z^2\,, \tag{1.30}$$

where

$$D \; = \; \frac{6\pi\epsilon_0 N r_0 \eta c^2}{\omega_s^2 \gamma C \hat{z}_0^3}S\,. \tag{1.31}$$

For the special case of space charge, we have

$$D \; = \; \frac{3 N r_0 \eta c^2}{2\omega_s^2 \gamma^3 \hat{z}_0^3}\left(1 + 2\ln\frac{b}{a}\right)\,.$$

Comparing this Hamiltonian with the ansatz (1.28) indicates that the stationary distribution must have the form

$$\psi \; = \; \frac{3\eta c N}{2\pi\omega_s \hat{z}_0^3}\sqrt{\hat{z}_0^2 - \frac{2}{\kappa\omega_s}H}\,,$$

and that self-consistency requires

$$\kappa^2 - 1 - D\kappa^{3/2} \; = \; 0\,,$$

or, in terms of the beam length $\hat{z}$, a fourth order equation,[12]

$$\left(\frac{\hat{z}}{\hat{z}_0}\right)^4 + D\frac{\hat{z}}{\hat{z}_0} - 1 \; = \; 0\,. \tag{1.32}$$

In the limit of weak beam intensity, $D = 0$, we have $\hat{z} = \hat{z}_0$. As the beam intensity increases, the beam shape remains parabolic, but its length changes — either shortens or lengthens according to the sign of D.

Figure 1.6 shows the dependence of $\frac{\hat{z}}{\hat{z}_0}$ on the wake strength parameter D.[13] The bunch lengthens below transition and shortens above transition if the impedance is capacitive like the space charge. The opposite holds if the impedance is inductive. In this model, for any given value of D, there is always a solution of $\hat{z}$.

Inductance wake — Elliptical beam for electron In the previous model, the beam area in phase space is kept constant as the beam intensity is varied. As we have just seen, this leads to a fourth order algebraic equation for the parabolic bunch length. The condition of constant phase space area applies to a proton beam when the accelerator operator carefully matches the injected

---

[12]A. Hofmann, Frontiers of Particle Beams, Lecture Notes in Physics 296, Springer-Verlag, p. 99 (1986).

[13]There are four solutions to Eq. (1.32). Figure 1.6 shows only one solution. The other three are unphysical; one is negative, the other two are complex.

Figure 1.6: Bunch length $\frac{\hat{z}}{\hat{z}_0}$ as a function of wake strength D for an elliptical beam with a pure inductance wake (1.26). The assumed beam distribution (1.28) applies to the case of a proton beam.

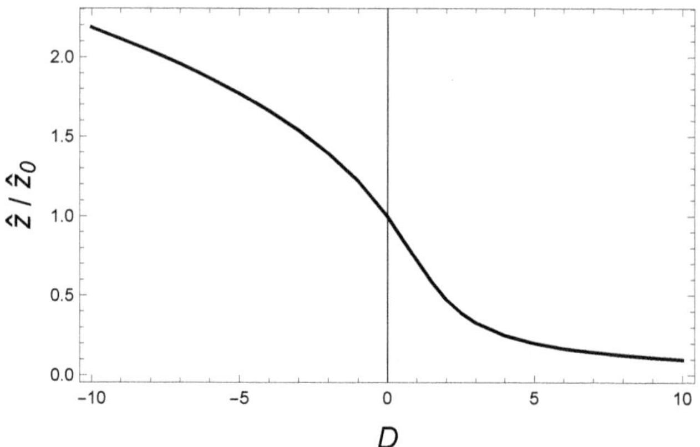

beam (a different beam intensity requires a different matching) to the distorted potential well so that there is no increase in emittance.

As mentioned, on the other hand for an electron beam, the stationary distribution has to assume a Haissinski form (1.21–1.23), and its δ-distribution is unperturbed by the potential well distortion. However, to obtain a qualitative illustration of potential well distortion of an electron beam, one could compromise the requirement of a Gaussian distribution and modify the ansatz distribution (1.28) slightly to read

$$\psi(z,\delta) = \begin{cases} \frac{3N|\eta|c\sqrt{\kappa}}{2\pi\omega_s \hat{z}_0^3}\sqrt{\hat{z}_0^2 - \left(\frac{\eta c}{\omega_s}\delta\right)^2 - \kappa z^2}, & \text{if } \left(\frac{\eta c}{\omega_s}\delta\right)^2 + \kappa z^2 < \hat{z}_0^2, \\ 0, & \text{otherwise.} \end{cases} \quad (1.33)$$

In this model, the distribution remains elliptical but now has a constant spread in δ, while the bunch length varies with the beam intensity. Unlike the previous model, its phase space area is no longer independent of beam intensity.

The line density $\rho(z)$ is still given by Eq. (1.29) with the total bunch length $\frac{2\hat{z}_0}{\sqrt{\kappa}}$. The total spread in δ is equal to the unperturbed value $\frac{2\hat{z}_0\omega_s}{|\eta|c}$. The Hamiltonian is still given by Eq. (1.30) with D given by Eq. (1.31).

Following a procedure like the one in the previous model, we obtain a self-consistency condition that leads to a cubic equation for $\frac{\hat{z}}{\hat{z}_0}$,

$$\left(\frac{\hat{z}}{\hat{z}_0}\right)^3 - \frac{\hat{z}}{\hat{z}_0} + D = 0. \quad (1.34)$$

Again, $\hat{z} = \hat{z}_0$ when $N = 0$, and the bunch lengthens or shortens according to $D < 0$ or $D > 0$. Bunch length as a function of wake strength D is shown

Figure 1.7: Bunch length $\frac{\hat{z}}{\hat{z}_0}$ as a function of wake strength D for an elliptical beam with a pure inductance wake (1.26). The assumed beam distribution (1.33) applies to the case of an electron beam. There is no physical solution when $D > \frac{2}{3^{3/2}}$.

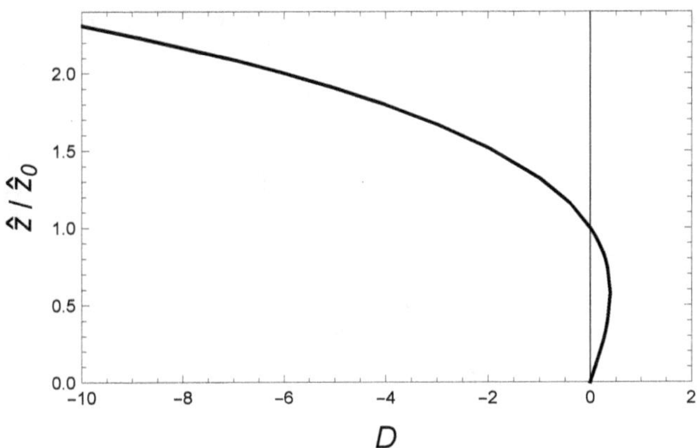

in Fig. 1.7. One difference from the proton model is that this electron model no longer allows a stationary solution for $D > \frac{2}{3^{3/2}}$. At $D = \frac{2}{3^{3/2}}$, the bunch assumes a length of $\frac{\hat{z}_0}{\sqrt{3}}$.

In the region $D > \frac{2}{3^{3/2}}$, the beam cannot maintain an unperturbed δ-spread in its stationary state. This can be one possible interpretation of the threshold of a microwave instability where stationary solution ceases to exist and collective modes become unstable in the longitudinal beam distribution.

Constant wake — Proton beam In this example, we consider a case of constant wake

$$W_0'(z < 0) = W_0. \tag{1.35}$$

Fundamental property on the wake function requires $W_0 > 0$. Unlike the previous examples, the impedance of this wake contains both real and imaginary parts,

$$Z_0^{\parallel}(\omega) = W_0 \left[\frac{i}{\omega} + \pi\delta(\omega) \right].$$

The solution we seek has to take into account both a centroid shift due to the real part of the impedance and a shape distortion due to the imaginary part. We need to take these into consideration when we design an ansatz for our model.

Consider the following ansatz distribution for a proton beam,

$$\psi(z, \delta) = \frac{N|\eta|c}{2\pi\omega_s \hat{z}_0} \frac{1}{\sqrt{\hat{z}_0^2 - \frac{1}{\kappa}\left(\frac{\eta c}{\omega_s}\delta\right)^2 - \kappa(z - \bar{z})^2}}, \tag{1.36}$$

where $\kappa > 0$ parametrizes the potential well distortion on the bunch shape and $\bar{z}$ gives the bunch centroid shift. The beam's phase space area in this ansatz is independent of the bunch intensity, so we are assuming a proton beam. To model an electron beam, we only have to replace $\kappa(z - \bar{z})^2$ by $(z - \bar{z})^2$ in Eq. (1.36) (Homework 1.10).

This beam distribution projects onto the z-axis with a constant line density,

$$\rho(z) \;=\; \begin{cases} \sqrt{\kappa}\,\frac{N}{2\hat{z}_0}, & \text{if } |z - \bar{z}| < \frac{\hat{z}_0}{\sqrt{\kappa}}, \\ 0, & \text{otherwise.} \end{cases}$$

Following the same procedure as before, a self-consistency condition leads to a fourth order equation for the bunch length,

$$\left(\frac{\hat{z}}{\hat{z}_0}\right)^4 - D\left(\frac{\hat{z}}{\hat{z}_0}\right)^3 - 1 \;=\; 0\,, \tag{1.37}$$

where

$$D \;=\; \frac{4\pi\epsilon_0 N r_0 \eta c^2 W_0}{2\omega_s^2 \gamma C \hat{z}_0}\,.$$

The same self-consistency condition also gives the bunch centroid shift,

$$\bar{z} \;=\; D\hat{z}_0\,. \tag{1.38}$$

Above transition, $D > 0$, the bunch centroid shifts forward and the bunch lengthens. Below transition, the opposite occurs.

Homework 1.5

(a) Follow the text and derive Eq. (1.25). To do so, you need to first show that it satisfies Eq. (1.24) and then show that it satisfies the normalization (1.19).

(b) Show that for a weak beam with $|\alpha N| \lesssim 1$, the peak beam density occurs at

$$\bar{z} \approx \frac{\alpha N}{\sqrt{2\pi}}\,\sigma_z\,.$$

Check this expected behavior against Fig. 1.5.

Homework 1.6

(a) For the inductance wake $W_0'(z) = S\delta'(z)$, show that the potential well distorted Haissinski beam distribution in the Gaussian form (1.23) satisfies

$$f(z) \;=\; f(0)\,e^{-z^2/2\sigma_z^2}\,,$$

where $f(z) = \alpha\rho(z)\,e^{-\alpha\rho(z)}$ with $\alpha = \frac{4\pi\epsilon_0 r_0 S}{\eta\sigma_\delta^2 \gamma C}$ and $\sigma_z = \frac{|\eta|c\sigma_\delta}{\omega_s}$.

(b) Given α, find $\alpha\rho(z)$ numerically as a function of $N = \int_{-\infty}^{\infty} dz\,\rho(z)$. Is there always a solution for any value of N? Figure 1.8 gives the result. Give an approximate expression for the rms bunch length when $|\alpha| \ll 1$.

Figure 1.8: A Gaussian beam potential well distorted by $W_0'(z) = S\delta'(z)$. The figure shows $\alpha\rho(z)$ as a function of $\frac{z}{\sigma_z}$ for five values of $\frac{\alpha N}{\sigma_z}$. The bunch lengthens if $\alpha < 0$, and shortens if $\alpha > 0$. There is no solution if $\alpha\rho(0) > 1$, corresponding to $\frac{\alpha N}{\sigma_z} > 1.53$.

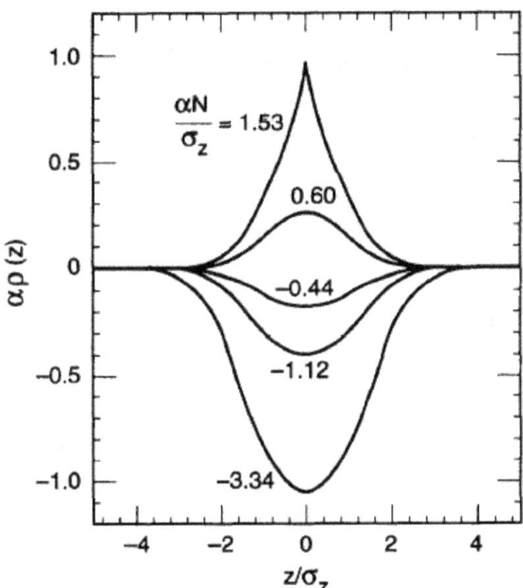

(c) Show that the bunch lengthens if $\alpha < 0$, and shortens if $\alpha > 0$. Give a physical reason why this is so. Show that $\rho(z) < \frac{1}{\alpha}$ for all z if $\alpha > 0$.

Homework 1.7 Repeat the study in Homework 1.6 for the resistance wake $W_0'(z) = S\delta(z)$. Observe that in this case there is a shift of the beam centroid, which is absent in the inductance case. You should know the physical origin of this shift.

Homework 1.8 Consider the beam distribution given by the ansatz form (1.28) and an inductance wake. Show the following:
(a) It gives a projection of line density $\rho(z)$ given by Eq. (1.29).
(b) It satisfies the normalization condition (1.19).
(c) Its self-consistent stationary bunch length is determined by Eq. (1.32).

Homework 1.9 Repeat Homework 1.8 for the case when the beam distribution is given by the ansatz form (1.33). In particular, derive Eq. (1.34).

Homework 1.10 The text analyzed the case of constant wake (1.35) for a proton beam. Repeat the analysis for an electron beam. Show that Eq. (1.38) still holds with the same definition of D as in the text, but Eq. (1.37) becomes

$$\left(\frac{\hat{z}}{\hat{z}_0}\right)^2 - D\left(\frac{\hat{z}}{\hat{z}_0}\right) - 1 = 0, \quad \text{or} \quad \frac{\hat{z}}{\hat{z}_0} = \sqrt{1 + \frac{D^2}{4}} + \frac{D}{2}.$$

1.2.4   Collective instability

In the previous section, we studied the time-independent stationary solutions of the Vlasov equation for a multiparticle system under the influence of wakefields. This led to the phenomenon of potential well distortion. In passing, we briefly mentioned a possible threshold of a microwave instability in the discussion of potential well distortion in Fig. 1.7.

It turns out that a wealth of beam dynamical properties can be studied by the corresponding time-dependent Vlasov equation,

$$\frac{\partial \psi}{\partial s} - \eta \delta \frac{\partial \psi}{\partial z} + \left[\frac{\omega_s^2}{\eta c^2} z - \frac{4\pi \epsilon_0 r_0}{\gamma C} \int_z^\infty dz' \, \rho(z') W_0'(z - z')\right] \frac{\partial \psi}{\partial \delta} = 0, \quad (1.39)$$

where $\psi(z, \delta, s)$ is the time-dependent beam distribution in the phase space $(q = z, p = -\frac{\eta c}{\omega} \delta)$ at time $t = \frac{s}{c}$, normalized by Eq. (1.19).

Equation (1.39) is the starting point of most literatures on collective beam instabilities. It is a partial-differential-integral equation, and even worse, it is nonlinear in ψ, the function we are solving for. General time-dependent solutions are very difficult to obtain and one relies on making drastic approximations and being content with limited success.

One drastic approximation that researchers often make is to assume the beam distribution is always close to some stationary solution ψ_0. Consequently, we have

$$\psi(z, \delta, s) = \psi_0(z, \delta) + \Delta\psi(z, \delta, s), \quad \text{where} \quad |\Delta\psi| \ll \psi_0. \quad (1.40)$$

Time dependence is contained only in the perturbation $\Delta\psi$.

Substituting Eq. (1.40) into (1.39) and keeping only to 1st order in $\Delta\psi$ give an equation that is linear in $\Delta\psi$. The problem is therefore shifted to the study of infinitesimal deviations from an assumed equilibrium. What we learn will be whether the assumed stationary distribution ψ_0 is stable against infinitesimal perturbations. This is a limited success because at the end we will learn whether the beam is stable or unstable if it starts with the one specific ψ_0 assumed, but yet there are an infinite number of possible stationary distributions to start the problem with. Nevertheless, this is an important research area which is outside the scope of this chapter.

1.3   Fokker–Planck equation

In the derivation of the Vlasov equation, we have assumed a conservative system without any diffusion or external damping effects. This is usually a good approximation for proton beams. For electron beams, synchrotron radiation contributes to both damping and diffusion,[14] and one needs to modify the Vlasov equation accordingly to obtain another equation called the *Fokker–Planck equation*.

Strictly speaking, our results obtained using the Vlasov equation apply only to protons and not electrons. On the other hand, on the subject of collective instabilities, when the instability occurs in a time shorter than the damping or diffusion times, the Vlasov treatment can apply also to electrons. The treatment of collective instabilities using the Fokker–Planck equation is beyond our scope here.[15]

1.3.1 Derivation

With diffusion due to some noise effect, the system is no longer deterministic. In the presence of noise, the single-particle equations of motion in the phase space (q, p) become

$$\begin{aligned}
\dot{q} &= f(q, p, t)\,, \\
\dot{p} &= g(q, p, t) + \sum_k \epsilon_k \delta(t - t_k)\,,
\end{aligned} \qquad (1.41)$$

where ϵ_k is the noise represented as a sudden and random change in momentum p of the particle at sudden and random time t_k. The noise is such that the sudden changes have an average value of zero. A nonzero average, if exists, can be transformed away by properly redefining the coordinates (see page 35 for a discussion).

We will derive the Fokker–Planck equation for a general system described by Eq. (1.41) — we are not specializing to synchrotron radiation unless specifically pointed out. In this system, we have assumed the noise occurs only in the momentum p.

A comment might be useful at this junction. There is no physical process that would give noise directly to q. We sometimes decompose the particle motion into a "betatron component" and a "synchrotron component". When a synchrotron radiation noise is applied, we then argue that there is a sudden noise delivered to x_β, which can be considered a noise applied to q, not to p. Although (approximately) correct in the limit when the synchrotron motion is slow, one needs to avoid a misconception here. The sudden change in x_β is not due to a sudden change in the physical coordinate x. The physical x of a particle does not change when a synchrotron photon is emitted. The reason x_β has a sudden change is because our *definition* of betatron component has changed. The change is in our artificial definition, not in the physics of the particle motion.

---

[14]M. Sands, The Physics of Electron Storage Rings, an Introduction, SLAC Report 121 (1970).

[15]See for example, A. Renieri, Frascati Lab. Report LNF-76/11(R) (1976); T. Suzuki, Part. Accel. 14, 91 (1983).

Let us make another word of caution here concerning the above statement of transforming away the nonzero average of noise. To transform away the average of the noise is not the same as replacing all the noise ϵ_k by $\epsilon_k - \langle\epsilon\rangle$. This is a subtle point, but an incorrect treatment here can lead to very wrong conclusions. We will return to this point on page 35.

One commonly encountered case of multiparticle system with noise is that of an electron beam in a storage ring. The noise comes mainly from emission of quantized photons as the electron circulates around the storage ring. In addition, the synchrotron radiation also gives rise to a radiation damping to the system. In contrast, the Vlasov equation is most relevant for a proton beam, whose synchrotron radiation is generally ignorable.

In Eq. (1.41), the functions f and g contain the information on the single-particle equations of motion, including the damping effect if any, and if included, we would have $\frac{\partial f}{\partial q} + \frac{\partial g}{\partial p} \neq 0$. Diffusion effect on the other hand is all contained in the noise term. Damping and diffusion, even when they come from the same source, e.g. synchrotron radiation, are described in different parts of the equation of motion (1.41); damping is deterministic, diffusion is stochastic.

Deterministic system   Let us first consider the system when f and g are given. This means the system is deterministic, including the damping effect if any. The diffusion effect will be considered momentarily. Consider a phase space element $\Delta q \Delta p$ in the (q, p) phase space as shown in Fig. 1.9. Represent the phase space element as a rectangular box,

$$A(q, p),$$
$$B(q + \Delta q, p),$$
$$C(q + \Delta q, p + \Delta p),$$
$$D(q, p + \Delta p).$$

The number of particles enclosed in the box is $\psi(q, p, t)\,\Delta q \Delta p$.

The phase space box evolves in time deterministically according to f and g. At time $t + dt$, the box has moved to $A'B'C'D'$ as shown in Fig. 1.9. We have used Δq and Δp (rather than dq and dp) to denote the dimensions of the box, but have used dt to denote the time increment. This is because we do not want the box size to be vanishingly small so that there are a sufficiently large number of statistical particles in it, but dt should be considered truly infinitesimal.

In general, the rectangular box deforms into a parallelogram. (The only case in which the rectangular box remains rigid in shape as time evolves is simple harmonic motion.) The vertices of the parallelogram are

$$A'[q + f(q, p, t)\,dt, p + g(q, p, t)\,dt],$$
$$B'[q + \Delta q + f(q + \Delta q, p, t)\,dt, p + g(q + \Delta q, p, t)\,dt],$$
$$C'[q + \Delta q + f(q + \Delta q, p + \Delta p, t)\,dt, p + \Delta p + g(q + \Delta q, p + \Delta p, t)\,dt],$$
$$D'[q + f(q, p + \Delta p, t)\,dt, p + \Delta p + g(q, p + \Delta p, t)\,dt].$$

Figure 1.9: A phase space element evolves from a rectangular box $ABCD$ at time t to a parallelogram $A'B'C'D'$ at time $t+dt$. The box shape is deformed. The area of the box shrinks if there is a damping effect.

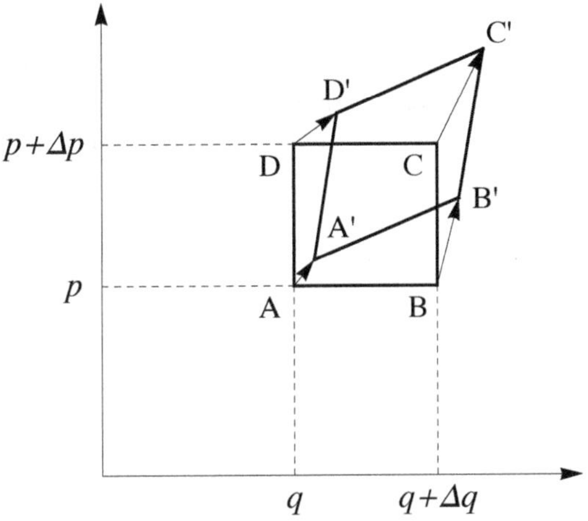

The condition that no particles leak into or out of the box gives

$$\psi(q,p,t)\,\text{area}(ABCD) \;=\; \psi(q+f\,dt, p+g\,dt, t+dt)\,\text{area}(A'B'C'D'). \quad (1.42)$$

In the present system,

$$\begin{aligned}
\text{area}(A'B'C'D') &= |\overrightarrow{A'B'} \times \overrightarrow{A'D'}| \\
&= \Delta q\,\Delta p\left[1 + \left(\frac{\partial f}{\partial q} + \frac{\partial g}{\partial p}\right)dt\right] \\
&= \text{area}(ABCD)\left[1 + \left(\frac{\partial f}{\partial q} + \frac{\partial g}{\partial p}\right)dt\right]. \quad (1.43)
\end{aligned}$$

The quantity $\left(\frac{\partial f}{\partial q} + \frac{\partial g}{\partial p}\right)$ is the contribution from the damping effect. A conservative system would have $\left(\frac{\partial f}{\partial q} + \frac{\partial g}{\partial p}\right) = 0$.

Equation (1.42) then gives

$$\psi(q,p,t) \;=\; \psi(q+f\,dt, p+g\,dt, t+dt)\left[1 + \left(\frac{\partial f}{\partial q} + \frac{\partial g}{\partial p}\right)dt\right] \quad (1.44)$$

or, to 1st order in dt,

$$\frac{\partial \psi}{\partial t} + f\frac{\partial \psi}{\partial q} + g\frac{\partial \psi}{\partial p} \;=\; -\left(\frac{\partial f}{\partial q} + \frac{\partial g}{\partial p}\right)\psi.$$

This is the equation we used earlier in Eq. (1.12).

Adding diffusion So far, the system we treated may not be conservative (if damping is included), but it still remains deterministic. The situation changes when diffusion noise, i.e. the term $\sum_k \epsilon_k \delta(t - t_k)$ in Eq. (1.41), is added.

Consider again the box and its evolution in the (q, p) phase space as shown in Fig. 1.9. In that description, dt has been regarded as truly infinitesimal. In the presence of noise, this can no longer be done. The time increment, now designated as Δt, has to be large enough that there are at least several ϵ_k kicks occurring within it. If we designate the total kick accumulated in Δt as E, and the probability density of having a particular accumulated kick E as $P(E)$, we have

$$E = \sum_k \epsilon_k \quad \text{from } t \text{ to } t + \Delta t,$$

with

$$\int P(E)\, dE = 1,$$

$$\int E P(E)\, dE = 0,$$

$$\int E^2 P(E)\, dE = \langle \dot{N} \epsilon^2 \rangle \Delta t,$$

where $\langle \dot{N} \epsilon^2 \rangle$ is the rate of increase of the second moment with $\dot{N}$ interpreted as the average occurring rate of the sudden kicks. The fact that the expected value of E^2 is proportional to Δt is a consequence of the noise being stochastic.

When noise is added, a moment's reflection shows that Eq. (1.44) needs to be modified to read

$$\int dE\, P(E)\psi(q, p - E, t) = \psi(q + f\,\Delta t, p + g\,\Delta t, t + \Delta t)\left[1 + \left(\frac{\partial f}{\partial q} + \frac{\partial g}{\partial p}\right)\Delta t\right].$$
$$(1.45)$$

The reason is because the particles in the $A'B'C'D'$ box come from the probability weighted integration of the earlier boxes, each displaced in momentum p by an amount of E. Note we have changed dt to Δt in this expression.

We now calculate

$$\int dE\, P(E)\psi(q, p - E, t) = \int dE\, P(E)\left[\psi - E\frac{\partial \psi}{\partial p} + \frac{E^2}{2}\frac{\partial^2 \psi}{\partial p^2} + \cdots\right]$$

$$= \psi(q, p, t) + \frac{1}{2}\langle \dot{N} \epsilon^2 \rangle \Delta t \frac{\partial^2 \psi}{\partial p^2} + \cdots.$$

When substituted into Eq. (1.45) and keep linear terms in Δt, we obtain the Fokker–Planck equation,

$$\frac{\partial \psi}{\partial t} + f\frac{\partial \psi}{\partial q} + g\frac{\partial \psi}{\partial p} = -\left(\frac{\partial f}{\partial q} + \frac{\partial g}{\partial p}\right)\psi + \frac{1}{2}\langle \dot{N} \epsilon^2 \rangle \frac{\partial^2 \psi}{\partial p^2} + \cdots. \qquad (1.46)$$

Compared with the Vlasov equation, the Fokker–Planck equation contains an additional damping term and a noise term on the right-hand-side. The damping term has appeared before in Eq. (1.12). In addition, don't forget the fact that f and g on the left-hand-side of the equation also contain the damping effect.

Note that it is the diffusion rate of the second moment of the noise that enters the Fokker–Planck equation. Different noise sources give the same average beam distribution, regardless of their detailed spectrum, as long as they have the same $\langle \dot{N} \epsilon^2 \rangle$. The fact that one and only one parameter, i.e. $\langle \dot{N} \epsilon^2 \rangle$, determines the beam distribution is a result of the central limit theorem in statistical mechanics.[16]

Additional terms represented by $(\cdots)$ on the right-hand-side of Eq. (1.46) involve higher moments $\langle \dot{N} \epsilon^n \rangle$ of the noise. These terms often do not contribute much and are ignored. We have a discussion of this in Sec. 1.3.4.

One more thing to note in Eq. (1.46) is the fact that it involves $\frac{\partial^2 \psi}{\partial p^2}$, and not $\frac{\partial^2 \psi}{\partial q^2}$, seemingly violating a symmetry between q and p in the Hamiltonian dynamics. The reason of this is because we have assumed the diffusion (as well as damping) occurs only in p and not in q [see Eq. (1.41)]. Had there been noise also in q, there will be a term involving $\frac{\partial^2 \psi}{\partial q^2}$ in Eq. (1.46). Furthermore, if the noise excites q and p in a correlated manner, then there will also be a term involving $\frac{\partial^2 \psi}{\partial q \partial p}$. Symmetry in Hamiltonian system is not in question. We will give the modified Fokker–Planck equation later when we analyze a general coupled system in Sec. 1.4, particularly Eq. (1.57).

For a damped simple harmonic oscillator, Eq. (1.10), the Fokker–Planck equation reads

$$\frac{\partial \psi}{\partial t} + \omega p \frac{\partial \psi}{\partial q} + (-\omega q - 2\alpha p)\frac{\partial \psi}{\partial p} = 2\alpha \psi + D_2 \frac{\partial^2 \psi}{\partial p^2}, \qquad (1.47)$$

where we have defined a diffusion coefficient $D_2 = \frac{1}{2}\langle \dot{N} \epsilon^2 \rangle$.

Homework 1.11 It would be instructive to derive Eq. (1.43). First convince yourself that the rectangle ABCD of Fig. 1.9 is deformed into a parallelogram when dt is infinitesimal. Then compute the area of the parallelogram to derive Eq. (1.43). Lastly, prove that if ψ satisfies Eq. (1.47), then once ψ is normalized initially, it will be normalized at all times.

1.3.2 Stationary solution of Fokker–Planck equation

The first thing to know about Eq. (1.47) is that it has an equilibrium stationary solution described by a bi-Gaussian distribution

$$\psi_0(q,p) = \frac{\alpha}{\pi D_2} \exp\left[-\frac{\alpha}{D_2}(q^2 + p^2)\right]. \qquad (1.48)$$

---

[16]It is indeed quite amazing that to a large degree no details of the noise spectrum matters. Tossing coins and synchrotron radiation clearly have very different noise spectra, but they both give strictly Gaussian distributions after sufficient noise kicks.

Unlike the Vlasov case, for which there are an infinite number of possible stationary solutions, here Eq. (1.48) is the unique bounded stationary solution to the Fokker–Planck equation.

The rms expectation values of q and p are

$$\sigma_q \;=\; \sigma_p \;=\; \left(\frac{D_2}{2\alpha}\right)^{1/2}. \tag{1.49}$$

We shall sometimes designate this quantity simply as σ in later developments. The dependence $\sigma_{p,q}^2 \propto D_2$ reflects a random walk process.

The equilibrium is reached as a result of balancing the damping and the diffusion effects and Eq. (1.48) is the only distribution that achieves such a balance at all points in the phase space. The fact that $\sigma_{q,p}$ depend on the ratio $\frac{D_2}{\alpha}$ is a direct consequence of this balancing act. As we will see repeatedly later, in attempts to solve the Fokker–Planck equation under more complex conditions, it is often convenient to start with an ansatz of a Gaussian form, or at least with a Gaussian factor, if the perturbation is weak. One exception to this general rule is described in Eq. (1.116).

The beam distribution in an electron storage ring therefore, unlike that of a proton case, tends to be Gaussian. The damping and diffusion coefficients depend on which dimension of motion is being considered. Table 1.1, extracted from standard textbooks on electron storage ring physics, gives a summary of the expressions for α, D_2 and the rms beam dimensions for the horizontal betatron x_β-motion, the vertical betatron y-motion, and the longitudinal synchrotron z-motion in terms of other storage ring parameters if the damping and diffusion come from synchrotron radiation. The table applies to an ideal planar storage ring for which there is no spurious coupling among these three dimensions, particularly no coupling between the x- and y-motions. It assumes synchrotron motion is slow compared with the particle revolution. It also assumes an iso-magnetic lattice design with uniform bending and uniform betatron focusing. The case with x-y coupling will be considered in Sec. 1.4.

In Table 1.1, for the x_β-motion, the displacement x_β is identified as the canonical momentum p, rather than the canonical coordinate q. This is because the synchrotron radiation noise occurs in x_β rather than x'_β (betatron components only). The same does not occur in the y-motion, where the synchrotron radiation excites y' due to radiation sideways and y' is considered the vertical momentum. The sideways radiation contributes a quantum noise spectrum with $\langle \dot{N} u^2 \theta_y^2 \rangle = \frac{23}{48\sqrt{3}} \frac{r_0 \hbar m c^4 \gamma^5}{\rho^3}$.

Homework 1.12 Show that $\psi_0(q,p)$ of Eq. (1.48) is a solution to the Fokker–Planck equation (1.47). Show that it is normalized according to $\int dq \int dp\, \psi_0 = 1$. Show that its rms moments are given by (1.49). A more adventurous reader may try to show that this ψ_0 is the unique solution to Eq. (1.47) that extends to the entire phase space and is normalizable.

Table 1.1: The phase space coordinates q, p and the damping and diffusion coefficients for the three dimensions of motion in an electron storage ring if damping and diffusion come from synchrotron radiation. Parameters used in the table are: $E_0 = \gamma mc^2 = $ beam energy, $\eta = $ phase slippage factor, $\omega_{x,y,s} = $ horizontal betatron, vertical betatron, synchrotron frequencies, $r_0 = $ classical radius of the beam particle, $\rho = $ bending magnet radius, $2\pi\rho = $ ring circumference, $T = \frac{2\pi\rho}{c} = $ revolution time, $\nu_{x,y} = \frac{\omega_{x,y}\rho}{c}$, averaged beta functions $\beta_{x,y} = \frac{\rho}{\nu_{x,y}}$, averaged dispersion function $= \frac{\rho}{\nu_x^2}$, synchrotron radiation energy loss per turn $U_0 = \frac{4\pi}{3} \frac{r_0 mc^2 \gamma^4}{\rho}$, $\mathcal{D}_u \equiv \frac{1}{2}\langle \dot{N}u^2 \rangle = \frac{55}{48\sqrt{3}} \frac{r_0 \hbar mc^4 \gamma^7}{\rho^3}$, $C_q \equiv \frac{55}{32\sqrt{3}} \frac{\hbar}{mc}$. Synchrotron radiation partition numbers are taken to be 1,1,2 for the x-, y- and z-motions.

	Horizontal betatron x_β	Vertical betatron y	Longitudinal synchrotron z
ω	ω_x	ω_y	ω_s
q	$-\frac{c}{\omega_x} x'_\beta$	y	z
p	x_β	$\frac{c}{\omega_y} y'$	$-\frac{\eta c}{\omega_s} \delta$
α	$\frac{U_0}{2E_0 T}$	$\frac{U_0}{2E_0 T}$	$\frac{U_0}{E_0 T}$
D_2	$\left(\frac{\rho}{E_0 \nu_x^2}\right)^2 \mathcal{D}_u$	$\frac{23}{110}\left(\frac{1}{\omega_y mc\gamma^2}\right)^2 \mathcal{D}_u$	$\left(\frac{\eta c}{\omega_s E_0}\right)^2 \mathcal{D}_u$
rms	$\sigma_{x\beta} = \frac{c}{\omega_x}\sigma_{x\beta'}$ $= \frac{1}{\nu_x^2}\sqrt{C_q \gamma^2 \rho}$	$\sigma_y = \frac{c}{\omega_y}\sigma_{y'}$ $= \frac{1}{\nu_y}\sqrt{\frac{23}{110} C_q \rho}$	$\frac{\omega_s}{|\eta| c}\sigma_z = \sigma_\delta$ $= \sqrt{\frac{C_q \gamma^2}{2\rho}}$

Homework 1.13 Consider a 2-D system with a separable Hamiltonian given by $H(x, p_x, y, p_y, t) = H_x(x, p_x, t) + H_y(y, p_y, t)$. With this Hamiltonian, the system is of course x-y decoupled. Show that the beam distribution in the 4-D phase space has the factorized form $\psi(x, p_x, y, p_y, t) = \psi_x(x, p_x, t) \times \psi_y(y, p_y, t)$, where ψ_x and ψ_y separately satisfy the 2-D Fokker–Planck equations of its own phase space.

This homework is the basis of the important conclusion that when additional degrees of freedom is taken into consideration in an existing system, what one does is to add the new Hamiltonian to the existing Hamiltonian by linear superposition.

Homework 1.14 Table 1.1 is a somewhat simplified version of the Fokker–Planck system for each of the three dimensions of beam motion in an ideal planar strong-focusing isomagnetic electron storage ring. Assuming some familiarity to basic electron storage ring physics, go over the table and check the approximations being made.

1.3.3　Haissinski solution

We mentioned earlier that a proton beam distribution satisfies the Vlasov equation, which allows an infinite number of stationary solutions. One particularly noteworthy family of these solutions is of the exponential form (1.18). Even among this family, however, the solution is not unique because the constant in the exponent is arbitrary.

An electron beam distribution, on the other hand, satisfying the Fokker–Planck equation (1.47), has a unique stationary solution, and it is of the type (1.18) with a specific value for the constant. For the case of a damped simple harmonic oscillator, for example, the unique solution is given by Eq. (1.48).

In this section, we will extend this unique solution to the cases when a potential well distortion is included. This solution has been referred to as the Haissinski distribution (see footnote 9) in the context of the Vlasov equation earlier. In the Fokker–Planck context, the solution has already been given by the Vlasov solution (1.22),

$$\psi(z,\delta) \;\propto\; \exp\left[-\frac{1}{\sigma_\delta^2}\,\frac{\omega_s H(z,\delta)}{\eta^2 c^2}\right],$$

except that the value of σ_δ can no longer be arbitrary and has to be calculated. Two example cases are discussed below, one with wakefield induced and one with RF bucket induced.

Wakefield induced potential well distortion　Substituting the Hamiltonian $H(z,\delta)$ from Eq. (1.20), we obtain

$$\psi(z,\delta) \;\propto\; \exp\left\{-\frac{1}{\sigma_\delta^2}\left[\frac{\delta^2}{2} + \frac{\omega_s^2 z^2}{2\eta^2 c^2} - \frac{4\pi\epsilon_0 r_0}{\eta\gamma C}\int_0^z dz'' \int_{z''}^\infty dz'\rho(z')W_0'(z''-z')\right]\right\},$$

which can be written in a factorized form

$$\psi(z,\delta) \;=\; \frac{1}{\sqrt{2\pi}\,\sigma_\delta}\,e^{-\frac{\delta^2}{2\sigma_\delta^2}}\,\rho(z)\,,$$

$$\rho(z) \;=\; \rho(0)\exp\left[-\frac{\omega_s^2 z^2}{2\eta^2 c^2\sigma_\delta^2} + \frac{4\pi\epsilon_0 r_0}{\eta\gamma C\sigma_\delta^2}\int_0^z dz'' \int_{z''}^\infty dz'\rho(z')W_0'(z''-z')\right].$$

The expression for $\rho(z)$ simply reproduces the Hassinski equation (1.23). Normalization is given by

$$\int_{-\infty}^\infty dz\,\rho(z) \;=\; N\,,$$

which determines $\rho(0)$.

We then substitute this expression into the Fokker–Planck equation (1.47) to obtain the unique value of σ_δ,

$$\sigma_\delta^2 \;=\; \frac{\omega_s^2 D_2}{2\alpha\eta^2 c^2}\,, \tag{1.50}$$

which can be recognized simply to be the Fokker–Planck value (1.49). The line density $\rho(z)$ satisfies the transcendental equation (1.23) with σ_δ given by the unique solution (1.50). This result has been applied to obtain Fig. 1.4.

RF bucket induced potential well distortion   In this illustration, consider the case when there is a superposition of two RF sources, one from a fundamental RF with frequency $\omega_{\rm rf}$ and the other from an m-th harmonic with frequency $m\omega_{\rm rf}$ (m = integer) and modulation voltage $V_{\rm mod}$, the total energy gain per turn is

$$eV_{\rm tot} = e\hat{V}_{\rm rf}\sin\left(\phi_s - \frac{\omega_{\rm rf}}{c}z\right) + e\hat{V}_{\rm mod}\sin\left(\phi_s - \frac{m\omega_{\rm rf}}{c}z\right) - e(\hat{V}_{\rm rf} + \hat{V}_{\rm mod})\sin\phi_s$$

where the last term comes from

$$e(\hat{V}_{\rm rf} + \hat{V}_{\rm mod})\sin\phi_s \;=\; U_0\,,$$

which is the energy gain per turn of a synchronous reference particle. We assume the RF voltage and the modulator voltage are in phase at position $z = 0$, where the synchronous particle is located and is a stable fixed point in phase space.

The Hamiltonian in the (q, p) phase space, where $q = z, p = -\frac{\eta c}{\omega_s}\delta$, has the form

$$H(z,\delta) = \frac{\eta^2 c^2}{\omega_s}\left[\frac{\delta^2}{2} + \frac{1}{\eta C E}\int_0^z eV_{\rm tot}(z')dz'\right].$$

When the acceleration voltage is small, meaning the longitudinal dynamics is slow so that adiabatic approximation (weak focusing) can be made, the longitudinal equilibrium beam distribution is given by the Haissinski solution

$$\psi(z,\delta) \;\propto\; \exp\left[-\frac{1}{\sigma_\delta^2}\frac{\omega_s H}{\eta^2 c^2}\right]$$

$$= \;\exp\left[-\frac{\delta^2}{2\sigma_\delta^2} - \frac{1}{\eta\sigma_\delta^2 C E}\int_0^z eV_{\rm tot}(z')dz'\right],$$

where σ_δ is given by Eq. (1.50).

Integrating over δ, we obtain

$$\rho(z) \propto \exp\left[-\frac{1}{\eta\sigma_\delta^2 C E}\int_0^z eV_{\rm tot}(z')dz'\right].$$

The proportionality constant is again determined by the normalization condition.

A side remark is in order. We know that in the Hamiltonian dynamics, there is a symmetry between q and p. It is possible to make a canonical transformation to switch their roles. In the Haissinski solution, however, it appears that this symmetry is lost; the distribution in δ is required to be Gaussian while the distribution in z can be generalized to include the wakefields. The reason behind this loss of symmetry, and it is important to keep in mind, is due to the fact that in the Haissinski model, all the damping, the noise excitation, and the wakefield effects occur in the δ'-equation. The z'-equation remains intact as

$$z' = -\eta\delta\,.$$

See Homework 1.16.

Homework 1.15 Go through the algebra to verify the conclusion made in the text that

(a) the unique stationary solution for σ_δ is given by Eq. (1.50) for the Fokker–Planck equation (1.47).

(b) the line density $\rho(z)$ then satisfies the Haissinski equation (1.23).

Homework 1.16 This homework can be considered an extension of Homework 1.4. The text points out the secret behind the Haissinki model lies in the fact that all the damping, the noise excitation, and the wakefield effects occur only in the δ'-equation. Question remains whether the model applies to the case when the z'-equation is perturbed by a nonlinearity such as a nonlinear phase slippage factor. Try to extend the Haissinski model to include the effect of a nonlinear phase slippage factor when η is a function of δ, i.e. when

$$z' = -\eta(\delta)\delta\,.$$

Find the condition on $\eta(\delta)$ for the Haissinski model to apply.

Solution As seen in Homework 1.4, stationary solution of the Vlasov equation can be obtained when both the phase slippage and the RF bucket are nonlinear. Unfortunately, stationary Haissinski solution of the Fokker–Planck equation applies only when phase slippage is linear, i.e. when $\eta(\delta)$ is independent of δ.

1.3.4   Distortion by higher moments in the noise spectrum

The nominal bi-Gaussian distribution (1.48) is a solution to the Fokker–Planck equation (1.47) only when we drop the contribution from all the higher order moments, i.e. when we drop all the terms in the $(\cdots)$ of Eq. (1.46). If included, we would have terms

$$\frac{1}{2}\langle \dot{N}\epsilon^2\rangle \frac{\partial^2\psi}{\partial p^2} + \cdots = \sum_{n=1}^{\infty} D_n \frac{\partial^n\psi}{\partial p^n}\,, \tag{1.51}$$

where

$$D_n = \frac{1}{n!}\langle \dot{N}\epsilon^n\rangle\,.$$

So far we have kept only the D_2 term. Indeed the central limit theorem says that one and only one parameter, i.e. D_2, dominates the stochastic system, regardless of the values of all the other moments of the noise spectrum. But now let us look closer at the effect due to the other terms.

Synchrotron radiation noise If the noise comes from synchrotron radiation (this does not have to be the case), the quantities $\langle \dot{N}\epsilon^n \rangle$ are related to the radiation spectral moment[17]

$$\langle \dot{N}u^n \rangle \;=\; \frac{5}{2\sqrt{3}}\,\frac{mc^2 r_0 \gamma}{\hbar\rho}\,\frac{3}{5\pi}\,\frac{2^n}{n+1}\,\Gamma\!\left(\frac{n}{2}+\frac{11}{6}\right)\Gamma\!\left(\frac{n}{2}+\frac{1}{6}\right)u_c^n\,,$$

where $\Gamma(x)$ is the Gamma function, and

$$u_c \;=\; \frac{3}{2}\,\frac{\hbar\gamma^3 c}{\rho}$$

is the critical photon energy of synchrotron radiation. For the case of synchrotron radiation noise, all coefficients D_n are positive.

To specify for the case of synchrotron radiation noise effect, we still need to connect the noise kick ϵ_k to the energy of the emitted photon u_k. This can be done using Table 1.1. Take the longitudinal motion for example, we have

$$\begin{aligned}
D_2 &= \frac{1}{2}\left(\frac{\eta c}{\omega_s E_0}\right)^2 \langle \dot{N}u^2 \rangle = \frac{55}{48\sqrt{3}}\,\frac{c^2\eta^2\gamma^5\hbar r_0}{m\rho^3\omega_s^2}\,,\\
D_4 &= \frac{1}{24}\left(\frac{\eta c}{\omega_s E_0}\right)^4 \langle \dot{N}u^4 \rangle = \frac{1309}{768\sqrt{3}}\,\frac{c^2\eta^4\gamma^9\hbar^3 r_0}{m^3\rho^5\omega_s^4}\,.
\end{aligned} \tag{1.52}$$

Effect of D_1 In the derivation of Eq. (1.46), we assumed $\int EP(E)\,dE = 0$. This means the noise excitations average to zero. This does not have to be the case. Consider the damped simple harmonic case (1.47) for example. If the D_1 term is included, we would have

$$\frac{\partial\psi}{\partial t} + \omega p\frac{\partial\psi}{\partial q} + (-\omega q - 2\alpha p)\frac{\partial\psi}{\partial p} \;=\; 2\alpha\psi + D_1\frac{\partial\psi}{\partial p} + D_2\frac{\partial^2\psi}{\partial p^2}\,.$$

The additional $D_1 = \langle \dot{N}\epsilon \rangle$ term can be absorbed into a redefinition of the coordinate q,

$$\bar{q} = q + \frac{D_1}{\omega}$$

$$\implies \quad \frac{\partial\psi}{\partial t} + \omega p\frac{\partial\psi}{\partial \bar{q}} + (-\omega\bar{q} - 2\alpha p)\frac{\partial\psi}{\partial p} \;=\; 2\alpha\psi + D_2\frac{\partial^2\psi}{\partial p^2}\,.$$

The effect of D_1 is removed by a shift of the coordinate q.

In case of the longitudinal synchrotron motion, this is the shift of the RF phase from the origin $\phi = 0$ to the synchronous phase ϕ_s. If we refer the longitudinal coordinate relative to the shifted RF phase, it is then equivalent to dropping D_1 from the Fokker–Planck equation.

We should make a subtle but important point of caution here. Mathematically, we have now clearly specified the effects of D_2 and D_1 in the Fokker–Planck

---

[17]A.W. Chao, Lectures on Accelerator Physics, World Scientific (2020), Chapter 6.

beam distribution — D_1 gives the distribution shift, D_2 gives the rms distribution spread. But what have we done really? Note particularly the fact that D_2 is still defined to be $\frac{1}{2}\langle \dot{N}\epsilon^2\rangle$. The nonzero average term D_1 only shifts the beam centroid position. It does not reduce the value of D_2, and as a result, it does not affect the overall Gaussian beam shape. It would, for example, be wrong if one considers the way to remove the noise average is to replace the noise ϵ_k by $\epsilon_k - \langle \epsilon\rangle$. Physically, this is because each of the noise excitations is by the full amount of ϵ_k, and the subsequent removal of the average D_1 term is done only after the excitation is fully smeared out by the rapid Hamiltonian flow.

It may be instructive to make an additional comment here on the physical origin of the reason that quantum diffusion is quantified by $\langle \dot{N}\epsilon^2\rangle$ instead of $\langle \dot{N}(\epsilon - \langle \epsilon\rangle)^2\rangle$.[18] Synchrotron radiation process, unlike a random walk process, has two sources of noise contributions, one coming from the random times of photon emissions, the other comes from the random energies of the photons at each emission. In synchrotron radiation, the former noise source dominates. Since the random radiation time dominates the diffusion, in a time period T long enough so that it contains many radiation events, the beam energy acquires a spread mainly due to a fluctuation of the number of radiation events among different electrons and not from a fluctuation of the many photon energies. As a result, the accumulated contribution to beam energy spread is determined by $\langle \dot{N}T\epsilon^2\rangle$ where $\dot{N}T$ is the expected number of photon emission events and ϵ is the full amount, without subtracting $\langle \epsilon\rangle$ from it, of the emitted photon energy. In particular, even when all emitted photons have the same energy ϵ, there will be a diffusion in the energy spread of the beam, and the same diffusion rate will apply.

In contrast, a random walk process contains a single noise source of the random step size. There is one random walk step at each time step and there is no noise introduced in the number of random walk steps. What dominates the diffusion process is the random step size. In this case, the net effect depends on whether the dynamics is in a 1-D physical space or a 2-D rotating phase space. In the 1-D physical space, the noise contribution to the dynamic variable is given by $\langle \dot{N}(\epsilon - \langle \epsilon\rangle)^2\rangle$. In a 2-D rotating phase space, since the oscillation amplitude depends on the total magnitude of ϵ regardless of $\langle \epsilon\rangle$, the diffusion is driven again by $\langle \dot{N}\epsilon^2\rangle$. In the present discussion, we emphasize that synchrotron radiation is a double-noise effect, not a random walk process which is a single-noise effect. The synchrotron radiation stochastic dynamics dictates that it is $\langle \dot{N}\epsilon^2\rangle$, not $\langle \dot{N}(\epsilon - \langle \epsilon\rangle)^2\rangle$, that determines the beam distribution.

Effect of D_4   As mentioned, after removing the D_1 effect by a centroid shift, the beam acquires a Gaussian distribution determined by the D_2 term. But we now want to see what the higher radiation spectral moments can do to the nominal Gaussian distribution.

---

[18]This point has been emphasized by M. Sands, Phys. Rev. 97, 470 (1955); J.M. Jowett, AIP Conf. Proc. 153, 864 (1987); X.J. Deng, et al., Phys. Rev. Accel. & Beams, 24, 094001 (2021); A.W. Chao, Lectures on Accelerator Physics, Chapter 6, World Scientific (2021).

Higher radiation moments such as D_4 are expected to move particles in the Gaussian core to the tails. Unlike D_1, they do affect the beam distribution and distorts it from Gaussian. As a result, the beam tail is enhanced in such a way that the higher distribution moments increase slightly. We will now demonstrate this effect.

If the D_4 term is included, we have

$$\frac{\partial \psi}{\partial t} + \omega p \frac{\partial \psi}{\partial q} + (-\omega q - 2\alpha p)\frac{\partial \psi}{\partial p} = 2\alpha\psi + D_2 \frac{\partial^2 \psi}{\partial p^2} + D_4 \frac{\partial^4 \psi}{\partial p^4}.$$

We look for a distorted stationary solution. We assume D_4 is small, and let the stationary solution be written as

$$\psi(q,p) = \frac{\alpha}{\pi D_2} \exp\left[-\frac{\alpha}{D_2}(q^2 + p^2)\right] \times e^{\xi(q,p)},$$

where $\xi(q,p)$ represents a small deviation with $|\xi| \ll 1$.

We look for an expression of $\xi(q,p)$ to 1st order in D_4. Let

$$q = a\sin\phi, \quad p = a\cos\phi.$$

For the effect of D_4, we look for a solution ξ that depends on a. This means both the unperturbed and the perturbed distributions depend only on a. The Fokker–Planck equation becomes, after some algebra,

$$-2\alpha a\psi'\cos^2\phi = 2\alpha\psi + D_2\left(\psi''\cos^2\phi + \frac{\psi'}{a}\sin^2\phi\right)$$
$$+ D_4\left[\cos^4\phi\,\psi'''' + 6\sin^2\phi\cos^2\phi\,\frac{\psi'''}{a} + 3\sin^2\phi(4\cos^2\phi - \sin^2\phi)\left(\frac{\psi'}{a^3} - \frac{\psi''}{a^2}\right)\right],$$

where a prime denotes a derivative with respect to a.

When $D_4 = 0$, it is easy to show that the solution is given by Eq. (1.48), i.e. $\xi(a) = 0$, as it should, and this solution is exact.

To solve for $\xi(a)$ when $D_4 \neq 0$, however, we need to average over ϕ. This yields

$$-\alpha a\psi' - 2\alpha\psi - \frac{D_2}{2}\left(\psi'' + \frac{\psi'}{a}\right) = D_4\left(\frac{3}{8}\psi'''' + \frac{3}{4}\frac{\psi'''}{a} - \frac{3}{8}\frac{\psi''}{a^2} + \frac{3}{8}\frac{\psi'}{a^3}\right).$$

Change variable from a to $x = \sqrt{\frac{\alpha}{D_2}}\,a$, and let

$$\psi(x) = \psi_0(x) + \Delta\psi(x), \quad \psi_0(x) = \frac{\alpha}{\pi D_2}e^{-x^2},$$

we have to 1st order in D_4,

$$-\alpha a\Delta\psi' - 2\alpha\Delta\psi - \frac{D_2}{2}\left(\Delta\psi'' + \frac{\Delta\psi'}{a}\right) \approx D_4\left(\frac{3}{8}\psi_0'''' + \frac{3}{4}\frac{\psi_0'''}{a} - \frac{3}{8}\frac{\psi_0''}{a^2} + \frac{3}{8}\frac{\psi_0'}{a^3}\right).$$

We also have

$$\Delta\psi(x) \approx \psi_0(x)\xi(x)\,.$$

After some simple algebra, we obtain (now with a prime denoting taking derivative with respect to x)

$$\frac{1}{2x}[(-1 + 2x^2)\xi' - x\xi''] \approx \frac{6\alpha^2 D_4}{D_2^2}\left(x^4 - 4x^2 + 2\right).$$

The solution is found to be

$$\xi(x) = \frac{6\alpha D_4}{D_2^2}\left(\frac{1}{2} - x^2 + \frac{x^4}{4}\right).$$

In terms of $\sigma = \sqrt{\frac{D_2}{2\alpha}}$, our final expression for the stationary beam distribution, to 1st order in D_4, is

$$\psi(q,p) = \frac{1}{2\pi\sigma^2}\exp\left[-\frac{q^2+p^2}{2\sigma^2} + \frac{3D_4}{\sigma^2 D_2}\left(\frac{1}{2} - \frac{q^2+p^2}{2\sigma^2} + \frac{(q^2+p^2)^2}{16\sigma^4}\right)\right]. \quad (1.53)$$

The validity of this expression requires

$$\frac{D_4}{\sigma^2 D_2} \ll 1\,.$$

If the noise comes from synchrotron radiation, and we are considering the longitudinal motion, then using expressions in (1.52), this validity criterion reads

$$\frac{D_4}{\sigma^2 D_2} = \frac{952}{275\sqrt{3}}\frac{u_c}{E_0} \ll 1\,, \quad (1.54)$$

which is easily fulfilled.

This distortion does not change the normalization, i.e. $\int dq \int dp\,\psi(q,p)$ remains unperturbed by D_4 to 1st order in D_4 (Homework 1.18). However, an inspection of Eq. (1.53) indicates that the beam is depleted around $q^2 + p^2 \sim \sigma^2$, while the beam center ($q^2 + p^2 \lesssim \sigma^2$) and beam tail ($q^2 + p^2 \gtrsim \sigma^2$) are enhanced. The overall rms of the beam distribution is unchanged. Higher distribution moments are increased, e.g. the fourth order and sixth order moments increase from the Gaussian values of $\langle(q^2 + p^2)^2\rangle = 8\sigma^4$ and $\langle(q^2 + p^2)^3\rangle = 48\sigma^6$ by the incremental amounts

$$\Delta\langle(q^2 + p^2)^2\rangle \approx 12\,\sigma^4\,\frac{D_4}{\sigma^2 D_2}\,,$$

$$\Delta\langle(q^2 + p^2)^3\rangle \approx 216\,\sigma^6\,\frac{D_4}{\sigma^2 D_2}\,. \quad (1.55)$$

The transfer of distribution from the beam core to the beam tail can have a significant effect on the quantum lifetime of the beam. A discussion of its effect on quantum lifetime is included on page 69. On the other hand, the quantity $\frac{D_4}{\sigma^2 D_2}$ is typically rather small.

Homework 1.17 Follow the derivation of the Fokker–Planck equation in the text, but this time fill in the steps to include the higher order correction terms and obtain Eq. (1.51). This straightforward exercise gives a feel of how the higher noise moments struggle to enter the stochastic equation when the leading D_2 term dominates. Validity of the central limit theorem is then physically connected to a convergence of a Taylor expansion series.

Homework 1.18

(a) Use Eq. (1.53) to show that, as stated in the text, to 1st order in $\frac{D_4}{\sigma^2 D_2}$, the normalization is unaffected by the distortion.

(b) Show that the rms beam sizes do not change either due to D_4.

(c) As the beam tail is enhanced, higher distribution moments are more seriously affected, and their changes will become increasingly positive toward higher orders. Verify Eq. (1.55).

Homework 1.19 For the case of synchrotron radiation noise, the text gives an expression (1.54) for the quantity $\frac{D_4}{\sigma^2 D_2}$ for the case of longitudinal motion. Follow the steps and Table 1.1 to give an expression of $\frac{D_4}{\sigma^2 D_2}$ for the horizontal betatron and the vertical betatron motions, and show that they too are typically very small quantities.

Homework 1.20 The text discussed the cases $D_1 \neq 0$ and $D_4 \neq 0$. We have not dealt with the case $D_3 \neq 0$. A more venturous reader might look into this case.

Solution This case cannot be dealt with by a $\Delta\psi$ that depends only on a as we did in the text for the case of D_4 effect. It will have to contain an angular modulation,

$$\frac{\Delta\psi}{\psi_0} = \xi(a) \begin{cases} \sin 3\phi, \\ \cos 3\phi. \end{cases}$$

1.4   Linear coupled system

The description so far is for a 1-D system. Fokker–Planck equation can be extended to 2- or 3-D systems following similar analysis. This is especially useful if two or more of the dimensions are coupled. The expectation is that, if the coupling is linear, the stationary solution to the coupled Fokker–Planck equation remains Gaussian in the multi-dimensional phase space. For the case of a 2-D decoupled system, consult Homework 1.13.

1.4.1   Fokker–Planck equation of a linear coupled system

In an electron storage ring, linear coupling can result from a skew quadrupole magnet, a solenoid, a crab cavity, or an RF cavity located at a dispersive location. In the following, we will assume the linear coupled equations of motion are known and can be written as

$$\dot{X} = CX + \sum_k E_k \delta(t - t_k).$$

For example, in the case of x- and y-motions, X is the column vector with canonical elements (x, p_x, y, p_y), C is the 4×4 coupling matrix, E_k is the vector describing the response in the four phase space coordinates to a single noise source ϵ_k that occurs at time $t = t_k$,

$$E_k = \epsilon_k \begin{bmatrix} \eta_1 \\ \eta_2 \\ \eta_3 \\ \eta_4 \end{bmatrix}.$$

The quantities $\eta_{1,2,3,4}$ specify the shock response of x, p_x, y, p_y to the single noise perturbation ϵ_k.

Near the coupling resonance $\omega_x \approx \omega_y$, C is parametrized by

$$C = \begin{bmatrix} 0 & \omega_x & -k_2 & 0 \\ -\omega_x & -2\alpha_x & -2k_1 & -k_2 \\ k_2 & 0 & 0 & \omega_y \\ -2k_1 & k_2 & -\omega_y & -2\alpha_y \end{bmatrix}, \tag{1.56}$$

where $\omega_{x,y}$ are the unperturbed horizontal and vertical betatron frequencies, $\alpha_{x,y}$ are the unperturbed damping rates, $k_{1,2}$ are two parameters that will later be connected to the real and imaginary parts of a coupling coefficient by Eq. (1.86). Consider only on-momentum particles; synchrotron motion is ignored in this section.

Derivation of the coupling matrix (1.56) is postponed to Sec. 1.4.2. Suffice to mention here that x and y here are the horizontal and vertical betatron displacements normalized by $\sqrt{\beta_x}$ and $\sqrt{\beta_y}$, respectively. As a result, the quantities $\langle x^2 \rangle, \langle y^2 \rangle$ are to be interpreted as the betatron emittances.

The diffusion effects are described by another symmetric matrix D whose elements are

$$D_{ij} = \frac{1}{2} \langle \dot{N} \epsilon^2 \rangle \eta_i \eta_j.$$

Let us consider the case when both the diffusion and the damping come from synchrotron radiation. In an ideal planar electron storage ring, the only non-vanishing elements of the D matrix are D_{22} and D_{44}, whose values have been given in Table 1.1 — D_{22} comes from quantum excitation through the horizontal dispersion; D_{44} comes from vertical sideways photon emission. In a general coupled system, however, a noise event simultaneously excites all four coordinates in a correlated manner, and all elements $D_{ij} \neq 0$. For example, photon emissions at a location with both a horizontal and a vertical dispersion would have $D_{22} \neq 0, D_{44} \neq 0$ and $D_{24} = D_{42} \neq 0$. In the remaining of this section, the general case is considered with an arbitrary diffusion matrix D.

The Fokker–Planck equation for this case can be derived similarly to that leading to Eq. (1.47). The resulting expression is

$$\frac{\partial \psi}{\partial t} + \sum_i \dot{x}_i \frac{\partial \psi}{\partial x_i} = 2(\alpha_x + \alpha_y)\psi + \sum_i \sum_j D_{ij} \frac{\partial^2 \psi}{\partial x_i \partial x_j},$$

or equivalently,

$$\frac{\partial \psi}{\partial t} + \sum_i \sum_j C_{ij} x_j \frac{\partial \psi}{\partial x_i} = -\sum_i C_{ii}\psi + \sum_i \sum_j D_{ij} \frac{\partial^2 \psi}{\partial x_i \partial x_j}. \tag{1.57}$$

Spectral higher moments of the radiation are ignored in Eq. (1.57).

Stationary solution Knowing the C and D matrices, the Fokker–Planck system is fully specified. We look for the stationary solution to Eq. (1.57). By declaring the solution shall be Gaussian, we have the ansatz

$$\psi_0 = Q \exp(-\tilde{X} A X), \tag{1.58}$$

where a tilde means taking the transpose of a matrix or a vector, A is a symmetric positive definite matrix yet to be found, and Q is the normalization constant not important to the following discussions.

Substituting the ansatz (1.58) into Eq. (1.57), and equating coefficients for terms containing the same $x_i x_j$, we obtain (Homework 1.22)

$$\widetilde{AC} + AC = -4\,ADA, \tag{1.59}$$

and

$$\text{trace}(C) = -2\,\text{trace}(DA). \tag{1.60}$$

There are $m(2m + 1)$ conditions and the same number of unknowns in Eq. (1.59), where m is the degree of freedom considered ($m = 2$ in the 2-D coupled case, 4-D phase space). The trace equation (1.60) is redundant since it follows simply by rewriting Eq. (1.59) as $A^{-1}\tilde{C}A + C = -4\,DA$ and then taking its trace. Solving Eq. (1.59) for A then yields the Gaussian beam distribution (1.58).

Second moment matrix Once the matrix A and the stationary beam distribution ψ_0 is obtained, the second-moment matrix of the beam distribution Σ can be obtained. The second moment Σ is a symmetric matrix with elements $\Sigma_{ij} = \langle x_i x_j \rangle$, and is related to the symmetric matrix A by

$$\Sigma = \frac{1}{2}A^{-1}. \tag{1.61}$$

We shall not prove this theorem here — it is actually not too difficult. Suffice it to mention that for the trivial 1×1 case with $\psi = e^{-x^2/2\sigma^2}$, this theorem predicts $\langle x^2 \rangle = \sigma^2$ as it should.

The second moment matrix Σ uniquely determines the Gaussian distribution in the phase space including all rms beam sizes and phase space orientations. The rms beam sizes are directly given by its diagonal elements. The tilt angle of the distribution projection onto the x_i-x_j plane is given by

$$\tan 2\theta_{ij} = \frac{2\langle x_i x_j \rangle}{\langle x_i^2 \rangle - \langle x_j^2 \rangle}. \tag{1.62}$$

Proof of Eq. (1.62) is also skipped here. One comment to make, however, is to be careful what the angle θ_{ij} means (or, perhaps even more importantly, what it does not mean) when x_i and x_j do not have the same physical dimensions.

Using expression (1.61), Eq. (1.59) can be rewritten as

$$\Sigma\tilde{C} + C\Sigma = -2D. \tag{1.63}$$

Although in principle it is straightforward to solve (1.63) for Σ_{ij}, it is useful and instructive to have a formal analytical solution of it. To do so, we diagonalize the matrix C by

$$C = V\lambda V^{-1}, \qquad \text{or} \qquad \tilde{C} = \tilde{V}^{-1}\lambda\tilde{V}, \tag{1.64}$$

where λ is the diagonal matrix whose elements are the eigenvalues of the matrix C and V is composed of the corresponding eigenvectors, i.e. the i-th column of V is given by the eigenvector corresponding to the eigenvalue λ_i. The real part of all eigenvalues $\mathrm{Re}(\lambda_i)$ must be negative (nonpositive, to be more accurate) for single particle motion to be stable.

Having established λ and V, the solution to Eq. (1.63) is found to be

$$\Sigma = VB\tilde{V}, \tag{1.65}$$

where

$$B_{ij} = -\frac{2}{\lambda_i + \lambda_j}[V^{-1}D\tilde{V}^{-1}]_{ij}. \tag{1.66}$$

The matrix B has the meaning of the symmetric second-moment matrix viewed in the eigenvector space.

The 1-D example　As a first application of this formalism, consider the 1-D case treated before with Eq. (1.48). We have

$$C = \begin{bmatrix} 0 & \omega \\ -\omega & -2\alpha \end{bmatrix},$$

$$D = \begin{bmatrix} 0 & 0 \\ 0 & D_2 \end{bmatrix}.$$

We then find

$$\lambda_{\pm} = -\alpha \pm i(\omega^2 - \alpha^2)^{1/2},$$

$$V = \begin{bmatrix} 1 & 1 \\ \frac{\lambda_+}{\omega} & \frac{\lambda_-}{\omega} \end{bmatrix},$$

$$B = \frac{D_2}{4}\frac{\omega^2}{\omega^2 - \alpha^2}\begin{bmatrix} \frac{1}{\lambda_+} & \frac{1}{\alpha} \\ \frac{1}{\alpha} & \frac{1}{\lambda_-} \end{bmatrix},$$

and the expected result

$$\Sigma \; = \; \frac{D_2}{2\alpha} \begin{bmatrix} 1 & 0 \\ 0 & 1 \end{bmatrix}. \tag{1.67}$$

The same result can of course also be obtained by algebraically solving the three simultaneous equations imbedded in Eq. (1.63).

RF phase jitter The damping and noise do not have to be limited to synchrotron radiation — for which the relevant parameters were listed in Table 1.1. For example, another source of noise could be due to a phase jitter of the RF system. Beam dynamics in the presence of RF noise is a rather extensive subject.[19] In one special case, however, when the RF noise has a white spectrum, the analysis simplifies.

A white noise in the RF phase translates to a white noise in the energy kicks as a particle passes by the RF system every turn,

$$\Delta\delta_i \; = \; \frac{eV_{\rm rf}\cos\phi_s}{E_0}\Delta\phi_i \qquad \Longrightarrow \qquad \Delta p_i \; = \; -\frac{\eta c}{\omega_s}\frac{eV_{\rm rf}\cos\phi_s}{E_0}\Delta\phi_i \, .$$

Since the RF phase jitter is uncorrelated to the synchrotron radiation noise, the RF phase jitter effect can be considered simply by adding to D_2 another contribution given by

$$(D_2)_{\rm RF\ noise} \; = \; \frac{1}{2T_0}\left(\frac{\omega_s C}{\omega_{\rm rf}}\right)^2 \langle\Delta\phi^2\rangle \, . \tag{1.68}$$

The rms beam energy spread and length will then be correspondingly larger due to the RF phase jitter. It is advisable to require $(D_2)_{\rm RF\ noise}$ be much smaller than the D_2 of Eq. (1.52).

Longitudinal 1-D case for an isochronous ring Consider the 1-D longitudinal dynamics of a storage ring whose momentum phase slippage factor η is near zero. The ring lattice is nearly isochronous. A conventional approach predicts an equilibrium bunch length approaching zero as $\sigma_z \propto \sqrt{|\eta|}$.[20] However, as pointed by Shoji, when $\eta \approx 0$, it is necessary to take into consideration of a partial-momentum-compaction effect.[21] The partial-momentum-compaction effect can be considered in terms of our formalism as follows.

We first note that the conventional result is readily obtained from Eq. (1.67) with

$$D_2 \; = \; \frac{1}{2}\langle\dot{N}\epsilon^2\rangle \; = \; \frac{1}{2}\langle\dot{N}u^2\rangle\frac{\eta^2 c^2}{\omega_s^2 E_0^2}$$

[19]G. Dôme, CERN Accel. School, Oxford (1985), p. 370; S. Krinsky and J.M. Wang, Part. Accel. 12, 107 (1982).

[20]M. Sands, SLAC report SLAC-121 (1970).

[21]Y. Shoji, H. Tanaka, M. Takao, and K. Soutome, Phys. Rev. E 54, R4556 (1996); K. Soutome, M. Takao, T. Tanaka, and Y. Shoji, Proc. 6th Euro. Part. Accel. Conf., Stockholm, 1998 (1998); X.J. Deng, et al., Phys. Rev. Accel. & Beams, 23, 044002 (2020); A.W. Chao, Lectures on Accelerator Physics, World Scientific (2020); Xiujie Deng, Thesis, Tsinghua Univ., Beijing (2021).

$$\implies \quad \sigma_z^2 = \frac{1}{4\alpha}\left\langle \dot{N}\frac{u^2}{E_0^2}\right\rangle \frac{\eta^2 c^2}{\omega_s^2}, \quad \sigma_\delta^2 = \frac{1}{4\alpha}\left\langle \dot{N}\frac{u^2}{E_0^2}\right\rangle. \tag{1.69}$$

The partial-momentum-compaction effect is included by adding an additional contribution D_{11} to the diffusion matrix, so that now

$$D = \begin{bmatrix} D_{11} & 0 \\ 0 & D_{22} \end{bmatrix}.$$

We then follow our formalism and obtain, when $\omega_s \gg \alpha$, (Homework 1.24)

$$\Sigma = \begin{bmatrix} \frac{D_{11}+D_{22}}{2\alpha} & 0 \\ 0 & \frac{D_{11}+D_{22}}{2\alpha} \end{bmatrix}. \tag{1.70}$$

The contribution from D_{22} comes from the conventional theory. The partial-momentum-compaction contributes to D_{11}. Taking both contributions into account, we obtain (see footnote 21)

$$\begin{aligned} \sigma_{z,\text{tot}}^2 &= \frac{\omega_s^2}{\eta^2 c^2}\sigma_{\delta,\text{tot}}^2 \\ &= \sigma_z^2 + \sigma_\delta^2\left(\Delta C^2 + \frac{1}{2}\mathcal{H}_1\mathcal{H}_2\right), \end{aligned} \tag{1.71}$$

where σ_z^2 and σ_δ^2 are given by Eq. (1.69);

$$\Delta C = \int_{s_1}^{s_2} ds\,\frac{D(s)}{\rho(s)}$$

is the partial momentum compaction at the observation position s_2 due to a photon emission at position s_1; and

$$\mathcal{H}_i = \frac{D_i^2 + (\alpha_i D_i + \beta_i D_i')^2}{\beta_i}, \quad i = 1, 2.$$

Observing at a fixed position s_2, we still need to average the result (1.71) over the emission position s_1. The result is to replace $\mathcal{H}_1$ of Eq. (1.71) by $\langle\mathcal{H}_1\rangle$, and ΔC^2 by $\langle\Delta C^2\rangle$, where

$$\langle\mathcal{H}_1\rangle = \frac{\oint ds_1\,\frac{\mathcal{H}_1(s_1)}{|\rho(s_1)|^3}}{\oint ds_1\,\frac{1}{|\rho(s_1)|^3}}, \qquad \langle\Delta C^2\rangle = \frac{\oint ds_1\,\frac{\Delta C(s_1)^2}{|\rho(s_1)|^3}}{\oint ds_1\,\frac{1}{|\rho(s_1)|^3}}.$$

This averaging over s_1 of course naturally takes into account of the radiation power by including the weighting function $\frac{1}{|\rho(s_1)|^3}$.

Taking into account of the additional contribution from the partial momentum compaction effect, Fig. 1.10 gives a comparison among the conventional calculation, the calculation applying Shoji's analysis, a calculation using the 6-D SLIM analysis, and computer simulation. Deviation from the conventional theory near the isochronous point $\eta = 0$ is rather drastic.

Figure 1.10: A comparison among the conventional calculation, the calculation applying Shoji's analysis, a calculation using the 6-D SLIM analysis, and a computer simulation. The isochronous lattice uses the MLS storage ring as an example. Parameters used include a beam energy of 630 MeV and an RF voltage of 50 MV. The observation point has $\mathcal{H}_2 = 0$. The upper panel shows the rms bunch length. The lower panel shows the rms energy spread. In contrast to the conventional theory, the bunch length does not drop to zero when $\eta = 0$, while the energy spread diverges there. [Courtesy Xiujie Deng (2020).]

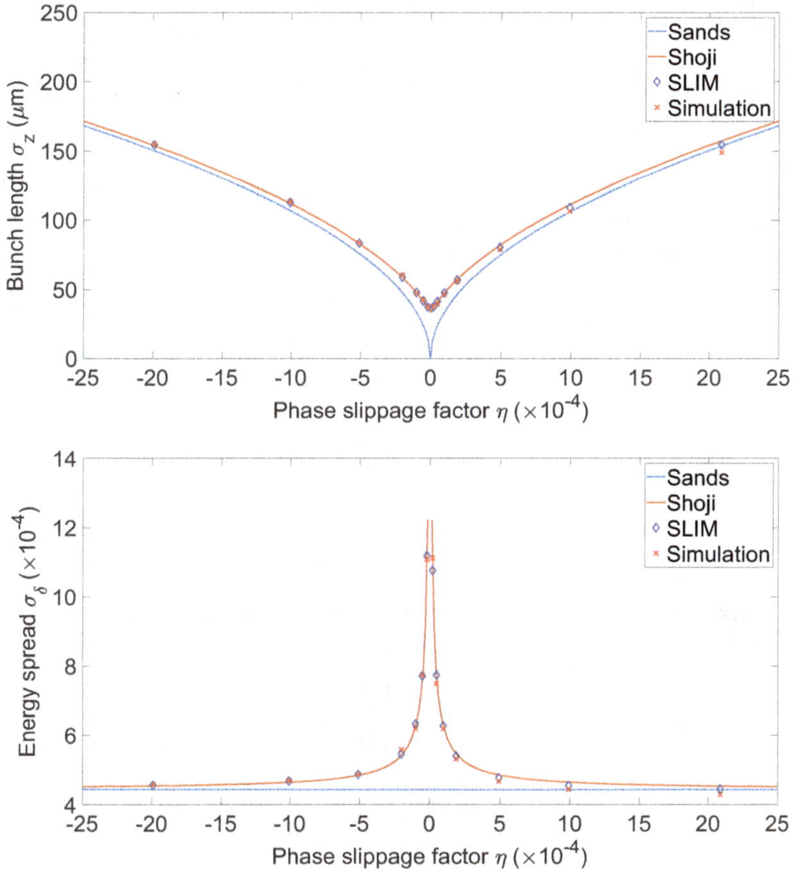

2-D x-y coupled example   A nontrivial 2-D application is for the case of x-y coupled motion when the horizontal and vertical betatron frequencies are close to each other.[22] The coupling matrix C has been given by Eq. (1.56). We

---

[22]Other 2-D coupled systems include, for example, synchrobetatron coupling when the RF system is installed at a dispersive location or when a crab cavity is installed.

consider a diffusion matrix given by

$$
D \;=\; D_2 \begin{bmatrix} 0 & 0 & 0 & 0 \\ 0 & \langle \eta_x^2 \rangle_B & 0 & \langle \eta_x \eta_y \rangle_B \\ 0 & 0 & 0 & 0 \\ 0 & \langle \eta_x \eta_y \rangle_B & 0 & \langle \eta_y^2 \rangle_B \end{bmatrix} ,
$$

where the angular brackets $\langle\ \rangle_B$ refer to averaging over the noise sources around the storage ring. We have assumed the one noise effect D_2 drives both p_x and p_y with different couplings η_x and η_y — but we also assume this noise source does not drive x or y. The subscript B is introduced to indicate averaging over bending magnets if the noise source is synchrotron radiation.[23]

Here we will ignore the nondiagonal correlation terms, $\langle \eta_x \eta_y \rangle_B \approx 0$, assuming they, unlike the diagonal terms, oscillate rapidly and average to zero around the storage ring — see Homework 1.25(b) to see what happens when they are included.

The analysis was given by a Chao–Lee formalism.[24] Given matrices C and D, we calculate the stationary beam distribution second moments. Skipping the algebra (Homework 1.25), we give the final result here,

$$
\langle x^2 \rangle \;=\; \frac{\langle x^2 \rangle_0}{1 + \kappa_y} + \frac{\alpha_y}{\alpha_x}\,\frac{\kappa_x \langle y^2 \rangle_0}{1 + \kappa_x} ,
$$

$$
\langle y^2 \rangle \;=\; \frac{\alpha_x}{\alpha_y}\,\frac{\kappa_y \langle x^2 \rangle_0}{1 + \kappa_y} + \frac{\langle y^2 \rangle_0}{1 + \kappa_x} , \tag{1.72}
$$

$$
\langle xy \rangle \;=\; \mathrm{Re}(\kappa')\left(\langle x^2 \rangle_0 - \langle y^2 \rangle_0 \right) ,
$$

where

$$
\langle x^2 \rangle_0 \;=\; \frac{D_2 \langle \eta_x^2 \rangle_B}{2\alpha_x} \qquad \text{and} \qquad \langle y^2 \rangle_0 \;=\; \frac{D_2 \langle \eta_y^2 \rangle_B}{2\alpha_y}
$$

give the horizontal and vertical betatron emittances in the absence of coupling; $\kappa_x, \kappa_y, \kappa'$ are coupling constants defined as

$$
\kappa_x \;=\; \frac{\frac{\alpha_x}{\alpha_y}(\alpha_x + \alpha_y)|k|^2}{(\alpha_x + \alpha_y)k^2 + \alpha_x \Delta\omega^2} ,
$$

$$
\kappa_y \;=\; \frac{\frac{\alpha_y}{\alpha_x}(\alpha_x + \alpha_y)|k|^2}{(\alpha_x + \alpha_y)k^2 + \alpha_y \Delta\omega^2} , \tag{1.73}
$$

$$
\kappa' \;=\; \frac{\alpha_x \alpha_y \Delta\omega}{(\alpha_x + \alpha_y)^2 |k|^2 + \alpha_x \alpha_y \Delta\omega^2}\,(k_1 - ik_2) ,
$$

where $|k|^2 = k_1^2 + k_2^2$, and $\Delta\omega = \omega_x - \omega_y$ is the small difference between the horizontal and vertical betatron frequencies. Quantities k_1 and k_2 are those that

---

[23]Be reminded that the noises are applied to p_x and p_y. Physically, no noise can give sudden kicks to x and y. In case one is considering the artificially defined "betatron" components only, and the noise is considered to be applied as kicks in the coordinates x_β and y_β and not their corresponding momenta, then a switching of roles of coordinates and momenta are required as done in Table 1.1.

[24]A.W. Chao and M.J. Lee, J. Appl. Phys. 47, 4453 (1976).

appeared in the coupling matrix C. We will give explicit expressions for them in the next Section 1.4.2 once a distribution of skew quadrupoles and solenoids are given. At a slight risk of confusion, here we have introduced three more coupling related constants κ_x, κ_y, and κ' by Eq. (1.73).

The result given by Eqs. (1.72) and (1.73) are derived under the condition

$$\omega_{x,y} \gg \left(|\omega_x - \omega_y|, |k|\right) \gg \alpha_{x,y}. \tag{1.74}$$

Physically, Eq. (1.74) requires the unperturbed x- and y-betatron motions be much faster than the coupling motion between them that yields the x-y exchange, while at the same time, it requires the coupling motion be much faster than the damping effects,

$$\begin{pmatrix}\text{Motion} \\ \text{within } x \text{ or } y\end{pmatrix} \gg \begin{pmatrix}\text{Motion exchang-} \\ \text{ing } x \text{ and } y\end{pmatrix} \gg \begin{pmatrix}\text{Damping and} \\ \text{diffusion effects}\end{pmatrix}.$$

Equation (1.72) gives the result of three second moments of the beam's Gaussian distribution. As to the remaining seven second moments, under the same approximation (1.74), they are given by[25]

$$\langle p_x^2 \rangle = \langle x^2 \rangle, \quad \langle p_y^2 \rangle = \langle y^2 \rangle, \quad \langle p_x p_y \rangle = \langle xy \rangle,$$
$$\langle x p_x \rangle = -\langle y p_y \rangle = 0, \quad \langle p_x y \rangle = -\langle x p_y \rangle = \mathrm{Im}(\kappa')(\langle x^2 \rangle_0 - \langle y^2 \rangle_0).$$

Coupling constant κ_y specifies the amount of beam emittance that is transferred from x- to y-dimension. Similarly, κ_x specifies the emittance transfer from y- to x-dimension. Coupling constant κ' is complex; its real and imaginary parts determine $(\langle xy \rangle, \langle p_x p_y \rangle)$ and $(\langle x p_y \rangle, \langle p_x y \rangle)$, respectively, which in turn are associated with the tilting of the beam distribution in the x-y and the x-p_y spaces. The other quantities $\langle x^2 \rangle, \langle y^2 \rangle, \langle p_x^2 \rangle, \langle p_y^2 \rangle$ involve only $|k|^2$.

It follows from Eqs. (1.72) and (1.73) that if the uncoupled beam has equal x- and y-emittances, the beam remains round and it does not tilt in the x-y plane due to linear coupling.

Equation (1.72) can be combined to yield an invariant condition,

$$\alpha_x \langle x^2 \rangle + \alpha_y \langle y^2 \rangle \ = \ \alpha_x \langle x^2 \rangle_0 + \alpha_y \langle y^2 \rangle_0. \tag{1.75}$$

The quantity on the left-hand-side is therefore an invariant, independent of the linear coupling. As mentioned, here x and y have been normalized by $\sqrt{\beta_x}$ and $\sqrt{\beta_y}$, respectively, so $\langle x^2 \rangle$ and $\langle y^2 \rangle$ represent the betatron emittances.

Special case when $\alpha_x = \alpha_y$ and $\eta_y = 0$ Equations (1.72) and (1.73) allows the calculation of the beam second moments when a distribution of skew

---

[25]Although both $\langle x p_x \rangle$ and $\langle y p_y \rangle$ are approximately zero, we have kept their relative signs in this expression because their next order terms have equal magnitude but opposite signs. Incidentally, the reason $\langle x p_x \rangle, \langle y p_y \rangle$ vanish here is because we normalized x and y by $\sqrt{\beta_{x,y}}$, so the Courant–Snyder phase space evolution has been removed.

quadrupole or solenoids are specified in a storage ring. One special case of interest is when $\alpha_x = \alpha_y$ and $\eta_y = 0$, which is approximately valid for strong focusing separate-function planar rings. In that case, we have

$$\begin{aligned}
\langle x^2 \rangle &= \frac{\langle x^2 \rangle_0}{1 + \kappa_y}\,, \\
\langle y^2 \rangle &= \frac{\kappa_y \langle x^2 \rangle_0}{1 + \kappa_y}\,, \\
\langle xy \rangle &= \mathrm{Re}(\kappa')\,\langle x^2 \rangle_0\,,
\end{aligned} \tag{1.76}$$

with

$$\begin{aligned}
\kappa_y &= \frac{2k^2}{2k^2 + \Delta\omega^2}\,, \\
\kappa' &= \frac{\Delta\omega(k_1 - ik_2)}{4k^2 + \Delta\omega^2}\,.
\end{aligned}$$

The invariant condition (1.75) becomes

$$\langle x^2 \rangle + \langle y^2 \rangle = \langle x^2 \rangle_0\,. \tag{1.77}$$

The tilt angle of the beam distribution in the x-y plane is given by

$$\tan 2\theta = \frac{2k_1}{\Delta\omega}\,. \tag{1.78}$$

This expression in turn should give an indication of the physical meaning of the coupling parameter k_1. In particular, we later will give an expression of k_1; it follows that the beam tilt in the x-y plane oscillates as k_1 oscillates around the storage ring. The beam tilts by $45°$ in the strong coupling limit $|k_1| \gg \Delta\omega$.

Equations (1.76) and (1.77) are familiar results where the coupling constant κ_y is regarded as phenomenological parameter. Application of Fokker–Planck equation provides a way of obtaining explicit expressions of these coupling constants $\kappa_x, \kappa_y, \kappa'$ and calculating them deterministically.

Homework 1.21 Equation (1.57) ignores all spectral moments other than the second moments. The higher moments are usually small unless one considers far tails in the beam distribution. The first moments are expected to be removed by a shift of the beam distribution centroid. This shift of beam centroid was discussed on page 35 for the 1-D case (2-D phase space). Follow a similar analysis to derive an expression of this centroid shift in the 2-D linearly coupled case (4-D phase space) when Eq. (1.57) is applied.

Homework 1.22 Derive Eqs. (1.59) and (1.60).

Homework 1.23 Follow the reasoning outlined in the text, derive Eq. (1.68) for the diffusion coefficient to due RF phase white noise.

Homework 1.24

(a) Follow the 1-D analysis of the text, derive Eq. (1.70).

(b) An alert reader may have noted that in the derivation we have not included $D_{12} = D_{21}$ in the diffusion matrix. Extend the calculation to include D_{12} and D_{21}. What are their effect on the beam distribution?

Solution (b) Follow the text and first show that

$$
\Sigma = \begin{bmatrix} \dfrac{4\alpha^2 D_{11} + 4\alpha\omega D_{12} + (D_{11}+D_{22})\omega_s^2}{2\alpha\omega_s^2} & -\dfrac{D_{11}}{\omega_s} \\[2ex] -\dfrac{D_{11}}{\omega_s} & \dfrac{D_{11}+D_{22}}{2\alpha} \end{bmatrix} .
$$

Then take the limit $\omega_s \gg \alpha$. In this limit, D_{12} and D_{21} do not contribute to the beam distribution.

Homework 1.25

(a) Readers willing to venture the algebra may derive Eqs. (1.72) and (1.73).

(b) When doing (a), include the contribution from the off-diagonal diffusion $\langle \eta_x \eta_y \rangle_B$.

(c) If skipping (a) and (b), at least verify the invariant condition (1.75).

Solution

(a) You will find that the formal solution (1.65) becomes cumbersome. It will be easier to solve directly Eq. (1.63). Also, don't forget to make the approximation (1.74).

(b) When $\langle \eta_x \eta_y \rangle_B$ terms are included in the diffusion matrix D, the result (1.72) becomes

$$
\langle x^2 \rangle = \frac{\langle x^2 \rangle_0}{1 + \kappa_y} + \frac{\alpha_y}{\alpha_x} \frac{\kappa_x \langle y^2 \rangle_0}{1 + \kappa_x} + \mathrm{Re}(\kappa') \frac{D_2}{\alpha_x} \langle \eta_x \eta_y \rangle_B ,
$$

$$
\langle y^2 \rangle = \frac{\alpha_x}{\alpha_y} \frac{\kappa_y \langle x^2 \rangle_0}{1 + \kappa_y} + \frac{\langle y^2 \rangle_0}{1 + \kappa_x} - \mathrm{Re}(\kappa') \frac{D_2}{\alpha_y} \langle \eta_x \eta_y \rangle_B ,
$$

$$
\langle xy \rangle = \langle p_x p_y \rangle = \mathrm{Re}(\kappa') \left[(\langle x^2 \rangle_0 - \langle y^2 \rangle_0) + \left(\frac{1}{\alpha_x} + \frac{1}{\alpha_y} \right) \frac{k_1 D_2}{\Delta\omega} \langle \eta_x \eta_y \rangle_B \right] ,
$$

$$
\langle p_x y \rangle = -\langle x p_y \rangle = \mathrm{Im}(\kappa') \left[(\langle x^2 \rangle_0 - \langle y^2 \rangle_0) + \left(\frac{1}{\alpha_x} + \frac{1}{\alpha_y} \right) \frac{k_1 D_2}{\Delta\omega} \langle \eta_x \eta_y \rangle_B \right] ,
$$

where the coupling constants $\kappa_x, \kappa_y, \kappa'$ are still those defined in Eq. (1.73).

1.4.2 Coupling matrix

Here we include a discussion as well as a derivation of the coupling matrix Eq. (1.56) for the case of x-y coupling due to skew quadrupoles and solenoids.

Symplecticity The system is Hamiltonian when we switch off the damping and diffusion effects. Let us examine this Hamiltonian system first. The equation of motion is given by

$$
\dot{X} = CX ,
$$

where C is the coupling matrix (1.56) with $\alpha_{x,y}$ set to zeroes,

$$C = \begin{bmatrix} 0 & \omega_x & -k_2 & 0 \\ -\omega_x & 0 & -2k_1 & -k_2 \\ k_2 & 0 & 0 & \omega_y \\ -2k_1 & k_2 & -\omega_y & 0 \end{bmatrix}. \tag{1.79}$$

It is easy to show that this equation of motion follows as the Hamilton equations with the Hamiltonian

$$H(x, p_x, y, p_y, t) = \frac{\omega_x}{2}(x^2 + p_x^2) + \frac{\omega_y}{2}(y^2 + p_y^2) + 2k_1 xy + k_2(xp_y - yp_x). \tag{1.80}$$

The solution to the equation of motion is given by

$$X(t) = M(t)X_0, \qquad \text{where} \qquad M(t) = e^{Ct},$$

with X_0 the initial condition at time $t = 0$. The matrix $M(t)$ is the transport map from $t = 0$ to t and is symplectic (Homework 1.26), satisfying

$$\widetilde{M(t)}SM(t) = S \quad \text{for all } t,$$

with S the symplectic form,

$$S = \begin{bmatrix} 0 & 1 & 0 & 0 \\ -1 & 0 & 0 & 0 \\ 0 & 0 & 0 & 1 \\ 0 & 0 & -1 & 0 \end{bmatrix}. \tag{1.81}$$

Derivation   Basically we need to prove that for the present linear x-y coupled system, its Hamiltonian is given by Eq. (1.80) when damping and diffusion effects are removed. This Hamiltonian is valid near the linear coupling difference resonance $\omega_x - \omega_y \approx n\omega_0$, or equivalently we define two reference tunes ν_{x0}, ν_{y0} such that $\nu_x \approx \nu_{x0}, \nu_y \approx \nu_{y0}$, and $\nu_{x0} - \nu_{y0} = n$ (n is an integer). The x-y coupling is driven by a distribution of skew quadrupole field $k(s) = \frac{1}{B\rho}\frac{\partial B_y}{\partial y}$ and we need to connect the parameters k_1 and k_2 to $k(s)$.

Following a *smooth approximation*,[26] the smoothed Hamiltonian is given by

$$K(\phi_x, J_x, \phi_y, J_y, s) = \frac{J_x}{R}(\nu_x - \nu_{x0}) + \frac{J_y}{R}(\nu_y - \nu_{y0})$$
$$+ \frac{\sqrt{J_x J_y}}{R}\left[ge^{i\phi_x - i\phi_y} + g^* e^{-i\phi_x + i\phi_y}\right],$$

where the canonical transformation is provided by

$$x = \sqrt{2\beta_x(s)J_x}\cos\left[\int_0^s \frac{ds'}{\beta_x(s')} - (\nu_x - \nu_{x0})\frac{s}{R} + \phi_x\right],$$
$$y = \sqrt{2\beta_y(s)J_y}\cos\left[\int_0^s \frac{ds'}{\beta_y(s')} - (\nu_y - \nu_{y0})\frac{s}{R} + \phi_y\right], \tag{1.82}$$

---

[26]For a derivation, see A.W. Chao, Lectures on Accelerator Physics, Chapter 3, World Scientific (2020).

with $\beta_{x,y}$ the beta-functions of the storage ring lattice design; $\phi_{x,y}$ and $J_{x,y}$ are slow variables now serving as the new canonical coordinates and momenta.

We have defined a dimensionless[27] complex coupling coefficient,[28]

$$g = \frac{1}{4\pi} \int_0^{2\pi R} ds \sqrt{\beta_x(s)\beta_y(s)}\, k(s)$$

$$\times \exp\left[i \int_0^s ds' \left(\frac{1}{\beta_x(s')} - \frac{1}{\beta_y(s')} \right) - i(\nu_x - \nu_y - n)\frac{s}{R} \right]. \quad (1.83)$$

For Eq. (1.83) to apply, we have assumed the observation point is at $s = 0$.

We shall designate $g = g_R + i g_I$. To avoid confusion, we shall refer to g as the "coupling coefficient", and $\kappa_x, \kappa_y, \kappa'$ in Eq. (1.73) as the "coupling constants". Our next job is to find k_1 and k_2 in terms of g_R and g_I.

Given the smoothed Hamiltonian, the new set of Hamilton equations gives

$$\frac{d\phi_x}{ds} = \frac{1}{R}(\nu_x - \nu_{x0}) + \frac{1}{2R}\sqrt{\frac{J_y}{J_x}}\left(g e^{i\phi_x - i\phi_y} + g^* e^{-i\phi_x + i\phi_y} \right),$$

$$\frac{d\phi_y}{ds} = \frac{1}{R}(\nu_y - \nu_{y0}) + \frac{1}{2R}\sqrt{\frac{J_x}{J_y}}\left(g e^{i\phi_x - i\phi_y} + g^* e^{-i\phi_x + i\phi_y} \right),$$

$$\frac{dJ_x}{ds} = -\frac{dJ_y}{ds} = -\frac{\sqrt{J_x J_y}}{R}\left(i g e^{i\phi_x - i\phi_y} - i g^* e^{-i\phi_x + i\phi_y} \right).$$

Keep in mind that a smooth approximation has been made, and the result is valid to 1st order in g, i.e. valid to 1st order in the skew quadrupole strength.

Consider the case of weak coupling when $\nu_x \approx \nu_y \approx \nu_0$ (i.e. when $n = 0$), normalize x by $\sqrt{\frac{\nu_x \beta_x}{R}}$ and similarly for y, and let $\theta = \frac{s}{R}$, the smoothed variables become

$$x \approx \sqrt{J_x}\cos(\nu_{x0}\theta + \phi_x), \quad y \approx \sqrt{J_y}\cos(\nu_{y0}\theta + \phi_y), \quad (1.84)$$

where $\phi_{x,y}, J_{x,y}$ are slowly varying with θ, as evidenced by their respective Hamilton equations given above — all terms on the right-hand-sides are small.

It then follows a straightforward algebra (Homework 1.27) that, near the resonance $\nu_x \approx \nu_y \approx \nu_0$ and to 1st order in the coupling strength,

$$\frac{d^2 x}{d\theta^2} + \nu_x^2 x \approx -2\nu_0 g_R y + 2g_I \frac{dy}{d\theta},$$

$$\frac{d^2 y}{d\theta^2} + \nu_y^2 y \approx -2\nu_0 g_R x - 2g_I \frac{dx}{d\theta},$$

---

[27] It actually has the dimensionality of the tune.

[28] For completeness, if there are solenoids $B_s(s)$ in addition to skew quadrupoles, a more general expression of g is (in this expression, $\alpha_{x,y}$ are the Courant–Snyder functions)

$$g = \frac{1}{4\pi} \int_0^{2\pi R} ds \sqrt{\beta_x(s)\beta_y(s)}\left[k(s) - \frac{B_s(s)}{2}\left(\frac{\alpha_x}{\beta_x} - \frac{\alpha_y}{\beta_y} \right) + i\frac{B_s}{2}\left(\frac{1}{\beta_x} + \frac{1}{\beta_y} \right) \right]$$

$$\times \exp\left[i \int_0^s ds' \left(\frac{1}{\beta_x(s')} - \frac{1}{\beta_y(s')} \right) - i(\nu_x - \nu_y - n)\frac{s}{R} \right].$$

See G. Guignard, CERN report CERN-ISR-MA/75-23 (1975).

or

$$\begin{aligned}
\ddot{x} + \omega_x^2 x &\approx -2\omega_0^2 \nu_0 g_R y + 2\omega_0 g_I \dot{y}, \\
\ddot{y} + \omega_y^2 y &\approx -2\omega_0^2 \nu_0 g_R x - 2\omega_0 g_I \dot{x},
\end{aligned}$$

$$(1.85)$$

where ω_0 is the revolution angular frequency.

This set of equations of motion in the (x, p_x, y, p_y) phase space is described by the Hamiltonian (1.80) where

$$k_1 = \omega_0 g_R, \qquad k_2 = -\omega_0 g_I,$$

or equivalently,

$$k_1 - i k_2 = \omega_0 g.$$

$$(1.86)$$

We have now the explicit expressions of k_1 and k_2 in terms of the coupling coefficient g.

We may substitute this result (1.86) into the coupling constants, Eq. (1.73), to obtain

$$\begin{aligned}
\kappa_x &= \frac{\frac{\bar{\alpha}_x}{\bar{\alpha}_y}(\bar{\alpha}_x + \bar{\alpha}_y)|g|^2}{(\bar{\alpha}_x + \bar{\alpha}_y)|g|^2 + \bar{\alpha}_x \Delta\nu^2}, \\
\kappa_y &= \frac{\frac{\bar{\alpha}_y}{\bar{\alpha}_x}(\bar{\alpha}_x + \bar{\alpha}_y)|g|^2}{(\bar{\alpha}_x + \bar{\alpha}_y)|g|^2 + \bar{\alpha}_y \Delta\nu^2}, \\
\kappa' &= \frac{\bar{\alpha}_x \bar{\alpha}_y g \Delta\nu}{(\bar{\alpha}_x + \bar{\alpha}_y)^2 |g|^2 + \bar{\alpha}_x \bar{\alpha}_y \Delta\nu^2},
\end{aligned}$$

where $\bar{\alpha}_{x,y} = \frac{\alpha_{x,y}}{\omega_0}$ specify the radiation damping rates.

Observing at a different position   The beam distribution we obtained so far refers to an observation point $s = 0$ with the coupling coefficient g given by Eq. (1.83). When observed at a different position $s = s^*$, by the smooth approximation, the coupling coefficient changes by a phase factor,

$$g\big|_{s=s^*} = g\big|_{s=0} \times \exp\left[i\int_0^{s^*} ds' \left(\frac{1}{\beta_x(s')} - \frac{\nu_x}{R}\right) - i\int_0^{s^*} ds' \left(\frac{1}{\beta_y(s')} - \frac{\nu_y}{R}\right) \right].$$

$$(1.87)$$

The magnitude $|g|^2$ does not change when changing the observation point. Only the phase of g changes. It follows from our result that $\langle xy \rangle$, $\langle p_x p_y \rangle$, $\langle x p_y \rangle$ and $\langle p_x y \rangle$ — the tilt angle related quantities — will depend on the point of observation, while $\langle x^2 \rangle$, $\langle y^2 \rangle$, $\langle p_x^2 \rangle$, $\langle p_y^2 \rangle$ — the emittance related quantities — are invariant around the storage ring. This is what one would expect.

As the beam circulates around the storage ring, its tilt angle in the x-y projection varies according to Eq. (1.78),

$$\tan 2\theta(s) = \frac{2 g_R(s)}{\nu_x - \nu_y - n}.$$

The reader is suggested to consider how $g_R(s)$ is expected to behave around the ring by observing Eq. (1.87). In particular, θ repeats itself one turn around the accelerator circumferences, i.e. $\theta(C) = \theta(0)$.

Homework 1.26 Consider the case when damping terms $\alpha_{x,y}$ are removed and the coupling matrix is given by Eq. (1.79).

(a) Show that the Hamilton equations of the Hamiltonian (1.80) is given by $\dot{X} = CX$.

(b) Write the Hamiltonian (1.80) as $\frac{1}{2}\tilde{X}HX$ with a symmetric matrix H, show that $C = SH$.

(c) Prove that the matrix $M(t) = e^{Ct}$ is symplectic for all time t.

Solution

(a)

$$H = \begin{bmatrix} \omega_x & 0 & 2k_1 & k_2 \\ 0 & \omega_x & -k_2 & 0 \\ 2k_1 & -k_2 & \omega_y & 0 \\ k_2 & 0 & 0 & \omega_y \end{bmatrix}.$$

(c) It needs only to be shown that $\tilde{C}S + SC = 0$.

Homework 1.27

(a) When x and y are normalized by $\sqrt{\beta_{x,y}}$ as in Eq. (1.84), show that the equation of motion (1.85) follows from the Hamiltonian $K(\phi_x, J_x, \phi_y, J_y, s)$ to 1st order in coupling strengths.

(b) In the derivation, note the interesting fact that the slowly varying ϕ_x and J_x each contributes exactly half of the right-hand-side of the $\ddot{x}$ equation, and similarly for the $\ddot{y}$ equation.

Homework 1.28

(a) Equation (1.85) defines a linearly coupled system with two modes. Find their two eigenfrequencies $\Omega_\pm$.

(b) These eigenfrequencies define the perturbed horizontal and vertical betatron frequencies in the coupled system. Consider a storage ring with given distribution of skew quadrupoles, and therefore given coupling coefficient g. Now measure the two perturbed betatron frequencies as the unperturbed horizontal betatron frequency is scanned through the coupling resonance $\omega_x = \omega_y$. Find an expression of the split $\Omega_+ - \Omega_-$ between the perturbed betatron frequencies. Show that during the scan this split goes through a minimum when exactly on resonance. Show that this minimum split is given by

$$(\Omega_+ - \Omega_-)_{\min} = 2\omega_0|g|.$$

Assume the coupling is weak.

The minimum tune split is simply given by twice the coupling coefficient. This offers a handy way to measure the coupling coefficient $|g|$. The phase of g, on the other hand, has to be measured by beam tilts at various locations around the storage ring.

Solution

$$\Omega_\pm^2 \;=\; \frac{1}{2}\left[\omega_x^2 + \omega_y^2 + 4\omega_0^2 g_I^2 \pm \sqrt{(\omega_x^2 - \omega_y^2)^2 + 8\omega_0^2 g_I^2(\omega_x^2 + \omega_y^2) + 16\omega_0^4 \nu_0^2 g_R^2}\;\right].$$

Homework 1.29 Given the Hamiltonian (1.80), or equivalently the coupling matrix (1.79), we know the problem is all analytically soluble. In particular, there are two invariants of the motion expressed in terms of quadratic forms,

$$g_i \;=\; \tilde{X} G_i X, \qquad i = 1, 2\,,$$

where $G_{1,2}$ are two symmetric positive definite matrices. The stationary beam distribution in the phase space can be expressed in terms of them by a Gaussian form.

Let C of Eq. (1.79) be diagonalized by Eq. (1.64) with eigenvector matrix V. The explicit expressions of $G_{1,2}$ are given by[29]

$$G_i \;=\; -iSVD_0H_iD_0\tilde{V}S\,, \qquad i \;=\; 1, 2, \tag{1.88}$$

where S is given by Eq. (1.81), and

$$D_0 = \begin{bmatrix} 1 & 0 & 0 & 0 \\ 0 & -i & 0 & 0 \\ 0 & 0 & 1 & 0 \\ 0 & 0 & 0 & -i \end{bmatrix}, \quad H_1 = \begin{bmatrix} 0 & 1 & 0 & 0 \\ 1 & 0 & 0 & 0 \\ 0 & 0 & 0 & 0 \\ 0 & 0 & 0 & 0 \end{bmatrix}, \quad H_2 = \begin{bmatrix} 0 & 0 & 0 & 0 \\ 0 & 0 & 0 & 0 \\ 0 & 0 & 0 & 1 \\ 0 & 0 & 1 & 0 \end{bmatrix}.$$

(a) Work out explicit expressions of these two invariants for the x-y coupled system in the presence of a distribution of skew quadrupoles.

(b) Express the equilibrium beam distribution in the phase space in terms of the two invariants g_1 and g_2.

1.5   Transient beam distribution

Another application of the Fokker–Planck equation is to find the transient beam behavior. Consider an electron beam injected into a storage ring with an initial distribution ψ_i that is different from the equilibrium Gaussian distribution ψ_0. As time goes on, it evolves towards the equilibrium form ψ_0 but it has to go through a transient. This transient evolution of beam distribution is discussed in this section.

Solve the Fokker–Planck equation   Consider the general Fokker–Planck equation (1.57) for a linear coupled system. We have shown that the equilibrium beam distribution of this system is a specific Gaussian form given by Eq. (1.58). The matrix A — or rather, the second-moment matrix Σ — is

[29] A.W. Chao, Lectures on Accelerator Physics, Sec. 6.5.7, World Scientific (2020).

solved by Eq. (1.63) or Eqs. (1.65–1.66). A similar technique will now be used to describe the transient.

At time $t = 0$, we assume the initial distribution is a displaced Gaussian,

$$\psi(0) \ = \ Q(0) \exp\left[-(\tilde{X} - \tilde{X}_0(0))A(0)(X - X_0(0))\right],$$

where $Q(0)$ is the normalization constant, $X_0(0)$ is the column vector specifying the initial beam center, and $A(0)$ is the symmetric distribution matrix at $t = 0$ — we are expecting time dependence on all these quantities Q, A and X_0. We assume $Q(0)$, $A(0)$, $X_0(0)$, and the matrices C and D are known. Here the matrix C contains the damping terms, i.e. $\alpha_{x,y}$ have not been set to zeroes.

It turns out that if the initial distribution is Gaussian, the beam will remain Gaussian at all times during the transient period. The distribution at time t can be written as

$$\psi(t) \ = \ Q(t) \exp\left[-(\tilde{X} - \tilde{X}_0(t))A(t)(X - X_0(t))\right]. \tag{1.89}$$

Substituting Eq. (1.89) into Eq. (1.57) and equating the coefficients containing the same $x_i x_j$ factors, we obtain

$$-\dot{A} \ = \ \widetilde{AC} + AC + 4ADA, \tag{1.90}$$

$$\dot{X}_0 \ = \ CX_0, \tag{1.91}$$

$$\frac{\dot{Q}}{Q} \ = \ \text{trace}(C) + 2\,\text{trace}(DA). \tag{1.92}$$

Equations (1.90) and (1.92) can be compared with the corresponding equations (1.59) and (1.60) for the equilibrium distribution. It is interesting to note that Eq. (1.91), which describes the motion of the beam distribution center, acts as if the beam is regarded as a single particle ignoring diffusion but take into account of the damping effects contained in the matrix C. The beam center thus executes a deterministic, damped, linearly coupled motion as dictated by the matrix C, starting with the initial value $X_0(0)$.

Like Eq. (1.59), there are $m(2m + 1)$ conditions and unknowns in Eq. (1.90). Equation (1.91) has $2m$ conditions for $2m$ unknowns. The two equations (1.90) and (1.91) for A and X_0 are decoupled from each other. Equation (1.92) can be shown to be redundant given Eq. (1.90).

Second moments Equation (1.90) can be rewritten in terms of the matrix $\Sigma = \frac{1}{2}A^{-1}$, yielding

$$\dot{\Sigma} \ = \ \Sigma\tilde{C} + C\Sigma + 2D. \tag{1.93}$$

We then diagonalize C by Eq. (1.64) to obtain

$$\dot{B} \ = \ B\lambda + \lambda B + 2V^{-1}D\tilde{V}^{-1}, \tag{1.94}$$

where B is related to Σ by Eq. (1.65), and is to be solved with the initial condition

$$B(0) \ = \ V^{-1}\Sigma(0)\tilde{V}^{-1}.$$

The solution is

$$B_{ij}(t) \;=\; B_{ij}(\infty) + [B_{ij}(0) - B_{ij}(\infty)]\exp[(\lambda_i + \lambda_j)t]\,, \qquad (1.95)$$

with the asymptotic values $B_{ij}(\infty)$ given in Eq. (1.66). A time $t = 0$, Σ is equal to the initial value $\Sigma(0)$. As $t \to \infty$, the exponential term diminishes and the beam distribution approaches that of the equilibrium.

Beam center The solution to Eq. (1.91) for the beam center can be written as

$$X_0(t) \;=\; VU(t)\,,$$

with

$$\begin{aligned}
u_i(t) &\;=\; u_i(0)e^{\lambda_i t}\,, \\
U(0) &\;=\; V^{-1}X_0(0)\,.
\end{aligned} \qquad (1.96)$$

The initial displacement is $X_0(0)$, while its asymptotic displacement vanishes, as expected.

The 1-D example, transient after injection The transient problem is now formally solved. As an example of its application, consider the earlier 1-D case of page 42. Let the injected beam have

$$\psi(0) \;=\; \delta(q - q_i)\delta(p - p_i)\,, \qquad (1.97)$$

i.e. the beam with vanishing size is injected off-centered by (q_i, p_i). The transient behavior found below for this initial condition is in fact the Green's function of the system. Knowing the Green's function, the transient behavior of arbitrary initial distributions can be obtained by superposition.

Distribution (1.97) can be described as a Gaussian with

$$X_0(0) \;=\; \begin{bmatrix} q_i \\ p_i \end{bmatrix},$$

$$\Sigma(0) \;=\; \begin{bmatrix} 0 & 0 \\ 0 & 0 \end{bmatrix}, \qquad \text{or} \qquad B(0) \;=\; \begin{bmatrix} 0 & 0 \\ 0 & 0 \end{bmatrix}.$$

With C, D, V, λ and $B(\infty)$ given on page 42 and following the prescription discussed above, the time-dependent solution is found to be

$$X_0(t) \;=\; \begin{bmatrix} q_0(t) \\ p_0(t) \end{bmatrix} = e^{-\alpha t} \begin{bmatrix} (\alpha q_i + \omega p_i)\frac{\sin \Omega t}{\Omega} + q_i \cos \Omega t \\ -(\alpha p_i + \omega q_i)\frac{\sin \Omega t}{\Omega} + p_i \cos \Omega t \end{bmatrix}, \qquad (1.98)$$

and

$$B(t) \;=\; \frac{\omega^2 D_2}{4\Omega^2} \begin{bmatrix} \dfrac{1-e^{2\lambda_+ t}}{\lambda_+} & \dfrac{1-e^{\lambda_+ t + \lambda_- t}}{\alpha} \\ \dfrac{1-e^{\lambda_+ t + \lambda_- t}}{\alpha} & \dfrac{1-e^{2\lambda_- t}}{\lambda_-} \end{bmatrix},$$

where $\Omega = \sqrt{\omega^2 - \alpha^2}$. The distribution moments are given by $\Sigma(t) = VB(t)\tilde{V}$, or explicitly,

$$\langle (q(t) - q_0(t))^2 \rangle = \Sigma_{11}(t)$$
$$= \frac{\omega^2 D_2}{2\alpha\Omega^2}(1 - e^{-2\alpha t}) - \frac{\alpha D_2}{2\Omega^2}(1 - 2^{-2\alpha t}\cos 2\Omega t) - \frac{D_2}{2\Omega}e^{2\Omega t}\sin 2\Omega t\,,$$
$$\langle (q(t) - q_0(t))(p(t) - p_0(t)) \rangle = \Sigma_{12}(t) = \Sigma_{21}(t)$$
$$= \frac{\omega D_2}{2\Omega^2}(1 - \cos 2\Omega t)e^{-2\alpha t}\,, \tag{1.99}$$
$$\langle (p(t) - p_0(t))^2 \rangle = \Sigma_{22}(t)$$
$$= \frac{\omega^2 D_2}{2\alpha\Omega^2}(1 - e^{-2\alpha t}) - \frac{\alpha D_2}{2\Omega^2}(1 - 2^{-2\alpha t}\cos 2\Omega t) + \frac{D_2}{2\Omega}e^{2\Omega t}\sin 2\Omega t\,.$$

As a check, note that $\Sigma_{11}(0) = \Sigma_{12}(0) = \Sigma_{21}(0) = \Sigma_{22}(0) = 0$, and $\Sigma_{11}(\infty) = \Sigma_{22}(\infty) = \frac{D}{2\alpha}$ and $\Sigma_{12}(\infty) = \Sigma_{21}(\infty) = 0.$[30]

The beam moments thus simply approach exponentially from the initial values to the equilibrium values. The second member of Eq. (1.99) is interesting to observe; it says that in order to reach the final state, the beam does not remain upright during the transient even though it is upright at $t = 0$ and again upright at $t \to \infty$; furthermore, its transient has a definitive sign.

Figure 1.11 illustrates the transient behavior of the injected beam using expressions (1.98) and (1.99). The beam has vanishing size and its center is displaced at time $t = 0$. The beam size diffuses towards its equilibrium value with rate 2α as the beam center damps out with rate α.

For the case of weak damping, we may keep only terms proportional to $\frac{1}{\alpha}$. Our results acquire simpler forms,

$$X_0(t) \approx e^{-\alpha t}\begin{bmatrix} p_i\sin\omega t + q_i\cos\omega t \\ -q_i\sin\omega t + p_i\cos\omega t \end{bmatrix},$$
$$\Sigma(t) \approx \frac{D_2}{2\alpha}\begin{bmatrix} 1 - e^{-2\alpha t} & 0 \\ 0 & 1 - e^{-2\alpha t} \end{bmatrix}. \tag{1.100}$$

In the more general case when the injected beam has finite size, the right-hand-side of Eq. (1.100) acquires additional terms $\Sigma_{ij}(0)e^{-2\alpha t}$ which indicate the initial knowledge of the beam size is exponentially damped out.

Although not included here, an extension to the case of beam transient in the presence of a 2-D linear coupling can be treated using the framework discussed above.

Homework 1.30

(a) Use Eq. (1.90) to prove Eq. (1.93).
(b) Show Eq. (1.94).
(c) Show Eq. (1.95).

---

[30] In general, chaining up an expression like $\Sigma_{11} = \Sigma_{12}\cdots$ is not allowed even if all terms are zero because all these terms do not have the same dimensionality. Here, however, it is allowed because we have chosen q and p, which have been normalized to have the same dimension.

Figure 1.11: Illustration of the transient behavior of a point-sized beam after its injection. The solid curve illustrates its centroid motion. The two dashed curves outline the beam size of $\pm\sqrt{\langle q^2 \rangle}$ around the centroid.

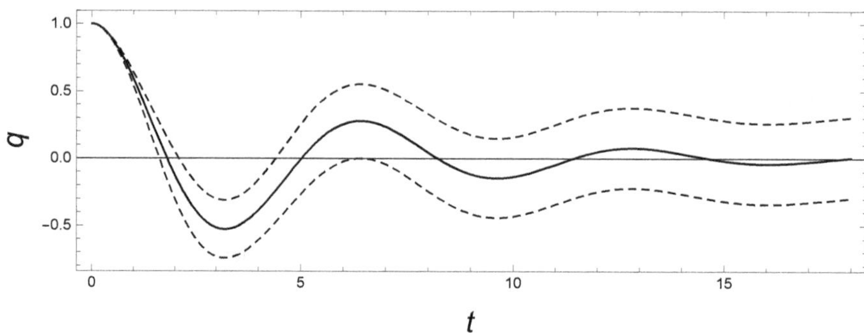

(d) Show Eq. (1.96).

(e) It may be instructive also to derive Eq. (1.99).

1.6   Quantum lifetime

We have shown the Fokker–Planck equation (1.47) has a bi-Gaussian equilibrium solution (1.48). However, this solution assumes that there is no physical limit on the value of q and p, which in reality means ignoring objects such as the vacuum chamber wall in the storage ring. In this section, we will take into consideration of the effect of this vacuum chamber wall.

Considering this bi-Gaussian equilibrium distribution, although the overall distribution looks stationary, individual particles are not. More specifically, they rapidly circulate around the phase angle ϕ and slowly change their amplitudes a due to diffusion and damping, while in doing so, keeping the overall beam distribution in a stationary state. In the presence of a vacuum chamber wall, the change in a makes it possible for particles to hit the wall and consequently be removed from the beam. As the process of particles' hitting the wall continues, even though slowly, the beam distribution cannot stay stationary; instead, it will decay with time. The corresponding beam lifetime is called the quantum lifetime.

In terms of the Fokker–Planck equation, the new ingredient is that we have to impose the extra condition that the distribution vanishes at the wall boundary. In the following, we will calculate the three quantum lifetimes for the three dimensions of motion. The simplest case is that for the vertical y-motion, so it will be discussed first. The more complicated cases of longitudinal and horizontal quantum lifetimes will be discussed subsequently. As we will see, an insightful heuristic physical argument can be used to derive the vertical quantum lifetime without invoking the Fokker–Planck equation.[31] The same argument does not apply to the more complex cases of longitudinal or horizontal limits or when there is x-y coupling. In these lectures, however, we shall discuss this heuristic argument and generalize it in Sec. 1.6.2, and then to apply it to the longitudinal case in Sec. 1.6.3 and to the horizontal-longitudinal coupled case in Sec. 1.6.4.

1.6.1 Vertical quantum lifetime

The rapid rotation in phase angle ϕ in particle motion provides a simplification of the problem. Figure 1.12 illustrates this point. Panel (a) shows the particle motion in phase space when there is no wall. Panel (b) shows a wall at the physical space $q = A$ removing all particles with $q > A$ without restriction on the momentum p. Panel (c) shows the removal of all particles with amplitude $a = \sqrt{q^2 + p^2} > A$ due to the rapid angular streaming in the phase space. The simplification is to consider the wall to be $a = A$ instead of $q = A$.

Figure 1.12: Particle motion in phase space with and without a wall boundary. The shaded regions indicate the wall.

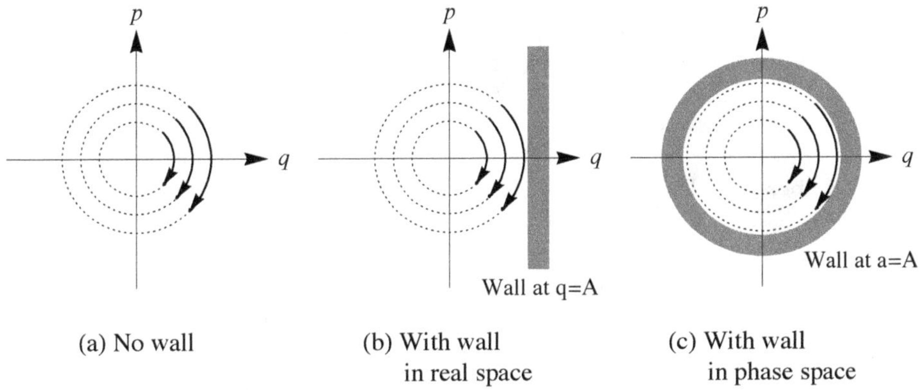

Fokker–Planck equation With this simplification, we first transform the problem from the Cartesian coordinates (q, p) to the polar coordinates (ϕ, a) according to Eq. (1.8). Expecting a quantum lifetime much longer than the phase space rotation period, the beam distribution is accurately given by a function of a and t only, i.e. $\psi = \psi(a, t)$, independent of phase angle ϕ. Substituting Eq. (1.8) into the Fokker–Planck equation (1.47) and averaging the resulting equation over ϕ give

$$\frac{\partial \psi}{\partial t} = 2\alpha\psi + \alpha a \frac{\partial \psi}{\partial a} + \frac{D_2}{2a} \frac{\partial}{\partial a}\left(a \frac{\partial \psi}{\partial a}\right). \tag{1.101}$$

---

[31]See M. Sands, The Physics of Electron Storage Rings, An Introduction, SLAC Report 121 (1970). See also A.W. Chao, Lectures on Accelerator Physics, World Scientific (2020).

The averaging over ϕ is valid only if we are interested in slow effects such as the effect of quantum lifetime. As will be seen later when we discuss the normal modes in Sec. 1.7, this simplification does not apply in general.

The boundary condition imposed by the wall is

$$\psi(a = A, t) \; = \; 0 \,, \tag{1.102}$$

for all time t. Equation (1.101) is to be solved together with (1.102).

Normal mode To find the quantum lifetime, we need to find the normal modes of the system. In this context, a normal mode is a particular pattern in phase space, each characterized by a particular value of lifetime. A beam injected into the storage ring at time $t = 0$ can be decomposed into a superposition of these normal modes. As time proceeds after injection, only the component in the mode with the longest lifetime remains. Quantum lifetime is thus identified to be the lifetime of the longest living mode.

Backtracking a bit, in the language of normal modes, our discussion of transient beam distribution in Sec. 1.5 can be viewed accordingly. Viewed in that context, the initial beam distribution is decomposed into the normal modes, and the longest living mode is the equilibrium mode with infinite lifetime, and all other modes die off as transient. In this case, although all the higher order modes die off quickly, the longest living mode lives forever, and the beam's quantum lifetime is identified to be infinite. The fundamental mode is the one with the stationary distribution.

Return to the case when a wall is imposed. Consider one normal mode with mode number n and lifetime τ_n. Let the normal mode be written as

$$\psi^{(n)}(a, t) \; = \; \psi_n(a)\, e^{-t/\tau_n} \,, \tag{1.103}$$

i.e. the mode has a fixed overall shape in phase space given by $\psi_n(a)$, but its overall magnitude diminishes with a lifetime τ_n. Substituting (1.103) into (1.101) gives

$$\psi_n'' + \left(\frac{1}{a} + \frac{2\alpha a}{D_2} \right) \psi_n' + \frac{2}{D_2} \left(2\alpha + \frac{1}{\tau_n} \right) \psi_n \; = \; 0 \,. \tag{1.104}$$

We need to look for the normal modes of the system by solving Eq. (1.104) together with the boundary condition $\psi_n(a = A) = 0$.

A more complete discussion of the normal modes, including the ϕ-dependencies, is delayed to Sec. 1.7. In this section, when ϕ is averaged over, we need only to look for the even modes with $n = $ even. With that in mind, at a slight risk of confusion, let us rewrite n as $2n$ from here on, and let

$$\psi_{2n}(a) \; = \; \exp\left(-\frac{a^2}{2\sigma^2} \right) g_{2n}(a) \,, \tag{1.105}$$

where $\sigma^2 = \frac{D_2}{2\alpha}$ is the natural rms beam size in the absence of the wall. Substituting into (1.104) leads to

$$g_{2n}'' + \left(\frac{1}{a} - \frac{a}{\sigma^2} \right) g_{2n}' + \frac{1}{\tau_{2n}\alpha\sigma^2} g_{2n} \; = \; 0 \,. \tag{1.106}$$

We further let

$$g_{2n}(a) = 1 + \sum_{k=1}^{\infty} c_k \left(\frac{a^2}{2\sigma^2} \right)^k . \qquad (1.107)$$

Note the expression of $g_{2n}(a)$ contains only even powers of a. The coefficients c_k's can be obtained by substituting (1.107) into (1.106) and identifying terms with equal powers in a. This leads to the recurrence equation,

$$c_{k+1} = \frac{1}{(k+1)^2} \left(k - \frac{1}{2\alpha\tau_{2n}} \right) c_k ,$$

which gives the solution

$$c_k = \frac{1}{(k!)^2} \left[(k-1) - \frac{1}{2\alpha\tau_{2n}} \right] \left[(k-2) - \frac{1}{2\alpha\tau_{2n}} \right] \cdots \left[-\frac{1}{2\alpha\tau_{2n}} \right] . \qquad (1.108)$$

The eigenvalue τ_{2n} is determined by the $(n+1)$-th solution for $g_{2n}(A) = 0$. There are an infinite number of solutions for τ_{2n}. The mode with the lowest value of τ_{2n} is identified to be the $n = 0$ mode. The quantum beam lifetime is identified to be τ_0.

The case without wall   In the absence of wall, Eq. (1.107) gives a spectrum of normal modes that have terminated power series. The first few are given below,

$$n = 0, \ \tau_0 = \infty, \ g_0(a) = 1 ,$$

$$n = 1, \ \tau_2 = \frac{1}{2\alpha}, \ g_2(a) = 1 - \frac{a^2}{2\sigma^2} ,$$

$$n = 2, \ \tau_4 = \frac{1}{4\alpha}, \ g_4(a) = 1 - \frac{a^2}{\sigma^2} + \frac{1}{2} \left(\frac{a^2}{2\sigma^2} \right)^2 ,$$

$$n = 3, \ \tau_6 = \frac{1}{6\alpha}, \ g_6(a) = 1 - \frac{3a^2}{2\sigma^2} + \frac{3}{2} \left(\frac{a^2}{2\sigma^2} \right)^2 - \frac{1}{6} \left(\frac{a^2}{2\sigma^2} \right)^3 . \qquad (1.109)$$

The case $n = 0$ is of course the stationary Gaussian distribution found earlier.

In general, $\tau_{2n} = \frac{1}{2n\alpha}$, and $g_{2n}(a)$ is an $2n$-th power polynomial of a. The phase space distribution ψ_{2n} of the n-th even mode is related to g_{2n} by Eq. (1.105). They satisfy the normalization condition

$$\int_0^{\infty} a \, da \, \psi_{2n}(a) = 0, \quad \text{except} \quad n = 0 . \qquad (1.110)$$

This normalization assures that as the n-th mode decays with rate $\tau_{2n} = \frac{1}{2n\alpha}$, the overall normalization is not affected. This normalization condition (1.110) is necessary for an internal consistency of our results.

As mentioned, the mode sequence (1.109) is an even sequence. There exists another sequence of odd modes which cannot be discussed by the phase-averaged equation (1.104). A general discussion can be found in Sec. 1.7.

Figure 1.13: Even ordered normal modes of the Fokker–Planck equation without boundary wall. The distributions ψ_{2n} are plotted versus $\frac{a}{\sigma}$.

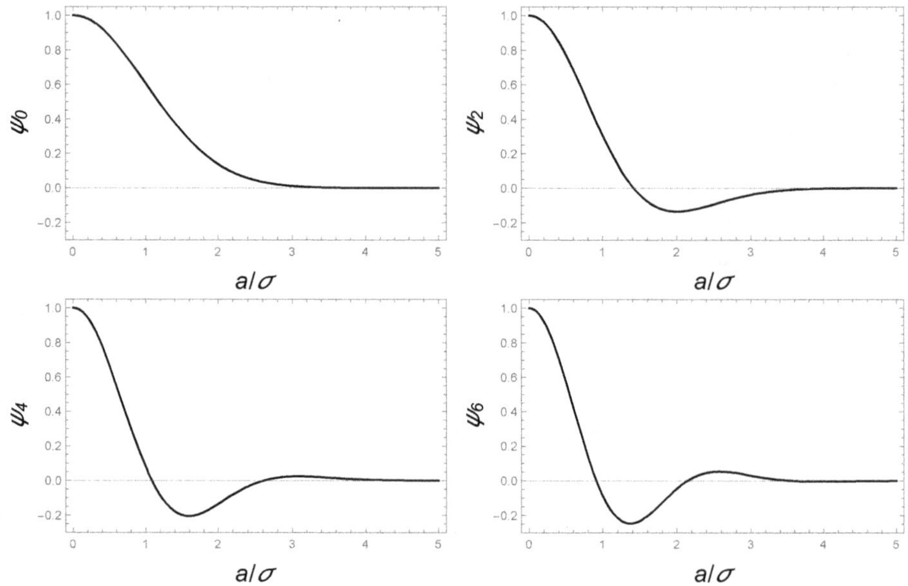

The lifetimes for $n \neq 0$ modes are due to the damping motion and not due to the wall boundary. The quantum lifetime, i.e. the lifetime for the $n = 0$ mode, is $\tau_0 = \infty$. Figure 1.13 shows the distributions ψ_{2n} as functions of $\frac{a}{\sigma}$ for a few lowest modes.

The case with wall   When the wall is imposed, we need to invoke the boundary condition $g_{2n}(A) = 0$. In most practical cases, the wall is far from the beam core, meaning $A \gg \sigma$. Under this condition, the quantum lifetime is much longer than the damping time $\frac{1}{\alpha}$, i.e. $\alpha\tau_0 \gg 1$. Equation (1.108) for the $n = 0$ mode becomes

$$c_k = \frac{1}{k\,(k!)}\left(\frac{-1}{2\alpha\tau_0}\right),$$

which gives

$$g_0(a) = 1 - \frac{1}{2\alpha\tau_0}h\left(\frac{a^2}{2\sigma^2}\right), \tag{1.111}$$

where we have defined a function

$$h(x) = \sum_{k=1}^{\infty}\frac{x^k}{k\,(k!)}.$$

The function $h(x)$ can also be written as

$$h(x) = \int_0^x dy \, \frac{e^y - 1}{y}. \tag{1.112}$$

For $x \gg 1$, it can be shown that $h(x) \approx \frac{e^x}{x}$.

The boundary condition $g_0(A) = 0$ gives, using Eq. (1.111),

$$\tau_0 = \frac{1}{2\alpha} h\left(\frac{A^2}{2\sigma^2}\right),$$

which in turn gives the familiar expression of quantum lifetime when $A \gg \sigma$ (see footnote 31),

$$\tau_0 \approx \frac{\sigma^2}{\alpha A^2} \exp\left(\frac{A^2}{2\sigma^2}\right). \tag{1.113}$$

The validity criterion that $\alpha\tau_0 \gg 1$ is safely assured by $A \gg \sigma$. For example, if $A = 6\,\sigma$, the quantum lifetime is 1.8×10^6 times the damping time $\frac{1}{\alpha}$.

Later in Eq. (8.91) of Chapter 8, we will give another expression of the quantum lifetime (1.113) with the factor $\frac{A^2}{\sigma^2}$ expressed in terms of the phase space area that is allowed by the aperture and measured in units of the rms beam emittance. The quantity $\frac{A^2}{\sigma^2}$ then acquires another important physical meaning.

Combining the results so far for the $n = 0$ mode, we have

$$\psi^{(0)}(a,t) = \exp\left(-\frac{t}{\tau_0} - \frac{a^2}{2\sigma^2}\right)\left[1 - \frac{h\left(\frac{a^2}{2\sigma^2}\right)}{h\left(\frac{A^2}{2\sigma^2}\right)}\right], \quad (a < A). \tag{1.114}$$

The quantity in the square brackets, for several values of $\frac{A}{\sigma}$, is plotted in Fig. 1.14.

From Fig. 1.14, it can be seen that the distribution is accurately given by the Gaussian form until a gets close to the wall $a = A$ where the distribution drops precipitously to zero. Note, however, that the drop approaching very close to the wall is not an abrupt cut-off as it might appear in Fig. 1.14; it follows a well behaved functional behavior according to Eq. (1.114). A magnified view of the distribution for the case $\frac{A}{\sigma} = 8$ near the wall is shown in Fig. 1.15.

In particular, the slope immediately in front of the wall has a well-defined value, not an abrupt slope approaching infinity. A further examination of the beam distribution at the wall shows that (Homework 1.33)

$$\psi^{(0)'}(A,t) \approx -\frac{A}{\sigma^2} \exp\left(-\frac{t}{\tau_0} - \frac{A^2}{2\sigma^2}\right). \tag{1.115}$$

Equation (1.115), together with the fact that the drop of the distribution immediately before the wall is not abrupt but gradual as dictated by it, is rather amazing. One immediately notes that, other than the overall damping factor e^{-t/τ_0}, it is the same as the slope of the original Gaussian distribution before

Figure 1.14: The beam distribution factor $1 - \dfrac{h(\frac{a^2}{2\sigma^2})}{h(\frac{A^2}{2\sigma^2})}$ plotted as function of $\frac{a}{\sigma}$ for several values of $\frac{A}{\sigma}$.

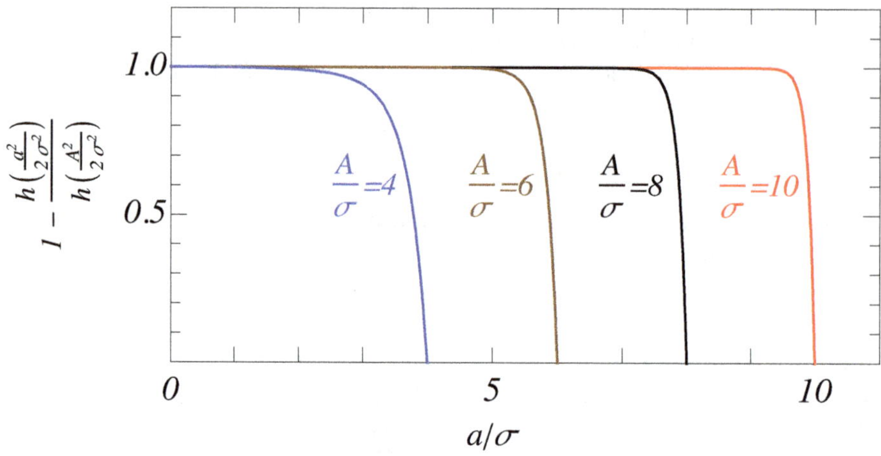

Figure 1.15: The beam distribution factor $1 - \dfrac{h(\frac{a^2}{2\sigma^2})}{h(\frac{A^2}{2\sigma^2})}$ plotted as function of $\frac{a}{\sigma}$ near the boundary wall for the case $\frac{A}{\sigma} = 8$. The distribution drops to zero at the wall in a well-defined manner with a very specific value of its slope.

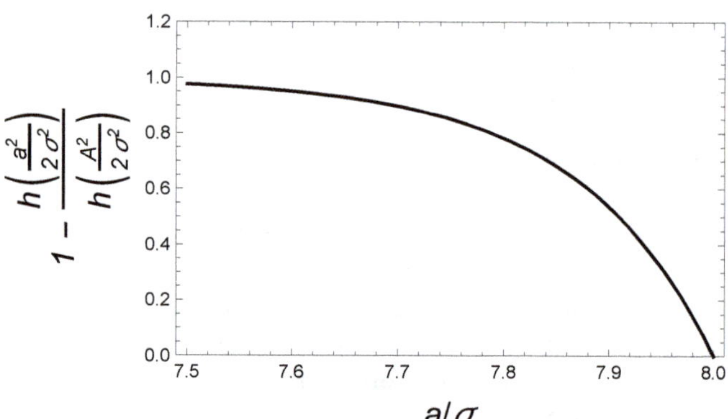

the wall was installed. The presence of the wall does not affect the *slope* of the beam distribution at the wall. In other words, as the distribution itself is so drastically cut by the wall, the cut is done in such a way that the slope at the wall is unaffected.

This is a very insightful observation, and as will be discussed in the following section, it is that physical insight that led to the heuristic argument used to derive the approximate quantum lifetime expression (1.113) without invoking the Fokker–Planck equation (see footnote 31).

Arbitrary wall position   Equation (1.113) is valid when $A \gg \sigma$ and $\tau_0 \gg \frac{1}{\alpha}$, i.e. when the wall is far away from the beam core. This is not too far wrong when A approaches σ but stays above it. On the other hand, it predicts that the quantum lifetime reaches a minimum at $A = \sqrt{2}\sigma$ with $\tau_0 = \frac{e}{2\alpha}$, which cannot be correct because smaller A is bound to give an even shorter lifetime. Equation (1.113) must break down when A approaches $\sim \sqrt{2}\sigma$.

So let us investigate the other extreme with the wall cuts into the beam core, $A \ll \sigma$. In this case, an approximate expression of quantum lifetime can be obtained by noting $\tau_0 \ll \frac{1}{\alpha}$, which implies, using Eq. (1.108),

$$c_k \approx \frac{1}{(k!)^2}\left(\frac{-1}{2\alpha\tau_0}\right)^k .$$

One notices that these coefficients are those appearing in the Taylor expansion of the Bessel function J_0. Indeed, substituting into Eq. (1.107), gives

$$g_0(a) = J_0\left(\frac{a}{\sigma\sqrt{\alpha\tau_0}}\right). \tag{1.116}$$

The boundary condition $g_0(A) = 0$ then gives

$$\tau_0 = \frac{1}{\alpha}\left(\frac{A}{\sigma\lambda_1}\right)^2, \tag{1.117}$$

where $\lambda_1 = 2.405$ is the first root of $J_0(x)$. The distribution (1.116) can also be written as $g_0(a) = J_0(\lambda_1 \frac{a}{A})$.

We conclude that the quantum lifetime is given by Eq. (1.113) when $A \gg \sigma$ and Eq. (1.117) when $A \ll \sigma$. One can actually calculate the exact quantum lifetime as a function of $\frac{A}{\sigma}$. A moment's reflection says that we have already solved a few special cases for the quantum lifetime when we wrote down Eq. (1.109), namely[32]

$$\tau_0 = \begin{cases} \frac{1}{2\alpha}, & \text{when } \frac{A}{\sigma} = \sqrt{2}, \\[4pt] \frac{1}{4\alpha}, & \text{when } \frac{A}{\sigma} = \sqrt{4 - 2\sqrt{2}} = 1.082, \\[4pt] \frac{1}{6\alpha}, & \text{when } \frac{A}{\sigma} = 0.912. \end{cases} \tag{1.118}$$

These special cases are just those given in Eq. (1.109) when the wall happens to be located at the first node of the corresponding eigen-distribution.

Without solving the problem for arbitrary $\frac{A}{\sigma}$, Fig. 1.16 shows the two asymptotic expressions, dashed curve for $A \gg \sigma$ and dotted curve for $A \ll \sigma$, as well as the few special cases as the circular dots. As mentioned, the dashed curve shows the existence of a minimum, which is unphysical.

---

[32]The value 0.912, for example, in this equation represents the first root of the equation $1 - \frac{3}{2}x^2 + \frac{3}{8}x^4 - \frac{1}{48}x^6 = 0$. The result is $\sqrt{6 - 4\sqrt{3}\sin\left[\frac{\pi}{6} + \frac{1}{3}\cos^{-1}\frac{1}{\sqrt{3}}\right]} = 0.912$.

Figure 1.16: Quantum lifetime plotted as a function of the wall aperture. The dashed and dotted curves show the asymptotic behavior for $A \gg \sigma$ and $A \ll \sigma$, respectively. The three circular dots are the exact solutions for the special cases of Eq. (1.118).

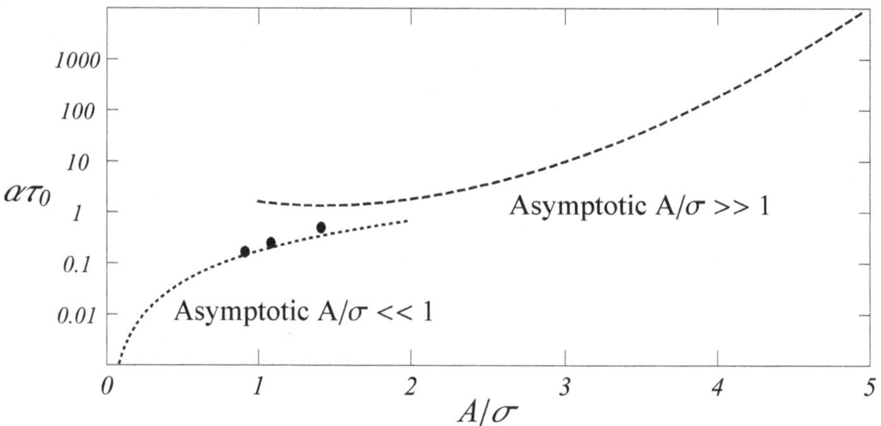

Homework 1.31 It may be instructive to go through the derivation in the text once. In particular, follow the procedure of finding the normal modes, each with a terminated power series in a.

(a) Derive the few modes listed in Eq. (1.109) and check against Fig. 1.13.

(b) Prove the normalization (1.110) for a general mode with arbitrary $n \neq 0$.

Homework 1.32 Derive Eq. (1.112). Show that

$$h(x) \approx e^x \left(\frac{1}{x} + \frac{1}{x^2} + \frac{2}{x^3} + \cdots \right), \qquad x \gg 1.$$

Solution Do this homework analytically, but as an aid, Fig. 1.17 demonstrates its asymptotic behavior numerically by plotting $xe^{-x}h(x)$ as a function of x.

Homework 1.33

(a) Use Eq. (1.114) to give an analytical derivation of Eq. (1.115).

(b) How precipitously does the distribution drop as it approaches the wall boundary? Show that the distribution drops towards zero in a distance $\sim \frac{\sigma^2}{A}$. The drop becomes sharper as A increases, as has been seen in Fig. 1.14.

Homework 1.34 Follow the text to obtain Eqs. (1.116) and (1.117) for the case when the wall boundary cuts into the beam core.

Homework 1.35 In the text as well as in Homework 1.34, we solved the quantum lifetime problem in the limit $A \ll \sigma$ and obtained an eigen-distribution

Figure 1.17: Function $xe^{-x}h(x)$ versus x.

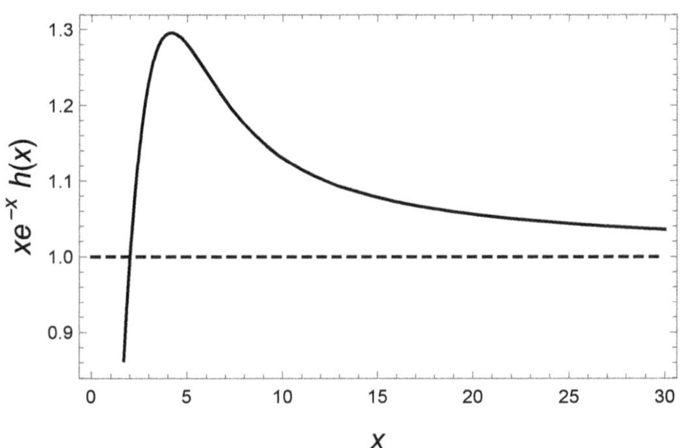

given by the Bessel function, as seen in Eqs. (1.116) and (1.117). A more general treatment can be obtained by solving Eq. (1.104) but with damping set to zero, $\alpha = 0$. Show that the solution for the n-th eigenmode ($n = 0, 1, 2, \cdots$) is given by

$$\psi_n(a) \;=\; J_0\left(\frac{\lambda_{n+1}a}{A}\right), \quad \text{with lifetime} \quad \tau_n \;=\; \frac{2A^2}{D_2\lambda_{n+1}^2},$$

where λ_n is the n-th root of $J_0(x) = 0$. The quantum lifetime corresponds to the case when $n = 0$ and $\lambda_1 = 2.405$. This result applies to a system with diffusion but no damping.

Homework 1.36

(a) Convince yourself of the few interesting special cases of quantum lifetime given by Eq. (1.118).

(b) A more venturous exercise would be to fill in the exact calculation to connect the three exact dots and approach the two asymptotes in appropriate limits.

Homework 1.37 Quantum lifetime is defined to be the lifetime of the lowest eigenmode with $n = 0$. When there is no wall, the $n = 0$ mode has infinite lifetime. When the wall is inserted, this mode acquires a quantum lifetime τ_0. The reason quantum lifetime is not defined for the other higher order modes is because those modes already have a short lifetime. When a beam is excited in those modes, the excitation damps out quickly with lifetimes $\frac{1}{2n\alpha}$ even without a wall boundary. This has been made clear by Eq. (1.109). However, this does not mean the wall does not have an effect on their damping times. When a wall is introduced, the mode damping rate will increase from the nominal value of $\frac{1}{2n\alpha}$ slightly.

Consider the next even eigenmode $2n = 2$ whose unperturbed mode distribution is $g_2(a) = 1 - \frac{a^2}{2\sigma^2}$ and damping time is $\tau_2 = \frac{1}{2\alpha}$. Follow the treatment in the text for the $n = 0$ case, show that when a wall is introduced at $a = A \gg \sigma$, the perturbed distribution becomes

$$\psi^{(2)}(a < A, t) = \exp\left(-\frac{t}{\tau_2} - \frac{a^2}{2\sigma^2}\right)\left[1 - \frac{a^2}{2\sigma^2} - \left(1 - \frac{A^2}{2\sigma^2}\right)\frac{h_2(\frac{a^2}{2\sigma^2})}{h_2(\frac{A^2}{2\sigma^2})}\right],$$

where

$$h_2(x) = \sum_{k=2}^{\infty} \frac{x^k}{k(k-1)(k!)},$$

with $h_2(x) \approx \frac{e^x}{x^2}$ if $x \gg 1$. Show that the perturbed damping time of this mode is slightly reduced from $\frac{1}{2\alpha}$,

$$\tau_2 = \frac{1}{2\alpha}\left[1 - \left(\frac{A^2}{2\sigma^2}\right)^3 \exp\left(-\frac{A^2}{2\sigma^2}\right)\right].$$

Solution Let $\tau_2 = \frac{1+\Delta}{2\alpha}$ with $|\Delta| \ll 1$, it follows that

$$c_1 \approx -1,$$

$$c_k \approx -\frac{\Delta}{k(k-1)(k!)} \qquad (k \geq 2).$$

$$\Longrightarrow \quad g_2(a) \approx 1 - \frac{a^2}{2\sigma^2} - h_2\left(\frac{a^2}{2\sigma^2}\right)\Delta.$$

The boundary condition $g_2(A) = 0$ then yields the value for Δ.

1.6.2   A heuristic argument and generalization

Before leaving the subject of vertical quantum lifetime, we will offer another shorter derivation of Eq. (1.113), provided one is not interested in detailed information such as the actual eigenmode distribution. The technique developed will be useful later in the more complex calculations of the horizontal and longitudinal quantum lifetimes. The physical meaning of this shorter derivation lies with the heuristic argument emphasized in the preceding section.

To proceed, we substitute Eq. (1.103) into Eq. (1.101) and integrate the resulting equation over $2\pi a\,da$ from 0 to A. An integration by parts gives a neat expression

$$-\frac{1}{\tau_n} = \pi D_2 A \psi_n'(A), \tag{1.119}$$

where we have assumed a normalization

$$\int_0^A 2\pi a\,da\,\psi_n(a) = 1.$$

Equation (1.119) relates the lifetime τ_n to the derivative of the beam distribution at the wall boundary. Without knowing ψ_n, as neat as it is, Eq. (1.119) is not very useful. However, for the lowest mode, we have previously explicitly solved the Fokker–Planck equation and derived Eq. (1.115). Substituting (1.115) into (1.119) indeed readily gives the quantum lifetime expression (1.113).

On the other hand, it can also be argued on a physical basis that the quantum lifetime is determined by the outward diffusion rate of the particle amplitudes across the wall. When there was no wall, in the Gaussian equilibrium, this diffusion rate is exactly equal to the inward flux of particle amplitudes due to damping. Since the wall is expected to completely stop the inward damping flux but basically does not affect the outgoing diffusion flux, we reach an insightful physical argument that (see footnote 31)

$$\text{the outward diffusion flux with the presence of the wall}$$
$$\approx \text{ the inward damping flux when there was no wall}. \qquad (1.120)$$

This heuristic argument is equivalent to replacing the distribution slope of Eq. (1.119) by that of the unperturbed Gaussian, namely replacing ψ_n by $\frac{1}{2\pi\sigma^2}e^{-a^2/2\sigma^2}$, and it is indeed what Eq. (1.115) says.

So we now have a shorter derivation in hand, i.e. we first establish an exact expression (1.119), and then approximate the eigen-distribution ψ_n by the unperturbed Gaussian to obtain an approximate expression of the quantum lifetime τ_0 of the lowest mode.

Equation (1.119) also has an important implication in practice. It gives a direct connection between the tail of the equilibrium beam distribution and the beam lifetime. In the analysis so far, we have considered a damped simple harmonic system whose natural distribution is Gaussian. In a practical storage ring, nonlinearities may cause the beam tail to be more populated than that expected of a Gaussian. Higher order moments of the noise spectrum as discussed in Sec. 1.3.4 would have a similar effect. The quantum lifetime predicted by (1.113) can easily be too optimistic by one or two orders of magnitude due to the sensitivity of quantum lifetime to details of the tail distribution. Directly applying Eq. (1.113) in those cases would be ill-advised. Conversely, measuring quantum lifetime is a sensitive method to obtain experimental information on the beam tail distribution using Eq. (1.119).

Effect of higher order noise spectrum   As just mentioned, one important application of Eq. (1.119) occurs when the beam distribution for some reason acquires an enhanced tail. Equation (1.119) allows an estimate of the reduced quantum lifetime.

Take for example the case when the higher order synchrotron radiation spectrum plays a role in the beam's longitudinal phase space distribution. Effects on the beam distribution due to higher order noise spectrum was discussed in Sec. 1.3.4. As mentioned then, the enhanced beam tail distribution can potentially affect the beam's quantum lifetime.

Consider an aperture limit at $a = A \gg \sigma$. For the longitudinal motion for example, $A = \frac{|\eta|c\hat{\delta}}{\omega_s}$ due to a momentum aperture limit $\hat{\delta}$.[33] It is clear that the quantum lifetime, as given by Eq. (1.119), depends sensitively on the beam tail distribution at $a = A$.

With beam distribution perturbed by D_4, we have found the equilibrium beam distribution given by (1.53). The parameter $\frac{D_4}{\sigma^2 D_2}$ is known to be very small, i.e. the perturbation on the beam core distribution is negligible. However, we are concerned for the distribution in the tail with $A \gg \sigma$. For example, we might consider the case when $\frac{D_4}{\sigma^2 D_2} \frac{A^2}{\sigma^2}$ is also small, but $\frac{D_4}{\sigma^2 D_2} \frac{A^4}{\sigma^4}$ may not be small. It is easy to show that the quantum lifetime effect becomes enhanced,

$$\frac{1}{\tau} = \alpha \frac{A^2}{\sigma^2} \exp\left(-\frac{A^2}{2\sigma^2}\right) \times \exp\left(\frac{3}{16} \frac{D_4}{\sigma^2 D_2} \frac{A^4}{\sigma^4}\right). \tag{1.121}$$

The additional enhancement factor $\exp\left(\frac{3}{16} \frac{D_4}{\sigma^2 D_2} \frac{A^4}{\sigma^4}\right)$ can be sizable. The case when α, D_2, D_4 are due to synchrotron radiation has been given by Table 1.1 and Eqs. (1.52) and (1.54).

Homework 1.38 Equation (1.119) is exact. It would be instructive to go through its derivation. As mentioned in the text, the challenge lies in finding the eigen-distribution $\psi_n(a)$ in the first place. However, the heuristic argument is made to approximate $\psi_n(a)$ simply by the unperturbed Gaussian distribution without wall. It is suggested to think through this argument, particularly Eq. (1.120), as some physical intuition is to be gained.

Homework 1.39 Although straightforward, it is useful to derive Eq. (1.121) to gain some feel of the effect of an enhanced beam tail on the quantum lifetime.

Homework 1.40 Although this section intends to discuss the vertical quantum lifetime, the text treated the effect of higher noise spectrum on the longitudinal quantum lifetime. Follow similar steps to find the effect of D_4 to the vertical quantum lifetime.

1.6.3   Longitudinal quantum lifetime

The vertical quantum lifetime calculation discussed above applies to a damped simple harmonic system. In the longitudinal dimension, this no longer applies because the RF voltage is in general not linear but is sinusoidal in the longitudinal coordinate, as discussed in Sec. 1.2.1. Furthermore, the RF bucket might be potential well distorted due to the wakefields. To cover those cases, an extension of the calculation is needed.

---

[33] Here we apply the result of vertical quantum lifetime to the longitudinal dimension as a simple illustration. A more complete treatment of longitudinal quantum lifetime is given in the following section.

Potential well distortion Let us consider a longitudinal system with Hamiltonian

$$H(q, p, t) \;=\; \frac{\omega_s}{2} p^2 + \int_0^q dq' K(q') \tag{1.122}$$
$$\implies \quad \dot{q} \;=\; \omega_s p, \qquad \dot{p} \;=\; -K(q).$$

This system describes a longitudinal synchrotron motion if we let (see Table 1.1)

$$q \;=\; z, \qquad p = -\frac{\eta c}{\omega_s}\delta.$$

As to the function $K(q)$, we know it depends only on q and is independent of p. In the absence of potential well distortion, we have a simple harmonic system with $K(q) = \omega_s q$. With potential well distortion, we have

$$K(q) = \begin{cases} \frac{\omega_s c}{\omega_{\rm rf}} \sin\left(\frac{\omega_{\rm rf}}{c} q\right), & \text{RF bucket,} \\[2mm] \omega_s q - \frac{4\pi\epsilon_0 \eta c^2 r_0}{\omega_s \gamma C} \int_q^\infty dq'\, \rho(q') W_0'(q - q'), & \text{wakefield,} \end{cases}$$

$$\implies \quad H = \begin{cases} \frac{\omega_s}{2} p^2 + \frac{\omega_s c^2}{\omega_{\rm rf}^2}\left[1 - \cos\left(\frac{\omega_{\rm rf}}{c} q\right)\right], & \text{RF bucket,} \\[2mm] \frac{\omega_s}{2}(q^2 + p^2) - \frac{4\pi\epsilon_0 \eta c^2 r_0}{\omega_s \gamma C} \int_0^q dq'' \int_{q''}^\infty dq'\, \rho(q') W_0'(q'' - q'), & \text{wakefield,} \end{cases}$$

where for the case of RF bucket, we have assumed the case of a nonaccelerating RF phase $\phi_s = 0$; and for the case of wakefield, $W_0'(q)$ is the longitudinal wake function of the storage ring and $\rho(q)$ is the line density of the particle distribution normalized by $\int_{-\infty}^\infty dq\, \rho(q) = N$.

Regardless of its origin, we will treat the general case for arbitrary $K(q)$ in this section. The requirement is that this potential well effect occurs in the $\dot{p}$-equation, not in the $\dot{q}$-equation. In addition, we will also assume that both the damping and the diffusion effects also occur only in p and not in q. When these conditions are fulfilled, the system offers a neat solution, as already demonstrated when we treated the Haissinski potential well distortion effect in Sec. 1.3.3, and will be shown again in our longitudinal quantum lifetime analysis below.

If the system is not distorted so that $K(q) = \omega_s q$, then our result will be the same as that of the vertical quantum lifetime. An external aperture limit might, for example, be imposed by the condition $\delta < A$, and the previous analysis applies straightforwardly. What we treat in this section, however, applies for arbitrary $K(q)$.

Aperture limit At the phase space origin ($q = 0, p = 0$), the value of the Hamiltonian (1.122) is zero and it increases as q or p increases. In a potential well distorted system, unlike the linear simple harmonic case, an aperture limit can be reached without introducing an external limit $\delta < A$. This is because in a nonlinear system, the stable region of phase space is limited to stay inside a bucket, the boundary of which is determined by the separatrix, and the separatrix now assumes the role of the physical aperture limit.

The phase space origin is the stable fixed point of the system. As q and p increase to approach the separatrix, the Hamiltonian reaches a value $\hat{H}$; all points on the separatrix have $H(q,p) = \hat{H}$. As far as the bucket is concerned, a particle with Hamiltonian larger than $\hat{H}$ is readily removed from the beam because it no longer executes bounded synchrotron oscillation within the bucket. This means there is equivalently a wall boundary specified by the condition $H = \hat{H}$. Associated with this particle loss mechanism is the longitudinal quantum lifetime.

A comment is in order here. Unlike a physical wall that strictly removes all particles hitting it, the separatirx boundary is not as strict. A particle accidentally crosses the separatrix still has a chance of crossing it again due to damping and survives. In the limit that phase space streaming is much faster than damping and diffusion, as we are assuming, it may be argued that the separatrix can still be considered a strict wall. It remains an issue, however, when the particle loss occurs in the neighborhood of the unstable fixed points, where phase space streaming slows down and ceases at the unstable fixed points. In our analysis below, we shall ignore these subtleties.

Take the case of RF bucket for example, the Hamiltonian on the separatrix has the value

$$\hat{H} \;=\; \frac{2\omega_s c^2}{\omega_{\rm rf}}\,. \tag{1.123}$$

The case of wakefield is more complex depending on the wake function $W_0'(q)$ and the beam distribution $\rho(q)$, but in principle can be obtained.

Smoothed Fokker–Planck equation   To calculate the longitudinal quantum lifetime with potential well distortion, we follow a procedure similar to the previous section except that the distribution is no longer a function simply of the amplitude a but of the Hamiltonian H. As mentioned earlier, this possibility requires

> (i) the potential well distortion appears only in $K(q)$, i.e. it appears only in the $\dot{p}$-equation and not in the $\dot{q}$-equation;
>
> (ii) the damping and the diffusion effects are applied only to p and not to q;
>
> (iii) the synchrotron motion is slow when observed on a turn-by-turn basis for the validity of the adiabatic approximation;
>
> (iv) in addition, as usual, for the Fokker–Planck equation to apply, we have also assumed the damping and the diffusion are slow compared with the Hamiltonian flow motion in the phase space. $\tag{1.124}$

In this case, the distribution function ψ has to conform to H and has to be a function of H, i.e.

$$\psi(q,p,t) \;=\; \psi(H,t)\,.$$

The separate dependencies on q and p are smoothed out due to the rapid phase space flow.

The Fokker–Planck equation reads

$$\frac{\partial \psi}{\partial t} + \frac{\partial H}{\partial p}\frac{\partial \psi}{\partial q} + \left(-\frac{\partial H}{\partial q} - 2\alpha p\right)\frac{\partial \psi}{\partial p} \; = \; 2\alpha\psi + D_2\frac{\partial^2 \psi}{\partial p^2}\,.$$

Now with ψ a function of H, it is easy to see that

$$\frac{\partial H}{\partial p}\frac{\partial \psi}{\partial q} - \frac{\partial H}{\partial q}\frac{\partial \psi}{\partial p} \; = \; 0\,.$$

Therefore,

$$\frac{\partial \psi}{\partial t} - 2\alpha p\frac{\partial \psi}{\partial p} \; = \; 2\alpha\psi + D_2\frac{\partial^2 \psi}{\partial p^2}\,.$$

Furthermore,

$$p\frac{\partial \psi}{\partial p} \; = \; \omega_s p^2 \psi'(H)\,,$$

$$\frac{\partial^2 \psi}{\partial p^2} \; = \; \omega_s\psi'(H) + \omega_s^2 p^2 \psi''(H)\,.$$

Under the conditions (1.124), we then make a critical step of averaging over the phase space angle. With the Hamiltonian given by a linear sum of a kinetic energy term and a potential well term, and with the kinetic term simply given by $\frac{\omega_s}{2}p^2$, as in Eq. (1.122), the two terms in Eq. (1.122) will equally share the Hamiltonian when the phase space angle is averaged over. In other words, we shall make the replacement

$$\frac{\omega_s}{2}p^2 \;\rightarrow\; \frac{1}{2}H\,,$$

as the result of averaging over the phase angle.

After making this replacement, we obtain a smoothed Fokker–Planck equation, valid when the Hamiltonian flow is much faster than the damping, the diffusion, and the quantum lifetime effects so that the beam distribution conforms to the Hamiltonian at all times. The smoothed Fokker–Planck equation reads

$$\frac{\partial \psi}{\partial t} - 2\alpha H\psi'(H) \; = \; 2\alpha\psi + D_2\omega_s[\psi'(H) + H\psi''(H)]\,. \tag{1.125}$$

This equation describes a slow evolution of the beam distribution as all remaining terms of the equation are slow.

Before discussing the quantum lifetime problem when an aperture limit $H < \hat{H}$ will be imposed, we need to find the unperturbed equilibrium solution to (1.125) without aperture limit. Designating this distribution by $\psi_0(H)$, the solution is found to be

$$\psi_0(H) \; = \; \frac{2\alpha}{D_2\omega_s}\exp\left(-\frac{2\alpha}{D_2\omega_s}H\right), \tag{1.126}$$

where we have normalized ψ_0 by $\int_0^\infty dH\psi_0(H) = 1$. Equation (1.126) is in fact just the Haissinski solution.

Quantum lifetime   To discuss the quantum lifetime problem, we look for an eigen-distribution

$$\psi(H,t) \; = \; e^{-t/\tau}\psi_1(H)\,,$$

which gives

$$-\frac{1}{\tau}\psi_1 \; = \; 2\alpha\psi_1 + 2\alpha H\psi_1' + \omega_s D_2(H\psi_1'' + \psi_1')\,. \tag{1.127}$$

Integrating Eq. (1.127) over dH from 0 to $\hat{H}$ and performing integration by parts as we did in deriving Eq. (1.119), we obtain the neat expression,

$$-\frac{1}{\tau} \; = \; \omega_s D_2 \hat{H}\psi_1'(\hat{H})\,, \tag{1.128}$$

which is a generalization of Eq. (1.119).

We now follow Sec. 1.6.2 of the shorter derivation of quantum lifetime based on a heuristic physical argument, and substitute the unperturbed distribution (1.126) — properly normalized — into the right-hand-side for $\psi_1'(\hat{H})$. This gives the final expression for the longitudinal quantum lifetime,

$$\frac{1}{\tau} \; = \; \frac{4\alpha^2}{D_2\omega_s}\hat{H}\,\exp\left(-\frac{2\alpha\hat{H}}{D_2\omega_s}\right)\,.$$

Defining an unperturbed (unperturbed by potential well distortion) rms beam size of q and p as

$$\sigma^2 \; = \; \sigma_q^2 \; = \; \sigma_p^2 \; = \; \sigma_z^2 \; = \; \frac{\eta^2 c^2}{\omega_s^2}\sigma_\delta^2 \; = \; \frac{D_2}{2\alpha}\,,$$

we have the final result

$$\frac{1}{\tau} \; = \; \frac{2\alpha\hat{H}}{\omega_s\sigma^2}\,\exp\left(-\frac{\hat{H}}{\omega_s\sigma^2}\right)\,. \tag{1.129}$$

RF bucket   For the case when the potential well is given by the RF bucket, substituting $\hat{H}$ from Eq. (1.123) gives an expression for the longitudinal quantum lifetime,

$$\frac{1}{\tau} \; = \; 2\alpha\left(\frac{2c^2}{\omega_{\mathrm{rf}}^2\sigma^2}\right)\exp\left(-\frac{2c^2}{\omega_{\mathrm{rf}}^2\sigma^2}\right)\,. \tag{1.130}$$

Validity of Eq. (1.130) requires

$$\sigma \; \ll \; \frac{c}{\omega_{\mathrm{rf}}}\,,$$

i.e. the unperturbed rms beam size must be much shorter than the RF wavelength divided by 2π.

Momentum aperture Another case occurs when the aperture limit is due to a limit in the momentum deviation $\delta < \hat{\delta}$ instead of the RF bucket — perhaps because the storage ring lattice can accept only particles with $|\delta| < \hat{\delta}$. In this case,

$$\hat{H} = \frac{\eta^2 c^2}{2\omega_s}\,\hat{\delta}^2\,.$$

Substituting into Eq. (1.129) then gives

$$\frac{1}{\tau} = 2\alpha\left(\frac{\eta^2 c^2 \hat{\delta}^2}{2\omega_s^2 \sigma^2}\right)\exp\left(-\frac{\eta^2 c^2 \hat{\delta}^2}{2\omega_s^2 \sigma^2}\right)\,. \tag{1.131}$$

In case of no potential well distortion, we can rewrite the momentum aperture limit $\hat{\delta}$ by an equivalent aperture limit $q < A$ by setting $A = \frac{|\eta| c \hat{\delta}}{\omega_s}$, we then obtain the result (1.113) as it should.

Homework 1.41

(a) Derive the expression of the longitudinal aperture limit Eq. (1.123) for the case of a nonaccelerating RF bucket.

(b) Extend the expression to the case with acceleration, $\phi_s \neq 0$.

Homework 1.42 Show that the smoothed Fokker–Planck equation (1.125) with appropriate normalization has the equilibrium solution given by (1.126).

1.6.4 Horizontal quantum lifetime

The quantum lifetime calculations discussed above applies to 1-D systems, either in the vertical or the longitudinal dimensions. A complication results if the aperture limit is in the horizontal dimension at a location with energy dispersion D_x. This is because the horizontal beam size at such a location contains both horizontal betatron and longitudinal synchrotron contributions and the aperture limitation is no longer 1-D.

The case when the aperture limit is nondispersive is straightforward; it is the same as the vertical case. In this section, we will concentrate on the case when there is a dispersion at the location of the aperture. Let A be the amplitude of the wall boundary, the aperture limitation can be written as

$$a_x + D_x a_\delta < A\,,$$

where a_x and a_δ are the oscillation amplitudes in the horizontal betatron coordinate x and synchrotron energy deviation δ, respectively. We assume D_x is positive; otherwise it is to be replaced by $|D_x|$.

The Fokker–Planck equation in the 2-D case is

$$\frac{\partial \psi}{\partial t} = \sum_{i=x,\delta}\left[2\alpha_i \psi + \alpha_i a_i \frac{\partial \psi}{\partial a_i} + \frac{D_{2i}}{2a_i}\frac{\partial}{\partial a_i}\left(a_i \frac{\partial \psi}{\partial a_i}\right)\right]\,. \tag{1.132}$$

As always, this equation describes the slow dynamics after smoothing over the rapid Hamiltonian flow in phase space. Both sides of the equation are slow.

To find the quantum lifetime, let the lowest eigen-distribution be

$$\psi(a_x, a_\delta, t) = e^{-t/\tau}\psi_1(a_x, a_\delta).\qquad(1.133)$$

If we denote the allowed region in the (a_x, a_δ) space by $\mathcal{R}$ and its boundary by $\mathcal{C}$, we require the normalization and boundary conditions,

$$\int_{\mathcal{R}} 4\pi^2 a_x da_x\, a_\delta da_\delta\, \psi_1(a_x, a_\delta) = 1,$$
$$[\psi_1(a_x, a_\delta)]_{\mathcal{C}} = 0.$$

In this example, $\mathcal{R} = \{a_x + D_x a_\delta < A\}, \mathcal{C} = \{a_x + D_x a_\delta = A\}$.

Substituting Eq. (1.133) into Eq. (1.132), integrating over the phase space region $\mathcal{R}$ and applying integration by parts yield

$$\frac{1}{\tau} = \alpha_x I_x + \alpha_\delta I_\delta,\qquad(1.134)$$

where

$$I_x = -4\pi^2\sigma_x^2 \int_{\mathcal{R}} a_\delta da_\delta \left[a_x \frac{\partial\psi_1}{\partial a_x}\right]_{\mathcal{C}},$$
$$I_\delta = -4\pi^2\sigma_\delta^2 \int_{\mathcal{R}} a_x da_x \left[a_\delta \frac{\partial\psi_1}{\partial a_\delta}\right]_{\mathcal{C}},$$

with

$$\sigma_x = \sqrt{\frac{D_{2x}}{2\alpha_x}}, \qquad \sigma_\delta = \sqrt{\frac{D_{2\delta}}{2\alpha_\delta}}.$$

These are the 2-D counterpart of Eqs. (1.119) and (1.128).

Following the by-now-familiar technique developed, we approximate ψ_1 by its natural unperturbed form

$$\psi_1(a_x, a_\delta) = \frac{1}{4\pi^2\sigma_x^2\sigma_\delta^2}\exp\left(-\frac{a_x^2}{2\sigma_x^2} - \frac{a_\delta^2}{2\sigma_\delta^2}\right).$$

This leads to the expressions

$$I_x = \frac{1}{\sigma_x^2\sigma_\delta^2}\int_0^{A/D_x} a_\delta da_\delta (A - D_x a_\delta)^2 \exp\left[-\frac{(A - D_x a_\delta)^2}{2\sigma_x^2} - \frac{a_\delta^2}{2\sigma_\delta^2}\right],$$
$$I_\delta = \frac{1}{\sigma_x^2 D_x^2\sigma_\delta^2}\int_0^A a_x da_x (A - a_x)^2 \exp\left[-\frac{a_x^2}{2\sigma_x^2} - \frac{(A - a_x)^2}{2D_x^2\sigma_\delta^2}\right].$$

By changing variables and defining

$$\sigma_T^2 = \sigma_x^2 + D_x^2\sigma_\delta^2, \qquad r = \frac{D_x^2\sigma_\delta^2}{\sigma_T^2},$$

it follows that

$$
\begin{aligned}
I_x &= \frac{A^4}{\sigma_x^2 D_x^2 \sigma_\delta^2} \exp\left(-\frac{A^2}{2\sigma_T^2}\right) \int_{-r}^{1-r} (y+r)(1-r-y)^2 dy \, \exp\left(-\frac{A^2 y^2}{2r\sigma_x^2}\right), \\
I_\delta &= \frac{A^4}{\sigma_x^2 D_x^2 \sigma_\delta^2} \exp\left(-\frac{A^2}{2\sigma_T^2}\right) \int_{r-1}^{r} (y+1-r)(r-y)^2 dy \, \exp\left(-\frac{A^2 y^2}{2r\sigma_x^2}\right).
\end{aligned}
\tag{1.135}
$$

The quantum lifetime is obtained once $I_{x,\delta}$ are calculated. Equations (1.134) and (1.135) are the solution for the general case. Simpler expressions are given below for two special cases.

Aperture with no dispersion The first special case applies when there is no dispersion at the aperture, i.e. $D_x = 0$, or $r = 0$. It is easy to show from Eq. (1.135) that in the limit $r \to 0$ (Homework 1.44),

$$
\begin{aligned}
I_x &= \frac{A^2}{\sigma_x^2} \exp\left(-\frac{A^2}{2\sigma_x^2}\right), \\
I_\delta &= 0,
\end{aligned}
\tag{1.136}
$$

as expected.

Aperture with moderate dispersion Another case with simpler result applies when σ_x and $D_x \sigma_\delta$ are comparable and are both not too close to A, or more precisely, if

$$
(\text{both } r \text{ and } 1-r) \ \gg \ \frac{\sigma_x D_x \sigma_\delta}{A\sigma_T}.
$$

The simplification results because the Gaussian factor in the integrands of I_x and I_δ limit significant contributions to the region $y^2 < \frac{r\sigma_x^2}{A^2}$, and as a result, the algebraic parts of the integrands can be approximated by setting $y = 0$ and the limits of the integral can be replaced by $\pm\infty$. After this process, we find

$$
\begin{aligned}
I_x &\approx \sqrt{2\pi}\, \frac{A^3 \sigma_x^3 D_x \sigma_\delta}{\sigma_T^7} \exp\left(-\frac{A^2}{2\sigma_T^2}\right), \\
I_\delta &\approx \sqrt{2\pi}\, \frac{A^3 \sigma_x D_x^3 \sigma_\delta^3}{\sigma_T^7} \exp\left(-\frac{A^2}{2\sigma_T^2}\right).
\end{aligned}
\tag{1.137}
$$

The quantum lifetime is then found to be

$$
\frac{1}{\tau} \approx \sqrt{2\pi}\, \frac{A^3 \sigma_x D_x \sigma_\delta}{\sigma_T^7} \exp\left(-\frac{A^2}{2\sigma_T^2}\right) \left(\alpha_x \sigma_x^2 + \alpha_\delta D_x^2 \sigma_\delta^2\right).
$$

For the case when the damping and diffusion come from synchrotron radiation, and if the storage ring is planar strong-focusing, separate-function design,

Figure 1.18: The 2-D quantum lifetime τ as a function of the aspect ratio parameter $r = \frac{D_x^2 \sigma_\delta^2}{\sigma_T^2}$. The lifetime τ is normalized by $\frac{e^{n^2/2}}{\alpha_x n^3}$. The curve applies when r stays away from 0 and 1.

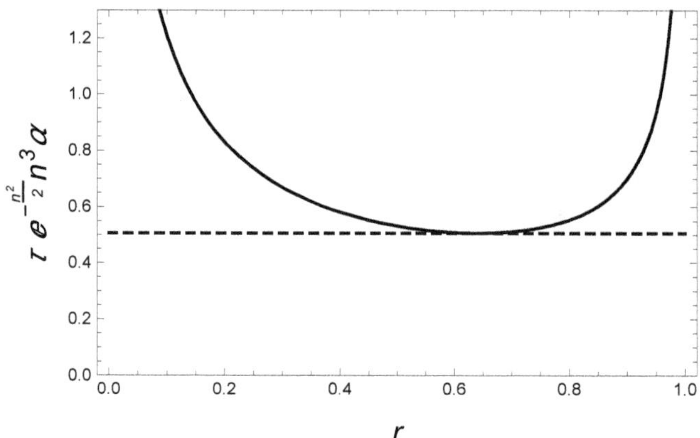

then we set $\alpha_\delta = 2\alpha_x$ (see also Table 1.1). We then obtain

$$\tau = \frac{e^{n^2/2}}{\sqrt{2\pi}\,\alpha_x n^3} \frac{1}{(1+r)\sqrt{r(1-r)}},$$

$$n = \frac{A}{\sigma_T},$$

$$r = \frac{D_x^2 \sigma_\delta}{\sigma_T^2},$$

$$\sigma_T^2 = \sigma_x^2 + D_x^2 \sigma_\delta^2.$$

For given σ_T, the shortest lifetime occurs when

$$r_{\min} = \frac{1+\sqrt{17}}{8} = 0.640, \quad \Longrightarrow \quad \begin{cases} D_x \sigma_\delta = 0.80\,\sigma_T, \\ \sigma_x = 0.60\,\sigma_T, \end{cases}$$

$$\Longrightarrow \tau_{\min} = 0.507 \frac{e^{n^2/2}}{\alpha_x n^3}. \tag{1.138}$$

Figure 1.18 shows the lifetime plotted as a function of r.

A comparison with the 1-D case indicates that the quantum lifetime in this 2-D case is shorter. In addition to a numerical factor, the denominator now contains n^3 instead of n^2, where $n \gg 1$.

Homework 1.43

(a) Derive Eq. (1.134).

(b) Derive Eq. (1.135).

Solution (b) For the expression of I_x, change variable from a_δ to y with $a_\delta = \frac{A}{D_x}(y+r)$. For I_δ, change $a_x = A(y-r+1)$.

Homework 1.44
    (a) Derive Eq. (1.136) in the limit $D_x \to 0$.
    (b) Follow the steps in the text to find the approximate expressions (1.137).

Homework 1.45 Verify Eq. (1.138).

Homework 1.46 The text treated the case of a horizontal aperture limit when the aperture location is dispersive, and the boundary condition appears as $a_x + D_x a_\delta < A$. A more venturous analysis can be extended to the case of an x-y coupled system.

    (a) Assuming linear coupling near the difference resonance $\nu_x - \nu_y \approx p$ and let $\alpha_x = \alpha_y$, the aperture limit can be written as

$$a_x^2 + a_y^2 \; < \; A^2 \, .$$

Follow a similar approach in this section to derive a 2-D x-y coupled quantum lifetime.

    (b) Extend (a) to the case with nonlinear x-y coupled beam near a difference resonance $m\nu_x - n\nu_y \approx p$. The aperture limit appears in the form

$$\frac{a_x^2}{n} + \frac{a_y^2}{m} \; < \; A^2 \, .$$

    Both (a) and (b) require modifications on the Fokker–Planck equation. Unlike the case treated in the text when only the aperture involves 2-D dynamics, here the Fokker–Planck equation is also 2-D.

1.7 Fokker–Planck normal mode

In the description of multiparticle systems using the Vlasov and Fokker–Planck equations, the collective beam motion is often analyzed in terms of normal modes. In the presence of wakefields, for example, the complex frequencies of these normal modes provide the criterion for their stability. The system is unstable if any one of its normal modes is found to be unstable. In the treatment of quantum lifetime, similarly, we identify the quantum lifetime to be the decay time of the lowest normal mode, as we did in Sec. 1.6.

Vlasov mode Due to the high degree of degeneracy of the Vlasov equation for a linear multiparticle system in the absence of potential well or wakefield effects, described by Eq. (1.7), its normal modes are not very interesting. The n-th normal mode, for example, can be written as

$$\psi^{(n)}(\phi, a, t) \; = \; e^{-i\Omega^{(n)}t} \, e^{in\phi} \, A_m^{(n)}(a) \, ,$$

where the normal mode frequency is $\Omega^{(n)} = n\omega$ and the radial eigen-function $A_m^{(n)}(a)$ could be the m-th member of some orthogonal polynomials. Each mode n in general degenerates into a set of modes, the index m specifying one of them, each specified by a radial eigen-function.

Complications occur when potential well distortion or wakefields are included as briefly discussed in Sec. 1.2.4. Depending on the wakefield pattern, specific solutions are needed to serve as the normal mode polynomials $A_m^{(n)}(a)$. The mode frequency $\Omega^{(n)}$ no longer is given simply by $n\omega$; it loses degeneracy and becomes dependent of both mode indices n and m. Furthermore, in general it becomes complex; the real part of $\Omega^{(n,m)}$ gives the mode frequency shift from the nominal value of $n\omega$; the imaginary part gives the instability growth rate (growth rate if positive, damping rate if negative) of the mode under consideration. Detailed study in this extensive subject is considered outside the present scope of this chapter.

Fokker–Planck mode   In the following, we will describe the normal modes of the Fokker–Planck equation in the ideal 1-D case without wall boundary, potential well distortion, or wakefields. We will start with the Fokker–Planck equation for a damped simple harmonic oscillator, Eq. (1.47), and recall that we then averaged over the phase angle and assumed no phase angle dependence of ψ to obtain Eq. (1.101). To find all the normal modes, however, the phase angle dependence must now be kept. We backtrack to Eq. (1.47).

Substituting Eq. (1.8) into (1.47), the complete Fokker–Planck equation in polar coordinates is found to be

$$
\frac{\partial\psi}{\partial t} + \omega\frac{\partial\psi}{\partial\phi} - 2\alpha a\cos^2\phi\,\frac{\partial\psi}{\partial a} + \alpha\sin 2\phi\,\frac{\partial\psi}{\partial\phi} \tag{1.139}
$$
$$
= 2\alpha\psi + D_2\cos^2\phi\,\frac{\partial^2\psi}{\partial a^2} - D_2\frac{\sin 2\phi}{a}\left[\frac{\partial^2\psi}{\partial a\partial\phi} - \frac{1}{a}\frac{\partial\psi}{\partial\phi}\right] + D_2\frac{\sin^2\phi}{a}\left[\frac{\partial\psi}{\partial a} + \frac{1}{a}\frac{\partial^2\psi}{\partial\phi^2}\right],
$$

where we have assumed ψ depends on both a and ϕ. Averaging over ϕ reduces to the smoothed Eq. (1.101), as it should.

Let the n-th normal mode be written as

$$
\psi^{(n)}(\phi, a, t) = e^{-i\Omega^{(n)}t}\sum_{m=-\infty}^{\infty} A_m^{(n)}(a)\,e^{-\alpha a^2/D_2}\,e^{im\phi}, \tag{1.140}
$$

where $\Omega^{(n)}$ is the normal mode frequency (it will be complex) and functions $A_m^{(n)}(a)$ are yet to be determined (they will be polynomials). The fact that $\psi^{(n)}(\phi, a, t)$ can be Fourier decomposed is because it is necessarily periodic in ϕ with period 2π.

Here m is not a mode index; it is a dummy index being summed over. Later we will see that there are multiple solutions of Eq. (1.140), each solution corresponds to a mode. To specify the individual solutions, there will be a second mode index ℓ to be assigned to each mode.

Substituting Eq. (1.140) into Eq. (1.139) and identifying the proper Fourier components, we obtain the iteration equation,

$$\left(-i\Omega^{(n)} + im\omega + \frac{m^2 D_2}{2a^2}\right) A_m^{(n)} + \left(\alpha a - \frac{D_2}{2a}\right) A_m^{(n)'} - \frac{D_2}{2} A_m^{(n)''}$$

$$-(m-2)\left(\frac{\alpha}{2} + \frac{m D_2}{4a^2}\right) A_{m-2}^{(n)} + \left(\frac{\alpha a}{2} + \frac{m D_2}{2a} - \frac{3 D_2}{4a}\right) A_{m-2}^{(n)'} - \frac{D_2}{4} A_{m-2}^{(n)''}$$

$$+(m+2)\left(\frac{\alpha}{2} - \frac{m D_2}{4a^2}\right) A_{m+2}^{(n)} + \left(\frac{\alpha a}{2} - \frac{m D_2}{2a} - \frac{3 D_2}{4a}\right) A_{m+2}^{(n)'} - \frac{D_2}{4} A_{m+2}^{(n)''} = 0,$$

$$m = -n, -n+2, -n+4, \cdots, n, \qquad (1.141)$$

where primes denote taking derivatives with respect to a.

Case $n = 0$ For the lowest mode with $n = 0$, there is only one normal mode. The solution is

$$\Omega^{(0)} = 0,$$

$$A_m^{(0)} = \begin{cases} 1, & \text{if } m = 0, \\ 0, & \text{if } m \neq 0. \end{cases}$$

This is just the stationary Gaussian solution of the Fokker–Planck equation.

Case $n = 1$ We expect

$$A_1^{(1)}(a) = f_1 a, \quad A_0^{(1)}(a) = 0, \quad A_{-1}^{(1)}(a) = f_{-1} a,$$

where coefficients $f_{\pm 1}$ are yet to be found, and all other $A_m^{(1)}(a) = 0$. Only two components in the iteration equation contribute, $m = \pm 1$, yielding

$$(-i\Omega^{(1)} + i\omega + \alpha) f_1 = -\alpha f_{-1},$$
$$(-i\Omega^{(1)} - i\omega + \alpha) f_{-1} = -\alpha f_1.$$

There are two solutions for the normal mode frequency Ω,

$$\Omega_\pm^{(1)} = -i\alpha \pm \sqrt{\omega^2 - \alpha^2},$$

meaning there are two normal modes, a "+ mode" and a "− mode". These modes damp with damping rate α. We have assumed $\omega > \alpha > 0$.

We next choose

$$f_1 = \alpha, \qquad f_{-1} = i\Omega_\pm^{(1)} - i\omega - \alpha.$$

The normal modes are therefore given by

$$\psi_\pm^{(1)}(\phi, a, t) = e^{-\alpha t \mp i\sqrt{\omega^2 - \alpha^2}\, t}\, a e^{-\alpha a^2 / D_2} \left[\alpha e^{i\phi} + (i\Omega_\pm^{(1)} - i\omega - \alpha) e^{-i\phi}\right].$$

Note that the normal modes found here are complex eigenmodes. Actual beam distribution of course is real. An actual beam distribution has to be decomposed into these eigenmodes, and after the superposition, it must be a real quantity. See Homework 1.48.

Case $n = 2$ There are three normal modes in this case, describing the damped quadrupolar motion of the beam distribution. Let

$$A_2^{(2)}(a) \; = \; a^2, \qquad A_0^{(2)}(a) \; = \; fa^2 + g, \qquad A_{-2}^{(2)}(a) \; = ha^2.$$

It follows from Eq. (1.141) that f, g, h have to satisfy the coupled equations,

$$
\begin{aligned}
-i\Omega^{(2)} + 2i\omega + 2\alpha + \alpha f &= 0, \\
-i\Omega^{(2)} f + 2\alpha f + 2\alpha h + 2\alpha &= 0, \\
-i\Omega^{(2)} g - 2D_2 f - 2D_2 h - 2D_2 &= 0, \\
-i\Omega^{(2)} h - 2i\omega h + 2\alpha h + f\alpha &= 0.
\end{aligned}
\qquad (1.142)
$$

Nontrivial solution of these equations requires $\Omega^{(2)}$ to be the eigenmode frequencies. There are three solutions corresponding to three normal modes,

$$
\begin{aligned}
\text{0-mode:} \quad \Omega_0^{(2)} &= -2i\alpha, \\
f &= -\frac{2i\omega}{\alpha}, \quad g = \frac{2i\omega D_2}{\alpha^2}, \quad h = -1, \\
\text{$\pm$-mode:} \quad \Omega_\pm^{(2)} &= -2i\alpha \pm 2\sqrt{\omega^2 - \alpha^2}, \\
f &= \frac{2i(-\omega \pm \sqrt{\omega^2 - \alpha^2})}{\alpha}, \\
g &= \frac{2iD_2(\omega \mp \sqrt{\omega^2 - \alpha^2})}{\alpha^2}, \\
h &= \frac{\alpha^2 - 2\omega^2 \pm 2\omega\sqrt{\omega^2 - \alpha^2}}{\alpha^2}.
\end{aligned}
\qquad (1.143)
$$

These modes damp with damping rate 2α. If the phase angle information is averaged over, these modes correspond to the $n = 2$ modes described by Eq. (1.109). This is easily verified because the 0-mode has

$$A_0^{(2)}(a) \; = \; fa^2 + g \; = \; -\frac{2i\omega}{\alpha}\left(a^2 - \frac{D_2}{\alpha}\right) \propto a^2 - 2\sigma^2,$$

as predicted by Eq. (1.109).

Case of general n   In general, there are $n+1$ modes with mode index n. If we designate them by another index ℓ which assumes values $\ell = n, n-2, \ldots, -n$, the eigen-frequencies are[34]

$$\Omega_\ell^{(n)} \; \approx \; -in\alpha + \ell\sqrt{\omega^2 - \alpha^2}. \qquad (1.144)$$

All these modes have damping rate $n\alpha$ when $\alpha \ll \omega$. Note that the mode frequencies are independent of the diffusion constant D_2. Note also that the higher modes have more complicated phase space structures, and they are damped

---

[34]A. Renieri, Frascati Laboratory report LNF-76/11 (R) (1976).

faster because the quantum diffusion smears out their subtle structures more easily.

Equation (1.144) holds when $\omega \gg \alpha$. Unlike the cases $n = 0, 1, 2$ when $\Omega_\ell^{(n)}$ can be solved exactly, cases when $n > 2$ are more complex; Eq. (1.144) is only approximate.

It should be noted that the mode frequency $\Omega_\ell^{(n)}$ is seen here to depend on two mode indices n and ℓ. The index n specifies the radial modes and defines the damping rate. The index ℓ specifies the azimuthal modes and defines the mode oscillation frequency.

Homework 1.47 Derive the iteration equation (1.141). It might be a bit tedious but straightforward.

Homework 1.48 Consider a beam with distribution $\psi = e^{-\alpha a^2/D_2}$ that is slightly displaced off-center by $q_0 \ll \frac{D_2}{\alpha}$ at time $t = 0$. The beam executes a subsequent transient damped dipole motion. Decompose the beam distribution in terms of the $n = 0$ and the two $n = 1$ eigenmodes. Give the beam distribution in its subsequent evolution when $t > 0$. Observe that the beam distribution remains real at all times.

Solution

$$
\begin{aligned}
\psi(q, p, t = 0) &= e^{-\alpha[(q-q_0)^2 + p^2]/D_2} \\
&\approx e^{-\alpha a^2/D_2} + \frac{2\alpha q_0 \sin\phi}{D_2} a e^{-\alpha a^2/D_2} .
\end{aligned}
$$

Decomposing it by

$$
\psi(q, p, t = 0) = \psi^{(0)}(a) + K_+ \psi_+^{(1)}(\phi, a, t = 0) + K_- \psi_-^{(1)}(\phi, a, t = 0) ,
$$

it can be shown that

$$
K_\pm = \frac{q_0}{2D_2} \left(-i \pm \frac{\alpha - i\omega}{\sqrt{\omega^2 - \alpha^2}} \right) .
$$

The subsequent motion is described by

$$
\psi(q, p, t > 0) = \psi^{(0)}(a) + K_+ \psi_+^{(1)}(\phi, a, t) + K_- \psi_-^{(1)}(\phi, a, t) .
$$

It is always real.

A continuation of this homework would ask for a comparison of your result with the result obtained using the transient Fokker–Planck solution of Sec. 1.5.

Homework 1.49 Derive Eq. (1.142) and then use it to verify Eq. (1.143) for the three quadrupole normal modes.

Chapter 2

Symplectification of Maps

2.1   Phase space

The symplecticity condition is an important and remarkable property of all Hamiltonian systems. It imposes a very strong constraint on the way we are allowed to formulate the dynamics of these systems. An intriguing observation to be made first is that this very strong constraint of the symplecticity condition follows from the mere fact that *there exists* a Hamiltonian that describes the particle motion. It does not matter what the Hamiltonian actually is.

To discuss the symplecticity condition, we must first enter the phase space. It is a profound observation that all equations of motion involve second order differential equations, as first evidenced by the Newton equation $m\ddot{x} = f$. As we know, to completely define the solution $x(t)$ to a second-order differential equation, we need to specify two initial conditions, i.e. the values of $x(0)$ and $\dot{x}(0)$. As a result, in order to describe the motion of a particle completely at all times t, we need to specify its initial conditions $(x(0), \dot{x}(0))$. If we define a space $(x, \dot{x})$, then the particle motion can be represented by a point moving in this plane starting with the position $(x(0), \dot{x}(0))$. This plane $(x, \dot{x})$ is called the *phase space*. The trajectory of the representative point in the phase space at all time t is uniquely determined by the differential equation of motion — or equivalently, the Hamiltonian — and the initial conditions.

In passing, we have arrived at another important conclusion, namely, once the initial conditions $(x(0), \dot{x}(0))$ are given, the subsequent motion of the particle is uniquely determined by the Hamiltonian of the system. It follows that if two particles occupy exactly the same phase space point at a given time, the only possibility is that these two particles coincide in their phase space trajectories at all times. Otherwise, two trajectories can never cross. All trajectories must move as laminar flows in phase space.

The phase space is an abstract construct. Motion of a representative point in phase space is different from and not to be mixed up with the motion of the actual particle in the real physical space. It should be appreciated that the concept of phase space has been a result of evolution in our understanding of dynamics accumulated over many decades by many pioneers, and was not available at the Galilean or even the Newtonian times.

As mentioned, it is a fundamental principle in dynamics that all equations of motion are described by second order differential equations. In case of a n-D system, the phase space is $2n$-D. Couplings among these dimensions, if any, are added on top and do not alter this fundamental property. Again, the state of a particle is represented by a point in the $2n$-D phase space; the motion of the particle is represented by its phase space representative point moving in time in the phase space starting initially from the initial conditions $(\vec{r}(0), \dot{\vec{r}}(0))$, and follows the equations of motion according to the Hamiltonian.

An additional comment should be made here concerning the legitimacy of even considering a "particle", i.e. a representative point, in phase space in the first place. This being legitimate is intimately due to a theorem called the Liouville theorem, which we will discuss momentarily. Without the Liouville theorem, the concept of a "particle" in phase space will not be applicable.

2.2 Symplecticity condition

Let us begin with the simplest dynamical system, i.e. a 1-D (2-D phase space) linear system. Let us consider the motion of the particle's coordinate y as a function of time, where the time is measured in a position coordinate s as we would do for an accelerator system. In such a dynamical system, evolution of particle motion is completely specified by the 2×2 transfer matrix $M_{s_1 \to s_2}$ from one position s_1 to another position s_2. For this system, the symplecticity condition reduces to something very simple, i.e. the determinant of $M_{s_1 \to s_2}$ must be equal to 1 regardless of s_1 or s_2 where M is evaluated. Indeed,

$$\text{1-D linear case}$$
$$\text{symplecticity} \quad \Longleftrightarrow \quad \det(M_{s_1 \to s_2}) = 1 \quad \text{for any } s_1, s_2 \, .$$

A neat proof is given below.

A dynamical system must have an equation of motion that describes the motion of its particles. For a 1-D system, the equation is a second order differential equation in $y(s)$,

$$y'' + K(s)y = 0 \, ,$$

where a prime means taking derivative with respect to s and $K(s)$ is an arbitrary focusing function along s. Consider two independent solutions $y_1(s)$ and $y_2(s)$ of this equation of motion. Form the *Wronskian*,

$$W(s) \equiv y_1 y_2' - y_2 y_1' \, . \tag{2.1}$$

It is easy to show that W is a constant of the motion, i.e. $W' = 0$. Note that (y, y') is the phase space.

One also observes, from the definition of the transfer matrix $M_{s_1 \to s_2}$, that

$$\begin{bmatrix} y_1 & y_2 \\ y_1' & y_2' \end{bmatrix}_{s_2} = M_{s_1 \to s_2} \begin{bmatrix} y_1 & y_2 \\ y_1' & y_2' \end{bmatrix}_{s_1} .$$

By taking the determinant on both sides, one sees $W(s_2) = \det(M_{s_1 \to s_2})\, W(s_1)$, which leads to $\det(M_{s_1 \to s_2}) = 1$ because $W(s)$ is a constant of the motion, and it cannot be zero because y_1 and y_2 are independent solutions to the equation of motion. Q.E.D.

For an n-D case ($2n$-D phase space), a linear motion can be described by a $2n \times 2n$ matrix. Symplecticity condition reads (we drop the $s_1 \to s_2$ subscript from M),

$$\tilde{M} S M = S, \tag{2.2}$$

where a tilde means taking the transpose of a matrix, and S is the matrix, sometimes called the *symplectic form*,

$$S = \begin{bmatrix} S_{2\times 2} & 0 & & & \\ 0 & S_{2\times 2} & & & \\ & & \ddots & & \\ & & & \ddots & \\ & & & & S_{2\times 2} \end{bmatrix},$$

consisting of a diagonal array of 2×2 matrix

$$S_{2\times 2} = \begin{bmatrix} 0 & 1 \\ -1 & 0 \end{bmatrix} .$$

It is easy to show that the condition (2.2) gives the condition $\det M = 1$ in a 1-D system, necessary and sufficient.

Symplecticity constraints　The symplecticity condition imposes a strong constraint on the n-D beam dynamical system through constraining its transfer maps. The general proof of symplecticity condition for n-D linear or nonlinear systems is omitted here. Additional comments, including its proof, can be found in Sec. 4.1 of Chapter 4.

Perhaps one would wonder where does the very strong symplecticity constraint originate from. To appreciate symplecticity, one should first appreciate the phase space. As an artifact and an abstract extension from 3-D space to 6-D, the degrees of freedom of the phase space are necessarily largely empty, and the motion of a particle's phase space coordinates must be heavily constrained. After all, all its momentum coordinates $p_1, p_2, \ldots$ are dictated to relate to the time derivatives of the spatial coordinates $q_1, q_2, \ldots$. The symplecticity condition is a reflection of this constraint and in a sense is a measure of the degree of emptiness in phase space. What is most amazing is that this property of unfilled degrees of freedom in phase space, as abstract as it may seem, has such an elegant, simple, and necessary-and-sufficient expression, the symplecticity condition (2.2), to describe it.

Homework 2.5 illustrates the symplecticity constraint more quantitatively. One might think of the $2n \times 2n$ matrix has $4n^2$ degrees of freedom in their matrix elements. Homework 2.5 shows however that only $n(2n+1)$ degrees of freedom are left when symplecticity is imposed. When n becomes big, it means only about half of the degrees of freedom are actually available. This is consistent with the fact that we are doubling the system dimensions when we went from real space to phase space. For the 3-D world we live in, only $\frac{21}{36}$ of our 6-D phase space's degrees of freedom are actually filled.

What is also remarkable is that the symplecticity condition (2.2) applies also to a nonlinear system if we identify M to be the *Jacobian matrix* of the nonlinear map, whose elements are defined as

$$M_{\alpha\beta} = \frac{\partial X_\alpha}{\partial (X_0)_\beta}, \tag{2.3}$$

where $(X_0)_\beta$ is the β-th component of the initial phase-space coordinates of a particle at $s = 0$, X_α is the α-th component of the final state X of the particle at an arbitrary final position s. The symplecticity condition (2.2) must hold for any s.

In a linear system, the Jacobian matrix is just the transfer matrix, and is independent of the particle's initial coordinates. In a nonlinear system, the Jacobian matrix M depends on the components of X_0, and the condition (2.2) must be satisfied for all X_0, thus imposing an *even stronger* constraint on the description of the nonlinear system.

Perhaps we can look at it another way. The system is Hamiltonian, i.e. it obeys the Hamilton equations,

$$\dot{\vec{q}} = \frac{\partial H}{\partial \vec{p}}, \qquad \dot{\vec{p}} = -\frac{\partial H}{\partial \vec{q}}.$$

This means the evolutions of both $\vec{q}(t)$ and $\vec{p}(t)$ are constrained by the one and the same Hamiltonian $H(\vec{q}, \vec{p}, t)$. It is therefore only reasonable that the description of the particle motion in terms of $(\vec{q}, \vec{p})$ must be constrained somehow.

Chain rule   Let us make a brief detour to discuss the Jacobian matrix. Note that the left-hand-side of Eq. (2.3) is specifically defined to be $M_{\alpha\beta}$, and not $M_{\beta\alpha}$. This convention is chosen so that the Jacobian matrices obey the chain rule

$$M(s_3|s_1) = M(s_3|s_2)M(s_2|s_1),$$

where $M(s_2|s_1)$ is the Jacobian matrix for the map from position s_1 to s_2, etc., i.e. the Jacobians of a string of maps can be multiplied sequentially to obtain the Jacobian of the total map.

The choice of $M_{\alpha\beta}$ on the right-hand-side of Eq. (2.3) is of course a result that we define matrix multiplication by multiplying rows into columns. Had we defined it as columns multiplied into rows, we would take the other definition, and the ordering of the matrices in the chain rule would reverse. The chain rule is of course valid, but the "reversed" ordering of matrix multiplications is only a convention; it does not have physical meaning.

Properties   The symplecticity condition resembles a "unitarity" condition due to the fact that the left-hand-side of (2.2) is *quadratic* in M, while the right-hand-side is almost a unit matrix. Such a condition obviously imposes a very strong constraint on the matrix M. For example, had the symplecticity condition read something such as $MS = S\tilde{M}$ with both sides of the equation linear in M, then M can be expanded or shrunk by some factor without consequences. If that were the case, beam dynamics would become very different — most likely much less interesting.

One expects that the unitarity nature tends to force the elements of transfer matrices to take on trigonometric or hyperbolic functions, noting the identities $\sin^2 + \cos^2 = 1$ and $\cosh^2 - \sinh^2 = 1$. The reason sines and cosines appear so often in the transport matrix elements is a consequence of the symplecticity condition. And that is also why particles in phase space naturally tend to *rotate*, although in real physical space they prefer to move in straight lines, according to Galileo.

Some of the consequences of the symplecticity condition follow simply from Eq. (2.2). In particular, one observes that

- the matrix S is symplectic; so is the identity matrix I;
- if M is symplectic, then $\det M = 1$;
- if M is symplectic, so are $-M, \tilde{M}$ and M^{-1};
- if both M are N are symplectic, then MN is symplectic;
- if λ is an eigenvalue with eigenvector E of a symplectic matrix M, then so is $\dfrac{1}{\lambda}$ with eigenvector SE. $\hspace{2cm}$ (2.4)

Liouville theorem   Another consequence of a symplectic map is that it obeys the *Liouville theorem*, i.e. the phase space volume is conserved as the system evolves according to the Hamiltonian map. Symplectic maps therefore are volume-preserving maps.[1] Liouville theorem follows because the Jacobian matrix, being symplectic, has unit determinant, which in turn assures that a volume element in phase space maintains its volume as it evolves with time. Note that the Liouville theorem is a consequence of the symplecticity condition, but the reverse is not necessarily true — it is true only in a 1-D system.

Let us continue a bit more with our 1-D system, with

$$\dot{x} = \frac{\partial H}{\partial p}, \qquad \dot{p} = -\frac{\partial H}{\partial x}.$$

Consider the map from time t to time $t + \Delta t$. The map is given by, to 1st order in Δt,

$$x(t + \Delta t) = x(t) + \frac{\partial H(x, p, t)}{\partial p} \Delta t,$$

---

[1]Volume here refers to the $2n$-D volume of phase space for a n-D system. For the case of 1-D system, this volume refers to an area.

$$p(t + \Delta t) \;=\; p(t) - \frac{\partial H(x,p,t)}{\partial x}\,\Delta t\,.$$

The Jacobian matrix is

$$J \;=\; \begin{bmatrix} \frac{\partial x(t+\Delta t)}{\partial x(t)} & \frac{\partial x(t+\Delta t)}{\partial p(t)} \\[4pt] \frac{\partial p(t+\Delta t)}{\partial x(t)} & \frac{\partial p(t+\Delta t)}{\partial p(t)} \end{bmatrix} \;=\; \begin{bmatrix} 1 + \frac{\partial^2 H}{\partial x \partial p}\Delta t & \frac{\partial^2 H}{\partial p^2}\Delta t \\[4pt] -\frac{\partial^2 H}{\partial x^2}\Delta t & 1 - \frac{\partial^2 H}{\partial x \partial p}\Delta t \end{bmatrix}.$$

It is easy to show that $\det J = 1$ to 1st order in Δt. Note that the Hamiltonian H here can be nonlinear in x and p (i.e. H may contain terms with orders equal or higher than cubic in x and p), and can depend on time t.

One might have a question how about the higher orders in Δt. For example, do we need to perform an additional step of symplectification — the subject of this chapter? Would higher orders in Δt destroy the Liouville theorem? The answer is that, in our discussion above, it is sufficient to keep to 1st order in Δt, as long as the Hamilton equation is applicable at all times t. In such a case, we can consider a truly infinitesimal Δt — unlike the numerical integrations to be discussed later in this chapter. After making the infinitesimal time step, the processes can be renewed and another time step can be initiated. By accumulating infinite number of infinitesimal time steps, one produces the evolution of the system in accordance with the Hamiltonian H.

Homework 2.1 The statement that the equation of motion is given by $y'' + K(s)y = 0$ for some function $K(s)$ is equivalent to saying the system is Hamiltonian for a linear system. Verify this equivalence. Find the Hamiltonian in terms of $K(s)$. Try to extend this problem to a 1-D nonlinear system.

Homework 2.2 If you know M, and you know that it is symplectic, how do you find M^{-1}?

Solution Use the identity

$$M^{-1} \;=\; -S\tilde{M}S\,.$$

This is a very useful expression because it would generally be time consuming to actually calculate M^{-1} by brute force, especially for the 2-D or 3-D cases. One consequence of this interesting formula is that M^{-1} will have the same matrix elements as M, just switched around positions and some flipped signs.

Homework 2.3 Prove the properties listed in (2.4).

Homework 2.4 Prove that, for the 1-D case, the symplecticity condition (2.2) is satisfied if and only if $\det(M) = 1$.

Homework 2.5
(a) Show that Eq. (2.2) imposes $n(2n-1)$ conditions on the matrix elements of M. The transfer matrix M therefore has only $n(2n+1)$ degrees of freedom in spite of the fact that it has $4n^2$ elements. For a 1-D system, there is one condition on the 4 matrix elements. The one condition is of course just $\det M = 1$.

n	$4n^2$	$n(2n-1)$	$n(2n+1)$
1	4	1	3
2	16	6	10
3	36	15	21

(b) For an n-D system with $n > 1$, convince yourself that the symplecticity condition implies $\det(M) = 1$, but $\det(M) = 1$ does not imply symplecticity.

Homework 2.6 Symplecticity condition is defined as $\tilde{M}SM = S$. Show that one might just equivalently define it as $MS\tilde{M} = S$.

Homework 2.7 Can symplectic matrices have $\det = -1$? This may not be as simple as it seems.

Homework 2.8 Show that the map constituted by successive symplectic maps is symplectic. This is valid even if the maps are nonlinear.

Solution Homework 2.3 was shown only for a linear system. Here, you need an additional step to show that combined Jacobian is equal to the product of the constituent Jacobians, i.e. prove the chain rule for the Jacobians.

Homework 2.9 Consider a 3-D dynamical system (6-D phase space). Let the equation of motion of the system have the form

$$\ddot{\vec{r}} = \vec{G}(\vec{r}),$$

for some vector function $\vec{G}(\vec{r})$.

(a) Is the system Hamiltonian?

(b) Show that this system is symplectic provided that $\nabla \times \vec{G} = \vec{0}$. This means if the force is derivable as a gradient of a potential, i.e. if $\vec{G} = -\nabla V$ for some potential function V, then the system is conservative and symplectic.

(c) If we replace the $\vec{G}(\vec{r})$ function by a more general case when it depends also on $\dot{\vec{r}}$, show that the system becomes nonsymplectic. For this system to be symplectic, the vector $\vec{G}$ must not have any dependence on momentum $\dot{\vec{r}}$.

We will return to this homework when we get to Homework 2.14.

Homework 2.10 Let M be a symplectic matrix, and another matrix N is related to M by $N = AM\tilde{A}$ with some matrix A. What is the condition on the matrix A for N to be also symplectic?

Solution It is OK if your answer is A should be symplectic, i.e. $\tilde{A}SA = S$. In this case $\det A = 1$. It is even better if your answer takes a note that $\tilde{A}SA = -S$ also works. In this case $\det A = -1$.

2.3 Symplectification of a linear map

We have considered solving the Hamilton equation of motion to obtain a transport matrix (linear) or a map (linear or nonlinear), that has the general form

$$x_f = x_f(x_i, x_i'), \qquad x_f' = x_f'(x_i, x_i').$$

We learned that the Jacobian matrix of the map is symplectic. In practical cases, however, the equation of motion is often not readily solvable analytically, and is to be solved by some numerical algorithm. There are many numerical algorithms, all of them aim for some good numerical accuracy to represent the map, but not all of them obey symplecticity. To perform numerical calculation of our Hamiltonian map, however, we will use only the symplectic ones. (The reason will be given later — and it is not only to maintain a long-term stability.) In fact, to assure this outcome, we will purposely look for algorithms with built-in symplecticity.

Let us first consider a 2-D linear system. Linear motion in (x, x', y, y') can be described by maps in matrix form. For example, the map for a horizontally focusing quadrupole magnet is

$$M = \begin{bmatrix} \cos kL & \frac{1}{k}\sin kL & 0 & 0 \\ -k\sin kL & \cos kL & 0 & 0 \\ 0 & 0 & \cosh kL & \frac{1}{k}\sinh kL \\ 0 & 0 & k\sinh kL & \cosh kL \end{bmatrix}, \tag{2.5}$$

where $k = \sqrt{\frac{G}{B_0\rho}}$ specifies the quadrupole strength.

Often a good approximation can be obtained by letting $L \to 0$, $\frac{G}{B_0\rho} \to \infty$, while holding $\frac{1}{f} \equiv \frac{GL}{B_0\rho}$ fixed. We then obtain the short-magnet approximation of Eq. (2.5),

$$M \approx \begin{bmatrix} 1 & 0 & 0 & 0 \\ -\frac{1}{f} & 1 & 0 & 0 \\ 0 & 0 & 1 & 0 \\ 0 & 0 & \frac{1}{f} & 1 \end{bmatrix}. \tag{2.6}$$

In the absence of x-y coupling, the matrix M degenerates into two 2×2 blocks. One can treat the x and y motions separately by studying their respective 2×2 blocks.

Because M satisfies the symplecticity condition, we know that

$$\det M = 1. \tag{2.7}$$

This condition is true whether M is 2×2, 4×4, or 6×6, and whether the system is stable or unstable. We also mentioned that for a linear 1-D case with 2×2 matrices, the symplecticity condition holds if and only if Eq. (2.7) holds. For higher dimensions, the symplecticity condition implies Eq. (2.7), but not vice versa. Obviously, Eqs. (2.5) and (2.6) satisfy Eq. (2.7).

Short magnet   Equation (2.6) is the map by taking the limit of zero magnet length. What do we do for a magnet with finite length, even though the length is short? Consider now a short quadrupole of length Δs, or a short slice of a quadrupole magnet from position s to position $s + \Delta s$. The map for the x motion is

$$M_{s \to s+\Delta s} = \begin{bmatrix} \cos k\Delta s & \frac{1}{k} \sin k\Delta s \\ -k \sin k\Delta s & \cos k\Delta s \end{bmatrix}.$$
(2.8)

One may be tempted to Taylor expand this matrix in terms of the small quantity Δs. This leads to

$$M_{s \to s+\Delta s} = \begin{bmatrix} 1 & 0 \\ 0 & 1 \end{bmatrix} + \Delta s \begin{bmatrix} 0 & 1 \\ -k^2 & 0 \end{bmatrix} + \Delta s^2 \begin{bmatrix} -\frac{k^2}{2} & 0 \\ 0 & -\frac{k^2}{2} \end{bmatrix} + \cdots .$$
(2.9)

Keeping only the leading term $\begin{bmatrix} 1 & 0 \\ 0 & 1 \end{bmatrix}$ is not a very useful thing to do because it loses all the physics introduced by the quadrupole. Obviously we will have to keep at least the terms of 1st order in Δs. Keeping up to the Δs terms while truncating all higher order terms, i.e.

$$M_{s \to s+\Delta s} \approx \begin{bmatrix} 1 & \Delta s \\ -k^2 \Delta s & 1 \end{bmatrix},$$
(2.10)

however, constitutes a problem because it violates the condition (2.7). This exercise says that, if truncations must be performed, then they must be performed carefully. A brute force truncation, like the one leading to Eq. (2.10), generally leads to violation of fundamental properties.

One might attempt to improve the symplecticity by including higher order terms in Δs. The table below shows the determinant as a function of the order kept ($\theta = k\Delta s$ with $|\theta| \ll 1$):

order kept	determinant
1	$1 + \theta^2$
2	$1 + \frac{1}{4}\theta^4$
3	$1 - \frac{1}{12}\theta^4 + \frac{1}{36}\theta^6$
4	$1 - \frac{1}{72}\theta^6 + \frac{1}{576}\theta^8$
5	$1 + \frac{1}{360}\theta^6 - \frac{1}{960}\theta^8 + \frac{1}{14400}\theta^{10}$

But as one can see, all we have achieved is increasing numerical accuracy by going to higher orders. All cases remain nonsymplectic.

The determinant of the map has the physical meaning of the magnification factor of the phase space area due to the mapping process (Homework 2.12). If $\det \neq 1$, phase space area is not conserved, violating Liouville theorem. If we take $n = 1$, for example, the phase space area gets magnified by a factor $(1 + \theta^2)$ in each application of the map. If not used with caution, it may lead to erroneous conclusions. In particular, repeatedly iterating map (2.10) could lead to wrong conclusions in terms of the long-term stability of the system.

One might ask why don't we just use the cosine and sine functions in the map and not to bother with the Taylor expansions running the risk of non-symplecticity. The answer is that the quadrupole case is being used only as an explicitly soluble example. The conclusions we will obtain using this example will be applied to cases when an analytical solution is not available, such as the case when the thick quadrupole is replaced by a thick sextupole or other nonlinear elements.

Note there is no objection to the fact that the map (2.10) is approximate. The objection is that it is nonsymplectic. So we need to explore ways to find approximations without sacrificing symplecticity, i.e. we need *symplectification* techniques.

We have been emphasizing the need of long-term tracking of particle motion as a main reason for requiring symplecticity of the beam dynamics maps. Perhaps it is worthwhile to mention another reason, of a more philosophical nature but at least equally important, namely symplecticity assures the integrity of the beam dynamics formulation being pursued. Violation of symplecticity casts doubt on whether the formulation is even correct if the system is known to be Hamiltonian. Insistence of symplecticity is advised to assure the integrity of the analysis and the formulation of the problem at hand. We consider the integrity of the formulation of the problem to be important.

Symplectification model   One trick to symplectify is to *artificially* add a $\mathcal{O}(\Delta s^2)$ term to Eq. (2.10) as follows,

$$M_{s \to s+\Delta s} \approx \begin{bmatrix} 1 & \Delta s \\ -k^2 \Delta s & 1 - k^2 \Delta s^2 \end{bmatrix}. \tag{2.11}$$

Note that truncating Eq. (2.9) to $\mathcal{O}(\Delta s^2)$ would give

$$M_{s \to s+\Delta s} \approx \begin{bmatrix} 1 - \frac{1}{2}k^2 \Delta s^2 & \Delta s \\ -k^2 \Delta s & 1 - \frac{1}{2}k^2 \Delta s^2 \end{bmatrix}, \tag{2.12}$$

which differs from (2.11). On the other hand, map (2.11) is symplectic while map (2.12) is not, in spite of the fact that (2.12) is numerically more accurate than (2.11).

Is there a physical meaning to Eq. (2.11)? The answer is yes. In fact, we are at this point launching onto the beginning of two important topics: map symplectification and canonical integration.[2] Both are active fields of research.

The physical meaning of Eq. (2.11) is seen as follows. One way to assure symplecticity is to model the quadrupole as a combination of interlacing drift spaces and lumped kicks. For example, one might model the quadrupole as sketched in Fig. 2.1(a). In the drift region, the map is given by

$$\begin{bmatrix} 1 & L \\ 0 & 1 \end{bmatrix}. \tag{2.13}$$

---

[2]See for example, H. Yoshida, Sec. 2.3.8, Handbook Accel. Phys. & Eng., 2nd edition, World Scientific (2013); H. Yoshida, Phys. Lett. A150, 262 (1990); E. Forest and R.D. Ruth, Physica D43, 105 (1990).

The lumped kick is obtained by concentrating the quadrupole strength to act as a δ-function kick to the passing particles. The matrix that describes that action is

$$\begin{bmatrix} 1 & 0 \\ -k^2 L & 1 \end{bmatrix}, \qquad (2.14)$$

as can be easily derived using the equation of motion $x'' = -k^2 x$.

The drift map (2.13) is symplectic — its determinant is equal to 1. What is important, however, is the general property that lumped kicks give symplectic maps — matrix (2.14) has determinant of 1. So the quadrupole map is decomposed into a product of two maps, one is drift map that keeps x' fixed while changing x, the other is a lumped *kick map* that keeps x fixed while changing x'. By modeling an accelerator element (linear or nonlinear) as combination of drifts and lumped kicks, one obtains a symplectic model of the element.

Incidentally, a kick map is a map that holds x fixed while changing x' by an amount that depends only on x. One could also consider the case when x' is held fixed while changing x by an amount that depends only on x'. Such is of course the case of a drift space. In that sense, there is no reason not to refer to a drift map also as a kick map. A drift map is symplectic just like a kick map is.

The observation that lumped kick maps, linear or nonlinear, are always symplectic can be proved easily as follows. Consider the kick map

$$\Delta x' \;=\; f(x) \,.$$

In general, $f(x)$ can be nonlinear. Consider a particle with initial coordinates (x_0, x_0') undergoing this map. Its final coordinates become

$$x_1 \;=\; x_0, \qquad x_1' \;=\; x_0' + f(x_0) \,.$$

The Jacobian matrix reads

$$M \;=\; \begin{bmatrix} 1 & 0 \\ f'(x_0) & 1 \end{bmatrix} \,.$$

Symplecticity is established because $\det M = 1$. The quadrupole lumped kick (2.14) is only a special case.

It needs to be noted however that this observation of kick map symplecticity holds only for 1-D systems. In general, kick maps in higher dimensions do not necessarily assure symplecticity. See Homework 2.14. Also a simple example in Homework 4.11 of Chapter 4 will show that kick maps are not the only way to symplectify a map. Generally speaking, the idea of universal symplectification by kick maps is applicable to the 1-D systems but requires additional care for higher dimensions.

We now return to our 1-D quadrupole example. If the quadrupole is modeled as a drift followed by a lumped kick, the total map for the quadrupole is

$$M \;=\; \begin{bmatrix} 1 & 0 \\ -k^2 L & 1 \end{bmatrix} \begin{bmatrix} 1 & L \\ 0 & 1 \end{bmatrix} = \begin{bmatrix} 1 & L \\ -k^2 L & 1 - k^2 L^2 \end{bmatrix}, \qquad (2.15)$$

which obviously has det $= 1$. But Eq. (2.15) has just reproduced Eq. (2.11). In other words, the artificial symplectification introduced in Eq. (2.11) could be a consequence of the particular *modeling* represented by Fig. 2.1(a).

This observation then opens up other ways to symplectify. For example, the model of Fig. 2.1(b) gives

$$
M = \begin{bmatrix} 1 & L \\ 0 & 1 \end{bmatrix} \begin{bmatrix} 1 & 0 \\ -k^2 L & 1 \end{bmatrix} = \begin{bmatrix} 1 - k^2 L^2 & L \\ -k^2 L & 1 \end{bmatrix}, \tag{2.16}
$$

which represents another way to symplectify map (2.10).

Both Figs. 2.1(a) and (b), although exactly symplectic, give maps that are accurate only up to $\mathcal{O}(L)$ [leading error term is of order $\mathcal{O}(L^2)$]. One way to improve the accuracy of the map while maintaining its symplecticity is to consider Fig. 2.1(c), the *thin-lens model*. It gives

$$
\begin{aligned}
M &= \begin{bmatrix} 1 & \frac{1}{2}L \\ 0 & 1 \end{bmatrix} \begin{bmatrix} 1 & 0 \\ -k^2 L & 1 \end{bmatrix} \begin{bmatrix} 1 & \frac{1}{2}L \\ 0 & 1 \end{bmatrix} \\
&= \begin{bmatrix} 1 - \frac{1}{2}k^2 L^2 & L - \frac{1}{4}k^2 L^3 \\ -k^2 L & 1 - \frac{1}{2}k^2 L^2 \end{bmatrix}.
\end{aligned} \tag{2.17}
$$

The thin-lens model is accurate up to order $\mathcal{O}(L^2)$, as can be shown by comparing (2.17) with the exact expression (2.8).

Note that, in this chapter, we have tried to avoid calling Eq. (2.6) the thin-lens model, and preferred to reserve the name for the much more accurate approximation (2.17).

We have chosen a horizontally focusing quadrupole to be our example application. All results are applicable to a horizontally defocusing (vertically focusing) quadrupole if we replace $k^2 \to -k^2$.

Numerical iterations of the maps (2.12), (2.15), (2.16), and (2.17) are shown respectively in Figs. 2.2(c) through (f) for $k\Delta s = 0.2$ (Δs is the step size, and the map is applied repeatedly to produce Fig. 2.2).

Figure 2.2(a) is obtained using the exact map (2.8), and the particle trajectory makes up a perfect circle in the phase space. Figure 2.2(f) is almost, but not quite, a circle. The deviation of Fig. 2.2(f) from a circle will be shown in Fig. 2.5(a) later. Figure 2.2(c) based on the nonsymplectic map (2.12), exhibits an outward spiral motion, although the spiraling is slower than that shown in Fig. 2.2(b) based on Eq. (2.10).

It is clear from Fig. 2.2 that for long-term stability purposes, enforcing the symplecticity condition is important. If a map is approximate but symplectic, the contours of particle trajectory in phase space may be slightly distorted, but the long-term stability of the motion is preserved, as shown in Figs. 2.2(d,e,f). This is no longer true if the symplecticity of the map is compromised, as shown in Figs. 2.2(b,c).

Homework 2.11 Follow the outline in the text to a more general case for a 1-D Hamiltonian system with equation of motion $y''(s) = f(y, s)$ for arbitrary force

Figure 2.1: Symplectification models.

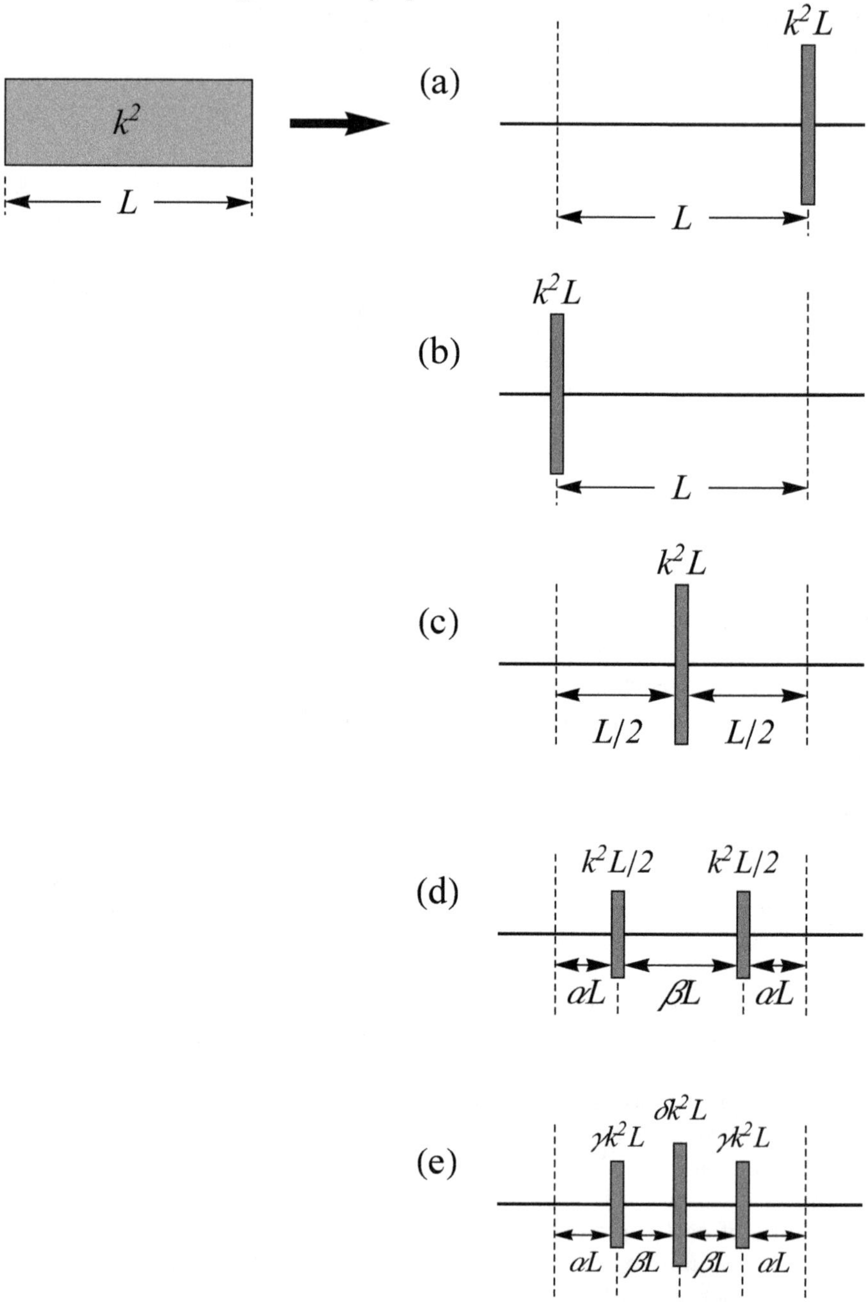

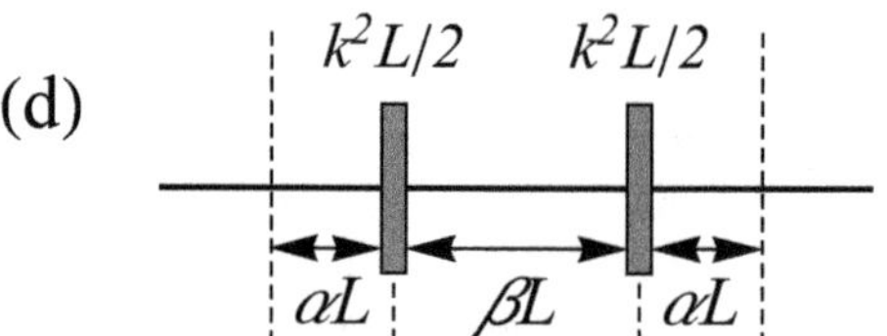

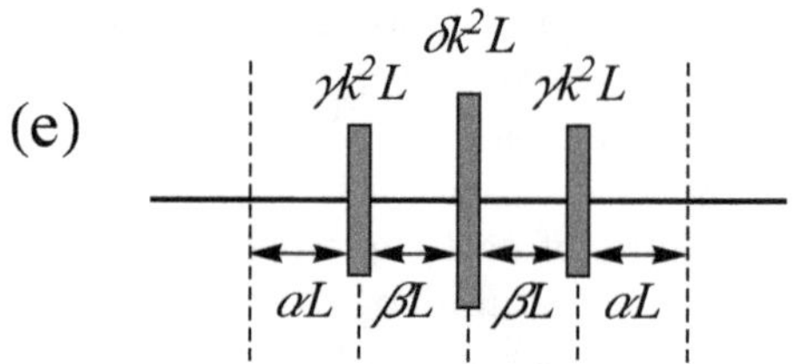

Figure 2.2: Phase space trajectories of a particle traversing a quadrupole as predicted using various tracking algorithms. (a) the exact map (2.8), (b) non-symplectic map (2.10), (c) nonsymplectic map (2.12), (d) symplectic map (2.15), (e) symplectic map (2.16), (f) symplectic thin-lens map (2.17).

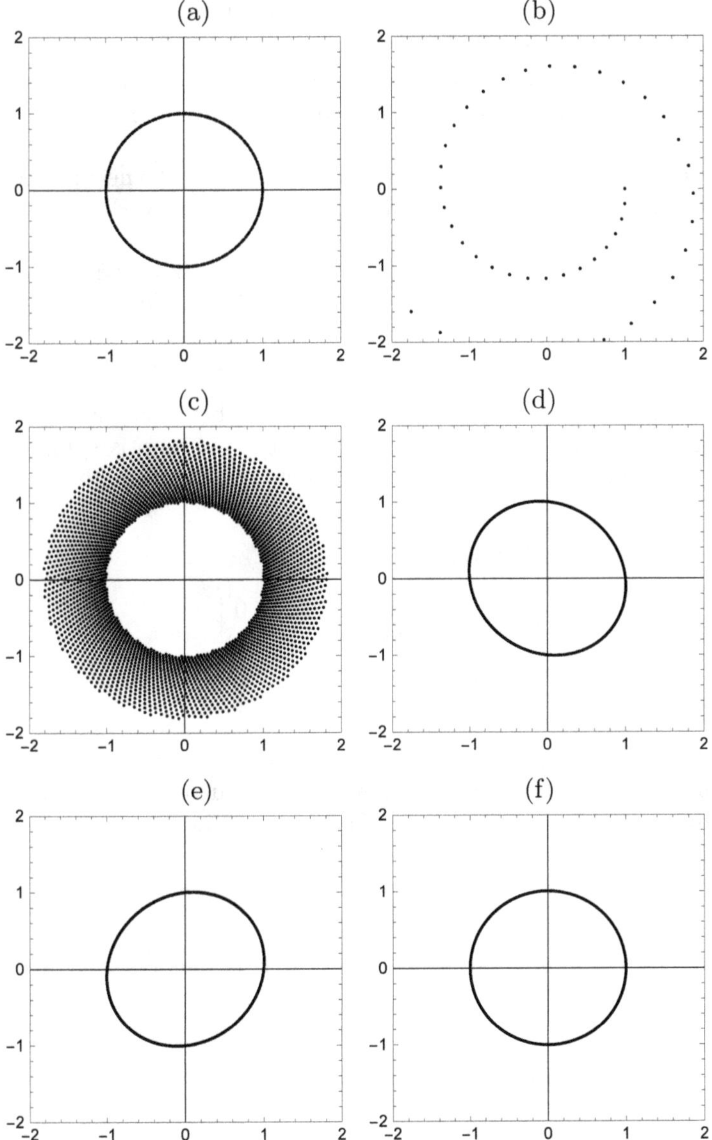

function $f(y, s)$. Show that the lumped kicks of this system, whether linear or nonlinear, always give symplectic maps.

Homework 2.12 Consider a system with equations of motion

$$\dot{q} \;=\; f(q,p,t), \qquad \dot{p} \;=\; g(q,p,t)\,.$$

Consider an infinitesimal time step dt and form the Jacobian matrix from time t to $t + dt$. Find the deviation of its determinant from 1.

It was shown in Eq. (1.43) of Chapter 1 that the quantity $\left(\frac{\partial f}{\partial q} + \frac{\partial g}{\partial p}\right)$ has the meaning of the changing rate of the phase space area element. Here we show that it is related to the determinant of the Jacobian matrix.

Homework 2.13 This is a simple exercise. Show that the lumped kick map for the system $x'' + k^2 x = 0$ over a distance L is given by Eq. (2.14).

Homework 2.14 We declared kick maps are symplectic in 1-D systems. This is not necessarily true for 2-D systems. Consider the map

$$\Delta x' \;=\; f(x,y), \qquad \Delta y' \;=\; g(x,y)\,,$$

for example. This arguably qualifies as a kick map because f and g depend only on x, y and do not depend on x', y'. Show that this map is symplectic only if $\frac{\partial f}{\partial y} = \frac{\partial g}{\partial x}$. A 3-D case was given in Homework 2.9 — although not revealed at the time.

Solution The Jacobian is given by

$$M \;=\; \begin{bmatrix} 1 & 0 & 0 & 0 \\ \frac{\partial f}{\partial x} & 1 & \frac{\partial f}{\partial y} & 0 \\ 0 & 0 & 1 & 0 \\ \frac{\partial g}{\partial x} & 0 & \frac{\partial g}{\partial y} & 1 \end{bmatrix}\,.$$

Homework 2.15 Still another symplectified form of a quadrupole, different from the model (2.11), valid to order $\mathcal{O}(L^2)$, is

$$\begin{bmatrix} 1 - \frac{1}{2}k^2 L^2 & L \\ -k^2 L + \frac{1}{4}k^4 L^3 & 1 - \frac{1}{2}k^2 L^2 \end{bmatrix}\,. \tag{2.18}$$

What is its corresponding physical model for this particular symplectification algorithm?

2.4   Higher order integrator

A symplectic model, such as (2.15), (2.16), or (2.17), is useful to model accelerator elements. But more importantly, when the accelerator element is nonlinear, we often do not have a closed form expression for its map as we do in the quadrupole example. These nonlinear maps have to be obtained by *ray tracing*. Performing ray tracing over many steps is also called *integrating*, thus the term "symplectic integrator".

For long-term stability considerations, it is impractical to trace with hundreds of steps per element due to computer limitations. One therefore searches for efficient ways to model an element, which are accurate and, for purpose of long-term tracking, are also symplectic. The quadrupole example is just an illustration of this search activity, as we shall continue in this section. Models obtained, some of them illustrated in Fig. 2.1, however, are all applicable to thick nonlinear elements.

Third order integrator One may proceed to consider Fig. 2.1(d) with the hope of finding a model that represents the quadrupole to order $\mathcal{O}(L^3)$. Obviously one has the condition $2\alpha + \beta = 1$. The transfer matrix, for the case of a thick quadrupole, is

$$
\begin{aligned}
M &= \begin{bmatrix} 1 & \alpha L \\ 0 & 1 \end{bmatrix} \begin{bmatrix} 1 & 0 \\ -\frac{1}{2}k^2 L & 1 \end{bmatrix} \begin{bmatrix} 1 & \beta L \\ 0 & 1 \end{bmatrix} \begin{bmatrix} 1 & 0 \\ -\frac{1}{2}k^2 L & 1 \end{bmatrix} \begin{bmatrix} 1 & \alpha L \\ 0 & 1 \end{bmatrix} \\
&= \begin{bmatrix} 1 - \frac{1}{2}k^2 L^2 + \frac{1}{4}\alpha\beta k^4 L^4 & L - \alpha(\alpha+\beta)k^2 L^3 + \frac{1}{4}\alpha^2\beta k^4 L^5 \\ -k^2 L + \frac{1}{4}\beta k^4 L^3 & 1 - \frac{1}{2}k^2 L^2 + \frac{1}{4}\alpha\beta k^4 L^4 \end{bmatrix}.
\end{aligned} \tag{2.19}
$$

Equation (2.19) does give $\det M = 1$. But this map cannot represent the quadrupole to order $\mathcal{O}(L^3)$. Compared with the exact map (2.8), in order to represent the quadrupole to order $\mathcal{O}(L^3)$, we need

$$
\begin{aligned}
\alpha(\alpha + \beta) &= \frac{1}{6}, \\
\frac{\beta}{4} &= \frac{1}{6}, \\
2\alpha + \beta &= 1,
\end{aligned} \tag{2.20}
$$

which, unfortunately, does not have a solution — 2 unknowns with 3 equations to satisfy.

Fourth order integrator Interestingly enough, it is possible to consider Fig. 2.1(e) to obtain a higher-order map correct to order $\mathcal{O}(L^4)$ [errors of order $\mathcal{O}(L^5)$]. We obviously have, to begin with, the conditions

$$
\begin{aligned}
2\alpha + 2\beta &= 1, \\
2\gamma + \delta &= 1.
\end{aligned} \tag{2.21}
$$

The map is given by

$$
\begin{aligned}
M &= \begin{bmatrix} 1 & \alpha L \\ 0 & 1 \end{bmatrix} \begin{bmatrix} 1 & 0 \\ -\gamma k^2 L & 1 \end{bmatrix} \begin{bmatrix} 1 & \beta L \\ 0 & 1 \end{bmatrix} \begin{bmatrix} 1 & 0 \\ -\delta k^2 L & 1 \end{bmatrix} \\
&\times \begin{bmatrix} 1 & \beta L \\ 0 & 1 \end{bmatrix} \begin{bmatrix} 1 & 0 \\ -\gamma k^2 L & 1 \end{bmatrix} \begin{bmatrix} 1 & \alpha L \\ 0 & 1 \end{bmatrix}
\end{aligned}
$$

$$
= \begin{bmatrix}
1 - \frac{1}{2}k^2L^2 + \beta\gamma(\alpha + \frac{\delta}{2})k^4L^4 & L - (\frac{\delta}{4} + \alpha\gamma + 2\alpha\beta\gamma)k^2L^3 \\
\quad -\alpha\beta^2\gamma^2\delta k^6L^6 & \quad +2\alpha\beta\gamma(\alpha\gamma + \frac{1}{2}\delta)k^4L^5 - \alpha^2\beta^2\gamma^2\delta k^6L^7 \\
\\
-k^2L + \beta\gamma(1+\delta)k^4L^3 & 1 - \frac{1}{2}k^2L^2 + \beta\gamma(\alpha + \frac{\delta}{2})k^4L^4 \\
\quad -\beta^2\gamma^2\delta k^6L^5 & \quad -\alpha\beta^2\gamma^2\delta k^6L^6
\end{bmatrix} .
$$

$$(2.22)$$

The conditions for (2.22) to represent the exact map to order $\mathcal{O}(L^4)$ are, in addition to (2.21),

$$
\begin{aligned}
\beta\gamma\left(\alpha + \frac{\delta}{2}\right) &= \frac{1}{24}, \\
\beta\gamma(1 + \delta) &= \frac{1}{6}, \\
\frac{\delta}{4} + \alpha\gamma + 2\alpha\beta\gamma &= \frac{1}{6}.
\end{aligned}
$$

$$(2.23)$$

Equations (2.21) and (2.23) — 4 unknowns, 5 equations — do have solutions! In fact, one finds

$$
\begin{aligned}
\beta &= \frac{1 - 2^{1/3}}{2(2 - 2^{1/3})} \approx -0.1756, \\
\alpha &= \frac{1}{2} - \beta = \frac{1}{2(2 - 2^{1/3})} \approx 0.6756, \\
\gamma &= \frac{1}{24\beta^2} = \frac{1}{2 - 2^{1/3}} \approx 1.3512, \\
\delta &= 1 - 2\gamma = -\frac{2^{1/3}}{2 - 2^{1/3}} \approx -1.7024.
\end{aligned}
$$

$$(2.24)$$

Note that β and δ are negative. This means the model involves seven computational steps as shown in Fig. 2.3. Following these 7 steps would yield a symplectic model of a thick quadrupole — or any thick nonlinear element — to order $\mathcal{O}(L^4)$.

The map for a quadrupole is then obtained by substituting Eq. (2.24) into Eq. (2.22),

$$
M = \begin{bmatrix}
1 - \frac{1}{2}k^2L^2 + \frac{1}{24}k^4L^4 & L - \frac{1}{6}k^2L^3 + \frac{1-2^{1/3}}{24(2-2^{1/3})^2}k^4L^5 \\
\quad + \frac{2^{1/3}}{48(2-2^{1/3})^3}k^6L^6 & \quad + \frac{2^{1/3}}{96(2-2^{1/3})^4}k^6L^7 \\
\\
-k^2L + \frac{1}{6}k^4L^3 & 1 - \frac{1}{2}k^2L^2 + \frac{1}{24}k^4L^4 \\
\quad + \frac{2^{1/3}}{24(2-2^{1/3})^2}k^6L^5 & \quad + \frac{2^{1/3}}{48(2-2^{1/3})^3}k^6L^6
\end{bmatrix} .
$$

$$(2.25)$$

It can be explicitly checked that the determinant of (2.25) is 1. This M is correct up to $\mathcal{O}(L^4)$. By introducing artificial terms of the orders of $\mathcal{O}(L^5)$, $\mathcal{O}(L^6)$, and $\mathcal{O}(L^7)$, therefore, we have symplectified the map.

Figure 2.3: Seven steps in the 4th order symplectic integration.

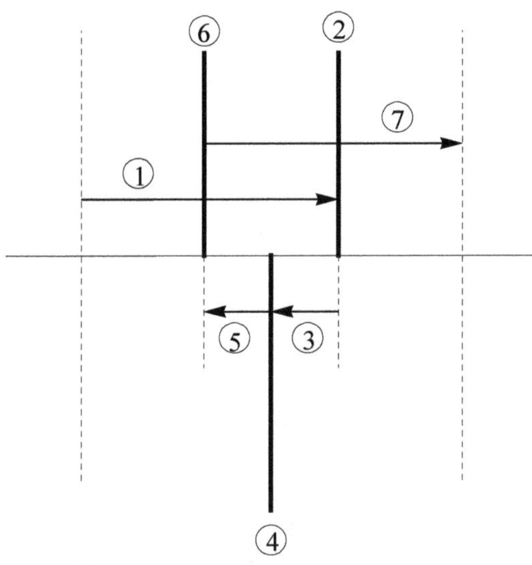

A way (not unique, see footnote 4 later) to quantify the error made by the map (2.25) is to compare its trace with that of the exact map. The exact map gives a trace of $2\cos kL$. The difference between the two traces is

$$\mathrm{tr}(M) - 2\cos kL \;=\; \frac{1}{360}\left(1 + \frac{15 \cdot 2^{1/3}}{(2 - 2^{1/3})^3}\right) k^6 L^6 + \mathcal{O}(k^8 K^8)$$

$$\approx \; 0.132286\, k^6 L^6\,.$$

Sixth order integrator Using the quadrupole as example to go to higher order integrators becomes cumbersome. One way to see the underlying mechanism in general is to apply Lie algebra to the analysis. We will not explore Lie algebra here, except to say that using Lie algebra not only gives the derivation of the above-mentioned integrators but also allows a systematic way to extend to higher order integrators.

For example, a sixth order integrator is found to be as follows (notation the same as in Table 2.1),[3]

$$(c_1 L)(d_1 SL)(c_2 L)(d_2 SL)(c_3 L)(d_3 SL)(c_4 L)(d_4 SL)(c_5 L)(d_5 SL)$$
$$\times (c_6 L)(d_6 SL)(c_7 L)(d_7 SL)(c_8 L)(d_8 SL)(c_9 L)(d_9 SL)(c_{10} L)\,, \qquad (2.26)$$

where

$$d_1 = d_3 = d_7 = d_9 = \frac{1}{(2 - 2^{1/3})(2 - 2^{1/5})} \approx 1.58722\,,$$

---

[3]H. Yoshida, Phys. Lett. A150, 262 (1990).

$$d_2 = d_8 = -\frac{2^{1/3}}{(2-2^{1/3})(2-2^{1/5})} \approx -1.99978\,,$$

$$d_4 = d_6 = -\frac{2^{1/5}}{(2-2^{1/3})(2-2^{1/5})} \approx -1.82324\,,$$

$$d_5 = \frac{2^{8/15}}{(2-2^{1/3})(2-2^{1/5})} \approx 2.29714\,,$$

$$c_1 = c_{10} = \frac{1}{2(2-2^{1/3})(2-2^{1/5})} \approx 0.793612\,,$$

$$c_2 = c_3 = c_8 = c_9 = \frac{1-2^{1/3}}{2(2-2^{1/3})(2-2^{1/5})} \approx -0.206277\,,$$

$$c_4 = c_7 = \frac{1-2^{1/5}}{2(2-2^{1/3})(2-2^{1/5})} \approx -0.118009\,,$$

$$c_5 = c_6 = -\frac{2^{1/5}(1-2^{1/3})}{2(2-2^{1/3})(2-2^{1/5})} \approx 0.23695\,. \tag{2.27}$$

One can check that $\sum_i c_i = \sum_i d_i = 1$.

We see that the 2nd order symplectic integrator (thin-lens approximation) requires one lumped kick. The 4th order integrator requires 3 lumped kicks. The 6th order integrator requires 9 lumped kicks. Together with the interlacing drift spaces, the 19-step algorithm is illustrated in Fig. 2.4.

A word of caution is to be issued here. Application of such high order map should be handled with some care because it involves accurate internal cancelation of numerical values. Concatenating all 19 steps into a single map, for example, as we did in Eq. (2.25) for the 4th order integrator, would be ill-advised; it is better to calculate numerically in 19 steps individually.

So far and particularly explicitly in Table 2.1, we have kicks interlaced by drifts. This is not a requirement. Instead of drifts, one can have any map which has an exact expression — exact, therefore symplectic. Most likely, this map with exact expression is a linear map. For example, in Homework 2.18, we have a combined-function magnet with quadrupole and sextupole fields. One can have sextupole kicks interlaced by quadrupole maps. The fact that the interlacing map does not have to be drifts is also evident if one uses Lie algebra to hunt for symplectic integrators.

Homework 2.16 In obtaining Eq. (2.24), we have first obtained an equation for β,

$$48\beta^3 - 24\beta^2 + 1 = 0\,. \tag{2.28}$$

Equation (2.24) is then the real solution of (2.28). But there are also two other complex solutions given by

$$\beta_\pm = \frac{1}{4}\left(\frac{1}{2-2^{1/3}} \pm \frac{2^{2/3}-1}{3^{1/2}2^{1/3}}i\right)$$
$$= 0.33780 \pm 0.06729i\,, \tag{2.29}$$

Figure 2.4: Nineteen steps in the 6th order symplectic integrator.

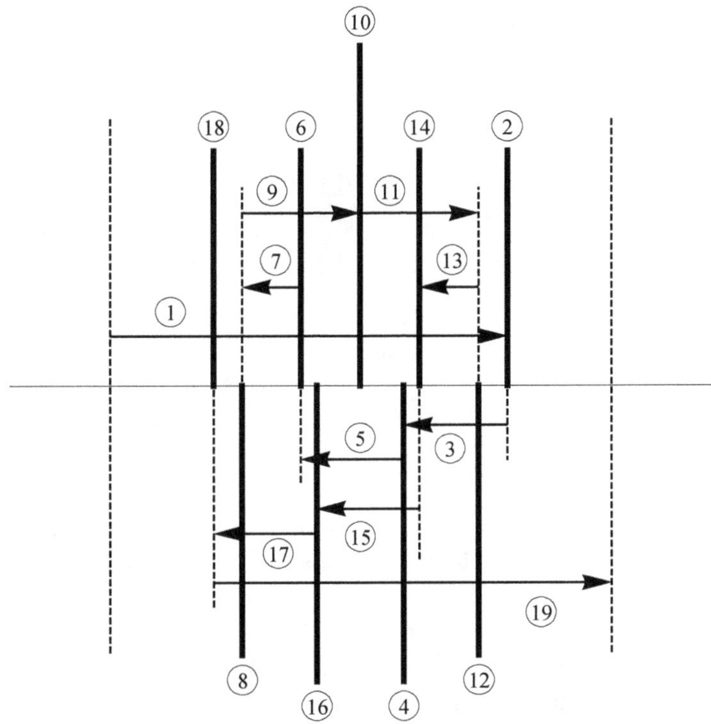

which in turn gives

$$\begin{aligned}
\alpha &= 0.16220 \mp 0.06729i\,, \\
\gamma &= 0.32440 \mp 0.13459i\,, \\
\delta &= 0.35121 \pm 0.26917i\,.
\end{aligned} \qquad (2.30)$$

Choosing these complex values are legitimate as far as symplectification is concerned. The total matrix, however, now contains higher order symplectifying terms which are complex.[4]

Homework 2.17 The 3rd order model Fig. 2.1(d) assumed a symmetry around the mid-point of the accelerator element and we showed there was no solution. An extra free parameter can be introduced if the symmetry is sacrificed. Is there a solution for this asymmetric 3rd order integrator?

---

[4]Would anyone try to use these complex symplectification in actual accelerator applications? Supposedly no one would. But it might still be considered for some purposes. For example, if used, the imaginary parts of your results might give an indication of the magnitude of the truncation error, even when, again, the symplecticity is not being questioned.

Solution Consider two kicks $-\delta k^2 L$ and $-\epsilon k^2 L$ spaced by three free spaces αL, βL and γL. For the system to describe a third order map, we find two solutions

$$
\begin{aligned}
\alpha &= \frac{1}{4} \mp i\frac{1}{4\sqrt{3}} = 0.25 \mp 0.14434i\,, \\
\beta &= \frac{1}{2}\,, \\
\gamma &= \alpha^* = 0.25 \pm 0.14434i\,, \\
\delta &= 2\alpha = 0.5 \mp 0.28868i\,, \\
\epsilon &= 2\alpha^* = 0.5 \pm 0.28868i\,.
\end{aligned}
\tag{2.31}
$$

There is no real solution and the resulting third order map is necessarily complex. The whole symplectified map becomes

$$
\begin{bmatrix}
1 - \frac{1}{2}k^2 L^2 + \frac{3-i\sqrt{3}}{72}k^4 L^4 & L - \frac{1}{6}k^2 L^3 + \frac{1}{72}k^4 L^5 \\
-k^2 L + \frac{1}{6}k^4 L^3 & 1 - \frac{1}{2}k^2 L^2 + \frac{3-i\sqrt{3}}{72}k^4 L^4
\end{bmatrix}.
$$

Homework 2.18 Consider the 1-D motion in a magnet with combined-function quadrupole and sextupole fields. Let the equation of motion be

$$
x'' + k^2 x = Sx^2\,.
\tag{2.32}
$$

Let L be the length of the magnet. Give the 2nd order symplectic integration (thin-lens approximation) for the map from the entrance to the exit of this magnet, considering three different approaches.

(a) Model both the quadrupole and the sextupole components as single lumped kicks in the middle of the drift L. Does the ordering of these two lumped kicks matter in your result?

(b) Model the quadrupole component as a thick-lens magnet, and the sextupole component as a lumped kick at the middle of the quadrupole.

(c) The results obtained in (a) and (b) are symplectic.
This exercise demonstrates the fact that the drifts in the models of Fig. 2.1 can be replaced by any map which has exact (symplectic) solution, quadrupole being one example.

Homework 2.19 Consider a horizontal sector dipole magnet of bending angle θ and bending radius ρ. The exact 4×4 map for an on-momentum particle is

$$
\begin{bmatrix}
\cos\theta & \rho\sin\theta & 0 & 0 \\
-\frac{1}{\rho}\sin\theta & \cos\theta & 0 & 0 \\
0 & 0 & 1 & \rho\theta \\
0 & 0 & 0 & 1
\end{bmatrix}.
$$

(a) Find the 2nd order thin-lens symplectic map of the magnet. Give your result in the form of a 4×4 matrix.

(b) Extend your result to give the thin-lens map of an off-momentum particle with $\frac{\Delta P}{P} = \delta_0$.

Table 2.1: Various techniques of explicit canonical integration for an accelerator element of strength S and length L. The symbol (L) means a drift length L. The symbol (SL) means a kick of integrated strength (SL). Values of $\alpha, \beta, \gamma, \delta$ are given by Eq. (2.24). See also Eqs. (2.26) and (2.27).

Integrator	Model	Error
First order	$(L)(SL)$	$\mathcal{O}(L^2)$
First order	$(SL)(L)$	$\mathcal{O}(L^2)$
Ray tracing	$(\frac{L}{n})(\frac{SL}{n}) \cdots$ repeat n times	$\mathcal{O}(\frac{L^2}{n})$
Second order (thin-lens)	$(\frac{L}{2})(SL)(\frac{L}{2})$	$\mathcal{O}(L^3)$
Ray tracing	$(\frac{L}{2n})(\frac{SL}{n})(\frac{L}{2n}) \cdots$ repeat n times	$\mathcal{O}(\frac{L^3}{n^2})$
Fourth order	$(\alpha L)(\gamma SL)(\beta L)(\delta SL)(\beta L)(\gamma SL)(\alpha L)$	$\mathcal{O}(L^5)$
Ray tracing	$(\frac{\alpha L}{n})(\frac{\gamma SL}{n})(\frac{\beta L}{n})(\frac{\delta SL}{n})(\frac{\beta L}{n})(\frac{\gamma SL}{n})(\frac{\alpha L}{n}) \cdots$ repeat n times	$\mathcal{O}(\frac{L^5}{n^4})$

2.5   Canonical integrator

These integration techniques are called explicit canonical integration techniques. This is because in the calculation process, all quantities are evaluated directly from previously known quantities and no implicit numerical inversion is required. An implicit scheme that assures symplecticity may use, for example, a generating function technique. A generating function by definition requires a mixing of the old and the new coordinates or momenta, and is therefore necessarily implicit. In this lecture, we do not cover the implicit schemes.

These explicit integration techniques give symplectic models of thick elements, and are most useful for particle tracking purposes. Most tracking programs use kick approximation of one kind or another. One often-used example is the thin-lens model, which is more technically known as a "second order explicit canonical integration". Another example is the ray tracing model, which is typically done with a first order integration but acquires accuracy by slicing the element into a large number of steps. As mentioned before, this requires extensive computer time and is not practical for long-term stability studies of large accelerators. Table 2.1 gives the various canonical integration techniques for an accelerator element (linear or nonlinear) of strength S and length L. For a quadrupole, we have $S = -k^2$.

These techniques apply to integration in general, not only to accelerator elements. Application of symplectic integration techniques goes beyond accelerator physics. Mirror symmetry as adopted in the thin-lens model helps in providing one order higher accuracy. Dividing an element into many steps also improves the accuracy. There are three ways to increase accuracy in an integrator algorithm,

Figure 2.5: Tracking results showing amplitude as a function of time using symplectic integrators, (a) 2nd order, (b) 4th order.

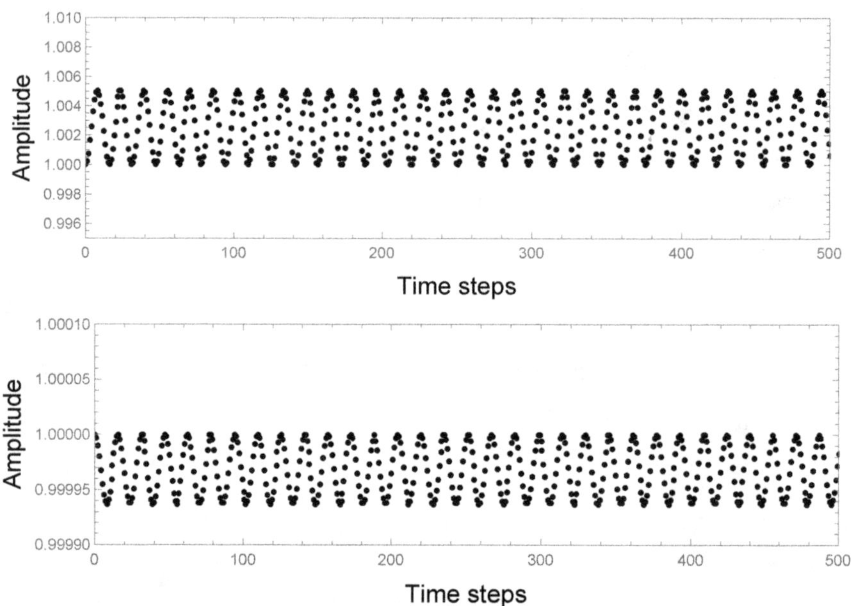

1. higher order integrator;

2. increase symmetry;

3. increase the number of integration steps.

Which one to use depends on the user's optimization considerations such as computing speed and his/her patience.

In Figs. 2.5(a) and (b), we compare the tracking results using the thin-lens map (2.17) and the 4th order map (2.25). Plotted in Fig. 2.5 is the quantity $A = \sqrt{x^2 + (\frac{x'}{k})^2}$ as a function of iteration number for a particle with initial conditions $(x = 1, \frac{x'}{k} = 0)$ and $kL = 0.2$. The exact map (2.5) would give a constant $A = 1$. Figure 2.5 exhibits the numerical accuracy of the two approximate maps by computing A. The reason $A > 1$ in Fig. 2.5(a) and $A < 1$ in Fig. 2.5(b) is due to the respective signs of the next order terms being dropped. Note the different vertical scales of the two graphs. Note also that in either maps — both being symplectic — the stability of the particle motion is not in question; only the numerical accuracy is being compared.

As mentioned, we derived the symplectic integrators in Table 2.1 using the linear quadrupole map. The same integrators apply also to other linear or nonlinear maps. It does not matter which map we use to derive the integrators.

In the remaining of this section, including a few homework problems, we consider a few explicit applications.

Sextupole As an illustration of symplectic integration of a nonlinear system, let us consider the 1-D motion in a thick sextupole with equation of motion

$$x'' = Sx^2 . \tag{2.33}$$

This problem can be solved exactly as follows. Integrating (2.33) gives a constant of the motion

$$C = \frac{1}{2}(x')^2 - \frac{1}{3}Sx^3 , \tag{2.34}$$

which gives

$$x' = \pm\sqrt{2C + \frac{2}{3}Sx^3} . \tag{2.35}$$

Quantity C is a constant. It is in fact the value of the Hamiltonian of the system. With the input coordinates (x_0, x_0'), we have $C = \frac{1}{2}(x_0')^2 - \frac{1}{3}Sx_0^3$. The exit coordinates of the particle are described by $(x(L), x'(L))$, where $x(L)$ satisfies

$$\int_{x_0}^{x(L)} \frac{du}{\pm\sqrt{(x_0')^2 - \frac{2}{3}S(x_0^3 - u^3)}} = L . \tag{2.36}$$

Equations (2.35) and (2.36) are the exact solution to the problem.

We now proceed to look for approximate algorithms perturbatively for short s. Hamilton equations (obtained from the Hamiltonian) can be used iteratively to obtain $x(s)$ and $x'(s)$ as Taylor expansion in s. Start with the zeroth order expressions $x(s) = x_0, x'(s) = x_0'$. Integrate to yield the 1st order expression $x(s) = x_0 + x_0's$. Inserting this 1st order expression into (2.33) then yields a 1st order expression $x'(s) = x_0' + Sx_0^2 s$, which can be integrated to give a 2nd order expression of $x(s)$. Repeating this process gives the approximate 4th order map,

$$x(s) = x_0 + x_0's + \frac{1}{2}Sx_0^2 s^2 + \frac{1}{3}Sx_0 x_0' s^3 + \frac{S}{12}(x_0'^2 + Sx_0^3)s^4 + \mathcal{O}(s^5) ,$$

$$x'(s) = x_0' + Sx_0^2 s + Sx_0 x_0' s^2 + \frac{S}{3}(x_0'^2 + Sx_0^3)s^3 + \frac{5}{12}S^2 x_0^2 x_0' s^4 + \mathcal{O}(s^5) . \tag{2.37}$$

Setting $s = L$ then gives expressions for $x(L)$ and $x'(L)$.

In comparison, the thin-lens approximation, which necessarily holds only to 1st order in S, gives

$$x(L) = x_0 + x_0'L + \frac{1}{2}Sx_0^2 L^2 + \frac{1}{2}Sx_0 x_0' L^3 + \frac{1}{8}Sx_0'^2 L^4 ,$$

$$x'(L) = x_0' + Sx_0^2 L + Sx_0 x_0' L^2 + \frac{1}{4}Sx_0'^2 L^3 , \tag{2.38}$$

which agrees with (2.37) up to order $\mathcal{O}(L^2)$. Map (2.38) is symplectic. If the $\mathcal{O}(s^5)$ terms are truncated, map (2.37) is not symplectic, even though it might

be numerically more accurate than Eq. (2.38). We will continue this thin-lens
sextupole discussion in Homework 4.115.

Applying the 4th order integrator in Table 2.1, we use Mathematica to obtain[5]

$$\begin{aligned}
x(L) &= x_0 + x_0's + \frac{1}{2}Sx_0^2 s^2 + \frac{1}{3}Sx_0 x_0' s^3 + \frac{S}{12}(x_0'^2 + Sx_0^3)s^4 \\
&\quad + 0.0188747 S^2 x_0' x_0^2 s^5 + \mathcal{O}(s^{6-22}), \\
x'(L) &= x_0' + Sx_0^2 s + Sx_0 x_0' s^2 + \frac{S}{3}(x_0'^2 + Sx_0^3)s^3 + \frac{5}{12}S^2 x_0^2 x_0' s^4 \\
&\quad + S^2 x_0(0.231125 x_0'^2 - 0.437566 S x_0^3)s^5 + \mathcal{O}(s^{6-21}).
\end{aligned} \tag{2.39}$$

These agree with the exact expressions to the appropriate orders respectively
but now higher orders (from 5th to 22nd order in s) are added artificially to
symplectify the map.

Runge–Kutta integrator   At this point, we shall point out that there are
various often-used numerical integration techniques that are nonsymplectic. As
an example, we mention the Runge–Kutta algorithm. Consider the differential
equation

$$x'' = f(x, x', s). \tag{2.40}$$

Given $x(0)$, $x'(0)$ at $s = 0$, Runge–Kutta gives approximate expressions of $x(L)$
and $x'(L)$ at $s = L$ as

$$\begin{aligned}
x(L) &\approx x(0) + Lx'(0) + \frac{1}{6}L(t_1 + t_2 + t_3), \\
x'(L) &\approx x'(0) + \frac{1}{6}(t_1 + 2t_2 + 2t_3 + t_4),
\end{aligned} \tag{2.41}$$

where

$$\begin{aligned}
t_1 &= Lf\left(x(0), x'(0), 0\right), \\
t_2 &= Lf\left(x(0) + \frac{1}{2}Lx'(0), x'(0) + \frac{1}{2}t_1, \frac{1}{2}L\right), \\
t_3 &= Lf\left(x(0) + \frac{1}{2}Lx'(0) + \frac{1}{4}Lt_1, x'(0) + \frac{1}{2}t_2, \frac{1}{2}L\right), \\
t_4 &= Lf\left(x(0) + Lx'(0) + \frac{1}{2}Lt_2, x'(0) + t_3, L\right).
\end{aligned} \tag{2.42}$$

Let us apply the Runge–Kutta technique to the x-motion in a quadrupole.
In this application, f is only a function of x. The result is

$$x(L) \approx x(0)\left(1 - \frac{1}{2}k^2 L^2 + \frac{1}{24}k^4 L^4\right) + x'(0)L\left(1 - \frac{1}{6}k^2 L^2\right),$$

---

[5]Explicit expressions of terms $\mathcal{O}(s^{6-22})$ and $\mathcal{O}(s^{6-21})$ are available but not listed in
Eq. (2.39). An important point, however, is the fact that Eq. (2.39) is not an infinite series. It terminates.

$$x'(L) \approx -k^2 L x(0)\left(1 - \frac{1}{6}k^2 L^2\right) + x'(0)\left(1 - \frac{1}{2}k^2 L^2 + \frac{1}{24}k^4 L^4\right). \qquad (2.43)$$

We can also apply it to the case of a sextupole, Eq. (2.33), to obtain

$$
\begin{aligned}
x(L) &\approx x_0 + x_0' L + \frac{1}{2}Sx_0^2 L^2 + \frac{1}{3}Sx_0 x_0' L^3 + \frac{S}{12}\left(x_0'^2 + Sx_0^3\right)L^4 \\
&\quad + \frac{1}{24}S^2 x_0^2 x_0' L^5 + \frac{1}{96}S^3 x_0^4 L^6 , \\
x'(L) &\approx x_0' + Sx_0^2 L + Sx_0 x_0' L^2 + \frac{S}{3}\left(x_0'^2 + Sx_0^3\right)L^3 + \frac{5}{12}S^2 x_0^2 x_0' L^4 \\
&\quad + S^2 x_0\left(\frac{5}{24}x_0'^2 + \frac{1}{16}Sx_0^3\right)L^5 + \frac{1}{12}S^2 x_0'\left(\frac{1}{2}x_0'^2 + x_0^3\right)L^6 \\
&\quad + \frac{1}{16}S^3 x_0^2 x_0'^2 L^7 + \frac{1}{48}S^3 x_0 x_0'^3 L^8 + \frac{1}{384}S^3 x_0'^4 L^9 . \qquad (2.44)
\end{aligned}
$$

The determinant of the Jacobian matrix for the transformation (2.43) is

$$1 - \frac{k^6 L^6}{72} + \frac{k^8 L^8}{576} \neq 1, \qquad (2.45)$$

and for the transformation (2.44), the determinant is

$$
\begin{aligned}
1 &- \frac{1}{72}\left(2x_0'^2 - 9Sx_0^3\right)S^2 L^6 + \frac{7}{36}x_0^2 x_0' S^3 L^7 + \frac{1}{144}x_0\left(7x_0'^2 + 15Sx_0^3\right)S^3 L^8 \\
&+ \frac{1}{288}x_0'\left(-x_0'^2 + 46Sx_0^3\right)S^3 L^9 + \frac{1}{576}x_0^2\left(45x_0'^2 + 16Sx_0^3\right)S^4 L^{10} \\
&+ \frac{1}{288}x_0 x_0'\left(4x_0'^2 + 13Sx_0^3\right)S^4 L^{11} + \frac{1}{576}x_0^3\left(15x_0'^2 + 2Sx_0^3\right)S^5 L^{12} \\
&+ \frac{1}{576}x_0^2 x_0'\left(4x_0'^2 + 3Sx_0^3\right)S^5 L^{13} + \frac{1}{1152}x_0 x_0'^2\left(x_0'^2 + 3Sx_0^3\right)S^5 L^{14} \\
&+ \frac{1}{2304}x_0^3 x_0'^3 S^6 L^{15} \neq 1. \qquad (2.46)
\end{aligned}
$$

In both cases, the Runge–Kutta integration is accurate to order $\mathcal{O}(L^4)$ but is not symplectic. We will continue this discussion later in Homework 4.117.

Note that in Eq. (2.40), the fact that $f(x, x', s)$ depends on x' already violates symplecticity. The system is non-Hamiltonian. Runge–Kutta is intended more generally than implied in our discussion emphasizing symplecticity.

Homework 2.20 The most straight ahead approach, without paying attention to the symplecticity or even the calculational efficiency, is the iteration approach mentioned before Eq. (2.37). Follow the text to verify Eq. (2.37).[6]

Homework 2.21 Consider an equation of motion

$$x'' = f(x). \qquad (2.47)$$

---

[6]The difference between this most straightforward and the various integrators mentioned in the text might be viewed as an illustration of how much progress we have made in developing our techniques. Reader will judge whether the sophistication is worthy of the effort.

Apply the thin-lens approximation to integrate the equation. Show that the result is

$$x'(L) \approx x_0' + Lf\left(x_0 + \frac{L}{2}x_0'\right),$$

$$x(L) \approx x_0 + Lx_0' + \frac{L}{2}f\left(x_0 + \frac{L}{2}x_0'\right). \tag{2.48}$$

Equation (2.48) is also called the leap-frog algorithm. Demonstrate explicitly the symplecticity of this map by calculating the determinant of the Jacobian matrix. To what order of L is the leap frog algorithm valid as an approximation to the exact answer? Give an estimate of the error.

Solution The exact Taylor map gives[7]

$$x(L) = x_0 + Lx_0' + \frac{1}{2}f(x_0)L^2 + \frac{1}{6}x_0'f'(x_0)L^3 + \cdots.$$

The leap-frog algorithm gives

$$x(L) = x_0 + Lx_0' + \frac{1}{2}f(x_0)L^2 + \frac{1}{4}x_0'f'(x_0)L^3 + \cdots.$$

The error terms are $\mathcal{O}(L^3)$.

Homework 2.22 This is a 2-D extension of the thin-lens algorithm. Consider a particle with initial coordinates (x_i, x_i', y_i, y_i') going through a sextupole magnet of strength S and length L. The equations of motion of this particle in the sextupole is given by

$$x'' = S(x^2 - y^2),$$
$$y'' = -2Sxy.$$

(a) Find the thin-lens map that relates the final coordinates (x_f, x_f', y_f, y_f') of the particle when exiting the sextupole in terms of the initial coordinates (x_i, x_i', y_i, y_i').

(b) Calculate the Jacobian matrix of your map.

(c) Demonstrate the symplecticity of your map. Note that $\det = 1$ is not sufficient in this case.

Solution

The first drift:

$$x_1 = x_i + \frac{L}{2}x_i', \qquad x_1' = x_i',$$

$$y_1 = y_i + \frac{L}{2}y_i', \qquad y_1' = y_i'.$$

---

[7]Be careful with the notation. Here, $f' = \frac{df}{dx}$, while $x' = \frac{dx}{ds}$. The two primes mean different things.

The lumped sextupole kick:

$$x_2 = x_1, \qquad x_2' = x_1' + SL(x_1^2 - y_1^2),$$
$$y_2 = y_1, \qquad y_2' = y_1' - 2SLx_1y_1.$$

The second drift:

$$x_3 = x_2 + \frac{L}{2}x_2', \qquad x_3' = x_2',$$
$$y_3 = y_2 + \frac{L}{2}y_2', \qquad y_3' = y_2'.$$

The whole map:

$$x_f = x_i + \frac{L}{2}x_i' + \frac{L}{2}\left\{x_i' + SL\left[\left(x_i + \frac{L}{2}x_i'\right)^2 - \left(y_i + \frac{L}{2}y_i'\right)^2\right]\right\},$$

$$x_f' = x_i' + SL\left[\left(x_i + \frac{L}{2}x_i'\right)^2 - \left(y_i + \frac{L}{2}y_i'\right)^2\right],$$

$$y_f = y_i + \frac{L}{2}y_i' + \frac{L}{2}\left[y_i' - \frac{1}{2}SL(2x_i + Lx_i')(2y_i + Ly_i')\right],$$

$$y_f' = y_i' - \frac{SL}{2}(2x_i + Lx_i')(2y_i + Ly_i').$$

The Jacobian matrix reads

$$\begin{bmatrix} 1+SL^2(x_i+\frac{L}{2}x_i') & L+\frac{SL^3}{2}(x_i+\frac{L}{2}x_i') & -SL^2(y_i+\frac{L}{2}y_i') & -\frac{SL^3}{2}(y_i+\frac{L}{2}y_i') \\ 2SL(x_i+\frac{L}{2}x_i') & 1+SL^2(x_i+\frac{L}{2}x_i') & -2SL(y_i+\frac{L}{2}y_i') & -SL^2(y_i+\frac{L}{2}y_i') \\ -\frac{SL^2}{2}(2y_i+Ly_i') & -\frac{SL^3}{4}(2y_i+Ly_i') & 1-\frac{SL^2}{2}(2x_i+Lx_i') & L-\frac{SL^3}{4}(2x_i+Lx_i') \\ -SL(2y_i+Ly_i') & -\frac{SL^2}{2}(2y_i+Ly_i') & -SL(2x_i+Lx_i') & 1-\frac{SL^2}{2}(2x_i+Lx_i') \end{bmatrix}.$$

It satisfies the symplectic condition for arbitrary (x_i, x_i', y_i, y_i').

Homework 2.23

(a) Apply Runge–Kutta integration to the simple example $x'' = a = $ constant over a distance L. The exact solution is trivial: $x = x_0 + x_0'L + \frac{a}{2}L^2$, $x' = x_0' + aL$. Show that Runge–Kutta algorithm reproduces the exact solution.

(b) Then do the same exercise for the nonlinear case $x'' = \frac{c}{x}$.

Homework 2.24

(a) The two nonsymplectic tracking in Figs. 2.2(b,c) show exponential growth of spiraling amplitude. Associate the growth rate to the deviation of the determinant of the transfer matrices from unity. Determinant of Eq. (2.10) is $1 + (k\Delta s)^2$. Determinant of Eq. (2.12) is $1 + \frac{(k\Delta s)^4}{4}$. The amplitude grows or damps exponentially depending on whether the determinant is > 1 or < 1. Give an expression of the growth rate in terms of $1 - \det(M)$. Compare your result quantitatively with an estimate of the growth rate obtained from Figs. 2.2(b,c).

(b) In case the map has to be truncated for some reason (due to computer capacity, or due to one's patience), how to set a tolerance level of the nonsymplecticity of the map used, or equivalently, given the degree of nonsymplecticity $\det(M) - 1$, what is the maximum number of map iterations allowed in your tracking simulation? This homework is recommended for its hands-on practical value.

(c) It was said in the text that symplecticity of a map is important for its application to long-term stability study. Does the long-term stability require the complete symplecticity condition, or does it require only the condition $\det(M) = 1$? This intriguing question is relevant only for 2-D or 3-D systems.

Homework 2.25 Consider a ray tracing model of a thick quadrupole. Let there be n steps, each of length $\frac{L}{n}$. The total map is given by

$$M = m^n,$$

where

$$m = \begin{bmatrix} 1 & 0 \\ -\frac{k^2 L}{n} & 1 \end{bmatrix} \begin{bmatrix} 1 & \frac{L}{n} \\ 0 & 1 \end{bmatrix} = \begin{bmatrix} 1 & \frac{L}{n} \\ -\frac{k^2 L}{n} & 1 - \frac{k^2 L^2}{n^2} \end{bmatrix}.$$

(a) Show that the matrix m can be written as

$$m = T \begin{bmatrix} e^{i\mu} & 0 \\ 0 & e^{-i\mu} \end{bmatrix} T^{-1}, \tag{2.49}$$

where $e^{\pm i\mu}$ are the eigenvalues of the matrix m with

$$\sin \frac{\mu}{2} = \frac{kL}{2n},$$

and

$$T = \begin{bmatrix} \frac{1}{\sqrt{2}} e^{-i\mu/2} & i\frac{L}{n} \frac{e^{i\mu/2}}{\sqrt{2} \sin \mu} \\ i\sqrt{2}\frac{n}{L} \sin \frac{\mu}{2} & \frac{1}{\sqrt{2} \cos \frac{\mu}{2}} \end{bmatrix}.$$

(b) Use the property (2.49) to show

$$M = \begin{bmatrix} \frac{\cos(n-\frac{1}{2})\mu}{\cos \frac{\mu}{2}} & \frac{L}{n} \frac{\sin n\mu}{\sin \mu} \\ -2\frac{n}{L} \sin n\mu \tan \frac{\mu}{2} & \frac{\cos(n+\frac{1}{2})\mu}{\cos \frac{\mu}{2}} \end{bmatrix}. \tag{2.50}$$

(c) Show that the determinant of (2.50) is 1. Show also that when $n = 1$ and when $n \to \infty$, expression (2.50) becomes the expected results (2.15) and (2.8), respectively.

(d) Show explicitly that for finite n, the error terms are of the order $\mathcal{O}(\frac{L^2}{n})$, as claimed in Table 2.1.

Solution Let m be eigenanalyzed as

$$m \begin{bmatrix} u_\pm \\ v_\pm \end{bmatrix} = e^{\pm i\mu} \begin{bmatrix} u_\pm \\ v_\pm \end{bmatrix}, \tag{2.51}$$

and choose normalization $u_+ v_- - u_- v_+ = 1$. Then

$$T = \begin{bmatrix} u_+ & u_- \\ v_+ & v_- \end{bmatrix}, \qquad \det T = 1. \tag{2.52}$$

Homework 2.26 Repeat the above Homework for the thin-lens model.

Solution Parameterize m as

$$m = \begin{bmatrix} a & b \\ c & a \end{bmatrix} \qquad \text{with} \qquad a^2 - bc = 1. \tag{2.53}$$

Eigenanalyze m according to Eq. (2.51), where the eigenvalues are $e^{\pm i\mu}$ with $\cos\mu = a$. The matrix T is found using Eq. (2.52) to be

$$T = \begin{bmatrix} \frac{1}{\sqrt{2}} & i\frac{b}{\sqrt{2}\sin\mu} \\ i\frac{\sin\mu}{\sqrt{2b}} & \frac{1}{\sqrt{2}} \end{bmatrix}, \qquad \det T = 1. \tag{2.54}$$

Matrix m can then be written as (2.49) and we find

$$M = m^n = \begin{bmatrix} \cos n\mu & \frac{b}{\sin\mu}\sin n\mu \\ -\frac{\sin\mu}{b}\sin n\mu & \cos n\mu \end{bmatrix}. \tag{2.55}$$

It follows that $\det M = 1$, and that M reduces to proper limits when $n = 1$ and $n \to \infty$. For the thin-lens model, parameters a, b, and c are given by (2.17). We have $\sin\frac{\mu}{2} = \frac{kL}{2n}$. Substituting this value for μ and $b = \frac{L}{n} - \frac{1}{4n^3}k^2 L^3$, it follows that model (2.55) agrees with the exact map with error terms of the order $\mathcal{O}(\frac{L^3}{n^2})$.

Alternative Solution A simpler derivation uses the fact that the map (2.55) can be written as a drift of length $-\frac{L}{2n}$ followed by the map (2.50) followed by a drift of length $\frac{L}{2n}$.

Homework 2.27 Repeat the above for the 4th order integrator.

Solution Results (2.53–2.55) still hold. The only difference is the values of a, b, and c. The error terms are of the order $\mathcal{O}(\frac{L^5}{n^4})$.

Homework 2.28 Equation (2.47) describes a conservative system. Its leap-frog map is symplectic. Consider a nonconservative system described by $x'' = f(x, x')$.

    (a) Develop a leap-frog algorithm for this system. Examine its symplecticity.

    (b) Observe how the extra dependency of f on x' makes the system non-symplectic.

Homework 2.29 Demonstrate explicitly that the exact sextupole map (2.35–2.36) is symplectic.

Solution This is suggested as an instructive exercise. You need to show

$$\frac{\partial x(L)}{\partial x_0}\frac{\partial x'(L)}{\partial x_0'} - \frac{\partial x(L)}{\partial x_0'}\frac{\partial x'(L)}{\partial x_0} = 1 \quad \text{exactly}. \tag{2.56}$$

This needs to be verified even though the functional dependencies are implicit. Clearly the same analysis applies if the sextupole equation of motion is replaced by any $x'' = f(x)$.

Chapter 3

Truncated Power Series Algebra

In the study of nonlinear dynamics in accelerators, one often needs high-order nonlinear maps. There are many examples of such studies and needs. Two of the more sophisticated applications might be:

- Given the maps of sufficiently high order and sufficiently symplectic for all beam-line components in the accelerator, one can use them to study the long-term behavior of the accelerator system by particle tracking;
- Given the nonlinear one-turn map of a circular accelerator, methods exist to extract various nonlinear dynamics quantities analytically.

But the purpose of this chapter is not to discuss the variety of reasons for these applications. For all these applications, question remains as to how to generate high order maps efficiently and accurately in the first place. Here we describe the truncated power series algebra (TPSA) technique, first introduced for accelerator applications by Berz in the 1980s.[1] Since then, a large number of efficient TPSA computer codes were developed.

3.1   Introducing TPSA

The TPSA technique is a powerful, practical calculational technique. It is useful not only for accelerators, but also for any calculational algorithm which relates some output quantities to some input quantities (see Fig. 3.1). Once the algorithm is given, TPSA allows one to generate power series expressions of the output quantities in terms of the input quantities. The order of the power series Ω is not limited (other than having sufficient memory space in a computer — see Homework 3.24) and can be specified by the user. One should note, however,

---

[1]M. Berz, Part. Accel. 24, 109 (1989); M. Berz, AIP Conf. Proc. 131, 77 (1992); M. Berz, Sec. 2.3.7, Handbook Accel. Phys. & Eng., 2nd ed., World Scientific (2013).

Figure 3.1: A numerical calculation algorithm with inputs and outputs. The inputs $x_1, x_2, \cdots, x_n$ and the outputs $y_1, y_2, \cdots, y_m$ consist of numerical numbers.

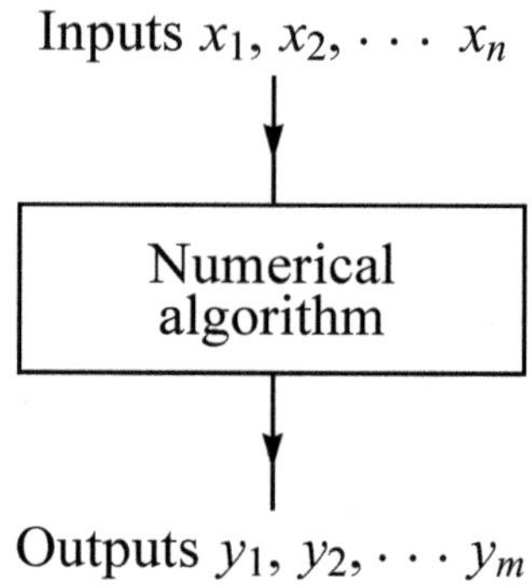

that TPSA does not contain physics; it is a calculational technique without physics contents.

What kind of algorithms are we talking about? The answer is any algorithm that numerically relates outputs to inputs in a well-defined manner. For example, one may have explicit functions

$$y_i \;=\; y_i(x_1, x_2, \cdots, x_n), \quad i = 1, 2, \cdots, m, \tag{3.1}$$

or one may have written a computer code with $X = (x_1, x_2, \cdots, x_n)$ as inputs, and $Y = (y_1, y_2, \cdots, y_m)$ as outputs. We are talking about any algorithm that allows one to calculate definitively the numerical values of Y once the numerical values of X are given. The algorithm defines a *map* from X to Y.

This algorithm, however, may be very complicated (the above computer code may contain 10,000 lines), and to carry it out may be very time consuming. In this case, one would want to approximate the map by something simpler such as a Taylor expansion up to order Ω, or symbolically

$$Y \;\approx\; \sum_{j}^{\Omega} C_j X^j + \mathcal{O}(X^{\Omega+1}), \tag{3.2}$$

which we recognize as a Taylor map from X to Y. Once we find this map (3.2), we can perform the calculation of the original algorithm much faster, albeit admittedly only approximately because higher order terms in the Taylor series beyond Ω-th order will be truncated. The TPSA technique is a way to calculate the coefficients C_j once the original algorithm is specified.

For some of the accelerator applications, one may consider X to be the coordinates $(x, x', y, y', z, \delta)$ of a particle at some starting position in a circular accelerator. In a tracking code, this X is followed element by element for one revolution and finally one obtains Y, which is the $(x, x', y, y', z, \delta)$ of the particle one revolution later. In a multiturn tracking, the result Y becomes the new input X for the next revolution, etc., and the process repeats for many revolutions as needed. Once this tracking program is written, TPSA allows one to extract the one-turn Taylor map relating Y to X up to some prespecified order Ω. Subsequent tracking study can then be performed using this Taylor map instead of the original element-by-element tracking. It is easy to see that for a large accelerator, the TPSA map can potentially be substantially more efficient than the one-turn element-by-element map (or numerical algorithm).

Note that if the original program contains a bug, the TPSA would still work, just that its resulting Taylor map contains the same bug. Note also that the original tracking program relates *numerically* a value of Y to a numerical value of X, while the TPSA converts the numerical algorithm to a map which relates Y to X *algebraically*. Although the values of X and Y changes each time, the TPSA map does not change.

Accelerator application of TPSA is of course not restricted to obtaining one-turn maps. Among many applications (some to be mentioned later in this chapter), TPSA allows one to obtain single-magnet maps, i.e. the high order Taylor maps (of order Ω) relating the exit coordinates $(x, x', y, y', z, \delta)$ to the entrance coordinates of single-magnet elements. Before the introduction of TPSA, these high order maps of magnet elements were obtained by analytically solving the particle's equation of motion. This is clearly a laborious procedure, ending with long expressions,[2] and the results are limited to relatively low orders. These pre-TPSA examples include TRANSPORT for second orders, MARYLIE for third orders, and COSY for fifth orders. Although analytic expressions allow more insight into the problem, TPSA makes this previous approach obsolete, at least as far as efficient generation of high order maps is concerned.

After truncation, the Taylor map is generally nonsymplectic. This Taylor map needs then to be symplectified. One approach of map symplectification, the explicit canonical integration, was the subject of Chapter 2. Another way to do that is to apply Lie algebra. In addition, Lie algebra also adds the analytical insights to the dry TPSA maps. The combined application of TPSA and Lie algebra results in a very powerful modern tool in the study of nonlinear dynamics. This has triggered a revolution on nonlinear dynamics research in accelerators in and after the 1980s. Lie algebra is the subject of Chapter 4.

The significance of the joint tool of TPSA and Lie algebra can perhaps be put in a historical note here. Nonlinear dynamics became an increasingly important topic since the invention of the strong focusing principle and when the stability of particle motion as well as the feasibility of circular accelerators — not to mention storage rings — was brought to question. In the 1960s and 1970s, this subject was studied intensely by applying nonlinear canonical perturbation theories in Hamiltonian dynamics. Transport maps were introduced and applied as just mentioned. The theory had made significant successes in some directions, but overloaded and cumbersome in some others. In the 1980s, two breakthroughs revolutionized the way nonlinear dynamics was carried out in accelerator physics. One of them was Lie algebra, the other being TPSA.

---

[2]The list of analytical expressions of 6-D 5th order Taylor maps for the basic accelerator beamline elements, when printed out, can be hundreds of pages long.

It should be noted that these two techniques maximize the total arsenal in accelerator physics particularly when they are used jointly. Used together, they revolutionized the field.

Issues with TPSA   We seem to have painted a rosy picture of TPSA so far. But in so doing, we have swept under the rug a rather serious problem. The catch lies in the truncation of the map. We mention three issues of concern.

1. In case the original map aims to transform the phase space coordinates $(x, x', y, y', z, \delta)$ for a Hamiltonian system, we know that the map is necessarily symplectic. The TPSA map, after the truncation, generally loses its symplecticity. Either the truncation error is known to be small enough, or an additional step is required to symplectify it.

2. There is no a priori handle on the convergence of the TPSA Taylor series and it is simply *assumed* that it converges. In reality, this is not always true, and in fact, most likely than not, it diverges. However, the divergence often occurs only in limited or even microscopic regions in phase space,[3] and we shall pretend that they do not cause any macroscopic problems.

3. In practice, there is still a question of how to choose the truncation order Ω. Intricate details such as chaos in phase space can be inadvertently affected by the choice of Ω. This TPSA truncation can sometimes be rather brutal by an inappropriate choice of Ω — it is not necessarily the higher the better. In the following, however, we shall not address this problem.

Calculation of derivatives   It is obvious from the definition of Taylor series that the coefficients C_j in Eq. (3.2) are related to the derivatives of Y with respect to X. So one may also say that the TPSA allows one to calculate the high order derivatives of the outputs with respect to the inputs, e.g. it calculates

$$\frac{\partial^{10} y_5}{\partial^3 x_1 \partial^3 x_2 \partial^4 x_8}$$

once the algorithm is known. If $Y = Y(X)$ is a simple analytic expression, one can find derivatives analytically by hand, but such case is unlikely for a 10,000-line tracking code.

Consider the case with a single input x and single output $y = f(x)$, where f may be given as an analytic expression or by a 10,000-line code. Let y be expressed as a truncated Taylor series around a reference point $x = a$,

$$y = f(x) = f(a) + (x-a)f'(a) + \frac{1}{2}(x-a)^2 f''(a) + \cdots + \frac{1}{\Omega!}(x-a)^\Omega f^{(\Omega)}(a), \quad (3.3)$$

---

[3]The reason the divergences typically occur only in a microscopic regions in phase space, in turn, is due to local detuning effects.

where $f^{(k)}(x)$ means the k-th derivative of $f(x)$. In the simplest application of the TPSA technique, we will illustrate how to obtain the coefficients $f(a)$, $f'(a)$, $f''(a)$, ... Both the order Ω of the power series and the reference point a are specified by the user.

We first note that to the extent that a function $f(x)$ can be represented (at least approximately) as a power series in x truncated to the Ω-th order, there is a one-to-one equivalence between the function $f(x)$ and the vector

$$(f(a), f'(a), \ldots, f^{(\Omega)}(a)). \tag{3.4}$$

In particular, this vector can be regarded as the *TPSA representation* of the function $f(x)$ around $x = a$. Once the coefficients in (3.4) are known using TPSA — there are $\Omega + 1$ of them — the value of $f(x)$ can be calculated using Eq. (3.3) for arbitrary values of x. All the coefficients involve taking derivatives of the function $f(x)$ at $x = a$.

One could imagine calculating the derivatives numerically instead of using TPSA. To do so, one first chooses a small number ϵ, and then computes

$$f'(a) \approx \frac{f(a + \epsilon) - f(a)}{\epsilon}. \tag{3.5}$$

This procedure, however, loses accuracy rapidly as one proceeds to high order derivatives. As we will see, TPSA is so remarkable because it does not calculate derivatives by subtracting two nearly equal numbers, and offers a way to calculate high order derivatives to computer accuracy![4]

Homework 3.1 Let $f(x)$ be represented by TPSA vector $(a_0, a_1, a_2, \ldots, a_\Omega)$ expanded at $x = a$. What can one say about the TPSA representation of $f(x)$ at $x = b$ in terms of the coefficients $a, b, a_0, a_1, \ldots$?

Solution $\quad b_k = a_k + a_{k+1}(b - a) + \frac{a_{k+2}}{2}(b - a)^2 + \cdots$.

Homework 3.2 Can you contemplate a TPSA representation of a random number generator?

3.2 TPSA

We have not yet described what TPSA does. To do so, let us consider first an example,

$$f(x) = \frac{1}{x + \frac{1}{x}},$$

---

[4]This is almost counterintuitive. We learned in school to compute derivatives like Eq. (3.5) — and that was what Newton did. It seems intrinsic, and a matter of definition, that a small quantity ϵ be invoked when one calculates derivatives numerically. It therefore seems inevitable that attempts to calculate high order derivatives numerically contain large numerical errors, but that intuition is evidently wrong.

and suppose we want to find the coefficient $f'(2)$. One way is to do it analytically, i.e. we find first that

$$f'(x) \;=\; -\frac{1 - \frac{1}{x^2}}{(x + \frac{1}{x})^2}\,.$$

Substituting $x = 2$ then gives

$$f'(2) \;=\; -\frac{3}{25}\,.$$

The second way is to find it numerically. For example,

$$f'(2) \;\approx\; \frac{f(2.1) - f(2)}{2.1 - 2} \;=\; \frac{0.38817 - 0.4}{2.1 - 2} \;=\; -0.1183\,.$$

By replacing the value 2.1 by something closer to 2, the accuracy of the calculation improves.

To compute $f'(2)$ using TPSA, let us first form a vector $v = (2,1)$ and try to find $f(v)$. The reason the first component of v is chosen to be 2 is because we want to compute $f'(2)$. As we shall explain later, the second component of v is always 1. We now have

$$f(v) \;=\; \frac{1}{v + \frac{1}{v}}\,. \tag{3.6}$$

As we will establish later, vectors are manipulated in such a way that

$$\frac{1}{(a_1, a_2)} \;=\; \left(\frac{1}{a_1}, -\frac{a_2}{a_1^2} \right), \tag{3.7}$$

$$(a_1, a_2) + (b_1, b_2) \;=\; (a_1 + b_1, a_2 + b_2)\,. \tag{3.8}$$

Thus,

$$
\begin{aligned}
f(v) \;&=\; \frac{1}{(2,1) + \frac{1}{(2,1)}} \;=\; \frac{1}{(2,1) + (\frac{1}{2}, -\frac{1}{4})} \\[2mm]
&=\; \frac{1}{(\frac{5}{2}, \frac{3}{4})} \;=\; \left(\frac{2}{5}, -\frac{3}{25} \right).
\end{aligned}
\tag{3.9}
$$

Now notice that the two components in the final vector are miraculously equal to $f(2)$ and $f'(2)$! In the above calculation, nowhere explicit expressions of $f'(x)$ was used, and nowhere subtraction of nearly-equal numbers was needed. The calculation is exact.

The simplest version of the TPSA technique is thus

$$f(v) \;=\; (f(a), f'(a)), \qquad \text{for any function } f(x) \text{ and } v = (a,1). \tag{3.10}$$

Note that in the calculation (3.9), only the input vector v is prespecified to be equal to $(2,1)$. In the intermediate steps of the calculation, the vector is of course changed from $(2,1)$. When the calculation is completed, we miraculously obtain $(f(a), f'(a))$.

Perhaps we can look at Eq. (3.10) in a different way like this: To evaluate $f(a)$, all we have to do is to insert $x = a$ in the expression of $f(x)$. But now we want to calculate $f'(a)$, so what do we do? The trick here is to enter a together with a small quantity that plays the role of ϵ in the numerical calculation of Eq. (3.5). In the TPSA technique, this "small quantity" is carried by the second element in the vector, so this time we enter not $x = a$ but $x = (a, 1)$ into $f(x)$, where the second element of the vector stands for one unit of the small quantity ϵ. Even though its magnitude is 1, it is still a small quantity (and it has one unit of ϵ at the input) because it is the vector's second element. By including this element, information of $f'(x)$ is carried along. As we will see later, we will extend the dimension of the vector, and the later elements of the vector will carry the higher orders of smallness — the later the element, the higher order its smallness, representing increasing powers of ϵ. By doing so, we will be able to calculate higher order derivatives. See also Homework 3.11.

The secret of TPSA is contained in the vector manipulation rules. How are those rules established? Let us examine it in 3 steps as follows.

1. Consider the simple identity function $f(x) = x$. We of course want to make sure (3.10) holds for this function. Suppose we take $v = (\alpha, \beta)$, then

$$f(v) = v = (\alpha, \beta).$$

   We want this to be equal to $(f(a), f'(a))$, but we know $f(a) = a$ and $f'(a) = 1$. This means we must choose $(\alpha, \beta) = (a, 1)$, i.e. the first component of v must be chosen to be equal to the value of x at which the derivative is to be taken, and the second component of v must be 1. This is of course what we did.

   We can also apply (3.10) to the constant function $f(x) = c$ (constant independent of x), i.e. we require $f(v) = c$ to be equal to $(f(a), f'(a)) = (c, 0)$. This gives the identification that

$$\text{constant } c = (c, 0). \tag{3.11}$$

2. Suppose we now have two functions $f(x)$ and $g(x)$, each satisfying (3.10). We now want to establish a rule which allows the new function $h(x) = f(x) + g(x)$ to satisfy (3.10) also. The left-hand-side of Eq. (3.10) reads

$$h(v) = f(v) + g(v) = (f(a), f'(a)) + (g(a), g'(a)),$$

   while the right-hand-side is

$$(f(a) + g(a), f'(a) + g'(a)).$$

   Obviously $h(x)$ satisfies (3.10) if and only if we establish the vector addition rule (3.8).

We know that the identity function and the constant function satisfy (3.10). By combining them, we now have established (3.10) for all functions of the type $f(x) = c + x$.

3. With $f(x)$ and $g(x)$ satisfying (3.10), we now want to find the rule for the function $h(x) = f(x)g(x)$ to satisfy (3.10). To do so, we note

$$h(v) = (f(a), f'(a))\,(g(a), g'(a)),$$
$$(h(a), h'(a)) = (f(a)g(a), f(a)g'(a) + f'(a)g(a)).$$

Thus (3.10) is assured if and only if we require the vector multiplication rule

$$(a_1, a_2)\,(b_1, b_2) = (a_1 b_1, a_2 b_1 + a_1 b_2). \tag{3.12}$$

If we take $f(x) = c$ and $g(x) = x$, we can use the vector multiplication rule to assure that the function $h(x) = cx$ satisfies (3.10).

If we take $f(x) = g(x) = x$, then the function $h(x) = x^2$ satisfies (3.10). Having established them, we repeat the rule to establish (3.10) for any integer power function $h(x) = x^n$. The vector addition rule then establishes the validity of (3.10) for any power series of x. We have thus established the TPSA for any arbitrary function that is expandable into a Taylor series just by two simple vector algebraic rules (3.8) and (3.12). Just substitute $v = (a, 1)$ into $f(v)$ and follow these simple rules to obtain $f(a)$ and $f'(a)$!

Now how about Eq. (3.7)? One can establish it by applying the vector multiplication rule as follows. Let

$$\frac{1}{(a_1, a_2)} = (x, y),$$

then

$$(a_1, a_2)\,(x, y) = 1 = (1, 0)$$
$$\implies (a_1 x, a_1 y + a_2 x) = (1, 0)$$
$$\implies a_1 x = 1, \qquad a_1 y + a_2 x = 0$$
$$\implies x = \frac{1}{a_1}, \qquad y = -\frac{a_2}{a_1^2}, \qquad \text{Q.E.D.}$$

Internal self-consistency requires

$$(a_1, a_2) + (b_1, b_2) = (b_1, b_2) + (a_1, a_2),$$
$$(a_1, a_2)(b_1, b_2) = (b_1, b_2)(a_1, a_2),$$

and indeed their validity is easily verified.

By applying the vector multiplication rule, we can also establish

$$c(a_1, a_2) = (c, 0)\,(a_1, a_2) = (ca_1, 0 \times a_1 + ca_2) = (ca_1, ca_2). \tag{3.13}$$

Homework 3.3 Show that

$$(a_1, a_2)^{-n} = \left(\frac{1}{a_1^n}, -\frac{na_2}{a_1^{n+1}} \right).$$

Solution Mathematical induction.

Homework 3.4 It is suggested to look at Homework 3.3 in a different way. Identify a function $f(x) \approx f(a) + f'(a)(x - a)$ to the vector $(a_1, a_2)^{-1} = (\frac{1}{a_1}, -\frac{a_2}{a_1^2})$. This means we return to look at the function $f(x)$ itself,

$$f(x) \approx \frac{1}{a_1} - \frac{a_2}{a_1^2}(x - a).$$

Raising $f(x)$ to n-th power,

$$f^n(x) \approx \left[\frac{1}{a_1} - \frac{a_2}{a_1^2}(x - a) \right]^n \approx \frac{1}{a_1^n} - \frac{na_2}{a_1^{n+1}}(x - a), \qquad \text{Q.E.D.}$$

In this Homework, however, n does not have to be an integer as mathematical induction requires. A useful special case when $n = -\frac{1}{2}$ is [see also Eq. (3.22)]

$$\sqrt{(a_a, a_2)} = \left(\sqrt{a_1}, \frac{a_2}{2\sqrt{a_1}} \right).$$

Homework 3.5 Use TPSA to calculate $f((a, b))$ with the function $(a > 0)$

$$f(x) = \frac{\sqrt{x} - 1}{\sqrt{x} + 1}.$$

Solution $\left(\frac{\sqrt{a}-1}{\sqrt{a}+1}, \frac{b}{\sqrt{a}(\sqrt{a}+1)^2} \right).$

Homework 3.6 Use TPSA to calculate $\frac{dF(x)}{dx}$, where $F(x) = f(f(x))$ with $f(x) = \frac{\sqrt{x}-1}{\sqrt{x}+1}$.

3.3   Higher order

To obtain higher derivatives (and thus higher order TPSA expansions), we need a larger dimension for the vectors. Consider again the case with one input variable and one output variable. Suppose we would like to obtain a power series like (3.3) to the Ω-th order. We first form the vector

$$v = (a, 1, 0, 0, \cdots, 0), \tag{3.14}$$

where the number of elements in the vector is $\Omega + 1$. The first component is the reference point a. The second component is 1 — we want the initial input to carry a pure and clean one unit of smallness. The remaining components are zeroes. We would like to have the property that when x is substituted by v in any function $f(x)$ — recall that $f(x)$ can represent the 10,000-line computer code — one would obtain

$$f(v) = (f(a), f'(a), f''(a), \cdots , f^{(\Omega)}(a)) . \tag{3.15}$$

As before, the vector (3.14) is chosen in such a way that (3.15) is automatically satisfied for the identity function $f(x) = x$. Indeed, in this case, $f(v) = (a, 1, 0, 0, \cdots 0)$, which is equal to the right-hand-side of Eq. (3.15) trivially.

Next we need to construct the vector manipulation rules so that (3.15) is satisfied for arbitrary functions $f(x)$ which can be expressed as a truncated power series. Consider first if we have two functions $f(x)$ and $g(x)$, each satisfying (3.15), and would like the functions $h(x) = f(x) + g(x)$ and $h(x) = cf(x)$ also to satisfy (3.15), we need the rules

$$c(a_0, a_1, a_2, \cdots , a_\Omega) = (ca_0, ca_1, ca_2, \cdots , ca_\Omega) , \tag{3.16}$$
$$(a_0, a_1, a_2, \cdots , a_\Omega) + (b_0, b_1, b_2, \cdots , b_\Omega)$$
$$= (a_0 + b_0, a_1 + b_1, a_2 + b_2, \cdots , a_\Omega + b_\Omega) . \tag{3.17}$$

If we consider the function $h(x) = f(x)g(x)$, validity of (3.15) requires

$$h(v) = f(v)g(v)$$
$$= (f(a), f'(a), f''(a), \cdots , f^{(\Omega)}(a)) \, (g(a), g'(a), g''(a), \cdots , g^{(\Omega)}(a))$$

to be equal to

$$(h(a), h'(a), h''(a), \cdots , h^{(\Omega)}(a))$$
$$= (f(a)g(a), f'(a)g(a) + f(a)g'(a), f''(a)g(a) + 2f'(a)g'(a) + f(a)g''(a),$$
$$\cdots , \sum_{k=0}^{\Omega} \frac{\Omega!}{k!(\Omega - k)!} f^{(k)}(a)g^{(\Omega-k)}(a)) .$$

This is assured if we choose the vector multiplication rule

$$(a_0, a_1, a_3, \cdots , a_\Omega) \, (b_0, b_1, b_3, ... , b_\Omega) = (c_0, c_1, c_3, \cdots , c_\Omega) ,$$
$$c_m = \sum_{k=0}^{m} \frac{m!}{k!(m - k)!} a_k b_{m-k} ,$$
$$c_0 = a_0 b_0, \quad c_1 = a_1 b_0 + a_0 b_1, \quad c_2 = a_2 b_0 + 2a_1 b_1 + a_0 b_2, \quad \cdots . \tag{3.18}$$

By identifying a constant c as the vector $(c, 0, 0, \ldots, 0)$, Eq. (3.16) follows from Eq. (3.18).

The remarkable thing, again, is that once a vector addition rule (3.17) and a vector multiplication rule (3.18) are established, high order derivatives of an arbitrary function $f(x)$ can be computed by substituting x by $v = (a, 1, 0, 0, \cdots)$. Once the high order derivatives are obtained, Taylor series expansion follows.

Homework 3.7 For the case with $\Omega = 2$, compute $(a_0, a_1, a_2)^{-1}$.

Homework 3.8

 (a) What is the TPSA representation of $f(x) = x^2$ for $\Omega = 4$?

 (b) What function does $(a^2, a, \frac{1}{2}, 0, 0)$ represent?

Solution

 (a) $(a^2, 2a, 2, 0, 0)$.

 (b) $\frac{1}{4}x^2$ expanded around $x = 2a$.

Homework 3.9 Consider the function $f(x) = \frac{1}{x + \frac{1}{x}}$.

 (a) Use TPSA to find an expression for $g(x) = \frac{d^4 f(x)}{dx^4}$.

 (b) The value of $g(2) = -\frac{912}{3125} = -0.29184$.

 (c) Calculate the approximate value by a Newton's algorithm,

$$g(2) \approx \frac{f(x + 2\epsilon) - 4f(x + \epsilon) + 6f(x) - 4f(x - \epsilon) + f(x - 2\epsilon)}{\epsilon^4}.$$

Solution (c) You obtain $g(2) \approx -0.157731$ if $\epsilon = 0.1$, and $g(2) \approx -0.274413$ if $\epsilon = 0.01$.

Homework 3.10 Suppose a function $f(x)$ has been identified with the vector $f(a, 1, 0, 0, 0, 0 \cdots) = (a_0, a_1, a_2, \cdots, a_\Omega)$, i.e. $f(a) = a_0, f'(a) = a_1, \cdots$, around $x = a$.

 (a) What is the TPSA representation of $f'(x)$ in terms of $a_0, a_1, a_2, \cdots, a_\Omega$?

 (b) Same question for $\int_a^x dx' f(x')$?

Homework 3.11 Suppose a function $f(x)$ has been identified with the vector $f(a, 1, 0, 0, 0, 0 \cdots) = (a_0, a_1, a_2, \cdots, a_\Omega)$, i.e. $f(a) = a_0, f'(a) = a_1, \cdots$, around $x = a$.

 (a) Show that

$$f((a, \lambda, 0, 0, \cdots)) = (a_0, \lambda a_1, \lambda^2 a_2, \cdots).$$

The cases $\lambda = 0$ and $\lambda = 1$ are the trivial cases. This exercise should reinforce the connection of the higher order components of the TPSA vector to represent higher order smallness in the intended incremental manner.

 (b) What is the vector that identifies $f(x + \lambda)$ around $x = a$, i.e. what is $f((a + \lambda, 1, 0, 0, 0, \cdots))$?

 (c) What is the vector representation of the function $f(\lambda x)$?

Solution You will note problem (b) is the same as Homework 3.1 in a different language and is asking for a different way to address it.

(b) $f((a + \lambda, 1, 0, 0, \cdots)) = (b_0, b_1, b_2, \cdots)$, where $b_k = \displaystyle\sum_{n=k}^{\Omega} a_n \frac{\lambda^{n-k}}{(n-k)!}$.

(c) $f((\lambda a, 0, 0, 0, \cdots)) = (b_0, b_1, b_2, \cdots)$, where $b_k = \lambda^k \displaystyle\sum_{n=k}^{\Omega} a_n \frac{(\lambda-1)^{n-k} a^{n-k}}{(n-k)!}$.

Homework 3.12 Consider the case with $\Omega = 3$. Show that

$$(1, 1, 0, 0)^2 = (1, 2, 2, 0), \quad (1, 1, 0, 0)^3 = (1, 3, 6, 6),$$
$$(0, 1, 0, 0)^2 = (0, 0, 2, 0), \quad (0, 1, 0, 0)^3 = (0, 0, 0, 6),$$
$$(0, 1, 0, 0)^n = (0, 0, 0, 0) \quad \text{if } n \geq 4.$$

Homework 3.13 Apply TPSA to calculate $f'(x), f''(x), f^{(3)}(x), \ldots$ for the function $f(x) = \frac{1}{x}$.

Solution Let

$$(x, 1, 0, 0, 0, \ldots)^{-1} = (b_0, b_1, b_2, b_3, \ldots),$$

we require

$$(1, 0, 0, 0, \ldots) = (x, 1, 0, 0, 0, \ldots)(b_0, b_1, b_2, b_3, \ldots)$$
$$= (xb_0, xb_1 + b_0, xb_2 + 2b_1, xb_3 + 3b_2, \ldots),$$
$$\implies \quad b_0 = \frac{1}{x}, \; b_1 = -\frac{1}{x}b_0, \; b_2 = -\frac{1}{x}2b_1, \; \cdots,$$
$$\implies \quad b_{k-1} = (-1)^k \frac{k!}{x^{k+1}}, \quad k = 1, 2, 3, \ldots,$$
$$\implies \quad (x, 1, 0, 0, 0, \ldots)^{-1} = \left(\frac{1}{x}, -\frac{1}{x^2}, \frac{2}{x^3}, -\frac{6}{x^4}, \cdots \right).$$

3.4   Special functions

One might question how useful TPSA is in practice because not all operations in the original calculational algorithm are additions and multiplications and (positive and negative) powers. What if we have functions like $e^x, \ln x, \sin x, \cos x$? Surely one can expand them into an infinite power series before we substitute v into them, but that would be computer time consuming and seems to defeat the purpose of TPSA. The answer to this question is that there are tricks that allow us to decompose the calculation of these special functions into a *finite* number of vector additions and multiplications. Some of these tricks are described below.

Since these special functions may appear somewhere in the midstream of the calculation algorithm, they operate on vectors of a general form $(a_0, a_1, a_2, \ldots)$, not only the form $(a, 1, 0, 0, 0, \ldots)$ at the initial step of the algorithm. In other words, we need to calculate the special functions for an arbitrary vector $(a_0, a_1, a_2, \ldots)$.

The trick that allows this calculation is to note again that the vector

$$v = (0, 1, 0, 0, \dots)$$

is a "smallness" vector in the sense that v^2 would have two zeroes in its first two elements, v^3 would have three zeroes, etc. By raising v beyond a certain power [more specifically, beyond the $(\Omega + 1)$-th power], all its elements vanish.

In fact, $(0, 1, 0, 0, \dots)$ is not the only vector that is small. Any vector whose first component is zero is small in the sense that

$$(0, \times, \times, \times, \dots)^k = (0, 0, 0, \dots \times, \times), \tag{3.19}$$

where each vector contains $\Omega + 1$ components, and an $\times$ means a nonzero component. The right-hand-side of Eq. (3.19) means that, if $k < \Omega + 1$, the leading k components vanishes, while if $k \geq \Omega + 1$, all components vanish.

Using Eq. (3.19), it then becomes straightforward to show that

$$e^{(a_0, a_1, a_2, \dots, a_\Omega)} = e^{a_0} \sum_{k=0}^{\Omega} \frac{1}{k!} (0, a_1, a_2, \dots a_\Omega)^k, \tag{3.20}$$

$$\ln(a_0, a_1, a_2, \dots, a_\Omega) = (\ln a_0, 0, 0, 0, \dots, 0)$$
$$+ \sum_{k=1}^{\Omega} (-1)^{k+1} \frac{1}{k} \left(0, \frac{a_1}{a_0}, \frac{a_2}{a_0}, \dots, \frac{a_\Omega}{a_0} \right)^k, \tag{3.21}$$

$$\sqrt{(a_0, a_1, a_2, \dots, a_\Omega)} = \sqrt{a_0} \left[(1, 0, 0, 0, \dots 0) + \frac{1}{2} (0, \frac{a_1}{a_0}, \frac{a_2}{a_0}, \dots, \frac{a_\Omega}{a_0}) \right.$$
$$\left. - \sum_{k=2}^{\Omega} (-1)^k \frac{(2k-3)!!}{(2k)!!} (0, \frac{a_1}{a_0}, \frac{a_2}{a_0}, \dots, \frac{a_\Omega}{a_0})^k \right], \tag{3.22}$$

$$\sin(a_0, a_1, a_2, \dots, a_\Omega) = \sin a_0 \sum_{k=0}^{\Omega} \frac{(-1)^k}{(2k)!} (0, a_1, a_2, \dots, a_\Omega)^{2k}$$
$$+ \cos a_0 \sum_{k=0}^{\Omega} \frac{(-1)^k}{(2k+1)!} (0, a_1, a_2, \dots, a_\Omega)^{2k+1}, \tag{3.23}$$

$$\cos(a_0, a_1, a_2, \dots, a_\Omega) = \cos a_0 \sum_{k=0}^{\Omega} \frac{(-1)^k}{(2k)!} (0, a_1, a_2, \dots, a_\Omega)^{2k}$$
$$- \sin a_0 \sum_{k=0}^{\Omega} \frac{(-1)^k}{(2k+1)!} (0, a_1, a_2, \dots, a_\Omega)^{2k+1}. \tag{3.24}$$

All series on the right-hand-sides terminate, and that they can all be evaluated readily by a finite number of applications of the vector addition and multiplication rules. Equations (3.21) and (3.22) require $a_0 \geq 0$.

Homework 3.14 For a first simple exercise,

  (a) use the special function rules to show

$$
\begin{aligned}
\sin(a, b) &= (\sin a, b \cos a)\,, \\
\cos(a, b) &= (\cos a, -b \sin a)\,, \\
e^{(a,b)} &= (e^a, b e^a)\,.
\end{aligned}
$$

  (b) use the result obtained in (a) to show

$$
\begin{aligned}
\sin((a, b) + (c, d)) &= \sin(a, b) \cos(c, d) + \cos(a, b) \sin(c, d)\,, \\
\cos((a, b) + (c, d)) &= \cos(a, b) \cos(c, d) - \sin(a, b) \sin(c, d)\,.
\end{aligned}
$$

Homework 3.15 Apply TPSA to calculate $f'(x), f''(x), f^{(3)}(x), \ldots$ for the function $f(x) = e^x$.

Solution This Homework is really asking you to find $e^{(x,1,0,0,0,\ldots)}$. There are more than one way to do this. For example, one may first note that

$$
\begin{aligned}
e^{(x,1,0,0,0,\ldots)} &= e^{(x,0,0,0,0,\ldots)+(0,1,0,0,0,\ldots)} = e^{(x,0,0,0,0,\ldots)} e^{(0,1,0,0,0,\ldots)} \\
&= \sum_{k=0}^{\infty} \frac{1}{k!}(x,0,0,0,0,\ldots)^k \sum_{n=0}^{\infty} \frac{1}{n!}(0,1,0,0,0,\ldots)^n\,.
\end{aligned}
$$

Then note that

$$
\begin{aligned}
(x,0,0,0,0,\ldots)^k &= (x^k,0,0,0,0,\ldots)\,, \\
(0,1,0,0,0,\ldots)^n &= (0,0,\ldots,n!,\ldots,0,0,\ldots)\,,
\end{aligned}
$$

where the term $n!$ on the right-hand-side of the second entry occurs at the $(n+1)$-th element of the vector. When $n \geq \Omega + 1$, the vector $(0,1,0,0,0,\ldots)^n$ vanishes. We then obtain, as expected,

$$
\begin{aligned}
e^{(x,1,0,0,\ldots)} &= (e^x,0,0,0,\ldots)(1,1,1,1,\ldots) \\
&= (e^x,e^x,e^x,e^x,\ldots)\,.
\end{aligned}
$$

Homework 3.16 Follow a path similar to the preceding homework to show that

$$
\begin{aligned}
\ln(x,1,0,0,0,\ldots) &= \left(\ln x, \frac{1}{x}, -\frac{1}{x^2}, \ldots\right)\,, \\
\sin(x,1,0,0,0,\ldots) &= (\sin x, \cos x, -\sin x, -\cos x, \ldots)\,, \\
\cos(x,1,0,0,0,\ldots) &= (\cos x, -\sin x, -\cos x, \sin x, \ldots)\,.
\end{aligned}
$$

You have guessed the results already, but the point of the homework is to practice the TPSA algebra.

Homework 3.17 Derive Eqs. (3.20–3.24).

Homework 3.18 Apply the result of this section to demonstrate the TPSA relationships

$$e^{\ln(x)} = x \qquad \text{and} \qquad \ln e^x = x.$$

Take $\Omega = 3$ for this demonstration.

Solution You may need to show these first,

$$e^{(a_0, a_1, a_2, a_3)} = \left(e^{a_0}, a_1 e^{a_0}, \left(a_2 + \frac{a_1^2}{2} \right) e^{a_0}, \left(a_3 + a_1 a_2 + \frac{a_1^3}{6} \right) e^{a_0} \right),$$

$$\ln(a_0, a_1, a_2, a_3) = \left(\ln a_0, \frac{a_1}{a_0}, \frac{a_2}{a_0} - \frac{a_1^2}{2a_0^2}, \frac{a_3}{a_0} - \frac{a_1 a_2}{a_0^2} + \frac{a_1^3}{3a_0^3} \right).$$

Homework 3.19 Derive the TPSA expressions for the special functions

$$\sin^{-1}(a_0, a_1, a_2, \ldots, a_\Omega),$$
$$\cos^{-1}(a_0, a_1, a_2, \ldots, a_\Omega).$$

Homework 3.20 Prove the expressions in TPSA representation,

$$\frac{d \sin^{-1} x}{dx} = \frac{1}{\sqrt{1 - x^2}},$$
$$\frac{d \cos^{-1} x}{dx} = -\frac{1}{\sqrt{1 - x^2}},$$
$$\frac{d \tan^{-1} x}{dx} = \frac{1}{1 + x^2}.$$

Homework 3.21 Find an expression for $(a_0, a_1, \ldots, a_\Omega)^\lambda$, where λ is a fractional number and $a_0 > 0$.

Solution Consider $x^\lambda = e^{\lambda \ln x}$.

Homework 3.22 Use TPSA to calculate $f'(4)$, where

$$f(x) = e^{2 \sin \left[\left(\frac{\sqrt{x} - 1}{\sqrt{x} + 1} \right) \pi \right]}.$$

Solution Enter $v = (4, 1)$ into $f(v)$. We have

$$\sqrt{v} = \left(2, \frac{1}{4} \right),$$

$$\frac{\sqrt{v} - 1}{\sqrt{v} + 1} = \frac{\left(1, \frac{1}{4} \right)}{\left(3, \frac{1}{4} \right)} = \left(1, \frac{1}{4} \right) \left(\frac{1}{3}, -\frac{1}{36} \right) = \left(\frac{1}{3}, \frac{1}{18} \right),$$

$$\sin \left[\left(\frac{\sqrt{v} - 1}{\sqrt{v} + 1} \right) \pi \right] = \sin \left[\left(\frac{\pi}{3}, \frac{\pi}{18} \right) \right] = \sin \frac{\pi}{3} (1, 0) + \cos \frac{\pi}{3} \left(0, \frac{\pi}{18} \right)$$

$$= \frac{\sqrt{3}}{2}(1,0) + \frac{1}{2}\left(0, \frac{\pi}{18}\right) = \left(\frac{\sqrt{3}}{2}, \frac{\pi}{36}\right),$$

$$e^{2\sin\left[\left(\frac{\sqrt{v}-1}{\sqrt{v}+1}\right)\pi\right]} = e^{\left(\sqrt{3}, \frac{\pi}{18}\right)} = e^{\sqrt{3}}\left[(1,0) + \left(0, \frac{\pi}{18}\right)\right] = e^{\sqrt{3}}\left(1, \frac{\pi}{18}\right),$$

$$\implies$$

$$f'(4) = \frac{e^{\sqrt{3}}\pi}{18}.$$

Homework 3.23 Consider the function $F(x)$ defined in the following computer code:

```
READ   X
X1 = X * *2 − 1.0
X2 = X1 * EXP(X1)
X3 = X1 * X2
F = COS(X1) * X3
OUTPUT   F
```

Use TPSA to compute F, $\frac{dF}{dx}$, $\frac{d^2 F}{dx^2}$ when $x = 1$.

3.5　Multiple inputs and outputs

So far we discussed the case with a single input variable x and a single output variable y. We will now generalize the algorithm to multiple variables. This is based on the fact that choice of inputs $x_1, x_2, \cdots, x_n$ contains quite some flexibility in the TPSA. All that is being asked is that these input variables are mutually independent.

It is trivial to generalize to multiple output variables. Each output variable can be treated independently of the other output variables. This means we can concentrate on just one of them each time and consider the case with only one output variable $y = y(x_1, x_2, \cdots, x_n)$.

To treat multiple input variables, we need vectors of larger dimension. Here for simplicity we will assume two input variables (x_1, x_2) and one output variable y below.

The TPSA requires that when input x_1 is substituted by vector v_1 and x_2 substituted by v_2, we want to obtain an output vector

$$y(v_1, v_2) = \left(y(a_0, b_0), \frac{\partial y}{\partial x_1}(a_0, b_0), \frac{\partial y}{\partial x_2}(a_0, b_0), \right.$$

$$\left. \frac{\partial^2 y}{\partial x_1^2}(a_0, b_0), \frac{\partial^2 y}{\partial x_1 \partial x_2}(a_0, b_0), \frac{\partial^2 y}{\partial x_2^2}(a_0, b_0), \cdots, \frac{\partial^\Omega y}{\partial x_2^\Omega}(a_0, b_0) \right), \quad (3.25)$$

where a_0, b_0 are prespecified reference positions for x_1, x_2 respectively. The first component on the right-hand-side is just y evaluated at the reference positions.

The next two terms are 1st order derivatives. The next three terms are second order derivatives, etc. The last group of $\Omega + 1$ terms are Ω-th order derivatives. All derivatives are evaluated at (a_0, b_0).

It is easy to show that in order for property (3.25) to hold for the identity functions $y(x_1, x_2) = x_1$ and $y(x_1, x_2) = x_2$, we must choose

$$
\begin{aligned}
v_1 &= (a_0, 1, 0, 0, 0, \cdots, 0), \\
v_2 &= (b_0, 0, 1, 0, 0, \cdots, 0).
\end{aligned}
\tag{3.26}
$$

The vector addition and multiplication rules are

$$
\begin{aligned}
& (a_{00}, a_{10}, a_{01}, a_{20}, a_{11}, a_{02}, \cdots, a_{0\Omega}) + (b_{00}, b_{10}, b_{01}, b_{20}, b_{11}, b_{02}, \cdots, b_{0\Omega}) \\
=\; & (a_{00} + b_{00}, a_{10} + b_{10}, a_{01} + b_{01}, \cdots, a_{0\Omega} + b_{0\Omega}), \\
& (a_{00}, a_{10}, a_{01}, a_{20}, a_{11}, a_{02}, \cdots, a_{0\Omega})\,(b_{00}, b_{10}, b_{01}, b_{20}, b_{11}, b_{02}, \cdots, b_{0\Omega}) \\
=\; & (c_{00}, c_{10}, c_{01}, c_{20}, c_{11}, c_{02}, \cdots, c_{0\Omega}),
\end{aligned}
\tag{3.27}
$$

$$
c_{mn} = \sum_{s=0}^{m} \sum_{t=0}^{n} a_{st} b_{m-s, n-t} \frac{m! n!}{s!(m-s)! t!(n-t)!}.
\tag{3.28}
$$

Once property (3.25) is established, to evaluate a high order derivative, we just have to substitute (3.26) into $y(x_1, x_2)$, processing the calculation using the vector manipulation rules (3.27–3.28), and the output vector contains the result. Once all derivatives are obtained, a Taylor map for y follows.

A point should be made about the highest order Ω. The Taylor expansion is made so that the series limits the sum of the powers of all the input variables to be equal or less than Ω. For example, if $\Omega = 10$, then the highest order terms include $x_1^{10}, x_1^7 x_2^3, x_2^{10}$, etc. A term like $x_1^7 x_2^6$ will be truncated even when neither x_1 nor x_2 is raised to power higher than 10.

On the other hand, in principle, it is possible to impose different maximum order for each input variable by careful bookkeeping. By properly controlling the highest order index Ω for different variables, some input variables can be studied with more or with less emphasis.

Applying TPSA to accelerator beam dynamics, we may consider $(x, x', y, y', z, \delta)$ at some entrance location of the accelerator beamline as inputs, and the same coordinates at the exit location as outputs. If the beamline is considered to be one turn around a circular accelerator, for example, the nonlinear Taylor one-turn map is then obtained by TPSA. This map can be used to track particles, or to calculate some nonlinear dynamics quantities (such as tune shifts with betatron amplitudes, or nonlinear resonance strengths) analytically.

There may be a case when one is interested in knowing the dependence of the one-turn nonlinear map on the strength S of some special magnet. In such a case, one can include S as one of the input variables, perhaps together with the phase space variables $(x, x', y, y', z, \delta)$. The map is then obtained as a Taylor expansion in terms of S in addition to all the other input variables. This can be useful for example in the study of sensitivity to magnet errors. One can also consider one of the inputs to be a fitting variable strength Q of a quadrupole in a lattice design. This allows varying Q to match the β-functions, betatron tunes, or nonlinear chromaticities if they are chosen as output variables.

Homework 3.24 Find the total number of elements needed in the TPSA vector to represent one output up to Ω-th order when there are M input variables.

Solution Consider M balls in $M + \Omega$ boxes to obtain $\frac{(M+\Omega)!}{M!\Omega!}$.

Homework 3.25 Find $f((x, 1, 0, 0, 0, 0, \ldots), (y, 0, 1, 0, 0, 0, \ldots))$ for the functions $f(x, y) = x + y, xy, \frac{1}{xy}$ and e^{x+y}.

Homework 3.26 Consider the linear map $X_f = MX_i$ where M is a 6×6 matrix, and X_i is given as a 6-D TPSA vector. Find the TPSA representation of X_f to second order $\Omega = 2$.

3.6   TPSA tool

TPSA does not have to be applied to accelerator problems. In general, let us say if you have a 10,000-line numerical code that calculates numerically the values of outputs $y_1, y_2, \cdots, y_m$ when given inputs $x_1, x_2, \cdots, x_n$, as sketched in Fig. 3.1, you may convert your numerical code into a TPSA Taylor map of the outputs in terms of the inputs. The TPSA converter is to be used as sketched in Fig. 3.2.

After the conversion, you have replaced a possibly tedious time-consuming numerical code into a simple straightforward Taylor map, an iteration of which now takes a fraction of a millisecond to perform. The TPSA map is an algebraic result, not a numerical result, connecting the outputs to inputs. Once obtained, it is applied the same way as in the original numerical code except that the outputs are calculated in a much shorter time.

But these numerical manipulations are not the only way to apply the TPSA map. Another application is to analyze the TPSA map to extract its analytical properties. The TPSA map by itself does not readily connect to analysis, e.g. to evaluate the high order chromaticities or resonance strengths for a storage ring lattice design. To do so, it requires connecting the TPSA map with analytical tools. A tool that allows nonlinear analysis in terms of a Taylor map is Lie algebra, a subject to be discussed in Chapter 4.

We mentioned on page 118 a few issues of the TPSA approach related to its truncation to the Ω-th order, including the fact that the TPSA maps are generally nonsymplectic after truncation. We did sweep these issues under the rug in our discussions. On the other hand, one can try to improve the accuracy of the TPSA by increasing the order Ω — assuming it converges.

The TPSA converter mentioned in Fig. 3.2 is a special processing program. Its inputs include the entire numerical code that is to be converted. In addition, it will require the user to specify which variables in the numerical code are to be considered the input variables and which are to be considered the output variables of the TPSA map. The user will also have to specify the maximum

Figure 3.2: A numerical calculation algorithm with inputs and outputs can be converted into a TPSA map.

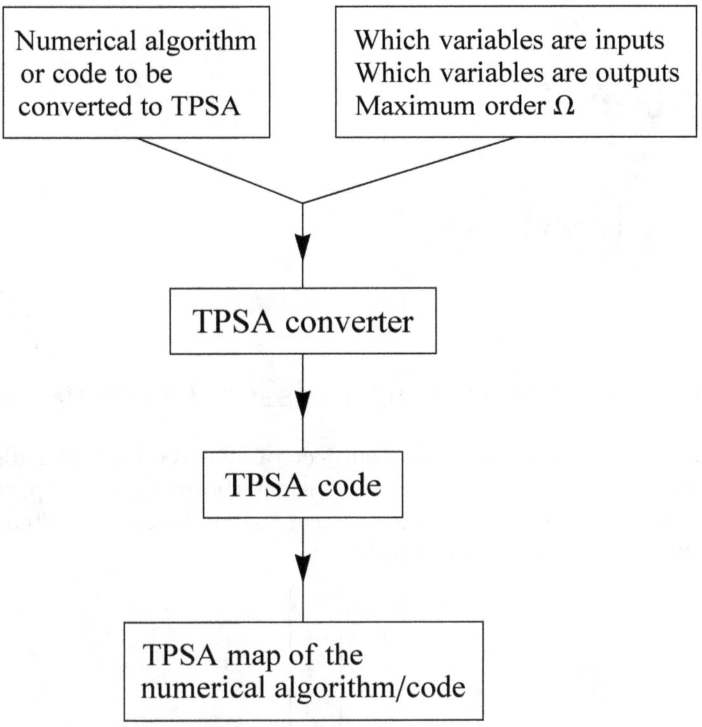

TPSA order Ω, as well as the reference points of all the input variables. The converter is a passive program in the sense of not affecting the original code being converted. If the original code contains a bug, it converts the same bug into the final TPSA map.

The principle of TPSA, as described above, is relatively straightforward. It might appear that the same can be expected of the TPSA converter. This is not so. The professionally prepared TPSA converter will have to consider the fact that the number of input and output variables vary from user to user. With many variables, and if accuracy is an issue, the maximum order Ω might be large, and in any case Ω needs to be flexible to be chosen by the user. This will lead to a demanding need of the computer's versatility and storage space. Sophisticated techniques for efficient and flexible data handling and storage will then be required, making the TPSA converter a high expertise.[5]

---

[5]M. Berz, et al., http://cosyinfinity.org; Y. Yan, AIP Conf. Proc. 297, 279 (1993); J. van Zeijts, AIP Conf. Proc. 297, 285 (1993).

Chapter 4

Lie Algebra

4.1 Symplecticity and Poisson bracket

Symplecticity — A recap The subject of symplecticity was discussed in Sec. 2.2. We give a brief recap here. Consider the motion of a particle in an n-dimensional ($2n$-dimensional phase space) linear dynamics system. Let the canonical coordinates of the particle be

$$X = \begin{bmatrix} q_1 \\ p_1 \\ q_2 \\ p_2 \\ \cdot \\ \cdot \\ \cdot \end{bmatrix}.$$

Let M be the $2n \times 2n$ matrix that describes the map that brings the canonical coordinates of the particle from the initial position $s = 0$ to the position of observation s in this linear dynamical system. The position coordinate s serves as the time variable of the system. Then M must satisfy the *symplecticity condition*

$$\tilde{M}SM = S, \tag{4.1}$$

where a tilde means taking the transpose of a matrix, and S is the matrix, sometimes called the *symplectic form*,

$$S = \begin{bmatrix} S_{2\times2} & 0 & & \\ 0 & S_{2\times2} & & \\ & & \cdots & \\ & & \cdots & \\ & & & S_{2\times2} \end{bmatrix},$$

that consists of a diagonal array of 2×2 matrix

$$S_{2\times2} = \begin{bmatrix} 0 & 1 \\ -1 & 0 \end{bmatrix}.$$

All symplectic matrices are necessarily even dimensional. Since $S^2 = -I$, S may be thought of as a matrix equivalent of the complex number $i = \sqrt{-1}$.

By its construction, the motion of a particle's phase space coordinates must be heavily confined. As discussed in Sec. 2.2, the symplecticity condition in a Hamiltonian system is a reflection of the fact that the degrees of freedom to describe the dynamics in the $2n$-dimensional phase space is severely restricted, and in this regard the phase space is an abstract space quite different from the n-D physical space. What is remarkable is that this property of unfilled degrees of freedom in the description of phase space dynamics has such an elegant simple expression, the symplecticity condition, to describe it.

In Sec. 2.2, we also introduced the Jacobian matrix of the map, whose elements are defined as

$$M_{\alpha\beta} \;=\; \frac{\partial X_\alpha}{\partial (X_0)_\beta}\,, \tag{4.2}$$

where $(X_0)_\beta$ is the β-th component of the initial coordinates of a particle at $s = 0$, X_α is the α-th component of the final state X of the particle at an arbitrary position s (even if $s < 0$). The Jacobian matrix M is a function of s and must be symplectic for all s and for all initial conditions X_0.

The symplecticity condition plays an important role in Hamiltonian dynamics. It imposes a strong constraint on the description of the Hamiltonian system. To be more specific, consider an n-D ($2n$-D phase space) linear system with matrix map M. The symplecticity condition (4.1) imposes a total of $n(2n - 1)$ conditions.[1] The $2n \times 2n$ matrix M has therefore $4n^2 - n(2n - 1) = n(2n + 1)$ independent elements. In particular, in a 2-D case, M contains 16 elements, but only 10 of them are independent.

The symplecticity condition imposes an even stronger constraint on nonlinear systems. In a linear system, the map is independent of X_0 or X; the symplecticity condition has only to hold for all s. In a nonlinear system, it has to hold for all s and all X_0.

Poisson bracket We will now introduce the *Poisson bracket*,

$$[f, g] \;=\; \sum_{i=0}^{n} \left(\frac{\partial f}{\partial q_i} \frac{\partial g}{\partial p_i} - \frac{\partial f}{\partial p_i} \frac{\partial g}{\partial q_i} \right)\,, \tag{4.3}$$

where f and g are arbitrary functions of s and the components of X. Note that although f and g depend on s in general, the Poisson bracket discriminates against the s-dependencies in that no explicit s-derivatives are included in its definition.

Using the notations already introduced, the Poisson bracket can also be written as

$$[f, g] \;=\; \sum_{\alpha,\beta} \frac{\partial f}{\partial X_\alpha} S_{\alpha\beta} \frac{\partial g}{\partial X_\beta}\,, \tag{4.4}$$

---

[1]To see this, first note that the matrix $\tilde{M}SM$ is necessarily antisymmetric. The diagonal elements of $\tilde{M}SM$ are necessarily zeroes. The number of free elements is therefore equal to the number of elements in the triangular upper (or lower) off-diagonal region.

or in matrix notation,

$$[f, g] = \widetilde{\frac{\partial f}{\partial X}} S \frac{\partial g}{\partial X}. \tag{4.5}$$

It is left for the reader to get a feel of the structure of the Poisson bracket here. In a sense, we defined a gradient operation here in the $2n$-D phase space. The Poisson bracket is then the dot product of two such gradients, except that a twisting operation by the matrix S is inserted before performing the dot product.

For what we plan to do, it is necessary to first get familiarized with the Poisson brackets. As a very first exercise, it follows immediately from the definition (4.3) that if f and g are both functions exclusively of the spatial coordinates, or exclusively of the momentum coordinates, then $[f, g] = 0$. In order for their Poisson bracket to be nonzero, both the spatial and the momentum coordinates must be invoked among f and g. Poisson bracket is a twist operation by construction.

As a second exercise, note that there are several properties of the Poisson brackets that are important. For example, let f, g, h be functions of X and s, and let a and b be some constants (independent of X, but can depend on s), we have

$$\begin{aligned}
[f, g] &= -[g, f], \\
[af + bg, h] &= a[f, h] + b[g, h], \\
[f, gh] &= [f, g]h + g[f, h], \\
[f, [g, h]] + [g, [h, f]] + [h, [f, g]] &= 0.
\end{aligned} \tag{4.6}$$

The last entry is called the *Jacobi identity*. The proofs of Eq. (4.6) are omitted here.

Proof of symplecticity condition   On the other hand, a fuller appreciation of the symplecticity condition (4.1) for the Jacobian matrix M can be found hidden in its derivation. We will now prove the symplecticity condition for a nonlinear system as follows. The proof for a linear system then follows as a special case.

The only information we have at this point is the fact that the system under consideration is Hamiltonian. This means there exists a Hamiltonian $H(X, s)$ so that the Hamilton equations of motion can be written as

$$X' = S \frac{\partial H}{\partial X}, \tag{4.7}$$

where a prime means taking derivative with respect to s. We are going to use Eq. (4.7) to calculate the derivative of $\tilde{M} S M$ with respect to s, where M is the Jacobian matrix (4.2). Before doing so, we establish the following,

$$M'_{\alpha\beta} = \left(\frac{\partial X_\alpha}{\partial X_{0\beta}} \right)' = \frac{\partial}{\partial X_{0\beta}} \left(S_{\alpha\gamma} \frac{\partial H}{\partial X_\gamma} \right)$$

$$\begin{aligned}
&= S_{\alpha\gamma}\frac{\partial^2 H}{\partial X_{0\beta}\partial X_\gamma} = S_{\alpha\gamma}\frac{\partial^2 H}{\partial X_\delta\partial X_\gamma}\frac{\partial X_\delta}{\partial X_{0\beta}} \\
&= S_{\alpha\gamma}H_{\delta\gamma}M_{\delta\beta} = (SHM)_{\alpha\beta} \\
\Longrightarrow \qquad M' &= SHM\,,
\end{aligned}$$
(4.8)

where we have adopted the notation that a repeated index means summation over the index, and we have defined a symmetric matrix H with elements

$$H_{\delta\gamma} = \frac{\partial^2 H}{\partial X_\delta\partial X_\gamma}\,.$$

Note that H is meant to contain X, not X_0, as its canonical coordinates. This requires that the *same* Hamiltonian H must hold throughout the dynamical process, and not only at its initial moment.

Equation (4.8) is rather intriguing and is of interest by its own right. But it is not what we are aiming to show here. Taking the derivative of the matrix $\tilde{M}SM$ with respect to s yields

$$\begin{aligned}
(\tilde{M}SM)' &= \tilde{M}'SM + \tilde{M}SM' \\
&= (\widetilde{M'})SM + \tilde{M}SM' \\
&= (\widetilde{SHM})SM + \tilde{M}SSHM \\
&= \tilde{M}\tilde{H}\tilde{S}SM - \tilde{M}HM \\
&= \tilde{M}HM - \tilde{M}HM = 0\,.
\end{aligned}$$
(4.9)

This means the matrix $\tilde{M}SM$ is an invariant, independent of s. In particular, its value can be obtained by evaluating it at $s = 0$. At $s = 0$, we have necessarily $M(s|0) = I$, i.e. the identity map. This then completes the proof of the symplecticity condition (4.1).

The use of the condition (4.7) demands the condition that the Jacobian matrix is symplectic only if we use *canonical* coordinates as the vector X. The transformation matrix for the vector (x, x', y, y'), for example, would not be symplectic in a solenoid, as x' and y' are not the canonical momenta in a solenoid. [See Homework 4.1, however.]

Algebra of Poisson bracket Consider any function f of s and X. Its value changes with time s either because of its explicit s dependence, or because it depends on X and changes because X changes. The total time derivative of f therefore can be written as

$$\begin{aligned}
f' &= \frac{\partial f}{\partial s} + \frac{\partial f}{\partial X_\alpha}X'_\alpha \\
&= \frac{\partial f}{\partial s} + \frac{\partial f}{\partial X_\alpha}S_{\alpha\beta}\frac{\partial H}{\partial X_\beta} \\
&= \frac{\partial f}{\partial s} + [f, H]\,.
\end{aligned}$$
(4.10)

Figure 4.1: Sophus Lie (1842–1899). [Courtesy Wikipedia.]

A quantity f is a constant of the motion if it is not explicitly s dependent, and that

$$[f, H] = 0. \tag{4.11}$$

Equations (4.10–4.11) are one of the reasons why Poisson brackets play an important role in dynamics; they are intimately related to the time evolution of phase space quantities.

At this point, the only relevant Poisson brackets seem to involve the Hamiltonian H; only the Poisson brackets with the Hamiltonian deserve attention. In the next section, however, we will see why the Poisson brackets in general are useful. In fact, we will see that, for our purpose, the Lie[2] algebra technique is basically an algebra of the Poisson brackets,[3]

$$\text{Lie algebra} \quad = \quad \text{algebra of Poisson brackets}. \tag{4.12}$$

In this context, we are saying that to apply Lie algebraic techniques to accelerator physics problems, it is required of us a proficient skill in manipulating the Poisson brackets, and we intend to develop it with these lectures. Suffice it to comment here that it is quite amazing that a purely mathematical construct — without introduction of any new physical principles — of a quantity like Eq. (4.3) can trigger so much substantial subsequent developments.

In later developments, we often compute the Poisson brackets of two Taylor series of X. It is easy to see that if f is an n-th order and g is an m-th order Taylor series, their Poisson bracket is another Taylor series of order $m + n - 2$. For example, the Poisson bracket of two quadratic forms is another quadratic form.

---

[2](Marius) Sophus Lie (1842–1899), Norwegian mathematician, noted for his work in differential equations, for which he developed the theory of continuous groups.

[3]The term *algebra* has a strict and rigorous — and complicated — definition in mathematics. We use the term as pedestrians.

Fundamental Poisson bracket The Poisson bracket of arbitrary functions f and g is given by Eqs. (4.3–4.5). There is one set of Poisson brackets that assumes particular significance, namely the *fundamental Poisson brackets*, Poisson brackets of the canonical coordinators themselves,

$$[X_\alpha, X_\beta] = S_{\alpha\beta} . \tag{4.13}$$

The proof is as follows. Consider a map from X_0 at $s = 0$ to X at s. The Jacobian matrix M of this map was defined in Eq. (4.2). The symplecticity of M implies that the fundamental Poisson brackets are preserved. To demonstrate this, consider the quantities X at position s as functions of the quantities X_0 at $s = 0$, and compute the Poisson brackets

$$\begin{aligned}
[X_\alpha, X_\beta] &= \frac{\partial X_\alpha}{\partial X_{0\gamma}} S_{\gamma\delta} \frac{\partial X_\beta}{\partial X_{0\delta}} = M_{\alpha\gamma} S_{\gamma\delta} M_{\beta\delta} \\
&= (M S \tilde{M})_{\alpha\beta} = S_{\alpha\beta} \\
&= [X_{0\alpha}, X_{0\beta}] ,
\end{aligned} \tag{4.14}$$

where again, repeated indices are summed over. In fact, Eq. (4.14) demonstrates that the transformation from X_0 to X is canonical if and only if all the fundamental Poisson brackets are preserved and are given by Eq. (4.13).

Our plan Perhaps it is helpful here to elucidate our plan to cover the subject of Lie algebra. We plan to divide the subject approximately into three parts. In the first part, we introduce the Poisson brackets, including enough material to introduce and to familiarize the reader with its operations. In the second part, we elevate the discussion and introduce a critical formula called Baker–Campbell–Hausdorff formula and its several variations, and include several applications of this formula to accelerator problems of a single-pass nature such as the cases encountered in a linac system. In the third part, we elevate the discussions again to include the cases when a particle moves in a repetitive system such as for a circular accelerator or a storage ring. The key quantity to learn in the third part is called normal form. In a sense, the BCH formula mainly applies to initial-value problems for single-pass systems or linacs, while normal forms apply to eigenvalue problems for repetitive systems or circular accelerators. These comments will become clearer as the subject develops.

	Subject	Analysis	System
part 1	Poisson bracket	algebra	single element
part 2	Baker–Campbell-Hausdorff formula	perturbation analysis	single passage
part 3	normal form	eigenanalysis	repetitive passages, storage ring

Homework 4.1 Let (q, p) be the canonical variables for a dynamical system with Hamiltonian $H(q, p, s)$. (a) Show that the variables $\bar{q} = aq$ and $\bar{p} = bp$ (a and b are constants) are not canonical unless $ab = 1$. (b) Define a Jacobi matrix $\bar{M}_{\alpha\beta} = \frac{\partial \bar{X}_\alpha}{\partial X_{0\beta}}$ with $X = (\bar{q}, \bar{p})$ and $X_0 = (q, p)$. Is the matrix $\bar{M}$ symplectic? The issue here is whether the use of noncanonical variables necessarily mean nonsymplecticity.

Homework 4.2 Show that, for a Hamiltonian system, if f and g are constants of the motion, so is $[f, g]$.

Solution Let H be the Hamiltonian governing the motion of the system. Use the Jacobi identity to show that if $[f, H] = 0$ and $[g, H] = 0$, then $[[f, g], H] = 0$.

Homework 4.3 As an illustration of Homework 4.2, consider a degenerate 2-D simple harmonic system described by the Hamiltonian

$$H(x, p_x, y, p_y, t) = \frac{1}{2}(\omega^2 x^2 + p_x^2 + \omega^2 y^2 + p_y^2).$$

(a) Show that

$$f_1 = \omega^2 x^2 + p_x^2 \qquad \text{and} \qquad f_2 = \omega^2 y^2 + p_y^2$$

are constants of the motion. Trying to find a third constant of the motion by forming the Poisson bracket $[f_1, f_2] = 0$ gives only a trivial case.

(b) However, show that there is another constant of the motion (the angular momentum)

$$g = x p_y - y p_x.$$

This third constant of the motion comes from the system's degeneracy due to rotational symmetry.

(c) By forming the Poisson bracket $[f_1, g]$ or $[f_2, g]$, one then finds another invariant

$$h = \omega^2 xy + p_x p_y.$$

(d) An alert reader would raise the question how could there be four invariants in this system, even degenerate. Indeed, the invariant h is not an independent invariant because it is related to the other three invariants. Show that they are related by $\omega^2 g^2 + h^2 = f_1 f_2$.

Homework 4.4 As another illustration of Homework 4.2, consider the 3-D system with Hamiltonian

$$H(x, p_x, y, p_y, z, p_z, t) = \frac{1}{2}(x^2 + p_x^2 + y^2 + p_y^2 + z^2 + p_z^2) + \epsilon(xy + yz + zx).$$

Show that

$$f_1 = (1-\epsilon)(x-y)^2 + (p_x - p_y)^2\,,$$
$$f_2 = (1-\epsilon)(y-z)^2 + (p_y - p_z)^2\,,$$
$$f_3 = (1-\epsilon)(z-x)^2 + (p_z - p_x)^2$$

are constants of the motion. By forming Poisson brackets among f_1, f_2, and f_3, show that

$$g = x(p_y - p_z) + y(p_z - p_x) + z(p_x - p_y)$$

is an invariant. By forming the Poisson brackets of $f_{1,2,3}$ and g, one finds more invariants

$$h_1 = (1-\epsilon)(x-y)(2z-x-y) + (p_x - p_y)(2p_z - p_x - p_y)\,,$$
$$h_2 = (1-\epsilon)(y-z)(2x-y-z) + (p_y - p_z)(2p_x - p_y - p_z)\,,$$
$$h_1 = (1-\epsilon)(z-x)(2y-z-x) + (p_z - p_x)(2p_y - p_z - p_x)\,.$$

Are these invariants all independent?

Homework 4.5

(a) Show that, if M is symplectic, so are $\tilde{M}$, M^{-1}, and

$$\bar{M} \equiv -S\tilde{M}S\,.$$

The last matrix $\bar{M}$ is called the *symplectic conjugate* of M.

(b) Show that if M_1 and M_2 are symplectic, then so is $M_1 M_2$.

Homework 4.6 Express a 4×4 symplectic matrix M in 2×2 block form as

$$M = \begin{bmatrix} A & B \\ C & D \end{bmatrix}.$$

(a) Show that

$$M^{-1} = \bar{M} = \begin{bmatrix} \bar{A} & \bar{C} \\ \bar{B} & \bar{D} \end{bmatrix} = \begin{bmatrix} (\det A)A^{-1} & (\det C)C^{-1} \\ (\det B)B^{-1} & (\det D)D^{-1} \end{bmatrix}.$$

(b) Show that the symplecticity condition gives six conditions

$$\det A + \det C = 1, \quad \det B + \det D = 1, \quad \bar{A}B + \bar{C}D = 0. \qquad (4.15)$$

(c) Since $\tilde{M}$ is also symplectic, applying the conditions (4.15) to $\tilde{M}$ gives

$$\det A + \det B = 1, \quad \det C + \det D = 1, \quad A\bar{C} + B\bar{D} = 0. \qquad (4.16)$$

The six conditions in Eq. (4.15) and the six conditions in Eq. (4.16) are algebraically equivalent. Combining the first two members of Eq. (4.15) and the first two members of Eq. (4.16) gives the necessary properties

$$\det A = \det D, \quad \det B = \det C\,.$$

Results in this exercise are useful when dealing with linearly coupled motions in an accelerator. In particular, you may want to check your 4×4 matrices against these properties to make sure they are symplectic as they should be.

Homework 4.7 Homework 4.6 can be extended to the 6×6 case. Show that if

$$
M \;=\; \begin{bmatrix} M_1 & M_2 & M_3 \\ M_4 & M_5 & M_6 \\ M_7 & M_8 & M_9 \end{bmatrix}
$$

is symplectic, then

$$
\begin{aligned}
\det M_1 + \det M_2 + \det M_3 &= \det M_4 + \det M_5 + \det M_6 \\
= \det M_7 + \det M_8 + \det M_9 &= \det M_1 + \det M_4 + \det M_7 \\
= \det M_2 + \det M_5 + \det M_8 &= \det M_3 + \det M_6 + \det M_9 = 1 \,.
\end{aligned}
$$

The reader is encouraged to test this interesting equation on the 6-D symplectic maps he/she learned in the past.

Homework 4.8 The text showed that a Hamiltonian system preserves the fundamental Poisson brackets as Eq. (4.13) described. The proof was given by Eq. (4.14). Show that the reverse is also true, i.e. if a system evolves in such a way that Eq. (4.13) holds at all times, then by slightly modifying Eq. (4.14), show that the system is symplectic.

4.2   Taylor and Lie map representations

4.2.1   The two representations

A map can be represented in various ways. Consider an n-D system and consider X the $2n$-component phase space vector. One example is the *Taylor map*, in which the final canonical variables are expressed as truncated power series in terms of the initial variables,

$$
X_\alpha \;=\; F_\alpha(X_0) \,, \tag{4.17}
$$

where X_α is the α-th component of the final X, and

$$
F_\alpha(X_0) \;=\; \Omega\text{-th order power series in the components of } X_0 \,. \tag{4.18}
$$

Since there are $2n$ variables, we have $2n$ polynomials like (4.18), each is of order Ω.

Application of Taylor maps to accelerators has a long history. In the linear case, the Taylor map is simply the linear map, which also has a matrix form, as adopted by Courant and Snyder.[4] An early version of a nonlinear (2nd order, i.e.

---

[4]E.D. Courant and H.S. Snyder, Ann. Phys. 3, 1 (1958).

$\Omega = 2$) Taylor map was adopted in the program TRANSPORT.[5] A more recent version of Taylor map scheme is developed in a 5th order program COSY.[6] A powerful tool to generate Taylor maps, introduced to accelerator dynamics by Berz,[7] uses the truncated power series algebra, the subject of Chapter 3.

Another way to represent a map is the *Lie map*, which is based on the Lie algebra techniques. Lie algebra techniques are an elegant and powerful tool in the study of nonlinear dynamics in accelerators. It was introduced by Dragt,[8] and extended and applied by many others.[9]

In the Courant–Snyder analysis of linear systems, we see how matrices are used extensively and effectively. The matrix analysis however is restricted to linear systems. The question is then how to generalize the Courant–Snyder analysis to nonlinear systems. Both the Taylor and the Lie representations can be regarded as ways to provide this generalization, and Courant–Snyder formalism is the basis for both.

It perhaps should be made a comment here that these formalisms, the Courant–Snyder, the TPSA, and the Lie algebra, are formal frameworks and tools to analyze the dynamical systems. They facilitate the study of physics and provide daily languages of communication but do not themselves contain physical meanings.

In the case of Taylor maps, this generalization to nonlinear systems is an obvious one. Equation (4.18) is the most straightforward way to represent a Taylor map. Another way to represent a Taylor map is in fact to cast it in a matrix form by extending the base vector to include higher order terms, which in turn allows the application of a range of mathematical tools to analyze the problem.[10] We shall not address this alternative Taylor analysis in our lectures other than a slight detour in Sec. 4.2.3.

To see the fact that Lie maps are also a way to generalize the Courant–Snyder analysis, the Courant–Snyder analysis must first be cast in the Lie formulation instead of a matrix (Taylor series) formulation. When done, the formulation then allows generalization to nonlinear systems in an elegant and natural manner. Details of these topics will be developed as we proceed.

In the Lie representation, an Ω-th order map is expressed as

$$e^{:G(X):} , \tag{4.19}$$

where

$$G(X) = (\Omega + 1)\text{-th order power series in the components of } X . \tag{4.20}$$

---

[5]K.L. Brown, SLAC Report 75 (1967).

[6]M. Berz, H.C. Hofmann, and H. Wollnick, Nucl. Instr. Meth. A258, 402 (1987).

[7]M. Berz, Part. Accel. 24, 109 (1989); M. Berz, Sec. 2.3.7, Handbook Accel. Phys. & Eng., 2nd ed., World Scientific (2013).

[8]A. Dragt, AIP Conf. Proc. 87, Phys. High Energy Accel., Fermilab, 1982, p. 147; see also A.J. Dragt, Lie Methods for Nonlinear Dynamics with Applications to Accelerator Physics, URL *http://www.physics.umd.edu/dsat/dsatliemethods.html*.

[9]See for example, E. Forest, SSCL Report 29 (1985); E. Forest, M. Berz, and J. Irwin, Part. Accel. 24, 91 (1989); J. Irwin, SSCL Report 228 (1989); Y. Yan, AIP Proc. 249, Phys. Part. Accel., 1993, p. 378.

[10]L.H. Yu, Phys. Rev. Accel. & Beams 20, 034001 (2017).

Note the subtle differences between Eq. (4.18) and (4.20),

- F_α is Ω-th order, while G is $(\Omega + 1)$-th order;

- there are $2n$ functions of F_α, while there is only one G;

- F_α is a function of X_0, while G is a function of X.

Equation (4.19) represents an operator. When applied to X, it gives the final coordinates in terms of the initial coordinates. In other words, the way the Lie map is to be used is[11]

$$X \;=\; e^{:G(X):}X\Big|_{X=X_0}. \tag{4.21}$$

Its detailed meaning and applications will become clear later.

Unless stated otherwise, we assume the map brings the origin $X_0 = 0$ to the origin $X = 0$. This means the lowest order leading terms of F_α are 1st order in X_0, and the leading terms in $G(X)$ are second order in X.

For a given accelerator system and a predetermined order Ω, if we compare the results of X obtained from the Taylor representation (4.17–4.18) and the Lie representation (4.19–4.20), we will find that they coincide up to the Ω-th order terms. There are no terms beyond the Ω-th order in the Taylor map (4.17). All terms beyond the Ω-th order are truncated. With this truncation, Taylor maps are not symplectic in general, although it does maintain its symplecticity up to the Ω-th order. In contrast, if expanded, the Lie map (4.21) contains higher order terms. These higher order terms, if all included without truncation, as we will see later, are going to make the Lie map strictly symplectic.

To study the long-term stability of the motion of a particle in an accelerator, one may encounter the situation when a high order Taylor map (say 12-th order) is needed, while a relatively low order (say 5th order) Lie map would suffice. The reason one needs a high order Taylor map is not so much to make the map extremely accurate. Rather, it is because the map needs to be extremely symplectic. If one uses Lie maps, which are always symplectic, a lower order map with less accuracy may suffice.

We shall return to more comparison between the Taylor and Lie maps on page 199.

It should be mentioned here that there are still other ways to represent a map. For example, one may choose to introduce a generating function, and then express it as a power series. However, we will discuss only the Taylor and the Lie representations.

---

[11] As indicated in Eq. (4.21), the substitution of X by X_0 is to be made *after* the application of the operator. It would be misleading, or at least ambiguous, to write for example $X = e^{:G(X):}X_0$ or $X = e^{:G(X_0):}X_0$. Before the last substitution of X_0 into the expression obtained, do not even think of inserting any numerical numbers in the process. All X's are considered dynamical variables, and we are just performing Poisson bracket algebra for a while!

4.2.2 Degree of freedom

Number of coefficients in a Taylor map The degree of freedom in the Taylor representation is given by how many coefficients are needed to describe a Taylor map. Consider the function $F_\alpha(X_0)$ which is a sum of k-th order terms where $k = 1, \cdots, \Omega$. Homework 4.9 shows that the total number of coefficients for the k-th order terms is

$$2n \times \frac{(2n + k - 1)!}{k!(2n - 1)!},\qquad(4.22)$$

where the factor $2n$ is because there are $2n$ functions of F_α. By adding the number of coefficients from $k = 1$ to Ω, we obtain Table 4.1.

Table 4.1: Total number of coefficients needed to describe a Taylor map of Ω-th order. The phase space is $2n$-dimensional.

	$\Omega = 1$	$\Omega = 2$	$\Omega = 3$	$\Omega = 4$	$\Omega = 5$	$\Omega = 6$
$n = 1$	4	10	18	28	40	54
$n = 2$	16	56	136	276	500	836
$n = 3$	36	162	498	1254	2766	5538

As a trivial example, consider the case $n = 1$ and $\Omega = 1$. This linear map can be written as a 2×2 matrix, with 4 elements. The Taylor map therefore has 4 coefficients. On the other hand, the symplecticity condition dictates that this matrix has unit determinant. So the total number of independent coefficients, its number of degrees of freedom, is 3. One of the four coefficients is redundant. Indeed, the Lie representation requires only 3 coefficients.

Taylor maps have the advantage that the final coordinates can be computed straightforwardly from the initial coordinates. However, as mentioned before, Taylor representation is in general nonsymplectic. In addition, it requires more coefficients than necessary to represent the map.

Number of coefficients in a Lie map Lie representation requires the minimum number of coefficients to represent a map up to a given order Ω. Lie maps have no redundancy of coefficients and are more concise than Taylor maps. The number of coefficients of the k-th order terms in the function $G(X)$ of Eq. (4.19) is

$$\frac{(2n + k - 1)!}{k!(2n - 1)!}.\qquad(4.23)$$

By adding the number of coefficients from $k = 2$ to $k = \Omega + 1$, we obtain Table 4.2. Taylor maps are more convenient for numerical tracking; Lie maps are better suited for analysis. The difference between Tables 4.1 and 4.2 is of course due to the symplecticity condition.

Homework 4.9 Prove the statement (4.22).

Table 4.2: Total number of coefficients needed to describe a Lie map of Ω-th order. The phase space is $2n$-dimensional.

	$\Omega = 1$	$\Omega = 2$	$\Omega = 3$	$\Omega = 4$	$\Omega = 5$	$\Omega = 6$
$n = 1$	3	7	12	18	25	33
$n = 2$	10	30	65	121	205	325
$n = 3$	21	77	203	455	917	1709

Solution Consider one of the functions $F_\alpha(X_0)$. To find the number of coefficients in F_α that are of k-th order, let us consider k "objects" and $2n - 1$ "partitions". Arrange the collection of objects and partitions in a certain order. Represent an object by a cross, and a partition by a vertical bar, one has for example the following pattern for $n = 2$ and $k = 4$,

$$\times \times \mid \times \mid\mid \times$$

One may identify this pattern with $x_1^2 x_2 x_4$. It is clear that each pattern is in one-to-one correspondence to a term in the polynomial. The total number of coefficients is then the same as the number of different patterns that can be made with k objects and $2n - 1$ partitions. The answer is

$$\frac{(2n + k - 1)!}{k!(2n - 1)!} \ .$$

Since there are $2n$ functions like F_α, we have proved Eq. (4.22).

Homework 4.10 The text mentioned a trivial case of a Taylor map with $n = 1$ and $\Omega = 1$. Consider here the first nontrivial nonlinear case $n = 1$, $\Omega = 2$. The Taylor map reads

$$
\begin{aligned}
x &= R_{11}x_0 + R_{12}p_0 + T_{111}x_0^2 + T_{112}x_0 p_0 + T_{122}p_0^2 , \\
p &= R_{21}x_0 + R_{22}p_0 + T_{211}x_0^2 + T_{212}x_0 p_0 + T_{222}p_0^2 .
\end{aligned}
\tag{4.24}
$$

There are ten coefficients in this representation — see Table 4.1. The Jacobian matrix of the map is

$$
M = \begin{bmatrix} R_{11} + 2T_{111}x_0 + T_{112}p_0 & R_{12} + T_{112}x_0 + 2T_{122}p_0 \\ R_{21} + 2T_{211}x_0 + T_{212}p_0 & R_{22} + T_{212}x_0 + 2T_{222}p_0 \end{bmatrix} .
$$

Due to symplecticity, the determinant of M has to be unity for arbitrary values of x_0 and p_0. This one condition $\det M = 1$ therefore actually yields three conditions,

$$
\begin{aligned}
R_{11}R_{22} - R_{12}R_{21} &= 1 , \\
R_{11}T_{212} + 2R_{22}T_{111} - 2R_{12}T_{211} - R_{21}T_{112} &= 0 , \\
2R_{11}T_{222} + R_{22}T_{112} - R_{12}T_{212} - 2R_{21}T_{122} &= 0 ,
\end{aligned}
\tag{4.25}
$$

where we have dropped terms second order in x_0 or p_0 because information is incomplete for these coefficients after the Taylor map is truncated to 2nd order in Eq. (4.24) — see Homework 4.11 to continue this discussion. We have thus obtained three conditions constraining the ten coefficients. The number of independent coefficients is then seven. Indeed, seven is the number of coefficients required in the corresponding Lie representation. See Table 4.2.

Homework 4.11 Equation (4.24) is a terminated power series because we have performed a truncation. If Eq. (4.24) turns out to be exact, i.e. the higher order terms all vanish exactly, what can we say about this map?

Solution In this case, we can impose the symplecticity condition exactly. Keep all terms including terms higher order in X_0, we have in addition to (4.25), three more conditions

$$
\begin{aligned}
T_{111}T_{212} - T_{112}T_{211} &= 0, \\
T_{111}T_{222} - T_{122}T_{211} &= 0, \\
T_{112}T_{222} - T_{122}T_{212} &= 0.
\end{aligned}
\tag{4.26}
$$

By solving Eqs. (4.25) and (4.26), it follows that the map $(x_0, p_0) \to (x, p)$ can be written as a two-step process, namely

$$
\begin{aligned}
x_1 &= R_{11}x_0 + R_{12}p_0, \\
p_1 &= R_{21}x_0 + R_{22}p_0,
\end{aligned}
\tag{4.27}
$$

followed by

$$
\begin{aligned}
x &= x_1 + T_{111}\left(\frac{T_{111}p_1 - T_{211}x_1}{R_{21}T_{111} - R_{11}T_{211}}\right)^2, \\
p &= p_1 + T_{211}\left(\frac{T_{111}p_1 - T_{211}x_1}{R_{21}T_{111} - R_{11}T_{211}}\right)^2.
\end{aligned}
\tag{4.28}
$$

Both steps (4.27) and (4.28) are symplectic. In case either $T_{111} = 0$ or $T_{211} = 0$, Eq. (4.28) is a kick map. Otherwise it is not a kick map and is symplectic nevertheless.

The trick introduced in this exercise, particularly Eq. (4.28), can be used to symplectify second order maps beyond those illustrated in Chapter 2.

4.2.3   Taylor invariant

For circular accelerators, a type of map of particular significance is the *one-turn map*, which gives the map for one revolution around the accelerator. It might be instructive at this point to consider how to construct a Taylor expression of an invariant from a Taylor one-turn map.

Suppose we are looking for an invariant of Ω-th order. Consider the vector

$$
Z = \begin{bmatrix} x \\ p \\ x^2 \\ xp \\ p^2 \\ \cdot \\ \cdot \\ \cdot \\ xp^{\Omega-1} \\ p^{\Omega} \end{bmatrix}.
$$

Let the one-turn map of the accelerator be written in a matrix form as

$$
Z_{\text{final}} = M Z_{\text{initial}}. \tag{4.29}
$$

In writing down Eq. (4.29), we have truncated the map M at the $(\Omega - 1)$-th order.

Let us make a side remark here. The Jacobian matrix of this nonlinear map will depend on X_0. It is a messy quantity because it involves both the accelerator properties and the particle properties. It does not factorize. On the other hand, when written in the form (4.29), it factorizes. The matrix M depends only on accelerator properties, while Z_{initial} depends only on particle properties. Potentially useful advances can be made to analyze this formulation of the system, particularly from the one-turn map M (see footnote 10). For our purposes here, however, we will be content with finding a Taylor invariant without exploring its further applications.

Let the invariant be expressed as an Ω-th order Taylor series,

$$
W = \tilde{V} Z, \tag{4.30}
$$

where V is a vector yet to be found, while the matrix M is assumed known.

Since W is an invariant, we must have

$$
\begin{aligned}
\tilde{V} Z &= \tilde{V} M Z \quad \text{for all } Z, \\
\implies \quad \tilde{V} &= \tilde{V} M, \\
\implies \quad \tilde{M} V &= V.
\end{aligned} \tag{4.31}
$$

This means V is just the eigenvector of $\tilde{M}$ with eigenvalue 1. This is the way V — and therefore the invariant W — can be found. It can happen that an eigenvector of eigenvalue 1 cannot be found, and if that happens, it means there is no invariant of the order being considered. Note that even in the linear case, Z must be kept up to the quadratic terms in order to find W.[12]

Incidentally, the reader might be curious what would be the expression for an invariant of a Lie map $e^{:G(X):}$. As will be clear later, the answer is simply $G(X)$.

---

[12]This is except for the trivial case when the map for (x, p) is the identity map, in which case x and p are invariants.

An example Let us apply our result to an example. Consider a one-turn map that represents a linear Courant–Snyder map with phase advance μ, followed by a thin-lens sextupole. The combined map is given by the Taylor form,

$$
\begin{aligned}
x &= x_0 \cos\mu + p_0 \sin\mu\,, \\
p &= -x_0 \sin\mu + p_0 \cos\mu + \epsilon x^2\,.
\end{aligned}
\tag{4.32}
$$

In the 1-D system, there can only be one invariant. Its 2nd order expression is $W_2 = x^2 + p^2$. Following the procedure outlined above, we find its Taylor expression to 3rd order in x, p. The result is

$$
W_3 = x^2 + p^2 - \epsilon \frac{\sin\mu\cos\mu}{(1+2\cos\mu)(1-\cos\mu)} x^3 - \epsilon x^2 p - \epsilon \frac{\sin\mu}{1+2\cos\mu} x p^2\,,
\tag{4.33}
$$

where x and p are values observed at the exit of the thin-lens sextupole. We will return to this expression in Homework 4.78. See also Homework 4.80.

This expression W_3 diverges when $\cos\mu = 1$ or $-\frac{1}{2}$, i.e. when $\frac{\mu}{2\pi} = \frac{n}{3}$ with n an integer. One may proceed to obtain a 4th order invariant W_4 and find that it diverges when $\frac{\mu}{2\pi} = \frac{n}{4}$ (Homework 4.13). It is clear that higher orders will incorporate more small-denominator terms involving increasingly more numerous higher order resonances. The series in fact generally does not converge.

One may perform a numerical test of the invariance of $W_2 = x^2 + p^2$ and W_3. Take for example a case with $\mu = 2\pi \times 0.23$ and $\epsilon = 0.05$. A particle with initial conditions $x_0 = 1$ and $p_0 = 0$ is tracked for 1000 iterations. The resulting values of x and p for each iteration are used to calculate the values of W_2 and W_3 and plotted as functions of iteration number in Fig. 4.2. It can be seen that W_2 is not a good invariant; its value fluctuates by about 10%. In comparison, W_3 fluctuates only by $< 0.1\%$. Round-off errors are negligible in these calculations.

Homework 4.12

(a) Consider the linear one-turn map

$$
\begin{aligned}
x &= x_0 \cos\mu + p_0 \sin\mu\,, \\
p &= -x_0 \sin\mu + p_0 \cos\mu\,.
\end{aligned}
$$

Find Taylor expressions of an invariant up to 2nd order, 3rd order, and 4th order in (x, p).

(b) Repeat the problem when the map is replaced by the one-turn Courant–Snyder matrix map. You should obtain the Courant–Snyder invariant as your result.

(c) Repeat the problem when the one-turn map is unstable with the linear one-turn map,

$$
\begin{aligned}
x &= x_0 \cosh\mu + p_0 \sinh\mu\,, \\
p &= x_0 \sinh\mu + p_0 \cosh\mu\,.
\end{aligned}
$$

Figure 4.2: Tracking results for the approximate invariants W_2 (upper) and W_3 (lower).

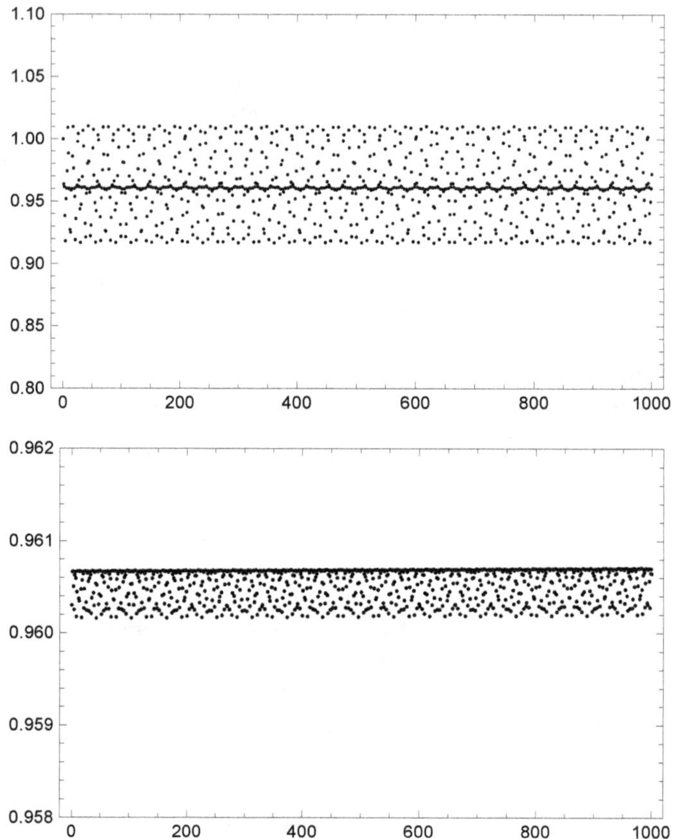

Solution (a) There are no 1st and 3rd order invariants. The 2nd order invariant is $x^2 + p^2$. There is only one 4th order invariant, $(x^2 + p^2)^2$. Note that in the derivation we must assume $\sin \mu \neq 0$. But then, what happens if $\sin \mu = 0$?

Homework 4.13

(a) Verify Eq. (4.33).

(b) Extend the expression to find the invariant to 4th order. Identify the resonance behavior mentioned in the text.

Solution (a) If you prefer to skip the way Eq. (4.33) is derived and just show that W_3, as given, is an invariant, you may simply substitute (4.32) into W_3 and show that it gives the same expression when you substitute x_0, p_0 into it. These two expressions should be equal up to the 3rd order in (x, p).

4.3 Algebra of operator

4.3.1 Lie operator

In this section, we will describe the techniques of Lie algebra, i.e. the algebra of Poisson brackets. For convenience we rewrite the Poisson bracket (4.3) in another notation, following Dragt,

$$:f:g \;\equiv\; [f,g]\,.$$

In this notation, the quantity $:f:$ is viewed as an operator — a Lie operator — which operates on the function g. Obviously we have $:f:g = -:g:f$. An identity map will be designated as 1.[13]

One can define functions of operators. For example, powers of an operator can be obtained by letting

$$
\begin{aligned}
(:f:)^2 g &= :f:(:f:g) = [f,[f,g]]\,,\\
(:f:)^3 g &= [f,[f,[f,g]]], \qquad \cdots \quad \text{etc.}
\end{aligned}
$$

It can be shown that

$$(:f:)^k(gh) \;=\; \sum_{n=0}^{k} \frac{k!}{n!(k-n)!}\,\bigl[(:f:)^n g\bigr]\bigl[(:f:)^{k-n}h\bigr]\,. \tag{4.34}$$

When $k=1$, it is just the third member of Eq. (4.6). The rest can be proved by induction.

It also follows from the Jacobi identity that the commutator of two operators $:f:$ and $:g:$, defined as

$$\{:f:,:g:\} \;\equiv\; :f:\,:g: - :g:\,:f:\,, \tag{4.35}$$

is equal to the operator $:[f,g]:$, i.e.

$$\{:f:,:g:\} \;=\; :[f,g]:\,. \tag{4.36}$$

Equation (4.36) has a useful variation, namely,

$$\{:f:,:g:\}h \;=\; :h::g:f\,. \tag{4.37}$$

Equation (4.37) is just another way of writing the Jacobi identity.

Equation (4.36) is the reason Poisson brackets play a prominent role in Lie algebra of operators. As we will see, whether two Lie operators commute or not is an important criterion in Lie algebra. The commutators of operators occur often — they occur as a residual remnant each time when we switch the

---

[13]Note that $:1:$ is not the identity map. In fact, one has

$$(\text{identity map})\; X = 1X = X \quad \text{while} \quad :1:X = 0\,.$$

ordering of two Lie operators, and Eq. (4.36) relates them to Poisson brackets. In particular, when $[f, g] = 0$, the operators $:f:$ and $:g:$ commute (however, see Homework 4.16). As a simple special case, if both f and g are functions exclusively of the spatial coordinates, or if both are exclusively of the momentum coordinates, then $[f, g] = 0$ and $:f:$ and $:g:$ commute.

Once powers of operators are defined, other functions of operators can then be constructed based on power series expansions. Of particular significance are the exponential operators,

$$e^{:f:} = \sum_{k=0}^{\infty} \frac{1}{k!} (:f:)^k . \tag{4.38}$$

The leading term on the right-hand-side of Eq. (4.38) is the identity map 1. Exponential operators are thus rooted as neighboring maps around the identity map. Homework 4.19 elaborates more on this important observation.

Homework 4.14

(a) Let $a(s)$ be a function of s and not a function of X. What is the operator $:a(s):$? What is $e^{:a(s):}$?

(b) Although $e^{\ln x} = x$ is an identity, show that the maps $e^{:\ln x:}$ and $:x:$ are quite different.

Solution (b) They are very different,

$$e^{:\ln x:} x = x, \qquad e^{:\ln x:} p = p + \frac{1}{x},$$
$$:x: x = 0, \qquad :x: p = 1 .$$

In particular, the map $e^{:\ln x:}$ is symplectic as its Jacobian matrix

$$\begin{bmatrix} 1 & 0 \\ -\frac{1}{x^2} & 1 \end{bmatrix}$$

has determinant of 1. The map $:x:$ is nonsymplectic.

Homework 4.15 Consider a function $f(x, p)$ that is a polynomial in x and p of the k-th order. What does the function $f(x, p)$ need to satisfy in order for the Lie operator $:f(x, p):$ to be symplectic?

Solution A hint of how to check the symplecticity of a Lie map can be obtained from Homework 4.14. Show that the only way for $:f(x, p):$ to be symplectic is for $f(x, p)$ to be a quadratic form with $k = 2$. Let $f(x, p) = ax^2 + bxp + cp^2$, it also has to satisfy the condition $b^2 - 4ac = -1$. In particular, the function $f(x, p)$ must be positive definite or negative definite.

Homework 4.16 It was mentioned that $:f:$ and $:g:$ commute if $[f, g] = 0$. In fact, $:f:$ and $:g:$ commute even if $[f, g] \neq 0$ but $=$ constant (independent of x and p but can depend on s).

(a) Although $[ax, bp] = ab \neq 0$, show that

$$:ax::bp: \ = \ :bp::ax:\,.$$

(b) Although $[ax^2, \frac{bp}{x}] = 2ab \neq 0$, show that

$$:ax^2::\frac{bp}{x}: \ = \ :\frac{bp}{x}::ax^2:\,.$$

Homework 4.17 Show that

$$:g::f:^2 g \ = \ -:f::g:^2 f\,. \tag{4.39}$$

This curious and somewhat sophisticated identity is of some use later.

Solution Define $h = :f:g$, then

$$
\begin{aligned}
:g::f:^2 g \ &= \ :g::f:h \ = \ -:f::h:g - :h::g:f \\
&= \ :f::g:h + :h::f:g \ = \ :f::g:h + :h:h \\
&= \ :f::g:h \ = \ :f::g::f:g \ = \ -:f::g:^2 f \quad \Longrightarrow \quad \text{Q.E.D.}
\end{aligned}
$$

4.3.2   Lie operator for accelerator element

The maps for accelerator elements can be represented as exponential Lie operators. For simplicity, let us consider first a 1-D system. A drift space of length L, for example, can be represented as the operator $\exp(: -\frac{1}{2}Lp^2:)$. To show that, let's first establish the following,

$$
\begin{aligned}
:p^2:x \ &= \ \frac{\partial p^2}{\partial x}\frac{\partial x}{\partial p} - \frac{\partial p^2}{\partial p}\frac{\partial x}{\partial x} \ = \ -2p\,, \\
:p^2:p \ &= \ \frac{\partial p^2}{\partial x}\frac{\partial p}{\partial p} - \frac{\partial p^2}{\partial p}\frac{\partial p}{\partial x} \ = \ 0\,, \\
(:p^2:)^2 x \ &= \ :p^2:(:p^2:x) \ = \ :p^2:(-2p) \ = \ 0\,, \\
(:p^2:)^2 p \ &= \ :p^2:(:p^2:p) \ = \ :p^2:(0) \ = \ 0\,.
\end{aligned}
$$

We then apply the operator $\exp(: - Lp^2/2:)$ to (x, p) to obtain

$$
\begin{aligned}
e^{:-Lp^2/2:}x \ &= \ x - \frac{1}{2}L:p^2:x + \frac{1}{8}L^2(:p^2:)^2 x + \cdots \\
&= \ x - \frac{1}{2}L(-2p) \ = \ x + Lp\,, \\
e^{:-Lp^2/2:}p \ &= \ p - \frac{1}{2}L:p^2:p + \cdots \ = \ p\,. \tag{4.40}
\end{aligned}
$$

If we identify the x and p on the right-hand-sides of Eq. (4.40) as the initial coordinate and momentum of the particle, we recognize that $\exp(: -\frac{1}{2}Lp^2:)x$ and $\exp(: -\frac{1}{2}Lp^2:)p$ give the final coordinate and momentum of the particle.

We therefore identify $\exp(:-\frac{1}{2}Lp^2:)$ as the Lie operator of a drift space, and the way the Lie map is to be applied is according to Eq. (4.21).

We have now established the fact that the operator $\exp(:-\frac{1}{2}Lp^2:)$ transforms x and p as a drift space does. A cautious reader may rightfully question whether it can represent the drift transformation for arbitrary functions of x and p — for example, whether it transforms x^2 to $(x+Lp)^2$, etc. If it does not, then its identification with a drift space map will remain questionable. To address this question, let us mention the important theorem which we will show later by Eqs. (4.43) and (4.45), that this map $\exp(:-\frac{1}{2}Lp^2:)$ indeed transforms an arbitrary function of x and p as a drift space does, and this important property applies to all exponential operators. See also Homework 4.20.

Similarly we can establish the Lie operators for other accelerator elements. A few simpler examples for a 1-D system are given in Table 4.3. We will elaborate more on these operators in the homework problems. For now, let us make the following observations on Table 4.3.

- The Lie map of a thick quadrupole differs from that of a thin-lens (taking the limit of zero length) quadrupole by a term that represents a drift space, as one might expect.

- The case of coordinate displacement is represented by a Lie map with linear exponents.

- The case of a coordinate scale change is symplectic if it conforms to the form $x = e^{-\lambda}x_0, p = e^{\lambda}p_0$.

- The table does not list a thick multipole or thick kick. This is because there is not an analytic Taylor map in general in that case. On the other hand, the Lie map for that case is straightforward and is given by

$$\exp\left(:-\frac{1}{2}Lp^2 + \int_0^x f(x')dx':\right).$$

A comment might be useful here concerning the function f that appears on the exponent of the Lie map $e^{:f:}$. It might appear at a first glance that it should be dimensionless because it appears on the exponent. A closer look says that what is dimensionless is $:f:$, not f. This requires that the function f must have a dimension of $(\text{coordinate}) \times (\text{momentum})$, i.e. it has the dimension of the phase space area, i.e. the dimension of the emittance. As we will see for example, for the linear Courant-Snyder system, f will be simply the Courant-Snyder emittance.

Another comment to be made is that not all symplectic maps are connected. Not all of them can be expressed by single exponential maps, or if insisted, requires imaginary exponent — when that happens, it means the effective Hamiltonian or effective emittance of the system is imaginary. In accelerator applications, however, we shall consider only exponential Lie maps with real exponents. They are all connected through the identity map I. [See the comment following Eq. (4.38) and Homework 4.19.] Those maps not expressible by such single Lie maps can be considered a product of two real exponential Lie maps.

Table 4.3: Taylor maps and Lie operators of some accelerator elements. The Lie operators are to be applied according to Eq. (4.21).

Element	Taylor map	Lie operator
Drift space	$x = x_0 + Lp_0$ $p = p_0$	$\exp\left(: -\tfrac{1}{2}Lp^2:\right)$
Thin-lens quad.	$x = x_0$ $p = p_0 - \tfrac{1}{f}x_0$	$\exp\left(: -\tfrac{1}{2f}x^2:\right)$
Thin-lens multipole	$x = x_0$ $p = p_0 + \lambda n x^{n-1}$	$\exp(:\lambda x^n:)$
Thin-lens kick	$x = x_0$ $p = p_0 + f(x)$	$\exp\left(: \int_0^x f(x')dx':\right)$
Thick focusing quad.	$x = x_0 \cos kL + \tfrac{p_0}{k}\sin kL$ $p = -kx_0 \sin kL + p_0 \cos kL$	$\exp\left[:-\tfrac{L}{2}(k^2x^2+p^2):\right]$
Thick defocus. quad.	$x = x_0 \cosh kL + \tfrac{p_0}{k}\sinh kL$ $p = kx_0 \sinh kL + p_0 \cosh kL$	$\exp\left[:\tfrac{L}{2}(k^2x^2 - p^2):\right]$
Coordinate shift	$x = x_0 - b$ $p = p_0 + a$	$\exp(:ax + bp:)$
Coordinate rotation	$x = x_0 \cos \mu + p_0 \sin \mu$ $p = -x_0 \sin \mu + p_0 \cos \mu$	$\exp\left[: -\tfrac{\mu}{2}(x^2 + p^2):\right]$
Scale change	$x = e^{-\lambda}x_0$ $p = e^{\lambda}p_0$	$\exp(:\lambda xp:)$

Changing the canonical coordinate scale   Table 4.3 assumes the canonical coordinates have been chosen to be x and $p = x'$. What if we want to scale these coordinates to $\bar{x} = \alpha x$ and $\bar{p} = \beta p$ and then use $\bar{x}$ and $\bar{p}$ as the new canonical coordinates? It is easy to see that the Poisson bracket changes according to

$$\frac{1}{\alpha\beta}\left[f(\alpha x, \beta p), g(\alpha x, \beta p)\right] = \left[f(x,p), g(x,p)\right],$$

which in turn gives

$$e^{:\frac{1}{\alpha\beta}f(\alpha x,\beta p):} = e^{:f(x,p):}.$$

As an example, the thick quadrupole map $e^{:-\frac{L}{2}(k^2x^2+p^2):}$ can be scaled to read

$$e^{:-\frac{kL}{2}(\bar{x}^2+\bar{p}^2):}$$

if we choose $\alpha = \tfrac{1}{k}, \beta = 1$.

This change of canonical coordinates is not to be confused with the scale map (telescope) mentioned in Table 4.3. In a telescope system, the phase space is not being changed, but here we are actually changing the phase space.

Homework 4.18

(a) Table 4.3 gives the map for a scale change, e.g. for a telescope system. The map is $e^{:\lambda xp:}$ to be applied to the 1-D coordinates (x,p). Prove that the

Taylor map is given by

$$x = e^{-\lambda} x_0, \qquad p = e^{\lambda} p_0 .$$

(b) Table 4.3 also gives the map $e^{:\frac{L}{2}(k^2 x^2 - p^2):}$. Derive the corresponding 2×2 matrix map.

(c) Now consider a 2-D system. A lumped sextupole (taking the limit of zero length) generates the map relating the final coordinates (x, x', y, y') to the initial coordinates (x_0, x'_0, y_0, y'_0) of a particle by

$$\begin{aligned}
x &= x_0 , \\
x' &= x'_0 + S(x_0^2 - y_0^2) , \\
y &= y_0 , \\
y' &= y'_0 - 2S x_0 y_0 .
\end{aligned}$$

Find the Lie representation of this lumped-sextupole map.

(d) This is another problem of a 2-D system. A particle going through a crab cavity can be represented by a matrix multiplication

$$X_{\text{out}} = M X_{\text{in}} ,$$

where

$$X = \begin{bmatrix} x \\ x' \\ z \\ \delta \end{bmatrix} \quad \text{and} \quad M = \begin{bmatrix} 1 & 0 & 0 & 0 \\ 0 & 1 & k & 0 \\ 0 & 0 & 1 & 0 \\ k & 0 & 0 & 1 \end{bmatrix} .$$

Find the Lie map for a crab cavity.

Solution

(a)
$$e^{:\lambda x p:} = 1 + (:\lambda x p:) + \frac{1}{2}(:\lambda x p:)^2 + \frac{1}{6}(:\lambda x p:)^3 + \cdots ,$$

$$\begin{aligned}
(:\lambda x p:)x &= [\lambda x p, x] = -\lambda x , \\
(:\lambda x p:)^2 x &= (-\lambda)^2 x , \\
(:\lambda x p:)^n x &= (-\lambda)^n x , \\
\Longrightarrow \quad e^{:\lambda x p:} x &= e^{-\lambda} x , \\
(:\lambda x p:)p &= [\lambda x p, p] = \lambda p , \\
(:\lambda x p:)^2 p &= \lambda^2 p , \\
(:\lambda x p:)^n p &= \lambda^n p , \\
\Longrightarrow \quad e^{:\lambda x p:} p &= e^{\lambda} p .
\end{aligned}$$

When $\lambda = 0$, it gives the identity map of course.

(b)
$$e^{:\frac{L}{2}(k^2 x^2 - p^2):} = 1 + \left(:\frac{L}{2}(k^2 x^2 - p^2):\right) + \frac{1}{2}\left(:\frac{L}{2}(k^2 x^2 - p^2):\right)^2$$
$$+ \frac{1}{6}\left(:\frac{L}{2}(k^2 x^2 - p^2):\right)^3 + \cdots ,$$

$$\left(:\frac{L}{2}(k^2x^2 - p^2):\right) x = Lp,$$

$$\left(:\frac{L}{2}(k^2x^2 - p^2):\right) p = Lk^2x,$$

$$\left(:\frac{L}{2}(k^2x^2 - p^2):\right)^2 x = \left(:\frac{L}{2}(k^2x^2 - p^2):\right) Lp = L^2k^2x,$$

$$\left(:\frac{L}{2}(k^2x^2 - p^2):\right)^2 p = \left(:\frac{L}{2}(k^2x^2 - p^2):\right) Lk^2x = L^2k^2p,$$

$$\left(:\frac{L}{2}(k^2x^2 - p^2):\right)^3 x = \left(:\frac{L}{2}(k^2x^2 - p^2):\right) L^2k^2x = L^3k^2p,$$

$$\left(:\frac{L}{2}(k^2x^2 - p^2):\right)^3 p = \left(:\frac{L}{2}(k^2x^2 - p^2):\right) L^2k^2p = L^3k^4x,$$

$$\implies \quad e^{:\frac{L}{2}(k^2x^2 - p^2):}x = (\cosh Lk)x + \frac{1}{k}(\sinh Lk)p,$$

$$e^{:\frac{L}{2}(k^2x^2 - p^2):}p = k(\sinh Lk)x + (\cosh Lk)p.$$

(b) — alternative: One can also use the formula to be established later in Eq. (4.63) that connects the f_2 quadratic form of the Lie map with the 2×2 matrix.

(c) Either by guessing, or establish a logic to derive the answer $e^{:S(\frac{x^3}{3} - xy^2):}$.

(c) — alternative: The later text will give a connection between f_3 and the 3rd order map, namely Eq. (4.87).

(d) Either by guessing, or establish a logic to derive the answer $e^{:kxz:}$.

(d) — alternative: One can also use the formula established later in Eq. (4.63).

Homework 4.19 In this homework, we introduce a map expressed as an exponential Lie map, but when the exponent is imaginary. This is equivalent to say the Hamiltonian of the system is imaginary. Consider a 4-D phase space (x, p_x, y, p_y). Consider a Lie map

$$e^{:\alpha(x-y)(p_x - p_y):}.$$

(a) Show that this is a linear map and its corresponding transfer matrix map is

$$T = \begin{bmatrix} \frac{1+e^{-2\alpha}}{2} & 0 & \frac{1-e^{-2\alpha}}{2} & 0 \\ 0 & \frac{1+e^{2\alpha}}{2} & 0 & \frac{1-e^{2\alpha}}{2} \\ \frac{1-e^{-2\alpha}}{2} & 0 & \frac{1+e^{-2\alpha}}{2} & 0 \\ 0 & \frac{1-e^{2\alpha}}{2} & 0 & \frac{1+e^{2\alpha}}{2} \end{bmatrix}.$$

(b) Show that the map T is symplectic.

(c) It is obvious that T is the identity map when $\alpha = 0$. Show that the identity map can also be given by choosing $\alpha = \pm i\pi$. Show that the x-y phase space exchange maps such as

$$\begin{bmatrix} 0 & 0 & 1 & 0 \\ 0 & 0 & 0 & 1 \\ 1 & 0 & 0 & 0 \\ 0 & 1 & 0 & 0 \end{bmatrix}$$

can be obtained by $\alpha = \pm i\frac{\pi}{2}$. Maps like this are not analytically connected to the identity map although they are symplectic.

Homework 4.20

(a) Calculate $e^{:-\frac{L}{2}(k^2 x^2 + p^2):}(xp)$.

(b) Calculate $\left(e^{:-\frac{L}{2}(k^2 x^2 + p^2):}x\right)\left(e^{:-\frac{L}{2}(k^2 x^2 + p^2):}p\right)$.

(c) Compare your results in (a) and (b). You will learn later of Eq. (4.43), the fact that (a) and (b) give the same result is not a coincidence. This property applies only when the operator is of an exponential form.

Homework 4.21

(a) Show that

$$e^{:\alpha\sqrt{x^2+p^2}:}x = x\cos\left(\frac{\alpha}{\sqrt{x^2+p^2}}\right) - p\sin\left(\frac{\alpha}{\sqrt{x^2+p^2}}\right),$$

$$e^{:\alpha\sqrt{x^2+p^2}:}p = x\sin\left(\frac{\alpha}{\sqrt{x^2+p^2}}\right) + p\cos\left(\frac{\alpha}{\sqrt{x^2+p^2}}\right).$$

(b) Use the result of (a) to show that the Lie map $e^{:\alpha\sqrt{x^2+p^2}:}$ is symplectic.

(c) Observe the fact that $x^2 + p^2$ is invariant as one would expect under the map $e^{:\alpha\sqrt{x^2+p^2}:}$.

(d) Try to generalize your result to the map $e^{:\alpha(x^2+p^2)^\beta:}$.

Solution (b) Show the fundamental Poisson bracket

$$\left[e^{:\alpha\sqrt{x^2+p^2}:}x, e^{:\alpha\sqrt{x^2+p^2}:}p\right] = 1.$$

4.3.3 Fundamental symplectic matrix

There are three fundamental symplectic 2×2 matrices and ten fundamental symplectic 4×4 matrices. All 2×2 and 4×4 symplectic matrices can be constructed from them as building blocks.[14] The three fundamental symplectic 2×2 matrices are

$$\begin{bmatrix} 1 & t_1 \\ 0 & 1 \end{bmatrix}, \quad \begin{bmatrix} 1 & 0 \\ t_2 & 1 \end{bmatrix}, \quad \begin{bmatrix} t_3 & 0 \\ 0 & \frac{1}{t_3} \end{bmatrix}.$$

---

[14]See, for example, A.W. Chao, Lectures on Accelerator Physics, Chapter 2, World Scientific (2020).

They correspond to a drift space, a quadrupole, and a telescope, respectively.

The ten fundamental symplectic 4×4 matrices are

$$
\begin{bmatrix} 1 & 0 & 0 & 0 \\ 0 & 1 & t_1 & 0 \\ 0 & 0 & 1 & 0 \\ t_1 & 0 & 0 & 1 \end{bmatrix}, \quad
\begin{bmatrix} 1 & 0 & 0 & t_2 \\ 0 & 1 & 0 & 0 \\ 0 & t_2 & 1 & 0 \\ 0 & 0 & 0 & 1 \end{bmatrix}, \quad
\begin{bmatrix} 1 & 0 & 0 & 0 \\ 0 & 1 & 0 & t_3 \\ -t_3 & 0 & 1 & 0 \\ 0 & 0 & 0 & 1 \end{bmatrix},
$$

$$
\begin{bmatrix} 1 & 0 & t_4 & 0 \\ 0 & 1 & 0 & 0 \\ 0 & 0 & 1 & 0 \\ 0 & -t_4 & 0 & 1 \end{bmatrix}, \quad
\begin{bmatrix} 1 & t_5 & 0 & 0 \\ 0 & 1 & 0 & 0 \\ 0 & 0 & 1 & 0 \\ 0 & 0 & 0 & 1 \end{bmatrix}, \quad
\begin{bmatrix} t_6 & 0 & 0 & 0 \\ 0 & \frac{1}{t_6} & 0 & 0 \\ 0 & 0 & 1 & 0 \\ 0 & 0 & 0 & 1 \end{bmatrix},
$$

$$
\begin{bmatrix} 1 & 0 & 0 & 0 \\ t_7 & 1 & 0 & 0 \\ 0 & 0 & 1 & 0 \\ 0 & 0 & 0 & 1 \end{bmatrix}, \quad
\begin{bmatrix} 1 & 0 & 0 & 0 \\ 0 & 1 & 0 & 0 \\ 0 & 0 & 1 & t_8 \\ 0 & 0 & 0 & 1 \end{bmatrix}, \quad
\begin{bmatrix} 1 & 0 & 0 & 0 \\ 0 & 1 & 0 & 0 \\ 0 & 0 & t_9 & 0 \\ 0 & 0 & 0 & \frac{1}{t_9} \end{bmatrix}, \quad
\begin{bmatrix} 1 & 0 & 0 & 0 \\ 0 & 1 & 0 & 0 \\ 0 & 0 & 1 & 0 \\ 0 & 0 & t_{10} & 1 \end{bmatrix}.
$$

It is clear that some symplectic maps can only be constructed from multiplying multiple fundamental matrices. In particular, all the fundamental matrices have diagonal elements equal to 1, which came from the fact that the leading term in the Taylor expansion of all exponential Lie maps start with 1.

If a symplectic matrix does not have its diagonal elements all equal to 1, this matrix can only be constructed from multiplying multiple fundamental matrices. For example, even in the 1-D case, the symplectic map

$$
-I = \begin{bmatrix} -1 & 0 \\ 0 & -1 \end{bmatrix}
$$

is not one of the fundamental maps. It however is just a π transformation, which can be constructed from multiple quadrupole and drift maps, for example,

$$
\begin{bmatrix} 1 & 0 \\ -\frac{1}{f} & 1 \end{bmatrix}
\begin{bmatrix} 1 & f \\ 0 & 1 \end{bmatrix}
\begin{bmatrix} 1 & 0 \\ -\frac{2}{f} & 1 \end{bmatrix}
\begin{bmatrix} 1 & f \\ 0 & 1 \end{bmatrix}
\begin{bmatrix} 1 & 0 \\ -\frac{1}{f} & 1 \end{bmatrix}
= \begin{bmatrix} -1 & 0 \\ 0 & -1 \end{bmatrix}.
$$

The art and challenge of constructing a desired map using the fundamental symplectic building blocks is the art and challenge of lattice design.

Since matrices deal only with linear maps, the corresponding Lie maps must be represented by $e^{:f(x,p):}$ with quadratic polynomials $f(x,p)$. In terms of Lie language, the inversion map $-I$ in the above lattice design is then expressed as a string of fundamental Lie maps,

$$
e^{:-\frac{x^2}{2f}:} \, e^{:-\frac{fx'^2}{2}:} \, e^{:-\frac{x^2}{f}:} \, e^{:-\frac{fx'^2}{2}:} \, e^{:-\frac{x^2}{2f}:} = e^{:-\frac{\pi}{2}\left(\frac{x^2}{\beta}+\beta p^2\right):},
$$

where the right-hand-side is a Lie exponential map representing $-I$ (β is an arbitrary β-function). If one insists to construct it using a fundamental map, then it requires an imaginary Hamiltonian,

$$
e^{:\pm i\pi xx':}.
$$

Lie map for fundamental matrix   One way to see why there are three fundamental symplectic 2×2 matrices is to consider the Lie map $e^{:f:}$, where f contains only a single term quadratic polynomial — a monomial — in x and x'. There are three degrees of freedom of linear Lie maps (the case of $n = 1$, $\Omega = 1$ in Table 4.2), with three fundamental f's, namely $f = x^2$, $f = xx'$ and $f = x'^2$. These Lie maps have the following corresponding matrix maps,

$$e^{:a_{11}x^2:} \quad \rightarrow \quad \begin{bmatrix} 1 & 0 \\ 2a_{11} & 1 \end{bmatrix},$$

$$e^{:a_{12}xx':} \quad \rightarrow \quad \begin{bmatrix} e^{-a_{12}} & 0 \\ 0 & e^{a_{12}} \end{bmatrix},$$

$$e^{:a_{22}x'^2:} \quad \rightarrow \quad \begin{bmatrix} 1 & -2a_{22} \\ 0 & 1 \end{bmatrix}.$$

It is easy to recognize that these are just those fundamental symplectic matrices mentioned earlier for the 2×2 case of a 1-D system.

From Table 4.3, the map $e^{:x^2:}$ is a thin-lens quadrupole, and $e^{:x'^2:}$ is a drift space. The map $e^{:xx':}$ is a scale change, i.e. a telescopic system. We therefore know the hardware to provide each of these cases. There does not seem to be a single accelerator hardware element that accomplishes the telescope, but it can be constructed by combining drifts and quadrupoles. These three fundamental maps can be constructed using two types of hardware, drifts and quadrupoles.

One can extend the consideration to a 2-D system. The number of fundamental matrices is ten. Take the coupled transverse case with phase space (x, x', y, y') for example. Table 4.4 gives the connection between the Lie maps and the 4×4 matrix maps. Six among these ten members are just the x and y 1-D building blocks. Only four actually invoke x-y coupling.

One can also consider the transverse-longitudinal coupled motion by replacing the phase space (x, x', y, y') by (x, x', z, δ).

One may ask if we have the hardware devices for each of these ten 2-D fundamental cases? Available hardware for the uncoupled x and y motions includes quadrupoles and drift spaces. For x-y coupling, there are skew quadrupoles and solenoids. For the uncoupled x and z motions, the available hardware includes RF cavities and momentum-compaction sections (which must contain dipoles). For x-z coupling, there are crab cavities and dipole magnets.

In the 3-D case(6-D phase space), it is easy to see that there are 21 fundamental symplectic matrices (see Homework 4.24).

Homework 4.22 Show that, as mentioned in the text,

(a) the Lie map $e^{:-\frac{\pi}{2}\left(\frac{x^2}{\beta}+\beta p^2\right):}$ gives the inversion $-I$ map with an arbitrary β (can be negative).

(b) the Lie map $e^{:\pm i\pi xp:}$ with an imaginary Hamiltonian also gives the inversion map.

Table 4.4: The ten fundamental symplectic building blocks of 2-D linear dynamics. Phase space is (x, x', y, y') but could also be, e.g. (x, x', z, δ).

Lie map	Matrix form	Lie map	Matrix form
$e^{:a_{11}x^2:}$	$\begin{bmatrix} 1 & 0 & 0 & 0 \\ 2a_{11} & 1 & 0 & 0 \\ 0 & 0 & 1 & 0 \\ 0 & 0 & 0 & 1 \end{bmatrix}$	$e^{:a_{12}xx':}$	$\begin{bmatrix} e^{-a_{12}} & 0 & 0 & 0 \\ 0 & e^{a_{12}} & 0 & 0 \\ 0 & 0 & 1 & 0 \\ 0 & 0 & 0 & 1 \end{bmatrix}$
$e^{:a_{22}x'^2:}$	$\begin{bmatrix} 1 & -2a_{22} & 0 & 0 \\ 0 & 1 & 0 & 0 \\ 0 & 0 & 1 & 0 \\ 0 & 0 & 0 & 1 \end{bmatrix}$	$e^{:a_{13}xy:}$	$\begin{bmatrix} 1 & 0 & 0 & 0 \\ 0 & 1 & a_{13} & 0 \\ 0 & 0 & 1 & 0 \\ a_{13} & 0 & 0 & 1 \end{bmatrix}$
$e^{:a_{23}x'y:}$	$\begin{bmatrix} 1 & 0 & -a_{23} & 0 \\ 0 & 1 & 0 & 0 \\ 0 & 0 & 1 & 0 \\ 0 & a_{23} & 0 & 1 \end{bmatrix}$	$e^{:a_{33}y^2:}$	$\begin{bmatrix} 1 & 0 & 0 & 0 \\ 0 & 1 & 0 & 0 \\ 0 & 0 & 1 & 0 \\ 0 & 0 & 2a_{33} & 1 \end{bmatrix}$
$e^{:a_{14}xy':}$	$\begin{bmatrix} 1 & 0 & 0 & 0 \\ 0 & 1 & 0 & a_{14} \\ -a_{14} & 0 & 1 & 0 \\ 0 & 0 & 0 & 1 \end{bmatrix}$	$e^{:a_{24}x'y':}$	$\begin{bmatrix} 1 & 0 & 0 & -a_{24} \\ 0 & 1 & 0 & 0 \\ 0 & -a_{24} & 1 & 0 \\ 0 & 0 & 0 & 1 \end{bmatrix}$
$e^{:a_{34}yy':}$	$\begin{bmatrix} 1 & 0 & 0 & 0 \\ 0 & 1 & 0 & 0 \\ 0 & 0 & e^{-a_{34}} & 0 \\ 0 & 0 & 0 & e^{a_{34}} \end{bmatrix}$	$e^{:a_{44}y'^2:}$	$\begin{bmatrix} 1 & 0 & 0 & 0 \\ 0 & 1 & 0 & 0 \\ 0 & 0 & 1 & -2a_{44} \\ 0 & 0 & 0 & 1 \end{bmatrix}$

Homework 4.23

(a) Design a telescopic system in the 1-D phase space (x, x') that gives the map $\begin{bmatrix} e^{-a_{12}} & 0 \\ 0 & e^{a_{12}} \end{bmatrix}$ using thin-lens quadrupoles and drift spaces.

(b) Design a telescopic system in the 2-D phase space (x, x', y, y') that gives the map $\begin{bmatrix} e^{-a_{12}} & 0 & 0 & 0 \\ 0 & e^{a_{12}} & 0 & 0 \\ 0 & 0 & 1 & 0 \\ 0 & 0 & 0 & 1 \end{bmatrix}$ using thin-lens quadrupoles and drift spaces.

Homework 4.24 Table 4.4 gives the ten fundamental symplectic matrices for a 2-D linear (x, x', y, y') system. It is straightforward — easier than it might appear to be if you include some guessing — to extend it to 21 matrices by including z-motion in the table for a linear $(x, x', y, y', z, \delta)$ system. Try several such matrices to know how this can be done.

4.3.4   Exponential Lie operator

Operator formulae   We will now establish a few frequently used formulae of the algebra of exponential operators. One thing to extract from this section is the fact that the exponential Lie operators are really special and have amazing properties.

First, we show that the inverse map of $\exp(:f:)$ is $\exp(-:f:)$, i.e.

$$(e^{:f:})^{-1} \;=\; e^{-:f:}. \tag{4.41}$$

Proof of Eq. (4.41):

$$
\begin{aligned}
e^{:f:}e^{-:f:} \;&=\; \sum_{m=0}^{\infty}\sum_{n=0}^{\infty}\frac{1}{m!\,n!}(:f:)^m(-:f:)^n \\[2mm]
&=\; \sum_{k=0}^{\infty}\sum_{n=0}^{k}\frac{1}{(k-n)!\,n!}(:f:)^{k-n}(-:f:)^n \\[2mm]
&=\; \sum_{k=0}^{\infty}\frac{1}{k!}\sum_{n=0}^{k}\frac{k!}{(k-n)!\,n!}(:f:)^{k-n}(-:f:)^n \,,
\end{aligned}
$$

and now since $:f:$ and $-:f:$ commute, we can rewrite[15]

$$
\begin{aligned}
e^{:f:}e^{-:f:} \;&=\; \sum_{k=0}^{\infty}\frac{1}{k!}(:f: - :f:)^k \\[2mm]
&=\; \sum_{k=0}^{\infty}\frac{1}{k!}(:f - f:)^k \;=\; \sum_{k=0}^{\infty}\frac{1}{k!}(:0:)^k \;=\; 1 \qquad \Longrightarrow \qquad \text{Q.E.D.}
\end{aligned}
$$

This derivation of Eq. (4.41) can be repeated to show that if $:f:$ and $:g:$ commute (i.e. if $:f: \, :g: = :g: \, :f:$, or if $[f,g] = $ constant depending only on s), then

$$e^{:f:}e^{:g:} \;=\; e^{:f+g:}. \tag{4.42}$$

In fact, Eq. (4.41), is just a special case of (4.42). When $:f:$ and $:g:$ do not commute, Eq. (4.42) is no longer valid. In that case, the right-hand-side of Eq. (4.42) will acquire additional residual terms that relate to the commutators of the operators. Furthermore, since the commutator of two operators are related to the operator formed by the Poisson bracket of the functions involved — see Eq. (4.36) — these residual terms will eventually be related to the Poisson brackets. A detailed analysis of this will be an important subject later, and the corresponding formula is called the Baker–Campbell–Hausdorff formula. The BCH formula can be considered a generalization of Eq. (4.42).

Next we establish the following,

$$e^{:f:}(gh) \;=\; (e^{:f:}g)(e^{:f:}h). \tag{4.43}$$

---

[15]If $:f:$ and $:g:$ do not commute, this step is not valid. For example, $(:f: + :g:)^2 = :f:^2 + :f: \, :g: + :g: \, :f: + :g:^2 \neq :f:^2 + 2:f: \, :g: + :g:^2$.

Note that Eq. (4.43) applies for exponential maps only. If one is interested in $:f:(gh)$, one should consider applying the third member of Eq. (4.6). In contrast, (4.43) is a quite different equation. The proof of Eq. (4.43) can be established using Eq. (4.34) and following similar steps as the proof of Eq. (4.41). See Homework 4.25. The reader should also recall the comment we emphasized following Eq. (4.40), and also Homework 4.20.

One useful variation of Eq. (4.43) is

$$e^{:f:}[g, h] \;=\; [e^{:f:}g, e^{:f:}h]\,. \tag{4.44}$$

It follows from Eq. (4.43) that, for an arbitrary function g of the components of X, one has

$$e^{:f:}g(X) = g(e^{:f:}X)\,. \tag{4.45}$$

The proof of Eq. (4.45) is as follows. Letting $g(X) = h(X) = X$ in Eq. (4.43) gives $e^{:f:}(X^2) = (e^{:f:}X)^2$, i.e. Eq. (4.45) holds if $g(X) = X^2$ including the cross products of the components of X. This can be extended to arbitrary powers of X. Since any function (of interest to us) can be expressed as power series (provided the series converges), Eq. (4.45) holds for arbitrary functions of X.

Equations (4.43–4.45) establish the fact that the most basic Lie operation is of the exponential type $e^{:f:}X$. This means we only need to know how the operators act on the components of the phase space; once these operations are found, exponential Lie operations on any complex functions of X follow. This is quite a remarkable property, and it applies only if the Lie operator is of the exponential type.

Equation (4.45) describes what happens when an exponential Lie operator is applied to a function. Lie operators can also be applied to other Lie operators. The counterpart of Eq. (4.45) in the operator case reads

$$e^{:f:}:g:e^{-:f:} \;=\; :(e^{:f:}g):\,. \tag{4.46}$$

Proof of Eq. (4.46) is given in Homework 4.29. Similar to Eq. (4.43), Eq. (4.46) is applicable only to exponential maps.

It follows from Eq. (4.46) that

$$e^{:f:}(:g:)^n e^{-:f:} \;=\; (e^{:f:}:g:e^{-:f:})^n \;=\; (:e^{:f:}g:)^n\,, \tag{4.47}$$

which in turn leads to the property that, for an arbitrary function G that can be expressed as power series, one has

$$e^{:f:}G(:g:)e^{-:f:} \;=\; G(:e^{:f:}g:)\,. \tag{4.48}$$

One particularly useful special case of Eq. (4.48) is

$$e^{:f:}e^{:g:}e^{-:f:} \;=\; \exp(:e^{:f:}g:)\,. \tag{4.49}$$

Table 4.5 lists several often-used formulae for Lie operators.

Note the difference between Eq. (4.48) — Lie operation on operators — and Eq. (4.45) — Lie operation on functions. In Eq. (4.45), one sees the powerful

Table 4.5: Some formulae for Lie operators. Here a is a constant and C is a constant vector, both independent of X, S is the matrix defined in Eq. (4.1), and f, g, and G are arbitrary functions. Square brackets mean Poisson brackets. Curly brackets mean commutators. A Poisson bracket is applied to two functions. A commutator is applied to two operators.

$$
\begin{aligned}
:a: &= 0, \quad e^{:a:} = 1 \\
:f:a &= 0, \quad e^{:f:}a = a \\
:f:f &= 0, \quad e^{:f:}f = f \\
\{:f:, :g:\} &= :[f,g]: \\
(e^{:f:})^{-1} &= e^{-:f:} \\
e^{:f:}e^{:g:} &= e^{:f+g:}, \quad \text{if } [f,g] = a \\
e^{:f:}g(X) &= g(e^{:f:}X) \\
e^{:\tilde{C}X:}g(X) &= g(X - SC) \\
e^{:f:}G(:g:)e^{-:f:} &= G(:e^{:f:}g:)
\end{aligned}
$$

property for exponential Lie operators that they *permeate* into the guts of the functions they apply to. In Eq. (4.48), one sees that the application assumes a sandwich form (similarity transformation), and that again, exponential Lie operators permeate into the guts of the operators being applied to.

The sandwich form in Eq. (4.48) has the important consequence that when a string of operators are being transformed, the result can be written as the same string of the transformed operators, i.e.

$$
\begin{aligned}
e^{:f:}(:g_1: :g_2: \dots :g_n:)e^{-:f:} &= (:e^{:f:}g_1:)(:e^{:f:}g_2:) \cdots (:e^{:f:}g_n:) \\
&= [:g_1(e^{:f:}X):][:g_2(e^{:f:}X):] \cdots [:g_n(e^{:f:}X):], \quad (4.50)
\end{aligned}
$$

where use has been made of Eq. (4.41). This sandwiching property has been used in Eq. (4.47) and will be used frequently later.

Equation (4.50) indicates that the sandwiching transformation is equivalent to substituting the dynamical coordinates X by $e^{:f:}X$ in the arguments of the operators being sandwiched. Here one sees that the operator $e^{:f:}$ first permeates into the component maps $:g_n:$'s, and then subsequently and individually permeates into the gut of each function g_n. Again, not all Lie maps have this remarkable property; only exponential ones do.

Perhaps a far-side comment can be made at this junction. We commented following Eq. (4.12) concerning the significance of introducing the Poisson brackets to the analysis of Hamiltonian beam dynamics, in spite of the apparent fact that they were no more than a mathematical construct designed to abbreviate long expressions. Our discussion up to this point facilitating exponential Lie operators follows from introducing the Lie operator $:f:$ to reexpress Poisson brackets. This step again seems to be a mathematical abstraction or perhaps merely an elevation of notation. One cannot help but admiring how these seemingly abstractions in reality have triggered and motivated so much progress.

Operator ordering   There is a subtlety concerning the ordering of Lie operators. Consider an accelerator section consisting of an element F (with operator $e^{:F:}$), followed by an element G (with operator $e^{:G:}$). Let X_0 be the coordinates of a particle entering element F. Let X_1 be the coordinates of the particle exiting element F and entering element G. Then we have [see Eq. (4.21)]

$$X_1 = \left. e^{:F(X):} X \right|_{X=X_0}. \tag{4.51}$$

Let X_2 be the coordinates exiting element G, then

$$\begin{aligned}
X_2 &= \left. e^{:G(X):} X \right|_{X=X_1} \\
&= \exp\left[:G\left(e^{:F(X):}X\right):\right] \left(e^{:F(X):}X\right)\Big|_{X=X_0} \\
&= \exp\left[:e^{:F(X):}G(X):\right] \left(e^{:F(X):}X\right)\Big|_{X=X_0} \\
&= \left(e^{:F:}e^{:G:}e^{:-F:}\right) e^{:F:} X \Big|_{X=X_0} \\
&= \left. e^{:F:}e^{:G:} X \right|_{X=X_0}. \tag{4.52}
\end{aligned}$$

In the derivation of Eq. (4.52), keep in mind that the substituting $X \to X_0$ is to be made only at the end after all Lie operations are completed.

Equation (4.52) indicates that, when a string of elements are connected into an accelerator section, the ordering of the operators is such that operators of the earlier elements appear to the *left* of operators of the later elements. This is opposite to what one might be used to when dealing with the linear systems with matrices!

It should be noted that the ordering of operators or matrices is a result of how their operations are precisely defined. The ordering by itself is an artifact. There is not an intrinsic reason of their orderings other than the definitions and conventions. For example, we have defined by convention that a matrix multiplication AB is to take the rows of A dot product into the columns of B. Had we defined the multiplication by taking the columns of A into rows of B, the ordering would be reversed in the matrix formalism. As to the Lie maps, its ordering follows from an accurate definition (4.21) or (4.51).

Symplecticity of exponential map   We are now in a position to show another key property of exponential operators, namely, an operator of exponential form $e^{:f:}$ is necessarily symplectic for an arbitrary function $f(X,s)$. To prove this, consider the map $X = \left. e^{:f:}X \right|_{X=X_0}$, and form the fundamental Poisson brackets

$$\begin{aligned}
[X_\alpha, X_\beta] &= \left. [e^{:f:}X_\alpha, e^{:f:}X_\beta] \right|_{X=X_0} \\
&= \left. \left(e^{:f:}[X_\alpha, X_\beta]\right) \right|_{X=X_0} \\
&= \left. \left(e^{:f:}S_{\alpha\beta}\right) \right|_{X=X_0} = S_{\alpha\beta}. \tag{4.53}
\end{aligned}$$

Equations (4.13–4.14) then prove that the operator $e^{:f:}$ is symplectic.

The fact that exponential operators are necessarily symplectic has an important practical meaning. In practice, one often has the exponential form $e^{:f:}$ of a symplectic map, and one needs to compute the effect of the map on the motion of a particle, i.e., one needs to find $e^{:f:}X$. For a general f, however, this is often not an easy task. The trick often adopted is to expand f in terms of a power series in the components of X,

$$e^{:f:} \;=\; e^{:(f_2+f_3+f_4+\cdots):}\,, \tag{4.54}$$

where f_k is a homogeneous power series of k-th order in the components of X. The exponential form allows the possibility that this power series be truncated at will without losing symplecticity — what one loses is accuracy but symplecticity is assured.

In contrast, had the map been in the form of $:f:$ instead of an exponential form, as would be the case of a Taylor map, a truncation would in general lead to a loss of the symplecticity of the map,

$$\begin{aligned}
\text{Taylor map} \;&=\; :(f_2 + f_3 + f_4 + \cdots):\,, \\
\text{Lie map} \;&=\; e^{:(f_2+f_3+f_4+\cdots):}\,.
\end{aligned}$$

Furthermore, the same map can sometimes also be written in a different form,

$$e^{:f:} \;=\; e^{:f_2:}e^{:f_3:}e^{:f_4:}\cdots\,. \tag{4.55}$$

As will become clear later, the f_2 in Eq. (4.55) is the same as the f_2 in Eq. (4.54), but all the higher order f_k's differ. Again in this form, exponential factors of higher orders can be truncated at will without losing symplecticity. In either the form (4.54) or (4.55), truncation down to the second order gives the linear map $e^{:f_2:}$. These two representations then coincide, and they both coincide with the Taylor linear map represented as a matrix.

Homework 4.25

(a) Prove Eq. (4.42), i.e. show that $e^{:f:}e^{:g:} = e^{:f+g:}$ if $[f, g] = 0$. Pay attention to the moment when you require the commutability of the two operators $:f:$ and $:g:$.

(b) Use similar steps to prove Eq. (4.43).

Homework 4.26 Equation (4.43) is a remarkable result. It applies when the Lie map being considered is an exponential map and, as pointed out in the text, does not apply to any arbitrary Lie map. Does Eq. (4.43) apply if the Lie map is symplectic but not an exponential form?

Solution Not all symplectic maps are connected to the identity map and, when that happens, it cannot represent a real single accelerator element in a beamline. Consider the map $:f(x, p):$ with $f(x, p) = \frac{1}{2}(x^2 + p^2)$ for example. It gives a linear map $\begin{bmatrix} 0 & -1 \\ 1 & 0 \end{bmatrix}$ which is symplectic. Show that $:f:(xp) \neq (:f:x)(:f:p)$.

In fact, $:\frac{1}{2}(x^2 + p^2):$ represents a linear Taylor map. It cannot represent a real accelerator element or beamline in the Lie algebra language. To represent $\begin{bmatrix} 0 & -1 \\ 1 & 0 \end{bmatrix}$, the Lie map has to be of an exponential form,

$$e^{:\frac{\pi}{4}(x^2+p^2):} .$$

The two maps $e^{:\frac{\pi}{4}(x^2+p^2):}$ and $:\frac{1}{2}(x^2 + p^2):$ are of course not the same.

Homework 4.27 Show that the quantity f is invariant under the map $e^{:f:}$.

Solution Consider the map $X = e^{:f:}X\big|_{X=X_0}$. We need to show that $f(X) = f(X_0)$, which is true because

$$f(X) = f\left(e^{:f:}X\big|_{X=X_0}\right) = e^{:f:}f(X)\big|_{X=X_0} = f(X)\big|_{X=X_0} = f(X_0) .$$

Admittedly this proof, like many of the Lie algebra proofs, is a bit abstract. If so, you might pretend a concrete function of $f(X)$ of your choice when you fill in the steps.

Homework 4.28 Show that if $[f, g] = 0$, then g is invariant under $e^{:f:}$. In particular, if f is the Hamiltonian, then $e^{:f:}$ describes the time evolution of the system, and this exercise says that if $[f, g] = 0$, then g is a constant of the motion.

Homework 4.29 Prove Eq. (4.46).

Solution It is suggested to try the proof yourself first. Lie algebra of operators often asks for a special kind of skills. This homework offers an initial glimpse of such skills to be acquired. Consider an arbitrary function $h(X)$, we have

$$\begin{aligned} e^{:f:}:g:h &= e^{:f:}[g, h] = [e^{:f:}g, e^{:f:}h] = :(e^{:f:}g):e^{:f:}h \\ \implies \quad e^{:f:}:g: &= :(e^{:f:}g):e^{:f:} , \end{aligned} \qquad (4.56)$$

where use has been made of Eq. (4.44). Multiplying both sides from the right by $e^{-:f:}$ and applying Eq. (4.41) proves Eq. (4.46).

4.3.5 Application to linear system

In this section, we shall try to connect the linear Lie maps with the linear maps described by transport matrices. As it will reveal, Lie maps are a clumsy way to describe linear systems. For linear systems, it is more efficient simply to stay with the transport matrices. The power of Lie maps is more to be revealed when we deal with nonlinear systems. On the other hand, before launching the subject of nonlinear systems, it remains necessary to establish the linear case first.

Connecting Lie map to transport matrix　Let us apply operator algebra to a linear system. Linear systems are described by Lie maps $e^{:f_2:}$ where f_2 is a quadratic polynomial of X. Consider a quadratic form

$$f_2 = -\frac{1}{2}\tilde{X}FX\,, \tag{4.57}$$

where F is a symmetric, positive definite matrix.[16] Suppose f_2 is given, we want to find the transport matrix of the linear system.

Given f_2 and F, we first observe that

$$:f_2:X = SFX\,, \tag{4.58}$$

where S is the symplectic form (4.1). The proof of Eq. (4.58) is as follows,

$$
\begin{aligned}
:f_2:X_\alpha &= \frac{\partial f_2}{\partial X_\beta}S_{\beta\gamma}\frac{\partial X_\alpha}{\partial X_\gamma} \\
&= -\frac{1}{2}\frac{\partial(F_{\ell m}X_\ell X_m)}{\partial X_\beta}S_{\beta\gamma}\delta_{\alpha\gamma} \\
&= -F_{\beta\ell}X_\ell S_{\beta\alpha} = (SFX)_\alpha, \qquad \text{Q.E.D.}
\end{aligned}
$$

It follows from Eq. (4.58) that

$$e^{:f_2:}X \leftrightarrow e^{SF}X, \quad \text{or} \quad e^{:f_2:} \leftrightarrow e^{SF}\,. \tag{4.59}$$

There is a reason we use $\leftrightarrow$ instead of $=$ in Eq. (4.59): although they physically perform the same operation, they are different objects. The left-hand-side of Eq. (4.59) is in a Lie operator form, while the right-hand-side is in a matrix form. The left-hand-side can be applied to the components of X individually, while the right-hand-side must be applied simultaneously to all components of X. The left-hand-side can be applied to any function of X, while the right-hand-side operates only on the vector X.

So the transport matrix is given by e^{SF}. But we are not done yet. Given the matrix F, we still need to calculate the matrix e^{SF}, i.e. it requires taking exponential of a matrix.

Conversely, in case we are given the transport matrix, we can calculate the quadratic form Lie map by taking its logarithm using Eq. (4.59). We are missing the step of how to calculate the exponential or the logarithm of a matrix.

Hamilton–Cayley theorem　Exponentiation of a matrix A is defined in a power series fashion as

$$e^A = \sum_{k=0}^{\infty}\frac{1}{k!}A^k\,.$$

---

[16]The condition of positive definiteness is for the particle motion to be stable. An example of exception can be found in Eq. (4.79). A symplectic linear map can have a corresponding F that is not positive definite, but one needs to pay attention to the system stability in that case. Being symplectic does not assure stability.

To be more specific, consider a 1-D system for which F is a symmetric 2×2 matrix. Let F be

$$F = \begin{bmatrix} a & b \\ b & c \end{bmatrix}. \tag{4.60}$$

The positive definiteness of F means the quantity $\tilde{X} F X \geq 0$ for arbitrary X. This leads to

$$b^2 \leq ac \qquad \text{and} \qquad a \geq 0.$$

We now need to compute the matrix e^{SF}. To do so, we apply a theorem called the *Hamilton–Cayley theorem*, which says that, for an $N \times N$ matrix A, and $f(A)$ being any function of A, we have

$$f(A) = \sum_{k=0}^{N-1} a_k A^k, \tag{4.61}$$

where a_k are a set of coefficients which satisfy the algebraic conditions

$$f(\lambda) = \sum_{k=0}^{N-1} a_k \lambda^k \qquad \text{with} \qquad \lambda = \text{eigenvalues of } A. \tag{4.62}$$

To stress the power of the Hamilton–Cayley theorem, note that the matrix $f(A)$ is going to be an $N \times N$ matrix; it has N^2 elements. However, Eq. (4.61) expresses it with only N (not N^2) "fitting parameters" a_k. Furthermore, these N fitting parameters can be found by solving Eq. (4.62) — N equations for N unknowns. The prove of the Hamilton–Cayley theorem is given in Homework 4.31.

As a consequence of this theorem, when A is raised to a power $\geq N$, the resulting matrix can always be expressed in terms of a summation over lower powers of A with order $\leq N - 1$. If $f(A)$ can be expressed as a power series in A, then it can be expressed as a power series with order $\leq N - 1$.

Take the case $N = 2$ for example. It is an intriguing observation, for example, that any function $f(A)$ of any 2×2 matrix A can be expressed in the form $a_0 + a_1 A$. Matrices are not just a display of an array of numbers laid out for convenient visualization. They are fascinating objects.

To apply the Hamilton–Cayley theorem to the present problem, we have $N = 2$ and

$$e^{SF} = \exp\left(\begin{bmatrix} b & c \\ -a & -b \end{bmatrix} \right) = a_0 + a_1 \begin{bmatrix} b & c \\ -a & -b \end{bmatrix},$$

where a_0 and a_1 are yet to be determined. The eigenvalues of the matrix $SF = \begin{bmatrix} b & c \\ -a & -b \end{bmatrix}$ are

$$\lambda_\pm = \pm i \sqrt{ac - b^2}.$$

The coefficients a_0 and a_1 therefore satisfy

$$e^{\lambda_+} = a_0 + a_1\lambda_+, \quad e^{\lambda_-} = a_0 + a_1\lambda_-$$

$$\implies \quad a_0 = \cos(\sqrt{ac - b^2}), \quad a_1 = \mathrm{Sgn}(a)\frac{\sin(\sqrt{ac - b^2})}{\sqrt{ac - b^2}}.$$

We thus obtain

$$e^{SF} = \cos(\sqrt{ac - b^2}) + \mathrm{Sgn}(a)\frac{\sin(\sqrt{ac - b^2})}{\sqrt{ac - b^2}}\begin{bmatrix} b & c \\ -a & -b \end{bmatrix}. \tag{4.63}$$

Equation (4.63) can be used to find the matrix form of the map when the Lie form of the map $e^{:f_2:} = e^{:-\frac{1}{2}(ax^2 + 2bxp + cp^2):}$ is known. One can also try to find the Lie form, given the matrix form, as follows. Suppose we know that the linear map has a matrix form

$$R = \begin{bmatrix} R_{11} & R_{12} \\ R_{21} & R_{22} \end{bmatrix} \quad \text{with} \quad \det R = 1. \tag{4.64}$$

By identifying the right-hand-side of Eq. (4.63) with Eq. (4.64), the coefficients a, b and c can be related to the R matrix elements according to

$$\cos(\sqrt{ac - b^2}) = \frac{1}{2}\mathrm{tr}R,$$

$$\frac{a}{-R_{21}} = \frac{2b}{R_{11} - R_{22}} = \frac{c}{R_{12}} = \frac{\sqrt{ac - b^2}}{\sin(\sqrt{ac - b^2})}. \tag{4.65}$$

Knowing a, b, c, the Lie form of the map is readily obtained.

For example, the R matrix may be a one-turn map of the Courant–Snyder form

$$R = \begin{bmatrix} \cos\mu + \alpha\sin\mu & \beta\sin\mu \\ -\gamma\sin\mu & \cos\mu - \alpha\sin\mu \end{bmatrix}, \tag{4.66}$$

where β and α are the β- and α-functions at the observation position in the circular accelerator, $\gamma = \frac{1+\alpha^2}{\beta}$, and μ is the betatron phase advance per turn. In terms of α, β, γ and μ, we find

$$a = \mu\gamma, \quad b = \mu\alpha, \quad \text{and} \quad c = \mu\beta,$$
$$ac - b^2 = \mu^2 \geq 0.$$

This gives, using Eq. (4.57),

$$f_2 = -\frac{\mu}{2}(\gamma x^2 + 2\alpha xp + \beta p^2)$$

$$= -\frac{\mu}{2} \times (\text{Courant–Snyder invariant}). \tag{4.67}$$

The Courant–Snyder map (4.66) has thus acquired a Lie operator form

$$e^{:f_2:} = e^{:-\frac{\mu}{2}(\gamma x^2 + 2\alpha xp + \beta p^2):}, \tag{4.68}$$

or, using the notation introduced in Eq. (4.59),

$$e^{:-\frac{\mu}{2}(\gamma x^2 + 2\alpha x p + \beta p^2):} \quad \leftrightarrow \quad \begin{bmatrix} \cos\mu + \alpha\sin\mu & \beta\sin\mu \\ -\gamma\sin\mu & \cos\mu - \alpha\sin\mu \end{bmatrix}. \qquad (4.69)$$

To recap, linear maps can be described by an exponential operator $\exp(:f_2:)$, where f_2 is a quadratic function of X. In particular, let f_2 be written as Eq. (4.57). We have shown that the map corresponding to the operator $\exp(:f_2:)$ can be written in a matrix notation as

$$X = M X_0, \qquad \text{where} \quad M = e^{SF}. \qquad (4.70)$$

The computation of the matrix e^{SF} can — although does not have to — proceed using the Hamilton–Cayley theorem.

Effective Hamiltonian   As we will see later, the one-turn map of a circular accelerator has the general form $e^{-C:H_{\text{eff}}:}$ where C is the accelerator circumference and H_{eff} is the effective Hamiltonian. Equation (4.68) says that in the 1-D linear case, the Courant–Snyder invariant (multiplied by $\frac{1}{2}$) has acquired the physical meaning of the effective Hamiltonian that describes the one-turn map, while the betatron phase advance per turn μ has the meaning of the total length of the accelerator (except that it is measured in radians instead of in meters).

We have asked f_2 to be positive definite because we want the system to be stable. A negative definite quadratic form also describes a stable system, but then the evolution (or phase advance) goes backward in time.

Concatenating linear maps   Consider two quadratic forms $f = -\frac{1}{2}\tilde{X}FX$ and $g = -\frac{1}{2}\tilde{X}GX$. Their corresponding maps can be concatenated into another linear map according to

$$e^{:h:} = e^{:f:}e^{:g:}, \qquad (4.71)$$

where $h = -\frac{1}{2}\tilde{X}HX$ is another quadratic form whose corresponding matrix H satisfies

$$e^{SH} = e^{SG}e^{SF}. \qquad (4.72)$$

The proof of Eq. (4.72) is as follows,

$$e^{:f:}e^{:g:}X_\alpha = e^{:f:}(e^{SG}X)_\alpha = e^{:f:}(e^{SG})_{\alpha\beta}X_\beta = (e^{SG})_{\alpha\beta}e^{:f:}X_\beta$$
$$= (e^{SG})_{\alpha\beta}(e^{SF}X)_\beta = (e^{SG}e^{SF}X)_\alpha \implies \text{Q.E.D.}$$

Comparing Eq. (4.71) in the Lie representation and (4.72) in the matrix representation, note that the ordering of the F and G components appear in reversed order as explained following Eq. (4.52).

2-D linear map   The map (4.68) can be extended to 2-D (4-D phase space). For example, we may include the synchrotron motion. Consider a 2-D linear system whose dynamical variables are (x, x', z, δ) where z is the longitudinal coordinate of a particle relative to an ideal reference particle, and $\delta = \frac{\Delta P}{P}$ is the momentum error relative to the reference particle with P the magnitude of the particle momentum. In the present consideration, we consider $\delta = $ constant in time, i.e., there are no RF acceleration devices and P is an invariant.

We consider only the case of an ideal planar storage ring here. In this special case, and when δ does not vary, the motion can be decomposed into the sum of a betatron component and a synchrotron component,

$$x = x_\beta + D\delta, \qquad x' = x'_\beta + D'\delta, \tag{4.73}$$

where D is the dispersion function. Since x and $p = x'$ in Eq. (4.68) actually are the betatron components x_β and x'_β, the relevant f_2 in the present system is

$$f_2 = -\frac{\mu}{2}(\gamma x_\beta^2 + 2\alpha x_\beta x'_\beta + \beta x'^2_\beta)$$

$$\rightarrow f_2 = -\frac{\mu}{2}\left[\gamma(x - D\delta)^2 + 2\alpha(x - D\delta)(x' - D'\delta) + \beta(x' - D'\delta)^2\right] + \frac{\alpha_c C}{2}\delta^2.$$

$$\tag{4.74}$$

In Eq. (4.74), x and x' refer to the total physical horizontal coordinates (4.73). An extra term $\frac{1}{2}C\alpha_c\delta^2$, where α_c is the momentum compaction factor and C is the accelerator circumference, has been added to f_2; its function will become clear momentarily.

Given the quadratic form (4.74), we then ask what is the 4×4 matrix R for the map $e^{:f_2:}$. This can be done by calculating the matrix e^{SF} (see Homework 4.39). It can also be done by calculating explicitly the quantities $e^{:f_2:}X$ and equating the results with the matrix form RX.

Skipping the algebra here, the result is

$$R = \begin{bmatrix} \cos\mu + \alpha\sin\mu & \beta\sin\mu & 0 & D - DR_{11} - D'R_{12} \\ -\gamma\sin\mu & \cos\mu - \alpha\sin\mu & 0 & D' - D'R_{22} - DR_{21} \\ D' - D'R_{11} + DR_{21} & -D + DR_{22} - D'R_{12} & 1 & (\gamma D^2 + 2\alpha DD' + \beta D'^2)\sin\mu - C\alpha_c \\ 0 & 0 & 0 & 1 \end{bmatrix},$$

$$\tag{4.75}$$

which is a familiar result from basic accelerator optics, with $R_{11} = \cos\mu + \alpha\sin\mu$, $R_{12} = \beta\sin\mu$, $R_{21} = -\gamma\sin\mu$, $R_{22} = \cos\mu - \alpha\sin\mu$, $\gamma = \frac{1+\alpha^2}{\beta}$.

It should be emphasized that the results (4.73)–(4.75) apply only in the case when $\delta = $ constant. Synchrotron oscillation has been excluded.

Homework 4.30 Consider Eq. (4.59).

(a) Show that for any nondegenerate matrix A,

$$\det(e^A) = e^{\mathrm{tr}A}.$$

Nondegeneracy means no two eigenvalues of A are equal.

(b) Use (a) to show that, with F any symmetric matrix, $\det(e^{SF}) = 1$.

(c) Show that the matrix e^{SF} is symplectic, where F is any symmetric matrix. The fact that its determinant equals 1, shown in (b), is not sufficient for it to be symplectic. Here you need to show $\widetilde{e^{SF}} S e^{SF} = S$.

Homework 4.31 Prove the Hamilton–Cayley theorem (4.61–4.62).

Solution Consider the characteristic polynomial $P(\lambda) = \det(A - \lambda I)$, which is an N-th order polynomial in λ. We have

$$P(\lambda) = (\lambda - \lambda_1)(\lambda - \lambda_2) \cdots (\lambda - \lambda_N),$$

where λ_j are the eigenvalues of A. Consider the function of A,

$$P(A) = (A - \lambda_1 I)(A - \lambda_2 I) \cdots (A - \lambda_N I).$$

If we apply $P(A)$ to V_j, which is an eigenvector of A, we obtain

$$P(A)V_j = 0. \tag{4.76}$$

The only way Eq. (4.76) can be true for all $j = 1, 2, \cdots N$ is when $P(A) = 0$ identically — assuming a nondegenerate system, i.e. no two of the eigenvalues are equal. The fact that $P(A) = 0$ is the Hamilton–Cayley theorem.

Homework 4.32 Consider the matrix

$$A = \begin{bmatrix} a & b \\ c & d \end{bmatrix}.$$

(a) Demonstrate the Hamilton–Cayley theorem explicitly, i.e. show that

$$P(A) = A^2 - (a + d)A + ad - bc = 0.$$

(b) Apply the Hamilton–Cayley theorem to obtain

$$f(A) = \left[\frac{\lambda_+ f(\lambda_-) - \lambda_- f(\lambda_+)}{\lambda_+ - \lambda_-} \right] + \left[\frac{f(\lambda_+) - f(\lambda_-)}{\lambda_+ - \lambda_-} \right] A, \tag{4.77}$$

where

$$\lambda_\pm = \frac{1}{2}(a + d) \pm \frac{1}{2}\sqrt{(a - d)^2 + 4bc}$$

are the eigenvalues of A. Given matrix A, Eq. (4.77) gives the matrix $f(A)$ for an arbitrary function $f(x)$.

(c) Observe from Eq. (4.77) that if the 2×2 matrix A is triangular (i.e. either $b = 0$ or $c = 0$), the matrix $f(A)$ is necessarily triangular for arbitrary $f(x)$. Similarly, if A is diagonal, so is $f(A)$.

(d) Find the matrices A^2, A^3, A^{-1}, $\sqrt{A}$, and e^A as special cases of Eq. (4.77).

Solution (d) All these matrices can be written in the form $a_0 + a_1 A$, as follows,

$$
\begin{aligned}
A^2 &= -(ad - bc) + (a + d)A\,, \\
A^3 &= -(a + d)(ad - bc) + (a^2 + ad + d^2 + bc)A\,, \\
A^{-1} &= \frac{1}{ad - bc}[(a + d) - A]\,, \\
\sqrt{A} &= \frac{1}{\sqrt{\lambda_+} + \sqrt{\lambda_-}}(\sqrt{ad - bc} + A)\,, \\
e^A &= e^{(a+d)/2}\left[\frac{\sinh\frac{\sqrt{(a-d)^2+4bc}}{2}}{\sqrt{(a-d)^2 + 4bc}}(-a-d+2A) + \cosh\frac{\sqrt{(a-d)^2+4bc}}{2}\right].
\end{aligned}
$$

Homework 4.33 It is possible to establish the connection (4.69) differently from the text.

(a) Show by brute force, without using the Hamilton–Cayley theorem, that the quantity $\exp(:f_2:)X$ with f_2 given by Eq. (4.68) gives a transformation according to the matrix (4.66).

(b) Another way is as follows. We have

$$
e^{:f_2:}X \;\leftrightarrow\; RX \qquad \Longrightarrow \qquad :f_2:X \;\leftrightarrow\; (\ln R)X\,.
$$

If R has the form (4.66), whose eigenvalues are $e^{\pm i\mu}$, an application of the Hamilton–Cayley theorem gives

$$
\ln R \;=\; \mu\begin{bmatrix} \alpha & \beta \\ -\gamma & -\alpha \end{bmatrix}.
$$

This map is to be identified with $:f_2:$. This can be done if f_2 is given by Eq. (4.67).

The analysis in (b) can be cast in a more crude statement by saying that the quadratic form of the Lie map is equal to the logarithm of the linear Taylor matrix map.

Homework 4.34 Given the 2×2 matrix maps, use these matrices to find the Lie operators representing (a) a drift space, (b) a thin-lens quadrupole, (c) a thick focusing quadrupole, and (d) a thick defocusing quadrupole. These results have been given in Table 4.3.

Solution

(a) Let the operator for a drift space L be written as $e^{:f_2:} = \exp(:-\frac{1}{2}\tilde{X}FX:)$. We need to find F. Let F be written as Eq. (4.60), then we have, using Eq. (4.63),

$$
\cos(\sqrt{ac - b^2}) + \frac{\sin(\sqrt{ac - b^2})}{\sqrt{ac - b^2}}\begin{bmatrix} b & c \\ -a & -b \end{bmatrix} = \begin{bmatrix} 1 & L \\ 0 & 1 \end{bmatrix}. \tag{4.78}
$$

The solution is found to be, using Eq. (4.65), $a = 0$, $b = 0$, $c = L$ which gives $f_2 = -\frac{1}{2}Lp^2$.

(b) Replace the right-hand-side of Eq. (4.78) by $\begin{bmatrix} 1 & 0 \\ -\frac{1}{f} & 1 \end{bmatrix}$ where $f =$ focal length. We have $a = \frac{1}{f}$, $b = 0$, $c = 0$, which gives $f_2 = -\frac{1}{2f}x^2$. For a defocusing thin-lens quadrupole, just let $f < 0$.

(c) Replace the right-hand-side of Eq. (4.78) by $\begin{bmatrix} \cos kL & \frac{1}{k}\sin kL \\ -k\sin kL & \cos kL \end{bmatrix}$. We find $a = k^2 L$, $b = 0$, $c = L$, which gives $f_2 = -\frac{L}{2}(k^2 x^2 + p^2)$.

(d) Replace k in the case above by ik to obtain $a = -k^2 L$, $b = 0$, $c = L$, which gives $f_2 = \frac{L}{2}(k^2 x^2 - p^2)$. Note that in this case, $ac - b^2 < 0$.

Homework 4.35 If the matrix R is not given by the Courant–Snyder representation (4.66) for a stable system, but is for an unstable system with

$$R = \begin{bmatrix} \cosh\mu + \alpha\sinh\mu & \beta\sinh\mu \\ \frac{1-\alpha^2}{\beta}\sinh\mu & \cosh\mu - \alpha\sinh\mu \end{bmatrix},$$

show that R can also be written as e^{SG}, where

$$G = \mu \begin{bmatrix} \frac{\alpha^2-1}{\beta} & \alpha \\ \alpha & \beta \end{bmatrix},$$

and the corresponding Lie representation of the map is

$$\exp\left[:-\frac{\mu}{2}\left(\frac{\alpha^2 - 1}{\beta}x^2 + 2\alpha xp + \beta p^2 \right): \right]. \tag{4.79}$$

The quadratic form in Eq. (4.79) is neither positive definite nor negative definite.

Homework 4.36 The Lie representation is very useful for generalization to nonlinear systems, but is awkward for a linear system. Linear systems are best studied using matrices. As a demonstration, try to concatenate a drift space followed by a thin-lens quadrupole. This can be easily done in the matrix language by multiplying two simple matrices. In the Lie language, one needs to concatenate

$$e^{:-\frac{1}{2}Lp^2:}e^{:-\frac{1}{2f}x^2:}$$

into a form of a single exponential. This can drive you crazy but can be done.

Homework 4.37 The Lie map (4.68) gives the one-turn Courant–Snyder map (4.66). The Courant–Snyder map from position 1 (with parameters α_1, β_1) to position 2 (with parameters α_2, β_2) is (ψ is the betatron phase advance from position 1 to position 2)

$$\begin{bmatrix} \sqrt{\frac{\beta_2}{\beta_1}}(\cos\psi + \alpha_1\sin\psi) & \sqrt{\beta_1\beta_2}\sin\psi \\ -\frac{1+\alpha_1\alpha_2}{\sqrt{\beta_1\beta_2}}\sin\psi + \frac{\alpha_1-\alpha_2}{\sqrt{\beta_1\beta_2}}\cos\psi & \sqrt{\frac{\beta_1}{\beta_2}}(\cos\psi - \alpha_2\sin\psi) \end{bmatrix}. \tag{4.80}$$

Find the Lie representation of this map.

Solution The Lie map is given by $\exp[-:\frac{1}{2}(ax^2 + 2bxx' + cx'^2):]$, where a, b, and c are determined by Eq. (4.65). In particular,

$$\sqrt{ac - b^2} = \cos^{-1}\left[\frac{1}{2}\left(\sqrt{\frac{\beta_2}{\beta_1}} + \sqrt{\frac{\beta_1}{\beta_2}}\right)\cos\psi + \frac{1}{2}\left(\alpha_1\sqrt{\frac{\beta_2}{\beta_1}} - \alpha_2\sqrt{\frac{\beta_1}{\beta_2}}\right)\sin\psi\right].$$

Again, one observes that the Lie map is much clumsier than the matrix map for linear problems. One special case occurs when $\alpha_1 = \alpha_2$ and $\beta_1 = \beta_2$. Then the Lie map has a simple form $\exp[-:\frac{1}{2}\psi(\gamma x^2 + 2\alpha xx' + \beta x'^2):]$.

Homework 4.38 A dog-leg beamline insertion has the matrix map

$$T = \begin{bmatrix} 1 & L & 0 & -L\theta \\ 0 & 1 & 0 & 0 \\ 0 & -L\theta & 1 & L\theta^2 \\ 0 & 0 & 0 & 1 \end{bmatrix},$$

where θ is the bending angle of each of the two dog-leg dipoles, assumed thin, and L is the distance between the two dipoles. This is not one of the ten fundamental maps, but since it is a linear system, there is in principle a Lie map $e^{:f_2(x,x',z,\delta):}$ that describes it. Find this quadratic form f_2.

Homework 4.39 Use the Hamilton–Cayley theorem to obtain the 4×4 matrix map (4.75) using the Lie exponential map with quadratic form (4.74).

Solution Due to its degeneracy (δ is constant), this is not as hard as it might seem. We first find from the expression of $f_2 = -\frac{1}{2}\tilde{X}FX$ that

$$F = \begin{bmatrix} \mu\gamma & \mu\alpha & 0 & -\mu(\gamma D + \alpha D') \\ \mu\alpha & \mu\beta & 0 & -\mu(\alpha D + \beta D') \\ 0 & 0 & 0 & 0 \\ -\mu(\gamma D + \alpha D') & -\mu(\alpha D + \beta D') & 0 & -\alpha_c C + \mu(\gamma D^2 + 2\alpha DD' + \beta D'^2) \end{bmatrix}.$$

The matrix SF has the eigenvalues

$$0, \quad 0, \quad i\mu, \quad -i\mu.$$

Hamilton–Cayley theorem then easily yields (4.75). Due to the degeneracy, however, the (34)-element of the resulting map contains a degree of freedom that can only be settled when RF is included.

Solution — alternative You may also explicitly calculate $e^{:f_2:}X$ and equating the result with the matrix form RX. This derivation does not show the additional degree of freedom due to the degeneracy. This may be considered an advantage or a disadvantage depending on your view.

4.3.6   Application to nonlinear system

So far we have been considering linear systems, which can be analyzed using matrices. As mentioned, using Lie technique for linear cases in fact is more cumbersome than using matrices. Lie technique is not very useful for linear cases.

The issue, however, is how to generalize the Courant–Snyder formalism — traditionally treated based on matrices — to nonlinear systems. A straightforward extension of matrices naturally leads to Taylor maps. But we already learned that Taylor maps have two shortcomings; it contains too much redundancy in its degrees of freedom, and it is not symplectic. Trying to extend the Courant–Snyder formalism to nonlinear systems, Taylor maps are not the best choice, or at least not the only one.

An alternative is the Lie maps. In fact, Lie algebra is arguably the most natural way to generalize the Courant–Snyder analysis to nonlinear systems; the straightforward Taylor extension is not the right answer.

Convert Lie map to Taylor map As a first application of the operator algebra to a nonlinear problem, consider a map which is known to have a second order Lie representation, accurate to order $\mathcal{O}(X^2)$, in the form

$$e^{:f_2:}e^{:f_3:}\,, \tag{4.81}$$

where f_2 and f_3 are homogeneous power series in X,

$$\begin{aligned}
f_2 &= -\frac{1}{2}\tilde{X}FX = -\frac{1}{2}F_{\ell m}X_\ell X_m\,,\\
f_3 &= C_{\ell mn}X_\ell X_m X_n\,.
\end{aligned} \tag{4.82}$$

The F and C coefficients are symmetric with respect to permutations on their respective subscript indices. As usual, repeated subscript indices are implicitly assumed to be summed over.

Let the Taylor representation of the 2nd order map be

$$X_\alpha = R_{\alpha\beta}X_{0\beta} + T_{\alpha\beta\gamma}X_{0\beta}X_{0\gamma} + \mathcal{O}(X_0^3)\,, \tag{4.83}$$

where the T-coefficients satisfy $T_{\alpha\beta\gamma} = T_{\alpha\gamma\beta}$. Our job is to find the R- and T-coefficients in terms of the F- and C-coefficients.

We first note that

$$\begin{aligned}
:f_3:X_\alpha &= \frac{\partial f_3}{\partial X_\delta}S_{\delta\gamma}\frac{\partial X_\alpha}{\partial X_\gamma}\\
&= \frac{\partial(C_{\ell mn}X_\ell X_m X_n)}{\partial X_\delta}S_{\delta\gamma}\delta_{\alpha\gamma}\\
&= C_{\ell mn}(\delta_{\delta\ell}X_m X_n + X_\ell\delta_{m\delta}X_n + X_\ell X_m\delta_{\delta n})S_{\delta\alpha}\\
&= -3S_{\alpha\delta}C_{\delta mn}X_m X_n\,,
\end{aligned}$$

and that

$$(:f_3:)^2 X_\alpha = \mathcal{O}(X^3)\,.$$

It follows that

$$
\begin{aligned}
e^{:f_3:}X_\alpha &= \left[1 + :f_3: + \frac{1}{2}:f_3:^2 + \cdots\right]X_\alpha \\
&= X_\alpha - 3S_{\alpha\delta}C_{\delta mn}X_m X_n + \mathcal{O}(X^3)\,.
\end{aligned}
$$

This then gives

$$
\begin{aligned}
e^{:f_2:}e^{:f_3:}X_\alpha &= e^{:f_2:}X_\alpha - 3S_{\alpha\delta}C_{\delta mn}e^{:f_2:}(X_m X_n) + \mathcal{O}(X^3) \\
&= (e^{SF}X)_\alpha - 3S_{\alpha\delta}C_{\delta mn}(e^{:f_2:}X_m)(e^{:f_2:}X_n) + \mathcal{O}(X^3)\,. \quad (4.84)
\end{aligned}
$$

Equation (4.84) then leads to

$$
R = e^{SF}\,, \tag{4.85}
$$

as expected, and

$$
T_{\alpha\beta\gamma} = -3S_{\alpha\delta}C_{\delta mn}R_{m\beta}R_{n\gamma}\,. \tag{4.86}
$$

Convert Taylor map to Lie map   Equations (4.85–4.86) express the R- and the T-coefficients in terms of the F- and C-coefficients. Their reverses can also be obtained. The procedure of finding F from R is given by Eqs. (4.65) and (4.60). The C-coefficients are found to be

$$
C_{\ell mn} = \frac{1}{3}S_{\ell\alpha}(R^{-1})_{\beta m}(R^{-1})_{\gamma n}T_{\alpha\beta\gamma}\,, \tag{4.87}
$$

as can be proven by back substitution into Eq. (4.86). This exercise shows an example of explicit transformation between a nonlinear Lie map and its corresponding Taylor map. It should be kept in mind that representations (4.83) and (4.81) agree with each other to 2nd order in X, but differ in higher orders.

Single sextupole   To be specific, consider a circular accelerator that is otherwise perfectly linear aside from a thin-lens sextupole at $s = 0$. The linear Lie map around $s = 0$ is (consider 1-D motion)

$$
e^{:f_2:} \qquad \text{where} \qquad f_2 = -\frac{\mu}{2}(\gamma x^2 + 2\alpha x p + \beta p^2)\,. \tag{4.88}
$$

Let the thin-lens sextupole be (see Table 4.3)

$$
e^{:f_3:} \qquad \text{where} \qquad f_3 = \lambda x^3\,. \tag{4.89}
$$

The one-turn map from $s = 0^+$, going through the accelerator, then the sextupole, and end up at $s = C^+$, is given by

$$
e^{:f_2:}e^{:f_3:}\,. \tag{4.90}
$$

As discussed earlier, note the ordering of the Lie operators.

The Taylor representation of the map is given by combining the two steps, i.e., passing through the accelerator,

$$
\begin{aligned}
x_1 &= x_0(\cos\mu + \alpha\sin\mu) + p_0\beta\sin\mu\,, \\
p_1 &= -x_0\gamma\sin\mu + p_0(\cos\mu - \alpha\sin\mu)\,,
\end{aligned}
\tag{4.91}
$$

followed by passing through the sextupole,

$$
x = x_1\,, \qquad \text{and} \qquad p = p_1 + 3\lambda x_1^2\,.
\tag{4.92}
$$

The combination of Eqs. (4.91–4.92) gives a Taylor map of the form (4.83) with all higher order terms terminated, and

$$
\begin{aligned}
R_{11} &= \cos\mu + \alpha\sin\mu\,, \\
R_{12} &= \beta\sin\mu\,, \\
R_{21} &= -\gamma\sin\mu\,, \\
R_{22} &= \cos\mu - \alpha\sin\mu\,, \\
T_{211} &= 3\lambda R_{11}^2\,, \\
T_{212} &= T_{221} = 3\lambda R_{12}R_{11}\,, \\
T_{222} &= 3\lambda R_{12}^2\,.
\end{aligned}
\tag{4.93}
$$

All unlisted T-coefficients vanish. This Taylor map is symplectic and is a special case of the map (4.27–4.28). The reason it can be symplectic is that (i) both the maps (4.91) and (4.92) are symplectic, and (ii) no truncation has been performed when concatenating them to obtain (4.93).

As to the Lie representation, the quadratic form f_2 is obtained by implementing Eq. (4.85) and is of course given by Eq. (4.88) as one would expect. The C-coefficients that describe f_3 according to Eq. (4.82) are given by Eq. (4.87). It is found that the only nonvanishing C-coefficient is

$$
C_{111} = \lambda\,.
$$

This leads to the expected Eq. (4.89) for f_3.

Monomial map   We now proceed to describe an interesting formula involving monomials. We have introduced them before in Sec. 4.3.3 as the fundamental linear symplectic maps with matrix representations. Here we will generalize on them to include nonlinear maps as well as fractional powers.

Consider a 1-D system. Let x and p be the canonical variables, a, α, and β be arbitrary constants. Consider the exponential operator $\exp(:ax^\alpha p^\beta:)$, where the exponent is a single term in the form of a power of x and p (α and β do not have to be integers). The formula reads

$$
\begin{aligned}
e^{a:x^\alpha p^\beta:}x &= \begin{cases} x\left[1 + a(\alpha-\beta)x^{\alpha-1}p^{\beta-1}\right]^{\beta/(\beta-\alpha)}, & \text{if } \alpha \neq \beta, \\ x\exp\left(-a\alpha x^{\alpha-1}p^{\alpha-1}\right), & \text{if } \alpha = \beta, \end{cases} \\[2ex]
e^{a:x^\alpha p^\beta:}p &= \begin{cases} p\left[1 + a(\alpha-\beta)x^{\alpha-1}p^{\beta-1}\right]^{\alpha/(\alpha-\beta)}, & \text{if } \alpha \neq \beta, \\ p\exp\left(a\alpha x^{\alpha-1}p^{\alpha-1}\right), & \text{if } \alpha = \beta. \end{cases}
\end{aligned}
\tag{4.94}
$$

The $\alpha = \beta$ case can be obtained by taking the limit $\alpha \to \beta$ in the $\alpha \neq \beta$ expressions. A discussion on Eq. (4.94) is mentioned in Homework 4.43.

To prove Eq. (4.94), one first notes that the map $e^{a:x^\alpha p^\beta:}$ describes a dynamical system with Hamiltonian [see Eq. (4.118) later],

$$H = -\frac{a}{T}x^\alpha p^\beta ,$$

where T is the total time period over which the Hamiltonian is being applied to the system. The equations of motion are

$$\dot{x} = \frac{\partial H}{\partial p} = -\frac{a\beta}{T}x^\alpha p^{\beta-1} ,$$

$$\dot{p} = -\frac{\partial H}{\partial x} = \frac{a\alpha}{T}x^{\alpha-1}p^\beta . \tag{4.95}$$

Since H is a constant of the motion, we have $x^\alpha p^\beta = x_0^\alpha p_0^\beta$. From the first line of Eq. (4.95), we have therefore

$$\dot{x} = -\frac{a\beta}{T}x_0^{\alpha(\beta-1)/\beta}p_0^{\beta-1}x^{\alpha/\beta}$$

$$\implies \quad x^{-\alpha/\beta}\dot{x} = -\frac{a\beta}{T}x_0^{\alpha(\beta-1)/\beta}p_0^{\beta-1}$$

$$\implies \quad [x(t)]^{(\beta-\alpha)/\beta} = x_0^{(\beta-\alpha)/\beta} - \frac{a(\beta-\alpha)}{T}tx_0^{\alpha(\beta-1)/\beta}p_0^{\beta-1} .$$

Setting x to be $x(T)$, we obtain

$$x = x_0\left[1 + a(\alpha-\beta)x_0^{\alpha-1}p_0^{\beta-1}\right]^{\beta/(\beta-\alpha)} ,$$

which proves the first line of Eq. (4.94).

Similarly,

$$\dot{p} = \frac{a\alpha}{T}x_0^{\alpha-1}p_0^{\beta(\alpha-1)/\alpha}p^{\beta/\alpha}$$

$$\implies \quad p = p_0\left[1 + a(\alpha-\beta)x_0^{\alpha-1}p_0^{\beta-1}\right]^{\alpha/(\alpha-\beta)} ,$$

which proves the second line in Eq. (4.94).

One particular case of Eq. (4.94) is when α and β are integers. Then the exponent is a monomial in x and p. Equation (4.94) then gives the exact result of the evolution of the dynamical variables due to the exponential monomial map. Special cases of monomial map includes several members listed in Table 4.3. They also provide the special cases of the fundamental symplectic matrices of Sec. 4.3.3.

Equation (4.94) is symplectic. To demonstrate that, we need only to show that the Poisson brackets of the two expressions in Eq. (4.94) is equal to 1, i.e.,

$$[x,p] = \frac{\partial x}{\partial x_0}\frac{\partial p}{\partial p_0} - \frac{\partial x}{\partial p_0}\frac{\partial p}{\partial x_0} = 1. \tag{4.96}$$

Perturbation small parameter   The Lie algebra we are developing is a perturbative technique. So far, we have been considering perturbation in powers of X, i.e. X is the small quantity in the perturbation treatment (assuming we are interested in the region of phase space close to the origin). This is not necessary; the perturbation expansion can be in other small quantities as well.

Consider a Taylor map which is 1st order in the strength ϵ of a certain linear or nonlinear perturbation in such a way that

$$X = X_0 + \epsilon G(X_0),\qquad(4.97)$$

where G is a vector function which could be nonlinear in X_0. Here, this map (4.97) is considered a 1st order map — 1st order in ϵ — even if $G(X_0)$ could be nonlinear in X_0. Our job is to find the corresponding Lie map to 1st order in ϵ. When $\epsilon = 0$, we have the identity map.

Let the Lie map be written as $\exp[:\epsilon f(X):]$, then we must have

$$:f(X):X = G(X),\qquad \text{or} \qquad [f(X), X] = G(X).$$

Using Eq. (4.4), this gives

$$-S_{ki}\frac{\partial f}{\partial X_i} = G_k, \qquad \text{or} \qquad -S\nabla f(X) = G(X).\qquad(4.98)$$

By inverting Eq. (4.98), we obtain

$$\nabla f(X) = SG(X)$$
$$\implies \qquad f(X) = \int_0^X dX'\, SG(X').\qquad(4.99)$$

The integral in Eq. (4.99) is along any path from 0 to X in the multidimensional phase space. This integral being independent of the path taken is a result of the dynamic system being conservative. Equation (4.98) can be used to obtain the Taylor map once the Lie map is known (knowing f, one can compute G). The fact that G is derived from the gradient of another function (and then multiplied by S) assures it to be conservative. Equation (4.99) can be used the other way around (knowing G, one can compute f as long as G is conservative).

Note that when we derive G from f, it is straightforward, while when we derive f from G, we must hurriedly add the condition that the system must be conservative. A Lie map is conservative automatically. A Taylor map must impose the conservative condition in addition.

The point that Lie algebra is a general perturbative technique and its small parameter does not have to be the components of X should be kept in mind. Let us consider an application to kick maps — 1st order in the kick strength but arbitrary order in X — below.

Consider a kick map described by Eq. (4.97) with

$$X = \begin{bmatrix} x \\ p \end{bmatrix}, \qquad G(X) = \begin{bmatrix} 0 \\ g(x) \end{bmatrix}.\qquad(4.100)$$

where $g(x)$ is a function of x and not p but otherwise arbitrary. Let us look for the corresponding exponential Lie map to 1st order in ϵ.

Before getting started, we need to be convinced first that the map (4.100) is symplectic, i.e. the system is conservative. Once that is established, Eq. (4.99) gives

$$\begin{bmatrix} \frac{\partial f(X)}{\partial x} \\ \frac{\partial f(X)}{\partial p} \end{bmatrix} = \nabla f(X) = \begin{bmatrix} 0 & 1 \\ -1 & 0 \end{bmatrix} \begin{bmatrix} 0 \\ g(x) \end{bmatrix} = \begin{bmatrix} g(x) \\ 0 \end{bmatrix}$$

$$\implies \quad f(X) = \int_0^x du\, g(u)\,, \tag{4.101}$$

independent of p. Conversely, if one is given the Lie map with $f(X)$ in the form (4.101), then

$$G(X) = -S\nabla f(X) = -\begin{bmatrix} 0 & 1 \\ -1 & 0 \end{bmatrix} \begin{bmatrix} g(x) \\ 0 \end{bmatrix} = \begin{bmatrix} 0 \\ g(x) \end{bmatrix}.$$

Homework 4.41 gives another more specific application.

Contact transformation   Consider the (x, p) phase space and a map $e^{:pf(x):}$ with some function $f(x)$. We can find the explicit expressions of

$$\begin{bmatrix} X \\ P \end{bmatrix} = e^{:pf(x):} \begin{bmatrix} x \\ p \end{bmatrix} \tag{4.102}$$

as follows.

The map $e^{:pf(x):}$ is a map from $t = 0$ to $t = T$ for a dynamical system with Hamiltonian $H = -\frac{pf(x)}{T}$. The Hamilton equations of motion are

$$\dot{x} = -\frac{1}{T}f(x), \qquad \dot{p} = \frac{1}{T}pf'(x)\,. \tag{4.103}$$

The first member of Eq. (4.103) gives

$$\int_{x_0}^x \frac{dx'}{f(x')} = -\frac{t}{T}\,, \tag{4.104}$$

where (x_0, p_0) are the initial coordinates at time $t = 0$.

Since the Hamiltonian is a constant of the motion, we have

$$pf(x) = p_0 f(x_0) \qquad \implies \qquad x = f^{-1}\left(\frac{p_0}{p}f(x_0)\right).$$

The second member of Eq. (4.103) then gives

$$\dot{p} = \frac{p}{T}f'\left[f^{-1}\left(\frac{p_0}{p}f(x_0)\right)\right]$$

$$\implies \quad \int_1^{p/p_0} \frac{du}{uf'[f^{-1}(\frac{1}{u}f(x_0))]} = \frac{t}{T}\,. \tag{4.105}$$

Knowing $f(x)$, Eq. (4.104) gives the solution for $x(t)$ as a function of t while Eq. (4.105) gives the solution for $p(t)$. When $t = T$, the solutions give explicit expressions of the final coordinates X and P.

If the map (4.102) is considered to be a change of phase space variables from (x, p) to (X, P), this type of coordinate change is called *contact transformation*. In particular, X and $\frac{P}{p_0}$ depend only on x_0 and not on p_0. Homework 4.45 gives a few explicit examples of such maps, or coordinate canonical transformations. Homework 4.46 gives a generalization of the contact maps to the case when the Lie exponent factorizes so that the Lie maps reads $e^{:f(x)g(p):}$.

Homework 4.40 As mentioned in the text, exponential Lie maps are always symplectic, while Taylor maps are generally not. However, Taylor maps should be symplectic up to the order it is truncated. Show that when Eqs. (4.85) and (4.86) are satisfied, the Taylor map (4.83) is symplectic to order $\mathcal{O}(X)$.

Solution The Jacobian matrix of the map (4.83) is

$$M_{\alpha\delta} = R_{\alpha\delta} + 2T_{\alpha\delta\beta}X_{0\beta} + \mathcal{O}(X^2)$$
$$\implies (\tilde{M}SM)_{\alpha\beta} = (\tilde{R}SR)_{\alpha\beta} + 2S_{\ell m}(T_{\ell\alpha n}R_{m\beta} - R_{\ell\alpha}T_{m\beta n})X_{0n} + \mathcal{O}(X^2).$$

You need to prove

$$\tilde{R}SR = S,$$
$$S_{\ell m}(T_{\ell\alpha n}R_{m\beta} - R_{\ell\alpha}T_{m\beta n}) = 0 \qquad \text{for all } \alpha, \beta, n.$$

Homework 4.41

(a) As an application of Eqs. (4.98–4.99), find the exponential Lie map that gives the Taylor map [a special case of the map (4.28)],

$$x = x_0 + \epsilon\alpha(x_0 - \alpha p_0)^2,$$
$$p = p_1 + \epsilon(x_0 - \alpha p_0)^2.$$

(b) Your result in (a) was obtained by applying the perturbation calculation and for that reason seemed valid only to 1st order in ϵ. Curiously, it is not so for this case. Show that the map

$$x = x_0 + bf(ax_0 + bp_0),$$
$$p = p_0 - af(ax_0 + bp_0) \qquad\qquad (4.106)$$

is symplectic for arbitrary values of a, b, and arbitrary function f. Find the Lie representation of the map. This map can be regarded as the generalized kick map. The usual kick map is obtained when either $a = 0$ or $b = 0$.

Solution

(a) Don't forget to check the symplecticity of the Taylor map first; otherwise the problem is not meaningful. The Lie map is

$$\exp\left[:\frac{\epsilon}{3}(x - \alpha p)^3:\right].$$

The result is meant to be only to 1st order in ϵ.

(b) Without making the perturbation approximation, the Lie map is

$$\exp\left[:-\int_0^{ax+bp} f(u)\,du:\right].$$

Homework 4.42 Factorize $\exp[:ax^2+2bxp+cp^2:]$ into a product of monomial map factors. Give physical meaning of each of the monomial factor maps. This factorization is exact. What happens if the case has $ax^2+2bxp+cp^2$ being negative definite with $a<0$ and $b^2-ac<0$?

Solution This factorization is not unique. One way is to factorize the map into a drift space of length L_1, followed by a thin-lens quadrupole of focal length f, followed by another drift space L_2, where

$$L_1 = \frac{\sqrt{ac-b^2}}{a}\tan\frac{\sqrt{ac-b^2}}{2}+\frac{b}{a},$$

$$f = \frac{\sqrt{ac-b^2}}{a\sin\sqrt{ac-b^2}},$$

$$L_2 = \frac{\sqrt{ac-b^2}}{a}\tan\frac{\sqrt{ac-b^2}}{2}-\frac{b}{a}.$$

This means we have factorized

$$e^{:ax^2+2bxp+cp^2:} = e^{:-\frac{1}{2}L_1 p^2:}e^{:-\frac{1}{2f}x^2:}e^{:-\frac{1}{2}L_2 p^2:}.$$

The total matrix map is

$$\begin{bmatrix} \cos\sqrt{ac-b^2}+\frac{b}{\sqrt{ac-b^2}}\sin\sqrt{ac-b^2} & \frac{c}{\sqrt{ac-b^2}}\sin\sqrt{ac-b^2} \\ -\frac{a}{\sqrt{ac-b^2}}\sin\sqrt{ac-b^2} & \cos\sqrt{ac-b^2}-\frac{b}{\sqrt{ac-b^2}}\sin\sqrt{ac-b^2} \end{bmatrix},$$

which, if we let $\sqrt{ac-b^2}=\psi, a=\gamma\psi, b=\alpha\psi, c=\beta\psi$ (which satisfies $\gamma=\frac{1+\alpha^2}{\beta}$), becomes

$$\begin{bmatrix} \cos\psi+\alpha\sin\psi & \beta\sin\psi \\ -\gamma\sin\psi & \cos\psi-\alpha\sin\psi \end{bmatrix}.$$

A generalization of this factorization into monomial maps to a nonlinear system can be found after we learn the Baker–Campbell–Hausdorff formula.

Homework 4.43 The text checked the monomial map Eq. (4.94) for its symplecticity, see Eq. (4.96). Another appreciation of Eq. (4.94) is provided by first considering the infinite series representation of the exponential map; application of this map on x or p then leads to infinite series. What Eq. (4.94) says is that the resulting infinite series can be summed exactly.

(a) Show this explicitly by first proving that, for $k \geq 0$,

$$(:x^\alpha p^\beta:)^k x = (\beta - \alpha)^k \frac{\Gamma(k + \frac{\beta}{\alpha-\beta})}{\Gamma(\frac{\beta}{\alpha-\beta})} x^{(\alpha-1)k+1} p^{(\beta-1)k},$$

$$(:x^\alpha p^\beta:)^k p = (\beta - \alpha)^k \frac{\Gamma(k - \frac{\alpha}{\alpha-\beta})}{\Gamma(-\frac{\alpha}{\alpha-\beta})} x^{(\alpha-1)k} p^{(\beta-1)k+1}. \qquad (4.107)$$

(b) Use Eq. (4.107) to sum the infinite series and prove Eq. (4.94).

Solution (a) Use mathematical induction.

Homework 4.44 The text resorted to the Hamiltonian dynamics to prove the monomial maps (4.94). Use the same method for a thick quadrupole magnet to obtain expressions for $e^{-\frac{L}{2}(k^2 x^2 + p^2):}x$ and $e^{-\frac{L}{2}(k^2 x^2 + p^2):}p$.

Homework 4.45 Apply the contact map (4.102) to the following special cases,
    (a) $f(x) = ax$;
    (b) $f(x) = ax^2$;
    (c) $f(x) = \alpha e^{ax}$;
    (d) $f(x) = \alpha \sin(ax)$.

Solution

$$(a) \qquad X = x_0 e^{-a}, \qquad P = p_0 e^a;$$

$$(b) \qquad X = \frac{x_0}{1 + ax_0}, \qquad P = p_0 (1 + ax_0)^2;$$

$$(c) \qquad X = -\frac{1}{a} \ln\left(e^{-ax_0} + \alpha a\right), \qquad P = p_0 \left(1 + \alpha a e^{ax_0}\right);$$

$$(d) \qquad \tan \frac{aX}{2} = \left(\tan \frac{ax_0}{2}\right) e^{-\alpha a},$$

$$P = \frac{p_0 e^{-\alpha a}}{2(1 + \cos ax_0)} \left[(1 + \cos ax_0)^2 e^{2\alpha a} + \sin^2 ax_0\right].$$

One may check that the Jacobians for these maps have unit determinants. As mentioned in the text, these contact maps give X and $\frac{P}{p_0}$ that depend on x_0 and not on p_0. One may also check some of them against Eq. (4.94) because they are special cases of monomial maps. The map for case (c), for example, is equivalent to a contact transformation with a generating function,

$$F(x, P) = -\frac{P}{a} \ln(e^{-ax} + \alpha a).$$

Homework 4.46 The discussion of contact maps in the text can be extended. Consider the map $e^{:f(x)g(p):}$.
    (a) Following similar steps as in the text, find expressions for

$$\begin{bmatrix} X \\ P \end{bmatrix} = e^{:f(x)g(p):} \begin{bmatrix} x \\ p \end{bmatrix}. \qquad (4.108)$$

(b) Apply the result obtained in (a) to find the explicit result when

$$f(x)g(p) \ = \ \alpha e^{ax+bp}.$$

(c) Work out the case

$$f(x)g(p) \ = \ \alpha x^a e^{bp}.$$

(d) Let us denote the coordinate x as ϕ and momentum p as A, and take

$$f(\phi)g(A) \ = \ \alpha\sqrt{A}\sin\phi.$$

Solution Following similar steps as in the text, we obtain

$$\int_{x_0}^{x} \frac{dx}{f(x)g'\left[g^{-1}\left(\frac{f(x_0)g(p_0)}{f(x)}\right)\right]} \ = \ -\frac{t}{T},$$

$$\int_{p_0}^{p} \frac{dp}{g(p)f'\left[f^{-1}\left(\frac{f(x_0)g(p_0)}{g(p)}\right)\right]} \ = \ \frac{t}{T}. \tag{4.109}$$

The map (4.108) is obtained by setting $t = T$ in Eq. (4.109). Equations (4.94), (4.104) and (4.105) are special cases of Eq. (4.109).

(b) For the case $f(x)g(p) = \alpha e^{ax+bp}$,

$$x(t) \ = \ x_0 - \alpha b \frac{t}{T} e^{ax_0+bp_0},$$

$$p(t) \ = \ p_0 + \alpha a \frac{t}{T} e^{ax_0+bp_0}. \tag{4.110}$$

(c) Taking $f(x)g(p) = \alpha x^a e^{bp}$ gives

$$x(t) \ = \ x_0 - \alpha b x_0^a e^{bp_0} \frac{t}{T},$$

$$p(t) \ = \ -\frac{a}{b}\ln\left(e^{-bp_0/a} - \alpha b x_0^{a-1} e^{b(a-1)p_0/a}\frac{t}{T}\right). \tag{4.111}$$

(d) There is nothing wrong when the phase space is polar coordinates. In this case, taking $f(\phi)g(A) = \alpha\sqrt{A}\sin\phi$ gives

$$\cot\phi(t) \ = \ \cot\phi_0 + \frac{\alpha}{2\sqrt{A_0}\sin\phi_0}\frac{t}{T},$$

$$A(t) \ = \ A_0\sin^2\phi_0 + \left(\sqrt{A_0}\cos\phi_0 + \frac{\alpha t}{2T}\right)^2. \tag{4.112}$$

Similarly, taking $f(\phi)g(A) = \alpha\sqrt{A}\cos\phi$ gives

$$\tan\phi(t) \ = \ \tan\phi_0 - \frac{\alpha}{2\sqrt{A_0}\cos\phi_0}\frac{t}{T},$$

$$A(t) \ = \ A_0\cos^2\phi_0 + \left(\sqrt{A_0}\sin\phi_0 + \frac{\alpha t}{2T}\right)^2. \tag{4.113}$$

One can check the symplecticity of the maps (4.110–4.113) by verifying the Jacobian matrices have unit determinants.

Homework 4.47 We learned in Homework 4.27 that if

$$e^{:f(x,p):}x = X(x,p), \qquad e^{:f(x,p):}p = P(x,p),$$

then

$$f(X(x,p), P(x,p)) = f(x,p). \qquad (4.114)$$

This means the quantity $f(x,p)$ is invariant when x and p are mapped forward by $e^{:f(x,p):}$.

(a) Prove that the quantity $f(x,p)$ is also invariant when x and p are mapped backward by $^{:-f(x,p):}$, i.e. show that

$$f(\bar{X}(x,p), \bar{P}(x,p)) = f(x,p), \qquad (4.115)$$

where

$$e^{:-f(x,p):}x = \bar{X}(x,p), \qquad e^{:-f(x,p):}p = \bar{P}(x,p).$$

(b) Check the validity of Eqs. (4.114) and (4.115) for the case of Eq. (4.94).

Solution (b) We have $f(x,p) = ax^\alpha p^\beta$, and $X, P, \bar{X}, \bar{P}$ are given by

$$
\begin{aligned}
X &= x[1 + a(\alpha - \beta)x^{\alpha-1}p^{\beta-1}]^{\beta/(\beta-\alpha)}, \\
P &= p[1 + a(\alpha - \beta)x^{\alpha-1}p^{\beta-1}]^{\alpha/(\alpha-\beta)}, \\
x &= \bar{X}[1 + a(\alpha - \beta)\bar{X}^{\alpha-1}\bar{P}^{\beta-1}]^{\beta/(\beta-\alpha)}, \\
p &= \bar{P}[1 + a(\alpha - \beta)\bar{X}^{\alpha-1}p\bar{P}^{\beta-1}]^{\alpha/(\alpha-\beta)}.
\end{aligned}
\qquad (4.116)
$$

Conditions (4.114) and (4.115) follow by observing

$$x^\alpha p^\beta = X^\alpha P^\beta = \bar{X}^\alpha \bar{P}^\beta. \qquad (4.117)$$

One can solve the last two expressions of Eq. (4.116) to obtain explicit expressions of $\bar{X}$ and $\bar{P}$ in terms of x and p. The result will be the same expressions as X and P except that a is replaced by $-a$, as one expects.

4.4 Baker–Campbell–Hausdorff formula

In the previous section, we tried to establish some familiarity with the Lie operators. Before we turn attention to accelerator applications in the following sections, we need to introduce a set of formulae — the Baker–Campbell–Hausdorff formula and its variations.[17] These variants of the BCH formulae, together with their inverses, are used repeatedly later in various circumstances. Familiarity with them and versatility in applying them are necessary skills to be acquired. A very brief introduction of the BCH formula, and why it is so much needed, was made following Eq. (4.42).

[17]J.E. Campbell, Proc. London Math. Soc., 29, 14 (1898); H.F. Baker, London Math. Soc., 34, 347 (1902); F. Hausdorff, Ber. Verhandl. Akad. Wiss. Leipzig, Math-nauturwiss, 58, 19 (1906); V.S. Varadarajan, Lie Groups, Lie Algebras, and Their Representations, Prentice-Hall, Englewood Cliffs, New Jersey, 1974.

4.4.1 Single accelerator element

An accelerator typically consists of a sequence of elements. Consider one particular element of length L. Let the particle motion in this element be described
by the Hamiltonian H which is independent of s inside the element. The map
for this element can be written as an exponential symplectic operator. Some
examples have been given in Table 4.3. For the element being considered here,
the map is[18]

$$e^{:-LH:} .$$
(4.118)

To prove Eq. (4.118), one only has to note that

$$\frac{dX}{ds} \;=\; -:H:X \qquad \Longrightarrow \qquad \frac{d^k X}{ds^k} \;=\; (-:H:)^k X \,,$$

where s, the distance along the accelerator, is chosen to be the time variable.
It follows that the map

$$X(s) \;=\; \sum_{k=0}^{\infty} \frac{s^k}{k!} \left(\frac{d^k X}{ds^k} \right)_{s=0}$$

can be represented by the Lie operator

$$\sum_{k=0}^{\infty} \frac{s^k}{k!} (-:H:)^k \;=\; e^{:-sH:} .$$

The operator (4.118) is obtained just by integrating the map to the end of the
element $s = L$. Again, we have assumed that the Hamiltonian H is independent
of s within this single element. Different elements of course can have different
Hamiltonians.

Given below are the Hamiltonians of a few typical accelerator elements,

$$H(x, p_x, y, p_y, z, \delta, s) \;=\; \begin{cases} -\frac{x\delta}{\rho} + \frac{x^2}{2\rho^2} + \frac{p_x^2 + p_y^2}{2(1+\delta)} & \text{dipole}\,, \\[2mm] \frac{K}{2}(x^2 - y^2) + \frac{p_x^2 + p_y^2}{2(1+\delta)}, & \text{quadrupole}\,, \\[2mm] \frac{S}{3}(x^3 - 3xy^2) + \frac{p_x^2 + p_y^2}{2(1+\delta)}, & \text{sextupole}\,, \\[2mm] \frac{\lambda}{4}(x^4 - 6x^2y^2 + y^4) + \frac{p_x^2 + p_y^2}{2(1+\delta)}, & \text{octupole}\,, \end{cases}$$
(4.119)

where K, S, and λ are the strengths of quadrupole, sextupole and octupole
elements, ρ is the dipole bending radius. The dynamical variables for these
Hamiltonians are $(x, p_x, y, p_y, z, \delta)$ — be reminded that they must be canonical.
In Eq. (4.119), all elements are considered thick-lens.

---

[18]We mentioned before that the exponent function on an exponential Lie operator must
have the dimension of the emittance. The reader is suggested to check that LH has the
correct dimension.

In general, for an n-th multipole other than dipole ($n = 1, 2, 3$ for quadrupole, sextupole, octupole, etc.), we have

$$H(x, p_x, y, p_y, z, \delta, s) = \frac{1}{n+1} \text{Re}[(\lambda_n + i\bar{\lambda}_n)(x + iy)^{n+1}] + \frac{p_x^2 + p_y^2}{2(1 + \delta)}, \quad (4.120)$$

where λ_n and $\bar{\lambda}_n$ are the normal and the skew components of the multipole field. Equation (4.119) contains only the normal components with $K = \lambda_1$, $S = \lambda_2$ and $\lambda = \lambda_3$. Explicit equations of motion for the Hamiltonians (4.119) can be found in Homework 4.48.

The nonlinear effects of δ through the term $\frac{p_x^2 + p_y^2}{2(1+\delta)}$ in the Hamiltonians (4.119–4.120), especially for the higher multipoles like sextupoles and octupoles, are often (not always) negligible. In those cases, we make the replacement,

$$\frac{p_x^2 + p_y^2}{2(1 + \delta)} \quad \rightarrow \quad \frac{p_x^2 + p_y^2}{2}. \quad (4.121)$$

In this approximation, the linear dispersion effects are contained by the $-\frac{x\delta}{\rho}$ term of the dipole magnets, while chromaticity effects are contained by the $\frac{p_x^2 + p_y^2}{2(1+\delta)}$ term of the quadrupoles.

The Hamiltonians (4.119–4.120) apply to the 6-D phase space ($x, p_x, y, p_y, z, \delta$). If one is interested only in the 4-D transverse phase space — even allowing the particle to be off-momentum — it is possible to simplify the Hamiltonians somewhat as follows. One first notes that for the elements described by (4.119–4.120), we have $x' = \frac{\partial H}{\partial p_x} = \frac{p_x}{1+\delta}$ and similarly $y' = \frac{p_y}{1+\delta}$. One then notes that without electric devices (e.g. an RF cavity), the Hamiltonians in (4.119–4.120) do not depend on z, and we have $\delta' = -\frac{\partial H}{\partial z} = 0$, i.e. δ is a constant of the motion and can be regarded as a numerical constant if one ignores the longitudinal dynamics. The Hamilton equations of motion then hold for the variables (x, x', y, y') with the new Hamiltonians,

$$H(x, x', y, y', s) = \begin{cases} \frac{1}{1+\delta}\left(-\frac{x\delta}{\rho} + \frac{1}{2\rho^2}x^2\right) + \frac{1}{2}(x'^2 + y'^2), & \text{dipole,} \\ \frac{K}{2(1+\delta)}(x^2 - y^2) + \frac{1}{2}(x'^2 + y'^2), & \text{quadrupole,} \\ \frac{S}{3(1+\delta)}(x^3 - 3xy^2) + \frac{1}{2}(x'^2 + y'^2), & \text{sextupole,} \\ \frac{\lambda}{4(1+\delta)}(x^4 - 6x^2y^2 + y^4) + \frac{1}{2}(x'^2 + y'^2), & \text{octupole.} \end{cases}$$
$$(4.122)$$

Hamiltonians (4.122) apply only if one ignores the longitudinal dynamics. In particular, if one wants to find the path length equation of motion by $z' = \frac{\partial H}{\partial \delta}$, he/she should use Eq. (4.119–4.120), not Eq. (4.122).

It might appear a bit casual when we write down the Hamiltonians in Eq. (4.119) without derivation and then drop the $(1+\delta)$ factor as in Eq. (4.121). If so, it is because the exact form of the Hamiltonians actually do not play too important a role in what is being discussed. As mentioned before, it is allowed in the Lie representation to make approximations on the exponent of the Lie map without disturbing the integrity — the symplecticity — of the system.

Approximating the Hamiltonian in the Lie form can be made as needed. In case a different better expression is to be used for some of the Hamiltonians in Eq. (4.119), for example, there is not an objection here of doing so. Once the Hamiltonian is chosen, the element map follows as given by Eq. (4.118).

The same cannot be said for a Taylor map. When a truncation or an approximation is made on the equations of motion, one needs to be careful to see if the possible loss of symplecticity affects the results.

Solenoid   We have not discussed the solenoids, to which we now turn to. We shall consider the case of fixed-momentum particles only ($\delta = 0$). What we need is its Hamiltonian. The Hamiltonian should be expressed in terms of the canonical coordinates (x, p_x, y, p_y),

$$H(x, p_x, y, p_y, s) = \frac{1}{2}\left(p_x + \frac{K}{2}y\right)^2 + \frac{1}{2}\left(p_y - \frac{K}{2}x\right)^2 , \tag{4.123}$$

where $K = \frac{B_s}{B\rho}$ is the solenoid strength; $p_x + \frac{K}{2}y$ and $p_y - \frac{K}{2}x$ are the mechanical components of the transverse momentum. The Lie map of the solenoid is $e^{:-LH:}$.

With the Hamiltonian (4.123), the Hamilton equations are

$$\frac{dx}{ds} = p_x + \frac{K}{2}y ,$$

$$\frac{dp_x}{ds} = -\frac{K^2}{4}x + \frac{K}{2}p_y ,$$

$$\frac{dy}{ds} = p_y - \frac{K}{2}x ,$$

$$\frac{dp_y}{ds} = -\frac{K^2}{4}y - \frac{K}{2}p_x .$$

This coupled equations can be solved to yield the 4×4 matrix map for the coordinates (x, p_x, y, p_y) over the solenoid of length L,

$$M_{\text{solenoid}} = \begin{bmatrix} \frac{1+C}{2} & \frac{S}{K} & \frac{S}{2} & \frac{1-C}{K} \\ -\frac{KS}{4} & \frac{1+C}{2} & -\frac{K(1-C)}{4} & \frac{S}{2} \\ -\frac{S}{2} & -\frac{1-C}{K} & \frac{1+C}{2} & \frac{S}{K} \\ \frac{K(1-C)}{4} & -\frac{S}{2} & -\frac{KS}{4} & \frac{1+C}{2} \end{bmatrix} , \tag{4.124}$$

with $C = \cos KL$, $S = \sin KL$.

With $KL = \pi$, for example, it is a phase space exchanger with

$$\begin{bmatrix} 0 & 0 & 0 & \beta \\ 0 & 0 & -\frac{1}{\beta} & 0 \\ 0 & -\beta & 0 & 0 \\ \frac{1}{\beta} & 0 & 0 & 0 \end{bmatrix} ,$$

where $\beta = \frac{2L}{\pi}$. Note that this phase space exchanger map is different from that discussed in Homework 4.19, which cannot be produced by a solenoid alone. When $KL = 2\pi$, the solenoid gives a unit map I.

It is important to note that we have been describing the motion in terms of the canonical coordinates (x, p_x, y, p_y). The equation of motion in a solenoid can also be described in terms of the mechanical coordinates (x, x', y, y'). If so, and if we are interested in knowing how the particle moves inside the solenoid, then we need to pay attention to the fringe fields. The effect of fringe fields is in fact exactly what makes the transformation from p_x, p_y coordinates to the x', y' coordinates as a particle enters the solenoid, and transforms back as it leaves.

Outside the solenoid, however, $K = 0$ and there is no difference between the mechanical momenta x', y' and the canonical momenta p_x, p_y. The map for (x, x', y, y') from the entrance of a solenoid to its exit, both considered outside of the solenoid, is given by the matrix M_{solenoid} of Eq. (4.124) without having to worry about the solenoid fringe fields.

Nonlinear map of drift space   We did not include drift space in Eqs. (4.119–4.120). As mentioned, in the context of Lie algebra, the exact Hamiltonian is not so critical an issue. As long as it gives the ingredients required for the investigation, it can be approximated without risking fundamental principles. With this in mind, we have somewhat casually adopted the Hamiltonians (4.119) for the 6-D dynamics and (4.122) for a reasonable approximation. One can of course also choose more accurate Hamiltonians to obtain more accurate results.

Incidentally, we stay close to the Hamiltonian dynamics and Lie algebra in our analysis to obtain Eq. (4.124) to avoid an error-prone process to establish the equations of motion from the kinematics without the help of the Hamiltonian. Drift space is a surprisingly tricky nonlinear beamline element. See also Homeworks 4.49 and 4.52 for comments on the nonlinear effects of a quadrupole magnet.

Take a drift space for example. A more accurate Hamiltonian can be written as

$$H(x, p_x, y, p_y, z, \delta, s) = \frac{\delta}{\beta_0} - \frac{1}{\beta_0}\sqrt{1 - \beta_0^2 + \beta_0^2(1 + \delta)^2 - \beta_0(p_x^2 + p_y^2)},$$

where $\delta = \frac{\Delta P}{P_0}$, and $P_0 = \frac{mc\beta_0}{\sqrt{1-\beta_0^2}}$. This Hamiltonian approaches Eqs. (4.119) and (4.122) in appropriate limits.

The Hamilton equations are nonlinear,

$$x' = \frac{p_x}{\sqrt{1 - \beta_0^2 + \beta_0^2(1 + \delta)^2 - \beta_0(p_x^2 + p_y^2)}},$$

$$p_x' = 0,$$

$$y' = \frac{p_y}{\sqrt{1 - \beta_0^2 + \beta_0^2(1 + \delta)^2 - \beta_0(p_x^2 + p_y^2)}},$$

$$p_y' = 0,$$

$$z' = \frac{1}{\beta_0} - \frac{(1 + \delta)\beta_0}{\sqrt{1 - \beta_0^2 + \beta_0^2(1 + \delta)^2 - \beta_0(p_x^2 + p_y^2)}},$$

$$\delta' = 0.$$

Note the difference between x', y' and p_x, p_y.

The three momenta are conserved in a drift space. One consequence is that the exponential Lie series terminates, and we obtain closed form result for the drift space map,

$$e^{:-LH:}x = x + \frac{p_x L}{\sqrt{1 - \beta_0^2 + \beta_0^2(1+\delta)^2 - \beta_0(p_x^2 + p_y^2)}},$$

$$e^{:-LH:}p_x = p_x,$$

$$e^{:-LH:}y = y + \frac{p_y L}{\sqrt{1 - \beta_0^2 + \beta_0^2(1+\delta)^2 - \beta_0(p_x^2 + p_y^2)}},$$

$$e^{:-LH:}p_y = p_y,$$

$$e^{:-LH:}z = z + L\left(\frac{1}{\beta_0} - \frac{(1+\delta)\beta_0}{\sqrt{1 - \beta_0^2 + \beta_0^2(1+\delta)^2 - \beta_0(p_x^2 + p_y^2)}}\right),$$

$$e^{:-LH:}\delta = \delta.$$

Taylor map expansion   As mentioned, given the Hamiltonian, the Lie map is simply $e^{:-LH:}$. Homeworks 4.49 and 4.50 give the matrix representations of the linear maps for quadrupoles and dipoles. On the other hand, although these two cases are applied to linear cases, the power of Lie technique lies really in nonlinear cases. Take the sextupole case in Eq. (4.119) as an illustration. With the approximation (4.121), we first work out the following, for $\delta = 0$,

$$:H:x = -\frac{\partial H}{\partial p_x} = -p_x,$$

$$:H:^2x = -:H:p_x = -\frac{\partial H}{\partial x} = -S(x^2 - y^2),$$

$$:H:^3x = -S\left(-\frac{\partial H}{\partial p_x}2x + \frac{\partial H}{\partial p_y}2y\right) = 2S(xp_x - yp_y),$$

$$:H:^4x = 2S\left(-\frac{\partial H}{\partial p_x}p_x + x\frac{\partial H}{\partial x} + \frac{\partial H}{\partial p_y}p_y - \frac{\partial H}{\partial y}y\right)$$

$$= 2S[-p_x^2 + p_y^2 + Sx(x^2 + y^2)],$$

$$:H:^5x = \mathcal{O}(S^2).$$

We then obtain the Taylor map

$$e^{:-LH:}x = x + p_x L - \frac{1}{2}SL^2(x^2 - y^2) - \frac{1}{3}SL^3(xp_x - yp_y)$$

$$+ \frac{1}{12}SL^4[-p_x^2 + p_y^2 + Sx(x^2 + y^2)] + \mathcal{O}(S^2L^5),$$

where $\mathcal{O}(S^2L^5)$ means terms same-or-higher order than S^2 in S *and* same-or-higher order than L^5 in L. Similarly, we have

$$
\begin{aligned}
e^{:-LH:}p_x &= p_x - SL(x^2-y^2) - SL^2(xp_x-yp_y) - \frac{1}{3}SL^3[p_x^2-p_y^2-Sx(x^2+y^2)] \\
&\quad + \frac{1}{12}S^2L^4(5x^2p_x - y^2p_x + 6xyp_y) + \mathcal{O}(S^2L^5)\,, \\
e^{:-LH:}y &= y + Lp_y + SL^2xy + \frac{1}{3}SL^3(xp_y + yp_x) \\
&\quad + \frac{1}{12}SL^4[2p_xp_y + Sy(x^2+y^2)] + \mathcal{O}(S^2L^5)\,, \\
e^{:-LH:}p_y &= p_y + 2SLxy + SL^2(xp_y+yp_x) + \frac{1}{3}SL^3[2p_xp_y+Sy(x^2+y^2)] \\
&\quad - \frac{1}{12}S^2L^4(x^2p_y - 6xyp_x - 5y^2p_y) + \mathcal{O}(S^2L^5)\,.
\end{aligned}
$$

So we have already obtained some useful results — in this case, a Taylor map for a thick sextupole magnet — on nonlinear dynamics! Extension to higher orders is straightforward. Note that no explicit use of the equations of motion has been made. These expressions are of course approximate and nonsymplectic if the higher order terms are truncated. Similar example applications can be found in Homeworks 4.52, 4.53 and 4.55.

Homework 4.48 Find the explicit 6-D equations of motion for the Hamiltonians given in Eqs. (4.119) and (4.120). We include this homework for completeness. As mentioned in the text, we try to apply Lie maps and avoid integrating the equations of motion.

Solution

(a) For a dipole,

$$
\begin{aligned}
x' &= \frac{\partial H}{\partial p_x} = \frac{p_x}{1+\delta}\,, \\
p_x' &= -\frac{\partial H}{\partial x} = \frac{\delta}{\rho} - \frac{x}{\rho^2}\,, \\
y' &= \frac{\partial H}{\partial p_y} = \frac{p_y}{1+\delta}\,, \\
p_y' &= -\frac{\partial H}{\partial y} = 0\,, \\
z' &= \frac{\partial H}{\partial \delta} = -\frac{x}{\rho} - \frac{p_x^2+p_y^2}{2(1+\delta)^2} = -\frac{x}{\rho} - \frac{1}{2}(x'^2 + y'^2)\,, \\
\delta' &= \frac{\partial H}{\partial z} = 0 \quad \Longrightarrow \quad \delta = \text{const}\,.
\end{aligned}
\tag{4.125}
$$

Note that under this Hamiltonian, $p_x \neq x'$ and $p_y \neq y'$. One may combine Eq. (4.125) to yield

$$
x'' + \frac{x}{(1+\delta)\rho^2} = \frac{\delta}{(1+\delta)\rho}, \qquad y'' = 0\,.
\tag{4.126}
$$

Closed form expressions of $x(s)$ and $y(s)$ can be found by solving Eq. (4.126).

(b) For a quadrupole,

$$x'' + \frac{K}{1+\delta}x \;=\; 0, \qquad y'' - \frac{K}{1+\delta}y \;=\; 0.$$

(c) For a sextupole,

$$x'' + \frac{S}{1+\delta}(x^2 - y^2) \;=\; 0, \qquad y'' - \frac{2S}{1+\delta}xy \;=\; 0.$$

(d) For an octupole,

$$x'' + \frac{\lambda}{1+\delta}(x^3 - 3xy^2) \;=\; 0, \qquad y'' + \frac{\lambda}{1+\delta}(y^3 - 3x^2 y) \;=\; 0.$$

(e) For a general multipole, we have the longitudinal component of the vector potential

$$A_s \;=\; -\frac{1}{n+1}\frac{P_0}{e}\,\mathrm{Re}[(\lambda_n + i\bar\lambda_n)(x + iy)^{n+1}],$$

which gives a magnetic field

$$B_y + iB_x \;=\; \frac{P_0}{e}(\lambda_n + i\bar\lambda_n)(x + iy)^n.$$

The Hamiltonian (4.120) is

$$H(x, p_x, y, p_y, z, \delta, s) \;=\; \frac{eA_s}{cP_0} + \frac{1}{2(1+\delta)}(p_x^2 + p_y^2),$$

and the equations of motion are

$$x'' + \frac{1}{1+\delta}\,\mathrm{Re}[(\lambda_n + i\bar\lambda_n)(x + iy)^n] \;=\; 0,$$

$$y'' - \frac{1}{1+\delta}\,\mathrm{Im}[(\lambda_n + i\bar\lambda_n)(x + iy)^n] \;=\; 0.$$

Homework 4.49 Given the Hamiltonian (4.119) of a quadrupole, find the matrix representation of the map $\exp(:-LH:)$ for the vector (x, p_x, y, p_y).

Solution Write the map as $\exp(:f_2:)$, where δ is regarded as a constant. Since x and y motions are decoupled, we can treat them separately with $f_2 = f_{2x} + f_{2y}$ — and note that $:f_{2x}:$ and $:f_{2y}:$ commute. For the x motion, f_{2x} can be written as Eqs. (4.57) and (4.60) with

$$a \;=\; KL, \qquad b \;=\; 0, \qquad c \;=\; \frac{L}{1+\delta}.$$

The matrix form can be obtained by applying Eq. (4.63). Similarly we obtain the matrix for the y motion. By combining the x and y results, we obtain the matrix representation of the map, assuming $K > 0$ and defining $\theta = \sqrt{\frac{KL^2}{1+\delta}}$,

$$\begin{bmatrix} \cos\theta & \frac{L}{(1+\delta)\theta}\sin\theta & 0 & 0 \\ -\frac{(1+\delta)\theta}{L}\sin\theta & \cos\theta & 0 & 0 \\ 0 & 0 & \cosh\theta & \frac{L}{(1+\delta)\theta}\sinh\theta \\ 0 & 0 & \frac{(1+\delta)\theta}{L}\sinh\theta & \cosh\theta \end{bmatrix}. \tag{4.127}$$

Note that had we used (x, x', y, y') as vector, the transformation matrix would be slightly different,

$$\begin{bmatrix} \cos\theta & \frac{L}{\theta}\sin\theta & 0 & 0 \\ -\frac{\theta}{L}\sin\theta & \cos\theta & 0 & 0 \\ 0 & 0 & \cosh\theta & \frac{L}{\theta}\sinh\theta \\ 0 & 0 & \frac{\theta}{L}\sinh\theta & \cosh\theta \end{bmatrix}. \tag{4.128}$$

Using the dynamical variables p_x and p_y, the map (4.127) is actually nonlinear in δ.

Note also that the path length variation, determined by the z' equation, is a nonlinear effect, which cannot be represented by a matrix. The same nonlinear effect appears in a drift space as discussed in the text. In this sense, drift spaces and quadrupoles are nonlinear elements. These nonlinear effects, however, are small, and it is often possible to ignore them by making the approximation (4.121).

Homework 4.50 Repeat the above Homework for dipole magnets for the vector $(x, p_x, y, p_y, z, \delta)$. In this 6-D problem, treat δ as a dynamical variable instead of a constant. Ignore nonlinear effects by making the approximation (4.121).

Solution The y-motion is separated from the x- and z-motions. To describe the y-motion, we have

$$f_{2y} = -\frac{L}{2}p_y^2, \tag{4.129}$$

which is just equivalent to a drift space. The remaining f_2 that describes the x- and z-motions is

$$f_2 = \frac{L}{\rho}x\delta - \frac{L}{2\rho^2}x^2 - \frac{L}{2}p_x^2. \tag{4.130}$$

Given Eqs. (4.129–4.130), the 6×6 transformation matrix is found to be

$$\begin{bmatrix} \cos\left(\frac{L}{\rho}\right) & \rho\sin\left(\frac{L}{\rho}\right) & 0 & 0 & 0 & \rho - \rho\cos\left(\frac{L}{\rho}\right) \\ -\frac{1}{\rho}\sin\left(\frac{L}{\rho}\right) & \cos\left(\frac{L}{\rho}\right) & 0 & 0 & 0 & \sin\left(\frac{L}{\rho}\right) \\ 0 & 0 & 1 & L & 0 & 0 \\ 0 & 0 & 0 & 1 & 0 & 0 \\ -\sin\left(\frac{L}{\rho}\right) & -\rho + \rho\cos\left(\frac{L}{\rho}\right) & 0 & 0 & 1 & -L + \rho\sin\left(\frac{L}{\rho}\right) \\ 0 & 0 & 0 & 0 & 0 & 1 \end{bmatrix}.$$

Homework 4.51 The Hamiltonian for a solenoid is given by Eq. (4.123).

(a) Show that

$$[H, xp_y - yp_x] = 0.$$

This result means that the angular momentum $xp_y - yp_x$ is conserved by the solenoid, i.e. its value is the same at the entrance, at the exit, as well as throughout its passage of the solenoid as long as you remember to set $p_x = x' - \frac{K}{2}y, p_y = y' + \frac{K}{2}x$. Note that the angular momentum is given by $xp_y - yp_x$, not by $xy' - yx'$.

(b) Show that

$$p_x^2 + p_y^2 + \frac{K^2}{4}(x^2 + y^2)$$

is also a constant of the motion.

Homework 4.52 Find the path length change Δz as a particle passes through a quadrupole of strength K and length L to order $\mathcal{O}(L^4)$.

Solution At the exit of the quadrupole, we have

$$
\begin{aligned}
z(L) &= e^{:-LH:}z\big|_{X=X_0} \\
&= \Big\{ z - L[H, z] + \frac{L^2}{2}[H, [H, z]] - \frac{L^3}{6}[H, [H, [H, z]]] \\
&\quad + \frac{L^4}{24}[H, [H, [H, [H, z]]]] + \mathcal{O}(L^5) \Big\}_{X=X_0}.
\end{aligned}
$$

With H given in Eq. (4.119), we find

$$
\begin{aligned}
[H, z] &= \frac{1}{2(1+\delta)^2}(p_x^2 + p_y^2), \\
[H, [H, z]] &= \frac{K}{(1+\delta)^2}(xp_x - yp_y), \\
[H, [H, [H, z]]] &= \frac{K^2}{(1+\delta)^2}(x^2 + y^2) - \frac{K}{(1+\delta)^3}(p_x^2 - p_y^2), \\
[H, [H, [H, [H, z]]]] &= -\frac{4K^2}{(1+\delta)^3}(xp_x + yp_y),
\end{aligned}
$$

$$
\begin{aligned}
\implies \Delta z = z(L) - z_0 &= -\frac{L}{2}(x_0'^2 + y_0'^2) + \frac{KL^2}{2(1+\delta)}(x_0 x_0' - y_0 y_0') \\
&\quad - \frac{K^2 L^3}{6(1+\delta)^2}(x_0^2 + y_0^2) + \frac{KL^3}{6(1+\delta)}(x_0'^2 - y_0'^2) \\
&\quad - \frac{4K^2 L^4}{24(1+\delta)^2}(x_0 x_0' + y_0 y_0') + \mathcal{O}(L^5).
\end{aligned}
\tag{4.131}
$$

Equations (4.128) and (4.131) constitute a Taylor map for the quadrupole.

Homework 4.53 Find the on-momentum Taylor map, to order $\mathcal{O}(\lambda)$, for (x, p_x, y, p_y) for a thick octupole of strength λ and length L.

Solution The problem asks for the map of the thick octupole. A thin-lens kick is not the right answer. The small parameter is λ; L is not necessarily small; all orders in L are to be kept. The Lie map is

$$e^{-:LH:}, \quad \text{where} \quad H = \frac{1}{2}(p_x^2 + p_y^2) + \frac{\lambda}{4}(x^4 - 6x^2y^2 + y^4).$$

The result is

$$
\begin{aligned}
e^{-:LH:}x &= x + Lp_x - \frac{L^2\lambda}{2}(x^3 - 3xy^2) + \frac{L^3\lambda}{2}[p_x(y^2 - x^2) + 2xyp_y] \\
&\quad + \frac{L^4\lambda}{4}(-xp_x^2 + xp_y^2 + 2yp_xp_y) + \frac{L^5\lambda}{20}(3p_xp_y^2 - p_x^3) + \mathcal{O}(\lambda^2), \\
e^{-:LH:}p_x &= p_x - L\lambda(x^3 - 3xy^2) + \frac{3L^2\lambda}{2}[p_x(y^2 - x^2) + 2xyp_y] \\
&\quad - L^3\lambda(xp_x^2 - xp_y^2 - 2yp_xp_y) + \frac{L^4\lambda}{4}(3p_xp_y^2 - p_x^3) + \mathcal{O}(\lambda^2).
\end{aligned}
\tag{4.132}
$$

Results for the y-dimension can be obtained by switching x and y in Eq. (4.132).

Homework 4.54 It often occurs when the Hamiltonian cleanly splits into components, e.g.,

$$H(x, p_x, y, p_y, t) = H(x, p_x, t) + H_y(y, p_y, t).$$

Convince yourself that in this case, the Lie map $e^{:-LH:}$ describes a x-y decoupled dynamics. This property is the basis for the fact — so far taken for granted — that when additional degrees of freedom is taken into account, all we need to do is to add additional terms linearly to the Hamiltonian, and the original dynamics can stay unaffected. See also Homework 1.13.

Homework 4.55 Find the on-momentum Taylor map, to order $\mathcal{O}(L^3)$, for (x, p_x, y, p_y) for a thick octupole of strength λ and length L.

4.4.2 Chain of elements

So far we have applied Lie operators for single elements. An accelerator typically consists of a sequence of elements. Let the elements be ordered according to the index $i = 1, 2, \ldots, N$ so that the beam passes sequentially through element 1, then element 2, etc. Let the i-th element have length L_i and Hamiltonian H_i. The position at the entrance to element 1 is identified as $s = 0$. The map for the i-th element is $e^{:-L_iH_i:}$. The one-trip map from the entrance of the first element to the exit of the last element is — again, note the ordering of the operators [see Eq. (4.52)],

$$e^{:-L_1H_1:}e^{:-L_2H_2:} \cdots e^{:-L_NH_N:}, \tag{4.133}$$

which we shall abbreviate as

$$\prod_{i=1}^{N} e^{:-L_iH_i:}. \tag{4.134}$$

Figure 4.3: The accelerator model viewed in Lie language.

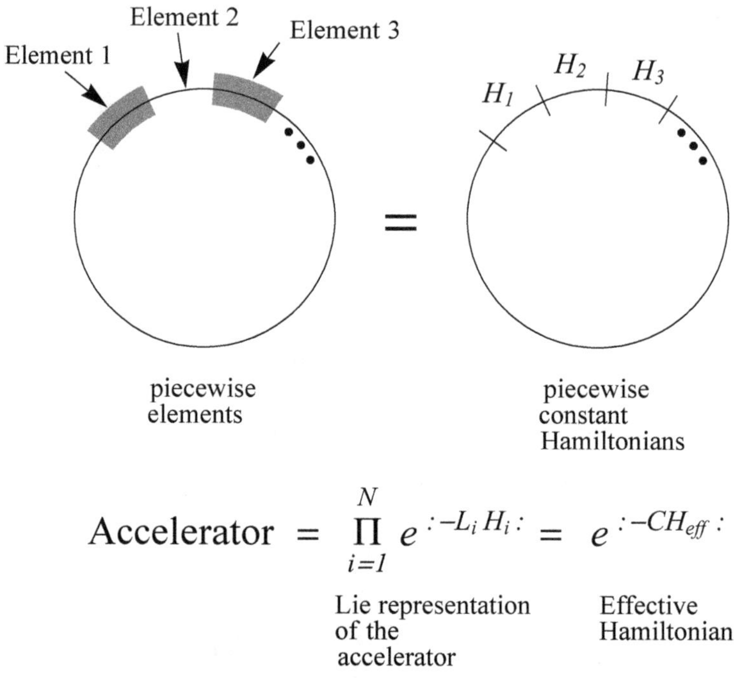

$$\text{Accelerator} \ = \ \prod_{i=1}^{N} e^{\,:-L_i H_i:} \ = \ e^{\,:-CH_{\text{eff}}:}$$

In case the N elements string into a circular accelerator, the map (4.133) becomes a one-turn map. The situation is illustrated in Fig. 4.3. The total one-turn map of the accelerator is obtained by multiplying a sequence of exponential operators. A simple example representing a single thin-lens sextupole in an otherwise perfectly linear accelerator was discussed following Eqs. (4.88–4.89).

Effective Hamiltonian The forms (4.133) or (4.134) are not yet very useful. To proceed, we would like to have the total map represented as a single exponential operator instead of a product of exponential operators. The process of combining exponential operators is called *concatenation*. After concatenation, the single exponential operator can be written as

$$\text{one turn map} \ = \ e^{:-CH_{\text{eff}}:}, \tag{4.135}$$

where C is the accelerator circumference.

Drawing an analogy of Eq. (4.135) to Eq. (4.118) for a single element allows the quantity H_{eff} to be identified as the *effective Hamiltonian* of the one-turn map. It describes the one-turn motion of particles in the system. It is a function of the dynamical variables X, but does not depend on s, i.e. it is time-independent — the same effective Hamiltonian applies turn after turn. As a consequence, H_{eff} is a constant of the motion when observed at the position

$s = 0$ turn after turn. When (or if) one has obtained the one-turn map in the form (4.135) by concatenation, it is impressive that a complex system of an accelerator, which consists of many pieces each having its own Hamiltonian, can be combined into a simple form (4.135) and a constant of the motion can be found this way. The jargon is that the system has been integrated.

In Fig. 4.3 and Eq. (4.135), we have assumed the total one-turn map is sufficiently closely connected to the identity map so that H_{eff} exists. In Homework 4.19 we introduced an example symplectic map (that of a phase space exchanger) that was disjoint from the identity map, so that the effective Hamiltonian was purely imaginary, and strictly speaking is outside of our description here. In case the one-turn map indeed contains a phase space exchanger, then the phase spaces get exchanged every turn. Our effective Hamiltonian description then has to be applied only when the particle motion is observed every two turns.

Another situation when H_{eff} does not exist is when the system is not integrable. In such cases, there is no constant of the motion, H_{eff} does not exist. This occurs notably when the phase space motion is chaotic. We shall not deal with the nonintegrable cases in our lectures. Underlying the Lie algebra technique in all our following developments is the assumption that the perturbative approach converges; the system is assumed integrable.

It is important to recognize that the quantity H_{eff} is not only a constant of the motion, but is also the Hamiltonian, which holds much more powerful implications. One can actually derive the equations of motion by applying the Hamilton equations to it. The condition is that the equations obtained this way apply only if one observes the motion at position $s = 0$. A third significance of H_{eff} is that it can be used to execute normal form analysis, which leads to analytic expressions of various physical quantities important to repetitive systems such as storage rings. To recap, H_{eff} has the following physical meanings.

- It is an invariant;
- It is a Hamiltonian;
- It permits normal form analysis.

Comparison of Lie and Taylor maps We compared the Lie maps and the Taylor maps before in Sec. 4.2. Perhaps we can add another comment here, as in Table 4.6.

In the Lie language, the single-element maps are easy — all we need to know is the Hamiltonian and the length of the element. The one-turn map, once obtained, also gives easy results because all is contained in the effective Hamiltonian, including the analyses of normal forms to be discussed later. What is difficult is the middle concatenation step. This middle step invokes the complicated BCH formula of various forms.

In contrast, in the Taylor language, single-element maps are difficult because they are to be obtained from explicitly solving the equation of motion inside the element. The final result is easily adopted for numerical calculations, but as a

Table 4.6: A comparison between the Lie and Taylor maps.

	Taylor map	Lie map
Single-element map	laborious integrating eq. of motion	easy $e^{-:L_iH_i:}$
Concatenation	straightforward	laborious BCH formula
Application	one-passage for linacs straightforward time domain nonsymplectic numerical tracking	one-turn for storage rings effective Hamiltonian frequency domain symplectic normal form analysis

map is nonsymplectic in general and does not allow analyses unless symplectified. On the other hand, the middle concatenation step is straightforward.

4.4.3   BCH formula

The basic formula that allows concatenation of exponential Lie operators is called the Baker–Campbell–Hausdorff (BCH) formula. The BCH formula is extensively used when the ordering of two Lie operators are to be switched. As mentioned before, when their ordering is to be switched, you expect to see Poisson brackets — see Eq. (4.36). Poisson brackets are the residual left-overs when two operators are forced to switch order. Two operators can be freely switched only if their Poisson bracket vanishes. BCH formula will involve lots of operations of Poisson brackets — Lie algebra at work.

We know that matrix multiplications cannot be switched around, they are not commutative, or not abelian. When applied to matrices, in fact, BCH formula can be used to calculate the residual of switching order in a matrix multiplication as well.

The BCH formula is not going to be trivial; it is not even easily intuitive. A first appreciation of this fact can be obtained by considering two 2×2 matrices A and B. The commutator $AB - BA$, the residual after their ordering is switched, can be worked out explicitly, and it will be found to be rather complicated and does not have an easy sign of a hidden formula. The fact however is that there is a hidden formula, the BCH formula — see Homework 4.68.

In this section, we shall introduce the BCH formula in four different forms, each is more convenient to be applied in different occasions although they come from the one and the same formula.

First form The first form of the BCH formula reads

$$e^{:f:}e^{:g:} \;=\; e^{:h:}\,, \tag{4.136}$$

where f and g are arbitrary functions of the dynamical variables X, and

$$\begin{aligned}
h \;=\;& f + g + \frac{1}{2}:f{:}g + \frac{1}{12}:f{:}^2 g + \frac{1}{12}:g{:}^2 f + \frac{1}{24}:f{::}g{:}^2 f \\
& - \frac{1}{720}:g{:}^4 f - \frac{1}{720}:f{:}^4 g + \frac{1}{360}:g{::}f{:}^3 g + \frac{1}{360}:f{::}g{:}^3 f \\
& + \frac{1}{120}:f{:}^2{:}g{:}^2 f + \frac{1}{120}:g{:}^2{:}f{:}^2 g + \mathcal{O}((f,g)^6)\,,
\end{aligned} \tag{4.137}$$

where $\mathcal{O}((f,g)^6)$ means terms of the order of $f^m g^n$ where $m + n \geq 6$.

It is easy to expect that the lowest order approximation of h is $f + g$. That explains the two leading terms on the right-hand-side of Eq. (4.137). The reader should note the fact that the residual of h from $f + g$ is fully consisting of the Poisson brackets of f and g. There are no noises from other higher order terms like f^2, fg, g^2, etc., which do not form Poisson brackets. This is one of the peculiarities that the BCH formula is telling us.

If we use the commutator notation of Eq. (4.35) and apply Table 4.5, we can rewrite Eqs. (4.136–4.137) as

$$\begin{aligned}
e^{:f:}e^{:g:} \;=\; \exp\Big[& :f: + :g: + \frac{1}{2}\{:f:, :g:\} + \frac{1}{12}\{:f:, \{:f:, :g:\}\} + \frac{1}{12}\{:g:, \{:g:, :f:\}\} \\
& + \frac{1}{24}\{:f:, \{:g:, \{:g:, :f:\}\}\} - \frac{1}{720}\{:g:, \{:g:, \{:g:, \{:g:, :f:\}\}\}\} \\
& - \frac{1}{720}\{:f:, \{:f:, \{:f:, \{:f:, :g:\}\}\}\} + \frac{1}{360}\{:g:, \{:f:, \{:f:, \{:f:, :g:\}\}\}\} \\
& + \frac{1}{360}\{:f:, \{:g:, \{:g:, \{:g:, :f:\}\}\}\} + \frac{1}{120}\{:f:, \{:f:, \{:g:, \{:g:, :f:\}\}\}\} \\
& + \frac{1}{120}\{:g:, \{:g:, \{:f:, \{:f:, :g:\}\}\}\} + :\mathcal{O}((f,g)^6):\Big]\,.
\end{aligned} \tag{4.138}$$

The function h is expressed in terms of the Poisson brackets of the functions f and g. The operator $:h:$ is expressed in terms of the commutators of the operators $:f:$ and $:g:$.

The proof of (4.136–4.138) is somewhat technical. To loosen the load on the text, it is moved to Homework 4.57. There does not seem to be an easy rule of the coefficients in Eqs. (4.137–4.138). Equations (4.136–4.138) are our first form of the BCH formula. It applies assuming the two functions f and g are small so that the power series converges.[19] Small f and g means the two operators $e^{:f:}$ and $e^{:g:}$ differ only slightly from identity maps.

When $:f:$ and $:g:$ commute (or when $[f, g] = \text{constant}$ depending only on s), we simply have $h = f + g$, as discussed in Eq. (4.42).

---

[19]There is no guarantee of convergence just because f and g are small, but f and g must be small so there is a chance of convergence.

Second form   The first form has an important variation, which gives the second form of the BCH formula. It is obtained by summing the infinite power series in the first BCH form over either the function f or the function g. If summed over g, then to 1st order in f, we have

$$e^{:f:}e^{:g:} \;=\; \exp\left[:g + \left(\frac{:g:}{e^{:g:}-1}\right)f + \mathcal{O}(f^2):\right]. \tag{4.139}$$

If we choose to sum over f, the summation can be done analytically to 1st order in g to yield

$$e^{:f:}e^{:g:} = \exp\left[:f + \left(\frac{:f:}{1-e^{-:f:}}\right)g + \mathcal{O}(g^2):\right]. \tag{4.140}$$

The proof of Eqs. (4.139–4.140) are given in Homeworks 4.58 and 4.59.

Equation (4.140) applies when the function g is small, but f does not have to be small. This expression is particularly useful when we try to concatenate the map for a small perturbation with the map for the rest of the accelerator. In that case, we would obviously choose g to be the perturbation, and f to be the rest of the accelerator. The price to pay in order for f not having to be small is that we have an analytic expression only to first order in g.

The operators $\frac{:g:}{e^{:g:}-1}$ and $\frac{:f:}{1-e^{-:f:}}$ in Eqs. (4.139) and (4.140) deserve some attention. Physically they introduce small denominators when $e^{:g:}-1$ and $1-e^{-:f:}$ get to operate on an eigenfunction of $:g:$ or $:f:$. If that happens, the factors $\frac{:g:}{e^{:g:}-1}$ and $\frac{:f:}{1-e^{-:f:}}$ will contribute terms with small denominators. As we will see later, these small denominators are the origin of resonances.

There is actually an analytic expression that extends Eq. (4.140) to second order in g. The proof is given in Homework 4.60. The term $\sim \mathcal{O}(g^2)$ in Eq. (4.140) is then given by

$$\frac{1}{2}\left(\frac{:f:}{1-e^{-:f:}}\right)^2 \int_0^1 du \int_0^u dv \left[e^{-v:f:}g, e^{-u:f:}g\right]. \tag{4.141}$$

A similar statement can be made with Eq. (4.139).

Map factorization   The BCH formula describes how to concatenate two exponential operators into one. It can be inverted to give a formula that describes how to factorize an exponential operator into a product of two exponential operators. In particular, Eq. (4.140) can be inverted to give

$$e^{:f+h:} \;=\; e^{:f:}\exp\left[:\left(\frac{1-e^{-:f:}}{:f:}\right)h:\right]e^{:\mathcal{O}(h^2):}. \tag{4.142}$$

In actual calculations, the operation $\frac{1-e^{-:f:}}{:f:}$ is not easy to perform — Taylor expansion in terms of $:f:$ gives an infinite series. In that case, one may prefer an alternative expression for Eq. (4.142) by noting

$$\left(\frac{1-e^{-:f:}}{:f:}\right)h(X) = \left(\int_0^1 du\, e^{-u:f:}\right)h(X) = \int_0^1 du\, h(e^{-u:f:}X)$$

$$\implies \quad e^{:f+h:} = e^{:f:}\exp\left[:\int_0^1 du\, e^{-u:f:}h:\right]e^{:\mathcal{O}(h^2):}. \tag{4.143}$$

What is the physical meaning of Eq. (4.143)? A kick by a slice of h at location $s = uL$ is equivalent to a kick $h(e^{-:uf:}X)$ at the beginning of the element. The effect of $e^{-:uf:}$ is just to transform the slice kick to the beginning of the element. The integration is then to sum up these kicks of the slices. This process is obviously valid only to 1st order in h. The factorization forms (4.142) and (4.143) are useful when a small perturbation term h is to be factored out from a large term f.

Third form A third form of the BCH formula takes a form similar to the first form except that it is more symmetric looking,

$$e^{:f:}e^{:g:}e^{:f:} \;=\; e^{:h:}, \tag{4.144}$$

where

$$
\begin{aligned}
h \;=\; & 2f + g + \frac{1}{6}{:}g{:}^2 f - \frac{1}{6}{:}f{:}^2 g + \frac{7}{360}{:}f{:}^4 g - \frac{1}{360}{:}g{:}^4 f \\
& + \frac{1}{90}{:}f{::}g{:}^3 f + \frac{1}{45}{:}g{::}f{:}^3 g - \frac{1}{60}{:}f{:}^2{:}g{:}^2 f + \frac{1}{30}{:}g{:}^2{:}f{:}^2 g + \mathcal{O}((f,g)^7).
\end{aligned}
\tag{4.145}
$$

The symmetry has the consequence that there are no even-order terms on the right-hand-side of Eq. (4.145). The symmetry simplifies the algebra. In practice, Eqs. (4.144–4.145) apply when the particle motion is observed at the middle of some element in the accelerator. The proof of Eqs. (4.144–4.145) is provided in Homework 4.61.

Fourth form Finally the fourth form of the BCH formula is when the right-hand-side of Eq. (4.145) is summed over g while keeping only to 1st order in f. This yields

$$
\begin{aligned}
e^{:f:}e^{:g:}e^{:f:} \;=\; & \exp\left[{:}g{:} + {:}\left(\frac{{:}g{:}}{e^{:g:} - 1}\right)f{:} + {:}\mathcal{O}(f^2){:}\right]e^{:f:} \\
\;=\; & \exp\left[{:}g{:} + {:}\left(\frac{{:}g{:}}{e^{:g:} - 1}\right)f{:} + {:}\left(\frac{{:}g{:}}{1 - e^{-:g:}}\right)f{:} + {:}\mathcal{O}(f^2){:}\right] \\
\;=\; & \exp\left[{:}g{:} + {:}\left({:}g{:}\coth(\tfrac{{:}g{:}}{2})f\right){:} + {:}\mathcal{O}(f^2){:}\right].
\end{aligned}
\tag{4.146}
$$

In other words, if we let

$$e^{:f:}e^{:g:}e^{:f:} \;=\; e^{:h:}, \tag{4.147}$$

then

$$h \;=\; g + {:}g{:}\coth\left(\frac{{:}g{:}}{2}\right)f + \mathcal{O}(f^2). \tag{4.148}$$

If g is also small, we can expand Eq. (4.148) in power series in g. The result agrees with Eq. (4.145) as it should. In the first and second lines of Eq. (4.146), we have applied Eqs. (4.139) and (4.140), respectively.

One can also apply the BCH formula to find h in Eq. (4.147) when f is not small but g is small. The result is

$$
\begin{aligned}
e^{:f:}e^{:g:}e^{:f:} &= e^{:2f:}\exp(:e^{-:f:}g:) \\
&= \exp\left[:2f:+:\left(\frac{:2f:}{1-e^{-:2f:}}\right)(e^{-:f:}g):+\mathcal{O}(g^2)\right] \\
&= \exp\left[:2f:+:\left(\frac{:f:}{\sinh:f:}\right)g:+\mathcal{O}(g^2)\right].
\end{aligned}
\tag{4.149}
$$

Again, expanding for small f gives an expression consistent with Eq. (4.145).

We have thus introduced four forms of the BCH formula. The first form is given by Eqs. (4.136–4.138), the second form by (4.139–4.140), the third form by (4.144–4.145), the fourth by (4.147–4.149). In passing, we have also given the map factorization formula, Eq. (4.143). These formulae are applied often in later sections.

Homework 4.56 Consider a map M that depends on a parameter α, and can be written as $M(\alpha) = e^{:h(\alpha):}$ with some function $h(\alpha)$ [$h(\alpha)$ is also a function of X]. Show that

$$
\frac{dM}{d\alpha}M^{-1} = :\left(\frac{e^{:h:}-1}{:h:}\right)\frac{dh}{d\alpha}:.
\tag{4.150}
$$

Equation (4.150) is a useful formula. It gives an expression of $\frac{dM}{d\alpha}$ when the expression of an exponential map $M(\alpha)$ is known.

Solution The trick used in this homework deserves some attention. Consider the map

$$
M = e^{:\beta h(\alpha):},
$$

where the dummy variable β will be set to 1 later. Define

$$
A(\beta) = \frac{dM}{d\alpha}M^{-1}.
$$

Then we have

$$
\begin{aligned}
\frac{dA}{d\beta} &= \left(\frac{d}{d\alpha}\frac{dM}{d\beta}\right)M^{-1}+\frac{dM}{d\alpha}\frac{dM^{-1}}{d\beta} \\
&= \left[\frac{d}{d\alpha}(:h:M)\right]M^{-1}-\frac{dM}{d\alpha}M^{-1}:h: \\
&= \frac{d:h:}{d\alpha}+:h:A-A:h: \\
&= :\frac{dh}{d\alpha}:-:Ah:,
\end{aligned}
\tag{4.151}
$$

where the last step used Eq. (4.36). The solution to (4.151) is

$$
A = :\left(\frac{e^{\beta:h:}-1}{:h:}\right)\frac{dh}{d\alpha}:,
\tag{4.152}
$$

as can be seen by back substitution. Having established Eq. (4.152), Eq. (4.150) follows by setting $\beta = 1$.

As a simple example of this formula, consider the case $M = e^{:f(\alpha)h(x,p):}$. The reader is suggested to derive

$$\frac{dM}{d\alpha} M^{-1} = f'(\alpha) {:} h {:} \,.$$

You may follow the technique to find an expression for $M^{-1}\frac{dM}{d\alpha}$. The answer is given by Eq. (4.162).

Homework 4.57 Prove the first form of the BCH formula, Eqs. (4.136–4.138).

Solution Consider

$$e^{:\alpha f:} e^{:\alpha g:} = e^{:h(\alpha):} = M(\alpha), \tag{4.153}$$

which gives

$$\frac{dM}{d\alpha} M^{-1} = {:}f{:} + M{:}g{:}M^{-1} = {:}f + Mg{:}. \tag{4.154}$$

Using Eq. (4.150), we obtain

$$f + Mg = \left(\frac{e^{:h:} - 1}{:h:} \right) \frac{dh}{d\alpha}. \tag{4.155}$$

Let

$$h = \alpha h_1 + \alpha^2 h_2 + \alpha^3 h_3 + \alpha^4 h_4 + \alpha^5 h_5 + \cdots . \tag{4.156}$$

Expand (4.155) in powers of α, the left-hand-side is given by

$$\begin{aligned}
\text{LHS} &= f + e^{:h:} g = f + g + {:}h{:}g + \frac{1}{2}{:}h{:}^2 g + \frac{1}{6}{:}h{:}^3 g + \frac{1}{24}{:}h{:}^4 g + \mathcal{O}(\alpha^5) \\[4pt]
&= f + g + \alpha(:h_1:g) + \alpha^2 \left({:}h_2{:}g + \frac{1}{2}{:}h_1{:}^2 g \right) \\[4pt]
&\quad + \alpha^3 \left({:}h_3{:}g + \frac{1}{2}{:}h_1{::}h_2{:}g + \frac{1}{2}{:}h_2{::}h_1{:}g + \frac{1}{6}{:}h_1{:}^3 g \right) \\[4pt]
&\quad + \alpha^4 \left({:}h_4{:}g + \frac{1}{2}{:}h_1{::}h_3{:}g + \frac{1}{2}{:}h_2{:}^2 g + \frac{1}{2}{:}h_3{::}h_1{:}g \right. \\[4pt]
&\quad \left. + \frac{1}{6}{:}h_1{:}^2{:}h_2{:}g + \frac{1}{6}{:}h_2{::}h_1{:}^2 g + \frac{1}{6}{:}h_1{::}h_2{::}h_1{:}g + \frac{1}{24}{:}h_1{:}^4 g \right) + \mathcal{O}(\alpha^5).
\end{aligned} \tag{4.157}$$

The right-hand-side of Eq. (4.155) is

$$\begin{aligned}
\text{RHS} &= \left(1 + \frac{1}{2}{:}h{:} + \frac{1}{6}{:}h{:}^2 + \frac{1}{24}{:}h{:}^3 + \frac{1}{120}{:}h{:}^4 \right) \left(h_1 + 2\alpha h_2 \right. \\[4pt]
&\quad \left. + 3\alpha^2 h_3 + 4\alpha^3 h_4 + 5\alpha^4 h_5 \right) \\[4pt]
&= h_1 + 2\alpha h_2 + \alpha^2 \left(\frac{1}{2}{:}h_1{:}h_2 + 3h_3 \right) + \alpha^3 \left(\frac{1}{6}{:}h_1{:}^2 h_2 + {:}h_1{:}h_3 + 4h_4 \right)
\end{aligned}$$

$$+\alpha^4\left(5h_5+\frac{3}{2}:h_1:h_4+\frac{1}{3}:h_1:^2h_3+\frac{1}{2}:h_2:h_3-\frac{1}{6}:h_2:^2h_1+\frac{1}{24}:h_1:^3h_2\right).$$

$$(4.158)$$

Equating the α coefficients of both sides gives

$$\mathcal{O}(\alpha^0)\implies\quad h_1=f+g\,,$$

$$\mathcal{O}(\alpha^1)\implies\quad h_2=\frac{1}{2}:f:g\,,$$

$$\mathcal{O}(\alpha^2)\implies\quad h_3=\frac{1}{3}\left(:h_2:g+\frac{1}{2}:h_1:h_2\right)=\frac{1}{12}:f:^2g+\frac{1}{12}:g:^2f\,,$$

$$\mathcal{O}(\alpha^3)\implies\quad h_4=\frac{1}{4}\left(:h_3:g+\frac{1}{2}:h_1::h_2:g+\frac{1}{6}:h_1:^3g-\frac{1}{6}:h_1:^2h_2-:h_1:h_3\right)$$

$$=\frac{1}{24}:f::g:^2f\,,$$

$$\mathcal{O}(\alpha^4)\implies\quad h_5=\frac{1}{5}\left(:h_4:g+\frac{1}{2}:h_1::h_3:g+\frac{1}{2}:h_2:^2g+\frac{1}{2}:h_3::h_1:g+\frac{1}{6}:h_1:^2:h_2:g\right.$$

$$+\frac{1}{6}:h_2::h_1:^2g+\frac{1}{6}:h_1::h_2::h_1:g+\frac{1}{24}:h_1:^4g-\frac{3}{2}:h_1:h_4$$

$$\left.-\frac{1}{3}:h_1:^2h_3-\frac{1}{2}:h_2:h_3+\frac{1}{6}:h_2:^2h_1-\frac{1}{24}:h_1:^3h_2\right)$$

$$=-\frac{1}{720}:f:^4g+\frac{1}{360}:g::f:^3g+\frac{1}{120}:g:^2:f:^2g+\frac{1}{120}:f:^2:g:^2f$$

$$+\frac{1}{360}:f:g:^3f-\frac{1}{720}:g:^4f\,.\qquad(4.159)$$

In the above derivation, we have used the identity (4.39) in Homework 4.17. This completes the proof of the first form of BCH formula when α is set to 1.

Homework 4.58 Prove the second form of the BCH formula, Eq. (4.139).

Solution Consider

$$e^{\alpha:f:}e^{:g:}\ =\ e^{:h(\alpha):}\ =\ M$$

$$\implies\quad\frac{dM}{d\alpha}M^{-1}\ =\ :f:\,.$$

With Eq. (4.150), we have

$$f\ =\ \left(\frac{e^{:h:}-1}{:h:}\right)\frac{dh}{d\alpha}\,.\qquad(4.160)$$

Define

$$h\ =\ g+\alpha h_1+\alpha^2 h_2+\cdots\,.$$

To the leading order in α, Eq. (4.160) gives

$$f\ =\ \left(\frac{e^{:g:}-1}{:g:}\right)h_1\,,$$

or

$$h_1 = \left(\frac{:g:}{e^{:g:} - 1}\right) f \qquad \Longrightarrow \qquad \text{Q.E.D.}$$

Homework 4.59 Prove the other second form of the BCH formula, Eq. (4.140).

Solution Let

$$e^{:f:}e^{\alpha:g:} = e^{:h(\alpha):} = M$$
$$\Longrightarrow \qquad M^{-1}\frac{dM}{d\alpha} = :g: . \tag{4.161}$$

At the same time, it can be shown (Homework 4.56) that

$$M^{-1}\frac{dM}{d\alpha} = :\left(\frac{1 - e^{-:h:}}{:h:}\right)\frac{dh}{d\alpha}:. \tag{4.162}$$

Combining Eqs. (4.161–4.162), we have

$$g = \left(\frac{1 - e^{-:h:}}{:h:}\right)\frac{dh}{d\alpha}. \tag{4.163}$$

Define

$$h = f + \alpha h_1 + \alpha^2 h_2 + \cdots.$$

To the leading order in α, Eq. (4.163) gives

$$g = \left(\frac{1 - e^{-:f:}}{:f:}\right) h_1, \tag{4.164}$$

or

$$h_1 = \left(\frac{:f:}{1 - e^{-:f:}}\right) g \qquad \Longrightarrow \qquad \text{Q.E.D.} \tag{4.165}$$

Solution — alternative Start with Eq. (4.139) which is valid for small f. Sandwich it by $e^{:g:}$ from the left and $e^{-:g:}$ from the right. This yields

$$\begin{aligned}
e^{:g:}e^{:f:} &= \exp\left\{:e^{:g:}\left[g + \left(\frac{:g:}{e^{:g:} - 1}\right) f\right] + \mathcal{O}(f^2):\right\} \\
&= \exp\left\{:g + \left[\left(\frac{:g:}{1 - e^{-:g:}}\right) f\right] + \mathcal{O}(f^2):\right\}.
\end{aligned} \tag{4.166}$$

Exchanging f and g then proves Eq. (4.140).

Homework 4.60 Prove the second order expression (4.141).

Solution To deal with the second order terms, instead of Eq. (4.164), we start with

$$g = \left(\frac{1 - e^{-:h:}}{:h:}\right)(h_1 + 2\alpha h_2). \tag{4.167}$$

The operator in Eq. (4.167) can be written as

$$\left(\frac{1 - e^{-:h:}}{:h:}\right) = \int_0^1 d\beta \; e^{-\beta:f+\alpha h_1:}.$$

If we denote the operator $e^{-\beta:f+\alpha h_1:}$ by M, then

$$\left(\frac{1 - e^{-:h:}}{:h:}\right) = \int_0^1 d\beta \left(M\Big|_{\alpha=0} + \alpha \frac{dM}{d\alpha}\Big|_{\alpha=0} \right) + \mathcal{O}(\alpha^2). \tag{4.168}$$

We need to compute $\frac{dM}{d\alpha}$. It follows from Eq. (4.150) that

$$\frac{dM}{d\alpha} = :\left(\frac{e^{-\beta:f+\alpha h_1:} - 1}{\beta:f + \alpha h_1:}\right) \beta h_1: \; e^{-\beta:f+\alpha h_1:}. \tag{4.169}$$

Substituting Eq. (4.169) into Eq. (4.168) yields

$$\left(\frac{1 - e^{-:h:}}{:h:}\right) = \int_0^1 d\beta \left[e^{-\beta:f:} + \alpha : \left(\frac{e^{-\beta:f:} - 1}{\beta:f:}\right) \beta h_1: \; e^{-\beta:f:} \right]. \tag{4.170}$$

Substituting Eq. (4.170) into Eq. (4.167) yields

$$g = \int_0^1 d\beta \left[e^{-\beta:f:} + \alpha : \left(\frac{e^{-\beta:f:} - 1}{\beta:f:}\right) \beta h_1: \; e^{-\beta:f:} \right] (h_1 + 2\alpha h_2). \tag{4.171}$$

Terms independent of α and terms 1st order in α must separately be equal on the two sides of Eq. (4.171). The terms constant in α gives simply Eq. (4.165). Terms linear in α give

$$0 = \int_0^1 d\beta \; e^{-\beta:f:} 2h_2 + \int_0^1 d\beta : \left(\frac{e^{-\beta:f:} - 1}{\beta:f:}\right) \beta h_1: \; e^{-\beta:f:} h_1. \tag{4.172}$$

Substituting h_1 from Eq. (4.165) then gives

$$0 = 2 \left(\frac{1 - e^{-:f:}}{:f:}\right) h_2 + \int_0^1 d\beta \left[\left(\frac{e^{-\beta:f:} - 1}{1 - e^{-:f:}}\right) g, e^{-\beta:f:} \left(\frac{:f:}{1 - e^{-:f:}}\right) g \right]$$

$$\implies h_2 = -\frac{1}{2} \left(\frac{:f:}{1 - e^{-:f:}}\right) \int_0^1 d\beta \left[\left(\frac{e^{-\beta:f:} - 1}{1 - e^{-:f:}}\right) g, e^{-\beta:f:} \left(\frac{:f:}{1 - e^{-:f:}}\right) g \right],$$

which in turn can be written in the form (4.141).

Homework 4.61 Prove the third form of the BCH formula (4.144–4.145).

Solution Consider

$$e^{:\alpha f:} e^{:\alpha g:} e^{:\alpha f:} = e^{:h(\alpha):} = M(\alpha).$$

We have

$$M^{-1}(\alpha) = e^{-:\alpha f:} e^{-:\alpha g:} e^{-:\alpha f:} = M(-\alpha)$$

$$\implies \quad e^{-:h(\alpha):} = e^{:h(-\alpha):}, \quad \text{or} \quad -h(\alpha) = h(-\alpha),$$

i.e. the function $h(\alpha)$ is an odd function of α.

Let

$$h(\alpha) \;=\; \alpha h_1 + \alpha^3 h_3 + \alpha^5 h_5 + \cdots . \tag{4.173}$$

We then consider

$$\frac{dM(\alpha)}{d\alpha} \;=\; :f\!:M + e^{:\alpha f:}\!:g\!:e^{:\alpha g:}e^{:\alpha f:} + M\!:f\!:$$

$$\implies \quad \frac{dM}{d\alpha}M^{-1} \;=\; :f\!: + e^{:\alpha f:}\!:g\!:e^{-:\alpha f:} + M\!:f\!:M^{-1}$$

$$\;=\; :f\!: + :e^{:\alpha f:}g\!: + :Mf\!:, \tag{4.174}$$

where use has been made of Table 4.5. We now combine Eqs. (4.150) and (4.174) to obtain

$$f + Mf + e^{:\alpha f:}g \;=\; \left(\frac{e^{:h:}-1}{:h:}\right)\frac{dh}{d\alpha}. \tag{4.175}$$

Expand both sides of Eq. (4.175) in powers of α. The left-hand-side reads

$$f + f + :h\!:f + \frac{1}{2}:h\!:^2 f + \frac{1}{6}:h\!:^3 f + \frac{1}{24}:h\!:^4 f$$

$$+ g + \alpha :f\!:g + \frac{1}{2}\alpha^2 :f\!:^2 g + \frac{1}{6}\alpha^3 :f\!:^3 g + \frac{1}{24}\alpha^4 :f\!:^4 g + \mathcal{O}(\alpha^5)$$

$$= \; 2f + g + \alpha(:h_1\!:f + :f\!:g) + \frac{1}{2}\alpha^2 \left(:h_1\!:^2 f + :f\!:^2 g\right)$$

$$+ \alpha^3 \left(\frac{1}{6}:h_1\!:^3 f + :h_3\!:f + \frac{1}{6}:f\!:^3 g\right)$$

$$+ \alpha^4 \left(\frac{1}{2}:h_3\!::h_1\!:f + \frac{1}{2}:h_1\!::h_3\!:f + \frac{1}{24}:h_1\!:^4 f + \frac{1}{24}:f\!:^4 g\right) + \mathcal{O}(\alpha^5).$$

The right-hand-side of Eq. (4.175) reads

$$\left(1 + \frac{1}{2}:h\!: + \frac{1}{6}:h\!:^2 + \frac{1}{24}:h\!:^3 + \frac{1}{120}:h\!:^4\right)\left(h_1 + 3\alpha^2 h_3 + 5\alpha^4 h_5\right) + \mathcal{O}(\alpha^5)$$

$$= \; h_1 + 3\alpha^2 h_3 + \alpha^3 \left(\frac{1}{2}:h_3\!:h_1 + \frac{3}{2}:h_1\!:h_3\right)$$

$$+ \alpha^4 \left(5h_5 + \frac{1}{2}:h_1\!:^2 h_3 + \frac{1}{6}:h_1\!::h_3\!:h_1\right) + \mathcal{O}(\alpha^5).$$

We then equate for each order of α on both sides of Eq. (4.175),

$$\mathcal{O}(\alpha^0) \implies \quad h_1 = 2f + g,$$

$$\mathcal{O}(\alpha^1) \implies \quad 0 = :h_1\!:f + :f\!:g,$$

$$\implies \quad 0 = :g\!:f + :f\!:g,\ \text{satisfied automatically},$$

$$\mathcal{O}(\alpha^2) \implies \quad 3h_3 = \frac{1}{2}:h_1\!:^2 f + \frac{1}{2}:f\!:^2 g,$$

$$\implies \quad h_3 = \frac{1}{6}:g\!:^2 f - \frac{1}{6}:f\!:^2 g,$$

$$\mathcal{O}(\alpha^3) \implies \quad \frac{1}{2}:h_3:h_1 + \frac{3}{2}:h_1:h_3 = \frac{1}{6}:h_1:^3 f + :h_3:f + \frac{1}{6}:f:^3 g \,,$$

$$\implies \quad \text{satisfied automatically}\,,$$

$$\mathcal{O}(\alpha^4) \implies \quad 5h_5 + \frac{1}{2}:h_1:^2 h_3 + \frac{1}{6}:h_1::h_3:h_1$$

$$= \frac{1}{2}:h_3::h_1:f + \frac{1}{2}:h_1::h_3:f + \frac{1}{24}:h_1:^4 f + \frac{1}{24}:f:^4 g$$

$$\implies \quad h_5 = \frac{1}{120}:h_1:^4 f + \frac{1}{120}:f:^4 g + \frac{1}{10}:h_3::g:f$$

$$+ \frac{1}{10}:h_1::h_3:f - \frac{1}{15}:h_1:^2 h_3 \,. \tag{4.176}$$

One still needs to substitute h_1 and h_3 into the h_5 expression above. Substituting into Eq. (4.176), using property (4.39), and simplifying give

$$h_5 \;=\; \frac{7}{360}:f:^4 g - \frac{1}{60}:f:^2:g:^2 f + \frac{1}{90}:f::g:^3 f + \frac{1}{45}:g::f:^3 g$$

$$+ \frac{1}{30}:g:^2:f:^2 g - \frac{1}{360}:g:^4 f \,. \tag{4.177}$$

Substituting h_1, h_3 and h_5 into Eq. (4.173) and setting $\alpha = 1$ then gives the result (4.145).

Homework 4.62 On the right-hand-side of the factorization formula Eq. (4.142), the ordering of the two factor maps are as shown. Show that one could choose the reversed ordering to obtain

$$e^{:f+h:} \;=\; e^{:\mathcal{O}(h^2):} \exp\left[:\left(\frac{e^{:f:} - 1}{:f:} \right) h: \right] e^{:f:}$$

$$=\; e^{:\mathcal{O}(h^2):} \exp\left[:\left(\int_0^1 du \; e^{u:f:} \right) h: \right] e^{:f:} \,. \tag{4.178}$$

A generalization of Eq. (4.178) and several applications can be found later with Eq. (4.275).

Homework 4.63 We now know well that Lie algebra does not obey the commutative law. But implicitly all along, we have assumed that Lie operators satisfy the associative law. As an explicit demonstration and as a practice of the algebra involved, check the internal self-consistency of the BCH formula (4.138) in this homework.

Consider for example three small functions $f(X), g(X), h(X)$, each of the order of $\sim \mathcal{O}(\epsilon)$. Use Eq. (4.138) to confirm the associative law

$$\left(e^{:f:} e^{:g:} \right) e^{:h:} \;=\; e^{:f:} \left(e^{:g:} e^{:h:} \right) \,,$$

up to and including order $\sim \mathcal{O}(\epsilon^3)$.

Solution Show explicitly that both sides are equal to

$$\exp\Big[:f: + :g: + :h: + \frac{1}{2}\{:g:,:h:\} + \frac{1}{2}\{:f:,:g:\} + \frac{1}{2}\{:f:,:h:\}$$

$$+\frac{1}{12}\{:f:,\{:f:,:g:\}\} + \frac{1}{12}\{:f:,\{:f:,:h:\}\} + \frac{1}{12}\{:g:,\{:g:,:f:\}\}$$

$$+\frac{1}{12}\{:g:,\{:g:,:h:\}\} + \frac{1}{12}\{:h:,\{:h:,:f:\}\} + \frac{1}{12}\{:h:,\{:h:,:g:\}\}$$

$$+\frac{1}{4}\{:f:,\{:g:,:h:\}\} + \frac{1}{12}\{:g:,\{:h:,:f:\}\} - \frac{1}{12}\{:h:,\{:f:,:g:\}\} + :\mathcal{O}(\epsilon^4): \Big].$$

In particular, if $h = f$, it reduces to

$$\exp\Big[2:f: + :g: - \frac{1}{6}\{:f:,\{:f:,:g:\}\} + \frac{1}{6}\{:g:,\{:g:,:f:\}\} + :\mathcal{O}(\epsilon^4): \Big],$$

which is in agreement to BCH the third form.

Homework 4.64 Use the BCH formulae to concatenate the map

$$e^{-:2f_2:}e^{:f_2+f_3:}e^{:f_2+g_3:}$$

into a form

$$e^{:h_3+\mathcal{O}(X^4):}.$$

Find h_3 in terms of f_2, f_3, and g_3.

Solution The intended physical meaning of the system should be clear. The result is

$$h_3 = \int_0^1 du\; e^{-u:f_2:}(e^{-:f_2:}f_3 + g_3).$$

Homework 4.65

(a) Concatenate the map

$$e^{:f+g:}e^{:-f+g:} \tag{4.179}$$

to 1st order in g, assuming g is small but f is not.

(b) Consider two short-but-strong sextupoles, each of the same length L but opposite polarity S and $-S$. Consider L to be short, but S strong in such a way that SL is held fixed. Arrange these two sextupoles back to back. The leading effect is that these two sextupoles will cancel each other, but the cancellation is not complete and one ends up with a residual error map. Apply the result in (a) to find this residual error map to leading order in L (but to all orders in LS). Consider on-momentum particles only.

(c) Repeat the problem this time holding LS fixed but let S to be weak. Find the map to 1st order in S but all orders in LS.

Solution

(a) The result is not $e^{:2g:}$ but

$$\exp\left[:2\left(\frac{e^{:f:}-1}{:f:}\right)g+\mathcal{O}(g^2):\right] = \exp\left[:2\left(\int_0^1 du\ e^{u:f:}\right)g+\mathcal{O}(g^2):\right]. \qquad (4.180)$$

(b) The map is given by Eq. (4.179) with

$$f = -\frac{LS}{3}(x^3-3xy^2), \qquad g = -\frac{L}{2}(p_x^2+p_y^2), \qquad (4.181)$$

and g is small and f is not small. Applying Eq. (4.180) gives the residual map

$$\exp\left[-:L\left(p_x^2+p_y^2-LSp_x(x^2-y^2)+2LSp_yxy+\frac{1}{3}L^2S^2(x^2+y^2)^2\right)+\mathcal{O}(L^2):\right].$$

Homework 4.66 Consider a thick-lens sextupole which is represented by the map [see Eq. (4.119)]

$$M = e^{:h_2+h_3:}, \qquad \text{where} \qquad \begin{cases} h_2 = -\frac{L}{2}(p_x^2+p_y^2), \\ h_3 = -\frac{L}{3}S(x^3-3xy^2). \end{cases}$$

The map M can be factorized to read

$$M = e^{:h_2:}e^{:g_3:}e^{:\mathcal{O}(X^4):}$$

where g_3 is a homogeneous 3rd order polynomial in X. Find g_3. Find the thin-lens limit of the result. The difference of this homework from Homework 4.65 is that this time the small parameter is not S or L, but X.

Solution Apply Eq. (4.142–4.143) to obtain

$$g_3 = \int_0^1 du\ h_3(e^{-u:h_2:}X)$$

$$= -\frac{1}{3}SL\int_0^1 du\left[(x-uLp_x)^3 - 3(x-uLp_x)(y-uLp_y)^2\right]$$

$$= -\frac{1}{3}SL\left[x^3 - \frac{3}{2}Lx^2p_x + L^2xp_x^2 - \frac{1}{4}L^3p_x^3 - 3xy^2 + \frac{3}{2}Lp_xy^2\right.$$

$$\left. +3Lxyp_y - 2L^2yp_xp_y - L^2xp_y^2 + \frac{3}{4}L^3p_xp_y^2\right].$$

Take the thin-lens limit $L \to 0$, $S \to \infty$, while holding LS fixed, this reduces to

$$g_3 \to -\frac{SL}{3}(x^3-3xy^2),$$

as it might have been guessed.

Homework 4.67 Factorize the map

$$\exp\left[:ax^3 + bx^2 p + cxp^2 + dp^3:\right] \tag{4.182}$$

into the form

$$e^{:Ax^3:}e^{:Bx^2 p:}e^{:Cxp^2:}e^{:Dp^3:}e^{:Ex^4:}e^{:Fx^3 p:}e^{:Gx^2 p^2:}e^{:Hxp^3:}e^{:Ip^4:}e^{:\mathcal{O}(X^5):}\,. \tag{4.183}$$

Apply both Eqs. (4.182–4.183) to x and p. Show that the results agree to order $\mathcal{O}(X^3)$. This factorization can be useful as one can apply the monomial map formula (4.94) to Eq. (4.183).

Solution

$$A = a, \quad B = b, \quad C = c, \quad D = d,$$
$$E = -\frac{3}{2}ab, \quad F = -3ac, \quad G = -\frac{3}{2}(3ad + bc),$$
$$H = -3bd, \quad I = -\frac{3}{2}cd\,.$$

Homework 4.68 The BCH formula can of course be applied to concatenate two linear maps. Take for example the first form, Eq. (4.136–4.137). Use what we learned from Lie algebra to see how two matrices A and B of the same dimension can be concatenated as

$$e^A e^B = e^C,$$

where

$$C = A + B + \frac{1}{2}\{A, B\} + \frac{1}{12}\{A, \{A, B\}\} + \frac{1}{12}\{B, \{B, A\}\} + \cdots. \tag{4.184}$$

We have used the curly brackets to mean the commutators of matrices.

Solution To demonstrate Eq. (4.184), start with the BCH formula. Consider the quadratic forms

$$f_2 = -\frac{1}{2}\tilde{X}FX, \qquad g_2 = -\frac{1}{2}\tilde{X}GX,$$

where F and G are symmetric matrices. Equations (4.71–4.72) and (4.136–4.137) say that

$$e^{SG}e^{SF} = e^{SH},$$

where H is a symmetric matrix for which the corresponding quadratic form $h_2 = -\frac{1}{2}\tilde{X}HX$ satisfies

$$h_2 = f_2 + g_2 + \frac{1}{2}[f_2, g_2] + \frac{1}{12}[f_2, [f_2, g_2]] + \frac{1}{12}[g_2, [g_2, f_2]] + \cdots,$$

where the square brackets are the Poisson brackets. We want to use this property to prove Eq. (4.184), at least for the special case when A and B can be written as SG and SF, respectively where F and G are symmetric matrices. To do that, note that

$$
\begin{aligned}
[f_2, g_2] &= \frac{1}{4}\frac{\partial(F_{\alpha\beta}X_\alpha X_\beta)}{\partial X_\ell} S_{\ell m}\frac{\partial(G_{\gamma\delta}X_\gamma X_\delta)}{\partial X_m} \\
&= F_{\alpha\ell}X_\alpha S_{\ell m}G_{\gamma m}X_\gamma = \tilde{X}FSGX \\
&= \frac{1}{2}\tilde{X}(GSF-FSG)X,
\end{aligned}
\tag{4.185}
$$

We have therefore demonstrated that $[f_2, g_2]$ is a quadratic form, and its corresponding matrix is given by

$$
GSF - FSG = -S\{SG, SF\}.
\tag{4.186}
$$

In the last step of Eq. (4.185), we have used the fact that

$$
FSG = -(\widetilde{GSF}),
\tag{4.187}
$$

to assure the symmetry of the matrix (4.186). Using Eq. (4.187), it follows that $[f_2, [f_2, g_2]]$ is also a quadratic form, and its corresponding matrix is

$$
-S\{S(-S\{SG, SF\}), SF\} = -S\{SF, \{SF, SG\}\}.
$$

The quadratic form h_2 therefore is given by the matrix

$$
\begin{aligned}
H = {} & F + G - \frac{1}{2}S\{SG, SF\} - \frac{1}{12}S\{SF, \{SF, SG\}\} \\
& -\frac{1}{12}S\{SG, \{SG, SF\}\} + \cdots,
\end{aligned}
$$

or

$$
\begin{aligned}
SH = {} & SF + SG + \frac{1}{2}\{SG, SF\} + \frac{1}{12}\{SF, \{SF, SG\}\} \\
& +\frac{1}{12}\{SG, \{SG, SF\}\} + \cdots,
\end{aligned}
\tag{4.188}
$$

which proves Eq. (4.184) if one identifies $A = SF, B = SG, C = SH$.

4.5   Localized radio-frequency cavity

Our plan of discussion   In the following sections, we will apply the Lie algebra techniques to several accelerator applications. Our plan is to cover the following subjects,[20]

---

[20]See footnotes 8 and 9. For additional reading, see for example A.J. Dragt and E. Forest, J. Math. Phys. 24(12), 2734 (1983); E. Forest, B. Leemann, and S. Chattopadhyay, SSC-N-118 (1986); E. Forest, D. Douglas, and B. Leemann, Nucl. Instr. Meth. in Phys. Res. A258, 355 (1987).

- localized RF cavity;
- a single localized sextupole;
- distribution of multipoles;
- multipole correction algorithm;
- higher order chromaticity;
- achromat;
- resonance strength.

To fully address these applications, it turns out there is one more important technique we still need to learn, namely the normal form technique to be covered in Sec. 4.8. But we will consider a few of the simpler applications first before we elevate our discussion to the normal forms.

The reason normal form provides a dividing line among the applications is that before the normal form, the physics is viewed based on a passage-by-passage, time domain, picture even when the analysis is applied to a circular accelerator. The normal form analysis views the physics as a repetitive system from a frequency domain picture. We touched upon this when we compared Taylor maps and Lie maps on page 199.

The localized RF system   As a first application of the techniques developed so far, consider the longitudinal particle motion whose dynamical variables are (z, δ). Consider an RF cavity located at $s = 0$, whose action on particle motion is given by the map

$$z = z_0, \qquad \delta = \delta_0 - V \sin k z_0 , \tag{4.189}$$

where V and k are related to the voltage and frequency of the RF cavity. Represent the rest of the accelerator by the simple map

$$z = z_0 + \alpha \delta_0, \qquad \delta = \delta_0 , \tag{4.190}$$

where α is related to the momentum compaction factor of the accelerator design. Here, $z > 0$ is beam head, $z < 0$ is beam tail, $\alpha > 0$ is below transition, $\alpha < 0$ is above transition.

Strictly speaking, the system described by Eqs. (4.189–4.190) is not integrable. This system is in fact filled with chaos when its phase space is examined by a powerful microscope. But to the extent that power series converge, it is at least approximately integrable when some appropriate parameters are "small", and in this sense, we sweep the chaos under the rug and insist to look for an invariant of the motion. In this first application of the Lie techniques, we will look for this invariant.

Lie algebra is a perturbative technique in our applications. To proceed, the first issue to address is what is the small perturbative parameter for the problem at hand. For the problem of localized RF, in this section below, we shall explore four different expressions of this invariant, each applicable in its own set of small parameters and in their respective ranges of validity.

When both α and V are small We will first find an expression of the invariant when both α and V are small — we will soon explain what does being small means. The two maps (4.189–4.190) are both close to the identity map. Let us first write them in their Lie algebraic forms (see Table 4.3),

$$M_{\mathrm{cav}} \;=\; \exp\left(:-\int_0^z dz'\, V\sin kz':\right) \;=\; \exp\left[:-\frac{V}{k}(1-\cos kz):\right],$$

$$M_{\mathrm{acc}} \;=\; \exp\left(:-\frac{1}{2}\alpha\delta^2:\right).$$

Observe the particle motion at the middle of the RF cavity, the one-turn map is

$$M_{\mathrm{cav}/2}M_{\mathrm{acc}}M_{\mathrm{cav}/2} \;=\; \exp\left[:-\frac{V}{2k}(1-\cos kz):\right]\exp\left(:-\frac{1}{2}\alpha\delta^2:\right)$$

$$\times\,\exp\left[:-\frac{V}{2k}(1-\cos kz):\right].$$

This map can be concatenated using the 3rd form of the BCH formula, Eqs. (4.144–4.145), if we identify

$$f \;=\; -\frac{V}{2k}(1-\cos kz)\,,$$

$$g \;=\; -\frac{1}{2}\alpha\delta^2\,. \qquad\qquad (4.191)$$

In the concatenation procedure, we note

$$:g:f \;=\; -\frac{1}{2}V\alpha\delta\sin kz\,,$$

$$:g:^2 f \;=\; -\frac{1}{2}V\alpha^2 k\delta^2\cos kz\,,$$

$$:g:^3 f \;=\; \frac{1}{2}V\alpha^3 k^2\delta^3\sin kz\,,$$

$$:g:^4 f \;=\; \frac{1}{2}V\alpha^4 k^3\delta^4\cos kz\,,$$

$$:f:g \;=\; \frac{1}{2}V\alpha\delta\sin kz\,,$$

$$:f:^2 g \;=\; -\frac{1}{4}V^2\alpha\sin^2 kz\,,$$

$$:f:^3 g \;=\; 0\,,$$

$$:f::g:^3 f \;=\; -\frac{3}{4}V^2\alpha^3 k^2\delta^2\sin^2 kz\,,$$

$$:f::g:^2 f \;=\; \frac{1}{4}V^2\alpha^2 k\delta\sin 2kz\,,$$

$$:f:^2:g:^2 f \;=\; -\frac{1}{8}V^3\alpha^2 k\sin kz\sin 2kz\,,$$

$$:g{::}f{:}^2 g \;=\; -\frac{1}{4}V^2\alpha^2 k\delta \sin 2kz\,,$$

$$:g{:}^2{:}f{:}^2 g \;=\; -\frac{1}{2}V^2\alpha^3 k^2\delta^2 \cos 2kz\,.$$

After concatenation, the one-turn map becomes $e^{:h:}$, where

$$
\begin{aligned}
h \;=\; & -\frac{1}{2}\alpha\delta^2 - \frac{V}{k}(1-\cos kz) \\
& -\frac{1}{12}V\alpha^2 k\delta^2\cos kz + \frac{1}{24}V^2\alpha\sin^2 kz \\
& -\frac{1}{720}V\alpha^4 k^3\delta^4\cos kz + \frac{1}{120}V^2\alpha^3 k^2\delta^2(3\sin^2 kz - 2) \\
& +\frac{1}{480}V^3\alpha^2 k\sin kz\sin 2kz + \mathcal{O}((V,\alpha)^7)\,.
\end{aligned}
\tag{4.192}
$$

Why are we interested in h? Because it is an invariant of the motion. That is, if we measure (z,δ) of a particle turn after turn as it passes by the middle of the RF cavity, and calculate the value of $h(z,\delta)$, we will find that the value of h does not vary from turn to turn.[21]

More importantly, we are interested in h not only because h is an invariant, but also because $-h$ is the effective Hamiltonian of the system. By finding h, we have replaced the piecewise-discrete system consisting of piecewise constant Hamiltonians by a smooth system which has an s-independent Hamiltonian $-h$. This smooth system is of course much simpler to handle than the original piecewise-discrete system. In fact, having found h, the problem has been "integrated". Its effective Hamilton equations read

$$\frac{dz}{dn} = \frac{\partial(-h)}{\partial\delta}\,, \qquad \frac{d\delta}{dn} = -\frac{\partial(-h)}{\partial z}\,,$$

where n is the turn number. We will pick up this topic of Hamilton equations using the effective Hamiltonian from time to time later.

For Eq. (4.192) to be an invariant, the series expansion must converge. This requires

$$|k\alpha V| \ll 1\,, \qquad |k\alpha\hat{\delta}| \ll 1\,, \tag{4.193}$$

where $\hat{\delta}$ is the peak value of δ during the evolution of the particle under consideration. For particles inside the RF bucket, $\hat{\delta} <$ bucket height. Since the bucket height $= \sqrt{\frac{4V}{k\alpha}}$, the second condition of (4.193) is automatically satisfied for particles inside the RF bucket if the first condition of (4.193) is satisfied. [For an idea of what does an RF bucket look like, see Fig. 4.4(a).]

---

[21] Provided h exists. If the power series expansion (4.192) does not converge, for example, then h may not exist. In that case, there may not be an invariant, and the system not integrable.

When only V is small One may also try to find an expression of the invariant in case $k\alpha V$ is small but $k\alpha\hat\delta$ is not necessarily small by applying the 4th BCH form (4.147–4.148). With f and g defined in Eq. (4.191), one obtains an invariant to 1st order in V which holds for particles of any δ including even outside the RF bucket. We note that

$$:g\!:\!f \;=\; \alpha\delta\frac{\partial f}{\partial z}$$

$$\Longrightarrow \qquad \left(:g\!:\coth\frac{:g\!:}{2}\right) f \;=\; -\frac{V}{2k}\left[\left(\alpha\delta\frac{\partial}{\partial z}\right)\coth\left(\frac{\alpha\delta}{2}\frac{\partial}{\partial z}\right)\right](1-e^{ikz})$$

$$=\; -\frac{V}{k} + \frac{V}{2k}(\alpha\delta ik)\coth\left(\frac{\alpha\delta}{2}ik\right)e^{ikz}$$

$$=\; -\frac{V}{k} + \frac{V}{2k}(\alpha\delta k)\cot\frac{\alpha\delta k}{2}\,e^{ikz}\,. \tag{4.194}$$

In the above expressions, only the real parts are meaningful, and we have used the fact that $i\coth(ix) = \cot x$. Substituting Eq. (4.194) into Eq. (4.148) gives the effective Hamiltonian $-h$ with

$$h \;=\; -\frac{1}{2}\alpha\delta^2 - \frac{V}{k} + \frac{1}{2}\alpha\delta V\cot\frac{\alpha\delta k}{2}\cos kz + \mathcal{O}(V^2)\,. \tag{4.195}$$

If $k\alpha\hat\delta$ is also small, Eq. (4.195) agrees with (4.192) to 1st order in V.

When only α is small The 4th BCH form also can be used to obtain an expression when $k\alpha\hat\delta$ is small, but $k\alpha V$ is not necessarily small. It follows from Eq. (4.149) that

$$h \;=\; 2f + \left(\frac{:f\!:}{\sinh:f\!:}\right)g + \mathcal{O}(\alpha^2)$$

$$=\; 2f + \left(1 - \frac{1}{6}:f\!:^2 + \frac{7}{360}:f\!:^4 + \cdots\right)g + \mathcal{O}(\alpha^2)$$

$$=\; 2f + \left(1 - \frac{1}{6}:f\!:^2\right)g + \mathcal{O}(\alpha^2)$$

$$=\; -\frac{V}{k}(1-\cos kz) - \frac{1}{2}\alpha\delta^2 + \frac{1}{24}V^2\alpha\sin^2 kz + \mathcal{O}(\alpha^2)\,. \tag{4.196}$$

Again, this expression is consistent with Eq. (4.192) in the appropriate limit.

As mentioned, $-h$ also has the meaning of being the one-turn effective Hamiltonian. For example, if one is interested in the one-turn equations of motion to 1st order in α, one may use Eq. (4.196) to obtain

$$\frac{dz}{dn} \;=\; -\frac{\partial h}{\partial\delta} \;=\; \alpha\delta + \mathcal{O}(\alpha^2)\,,$$

$$\frac{d\delta}{dn} \;=\; \frac{\partial h}{\partial z} \;=\; -V\left[\sin kz - \frac{1}{24}\alpha kV\sin 2kz\right] + \mathcal{O}(\alpha^2)\,, \tag{4.197}$$

where n is a turn number index which increases by 1 per turn.[22] What we have shown is that the discrete maps (4.189–4.190) are approximately equivalent to the continuous evolution (4.197), provided the particle motion is observed only at discrete times at the middle of the RF cavity when $n =$ integers. Equation (4.197) however can be solved as if the time variable n assumes continuous values provided that we observe the solution only when $n =$ integers.

When z and δ are small Sometimes it is more useful to find an invariant (and the effective Hamiltonian) that is valid when the dynamical variables z and δ are small, i.e. when the synchrotron motion is close to the origin of the phase space. In this representation, V and α do not have to be small. How do we proceed in this case? To proceed, let us write the map of half of the RF cavity as

$$M_{\mathrm{cav}/2} \;=\; \exp\left[: -\frac{V}{2k}(1 - \cos kz):\right] \;=\; e^{:f_2+f_{\mathrm{NL}}:},$$

where f_2 is the term in f that is quadratic in the dynamical variables, and f_{NL} contains the remaining, higher order nonlinear terms of f, i.e.

$$f_2 \;=\; -\frac{V}{4}kz^2,$$

$$f_{\mathrm{NL}} \;=\; -\frac{V}{2k}\left(1 - \cos kz - \frac{1}{2}k^2z^2\right).$$

Since $:f_2:$ and $:f_{\mathrm{NL}}:$ commute, we have

$$M_{\mathrm{cav}/2} \;=\; e^{:f_2:}e^{:f_{\mathrm{NL}}:} \;=\; e^{:f_{\mathrm{NL}}:}e^{:f_2:}.$$

The one-turn map then reads

$$M \;=\; e^{:f_{\mathrm{NL}}:}e^{:f_2:}e^{:g_2:}e^{:f_2:}e^{:f_{\mathrm{NL}}:}, \tag{4.198}$$

where $g = -\frac{1}{2}\alpha\delta^2$ has been designated as g_2 because it is quadratic in δ.

The three operators in the middle of Eq. (4.198) describe the linearized synchrotron motion, which we designate as

$$M_2 \;=\; e^{:f_2:}e^{:g_2:}e^{:f_2:}.$$

The action of M_2 on the vector $X = (z,\delta)$ can be described most conveniently by a matrix [linearizing Eqs. (4.189–4.190)],

$$M_2 \;=\; \begin{bmatrix} 1 & 0 \\ -\frac{1}{2}kV & 1 \end{bmatrix}\begin{bmatrix} 1 & \alpha \\ 0 & 1 \end{bmatrix}\begin{bmatrix} 1 & 0 \\ -\frac{1}{2}kV & 1 \end{bmatrix}$$

$$\;=\; \begin{bmatrix} 1 - \frac{1}{2}kV\alpha & \alpha \\ -kV + \frac{1}{4}k^2V^2\alpha & 1 - \frac{1}{2}kV\alpha \end{bmatrix}. \tag{4.199}$$

---

[22]Usually we have the expression that the one-turn map is given by $e^{:-Lh:}$, but here we have the expression $e^{:-h:}$. Obviously this is because we used the number of turns n to be our length unit, and the total length for one turn is 1.

For sufficiently small z and δ, we can ignore the nonlinear effects described by f_{NL}. The one-turn map is just the linear map M_2. Let us first find the "Courant–Snyder" invariant of this linear system. This means we are looking for a quadratic invariant, that is solely determined by M_2.

Let the Lie representation of M_2 be written as $\exp(:F_2:)$. Let F_2 be written as $-\frac{1}{2}\tilde{X}FX$, where F is a symmetric matrix parametrized as Eq. (4.60). By applying the Hamilton–Cayley technique using Eq. (4.65), we obtain

$$a = \frac{\mu}{\alpha}\sin\mu, \qquad b = 0, \qquad c = \frac{\alpha\mu}{\sin\mu},$$

where μ is determined by

$$\cos\mu = 1 - \frac{kV\alpha}{2}. \tag{4.200}$$

For stability of the system, we must have $4 \geq kV\alpha \geq 0$. We choose μ to be between 0 and π.

The quadratic invariant — the Courant–Snyder invariant — is given by

$$\begin{aligned} F_2 &= -\frac{1}{2}\left(az^2 + 2bz\delta + c\delta^2\right) \\ &= -\frac{\mu}{2}\left(\frac{\sin\mu}{\alpha}z^2 + \frac{\alpha}{\sin\mu}\delta^2\right). \end{aligned} \tag{4.201}$$

It goes without saying by now that the quantity

$$\frac{\sin\mu}{\alpha}z^2 + \frac{\alpha}{\sin\mu}\delta^2$$

is just the Courant–Snyder invariant in the longitudinal dimension. The reason the phase space ellipse is upright is due to the mirror symmetry when we observe the motion at the middle of the RF cavity.

We next look for the higher order terms of the invariant. To do this, we will apply Eq. (4.148) to the one-turn map

$$M = e^{:f_{\mathrm{NL}}:}e^{:F_2:}e^{:f_{\mathrm{NL}}:}.$$

To 1st order in f_{NL}, the invariant is given by

$$h = F_2 + :F_2:\coth\left(\frac{:F_2:}{2}\right)f_{\mathrm{NL}}. \tag{4.202}$$

To calculate the invariant, therefore, we need to find $:F_2:\coth(\frac{:F_2:}{2})f_{\mathrm{NL}}$.

We learned before that an exponential operator operating on a complicated function can be simplified by using (see Table 4.5) $e^{:f:}g(X) = g(e^{:f:}X)$. But here we have a complicated operator operating on a complicated function, and the question is how to proceed. The answer lies in the idea of *eigenmodes*. To see this, note that the complicated operator is made up of $:F_2:$, which is a much simpler object. If the function f_{NL} to be operated on can be decomposed into some eigenmodes of the operator $:F_2:$, then the calculation will simplify considerably. So our first job is to find these eigenmodes of $:F_2:$.[23]

We first simplify the expression of F_2 by introducing the dynamical variables

$$\bar{z} = z\sqrt{\frac{\sin\mu}{|\alpha|}}, \qquad \bar{\delta} = \delta\sqrt{\frac{|\alpha|}{\sin\mu}}. \tag{4.203}$$

This is a canonical transformation because the fundamental Poisson brackets are preserved (see Homework 4.1),

$$[\bar{z}, \bar{\delta}] = [z, \delta] = 1.$$

This means we can treat $(\bar{z}, \bar{\delta})$ as the new dynamical variables. The new F_2 reads

$$F_2 = -\frac{\mu}{2}(\bar{z}^2 + \bar{\delta}^2).$$

Equation (4.203) is just the longitudinal equivalent of the Courant–Snyder transformation to normalized coordinates.

Applying $:F_2:$ on $\bar{z}$ and $\bar{\delta}$ gives

$$:F_2:\begin{bmatrix} \bar{z} \\ \bar{\delta} \end{bmatrix} = \mu\,\mathrm{sgn}(\alpha)\begin{bmatrix} \bar{\delta} \\ -\bar{z} \end{bmatrix}.$$

One then notes that if we apply $:F_2:$ to $\bar{z} \pm i\bar{\delta}$, we obtain

$$:F_2:(\bar{z} \pm i\bar{\delta}) = \mp i\,\mathrm{sgn}(\alpha)\mu\,(\bar{z} \pm i\bar{\delta}). \tag{4.204}$$

This means $(\bar{z} \pm i\bar{\delta})$ are eigenmodes of $:F_2:$ with eigenvalues $\mp i\,\mathrm{sgn}(\alpha)\mu$. We have thus found two eigenmodes of $:F_2:$. But we need to find more such modes, as is done next.

To simplify the consideration, below we will consider the case below transition only, with $\alpha > 0$. Make another canonical transformation from $(\bar{z}, \bar{\delta})$ to (ϕ, A) according to[24]

$$\bar{z} = \sqrt{2A}\sin\phi, \qquad \bar{\delta} = \sqrt{2A}\cos\phi.$$

or

$$A = \frac{1}{2}(\bar{z}^2 + \bar{\delta}^2), \qquad \phi = \tan^{-1}\frac{\bar{z}}{\bar{\delta}}.$$

This is a canonical transformation because

$$[\phi, A] = 1.$$

---

[23]Beware. We are touching upon the topic of normal form.

[24]Remember that ϕ is the coordinate, A is the momentum, not the other way around. Keep a good habit of writing (ϕ, A), not (A, ϕ). Unfortunately, we have been calling this an "action-angle" transformation. It really should be called an "angle-action" transformation.

The new F_2 is

$$F_2 = -\mu A. \tag{4.205}$$

This new form of F_2 deserves a closer look. We know that the final one-turn map we aim for has the form $e^{:-CH_{\mathrm{eff}}:}$. The end product on the exponent of this total Lie map contains three factors: one minus sign, one total circumference, and one effective Hamiltonian. Here this new form of F_2, obtained after a series of transformations, we managed it to contain three factors, and only three factors, exactly as required. In fact, the end result cannot be simpler. We have managed to make transformations to make the final map "as simple as possible". It is not possible to do better. As we will see later, this will lead to the way to proceed to the normal form formalism.

It follows now from Eq. (4.205) that

$$:F_2:A = 0, \qquad :F_2:\phi = \mu,$$

and that $e^{in\phi}$ is an eigenmode of $:F_2:$ with eigenvalue $in\mu$ for any integer n, i.e.,

$$:F_2:e^{in\phi} = in\mu e^{in\phi}. \tag{4.206}$$

Equation (4.204) is just the special case of Eq. (4.206) with $n = \pm 1$ and $\alpha > 0$.

We have thus found a host of eigenmodes of the operator $:F_2:$. Furthermore, $e^{in\phi}$ is also an eigenmode of any function of $:F_2:$, i.e.,

$$G(:F_2:)e^{in\phi} = G(in\mu)e^{in\phi}.$$

We are now ready to calculate $:F_2:\coth(\frac{:F_2:}{2})f_{\mathrm{NL}}$. We first note that

$$
\begin{aligned}
f_{\mathrm{NL}}(z) &= \frac{1}{48}Vk^3 z^4 + \mathcal{O}(z^6) \\
&= \frac{1}{48}Vk^3 \frac{\alpha^2}{\sin^2 \mu}\bar{z}^4 + \mathcal{O}(\bar{z}^6) \\
&= \frac{1}{12}Vk^3 \frac{\alpha^2}{\sin^2 \mu}A^2 \sin^4 \phi + \mathcal{O}(A^3).
\end{aligned}
$$

We then decompose f_{NL} into a linear combination of the eigenmodes of $:F_2:$ as

$$f_{\mathrm{NL}} = \frac{1}{192}Vk^3 \frac{\alpha^2}{\sin^2 \mu}A^2 \left(e^{i4\phi} - 4e^{i2\phi} + 6 - 4e^{-i2\phi} + e^{-i4\phi}\right) + \mathcal{O}(A^3).$$

We then find

$$
\begin{aligned}
:F_2:\coth\left(\frac{:F_2:}{2}\right)f_{\mathrm{NL}} &= \frac{1}{192}Vk^3 \frac{\alpha^2}{\sin^2 \mu}A^2 \Big[i4\mu \coth(i2\mu)e^{i4\phi} \\
&\qquad -i8\mu \coth(i\mu)e^{i2\phi} + 12 + i8\mu \coth(-i\mu)e^{-i2\phi} \\
&\qquad -i4\mu \coth(-i2\mu)e^{-i4\phi}\Big] + \mathcal{O}(A^3) \\
&= \frac{1}{48}Vk^3 \frac{\alpha^2}{\sin^2 \mu}A^2 \left(2\mu \cot 2\mu \cos 4\phi - 4\mu \cot \mu \cos 2\phi + 3\right), \tag{4.207}
\end{aligned}
$$

where we have used the fact that $x \coth x = 2$ as $x \to 0$ and $i \coth(ix) = \cot x$. Adding up Eqs. (4.205) and (4.207) yields h according to Eq. (4.202).

Equation (4.207) can be expressed in terms of $(\bar{z}, \bar{\delta})$ as

$$
:\!F_2\!: \coth\left(\frac{:\!F_2\!:}{2}\right) f_{\mathrm{NL}} = \frac{1}{48} V k^3 \frac{\alpha^2}{\sin^2 \mu} \left[\bar{z}^4 \left(\frac{\mu}{2} \cot 2\mu + \mu \cot \mu + \frac{3}{4} \right) \right.
$$
$$
\left. + \bar{z}^2 \bar{\delta}^2 \left(-3\mu \cot 2\mu + \frac{3}{2} \right) + \bar{\delta}^4 \left(\frac{\mu}{2} \cot 2\mu - \mu \cot \mu + \frac{3}{4} \right) \right].
$$

It can also be expressed in terms of the original coordinates (z, δ). When substituted into the invariant expression (4.202), one obtains finally

$$
\begin{aligned}
h \;=\; &-\frac{\mu}{2}\left(\frac{\sin \mu}{\alpha} z^2 + \frac{\alpha}{\sin \mu} \delta^2 \right) \\
&+ \frac{1}{48} V k^3 \frac{\alpha^2}{\sin^2 \mu} \left[\frac{\sin^2 \mu}{\alpha^2} z^4 \left(\frac{\mu}{2} \cot 2\mu + \mu \cot \mu + \frac{3}{4} \right) \right. \\
&\left. + z^2 \delta^2 \left(-3\mu \cot 2\mu + \frac{3}{2} \right) + \frac{\alpha^2}{\sin^2 \mu} \delta^4 \left(\frac{\mu}{2} \cot 2\mu - \mu \cot \mu + \frac{3}{4} \right) \right] + \mathcal{O}((z,\delta)^6).
\end{aligned}
\tag{4.208}
$$

Equation (4.207) gives the effective Hamiltonian in the (ϕ, A) phase space. This expression can be used to calculate the dependence of the synchrotron tune on the longitudinal action A. See Homework 4.107.

Tracking result   We can check the various invariants obtained in this section numerically as illustrated in Fig. 4.4. Figure 4.4(a) shows the turn-by-turn tracking result of the system obtained by iterating the exact maps (4.189–4.190). The motion of a particle is observed as it passes by the middle of the RF cavity turn after turn. The observed coordinates (z, δ) are plotted for 500 turns for three particles whose initial conditions are[25]

$$
\text{first particle} : \left(kz = 3.14 \times \frac{1}{3}, \delta = 0 \right),
$$
$$
\text{second particle} : \left(kz = 3.14 \times \frac{2}{3}, \delta = 0 \right),
$$
$$
\text{third particle} : \left(kz = 3.14, \delta = 0 \right).
\tag{4.209}
$$

The parameters used are $k = 1$, $V = 0.3$, $\alpha = 0.3$. The small-amplitude synchrotron tune, determined by Eq. (4.200), is $\frac{\mu}{2\pi} = 0.04793$. The trajectory of the third particle approximately traces out the boundary of the RF bucket, within which particle motion is to be confined.

Figure 4.4(b) shows the numerical value of the approximate invariant given by the first two terms [terms linear in (α, V)] of Eq. (4.192) as a function of turn number as the third particle circulates around the accelerator. If it is a true invariant, this quantity would be constant. Figure 4.4(b) shows that

---

[25]In Fig. 4.4(a), we are not exploring chaotic motions. Had we looked in great detail in the phase space, there will be small regions where particle motion is chaotic. See Homework 4.73.

Figure 4.4: Tracking results of various invariants for the case of a localized RF.

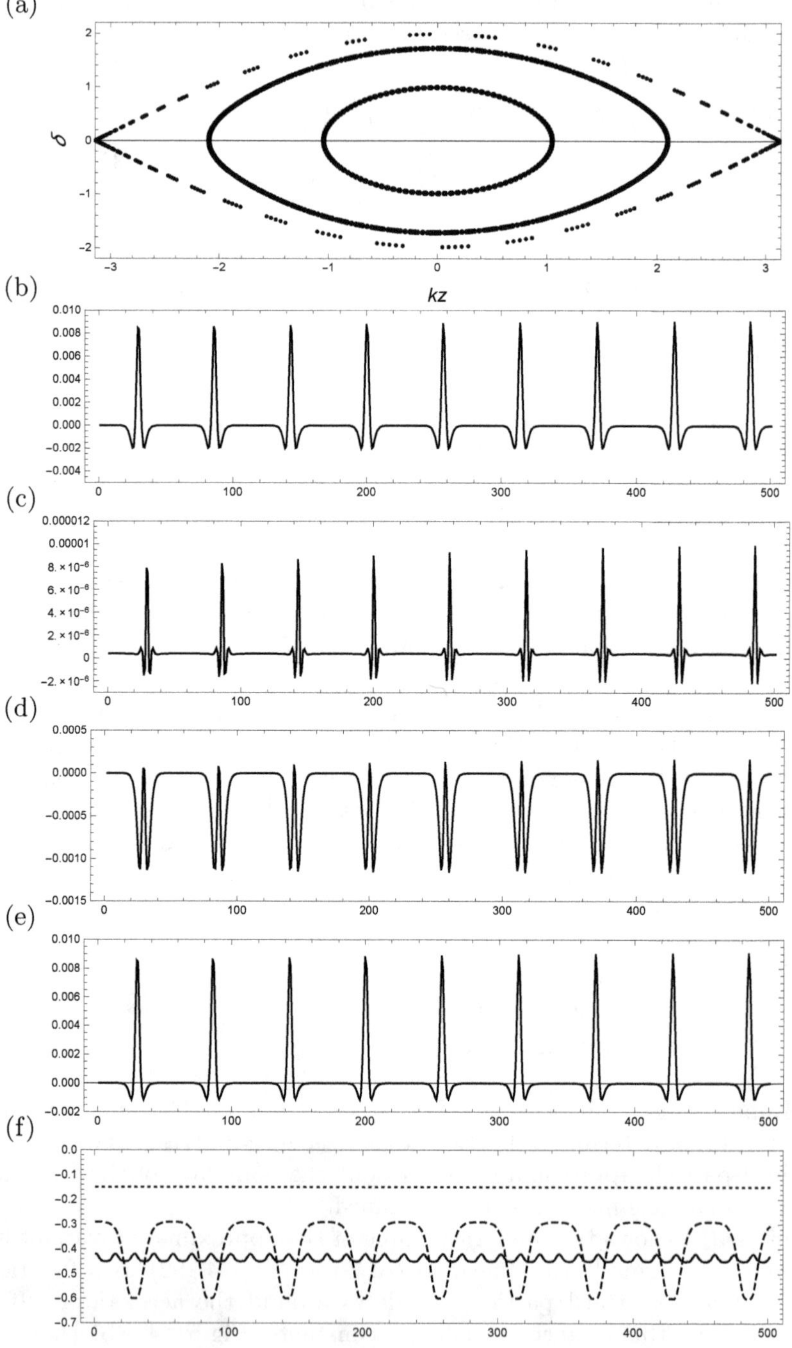

it is not an accurate invariant, although the vertical scale is rather expanded. Figure 4.4(c) is the same as Fig. 4.4(b) for the third particle, except that all terms in Eq. (4.192) are included, i.e. it is a plot of the 6th order invariant. For convenience, in Figs. 4.4(b) and (c), the vertical coordinates refer to $h + 0.6$, instead of the invariant h. It can be seen that the invariant plotted in Fig. 4.4(c) is indeed much more accurate than that of Fig. 4.4(b). For these figures, conditions (4.193) are reasonably satisfied [$k\alpha V = 0.09$, $k\alpha\hat{\delta} = 0.6$ for the third particle].

Figure 4.4(d) plots the time evolution of $h + 0.6$ for the third particle where the invariant h is as calculated using Eq. (4.195), an invariant for small $k\alpha V$. Figure 4.4(e) shows the same using Eq. (4.196), an invariant for small $k\alpha\hat{\delta}$.

One can also plot the invariant h when it is calculated using Eq. (4.208) for the three particles. This is shown in Fig. 4.4(f), dotted line for the first particle, solid and dashed lines for the second and third particles, respectively. In this figure, note that the small-amplitude synchrotron period is given by

$$\frac{2\pi}{\cos^{-1}(1 - \frac{1}{2}k\alpha V)} \approx 21 \text{ turns}.$$

One observes that the invariance holds for small amplitude particles better. The second particle with intermediate amplitude executes synchrotron oscillation with a period slightly slower than 21 turns. Observation of the dashed curve indicates the synchrotron oscillation of the third particle has slowed down considerably from 21 turns. A particle even closer to the RF bucket has its synchrotron oscillation period $\to \infty$.

Recap This section on localized RF cavity has the goal, in addition to give an example of how to apply the BCH formula, to introduce the importance of knowing what small parameters are in the consideration. In particular, we found the expressions of the effective Hamiltonian for various cases with small momentum compaction, small RF voltage, or small deviations from the phase space origin.

Later, we will have another example application when the length of the nonlinear element under consideration, ℓ, is short. In that case, ℓ becomes the small parameter. Some examples will be given in Sec. 4.7.3 under the subject of correction schemes for nonlinear perturbations.

Perhaps let us summarize what we have learned from this example of a localized RF cavity below.

- The first thing to do is always to identify the small parameter in the study. The choice of small parameters will be different in different situations.

- Find a way to expand in terms of power series in the small parameter identified.

- Apply BCH formula of an appropriate form. Some algebra will be required, but mostly straightforwardly. In some cases, normal mode techniques will be needed.

Homework 4.69 Show that the invariant (4.208) agrees with the invariant (4.192) for small z, δ, V, α with a 9th order error $\sim \frac{1}{280} \alpha^4 k^3 V^3 \delta^2$.

Homework 4.70

(a) Consider the invariant (4.201) as the effective Hamiltonian. Derive and solve the effective one-turn equations of motion for z and δ.

(b) Derive the equations of motion for the more accurate effective Hamiltonian (4.208).

Homework 4.71 Find the Courant–Snyder parameters for the linear motion described by Eq. (4.199). Plot $\beta(s)$ and $\alpha(s)$ as functions of s around the accelerator. Note that these are *longitudinal* β- and α-functions.

Solution Assuming the momentum compaction factor $\alpha > 0$ and assume the momentum compaction is evenly distributed over the storage ring of circumference C, the β-function in the ring is given by

$$\beta(s) = \frac{\alpha}{\sin \mu} \left[1 - \alpha k V \frac{s}{C} \left(1 - \frac{s}{C} \right) \right].$$

Stability requires $\beta(s) > 0$ for all s.

Homework 4.72 Consider the 1-D motion in a circular accelerator with a weak thin quadrupole of strength q inserted locally at position $s = 0$. Apply what you learned in this section to analyze the particle motion for small q. Compare your result with an exact solution using matrix formalism.

Homework 4.73 If one looks closely into Fig. 4.4(a), one might notice that there are particles in the neighborhood of the third particle whose turn-to-turn trajectory behaves chaotically. To see this more clearly, consider a case with $V = 0.3$, $\alpha = 0.3$, and $k = 8$. We now have $k\alpha V = 0.72$, which is no longer $\ll 1$, and the series expansions may no longer converge, especially for particles very close to the unstable fixed points. Perform numerical tracking for a particle with initial conditions $(kz = 3.141, \delta = 0)$ to observe this chaotic trajectory.

Solution The portion of the trajectory between $kz = -4$ to $kz = 4$ (modulus 2π) is shown in Fig. 4.5. A bucket without chaos was shown in Fig. 1.1.

Homework 4.74 Consider a 1-D motion of a particle going through an accelerator element that has a length L and is described by the equation of motion

$$x'' + \lambda \sin \alpha x = 0.$$

(a) Find the Lie map that describes the motion of a particle through the element.

One may look for Taylor series approximations of this map in different forms. Each form is valid when a certain parameter is small.

(b) Find the Taylor map up to third order in L. [Error terms are $\mathcal{O}(L^4)$.] This expression is valid when the element is short.

Figure 4.5: Tracking results showing chaos.

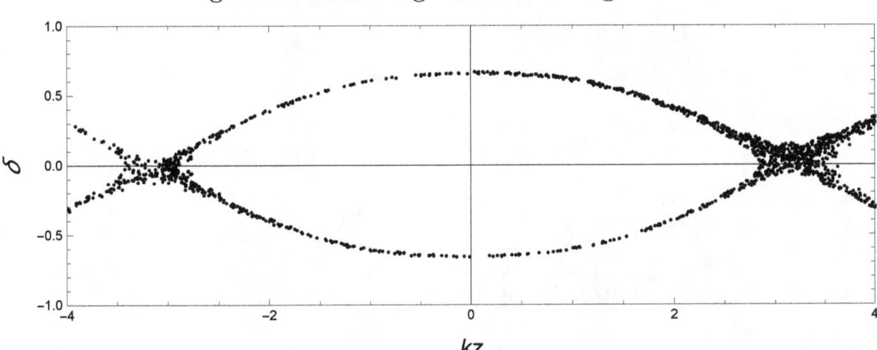

(c) Find the Taylor map up to third order in X for arbitrary values of L and λ. This expression holds when particles considered are close to the design orbit.

Solution

(a)

$$M = \exp\left[:-\frac{L}{2}p^2 - \frac{\lambda L}{\alpha}(1 - \cos\alpha x):\right]. \tag{4.210}$$

(b)

$$x = x_0 + Lp_0 - \frac{\lambda L^2}{2}\sin\alpha x_0 - \frac{\lambda\alpha L^3}{6}p_0\cos\alpha x_0 + \mathcal{O}(L^4),$$

$$p = p_0 - \lambda L\sin\alpha x_0 - \frac{\lambda\alpha L^2}{2}p_0\cos\alpha x_0$$

$$+\frac{\lambda\alpha L^3}{6}(\lambda\cos\alpha x_0 + \alpha p_0^2)\sin\alpha x_0 + \mathcal{O}(L^4). \tag{4.211}$$

(c) To 4th order in X, the map (4.210) can be written as

$$M = e^{:f_2:}\exp\left[\frac{L\lambda\alpha^3}{24}:\left(\int_0^1 due^{-u:f_2:}x^4\right):\right],$$

where

$$f_2 = -\frac{L}{2}p^2 - \frac{L\lambda\alpha}{2}x^2$$

describes the linearized motion. With

$$e^{-u:f_2:}\begin{bmatrix}x\\p\end{bmatrix} = \begin{bmatrix}x\cos u\theta - \frac{p}{\sqrt{\lambda\alpha}}\sin u\theta\\\sqrt{\lambda\alpha}x\sin u\theta + p\cos u\theta\end{bmatrix},$$

where $\theta = \sqrt{\lambda\alpha}L$ is the betatron phase advance if only the linear terms are kept, we have

$$M = e^{:f_2:}\exp\left[\frac{L\lambda\alpha^3}{24}:\int_0^1 du\left(x\cos u\theta - \frac{p}{\sqrt{\lambda\alpha}}\sin u\theta\right)^4:\right]$$

$$
= e^{:f_2:} \exp\left[: \frac{\theta p^4}{768\lambda^2 L}(12\theta - 8\sin 2\theta + \sin 4\theta)\right.
$$

$$
+ \frac{\alpha x p^3}{192\lambda}(-3 + 4\cos 2\theta - \cos 4\theta)
$$

$$
+ \frac{\alpha \theta x^2 p^2}{128\lambda L}(4\theta - \sin 4\theta) + \frac{\alpha^2 x^3 p}{192}(-5 + 4\cos 2\theta + \cos 4\theta)
$$

$$
\left. + \frac{\theta \alpha^2 x^4}{768 L}(12\theta + 8\sin 2\theta + \sin 4\theta) :\right].
$$

The 3rd order Taylor map is found to be

$$
x = x_0 \cos\theta + \frac{p_0}{\sqrt{\alpha\lambda}}\sin\theta + \frac{\alpha^2 x_0^3}{96}\sin\theta(6\theta + 7\sin 2\theta + \sin 4\theta)
$$

$$
+ \frac{\theta \alpha x_0^2 p_0}{64\lambda L}(-4\theta\cos\theta - 7\sin\theta + 2\sin 3\theta + \sin 5\theta)
$$

$$
+ \frac{\alpha x_0 p_0^2}{32\lambda}\sin\theta(2\theta + \sin 2\theta - \sin 4\theta)
$$

$$
- \frac{\theta p_0^3}{192\lambda^2 L}(12\theta\cos\theta + \sin\theta - 6\sin 3\theta + \sin 5\theta),
$$

$$
p = -x_0\sqrt{\alpha\lambda}\sin\theta + p_0\cos\theta + \frac{\theta \alpha^2 x_0^3}{192 L}(12\theta\cos\theta - 3\sin\theta + 6\sin 3\theta + \sin 5\theta)
$$

$$
+ \frac{\alpha^2 x_0^2 p_0}{32}\sin\theta(2\theta - 3\sin 2\theta - \sin 4\theta)
$$

$$
+ \frac{\theta \alpha x_0 p_0^2}{64\lambda L}(4\theta\cos\theta - 5\sin\theta + 2\sin 3\theta - \sin 5\theta)
$$

$$
+ \frac{\alpha p_0^3}{96\lambda}\sin\theta(6\theta - 5\sin 2\theta + \sin 4\theta). \tag{4.212}
$$

Equation (4.211) in the limit of small L and Eq. (4.212) in the limit of small x_0 and p_0 can be shown to be consistent.

4.6   Single sextupole

The previous section considered the synchrotron motion of particles in an accelerator containing a single nonlinear thin-lens element, i.e. a localized RF cavity. The rest of the accelerator is regarded as perfectly linear. A similar situation occurs in the transverse x- and y-dimensions. In this section, we will consider a single thin sextupole in a circular accelerator whose beam dynamics is otherwise perfectly linear. One difference, however, is that in this section, we are dealing with a 2-D system (4-D phase space). We consider on-momentum particles, so the longitudinal motion is ignored.

The thin-lens sextupole has the map

$$
M_{\text{sext}} = e^{:-LH:} = \exp[:\lambda(x^3 - 3xy^2):], \tag{4.213}
$$

where $\lambda = -\frac{1}{3}SL$ is the integrated strength of the thin-lens sextupole with S the quantity that appears in the Hamiltonian (4.119),

$$\lambda = -\frac{1}{6}\frac{L}{B\rho}\frac{\partial^2 B_y}{\partial x^2}.$$
(4.214)

One might compare Eq. (4.213) with Eq. (4.89), which applies when the y-motion is ignored. If a particle has $y = 0$ and $y' = 0$ initially, it will stay in the horizontal plane at all times. For those particles, ignoring y-motion, applying Eq. (4.89) is legitimate, but we want to include y-motion in this section.

The rest of the accelerator can be described by the map

$$M_{\mathrm{acc}} = e^{:f_2:},$$

where

$$f_2 = -\frac{\mu_x}{2}(\gamma_x x^2 + 2\alpha_x x p_x + \beta_x p_x^2) - \frac{\mu_y}{2}(\gamma_y y^2 + 2\alpha_y y p_y + \beta_y p_y^2).$$
(4.215)

The quantities $\alpha_{x,y}$, $\beta_{x,y}$ and $\gamma_{x,y}$ are the unperturbed Courant–Snyder parameters evaluated at the position of the sextupole.

We will observe the motion of a particle at the exit end of the sextupole. The one-turn Lie map is

$$M_{\mathrm{acc}}M_{\mathrm{sext}}.$$
(4.216)

Effective Hamiltonian away from resonance   We now concatenate the two factor maps in Eq. (4.216) to become $\exp(:-H_{\mathrm{eff}}:)$, where H_{eff} is an invariant and the effective Hamiltonian when the time variable is taken to be the number of turns n. To 1st order in the sextupole strength, we apply the BCH form (4.140) to obtain the invariant as

$$h = f_2 + \left(\frac{:f_2:}{1 - e^{-:f_2:}}\right)\lambda(x^3 - 3xy^2) + \mathcal{O}(\lambda^2).$$
(4.217)

The $\mathcal{O}(\lambda^2)$ terms in Eq. (4.217) are small for two reasons. First, they are of the order of λ^2, and are therefore small when the sextupole is weak. Second, these terms are also of order $(x, y)^4$, and are small near the origin of the phase space. In what follows, we will simply drop these terms.[26]

We learned in the previous section that, to evaluate (4.217), we need to find the eigenmodes of the operator $:f_2:$. To do so, we also learned to make two successive canonical transformations. The first is

$$\bar{x} = \frac{x}{\sqrt{\beta_x}}, \qquad \bar{p}_x = \frac{\alpha_x x + \beta_x p_x}{\sqrt{\beta_x}},$$
(4.218)

or

$$x = \sqrt{\beta_x}\,\bar{x}, \qquad p_x = \frac{-\alpha_x \bar{x} + \bar{p}_x}{\sqrt{\beta_x}},$$
(4.219)

---

[26]Be reminded that there are potentially subtle nonlinear dynamics effects, such as higher order resonances and chaotic effects, that feed on these higher order terms.

and similar pairs of expressions with x replaced by y. The function f_2 becomes

$$f_2 \;=\; -\frac{\mu_x}{2}(\bar{x}^2 + \bar{p}_x^2) - \frac{\mu_y}{2}(\bar{y}^2 + \bar{p}_y^2)\,.$$

The second transformation is

$$\bar{x} \;=\; \sqrt{2A_x}\,\sin\phi_x, \qquad \bar{p}_x \;=\; \sqrt{2A_x}\,\cos\phi_x\,, \tag{4.220}$$

and a similar pair of expressions with x replaced by y. Both transformations (4.218–4.219) and (4.220) are canonical. The new expression of f_2 is

$$f_2 \;=\; -\mu_x A_x - \mu_y A_y\,.$$

The eigenmodes of $:f_2:$ are, as we found in the previous section, $e^{in_x\phi_x + in_y\phi_y}$ with eigenvalues $in_x\mu_x + in_y\mu_y$. That is,

$$:f_2:A_x \;=\; 0, \qquad :f_2:A_y \;=\; 0\,,$$
$$:f_2:e^{in_x\phi_x + in_y\phi_y} \;=\; (in_x\mu_x + in_y\mu_y)e^{in_x\phi_x + in_y\phi_y}\,. \tag{4.221}$$

We now have

$$\left(\frac{:f_2:}{1 - e^{-:f_2:}}\right)\lambda(x^3 - 3xy^2) \tag{4.222}$$

$$= \lambda\left(\frac{:f_2:}{1 - e^{-:f_2:}}\right)\left[(2\beta_x A_x)^{3/2}\sin^3\phi_x - 3(2\beta_x A_x)^{1/2}(2\beta_y A_y)\sin\phi_x \sin^2\phi_y\right].$$

We then note that

$$\left(\frac{:f_2:}{1 - e^{-:f_2:}}\right)\sin^3\phi_x$$

$$= \frac{i}{8}\left(\frac{:f_2:}{1 - e^{-:f_2:}}\right)\left(e^{i3\phi_x} - 3e^{i\phi_x} + 3e^{-i\phi_x} - e^{-i3\phi_x}\right)$$

$$= \frac{i}{8}\left(\frac{i3\mu_x}{1 - e^{-i3\mu_x}}e^{i3\phi_x} - \frac{i3\mu_x}{1 - e^{-i\mu_x}}e^{i\phi_x} + \frac{-i3\mu_x}{1 - e^{i\mu_x}}e^{-i\phi_x} - \frac{-i3\mu_x}{1 - e^{i3\mu_x}}e^{-i3\phi_x}\right)$$

$$= -\frac{3}{8}\mu_x\left[\frac{\sin(3\phi_x + \frac{3\mu_x}{2})}{\sin\frac{3\mu_x}{2}} - \frac{\sin(\phi_x + \frac{\mu_x}{2})}{\sin\frac{\mu_x}{2}}\right], \tag{4.223}$$

and that

$$\left(\frac{:f_2:}{1 - e^{-:f_2:}}\right)\sin\phi_x \sin^2\phi_y$$

$$= \frac{i}{8}\left(\frac{:f_2:}{1 - e^{-:f_2:}}\right)\left(e^{i\phi_x + i2\phi_y} - 2e^{i\phi_x} + e^{i\phi_x - i2\phi_y}\right.$$

$$\left. - e^{-i\phi_x + i2\phi_y} + 2e^{-i\phi_x} - e^{-i\phi_x - i2\phi_y}\right)$$

$$= -\frac{1}{8}(\mu_x + 2\mu_y)\frac{\sin(\phi_x + 2\phi_y + \frac{\mu_x}{2} + \mu_y)}{\sin(\frac{\mu_x}{2} + \mu_y)} + \frac{1}{4}\mu_x\frac{\sin(\phi_x + \frac{\mu_x}{2})}{\sin\frac{\mu_x}{2}}$$

$$- \frac{1}{8}(\mu_x - 2\mu_y)\frac{\sin(\phi_x - 2\phi_y + \frac{\mu_x}{2} - \mu_y)}{\sin(\frac{\mu_x}{2} - \mu_y)}\,. \tag{4.224}$$

Substituting Eqs. (4.222–4.224) into (4.217) gives an expression of the effective Hamiltonian to 1st order in λ [and to order $\mathcal{O}(A_{x,y}^{3/2})$ in amplitudes],

$$
\begin{aligned}
h \;=\; &-\mu_x A_x - \mu_y A_y - \frac{3}{8}\lambda\mu_x(2\beta_x A_x)^{3/2}\left[\frac{\sin(3\phi_x + \frac{3\mu_x}{2})}{\sin\frac{3\mu_x}{2}} - \frac{\sin(\phi_x + \frac{\mu_x}{2})}{\sin\frac{\mu_x}{2}}\right] \\
&-3\lambda(2\beta_x A_x)^{1/2}(2\beta_y A_y)\left[-\frac{1}{8}(\mu_x + 2\mu_y)\frac{\sin(\phi_x + 2\phi_y + \frac{\mu_x}{2} + \mu_y)}{\sin(\frac{\mu_x}{2} + \mu_y)}\right. \\
&\left.+\frac{1}{4}\mu_x\frac{\sin(\phi_x + \frac{\mu_x}{2})}{\sin\frac{\mu_x}{2}} - \frac{1}{8}(\mu_x - 2\mu_y)\frac{\sin(\phi_x - 2\phi_y + \frac{\mu_x}{2} - \mu_y)}{\sin(\frac{\mu_x}{2} - \mu_y)}\right].
\end{aligned}
\tag{4.225}
$$

We shall continue this discussion. However, Eq. (4.225) contains a disturbing aspect that we need to discuss first, i.e. the appearance of resonances.

Resonance The expression (4.225) contains divergences when one of the following *resonance* conditions is satisfied

$$
\begin{aligned}
\nu_x &= \text{integer}, \\
3\nu_x &= \text{integer}, \\
\nu_x + 2\nu_y &= \text{integer}, \\
\nu_x - 2\nu_y &= \text{integer},
\end{aligned}
\tag{4.226}
$$

where we have defined the unperturbed tunes

$$
\nu_{x,y} \;=\; \frac{\mu_{x,y}}{2\pi}.
$$

Away from resonances (4.226), Eq. (4.225) gives the expression of the invariant. Near resonances, the above analysis leading to Eq. (4.225) breaks down. We need to treat the problem differently, which we will do momentarily.

Note that had we kept some higher order terms in λ, we would have found terms that diverge at higher order resonances. For example, had we kept a term of order X^4, the expression for h would contain a term of the type

$$
\left(\frac{:f_2:}{1 - e^{-:f_2:}}\right)\sin 4\phi_x,
\tag{4.227}
$$

which diverges when $\nu_x = \frac{1}{4} \times$ integer. When dropping these higher order terms in Eqs. (4.217) and (4.225), therefore, we have inadvertently dropped effects of higher order resonances. This is of course quite serious. As a result, if one is interested in knowing the effect of a higher order resonance, it is necessary to keep terms of sufficiently high order for the resonance of interest to show its effect.

Perhaps we illustrate this same comment in a different way. Take for example a lattice design with $\nu_y = 10.17$. It is clear that it can be driven (divergent) by a 100-th order resonance because $100 \times \nu_y = $ integer. To answer this question properly, in principle, we need to have a Hamiltonian expanded to 100th order in X. (See Homework 4.75.)

Even worse, the question arises as to whether the expansion converges at all since there is always some high order resonance of interest even if ν_x and ν_y are irrational numbers. We will not pursue these subtleties however, assuming that the higher order resonances are sufficiently weak to be ignored.[27]

One more comment is in order concerning the form of Eq. (4.225). Note that each resonance of Eq. (4.226) appears in Eq. (4.225) separately, each responsible for a distinct term. This is indication that, to 1st order in λ, each resonance has a clear driving term responsible for its nonlinear dynamical effects. The resonance $3\nu_x =$ integer, for example, would have a driving term proportional to

$$\lambda \beta_x^{3/2} A_x^{3/2} \sin\left(3\phi_x + \frac{3\mu_x}{2}\right).$$

Driving terms of the other resonances can be identified similarly. When there are multiple sextupoles, resonance driving terms, to 1st order in sextupole strengths, can be found by superposition.

Away from resonance, 1-D case   Before we proceed to treat the case near a resonance, for simplicity, we now consider a 1-D case, again away from resonances. This can be obtained by considering only those particles with no y-motion $(y = 0, y' = 0)$. We will further drop the subscripts x. The one-turn map then becomes

$$e^{:-\mu A:}e^{:\lambda x^3:}. \tag{4.228}$$

This map (4.228) is also called a *Hénon map* or a *standard map* in the literature.[28]

The invariant, to 1st order in λ and away from resonances, is

$$
\begin{aligned}
h &= -\mu A + \left(\frac{:-\mu A:}{1 - e^{:\mu A:}}\right)\lambda x^3 \\
&= -\mu A - \frac{3}{8}\mu\lambda(2\beta A)^{3/2}\left[\frac{\sin(3\phi + \frac{3\mu}{2})}{\sin\frac{3\mu}{2}} - \frac{\sin(\phi + \frac{\mu}{2})}{\sin\frac{\mu}{2}}\right]. \tag{4.229}
\end{aligned}
$$

The effective Hamiltonian is of course given by $H_{\text{eff}} = -h$. Equation (4.229) is just Eq. (4.225) when $A_y = 0$.

Particles move in the phase space along contours of constant h. To study the phase space topology, let us go from the (ϕ, A) back to the $(\bar{x}, \bar{p})$ coordinates according to Eq. (4.220). Then Eq. (4.229) becomes

$$h = -\frac{\mu}{2}(\bar{x}^2 + \bar{p}^2) - \frac{3}{8}\mu\lambda\beta^{3/2}\bar{x}\left[(3\bar{p}^2 - \bar{x}^2)\cot\frac{3\mu}{2} - (\bar{x}^2 + \bar{p}^2)\cot\frac{\mu}{2} - 4\bar{x}\bar{p}\right]. \tag{4.230}$$

---

[27]Or more accurately, assuming they are sufficiently weak so that a tiny detuning effect can suppress them.

[28]See for example, A. Dragt, Sec. 2.4.6, Handbook Accel. Phys. & Eng., 2nd ed., World Scientific (2013).

Figure 4.6: Contour plots in the $(\bar{x}, \bar{p})$ phase space with a sextupole perturbation, away from resonances.

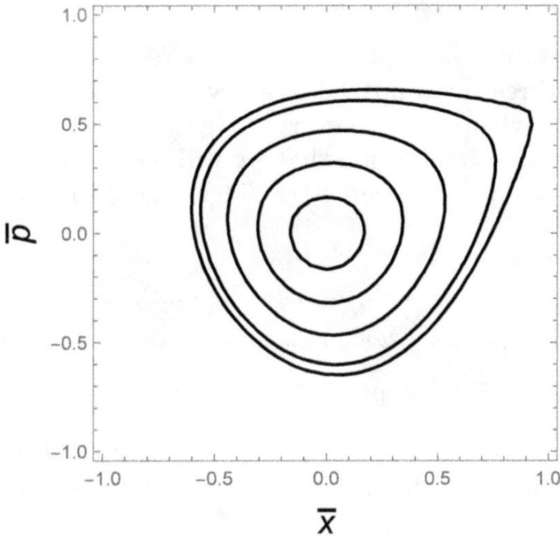

Figure 4.6 shows these contours in the $(\bar{x}, \bar{p})$ space for the case with $\nu = \frac{\mu}{2\pi} = 0.17$ and $\frac{3}{8}\lambda\beta^{3/2} = 0.1$. The five contours correspond to $-\frac{h}{\mu} = 0.0137, 0.0515, 0.109, 0.180$ and 0.209. As seen in Fig. 4.6, a small-amplitude particle traces out a circular contour. As the amplitude increases, the contour distorts from a circle. The reason the contours are not plotted further out in Fig. 4.6 is because the invariant expression (4.230) breaks down there as a resonance is approaching.

Effective Hamiltonian near an isolated resonance   The invariance of (4.229), and therefore the validity of Fig. 4.6, requires ν and 3ν to be away from an integer, i.e. away from resonances. It turns out that being away from all resonances is not the only case when our analysis can be applied to obtain useful results. It is quite amazing that another case to the other extreme, i.e. the case when a resonance is very close by, can also be treated analytically. The condition is that there must be one and only one resonance that plays the dominating role. If there are multiple comparable resonances competing, most likely chaos emerges,[29] and our analysis breaks down.

To illustrate how to deal with the situation when there is one and only one single resonance nearby, consider the case when

$$\nu \approx \frac{p}{3},$$
(4.231)

---

[29] It is as if the particle loses track of which resonance contour to follow, so it ends up with random walking.

where p is some integer. Let d be the distance of ν from the resonance with

$$\nu = \frac{p+d}{3}, \qquad |d| \ll 1. \tag{4.232}$$

The clever trick of treating the case near a resonance (4.231) is to observe the system every three turns instead of every turn as we have been doing. The motion of particles, when observed every three turns, will appear to move slowly. And it is this slow (strobe) motion that we will now study — slow dynamics is what our analysis tend to handle better. Thus, the 3-turn map is

$$\begin{aligned}
M^3 &= (e^{:-\mu A:}e^{:\lambda x^3:})^3 \\
&= e^{:-\mu A:}e^{:\lambda x^3:}e^{:-\mu A:}e^{:\lambda x^3:}e^{:-\mu A:}e^{:\lambda x^3:} \\
&= e^{:-3\mu A:}e^{:2\mu A:}e^{:\lambda x^3:}e^{:-2\mu A:}e^{:\mu A:}e^{:\lambda x^3:}e^{:-\mu A:}e^{:\lambda x^3:} \\
&= e^{:-3\mu A:}\exp(:e^{:2\mu A:}\lambda x^3:)\exp(:e^{:\mu A:}\lambda x^3:)e^{:\lambda x^3:}.
\end{aligned} \tag{4.233}$$

The first operator $e^{:-3\mu A:}$ in the last line of Eq. (4.233) takes on a different form near the resonance. In fact, this is where the strobe action is taking place, i.e.

$$e^{:-3\mu A:} = e^{:-6\pi\nu A:} = e^{:-2\pi d A:},$$

where the last step is because

$$e^{:2\pi p A:} = 1 \tag{4.234}$$

identically for any integer p.

The remaining three operators in Eq. (4.233) can be combined to first order in λ easily,

$$\begin{aligned}
\exp(:e^{:2\mu A:}&\lambda x^3:)\exp(:e^{:\mu A:}\lambda x^3:)e^{:\lambda x^3:} \\
&= \exp[:(e^{:2\mu A:}+e^{:\mu A:}+1)\lambda x^3 + \mathcal{O}(\lambda^2):] \\
&= \exp\left[:\left(\frac{1-e^{:3\mu A:}}{1-e^{:\mu A:}}\right)\lambda x^3 + \mathcal{O}(\lambda^2):\right] \\
&= \exp\left[:\left(\frac{1-e^{:2\pi d A:}}{1-e^{:\mu A:}}\right)\lambda x^3 + \mathcal{O}(\lambda^2):\right].
\end{aligned}$$

We then apply the BCH formula to obtain

$$M^3 = e^{:3h:},$$

where $3h$ is a new expression of the invariant near the resonance (4.231) and is given by

$$\begin{aligned}
3h &= -2\pi d A + \left(\frac{:-2\pi d A:}{1-e^{:2\pi d A:}}\right)\left(\frac{1-e^{:2\pi d A:}}{1-e^{:\mu A:}}\right)\lambda x^3 \\
&= -2\pi d A + \left(\frac{:-2\pi d A:}{1-e^{:\mu A:}}\right)\lambda x^3.
\end{aligned}$$

The reason it is $3h$ instead of h that appears in the concatenated total map is that the particle motion is now observed every three turns and the natural time unit is 3. This leads to, after a few steps of algebra,

$$h = -\frac{2\pi d}{3}A - \frac{\pi d\lambda}{4}(2\beta A)^{3/2}\left[\frac{\sin\left(3\phi + \frac{3\mu}{2}\right)}{\sin\frac{3\mu}{2}} - \frac{\sin\left(\phi + \frac{\mu}{2}\right)}{\sin\frac{\mu}{2}}\right]. \tag{4.235}$$

The expression (4.235), unlike Eq. (4.229), is well-behaved near the resonance as $d \to 0$. Interestingly, it is very similar to Eq. (4.229), provided one pretends that μ in the one-turn map can be replaced by $\frac{2\pi d}{3}$, the phase advance by the strobe action. To see this, let us consider the invariant (4.229), and insist on applying it even near a 3rd order resonance $\nu \approx \frac{1}{3}$, we may keep only the $\sin\frac{3\mu}{2}$ term in Eq. (4.229), i.e.

$$h \approx -\mu A - \frac{3}{8}\mu\lambda(2\beta A)^{3/2}\frac{\sin\left(3\phi + \frac{3\mu}{2}\right)}{\sin\frac{3\mu}{2}}. \tag{4.236}$$

Multiplying h by the factor $\frac{2\pi d}{3\mu}$ is still an invariant — although if we call it the effective Hamiltonian, then we are slowing down the motion from the regular time to the strobing time. Thus we write

$$h \approx -\frac{2\pi d}{3}A - \frac{\pi d\lambda}{4}(2\beta A)^{3/2}\frac{\sin\left(3\phi + \frac{3\mu}{2}\right)}{\sin\frac{3\mu}{2}}, \tag{4.237}$$

which is the same as Eq. (4.235) when $d \to 0$.

We further note that

$$\frac{\sin\left(3\phi + \frac{3\mu}{2}\right)}{\sin\frac{3\mu}{2}} \approx \frac{\sin 3\phi}{\pi d}.$$

The invariant then becomes

$$h \approx -\frac{2\pi d}{3}A - \frac{\lambda}{\sqrt{2}}(\beta A)^{3/2}\sin 3\phi. \tag{4.238}$$

We can now study the phase space topology based on (4.238) near the resonance (4.231). Again make the transformation to the Cartesian coordinates by Eq. (4.220),[30]

$$h \approx -\frac{\pi d}{3}(\bar{x}^2 + \bar{p}^2) - \frac{\lambda\beta^{3/2}}{4}\bar{x}(3\bar{p}^2 - \bar{x}^2). \tag{4.239}$$

Because the invariant (4.239) has the significance of being the effective Hamiltonian observing the strobe motion, it gives the fixed points[31] according to

$$\frac{\partial h}{\partial \bar{x}} = -\frac{2\pi d}{3}\bar{x} - \frac{\lambda\beta^{3/2}}{4}(3\bar{p}^2 - 3\bar{x}^2) = 0,$$

---

[30]Note in this case, however, $\bar{x}$ and $\bar{p}$ are not related to the original x and p by Eq. (4.218) because of the strobing.

[31]Fixed points are special points in phase space. A particle located at a fixed point will stay there turn after turn.

$$\frac{\partial h}{\partial \bar{p}} = -\frac{2\pi d}{3}\bar{p} - \frac{\lambda\beta^{3/2}}{4}6\bar{x}\bar{p} = 0. \qquad (4.240)$$

There are four fixed points corresponding to the four solutions to Eq. (4.240). One of them is the origin $(\bar{x},\bar{p}) = (0,0)$. The other three are

$$(\bar{x},\bar{p}) = \begin{cases} (\epsilon,0), \\ (-\frac{1}{2}\epsilon, \frac{\sqrt{3}}{2}\epsilon), \\ (-\frac{1}{2}\epsilon, -\frac{\sqrt{3}}{2}\epsilon), \end{cases} \qquad (4.241)$$

where we have introduced a new parameter

$$\epsilon = \frac{8\pi d}{9\lambda\beta^{3/2}}. \qquad (4.242)$$

At the three fixed points (4.241), the value of h is given by

$$h = -\frac{\pi}{9}d\epsilon^2. \qquad (4.243)$$

Particles move in the phase space along contours of constant Hamiltonian $-h$. The constant-h contour that goes through the fixed points are called the separatrix. To find the separatrix, we set h of Eq. (4.239) to be equal to (4.243). This gives

$$\left(\bar{x}+\frac{1}{2}\epsilon\right)\left(\bar{x}-\epsilon+\sqrt{3}\bar{p}\right)\left(\bar{x}-\epsilon-\sqrt{3}\bar{p}\right) = 0. \qquad (4.244)$$

The separatrix of the system (4.239), therefore, consists of three straight lines, as shown in Fig. 4.7(a).

Comparing Fig. 4.7(a), the phase space contours near resonance, with Fig. 4.6, the contours away from resonances, one notes that the contours near resonances are more or less triangular in shape, while the contours away from resonances are more approximately like circles.

Figure 4.7(a) shows the contour plot of the function $\frac{\xi}{\epsilon^3}$ in the $(\frac{\bar{x}}{\epsilon}, \frac{\bar{p}}{\epsilon})$ space, where

$$\xi = -\frac{8}{3\lambda\beta^{3/2}}\left(h+\frac{\pi d\epsilon^2}{9}\right) = \epsilon(\bar{x}^2+\bar{p}^2)+\frac{2}{3}\bar{x}(3\bar{p}^2-\bar{x}^2)-\frac{1}{3}\epsilon^3.$$

The dashed straight lines are the separatrices. The three intersection points are the fixed points (4.241). The solid curves are the constant-h contours. The numerical values for the contours shown are when $\frac{\xi}{\epsilon^3} = -0.31, -0.18, 0, 0.5$, respectively.

The topology scales with the parameter ϵ, i.e. essentially the ratio of distance to resonance d to the sextupole strength λ. The topology for the case with negative ϵ is given by the mirror reflection of Fig. 4.7(a) with respect to the vertical $\bar{x} = 0$ line.

When far away from resonances or when the sextupole strength vanishes, $\epsilon = \infty$, we have the unperturbed case where the topology consists of concentric circles.

Figure 4.7: Phase space contour plots, (a) near a third order resonance, (b) exactly on resonance.

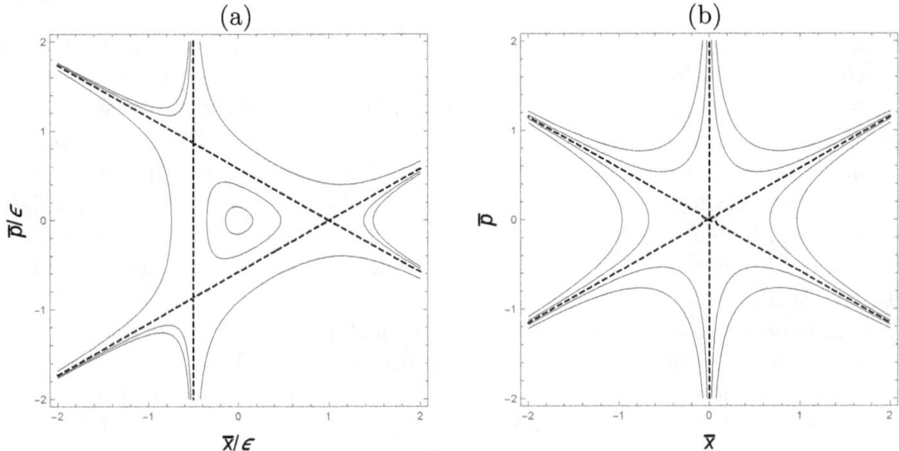

Right on resonance, when $\epsilon = 0$, scaling with ϵ as done in Fig. 4.7(a) is no longer applicable, the phase space looks like Fig. 4.7(b), the contour plot of ξ in the $(\bar{x}, \bar{p})$ space. The three fixed points and the origin coincide. Values of ξ for the contours are $-0.6, -0.2, 0, 0.2, 0.6$, respectively. Exactly on resonance, there is no region in the phase space that provides stable motion for the particles.

Effective Hamiltonian away from resonance — 2-D case continued
We now return to the 2-D case starting with the effective invariant (4.225) away from resonances. The effective Hamiltonian is of course one of the invariants. We first rewrite it as

$$-h \;=\; \mu_x W_x + \mu_y W_y \,, \tag{4.245}$$

where

$$
\begin{aligned}
W_x \;=\;& A_x + \frac{3}{8}\lambda(2\beta_x A_x)^{3/2}\left[\frac{\sin(3\phi_x + \frac{3\mu_x}{2})}{\sin\frac{3\mu_x}{2}} - \frac{\sin(\phi_x + \frac{\mu_x}{2})}{\sin\frac{\mu_x}{2}}\right]\\
&+ \frac{3}{8}\lambda(2\beta_x A_x)^{1/2}(2\beta_y A_y)\left[-\frac{\sin(\phi_x + 2\phi_y + \frac{\mu_x}{2} + \mu_y)}{\sin(\frac{\mu_x}{2} + \mu_y)}\right.\\
&\left.+2\frac{\sin(\phi_x + \frac{\mu_x}{2})}{\sin\frac{\mu_x}{2}} - \frac{\sin(\phi_x - 2\phi_y + \frac{\mu_x}{2} - \mu_y)}{\sin(\frac{\mu_x}{2} - \mu_y)}\right],\\
W_y \;=\;& A_y + \frac{3}{4}\lambda(2\beta_x A_x)^{1/2}(2\beta_y A_y)\left[-\frac{\sin(\phi_x + 2\phi_y + \frac{1}{2}\mu_x + \mu_y)}{\sin(\frac{\mu_x}{2} + \mu_y)}\right.\\
&\left.+ \frac{\sin(\phi_x - 2\phi_y + \frac{\mu_x}{2} - \mu_y)}{\sin(\frac{\mu_x}{2} - \mu_y)}\right].
\end{aligned}
\tag{4.246}
$$

We then make the physical observation that if we make the artificial replacements

$$\mu_x \to \mu_x + 2\pi n_x \quad \text{and} \quad \mu_y \to \mu_y + 2\pi n_y \tag{4.247}$$

for arbitrary integers of $n_{x,y}$, the quantity h in Eq. (4.245) must still be an invariant. This is because the sextupole is located at a fixed location, and it will not know the integer part of the x- and y-tunes. Since the expressions of W_x and W_y in Eq. (4.246) are unchanged by the replacements (4.247), an inspection of Eq. (4.245) shows that the only way for h to be an invariant is that W_x and W_y are *separately* invariants by their own. That is, there are actually two invariants W_x and W_y, instead of only one invariant h. The important difference between $W_{x,y}$ and h, however, is that h maintains the unique physical meaning of the effective Hamiltonian of the system.

One may try to apply this argument to the 1-D case, but that does not lead to any information of new invariants. The quantity h in the 1-D case contains a multiplicative factor μ, which can be simply scaled away, as we did in the step from (4.236) to (4.237). As mentioned, the rescaling from (4.236) to (4.237) only slows down the Hamiltonian motion from the regular time to the stroboscopic time.

One can actually prove mathematically the invariance of W_x and W_y using the fact that $-h$ is the effective Hamiltonian. One first note that the rates of change of W_x per turn is given by

$$\frac{dW_x}{dn} = \frac{\partial W_x}{\partial A_x}\frac{dA_x}{dn} + \frac{\partial W_x}{\partial A_y}\frac{dA_y}{dn} + \frac{\partial W_x}{\partial \phi_x}\frac{d\phi_x}{dn} + \frac{\partial W_x}{\partial \phi_y}\frac{d\phi_y}{dn}, \tag{4.248}$$

and a similar expression for $\frac{dW_y}{dn}$. It then follows straightforwardly from

$$\frac{dA_x}{dn} = \frac{\partial h}{\partial \phi_x}, \qquad \frac{d\phi_x}{dn} = -\frac{\partial h}{\partial A_x},$$
$$\frac{dA_y}{dn} = \frac{\partial h}{\partial \phi_y}, \qquad \frac{d\phi_y}{dn} = -\frac{\partial h}{\partial A_y}, \tag{4.249}$$

and the expression of h that $\frac{dW_x}{dn} = \frac{dW_y}{dn} = 0$, i.e., W_x and W_y are invariants.

We will return to the topic of single sextupoles later after we introduce the normal form technique. We will discuss the case of a general distribution of sextupoles. The study of strategically arranged distributions of sextupoles will also be discussed under the subject of achromats.

Homework 4.75 Consider vertical y-motion of a perfectly linear storage ring with tune $\nu_y = 10.17$ exactly. The Lie map of the ring is given by $e^{:-2\pi\nu_y A:}$. In this storage ring, we now insert a small thin perturbation $e^{:\epsilon y^{100}:}$. The system is exactly on a resonance $\nu_y = \frac{1017}{100}$. Follow the text to analyze its phase space topology to 1st order in ϵ.

Homework 4.76 Following Eq. (4.228), the text analyzed the system when observed at the exit of the thin sextupole. Repeat the analysis this time observing the system before the sextupole. Examine its phase space away from resonances and compare it with Fig. 4.6.

Homework 4.77 Figure 4.6 gives the trajectories of particles when Eq. (4.230) is the effective Hamiltonian. Write a numerical tracking program for the exact system (4.216). Compare the results with those shown in Fig. 4.6.

Homework 4.78 We worked out an expression of invariant W_3 in Eq. (4.33). Show that it agrees with the invariant h of Eq. (4.230).

Solution Away from resonances, Eqs. (4.230) and (4.33) are identical if one notes $\epsilon = 3\lambda\beta^{3/2}$, and the trigonometric identities

$$\cot\frac{3\mu}{2} + \cot\frac{\mu}{2} = \frac{4\sin\mu\cos\mu}{(1 + 2\cos\mu)(1 - \cos\mu)},$$

$$3\cot\frac{3\mu}{2} - \cot\frac{\mu}{2} = -\frac{4\sin\mu}{1 + 2\cos\mu}.$$

Homework 4.79 Equations (4.239–4.242) studied the phase space topology in the Cartesian coordinates. One can do the same in the polar coordinates (ϕ, A) using Eq. (4.238). Show that the fixed points determined by $\frac{\partial h}{\partial A} = 0$ and $\frac{\partial h}{\partial \phi} = 0$ in the (ϕ, A) space give the same result as Eq. (4.241).

Solution Take the case $\epsilon > 0$ for example. The solutions are given by $A = \frac{\epsilon^2}{2}$ and $\phi = 90°, 210°, 330°$.

Homework 4.80 Consider the Taylor map (4.32). Find the fixed points in the (x, p) space. What is the connection, if any, between these fixed points and the fixed points found using the effective Hamiltonian (4.230)?

Solution To be a fixed point for the map (4.32), it is necessary that

$$x = x\cos\mu + p\sin\mu,$$
$$p = -x\sin\mu + p\cos\mu + \epsilon x^2.$$

There are two solutions. One is the trivial case of the origin. The other is located at

$$x = \frac{2}{\epsilon}\tan\frac{\mu}{2}, \qquad p = \frac{2}{\epsilon}\tan^2\frac{\mu}{2}. \tag{4.250}$$

The fixed point (4.250) is not the fixed point for Eq. (4.230). The fixed point in Fig. 4.6 for example is located at $(\bar{x}, \bar{p}) = (0.95, 0.56)$ while Eq. (4.250) gives $(1.48, 0.87)$ using the same parameters $(\mu = 2\pi \times 0.17, \epsilon = 3\lambda\beta^{3/2} = 0.8)$. Give a reason why they differ.

Homework 4.81 Verify the invariance of W_x and W_y in Eq. (4.246) by carrying out the steps (4.248–4.249).

Homework 4.82 Consider a thin-lens octupole in an otherwise perfectly linear optics of a circular accelerator. Consider 1-D motion. The Lie map is $e^{:f_2:}e^{:\lambda x^4:}$. Find the effective Hamiltonian to 1st order in λ. First do this away from resonances. Then repeat the analysis close to a resonance $\nu_x \approx \frac{1}{4}$.

Homework 4.83 Extend the calculation in the text to find an expression of the effective Hamiltonian for a single sextupole to 2nd order in the sextupole strength. Do this for the 1-D case. Results obtained in Homeworks 4.82 and 4.83 can be used to calculate the tune shift with betatron amplitude, as will be described later in Secs. 4.8.4 and 4.9.3.

4.7   Distribution of multipole

In the preceding section, we considered the effects of a single sextupole in an otherwise perfectly linear accelerator. In this section, we will consider the effects of a distribution of multipoles — of which sextupoles are a special case. The topic of achromats — special distribution of multipoles that minimizes optical aberrations — is postponed to a later section 4.10.5.

4.7.1   One-turn map

As illustrated in Fig. 4.3, an accelerator typically consists of elements of piecewise sections, each given by its specific Hamiltonian. Let the i-th element have Hamiltonian H_i and length L_i, then we have the total Lie map M_{tot} of the accelerator given by Eq. (4.134), where N is the total number of elements in the accelerator. For a circular accelerator, this total map gives the one-turn map around the accelerator starting from the entrance of element $i = 1$.

Element map   These Hamiltonians H_i are in general nonlinear. One way to proceed is as follows. First, for each element, the Hamiltonian H_i is Taylor expanded as

$$H_i = H_{2i} + H_{3i} + H_{4i} + \cdots , \qquad (4.251)$$

where H_{ki} is a homogeneous polynomial that contains terms k-th order in the components of the dynamic variable X. The leading order is taken to be quadratic in X. If H_i contains only second order terms, i.e. if $H_i = H_{2i}$, we have a linear element. We regard the nonlinear terms to be small, which would be the case if we restrict our attention to phase space regions close to the origin.

We will consider Hamiltonians up to $(\Omega+1)$-th order in X. This means maps are considered up to the Ω-th order. Higher order terms are truncated. The analysis below applies exactly for Hamiltonians quadratic in X, but represents a perturbation analysis as far as the higher order terms are concerned.

Having made the Taylor expansion (4.251), the map of the i-th element reads

$$e^{-:L_i H_i:} = e^{-:L_i[H_{2i}+H_{3i}+\cdots+H_{(\Omega+1)i}]:} . \qquad (4.252)$$

The BCH formula can be applied to factorize (4.252) to become

$$e^{-:L_iH_i:} \;=\; e^{:f_{2i}:}e^{:f_{3i}:}\cdots e^{:f_{(\Omega+1)i}:}\,, \tag{4.253}$$

where the function f_{ki} is a k-th order homogeneous polynomial in the components of X. Knowing $H_{2i},\dots,H_{(\Omega+1)i}$, we are looking for expressions for $f_{2i},\dots,f_{(\Omega+1)i}$.

The expression for f_{2i} is simply

$$f_{2i} \;=\; -L_iH_{2i}\,. \tag{4.254}$$

The next order term f_{3i} follows from Eq. (4.143),

$$f_{3i} \;=\; -L_i\int_0^1 du\; e^{:uL_iH_{2i}:}H_{3i}\,. \tag{4.255}$$

The reader is reminded of Eq. (4.87), which gives f_3 when the Taylor map is known.

The higher order factors can be obtained by iteration. For example,

$$\begin{aligned}
f_4 \;=\;& -L_i\int_0^1 du\; e^{:uL_iH_{2i}:}H_{4i}\\
& -\frac{L_i^2}{2}\int_0^1 du\int_0^u dv\;[e^{:vL_iH_{2i}:}H_{3i},\, e^{:uL_iH_{2i}:}H_{3i}]\,.
\end{aligned} \tag{4.256}$$

Homework 4.84 illustrates the procedure of Eqs. (4.253–4.256) by a simple example. Derivation of Eq. (4.256) is left to Homework 4.85.

For later convenience, we summarize the factorization Eqs. (4.254–4.256) as follows,

$$e^{:h_2+h_3+h_4:} \;=\; e^{:f_2:}e^{:f_3:}e^{:f_4:}\cdots$$

$$\implies \begin{cases}
f_2 \;=\; h_2\,,\\
f_3 \;=\; \int_0^1 du\, e^{:-uh_2:}h_3\,,\\
f_4 \;=\; \int_0^1 du\, e^{:-uh_2:}h_4 - \frac{1}{2}\int_0^1 du\int_0^u dv\,[e^{:-vf_2:}h_3,\, e^{:-uf_2:}h_3]\,.
\end{cases} \tag{4.257}$$

Total map Having obtained the maps of all individual elements in the form (4.253), the total map M_{tot} is expressed as the product of a long string of factor maps. A procedure is followed using the BCH formula to commute the factor maps in M_{tot} in such a way that at the end of the procedure it reads

$$M_{\mathrm{tot}} \;=\; e^{:f_2:}e^{:f_3:}\cdots e^{:f_{\Omega+1}:}\,, \tag{4.258}$$

where again f_k is a k-th order homogeneous polynomial in X. The Courant–Snyder linear one-turn map is just $e^{:f_2:}$. In fact, f_2 is the Courant–Snyder invariant as given by Eq. (4.67) in the 1-D case, and by Eq. (4.215) in the 2-D case.

After M_{tot} has been obtained in the form of Eq. (4.258), the BCH formula is applied to concatenate all the factors into a single factor,

$$M_{\text{tot}} \;=\; e^{:F:}\,. \tag{4.259}$$

where F is related to the effective Hamiltonian, as was mentioned in Eq. (4.135), by $F = -CH_{\text{eff}}$. This is how the effective Hamiltonian can be obtained explicitly as a Taylor expansion in X. The quantity F obtained here is an $(\Omega+1)$-th order Taylor series in X. It is an invariant because (see Homework 4.27)

$$M_{\text{tot}}F \;=\; F\,.$$

The effective Hamiltonian, once obtained, contains a wealth of analytical information on the nonlinear dynamics of the circular accelerator system. The procedure described here is a natural way to generalize the Courant–Snyder analysis to a nonlinear system. In particular, the leading quadratic terms of F give the Courant–Snyder invariant. Including higher order terms, F then acquires the physical meaning of the generalized nonlinear Courant–Snyder invariant to the $(\Omega+1)$-th order.

Particle tracking using Lie map   In passing, Lie maps can also be used for numerical particle tracking, although not its main strength. When doing so, it is often necessary to make sure the numerical application keeps the symplecticity of the map. For example, one may consider applying the one-turn map (4.259) for particle tracking. One notes first that F, being truncated to become an $(\Omega+1)$-th order Taylor series, is only an approximate expression. Nevertheless, the map (4.259) remains exactly symplectic. To perform particle tracking, one would attempt to compute

$$(X_j)_{\text{final}} \;=\; \left. e^{:F:}X_j \right|_{X=X_{\text{initial}}}, \tag{4.260}$$

where X_j represents the j-th component of X, and one needs to compute (4.260) for all the components. To do that, one expands the exponential operator into an infinite series,

$$(X_j)_{\text{final}} \;=\; \sum_{n=0}^{\infty} \frac{1}{n!}:F:^{n}X_j \bigg|_{X=X_{\text{initial}}}. \tag{4.261}$$

Each of the terms in Eq. (4.261) can be evaluated exactly because F is a Taylor series that terminates. Application of a finite power of $:F:$ on X is therefore a terminated Taylor series, which can be evaluated exactly. However, Eq. (4.261) is an infinite series. To perform numerical tracking, it is necessary to truncate it to a certain order. This truncation, unlike the one that truncates F to $(\Omega+1)$-th order, is in general nonsymplectic, and must be done at a sufficiently high order that the nonsymplecticity introduced is negligible numerically.

This inconvenience of having to check symplecticity numerically for tracking can be avoided in several ways. As an example, we mention one way as follows. Instead of tracking with (4.259) or (4.258), one could factorize the map as

$$M_{\text{tot}} = \text{product of } (e^{:\text{monomial}:}). \tag{4.262}$$

Each factor in Eq. (4.262) is a Lie map whose exponent contains a single monomial term in X. The highest order monomial in the representation (4.262) is $(\Omega + 1)$. The number of factors in (4.262) is therefore finite. If all factors in Eq. (4.262) are concatenated into one exponential map, and the exponent is truncated to $(\Omega + 1)$-th order, one regains (4.259). An example of the factorization (4.262) was given in Homework 4.67.

The factorized form (4.262) now allows exactly symplectic tracking because one can now apply the exact formulae (4.94). The price to pay is that there is a large number of factors in Eq. (4.262) if a high order map is required, which slows down the tracking program.

An alert reader would have noted that the effective Hamiltonian F is not so streamlined for particle tracking and wonder why we care to calculate it in the first place. This is indeed the case. The usefulness of $F = -CH_{\text{eff}}$ lies in the analysis of the beam dynamics, the normal form analysis that we shall discuss soon.

Concatenating nonlinear maps We have thus described how a one-turn Lie map can be obtained once the accelerator design is given, and how the one-turn Lie map might be used for particle tracking. To illustrate the concatenation procedure more explicitly, consider the following. Suppose we are interested in a 3rd order map ($\Omega = 3$), and we have two element-maps which have been factorized into the form (4.253), i.e. we have

$$e^{:f:} = e^{:f_2:}e^{:f_3:}e^{:f_4:} \qquad \text{and} \qquad e^{:g:} = e^{:g_2:}e^{:g_3:}e^{:g_4:}.$$

Our job is to concatenate these two maps into the form (4.258). To do so, let us write the total map as

$$M = e^{:f:}e^{:g:}.$$

We then note that

$$\begin{aligned}
M &= e^{:f_2:}e^{:f_3:}e^{:f_4:}e^{:g_2:}e^{:g_3:}e^{:g_4:} \\
&= \left(e^{:f_2:}e^{:g_2:}\right)\left(e^{-:g_2:}e^{:f_3:}e^{:g_2:}\right)\left(e^{-:g_2:}e^{:f_4:}e^{:g_2:}\right)e^{:g_3:}e^{:g_4:}. \tag{4.263}
\end{aligned}$$

The idea of the second line is to commute the lower order maps to the left of the expression. The first pair of brackets describes the linear map of the combined map, i.e.

$$e^{:f_2:}e^{:g_2:} = e^{:h_2:},$$

where the quadratic h_2 is such that the matrix relation [see Eqs. (4.71–4.72)]

$$e^{SH} = e^{SG}e^{SF}$$

holds, where the symmetric matrices F, G and H are related to f_2, g_2 and h_2 by $f_2 = -\frac{1}{2}\tilde{X}FX$, etc.

The second pair of brackets in Eq. (4.263) can be written as

$$e^{-:g_2:}e^{:f_3:}e^{:g_2:} = \exp\left(:e^{:-g_2:}f_3:\right) = \exp\left[:f_3(e^{:-g_2:}X):\right] = \exp\left[:f_3(e^{-SG}X):\right],$$

i.e. it is a second order map described by $e^{:f_3:}$, except that the arguments of f_3 are transformed from the old coordinates X to the new coordinates according to $e^{:-g_2:}$. Note the minus sign, i.e. it is the inverse of $e^{:g_2:}$ that appears in this scheme. The reader is suggested to associate it with a physical explanation.

Similarly, maps in the third pair of brackets in Eq. (4.263) can be combined into

$$\exp\left[:f_4(e^{-SG}X):\right].$$

We therefore have

$$
\begin{aligned}
M &= e^{:h_2:}\exp\left(:f_3(e^{-SG}X):\right)\exp\left(:f_4(e^{-SG}X):\right)e^{:g_3:}e^{:g_4:} \qquad (4.264)\\
&= e^{:h_2:}\left[\exp\left(:f_3(e^{-SG}X):\right)e^{:g_3:}\right]\left[e^{-:g_3:}\exp\left(:f_4(e^{-SG}X):\right)e^{:g_3:}\right]e^{:g_4:}.
\end{aligned}
$$

The two maps in the first pair of square brackets of Eq. (4.264) are both second order. They can be concatenated using the BCH formula of the first form (4.137) to read

$$
\begin{aligned}
\exp\left(:f_3(e^{-SG}X):\right)e^{:g_3:} &= \exp\left(:f_3(e^{-SG}X)+g_3+\frac{1}{2}[f_3(e^{-SG}X),g_3]+\mathcal{O}(X^5):\right)\\
&= \exp\left(:f_3(e^{-SG}X)+g_3:\right)\exp\left(:\frac{1}{2}[f_3(e^{-SG}X),g_3]:\right),
\end{aligned}
$$

where the square brackets are the Poisson brackets. The term represented by the Poisson brackets is 4th order in X. Terms equal or higher ordered than 5th are dropped in the last line.

The second pair of square brackets in Eq. (4.264), to 4th order in X, is simply

$$
\begin{aligned}
e^{-:g_3:}\exp\left(:f_4(e^{-SG}X):\right)e^{:g_3:} &= \exp\left(:e^{-:g_3:}f_4(e^{-SG}X):\right)\\
&= \exp\left(:f_4(e^{-SG}X)+\mathcal{O}(X^5):\right).
\end{aligned}
$$

Combining the results so far gives

$$
\begin{aligned}
M &= e^{:h_2:}\exp\left(:f_3(e^{-SG}X)+g_3:\right)\exp\left(:\frac{1}{2}[f_3(e^{-SG}X),g_3]:\right)\\
&\quad\times\exp\left(:f_4(e^{-SG}X):\right)e^{:g_4:}.
\end{aligned}
$$

If we then write

$$M = e^{:h_2:}e^{:h_3:}e^{:h_4:},$$

we have the explicit final results,

$$
\begin{aligned}
h_3 &= f_3(e^{-SG}X)+g_3,\\
h_4 &= \frac{1}{2}[f_3(e^{-SG}X),g_3]+f_4(e^{-SG}X)+g_4.
\end{aligned}
$$

Homework 4.84 As an illustration of the factorization procedure described from Eq. (4.253) to Eq. (4.256), factorize the 3rd order map for a thick magnet with combined sextupolar and octupolar fields,

$$M \;=\; \exp\left(: -\frac{L}{2}p^2 - Sx^3 - \epsilon x^4 : \right),$$

into a form

$$M \;=\; e^{:f_2:}\,e^{:f_3:}\,e^{:f_4:}\,e^{:\mathcal{O}(X^5):}\,.$$

Find expressions for $f_{2,3,4}$. This exercise can be compared with Homework 4.67.

Solution

$$f_2 \;=\; -\frac{L}{2}p^2\,,$$

$$
\begin{aligned}
f_3 \;&=\; -S\int_0^1 du\,\left(e^{-u:f_2:}x\right)^3 \;=\; -S\int_0^1 du\,(x - uLp)^3 \\
&=\; -S\left(x^3 - \frac{3}{2}Lx^2p + L^2xp^2 - \frac{1}{4}L^3p^3 \right),
\end{aligned}
$$

$$
\begin{aligned}
f_4 \;&=\; -\epsilon\int_0^1 du\,\left(e^{-u:f_2:}x\right)^4 - \frac{S^2}{2}\int_0^1 du\int_0^u dv[(e^{-v:f_2:}x)^3,(e^{-u:f_2:}x)^3] \\
&=\; -\epsilon\int_0^1 du(x - uLp)^4 - \frac{S^2}{2}\int_0^1 du\int_0^u dv[(x-vLp)^3,(x-uLp)^3] \\
&=\; -\epsilon\left(x^4 - 2Lx^3p + 2L^2x^2p^2 - L^3xp^3 + \frac{1}{5}L^4p^4 \right) \\
&\qquad - \frac{S^2L}{2}\left(-\frac{3}{28}L^4p^4 + \frac{3}{4}L^3xp^3 - \frac{9}{4}L^2x^2p^2 + 3Lx^3p - \frac{3}{2}x^4 \right).
\end{aligned}
$$

Homework 4.85 Prove Eq. (4.256).

Solution Instead of iteration as suggested in the text, we could show by back substitution using the second form of the BCH formula, Eqs. (4.140) and (4.141). It suffices to show that

$$e^{:-L_iH_{2i}-L_iH_{3i}-L_iH_{4i}:} \;=\; e^{:f_2:}e^{:f_3:}e^{:f_4:} \;=\; e^{:f_2:}e^{:f_3+f_4:}e^{:\mathcal{O}(X^5)} \qquad (4.265)$$

holds when $f_{2,3,4}$ are given by Eqs. (4.254–4.256). Equation (4.257) is also obtained in the process.

Homework 4.86 This homework intends to illustrate a point that there is a wide range of possible applications at this stage. Find factorized 2-D Lie maps to $\mathcal{O}(X^4)$ for thick combined-function magnets, e.g., (a) combined quadrupole and sextupole, (b) combined quadrupole and solenoid, (c) combined dipole and quadrupole, (d) combined dipole and sextupole. Apply the results to obtain their corresponding 4th order Taylor maps.

4.7.2 A perturbation theory

As another illustration of the Lie techniques, we will develop a particular perturbation theory. Consider the dynamical system described by the Hamilton equation

$$X' = -:H:X,\qquad(4.266)$$

where a prime means derivative with respect to the time variable s, and $H(X, s)$ is the Hamiltonian of the system. Let the map that describes the evolution of X be designated formally as M, so that

$$X(s) = MX(0),\qquad(4.267)$$

and, knowing that M is going to be an exponential map,

$$f(X(s)) = Mf(X(0)),\qquad(4.268)$$

for any function f of X.

The map M is now considered as a function of s. We have, by substituting Eq. (4.267) into Eq. (4.266), that

$$\begin{aligned}
M'X(0) &= -[H(X, s), MX(0)] \\
&= -M[H(X(0), s), X(0)],
\end{aligned}\qquad(4.269)$$

where the square brackets are the Poisson brackets and use has been made of Eq. (4.268) in the second step. Equation (4.269) can be written symbolically as

$$M' = -M:H:.\qquad(4.270)$$

In particular, if the Hamiltonian is independent of s, then we have

$$M' = -M:H: \quad\Longrightarrow\quad M = e^{-s:H:},$$

which is what one expects. But the form (4.270) now has the ability to treat s-dependent Hamiltonians.

Consider a system described by a Hamiltonian that is basically given by an unperturbed Hamiltonian $H_0(X, s)$, but contains in addition a small perturbation,

$$H = H_0(X, s) + \epsilon V(X, s),$$

where ϵ is considered to be small. Let us assume the unperturbed Lie map, $M_0(X, s)$, from position $s = 0$ to position s is known. We will now find an expression for the evolution map $M(s)$ from position $s = 0$ to position s for this slightly perturbed system to 1st order in ϵ.

We know M is approximately given by M_0. Let M be written in the form

$$M = NM_0,\qquad(4.271)$$

where N is approximately equal to the identity map. To first order in ϵ, we let

$$N = e^{:\epsilon f:}.\qquad(4.272)$$

We need to find an expression for $f(X, s)$.

Using (4.271), the left-hand-side of Eq. (4.270) reads

$$M' = N'M_0 + NM_0' = N'M_0 - NM_0{:}H_0{:}.$$

The right-hand-side of Eq. (4.270) is given by

$$-M{:}H{:} = -NM_0({:}H_0{:} + \epsilon{:}V{:}).$$

Substituting into Eq. (4.270) gives

$$N' = -\epsilon NM_0{:}V{:}M_0^{-1} = -\epsilon N{:}(M_0 V){:}. \tag{4.273}$$

We need an expression for N' when N is given by Eq. (4.272). The result is obtained using Eq. (4.150), i.e.

$$N' = {:}\left(\frac{e^{{:}\epsilon f{:}} - 1}{{:}\epsilon f{:}}\right)\epsilon f'{:}N\,.$$

Note that

$$N' \neq \epsilon{:}f'{:}N, \qquad \text{and} \qquad N' \neq N\epsilon{:}f'{:}, \tag{4.274}$$

as one might be tempted to write using Eq. (4.272). Expressions (4.274) hold only if the operators ${:}f{:}$ and ${:}f'{:}$ commute, which is not true in general. The correct expression for N' has to be obtained by Eq. (4.150).

To 1st order in ϵ, however, either one of the expressions in Eq. (4.274) actually applies. We will take the second of the pair. We then obtain from Eq. (4.273), to 1st order in ϵ, that

$$\epsilon f' = -\epsilon M_0 V\,.$$

Solving this equation gives the result we are looking for,

$$\begin{aligned}
f(X, s) &= -\int_0^s ds'\, M_0(X, s')V(X, s') \\
&= -\int_0^s ds'\, V(M_0(X, s')X, s')\,. \tag{4.275}
\end{aligned}$$

The physical meaning of Eq. (4.275) has been described following Eq. (4.143). To 1st order in ϵ, the effect of errors at position s' can be transformed to the observation position s by the map M_0.

If $H_0 = H_0(X)$ is independent explicitly on s, then $M_0 = e^{-s{:}H_0{:}}$, and

$$f(X, s) = -\int_0^s ds'\, e^{-s'{:}H_0{:}}V(X, s')\,. \tag{4.276}$$

Equation (4.276) is just a generalization of the BCH formula (4.178). This is seen as follows. Consider $H = H_0 + \epsilon V$ where H_0 and V are s-independent.

Then we have

$$
\begin{aligned}
M &= e^{-:sH_0 + s\epsilon V:} \\
&= \exp\left[:\int_0^1 du\, e^{-:usH_0:}(-s\epsilon V):\right] e^{-:sH_0:} \\
&= \exp\left[-\epsilon:\left(\int_0^s ds'\, e^{-:s'H_0:}V\right):\right] e^{-:sH_0:},
\end{aligned}
\tag{4.277}
$$

which confirms Eq. (4.276).

Weak multipole in a drift space　As an application of Eqs. (4.275) and (4.276), consider a weak error perturbation uniformly distributed over a drift distance L. Suppose we care only about the optics in 1-D (e.g. we care only about not messing up the x-emittance and don't care much about the y-emittance in this application). Let the perturbed Hamiltonian be

$$
H = \frac{1}{2}p^2 + \epsilon v(x).
$$

This is a case when both H_0 and V are independent of s. Equation (4.276) gives

$$
f = -\int_0^s ds'\, e^{-:\frac{s'}{2}p^2:}v(x) = -\int_0^s ds'\, v(x + s'p).
$$

For example, the perturbation may be a multipole. From Eq. (4.120), we have

$$
\epsilon v(x) = \frac{\lambda}{n+1}\, x^{n+1}.
$$

Then we have

$$
\epsilon f = -\frac{\lambda}{(n+1)(n+2)}\frac{(x+sp)^{n+2} - x^{n+2}}{p}.
$$

For a focusing quadrupole error of strength $\lambda = k^2$, we have $n = 1$ and

$$
\epsilon f = -\frac{k^2}{6}(3sx^2 + 3s^2 xp + s^3 p^2).
\tag{4.278}
$$

Once f is obtained, the map M has been decomposed as $M = e^{:\epsilon f:}M_0$. The system map from $s = 0$ to s is then approximated (to 1st order in k^2, no limit on s) by a lumped kick $e^{:\epsilon f:}$ followed by a free space drift.

Weak multipole in a long linear channel　Equation (4.275) can also be applied to a linear channel instead of a drift space as in the preceding example. In this case, H_0 depends on s, the map $M_0(X, s')$ is to be replaced by the Courant–Snyder map from $s = 0$ to s'. Similar technique can be applied to the algorithm for corrector compensation, an illustrative example is given in Homework 4.97.

Homework 4.87 The text gives an example of a weak multipole uniformly distributed over a long drift space. Extend the analysis to the case of a combined-function quadrupole-sextupole. Let the sextupole be weak but the quadrupole not necessarily so. Decompose the original map

$$e^{:-\frac{L}{2}(k^2 x^2 + p^2) - \frac{SL}{3} x^3:}$$

to read, to 1st order in SL,

$$e^{:SLf:} e^{:-\frac{L}{2}(k^2 x^2 + p^2):} \, .$$

Find the function f. You may equally consider this an exercise on the BCH formula.

Homework 4.88 Equations (4.275) to (4.277) apply when the perturbed map M is written in the form (4.271). Repeat the procedure if M assumes a different form $M = M_0 N$. Give the physical meaning of your result as was done following Eq. (4.275).

Homework 4.89 The text gave an example application of the perturbation theory, Eqs. (4.275) and (4.276), to the simple case of a long weak quadrupole. The quadrupole map was approximated to read $e^{:\epsilon f:} e^{:-\frac{1}{2}p^2:}$ to 1st order in k^2, where ϵf is given by Eq. (4.278). Since ϵf is a quadratic form, the map $e^{:\epsilon f:}$ is linear and can be written in a matrix form.

Calculate explicitly the matrix map for $e^{:\epsilon f:} e^{:-\frac{1}{2}p^2:}$. Compare your result with the exact map of a long focusing quadrupole to check the validity of the perturbation theory.

4.7.3   Error multipole correction algorithm

Equations (4.275) and (4.276) have wide applications. We mentioned one application to weak multipole in a long drift space. In this section, we consider their applications to error correction algorithms in accelerators.

An example Suppose we have an error perturbation $v(x)$ uniformly distributed from position $s = 0$ to $s = L$, and we want to make a correction so that, when observed downstream from position $s = L$, its effect is compensated, at least to 1st order in the perturbation strength. Suppose we introduce a set of N_c thin-lens multipole correctors (of the same type of multipole as the error multipole). Let the i-th corrector ($i = 1, 2, \ldots, N_c$) have location $s = s_i$ and strength α_i. How should we choose the corrector strengths α_i, assuming the corrector positions are given?

Let the Hamiltonian, including the error perturbation and the correctors, be

$$H = \frac{1}{2}p^2 + \epsilon v(x) + \sum_{i=1}^{N_c} \alpha_i \delta(s - s_i) v(x) \, .$$

The first term represents the drift space. The second term represents the weak perturbation uniformly distributed along s. The third term represents the correctors. The correctors and the perturbation are multipoles of the same type represented by $v(x)$. The first term is considered the unperturbed leading term, the other two terms are perturbations.

From our analysis of Eqs. (4.275) and (4.276), we know that to 1st order in the perturbation strength, the correction requires

$$\int_0^L ds'\, e^{-:\frac{s'}{2}p^2:}\, v(x) \left[\epsilon + \sum_{i=1}^{N_c} \alpha_i \delta(s' - s_i)\right]$$

$$= \epsilon \int_0^L ds'\, v(x + s'p) + \sum_{i=1}^{N_c} \alpha_i v(x + s_i p) = 0. \qquad (4.279)$$

Physical meaning of Eq. (4.279) should be apparent.

To apply Eq. (4.279), we consider s_i and $v(x)$ to be given, and we need to compute α_i so that Eq. (4.279) is satisfied for all values of x and p. In case the perturbation is an n-th order multipole, $\epsilon v(x) = \frac{\lambda}{n+1} x^{n+1}$, this correction scheme can be accomplished by having a finite number of correctors. If only 1-D is of interest, there are $n + 2$ coefficients to vanish in Eq. (4.279). This requires $n + 2$ correctors. A distributed quadrupole generally requires three lumped quadrupoles to correct, and a distributed sextupole requires four sextupole correctors, etc.

Take the case of an error focusing quadrupole for example. Suppose we introduce three thin-lens quadrupoles of strength $\alpha_{1,2,3}$ at locations $s = 0, \frac{L}{2}$, and L, what should the corrector strengths be?

The correctors Hamiltonian is

$$\frac{\alpha_1}{2} x^2 \delta(s) + \frac{\alpha_2}{2} x^2 \delta\left(s - \frac{L}{2}\right) + \frac{\alpha_3}{2} x^2 \delta(s - L). \qquad (4.280)$$

To 1st order in k^2, the correction condition (4.279) requires

$$-\frac{k^2}{6}(3Lx^2 + 3L^2 xp + L^3 p^2) + \frac{\alpha_1}{2} x^2 + \frac{\alpha_2}{2}\left(x + \frac{L}{2}p\right)^2 + \frac{\alpha_3}{2}(x + Lp)^2 = 0, \qquad (4.281)$$

where the first term is from Eq. (4.278). Note that this result applies when $s > L$, so the integration covers the whole length of L. Once accomplished, however, the correction has been made for all positions downstream of $s > L$.

Equation (4.281) is to hold for all values of x and p. This means the coefficients of x^2, xp and p^2 must vanish separately. This gives three conditions to determine the three corrector strengths. The solution is found to be

$$\alpha_1 = \alpha_3 = \frac{1}{6} kL, \quad \alpha_2 = \frac{2}{3} kL. \qquad (4.282)$$

The relative weights of the three corrector strengths are just the Simpson's rule of integral approximation.

In case the perturbation is not a pure multipole, or if the perturbation is a multipole but the number of correctors available is less than $n + 2$, then Eq. (4.279) cannot be fulfilled exactly. One can still optimize the cancellation approximately if the dependencies of the function $v(x + s'p)$ on x and p do not vary much in the range $0 < s' < L$. To do so, let us rewrite the corrector condition,

$$\epsilon \int_0^1 du\, v(x + uLp) \;=\; -\frac{1}{L} \sum_i \alpha_i v(x + u_i Lp)\,,$$

and approximate the function $v(x + uLp)$ as $G(u)F(x,p)$. The function $F(x,p)$ cancels out in the condition and it becomes clear that we are just looking for a set of values α_i and s_i which best approximates the integral

$$\epsilon \int_0^1 du\, G(u) \;=\; -\frac{1}{L} \sum_i \alpha_i G(u_i)\,.$$

for any reasonably smooth function $G(u)$. Simpson's rule is then just one way to make this approximation. Simpson's rule is exact if $v(x)$ and H_0 are quadratic in x.

Another example   As another example, consider an accelerator section from $s = 0$ to $s = L$ described by an unperturbed Hamiltonian $H_0(X)$. Now consider some perturbation in this section. Let the perturbation strength distribution be $\alpha(s)$. We consider $\alpha(s)$ to be small. Let the perturbation Hamiltonian be $\alpha(s)f(X)$. The Hamiltonian of the system is then

$$H(X, s) \;=\; H_0(X) + \alpha(s)f(X)\,.$$

Suppose we now regard the perturbation to be due to error sextupoles — which specializes $f(X) = \frac{S}{3}x^3$, and suppose we want to insert a single thin-lens correction sextupole at the middle of the drift space ($s = \frac{L}{2}$) to correct for the effect of this error sextupole distribution, what would be the optimal setting of this thin-lens correction sextupole?

When the correction sextupole is included, the Hamiltonian becomes

$$H(X, s) \;=\; H_0(X) + \alpha(s)f(X) + \beta\delta\left(s - \frac{L}{2}\right)f(X)\,,$$

where β, yet to be chosen, characterizes the strength of the correction sextupole. Our analysis says the choice is to be made according to

$$\epsilon f \;=\; -\int_0^L ds\, e^{-s:H_0:} \left[\alpha(s) + \beta\delta\left(s - \frac{L}{2}\right)\right] f(X) = 0\,. \tag{4.283}$$

Equation (4.283) is to be valid for all X. The function $f(X)$ describes the action of the specific perturbation. The operation $e^{-s:H_0:}$ is to transform the perturbation effect to the observation position $s = L$. Clearly, the effect of a sextupole at position s is in general not described by the function $f(X)$ when observed at a position $s \neq L$. This means that, in general, the condition (4.283) is impossible to be met exactly, and we do not have a single corrector that entirely removes the 1st order error effects. But if the drift space is short enough, this can be done because $e^{-s:H_0:} \approx 1$ — which is equivalent to ignoring the difference made on $f(X)$ due to the transport from s to L — and we have the approximate condition

$$\int_0^L ds \left[\alpha(s) + \beta\delta\left(s - \frac{L}{2}\right)\right] = 0, \qquad \text{or} \qquad \beta = -\int_0^L ds\,\alpha(s). \qquad (4.284)$$

In other words, the optimal correction is achieved by — not surprisingly — choosing the corrector to be of the same perturbation type [reflected by the fact that the corrector and the error have the same $f(X)$], and choosing β to be negative of the total integral of the error perturbation strength. It is easy to see that this choice of β is independent of the location of the corrector sextupole.

If the beamline section length is not that short, to improve the accuracy of the correction, we can insert two correctors. For example, one might insert one corrector at the entrance ($s = 0$) and another at the exit ($s = L$) of the section. The condition for correction becomes

$$\int_0^L ds\,[1 - s:H_0:][\alpha(s) + \beta_1\delta(s) + \beta_2\delta(s - L)]f(X) = 0, \qquad (4.285)$$

where we have kept one more term in $e^{-s:H:}$, i.e., we now use $e^{-s:H_0:} \approx 1 - s:H_0:$. Regardless of the expression of H_0, Eq. (4.285) can be satisfied if

$$\int_0^L ds\,[\alpha(s) + \beta_1\delta(s) + \beta_2\delta(s - L)] = 0,$$

$$\int_0^L ds\,s[\alpha(s) + \beta_1\delta(s) + \beta_2\delta(s - L)] = 0,$$

or

$$\beta_1 = -\int_0^L ds\,\left(1 - \frac{s}{L}\right)\alpha(s),$$

$$\beta_2 = -\frac{1}{L}\int_0^L ds\,s\alpha(s). \qquad (4.286)$$

The solution (4.286) can be combined to yield

$$\beta_1 + \beta_2 + \int_0^L ds'\alpha(s') = 0,$$

$$\beta_1 L + \int_0^L ds'(L - s')\alpha(s') = 0. \qquad (4.287)$$

The condition (4.284) and the first condition of Eq. (4.287) have the physical meaning that the total angular kick of the perturbation (errors plus correctors) observed at $s = L$ is zero. The physical meaning of the second condition of Eq. (4.287) is that the total perturbation on the displacement of the particle's trajectory is zero when observed at $s = L$.

One can further improve the accuracy of correction when the drift space is not too short by having three correctors, located at $s = 0$, $s = \frac{L}{2}$ and $s = L$. The conditions for correction is then

$$\int_0^L ds \left[\alpha(s) + \beta_1\delta(s) + \beta_2\delta\left(s - \frac{L}{2}\right) + \beta_3\delta(s - L)\right] = 0,$$

$$\int_0^L ds\, s \left[\alpha(s) + \beta_1\delta(s) + \beta_2\delta\left(s - \frac{L}{2}\right) + \beta_3\delta(s - L)\right] = 0,$$

$$\int_0^L ds\, s^2 \left[\alpha(s) + \beta_1\delta(s) + \beta_2\delta\left(s - \frac{L}{2}\right) + \beta_3\delta(s - L)\right] = 0,$$

or

$$\beta_1 = \int_0^L ds\, \alpha(s) \left(-1 + \frac{3s}{L} - \frac{2s^2}{L^2}\right),$$

$$\beta_2 = \int_0^L ds\, \alpha(s) \left(-\frac{4s}{L} + \frac{4s^2}{L^2}\right),$$

$$\beta_3 = \int_0^L ds\, \alpha(s) \left(\frac{s}{L} - \frac{2s^2}{L^2}\right). \tag{4.288}$$

Equations (4.284), (4.286) and (4.288) give various error correction algorithms for short accelerator sections. In this algorithm, it does not matter which type of errors are being considered, sextupole or otherwise. It is interesting to note that it also does not matter which unperturbed H_0 is being considered. Whether the unperturbed section is in a drift space or inside a long sextupole will not matter as long as the section length is short enough. The same correction algorithm applies.

Homeworks 4.96 and 4.97 give two more explicit examples applying this correction algorithm. Compensation for the beam dynamics effects of error multipoles is an important topic for accelerator designs. Depending on the problem at hand, the most effective way to provide the compensation varies. It should be mentioned here that the compensation discussed in this section applies only for the case when the effects of error multipoles are to be compensated locally (or almost locally within the beamline section). In particular, if it is the effect accumulated over one turn of the accelerator that is to be compensated, i.e., if the errors are to be corrected globally, the compensation scheme used would be quite different — we then resort to the normal form analysis.

Optimization of corrector location   Equation (4.288) describes how to correct for some known nonlinear effects of a short accelerator section from $s = 0$ to $s = L$ using three correctors located at $s = 0$, $\frac{L}{2}$, and L. The accuracy of this correction is up to order $\mathcal{O}(L^3)$ in the Lie exponent. One might try to improve the correction accuracy by freeing up the corrector locations.

To simplify the algebra, change coordinates so that the region of interest is from $s = -L$ to L. Let the three correctors be located at $s = -s_0, 0, +s_0$, where s_0 is to be chosen for optimization of the correction accuracy. As before, the conditions for the correction are given by

$$\int_{-L}^{L} ds\,[\alpha(s) + \beta_1\delta(s + s_0) + \beta_2\delta(s) + \beta_3\delta(s - s_0)] = 0,$$

$$\int_{-L}^{L} ds\,s\,[\alpha(s) + \beta_1\delta(s + s_0) + \beta_2\delta(s) + \beta_3\delta(s - s_0)] = 0,$$

$$\int_{-L}^{L} ds\,s^2\,[\alpha(s) + \beta_1\delta(s + s_0) + \beta_2\delta(s) + \beta_3\delta(s - s_0)] = 0,$$

or

$$\beta_1 = \frac{I_1 s_0 L - I_2 L^2}{2s_0^2},$$

$$\beta_2 = -I_0 + \frac{I_2 L^2}{s_0^2},$$

$$\beta_3 = \frac{-I_1 s_0 L - I_2 L^2}{2s_0^2}, \tag{4.289}$$

where

$$I_k = \frac{1}{L^k}\int_{-L}^{L} ds\,\alpha(s)s^k. \tag{4.290}$$

When $s_0 = L$, this reduces to the result (4.288).

To improve the accuracy, we can choose s_0 so that the next order perturbation is canceled, i.e.

$$\int_{-L}^{L} ds\,s^3\,[\alpha(s) + \beta_1\delta(s + s_0) + \beta_2\delta(s) + \beta_3\delta(s - s_0)] = 0.$$

This is fulfilled when

$$s_0 = \sqrt{\frac{I_3}{I_1}}\,L.$$

Equation (4.289) then gives the corrector strengths

$$\beta_1 = -\frac{I_1(I_2 - \sqrt{I_1 I_3})}{2I_3},$$

$$\beta_2 = -I_0 + \frac{I_1 I_2}{I_3},$$

$$\beta_3 = -\frac{I_1(I_2 + \sqrt{I_1 I_3})}{2I_3}. \tag{4.291}$$

The residual effect of the errors after correction is given by

$$I_4 - \frac{I_2 I_3}{I_1} = \mathcal{O}(L^5).$$

For the special case when $\alpha(s)$ is an even function of s so that $I_1 = 0$ and $I_3 = 0$, a $\mathcal{O}(L^5)$ correction [error $\sim \mathcal{O}(L^6)$] can be achieved with three correctors,

$$s_0 = \sqrt{\frac{I_4}{I_2}}, \qquad \beta_1 = \beta_3 = -\frac{I_2^2}{2I_4}, \qquad \beta_2 = -I_0 + \frac{I_2^2}{I_4}. \qquad (4.292)$$

Numerical algorithm of integration   We alluded to the Simpson's rule of numerical integration in Eq. (4.282). Somewhat unexpectedly, Lie algebra also yields various algorithms to perform numerical integrations. We make a small detour on this subject here. Techniques developed in this section should also be related to the map symplectification techniques developed in Chapter 2.

To illustrate the connection, let us consider Eq. (4.285) for example. Identify $e^{-s:H_0:}$ as a function of s and call it $f(s)$. If it behaves sufficiently smoothly in the region between $s = 0$ and $s = 1$, we have shown that the best choice of β_1 and β_2 that makes

$$\int_0^1 ds\, f(s)[\alpha(s) + \beta_1 \delta(s) + \beta_2 \delta(s - 1)] \approx 0 \qquad (4.293)$$

is given by Eq. (4.286) regardless of the exact form of $f(s)$. Now if we take $\alpha(s) = 1$, Eq. (4.286) gives

$$\beta_1 = \beta_2 = -\frac{1}{2}.$$

Substituting back into Eq. (4.293) yields the trapezoidal rule of numerical integration,

$$\int_0^1 dx\, f(x) \approx \frac{1}{2}f(0) + \frac{1}{2}f(1), \qquad (4.294)$$

where $f(x)$ is any smooth function of x.

In fact, Eq. (4.284) is simply the cruder approximation

$$\int_0^1 dx\, f(x) \approx f\left(\frac{1}{2}\right).$$

Homework 4.92 asks to show that Eq. (4.288) is related to the Simpson's rule of numerical integration,

$$\int_0^1 dx\, f(x) \approx \frac{1}{6}f(0) + \frac{2}{3}f\left(\frac{1}{2}\right) + \frac{1}{6}f(1). \qquad (4.295)$$

Homeworks 4.91 and 4.92 also ask for generalization of the integration technique to include weight functions.

One can improve the numerical integration accuracy by applying the correction algorithm with optimized corrector locations. Considering the case $\alpha(x) = 1$. In this case, we apply Eq. (4.292). This leads to

$$\int_{-1}^1 dx\, f(x) \approx \frac{5}{9}f\left(-\sqrt{\frac{3}{5}}\right) + \frac{8}{9}f(0) + \frac{5}{9}f\left(\sqrt{\frac{3}{5}}\right). \qquad (4.296)$$

Equation (4.296) is exact if $f(x)$ is any 5th order Taylor series in x. In contrast, Simpson's rule (4.295) is exact if $f(x)$ is a 3rd order Taylor series. One may also compare this with Legendre integration involving roots of $P_3(x) = 0$.

Weight function   One can also include a weight function by considering $\alpha(x) = W(x)$, as considered in Homeworks 4.91 and 4.92. The same extension can be considered based on the scheme with optimized correction locations. If $W(x)$ is not an even function of x, Eq. (4.291) is to be used. It gives

$$\int_{-1}^{1} dx\, W(x) f(x) \approx -\beta_1 f(-s_0) - \beta_2 f(0) - \beta_3 f(s_0)\,,$$

where $s_0 = \sqrt{\frac{I_3}{I_1}}$, and $\beta_{1,2,3}$ are given by Eq. (4.291) with

$$I_k = \int_{-1}^{1} dx\, W(x) x^k\,.$$

For example, when $W(x) = x$, we find

$$\int_{-1}^{1} dx\, x f(x) \approx \frac{1}{3}\sqrt{\frac{5}{3}} \left[f\left(\sqrt{\frac{3}{5}}\right) - f\left(-\sqrt{\frac{3}{5}}\right) \right].$$

which is consistent with Eq. (4.296).

In case $W(x)$ is an even function of x, $I_1 = 0$ and $I_3 = 0$, we apply Eq. (4.292) to obtain

$$\int_{-1}^{1} dx\, W(x) f(x) \approx I_0 f(0) + \frac{I_2^2}{2 I_4} \left[f\left(\sqrt{\frac{I_4}{I_2}}\right) + f\left(-\sqrt{\frac{I_4}{I_2}}\right) - 2 f(0) \right],$$

where $I_k = \int_{-1}^{1} dx\, W(x) x^k$. As special cases, we have

$$\int_{-1}^{1} dx\, f(x)(1 - x^2) \approx \tfrac{14}{45} f\left(-\sqrt{\tfrac{3}{7}}\right) + \tfrac{32}{45} f(0) + \tfrac{14}{45} f\left(\sqrt{\tfrac{3}{7}}\right),$$

$$\int_{-1}^{1} dx\, f(x)\sqrt{1 - x^2} \approx \tfrac{\pi}{8} f\left(-\sqrt{\tfrac{1}{2}}\right) + \tfrac{\pi}{4} f(0) + \tfrac{\pi}{8} f\left(\sqrt{\tfrac{1}{2}}\right),$$

$$\int_{-1}^{1} dx\, \frac{f(x)}{\sqrt{1 - x^2}} \approx \tfrac{\pi}{3} f\left(-\sqrt{\tfrac{3}{4}}\right) + \tfrac{\pi}{3} f(0) + \tfrac{\pi}{3} f\left(\sqrt{\tfrac{3}{4}}\right).$$

One may compare this with Tsybechev integration.

Homework 4.90 Equation (4.284) can be applied to the case of a sextupole perturbation, and it then gives the optimal setting of a single sextupole corrector to compensate for a distribution of error sextupoles. Consider the case when the error sextupoles are a weak uniform distribution from $s = 0$ to $s = L$. Estimate the residual effect of the sextupoles after correction by computing the net Lie or Taylor map.

Homework 4.91 The trapezoidal rule (4.294) applies when $f(x)$ is a smooth function of x between 0 and 1. If one is interested in calculating the integral of $f(x)W(x)$ where $f(x)$ is smooth but $W(x)$ is not, we can reconsider Eq. (4.293) but this time take $\alpha(s) = W(s)$. Show that this leads to a generalized trapezoidal rule,

$$\int_0^1 dx\, f(x)W(x) \approx \beta_1 f(0) + \beta_2 f(1)\,,$$

where

$$\beta_1 = \int_0^1 dx\,(1-x)W(x)\,,$$

$$\beta_2 = \int_0^1 dx\, xW(x)\,.$$

The rule can be rewritten over the range $1 > x > -1$,

$$\int_{-1}^1 dx\, f(x)W(x) \approx \beta_1 f(-1) + \beta_2(1)\,,$$

$$\beta_1 = \frac{1}{2}\int_{-1}^1 dx\,(1-x)W(x)\,,$$

$$\beta_2 = \frac{1}{2}\int_{-1}^1 dx\,(1+x)W(x)\,. \tag{4.297}$$

As applications of Eq. (4.297), show that

$$\int_{-1}^1 dx\, f(x)(1-x^2) \approx \tfrac{2}{3}[f(-1)+f(1)]\,,$$

$$\int_{-1}^1 dx\, f(x)\sqrt{1-x^2} \approx \tfrac{\pi}{4}[f(-1)+f(1)]\,,$$

$$\int_{-1}^1 dx\, \frac{f(x)}{\sqrt{1-x^2}} \approx \tfrac{\pi}{2}[f(-1)+f(1)]\,.$$

Homework 4.92 Follow the text showing the trapezoidal rule to prove the Simpson's rule (4.295). Follow Homework 4.91 to find the Simpson's rule with a weight function.

Solution The Simpson's rule with weight function is

$$\int_0^1 dx f(x)W(x) \approx \beta_1 f(0) + \beta_2 f\left(\frac{1}{2}\right) + \beta_3 f(1)\,,$$

where

$$\beta_1 = \int_0^1 dx \, (1-x)(1-2x)W(x) \,,$$

$$\beta_2 = 4\int_0^1 dx \, x(1-x)W(x) \,,$$

$$\beta_3 = -\int_0^1 dx \, x(1-2x)W(x) \,.$$

Rewriting this results in the range $1 > x > -1$, we obtain

$$\int_{-1}^1 dx \, f(x)W(x) \approx \beta_1 f(-1) + \beta_2 f(0) + \beta_3 f(1) \,,$$

$$\beta_1 = -\frac{1}{2}\int_{-1}^1 dx \, W(x)x(1-x) \,,$$

$$\beta_2 = \int_{-1}^1 dx \, W(x)(1-x^2) \,,$$

$$\beta_3 = \frac{1}{2}\int_{-1}^1 dx \, W(x)x(1+x) \,. \tag{4.298}$$

As applications of Eq. (4.298), we have

$$\int_{-1}^1 dx \, f(x)(1-x^2) \approx \frac{2}{15}f(-1) + \frac{16}{15}f(0) + \frac{2}{15}f(1) \,,$$

$$\int_{-1}^1 dx \, f(x)\sqrt{1-x^2} \approx \frac{\pi}{16}f(-1) + \frac{3\pi}{8}f(0) + \frac{\pi}{16}f(1) \,,$$

$$\int_{-1}^1 dx \, \frac{f(x)}{\sqrt{1-x^2}} \approx \frac{\pi}{4}f(-1) + \frac{\pi}{2}f(0) + \frac{\pi}{4}f(1) \,.$$

Homework 4.93 Concerning the optimization of the corrector locations,
(a) verify Eqs. (4.289–4.291).
(b) continue to verify Eq. (4.292).

Homework 4.94
(a) Use Lie algebraic technique developed in this section to show that

$$\int_{-\infty}^\infty dx \, e^{-x^2/2}f(x) \approx \sqrt{2\pi}\left[\frac{1}{6}f(-\sqrt{3}) + \frac{2}{3}f(0) + \frac{1}{6}f(\sqrt{3})\right].$$

This formula is exact if $f(x)$ is any polynomial in x up to 5th order. You need to think of a way to extend the integration range to infinities.
(b) Similarly show that

$$\int_0^\infty dx \, e^{-x}f(x) \approx \frac{1}{4}\left[(2+\sqrt{2})f(2-\sqrt{2}) + (2-\sqrt{2})f(2+\sqrt{2})\right].$$

Homework 4.95

(a) Determine the strengths of four correctors located at $s = 0, \frac{L}{3}, \frac{2L}{3}$, and L to provide a correction accuracy of $\mathcal{O}(L^4)$ in the Lie exponent.

(b) Free up the locations of these four correctors to achieve a higher correction accuracy. Compare this with Legendre integration involving the roots of $P_4(x) = 0$.

Homework 4.96 Consider an accelerator beamline consisting of a thin focusing quadrupole at location $s = 0$, followed by two dipoles, each of length D, followed by a thin defocusing quadrupole at location $s = 2D$, followed by another two dipoles, each of length D, followed by a thin focusing quadrupole at $s = 4D$. Suppose that the four dipoles are known to have weak error sextupole fields with strengths of $\alpha_i, i = 1, \ldots, 4$, and that it is possible to insert some thin correctors next to each of the three quadrupoles. Assume the sextupole field within each dipole is uniform over the length of the dipole.

(a) Consider only the horizontal motion and only the on-momentum particles. Design a single-pass correction scheme and determine the corrector strengths for this transport line.

(b) Consider the 2-D x- and y on-momentum motion. Give expressions of the Lie and Taylor maps before and after correction to the leading order in α and D.

Solution (a) Three thin sextupoles at the locations next to the quadrupoles serve as the correctors. This homework requires you to know that this is a situation to apply Eq. (4.288). The result is

$$\beta_1 = -\frac{2}{3}\alpha_1 D - \frac{1}{6}\alpha_2 D + \frac{1}{12}\alpha_3 D + \frac{1}{12}\alpha_4 D,$$

$$\beta_2 = -\frac{5}{12}\alpha_1 D - \frac{11}{12}\alpha_2 D - \frac{11}{12}\alpha_3 D - \frac{5}{12}\alpha_4 D,$$

$$\beta_3 = \frac{1}{12}\alpha_1 D + \frac{1}{12}\alpha_2 D - \frac{1}{6}\alpha_3 D - \frac{2}{3}\alpha_4 D.$$

These corrector settings are independent of the strengths of the quadrupoles or the dipoles. The length D must not be too long for this correction scheme to apply.

Homework 4.97 The text considered the case of a uniform error quadrupole and gave a scheme, Eq. (4.282), of using three thin quadrupoles to make an exact correction to 1st order of the error quadrupole strength.

(a) Follow a similar analysis to extend the example to a uniform sextupole. Consider 1-D x-motion only. You need four thin sextupole correctors.

(b) How many corrector sextupoles are needed if y-motion is also to be corrected?

Homework 4.98 The text illustrates situations of error correction in typical simple cases. This homework intends to illustrate a somewhat more involved example. Equations (4.280–4.282) are for the case to correct a weak uniform error quadrupole with three lumped quadrupole correctors. Consider the case to correct a weak uniform error quadrupole in a not-so-weak, unperturbed quadrupole. Analyze this problem. Show that again three correctors are needed and give their required corrector settings. With this setting, you are supposed to correct the error effect exactly, at least to 1st order in the error quadrupole strength.

Solution Let K and ΔK be the unperturbed and error quadrupole strengths respectively. The Hamiltonian is

$$H \;=\; \frac{1}{2}p^2 + \frac{K}{2}x^2 + \frac{\Delta K}{2}x^2 + \sum_{i=1}^{3} \frac{\alpha_i}{2}x^2\delta(s-s_i)\,.$$

We want

$$\int_0^L ds'\, e^{-:s'\left(\frac{1}{2}p^2 + \frac{1}{2}Kx^2\right):}\left[\Delta K x^2 + \sum_i \alpha_i x^2\delta(s-s_i)\right] \;=\; 0\,,$$

$$\Longrightarrow \quad \Delta K \int_0^L ds'\left(x\cos\sqrt{K}s' + \frac{p}{\sqrt{K}}\sin\sqrt{K}s'\right)^2$$

$$+\sum_i \alpha_i\left(x\cos\sqrt{K}s_i + \frac{p}{\sqrt{K}}\sin\sqrt{K}s_i\right)^2 \;=\; 0\,,$$

$$\Longrightarrow \quad \begin{cases} \Delta K \int_0^L ds'\cos^2\sqrt{K}s' + \sum_i \alpha_i \cos^2\sqrt{K}s_i = 0\,, \\[4pt] \Delta K \int_0^L ds'\sin\sqrt{K}s'\cos\sqrt{K}s' + \sum_i \alpha_i \sin\sqrt{K}s_i\cos\sqrt{K}s_i = 0\,, \\[4pt] \Delta K \int_0^L ds'\sin^2\sqrt{K}s' + \sum_i \alpha_i \sin^2\sqrt{K}s_i = 0\,. \end{cases}$$

Let the three correctors be located at $s_{1,2,3} = 0, \frac{L}{2}$ and L. The corrector strengths are found to be (with $q = \sqrt{K}L$)

$$\alpha_1 \;=\; \alpha_3 \;=\; \frac{\Delta K}{4\sqrt{K}\sin^2\frac{q}{2}}(-q + \sin q)\,,$$

$$\alpha_2 \;=\; \frac{\Delta K}{2\sqrt{K}\sin^2\frac{q}{2}}(q\cos q - \sin q)\,.$$

When $K=0$ or when L is short, this reduces to the Simpson's rule $\alpha_1 = \alpha_3 = -\frac{\Delta KL}{6}$, $\alpha_2 = -\frac{2\Delta KL}{3}$.

Homework 4.99 Consider a long superconducting dipole magnet that contains a uniform error octupole along its length L. To correct for its nonlinear effects, ideally one introduces a corrector octupole with the negative of the error octupole uniformly along the dipole magnet. Such would be the case of a *bore tube winding corrector* considered for example in some superconducting collider storage rings. On the other hand, a lesser compensation scheme would be to introduce lumped correctors at the two ends of the magnet where spool pieces can be more easily available. Use Lie algebra to analyze the net beam nonlinear dynamical effect when the long magnet's error is compensated by the lumped correctors at its ends. For simplicity, consider 1-D horizontal dynamics only.

4.8 Normal form

4.8.1 Nonlinear map

In our discussions so far, linear maps have played an important role. This is not unexpected. We can learn and have learned much from linear maps. Linear maps can simultaneously be viewed as Taylor maps and Lie maps and therefore have the advantages of both. In particular, we learned that much of the linear system analysis can be handled by matrices, the lowest order Taylor maps. On the other hand, we have also learned that an elegant treatment of linear systems is provided by a Courant–Snyder analysis, and the Courant–Snyder analysis is intimately related to the Lie representation of linear maps. [See for example Eqs. (4.68) and (4.215).]

In fact, even the very concept of maps is learned from analyzing the linear systems. In a linear system, one speaks of matrices and Courant–Snyder transformation of these matrices without having to think of them as maps. But when considering a nonlinear system, the map concept becomes inevitable. There are more reasons why maps are useful.

1. We are interested in predicting the particle's motion at an observation position when its motion is given at an entrance position upstream. This implies the need of a map. In particular, for a circular accelerator, we are then talking about one-turn maps.

2. Maps are easier to handle mathematically — maps can be multiplied; Hamiltonians cannot.[32]

3. A map allows an analytic expression of the final coordinates as functions of the initial coordinates, while a tracking code only establishes their numerical connection. There is an essential difference.

One then realizes that representing nonlinear maps in their Lie forms provides a natural way to generalize the Courant–Snyder analysis to nonlinear systems. A specific generalization is seen when the effective Hamiltonian takes on the role of the Courant–Snyder invariant when the system goes nonlinear. In Lie language, the one-turn map $e^{:-\frac{\mu}{2}(\gamma x^2 + 2\alpha xp + \beta p^2):}$ is generalized to $e^{:-CH_{\text{eff}}:}$ in a natural manner.

But being able to write the nonlinear maps in Lie forms, such as $e^{:-LH:}$ for single elements or even after concatenating into $e^{:-CH_{\text{eff}}:}$ for a storage ring, constitutes only a first step toward completing the analogy to linear analysis — in particular, the analogy to the introduction of the Courant-Snyder α, β, γ and dispersion functions and the tunes is yet to be made. Lie maps for the nonlinear systems are the equivalent of matrices for the linear systems. What is missing is the equivalent of the Courant–Snyder transformation and the analyses of the properties of the linear system subsequent to the transformation. In this section,

---

[32]Lie algebra can be viewed as a technique to express maps in terms of Hamiltonians. To multiply Lie maps together, however, requires heavy use of the BCH formula.

we will introduce a *normal form* technique which is the nonlinear extension of the Courant–Snyder formalism.

Indeed, having obtained the one-turn Lie map $e^{:-CH_{\text{eff}}:}$ with the effective Hamiltonian H_{eff}, and transformed it into its normal form, many physical quantities related to the nonlinear dynamics of the system can be extracted. These physical quantities include

closed orbit distortion in the presence of nonlinearities;

distorted Courant–Snyder functions for off-momentum particles;

strengths for various resonances;

smear or distortion functions;

betatron tunes as functions of momentum deviation;

betatron tunes as functions of betatron amplitudes;

KAM tori (or KAM invariants).

When completed, the end result is that these quantities are expressed as explicit Taylor expressions in terms of the perturbations up to the order specified by the map. These Taylor expressions require them to be expanded in terms of perturbation quantities small enough for the series to converge. If the map treats the dynamic variables (components of X) as perturbations, these quantities are expressed to the specified order in terms of the components of X. If the map treats the strength of a certain nonlinear element as the perturbation, explicit expressions of these quantities as functions of this nonlinear element strength can be obtained. It is also possible to simultaneously use X and element strengths as perturbations.

Once one recognizes the significant role maps play, the questions are then (a) how to generate maps efficiently, and (b) once generated, how to analyze those maps. The subject of high-order map generation is addressed for example by the technique of truncated power series algebra (TPSA, or differential algebra), the subject of Chapter 3. In this section, we address question (b) by introducing the normal form technique.

4.8.2　1-D linear system

To introduce the normal forms, let us first consider the nonlinear map for a 1-D system,

$$M = e^{:f_2(X):}e^{:f_3(X):}\cdots,\tag{4.299}$$

where X stands for the physical coordinates $X = (x, p_x)$ with $p_x = x'$. If the system is linear, we have $f_k = 0$ for $k \geq 3$. From the Courant–Snyder analysis, we know it is convenient to transform the physical coordinates to the normalized coordinates

$$U = \begin{bmatrix} \bar{x} \\ \bar{p}_x \end{bmatrix} = \begin{bmatrix} \frac{1}{\sqrt{\beta}} & 0 \\ \frac{\alpha}{\sqrt{\beta}} & \sqrt{\beta} \end{bmatrix} \begin{bmatrix} x \\ p_x \end{bmatrix},\tag{4.300}$$

where α, β are the Courant-Snyder functions. The first question we ask is how to introduce this transformation in the Lie language.

It is important to keep in mind that the Courant–Snyder transformation (4.300) is not the unique way to formulate the 1-D linear system in normal form. In that sense, Courant–Snyder formalism is not unique; it is a convention, an artifact without its own physical meaning. However, it is an elegant formalism and it is the formalism that our community has adopted as the universal language of communication.[33] Once we choose the particular form of Courant–Snyder formalism, our following normal form analysis is chosen conforming to that choice.

To proceed, consider the map in Lie representation,

$$N = A_2 M A_2^{-1}, \tag{4.301}$$

where A_2^{-1} is the linear symplectic map whose matrix representation $\mathbf{A_2}^{-1}$ is such that[34]

$$\mathbf{A_2}^{-1} X = U, \quad \text{or } \mathbf{A_2}^{-1} = \begin{bmatrix} \frac{1}{\sqrt{\beta}} & 0 \\ \frac{\alpha}{\sqrt{\beta}} & \sqrt{\beta} \end{bmatrix}, \quad \text{or } \mathbf{A_2} = \begin{bmatrix} \sqrt{\beta} & 0 \\ -\frac{\alpha}{\sqrt{\beta}} & \frac{1}{\sqrt{\beta}} \end{bmatrix}.$$

The map N constitutes a similarity transformation from M and it operates on coordinates U. We have

$$\begin{aligned} N &= A_2 e^{:f_2(X):} e^{:f_3(X):} \cdots A_2^{-1} \big|_{X=U} \\ &= A_2 e^{:f_2(X):} A_2^{-1} A_2 e^{:f_3(X):} A_2^{-1} \cdots \big|_{X=U} \\ &= e^{:A_2 f_2(X):} e^{:A_2 f_3(X):} \cdots \big|_{X=U} \\ &= e^{:f_2(A_2 U):} e^{:f_3(A_2 U):} \cdots, \end{aligned} \tag{4.302}$$

where $f_n(A_2 U)$ just means $f_n(X)$ when it is reexpressed in terms of $U = A_2^{-1} X$. Take Eq. (4.68) for example,

$$f_2(X) = -\frac{\mu}{2}(\gamma x^2 + 2\alpha x p_x + \beta p_x^2) = -\frac{\mu}{2}\left[\left(\frac{x}{\sqrt{\beta}}\right)^2 + \left(\frac{\alpha x + \beta x'}{\sqrt{\beta}}\right)^2\right]$$

$$\implies \quad f_2(A_2 U) = -\frac{\mu}{2}(\bar{x}^2 + \bar{p}_x^2).$$

The map N describes the dynamics of the same physical system as described by M, and it has the same form as M except that all coordinates X are transformed into U. If the system is linear, one can use a matrix language. The matrix representation of Eq. (4.301) is, recalling the reversed ordering,

$$\mathbf{N} = \mathbf{A_2}^{-1}\mathbf{M}\mathbf{A_2}.$$

---

[33]For more discussion, see A.W. Chao, Lectures on Accelerator Physics, World Scientific (2020).

[34]In Secs. 4.8.2 and 4.8.3 (but not elsewhere in these lectures), since we will often switch between matrix and Lie representations of the same maps, to avoid possible confusion, we shall designate the matrix representations in bold letters. There is only one and the same map; just that they have different representations.

264   4. Lie Algebra

If the map M is applied to X, we have

$$\mathbf{M}X \;=\; \mathbf{A_2 N A_2}^{-1}X \;=\; \mathbf{A_2 N}U \,.$$

It is also easy to see that applying the map k times gives

$$\mathbf{M}^k X \;=\; \mathbf{A_2 N}^k U \,. \tag{4.303}$$

This means that to study the multiturn beam dynamics, one can study the effects of the map M on the physical coordinates X, or equivalently, one can study the effects of the map N on the normalized coordinates U. The operation $\mathbf{A_2}$ on the right-hand-side of Eq. (4.303) simply means that after the multiturn study, one needs to remember to transform the results back to the physical space, but such transformation needs to be done only once at the end. The *dynamics* of the map M in terms of the X-coordinates and the dynamics of the map N in terms of the U-coordinates are identical.

4.8.3   3-D linear system

Let us now consider a 3-D case. More specifically let us consider a section of accelerator design that contains only static magnetic devices and no electric devices so that $\delta = \frac{\Delta P}{P}$ is an invariant. For example, if there is only one RF cavity in a circular accelerator, and the cavity is considered to be a thin-lens element located at position $s = 0$, the accelerator section being considered could be the map from $s = 0^+$ to $s = C^-$ for one turn around the accelerator excluding the RF cavity. The RF map then has to be taken into account separately in this description.

We will adopt the dynamic variables $X = (x, x', y, y', z, \delta)$. Let the map through the section be M, which has the form (4.299). Because δ is a constant of the motion, all functions $f_n(X)$ do not depend on z, as one would also expect physically — particles with different z's all get the same kicks from the static magnetic devices. For a linear system, $f_3(X) = 0$, etc.

We then make a transformation according to Eq. (4.301), where A_2 is chosen such that the function $f_2(A_2 U)$ is given by

$$f_2(A_2 U) \;=\; -\frac{\bar{\mu}_x}{2}(\bar{x}^2 + \bar{p}_x^2) - \frac{\bar{\mu}_y}{2}(\bar{y}^2 + \bar{p}_y^2) - \frac{\bar{\alpha}_c}{2}\delta^2 \,, \tag{4.304}$$

where

$$U \;=\; \begin{bmatrix} \bar{x} \\ \bar{p}_x \\ \bar{y} \\ \bar{p}_y \\ \bar{z} \\ \delta \end{bmatrix} \;=\; \mathbf{A_2}^{-1} X \,, \tag{4.305}$$

and $\bar{\mu}_{x,y}$ (the normal mode frequencies) and $\bar{\alpha}_c$ (the normal mode momentum compaction factor) are constants determined by the system. Note that the variable δ is the same before and after the transformation and that $f_2(A_2U)$ [and $f_n(A_2U)$ if the system is nonlinear] do not depend on $\bar{z}$.

How to find $\mathbf{A_2}^{-1}$ and to obtain the normalized coordinates U knowing the one-turn linear map $\mathbf{M}$ of a circular accelerator? We want U (and the corresponding coordinate transformation map A) to be chosen so that the f_2 function of the linear Lie map has the form (4.304).

Note that we aim to make this normal form transformation even when there are general couplings of the 3-D motion. The matrix of the linear map can be written in general as

$$\mathbf{M} = \begin{bmatrix} & & & & 0 & R_{16} \\ & & \mathbf{R} & & 0 & R_{26} \\ & & & & 0 & R_{36} \\ & & & & 0 & R_{46} \\ R_{51} & R_{52} & R_{53} & R_{54} & 1 & R_{56} \\ 0 & 0 & 0 & 0 & 0 & 1 \end{bmatrix}, \tag{4.306}$$

where $\mathbf{R}$ is a 4×4 matrix. As written, Eq. (4.306) contains 25 yet-unspecified elements, but symplecticity of $\mathbf{M}$ imposes strong constraints on these elements. First, it requires that $\mathbf{R}$ be symplectic. This reduces the number of independent elements of $\mathbf{R}$ from 16 to 10. Second, it also requires that the R-elements satisfy

$$\tilde{\mathbf{R}}\mathbf{S} \begin{bmatrix} R_{16} \\ R_{26} \\ R_{36} \\ R_{46} \end{bmatrix} + \begin{bmatrix} R_{51} \\ R_{52} \\ R_{53} \\ R_{54} \end{bmatrix} = 0, \tag{4.307}$$

where $\mathbf{S}$ is the 4×4 member of the symplectic form

$$\mathbf{S} = \begin{bmatrix} 0 & 1 & 0 & 0 \\ -1 & 0 & 0 & 0 \\ 0 & 0 & 0 & 1 \\ 0 & 0 & -1 & 0 \end{bmatrix},$$

and a tilde means taking the transpose of a matrix. Equation (4.307) constitutes four more conditions. The total number of independent elements in Eq. (4.306) is fifteen. We now assume the given matrix $\mathbf{M}$ satisfies all the constraints mentioned above but is otherwise arbitrary, and we need to find the matrix $\mathbf{A_2}^{-1}$.

We will make a succession of three canonical transformations. These transformations follow the Courant–Snyder transformation but is its generalization to 3-D linear system. We will obtain a "normal form" result at the end. This discussion then serves as the background to discuss normal forms for nonlinear systems later.

First transformation   We first make a transformation of coordinates,

$$\text{from} \quad X = \begin{bmatrix} x \\ x' \\ y \\ y' \\ z \\ \delta \end{bmatrix} \quad \text{to} \quad X_1 = \mathbf{B}X\,, \tag{4.308}$$

where $\mathbf{B}$ is the symplectic matrix

$$\mathbf{B} = \begin{bmatrix} 1 & 0 & 0 & 0 & 0 & -\eta_x \\ 0 & 1 & 0 & 0 & 0 & -\eta_x' \\ 0 & 0 & 1 & 0 & 0 & -\eta_y \\ 0 & 0 & 0 & 1 & 0 & -\eta_y' \\ \eta_x' & -\eta_x & \eta_y' & -\eta_y & 1 & 0 \\ 0 & 0 & 0 & 0 & 0 & 1 \end{bmatrix}.$$

The four components $(\eta_x, \eta_x', \eta_y, \eta_y')$ are called the dispersion functions. They form a vector that satisfies a fixed point condition,

$$\mathbf{R}\begin{bmatrix} \eta_x \\ \eta_x' \\ \eta_y \\ \eta_y' \end{bmatrix} + \begin{bmatrix} R_{16} \\ R_{26} \\ R_{36} \\ R_{46} \end{bmatrix} = \begin{bmatrix} \eta_x \\ \eta_x' \\ \eta_y \\ \eta_y' \end{bmatrix}, \tag{4.309}$$

or

$$\begin{bmatrix} \eta_x \\ \eta_x' \\ \eta_y \\ \eta_y' \end{bmatrix} = -(\mathbf{R} - 1)^{-1} \begin{bmatrix} R_{16} \\ R_{26} \\ R_{36} \\ R_{46} \end{bmatrix}.$$

The condition (4.309) allows expressions of the elements $R_{16}, R_{26}, R_{36}, R_{46}$ in terms of $\eta_x, \eta_x', \eta_y, \eta_y'$. The symplecticity condition (4.307) then allows expressions of $R_{51}, R_{52}, R_{53}, R_{54}$ in terms of $\eta_x, \eta_x', \eta_y, \eta_y'$ as

$$\begin{bmatrix} R_{51} \\ R_{52} \\ R_{53} \\ R_{54} \end{bmatrix} = (1 - \tilde{\mathbf{R}})\mathbf{S} \begin{bmatrix} \eta_x \\ \eta_x' \\ \eta_y \\ \eta_y' \end{bmatrix}. \tag{4.310}$$

It is easy to show that the transformation matrix for the new coordinates (4.308) is given by

$$\mathbf{BMB}^{-1} = \begin{bmatrix} & & & & 0 & 0 \\ & & \mathbf{R} & & 0 & 0 \\ & & & & 0 & 0 \\ & & & & 0 & 0 \\ 0 & 0 & 0 & 0 & 1 & \bar{\alpha}_c C \\ 0 & 0 & 0 & 0 & 0 & 1 \end{bmatrix}, \tag{4.311}$$

where (C represents the storage ring circumference)

$$\bar{\alpha}_c C \;=\; R_{56} + \eta'_x R_{16} - \eta_x R_{26} + \eta'_y R_{36} - \eta_y R_{46}\,,$$

and use has been made of Eqs. (4.309) and (4.310). The matrix (4.311) is obviously simpler than the matrix (4.306).

Second transformation We can make the matrix even simpler by introducing a second transformation which deals with the matrix $\mathbf{R}$ in Eq. (4.311). Assuming the betatron motion is stable, $\mathbf{R}$ can be block diagonalized by a 4×4 matrix $\mathbf{T}$ so that

$$\mathbf{TRT}^{-1} = \begin{bmatrix} \cos\bar{\mu}_x & \sin\bar{\mu}_x & 0 & 0 \\ -\sin\bar{\mu}_x & \cos\bar{\mu}_x & 0 & 0 \\ 0 & 0 & \cos\bar{\mu}_y & \sin\bar{\mu}_y \\ 0 & 0 & -\sin\bar{\mu}_y & \cos\bar{\mu}_y \end{bmatrix}.$$

We shall assume that the matrix $\mathbf{T}$ has been found.

The 6×6 matrix $\mathbf{A}_2^{-1}$ of Eq. (4.305) is given by

$$\mathbf{A}_2^{-1} \;=\; \underbrace{\begin{bmatrix} & & & & 0 & 0 \\ & \mathbf{T} & & & 0 & 0 \\ & & & & 0 & 0 \\ & & & & 0 & 0 \\ 0 & 0 & 0 & 0 & 1 & 0 \\ 0 & 0 & 0 & 0 & 0 & 1 \end{bmatrix}}_{\mathbf{B}}$$

$$=\; \begin{bmatrix} & & & & 0 & (A_2^{-1})_{16} \\ & \mathbf{T} & & & 0 & (A_2^{-1})_{26} \\ & & & & 0 & (A_2^{-1})_{36} \\ & & & & 0 & (A_2^{-1})_{46} \\ \eta'_x & -\eta_x & \eta'_y & -\eta_y & 1 & 0 \\ 0 & 0 & 0 & 0 & 0 & 1 \end{bmatrix}, \qquad (4.312)$$

where

$$\begin{bmatrix} (A_2^{-1})_{16} \\ (A_2^{-1})_{26} \\ (A_2^{-1})_{36} \\ (A_2^{-1})_{46} \end{bmatrix} = -\mathbf{T} \begin{bmatrix} \eta_x \\ \eta'_x \\ \eta_y \\ \eta'_y \end{bmatrix}.$$

One can also take the inverse of Eq. (4.312) to obtain

$$\mathbf{A}_2 \;=\; \begin{bmatrix} & & & & 0 & \eta_x \\ & \mathbf{T}^{-1} & & & 0 & \eta'_x \\ & & & & 0 & \eta_y \\ & & & & 0 & \eta'_y \\ (A_2)_{51} & (A_2)_{52} & (A_2)_{53} & (A_2)_{54} & 1 & 0 \\ 0 & 0 & 0 & 0 & 0 & 1 \end{bmatrix},$$

where

$$
\begin{bmatrix} (A_2)_{51} \\ (A_2)_{52} \\ (A_2)_{53} \\ (A_2)_{54} \end{bmatrix} = -\tilde{\mathbf{T}}^{-1}\mathbf{S} \begin{bmatrix} \eta_x \\ \eta_x' \\ \eta_y \\ \eta_y' \end{bmatrix} .
$$

In terms of the new coordinates (4.305), finally, the transformation matrix assumes a simple form

$$
\begin{aligned}
\mathbf{N} \;=\;& \mathbf{TBM(TB)}^{-1} \\[4pt]
=\;& \begin{bmatrix}
\cos\bar{\mu}_x & \sin\bar{\mu}_x & 0 & 0 & 0 & 0 \\
-\sin\bar{\mu}_x & \cos\bar{\mu}_x & 0 & 0 & 0 & 0 \\
0 & 0 & \cos\bar{\mu}_y & \sin\bar{\mu}_y & 0 & 0 \\
0 & 0 & -\sin\bar{\mu}_y & \cos\bar{\mu}_y & 0 & 0 \\
0 & 0 & 0 & 0 & 1 & \bar{\alpha}_c C \\
0 & 0 & 0 & 0 & 0 & 1
\end{bmatrix} .
\end{aligned}
\tag{4.313}
$$

We have thus transformed the complicated matrix (4.306) to a much simpler form (4.313) by two coordinate transformations. Application to a special case of 3-D linear system is given in Homework 4.100.

Each of the maps $\mathbf{B}, \mathbf{T}, \mathbf{N}$ plays a specific role. The map $\mathbf{B}$ contains the dispersion function information of the system, $\mathbf{T}$ contains the betatron functions and $\mathbf{N}$ contains the normal mode frequencies. The dispersion and the betatron functions, and therefore $\mathbf{B}$ and $\mathbf{T}$, depend on the location s of where the one-turn map (4.306) is taken. On the other hand, the normal mode frequencies, and therefore $\mathbf{N}$, are s-independent. The formalism is truly elegant. See Homework 4.101.[35]

Third transformation   The components of U are the normalized coordinates which transform the linear part of the map into $e^{:f_2(A_2U):}$ with $f_2(A_2U)$ given by Eq. (4.304). To complete the normal form representation, we still need to introduce a third coordinate transformation. Following what we learned with Eq. (4.220), we introduce the coordinates $(\phi_x, A_x, \phi_y, A_y)$ according to

$$
\begin{aligned}
\bar{x} &= \sqrt{2A_x}\sin\phi_x, & \bar{p}_x &= \sqrt{2A_x}\cos\phi_x , \\
\bar{y} &= \sqrt{2A_y}\sin\phi_y, & \bar{p}_y &= \sqrt{2A_y}\cos\phi_y .
\end{aligned}
\tag{4.314}
$$

This transformation (4.314) is of course canonical. We have then finally, after incorporating the third transformation into A_2,

$$
f_2(A_2U) = -\bar{\mu}_x A_x - \bar{\mu}_y A_y - \frac{1}{2}\bar{\alpha}_c C\delta^2 .
\tag{4.315}
$$

After three successive canonical transformations, we thus obtain a linear Lie map with f_2 of a very simple form, Eq. (4.315). The eigenmodes of $:f_2:$ are given by Eq. (4.221). The simple additive nature of the three terms in

---

[35] On the other hand, it should be pointed out again that it is also not unique.

the Hamiltonian (4.315), each representing one degree of freedom of the 3-D dynamics, is a consequence as discussed in Homeworks 1.13 and 4.54.

What has been described so far since Eq. (4.300) is the normal form formalism for a linear system. Given a general linear map $M = e^{:f_2:}$, we have found a transformation A (or a succession of transformations whose net result is a transformation A) such that the transformed map $N = AMA^{-1}$ has the very simple form $e^{:h_2:}$, where $h_2 = f_2(A_2U)$ is a function only of A_x, A_y, and δ. In contrast, note that in the original map $e^{:f_2:}$, f_2 is a function of 5 variables x, x', y, y', and δ.

An additional comment is obligatory here. The reduction of f_2 from five variables to three variables is not only simplification in appearance. The new f_2, being the effective Hamiltonian, now predicts the three variables are constants of the motion. The system has been integrated.

Homework 4.100 In the absence of x-y coupling, Eqs. (4.306–4.313) have simpler forms. Let the linear part of the accelerator section be represented by the transformation matrix [see Eq. (4.75)]

$$
\mathbf{M} = \begin{bmatrix}
R_{11} & R_{12} & 0 & 0 & 0 & \eta - \eta R_{11} - \eta' R_{12} \\
R_{21} & R_{22} & 0 & 0 & 0 & \eta' - \eta' R_{22} - \eta R_{21} \\
0 & 0 & R_{33} & R_{34} & 0 & 0 \\
0 & 0 & R_{43} & R_{44} & 0 & 0 \\
\eta' - \eta' R_{11} + \eta R_{21} & -\eta + \eta R_{22} - \eta' R_{12} & 0 & 0 & 1 & A_\eta \sin\mu_x - \alpha_c C \\
0 & 0 & 0 & 0 & 0 & 1
\end{bmatrix},
$$

$$(4.316)$$

where

$$
\begin{aligned}
R_{11} &= \cos\mu_x + \alpha_x \sin\mu_x, \\
R_{12} &= \beta_x \sin\mu_x, \\
R_{21} &= -\gamma_x \sin\mu_x, \\
R_{22} &= \cos\mu_x - \alpha_x \sin\mu_x, \\
R_{33} &= \cos\mu_y + \alpha_y \sin\mu_y, \\
R_{34} &= \beta_y \sin\mu_y, \\
R_{43} &= -\gamma_y \sin\mu_y, \\
R_{44} &= \cos\mu_y - \alpha_y \sin\mu_y, \\
A_\eta &= \gamma_x \eta^2 + 2\alpha_x \eta\eta' + \beta_x \eta'^2,
\end{aligned}
$$

with $\gamma_{x,y} = \frac{1+\alpha_{x,y}^2}{\beta_{x,y}}$. Find the matrix $\mathbf{A_2}$ and $f_2(A_2U)$.

Solution One should first check that the symplectic condition (4.307) is satisfied (and it is) by the matrix $\mathbf{M}$. The matrix map (4.316) has a Lie form $e^{:f_2(X):}$ with [see Eq. (4.74)]

$$
\begin{aligned}
f_2(X) &= -\frac{\mu_x}{2}\left[\gamma_x(x-\eta\delta)^2 + 2\alpha_x(x-\eta\delta)(x'-\eta'\delta) + \beta_x(x'-\eta'\delta)^2\right] \\
&\quad -\frac{\mu_y}{2}\left(\gamma_y y^2 + 2\alpha_y yy' + \beta_y y'^2\right) - \frac{1}{2}\alpha_c C\delta^2.
\end{aligned}
$$

$$(4.317)$$

Following the procedure outlined in the text, we find

$$U = \mathbf{A_2}^{-1}X = \begin{bmatrix} \dfrac{x-\eta\delta}{\sqrt{\beta_x}} \\[2ex] \dfrac{\alpha_x(x-\eta\delta)+\beta_x(x'-\eta'\delta)}{\sqrt{\beta_x}} \\[2ex] \dfrac{y}{\sqrt{\beta_y}} \\[2ex] \dfrac{\alpha_y y+\beta_y y'}{\sqrt{\beta_y}} \\[2ex] z+\eta'x-\eta x' \\[1ex] \delta \end{bmatrix} = \begin{bmatrix} \bar{x} \\ \bar{p}_x \\ \bar{y} \\ \bar{p}_y \\ \bar{z} \\ \delta \end{bmatrix},$$

$$\bar{\mu}_x = \mu_x, \qquad \bar{\mu}_y = \mu_y, \qquad \bar{\alpha}_c = \alpha_c,$$

$$f_2(A_2 U) = -\frac{\mu_x}{2}(\bar{x}^2+\bar{p}_x^2) - \frac{\mu_y}{2}(\bar{y}^2+\bar{p}_y^2) - \frac{1}{2}\alpha_c C\delta^2, \qquad (4.318)$$

and

$$\mathbf{A_2}^{-1} = \begin{bmatrix} \dfrac{1}{\sqrt{\beta_x}} & 0 & 0 & 0 & 0 & -\dfrac{\eta}{\sqrt{\beta_x}} \\[2ex] \dfrac{\alpha_x}{\sqrt{\beta_x}} & \sqrt{\beta_x} & 0 & 0 & 0 & -\dfrac{\alpha_x\eta+\beta_x\eta'}{\sqrt{\beta_x}} \\[2ex] 0 & 0 & \dfrac{1}{\sqrt{\beta_y}} & 0 & 0 & 0 \\[2ex] 0 & 0 & \dfrac{\alpha_y}{\sqrt{\beta_y}} & \sqrt{\beta_y} & 0 & 0 \\[2ex] \eta' & -\eta & 0 & 0 & 1 & 0 \\[1ex] 0 & 0 & 0 & 0 & 0 & 1 \end{bmatrix},$$

$$\mathbf{A_2} = \begin{bmatrix} \sqrt{\beta_x} & 0 & 0 & 0 & 0 & \eta \\[2ex] -\dfrac{\alpha_x}{\sqrt{\beta_x}} & \dfrac{1}{\sqrt{\beta_x}} & 0 & 0 & 0 & \eta' \\[2ex] 0 & 0 & \sqrt{\beta_y} & 0 & 0 & 0 \\[2ex] 0 & 0 & -\dfrac{\alpha_y}{\sqrt{\beta_y}} & \dfrac{1}{\sqrt{\beta_y}} & 0 & 0 \\[2ex] -\dfrac{\alpha_x\eta+\beta_x\eta'}{\sqrt{\beta_x}} & \dfrac{\eta}{\sqrt{\beta_x}} & 0 & 0 & 1 & 0 \\[1ex] 0 & 0 & 0 & 0 & 0 & 1 \end{bmatrix}. \qquad (4.319)$$

The matrix representation of the map N is

$$\mathbf{N} = \mathbf{A_2}^{-1}\mathbf{M}\mathbf{A_2} = \begin{bmatrix} \cos\mu_x & \sin\mu_x & 0 & 0 & 0 & 0 \\ -\sin\mu_x & \cos\mu_x & 0 & 0 & 0 & 0 \\ 0 & 0 & \cos\mu_y & \sin\mu_y & 0 & 0 \\ 0 & 0 & -\sin\mu_y & \cos\mu_y & 0 & 0 \\ 0 & 0 & 0 & 0 & 1 & -\alpha_c C \\ 0 & 0 & 0 & 0 & 0 & 1 \end{bmatrix}.$$

Equation (4.316) gives the 6×6 matrix for one-turn map around a certain position in the circular accelerator. For completeness and later use, we give below the 6×6 matrix from position 1 to position 2,

$$M = \begin{bmatrix} R_{11} & R_{12} & 0 & 0 & 0 & \eta_2 - R_{11}\eta_1 - R_{12}\eta_1' \\ R_{21} & R_{22} & 0 & 0 & 0 & \eta_2' - R_{21}\eta_1 - R_{22}\eta_1' \\ 0 & 0 & R_{33} & R_{34} & 0 & 0 \\ 0 & 0 & R_{43} & R_{44} & 0 & 0 \\ A & B & 0 & 0 & 1 & \alpha_{1\to2}C - A\eta_1 - B\eta_1' \\ 0 & 0 & 0 & 0 & 0 & 1 \end{bmatrix},$$

$$R_{11} = \sqrt{\frac{\beta_{x2}}{\beta_{x1}}}\left[\cos\psi_x + \alpha_{x1}\sin\psi_x\right],$$

$$R_{12} = \sqrt{\beta_{x1}\beta_{x2}}\,\sin\psi_x,$$

$$R_{21} = -\frac{1}{\sqrt{\beta_{x1}\beta_{x2}}}\left[(1 + \alpha_{x1}\alpha_{x2})\sin\psi_x - (\alpha_{x1} - \alpha_{x2})\cos\psi_x\right],$$

$$R_{22} = \sqrt{\frac{\beta_{x1}}{\beta_{x2}}}\left[\cos\psi_x - \alpha_{x2}\sin\psi_x\right],$$

$$R_{33} = \sqrt{\frac{\beta_{y2}}{\beta_{y1}}}\left[\cos\psi_y + \alpha_{y1}\sin\psi_y\right],$$

$$R_{34} = \sqrt{\beta_{y1}\beta_{y2}}\,\sin\psi_y,$$

$$R_{43} = -\frac{1}{\sqrt{\beta_{y1}\beta_{y2}}}\left[(1 + \alpha_{y1}\alpha_{y2})\sin\psi_y - (\alpha_{y1} - \alpha_{y2})\cos\psi_y\right],$$

$$R_{44} = \sqrt{\frac{\beta_{y1}}{\beta_{y2}}}\left[\cos\psi_y - \alpha_{y2}\sin\psi_y\right],$$

$$A = R_{21}\eta_2 - R_{11}\eta_2' + \eta_1',$$

$$B = -R_{12}\eta_2' + R_{22}\eta_2 - \eta_1,$$

$$\alpha_{1\to2}C = -\int_{s_1}^{s_2} ds\,\frac{\eta(s)}{\rho(s)}, \tag{4.320}$$

where ψ_x and ψ_y are the horizontal and vertical betatron phase advances from position 1 to position 2, and $\alpha_{1\to2}$ is the corresponding partial momentum compaction factor. As mentioned in Homework 4.36, an explicit expression of the corresponding Lie representation can only be done transcendentally.

4.8.4 Nonlinear system

We are now ready to introduce the idea of normal form for a nonlinear system. For a nonlinear map M such as (4.299), the idea of normal form is to look for a nonlinear symplectic transformation A such that the map after making the transformation,

$$N = AMA^{-1}, \tag{4.321}$$

is "as simple as possible". Drawing an analogy to the linear analysis, this means the transformed map N — which has the "simplest possible" form — depends only on A_x, A_y and δ even though the original map M depends on five dynamical variables x, x', y, y', and δ.[36]

It should be emphasized that, recalling the discussion following Eq. (4.303), A is a coordinate transformation that can be detached from the multiturn dynamics. All dynamical effects that involve multiple turns — such as the beam dynamics encountered in circular accelerators — are contained in the map N, and that is why we would like N to be simple.

As mentioned before, normal forms are not as useful for single pass devices such as a transport line or a linac. This comment applies in the linear case as well; there is no advantage to be gained by applying the Courant–Snyder formalism to linacs.

The map A and A^{-1} are applied once and only once before and after the study of the multiturn effects. We shall see that A has the physical meaning of the Courant–Snyder parameters generalized to the nonlinear systems, that N contains information about tune dependence on the dynamical variables A_x, A_y, δ, and that N also contains information about the resonance strengths. In short, A contains a static coordinate transformation and is a kinematic quantity, while N contains all the real beam dynamics.[37]

We want N to be simple by making a clever choice of A. Note that A is not unique. This is true even in the linear case; the Courant–Snyder formalism is not unique either. Now for the nonlinear case, we shall try below to make one choice based on Lie algebra as cleverly as we can make it.

Second order   To demonstrate, consider a second order map

$$M = e^{:f_2:}e^{:f_3:} . \tag{4.322}$$

Again, we are looking for a map N and a transformation A in the form of (4.321) so that N is simple. Let us first consider the second order transformation

$$A = e^{:F_3:}A_2 , \tag{4.323}$$

where A_2 is the linear transformation (4.305) which transforms the X-coordinates to the U-coordinates. Note the specific ordering of the two factor maps in Eq. (4.323). The function F_3 is yet to be found.

After the transformation, we have

$$N = AMA^{-1} = e^{:F_3(X):}A_2 e^{:f_2(X):}e^{:f_3(X):}A_2^{-1}e^{-:F_3(X):}\Big|_{X=U}$$

$$= e^{:F_3(U):}e^{:f_2(A_2 U):}e^{:f_3(A_2 U):}e^{-:F_3(U):} ,$$

---

[36] For a nonlinear system, this ambition requires an additional condition that the system is away from resonances. When there is a resonance near by, there are complications. See later. For a linear system, this additional condition was avoided because we have categorically assumed the linear system is stable.

[37] Perhaps let us put it in another way. The map A^{-1} is like putting on eye glasses. The map A is to take off the glasses. You do them only once. The movie you are watching, through the eye glasses, is given by the map N.

where use has been made of Eq. (4.302). The map $e^{:f_2(A_2U):}$ is a simple rotation map corresponding to Eq. (4.313). The function $f_2(A_2U)$ will become a function only of A_x, A_y, and δ if one makes a further transformation (4.314).

Considering the U coordinates, although at a slight risk of confusion, we write $f_n(A_2U)$ as $f_n(U)$, or simply as f_n to 2nd order in U,

$$
\begin{aligned}
N &= e^{:f_2:}e^{-:f_2:}e^{:F_3:}e^{:f_2:}e^{:f_3:}e^{-:F_3:} \\
&= e^{:f_2:}\exp(:e^{-:f_2:}F_3:)e^{:f_3:}e^{-:F_3:} \\
&= e^{:f_2:}\exp[:(e^{-:f_2:}-1)F_3 + f_3 + \mathcal{O}(U^4):].
\end{aligned}
\tag{4.324}
$$

An inspection of Eq. (4.324) indicates that, at this point in an effort to make N as simple as possible, one would be tempted to choose F_3 — recalling that F_3 is still at our disposal — as

$$
F_3 = \left(\frac{1}{1-e^{-:f_2:}}\right)f_3,
\tag{4.325}
$$

so that the map N, to 2nd order in U, is simply the linear map $e^{:f_2:}$. It however turns out that this cannot be done in general.[38] On the other hand, we can try to approach Eq. (4.325) as closely as possible as follows.

Consider a function $f_n(U)$, which is an n-th order homogeneous polynomial in the components of U (except $\bar{z}$). It is possible to express $f_n(U)$ as

$$
f_n = \sum_{\substack{a,b,c,d,e=0 \\ a+b+c+d+e=n}}^{n} C^{(n)}_{abcd,e}|abcd,e\rangle,
\tag{4.326}
$$

where

$$
\begin{aligned}
|abcd,e\rangle &\equiv \left(\sqrt{A_x}e^{i\phi_x}\right)^a\left(\sqrt{A_x}e^{-i\phi_x}\right)^b\left(\sqrt{A_y}e^{i\phi_y}\right)^c\left(\sqrt{A_y}e^{-i\phi_y}\right)^d\delta^e \\
&= A_x^{(a+b)/2}A_y^{(c+d)/2}e^{i(a-b)\phi_x}e^{i(c-d)\phi_y}\delta^e
\end{aligned}
\tag{4.327}
$$

are the eigenmodes (sometimes referred to as the resonance basis) of $:f_2:$ with

$$
:f_2:|abcd,e\rangle = i[(a-b)\bar{\mu}_x + (c-d)\bar{\mu}_y]\,|abcd,e\rangle.
\tag{4.328}
$$

We have used the fact that $:f_2:$ leaves δ untouched. Equation (4.326) simply decomposes f_n as a linear superposition of eigenmodes of $:f_2:$.

It follows from the fact that $f_n(U)$ is a real function that

$$
{C^{(n)}_{abcd,e}}^{*} = C^{(n)}_{badc,e}.
\tag{4.329}
$$

In what follows, we will simplify the notations by dropping the bars on the normal mode frequencies $\bar{\mu}_{x,y}$. We will also drop the bars on $\bar{\alpha}_c$ and $\bar{z}$.

---

[38]If this could be done, all of the nonlinear dynamics are contained simply in the boring coordinate transformation A. There would be no multiturn beam dynamics to speak of any more.

For f_3 we have

$$f_3 = \sum_{a,b,c,d,e=0}^{3} C^{(3)}_{abcd,e}|abcd,e\rangle\,. \tag{4.330}$$

Using Eq. (4.328) the right-hand-side of Eq. (4.325) reads

$$\left(\frac{1}{1-e^{-:f_2:}}\right)f_3 = \sum_{a,b,c,d,e} C^{(3)}_{abcd,e}\left[\frac{1}{1-e^{-i(a-b)\mu_x-i(c-d)\mu_y}}\right]|abcd,e\rangle\,. \tag{4.331}$$

The expression (4.331) is fine except for those terms in the summation when one finds zero denominators,

$$e^{-i(a-b)\mu_x-i(c-d)\mu_y} = 1\,,$$

$$\text{or}\qquad (a-b)\frac{\mu_x}{2\pi} + (c-d)\frac{\mu_y}{2\pi} = \text{integer}\,.$$

There are two circumstances when this condition is met. One occurs when

$$a = b \quad\text{and}\quad c = d\,. \tag{4.332}$$

The other occurs only when a "resonance" condition is fulfilled,[39] i.e.

$$m\frac{\mu_x}{2\pi} + n\frac{\mu_y}{2\pi} = p\,, \tag{4.333}$$

where m, n, and p are integers without common denominators. When Eq. (4.333) is satisfied, terms with

$$a - b = km, \qquad\text{and}\quad c - d = kn \tag{4.334}$$

must also be avoided, where k is any positive or negative integer.

Since the choice of F_3 is at our disposal, we will choose it to be given by Eq. (4.331) *excluding* terms that give us trouble. In other words, we certainly exclude from the summation those terms corresponding to Eq. (4.332) and, when a resonance is uncomfortably close by, also those terms corresponding to Eq. (4.334).

We postpone the discussion of the resonance terms till later. For now, let us consider the case away from all resonances. In this case, we have

$$F_3 = \sum_{\begin{cases} a,b,c,d,e=0 \\ a+b+c+d+e=3 \\ \text{unless } a=b \text{ and } c=d \end{cases}}^{3} C^{(3)}_{abcd,e}\left[\frac{1}{1-e^{-i(a-b)\mu_x-i(c-d)\mu_y}}\right]|abcd,e\rangle\,, \tag{4.335}$$

---

[39] Recall that $\mu_{x,y}$ here are really $\bar\mu_{x,y}$, the normal mode frequencies, which take into account the linear perturbations of the accelerator. Skew quadrupoles effects on the tunes, for example, are to be contained in $\mu_{x,y}$.

which leads to

$$(e^{-:f_2:} - 1)F_3 + f_3 = \sum_{a,c,e} C^{(3)}_{aacc,e}|aacc,e\rangle$$

$$= \sum_{a,c,e} C^{(3)}_{aacc,e} A_x^a A_y^c \delta^e$$

$$= C^{(3)}_{0000,3}\delta^3 + C^{(3)}_{1100,1}A_x\delta + C^{(3)}_{0011,1}A_y\delta \equiv h_3 . \tag{4.336}$$

We have designated the above function as h_3.

Substituting Eq. (4.336) into Eq. (4.324) gives the simplest form of N, to second order in U, as

$$N = e^{:f_2:}e^{:h_3:} . \tag{4.337}$$

Both f_2 and h_3 are functions only of A_x, A_y, and δ. Note that h_3 contains only three terms in the summation, each proportional to $A_x\delta, A_y\delta$, and δ^3, respectively. Since $:f_2:$ and $:h_3:$ commute, we can write

$$N = e^{:f_2+h_3:} . \tag{4.338}$$

An inspection of Eqs. (4.322) and (4.337) indicates that the normal form transformation basically means dropping the oscillating terms — terms depending on the phases ϕ_x and ϕ_y — from f_3, and keeping only the zeroth harmonic terms. The reason these zeroth harmonic terms cannot be dropped is because it is impossible to find F_3 to transform them away.

Higher order   The normal form procedure from Eq. (4.322) to Eq. (4.338) can be extended to higher orders. Consider a third order map

$$M = e^{:f_2(X):}e^{:f_3(X):}e^{:f_4(X):} . \tag{4.339}$$

By making the transformation with

$$A = e^{:F_4(U):}e^{:F_3(U):}A_2 , \tag{4.340}$$

we obtain

$$N = AMA^{-1} = e^{:F_4:}e^{:F_3:}e^{:f_2:}e^{:f_3:}e^{:f_4:}e^{-:F_3:}e^{-:F_4:} . \tag{4.341}$$

All functions in Eq. (4.341) are expressed as functions of U. We choose F_3 to be given by Eq. (4.335).

To proceed, we need to go back to Eq. (4.324) and keep terms $\mathcal{O}(U^3)$. Consider the following,

$$e^{:F_3:}e^{:f_2:}e^{:f_3:}e^{-:F_3:} = e^{:f_2:}\left(e^{-:f_2:}e^{:F_3:}e^{:f_2:}\right)\left(e^{:f_3:}e^{-:F_3:}\right)$$

$$= e^{:f_2:}\exp\left(:e^{-:f_2:}F_3:\right)\exp\left(:f_3 - F_3 - \frac{1}{2}:f_3:F_3 + \mathcal{O}(U^5):\right)$$

$$= e^{:f_2:}\exp\left[:e^{-:f_2:}F_3 + f_3 - F_3 - \frac{1}{2}:f_3:F_3\right.$$

$$-\frac{1}{2}:(f_3 - F_3):e^{-:f_2:}F_3:\Big] \, e^{:\mathcal{O}(U^5):}$$

$$= \; e^{:f_2:}e^{:h_3:}e^{:g_4:}e^{:\mathcal{O}(U^5):} \, , \tag{4.342}$$

where use has been made of the BCH formula (4.137), h_3 is given by Eq. (4.336), and we have introduced

$$g_4 \; = \; -\frac{1}{2}\left(:f_3: + :f_3:e^{-:f_2:} - :F_3:e^{-:f_2:}\right) F_3 \, . \tag{4.343}$$

By definition of h_3 in Eq. (4.336), Eq. (4.343) can also be written as

$$g_4 \; = \; -\frac{1}{2}\left[:f_3:F_3 + :(f_3 - F_3):h_3\right] \, . \tag{4.344}$$

Having obtained Eq. (4.342), we have, after dropping the high order contributions from $e^{:\mathcal{O}(U^5):}$,

$$\begin{aligned}
e^{:F_3:}e^{:f_2:}e^{:f_3:}e^{:f_4:}e^{-:F_3:} \; &= \; \left(e^{:F_3:}e^{:f_2:}e^{:f_3:}e^{-:F_3:}\right)\left(e^{:F_3:}e^{:f_4:}e^{-:F_3:}\right) \\
&= \; \left(e^{:f_2:}e^{:h_3:}e^{:g_4:}\right)\left(e^{:f_4:}\right) \\
&= \; e^{:f_2:}e^{:h_3:}e^{:f_4+g_4:} \, . \tag{4.345}
\end{aligned}$$

Substituting Eq. (4.345) into Eq. (4.341) gives

$$\begin{aligned}
N \; &= \; e^{:F_4:}e^{:f_2:}e^{:h_3:}e^{:f_4+g_4:}e^{-:F_4:} \\
&= \; e^{:f_2:}\exp(:e^{-:f_2:}F_4:)e^{:h_3:}e^{:f_4+g_4:}e^{-:F_4:} \\
&= \; e^{:f_2:}e^{:h_3:}\exp\left[:(e^{-:f_2:} - 1)F_4 + f_4 + g_4:\right] \, .
\end{aligned}$$

Following the steps which are by now familiar, we decompose

$$f_4 + g_4 \; = \; \sum_{a,b,c,d,e}^{4} C^{(4)}_{abcd,e}|abcd,e\rangle \, ,$$

and choose F_4 to be (away from resonances)

$$F_4 \; = \; \sum_{\substack{a,b,c,d,e=0 \\ a+b+c+d+e=4 \\ \text{unless } a=b \text{ and } c=d}}^{4} C^{(4)}_{abcd,e}\left(\frac{1}{1 - e^{-i(a-b)\mu_x - i(c-d)\mu_y}}\right)|abcd,e\rangle \, .$$

Then, to third order, we have

$$\begin{aligned}
N \; &= \; e^{:f_2:}e^{:h_3:}e^{:h_4:} \\
&= \; e^{:f_2+h_3+h_4:} \, , \tag{4.346}
\end{aligned}$$

where

$$h_4 \; = \; \sum_{a,c,e} C^{(4)}_{aacc,e}A_x^a A_y^c \delta^e \tag{4.347}$$

is a function of A_x, A_y, and δ. In Eq. (4.346), we have used the fact that f_2, h_3 and h_4 commute because they are all only functions of A_x, A_y, δ which are constants of the motion.

In general, h_4 contains six terms, proportional to $\delta^4, A_x\delta^2, A_y\delta^2, A_x^2, A_xA_y$, and A_y^2, respectively. Knowing $A_2, F_3(U)$, and $F_4(U)$, the coordinate transformation A, given by Eq. (4.340), is obtained. Note that although the normal form has made N simple, A_2, F_3 and F_4, and consequently also A and A^{-1}, remain complicated. Fortunately, we need to apply them only once before and once after the multiturn dynamics study.

One can extend the procedure to higher orders and in general obtains an Ω-th order representation

$$
\begin{aligned}
M &= e^{:f_2(X):}e^{:f_3(X):}\cdots e^{:f_{\Omega+1}(X):}\,, \\
A &= e^{:F_{\Omega+1}(U):}\cdots e^{:F_3(U):}A_2\,, \\
N &= e^{:f_2+h_3+h_4+\cdots\, h_{\Omega+1}:}\,,
\end{aligned}
\tag{4.348}
$$

where $f_2, h_3, \cdots, h_{\Omega+1}$ are functions of A_x, A_y, and δ only. Equation (4.348) is the normal form representation of the map M away from resonances.

Applications of normal forms as well as the effect of resonances are subjects in the following sections.

Homework 4.101 Prove that in the normal form transformation, Eqs. (4.321) and (4.348), of a one-turn map $M(s)$ around the location s, the map N is s-independent. The s-dependence is contained in the coordinate transformation map A.

Homework 4.102

(a) Calculate $:A_{x,y}:|abcd, e\rangle$. The result should reconfirm what you already know, i.e. $|abcd, e\rangle$ is an eigenmode of the operators $:A_{x,y}:$.

(b) Calculate $e^{:\pm i\phi_{x,y}:}|abcd, e\rangle$.

(c) Calculate the Poisson bracket $\left[|a_1b_1c_1d_1, e_1\rangle, |a_2b_2c_2d_2, e_2\rangle\right]$.

Solution

(b)
$$
e^{:i\phi_x:}|abcd, e\rangle = \frac{i}{2}(a+c)\left|a-\frac{1}{2}, b, c-\frac{3}{2}, d, e\right\rangle,
$$

$$
e^{:-i\phi_x:}|abcd, e\rangle = -\frac{i}{2}(a+c)\left|a-\frac{3}{2}, b, c-\frac{1}{2}, d, e\right\rangle.
$$

(c)
$$
\left[|a_1b_1c_1d_1, e_1\rangle, |a_2b_2c_2d_2, e_2\rangle\right]
$$
$$
= i(a_1c_2-a_2c_1)\,|a_1+a_2-1, b_1+b_2, c_1+c_2-1, d_1+d_2, e_1+e_2\rangle
$$
$$
+i(b_1d_2-b_2d_1)\,|a_1+a_2, b_1+b_2-1, c_1+c_2, d_1+d_2-1, e_1+e_2\rangle.
$$

Homework 4.103 Equations (4.326–4.329) are when $f_2(A_2U)$ describes a stable motion given by Eq. (4.315). If the system is linearly unstable (see Homework 4.35), how are $f_2(A_2U)$ and Eqs. (4.326–4.329) modified?

Solution Even though unstable, the system is symplectic. Also, unlike the stable case, the corresponding eigenmodes do not involve complex quantities.

4.9 Application away from resonance

4.9.1 Invariant of motion

In the previous section, we transformed the nonlinear map M by a map A into a normal form N according to

$$N = AMA^{-1} \quad \text{or} \quad M = A^{-1}NA.$$

We noted that, away from resonances, N has the form $e^{:h:}$ where h depends only on A_x, A_y, and δ out of the phase space coordinates $(\phi_x, A_x, \phi_y, A_y, z, \delta)$ [see Eq. (4.348)]. It follows that the quantities A_x, A_y and δ are invariant under the map N, i.e.,

$$NA_x = A_x, \qquad NA_y = A_y, \qquad N\delta = \delta.$$

The quantities A_x, A_y, and δ are the three constants of the motion up to order $\mathcal{O}(X^\Omega)$ in the 3-D motion under the map M. For this 3-D system (6-D phase space), there are at most three invariants. Away from resonances, we have now found them all, and the system has been fully integrated.

Needless to say, since A_x, A_y, δ are constants of the motion, so is the function h on the exponent of the Lie map M. However, as pointed out before, $-h$ has the additional physical meaning of being the effective Hamiltonian.

Three invariants Let us consider the quantities

$$W_x = A^{-1}A_x \quad \text{and} \quad W_y = A^{-1}A_y.$$

The quantities W_x and W_y are just the invariants A_x and A_y observed back in the X coordinates. To obtain an expression of W_x, one first writes $A_x = \frac{1}{2}(\bar{x}^2 + \bar{p}_x^2)$, and then expresses $\bar{x}$ and $\bar{p}_x$ (components of the normalized coordinates U) in terms of the physical X components using $U = A^{-1}X$; the resulting expression in terms of the X coordinates is W_x. Similar operation applies for W_y.

The two quantities $W_{x,y}$ are invariant under the map M, as can also be seen as follows,

$$MW_x = A^{-1}NAW_x = A^{-1}NA_x = A^{-1}A_x = W_x,$$

and similarly for W_y. As a particle's motion is observed turn after turn, the physical coordinates change but the values of W_x and W_y remain constant.

There is a third invariant, which is simply δ. Its invariance follows from the fact that M is independent of z. Particle motion in the 6-D $(x, p_x, y, p_y, z, \delta)$ phase space must stay on contours of constant W_x, W_y and δ.

4.9.2 Effective Hamiltonian

We also look for an expression of the effective Hamiltonian of the nonlinear system for the one-turn motion in the $(x, p_x, y, p_y, z, \delta)$ phase space. This is obtained by first writing $N = e^{:h(A_x, A_y, \delta):}$ and then noting

$$
\begin{aligned}
M &= A^{-1}NA = A^{-1}e^{:h(A_x, A_y, \delta):}A \\
&= \exp\left[:A^{-1}h(A_x, A_y, \delta):\right] = \exp\left[:h(A^{-1}A_x, A^{-1}A_y, A^{-1}\delta):\right] \\
&= e^{:h(W_x, W_y, \delta):}.
\end{aligned}
$$

This means $-h(W_x, W_y, \delta)$ is the effective Hamiltonian in the X phase space away from all resonances. The time variable for this Hamiltonian is turn of revolutions. The case when a resonance is close by will be addressed later — suffice it to say here that then the effective Hamiltonian would contain additional terms.

1-D Courant–Snyder system Take the 1-D Courant–Snyder system for example. We find explicit expressions of the invariant W_x and the effective one-turn Hamiltonian as follows.

The invariant W_x is obtained by substituting

$$
\bar{x} = \frac{1}{\sqrt{\beta}}x, \qquad \bar{p}_x = \frac{1}{\sqrt{\beta}}(\alpha x + \beta p_x)
$$

into the expression

$$
A_x = \frac{1}{2}(\bar{x}^2 + \bar{p}_x^2).
$$

The result is, as expected,

$$
W_x = \frac{1}{2\beta}[x^2 + (\alpha x + \beta p_x)^2].
$$

The one-turn map for the system is given by Eq. (4.68). It follows that the effective Hamiltonian is

$$
\begin{aligned}
H(x, p_x, n) &= -f_2 = \mu W_x = \frac{\mu}{2\beta}\left[x^2 + (\alpha x + \beta p_x)^2\right] \\
&= \frac{\mu}{2}\left(\gamma x^2 + 2\alpha x p_x + \beta p_x^2\right),
\end{aligned}
$$

with n the turn number as the time variable.

The effective Hamiltonian can be used to describe the one-turn behavior of the variables x and p_x by the Hamilton equations,

$$
\begin{aligned}
\frac{dx}{dn} &= \frac{\partial H}{\partial p_x} = \mu(\alpha x + \beta p_x), \\
\frac{dp_x}{dn} &= -\frac{\partial H}{\partial x} = -\mu(\gamma x + \alpha p_x),
\end{aligned}
\tag{4.349}
$$

where n runs increments of one unit for each complete turn. In this effective Hamiltonian language, however, n is assumed to extend to a continuum. Although the Courant–Snyder functions α, β, and γ depend on the location around the accelerator, in the effective Hamiltonian description here, they are taken as constants in the turn index n. The result obtained applies only when the particle motion is observed at this particular location.

Equation (4.349) can be solved to yield

$$\begin{aligned}
x(n) &= x(0)(\cos n\mu + \alpha \sin n\mu) + p(0)\beta \sin n\mu, \\
p_x(n) &= -x(0)\gamma \sin n\mu + p(0)(\cos n\mu - \alpha \sin n\mu).
\end{aligned} \tag{4.350}$$

When n is set to 1, one recovers map (4.66).

Fixed point for an off-momentum particle Consider now a 3-D linear system in the absence of RF devices. For an on-momentum particle, the fixed point of the one-turn map in the X space is simply the origin $X = 0$ of the 6-D phase space. Consider an off-momentum particle with $\delta = \delta_0 = \text{constant}$. Its fixed point will be identified with the dispersion function. First note that the vector

$$U_0 = \begin{bmatrix} 0 \\ 0 \\ 0 \\ 0 \\ 0 \\ \delta_0 \end{bmatrix} \tag{4.351}$$

is transformed by N according to

$$NU\big|_{U=U_0} = \begin{bmatrix} 0 \\ 0 \\ 0 \\ 0 \\ z_1 \\ \delta_0 \end{bmatrix} \equiv U_1.$$

The vector U_1 differs from U_0 only in its z-component, and z_1 is some quantity whose exact form does not matter here [see Homework 4.104]. The first four components of U_1 remain zero because N does not couple the three normal modes of U.

Consider the vector

$$X_0 = AU\big|_{U=U_0}. \tag{4.352}$$

We have

$$\begin{aligned}
MX\big|_{X=X_0} &= AMU\big|_{U=U_0} = AA^{-1}NAU\big|_{U=U_0} \\
&= NAU\big|_{U=U_0} = AU\big|_{U=U_1}.
\end{aligned} \tag{4.353}$$

In writing down the first step in Eq. (4.353), recall that the earlier Lie operators apply from the left.

Since the first four components of the vector on the right-hand-side of Eq. (4.353) are the same as those of X_0, Eq. (4.353) says that the first four components of X_0 form a fixed point, i.e. its position does not change turn after turn, in the (x, p_x, y, p_y) space for an off-momentum particle.

3-D linear system   Consider the 3-D idealized uncoupled linear system described in Sec. 4.8.3 and Homework 4.100. We may find expressions of the invariants W_x, W_y, the effective Hamiltonian, and the off-momentum fixed point as follows.

The connection between the physical coordinates $(x, p_x = x', y, p_y = y', z, \delta)$ and the normalized coordinates $(\bar{x}, \bar{p}_x, \bar{y}, \bar{p}_y, \bar{z}, \delta)$ is provided by Eq. (4.318). The two betatron invariants then have the expressions

$$W_x = \frac{1}{2\beta_x} \left\{ (x - \eta\delta)^2 + [\alpha_x(x - \eta\delta) + \beta_x(p_x - \eta'\delta)]^2 \right\},$$

$$W_y = \frac{1}{2\beta_y} \left[y^2 + (\alpha_y y + \beta_y p_y)^2 \right].$$

The effective Hamiltonian is

$$H = \frac{\mu_x}{2\beta_x} \left\{ (x - \eta\delta)^2 + [\alpha_x(x - \eta\delta) + \beta_x(p_x - \eta'\delta)]^2 \right\}$$
$$+ \frac{\mu_y}{2\beta_y} \left[y^2 + (\alpha_y y + \beta_y p_y)^2 \right] + \frac{1}{2}\alpha_c\delta^2.$$

A somewhat lengthy but straightforward algebra similar to that of Eqs. (4.349–4.350) verifies that the Hamilton equations using this effective Hamiltonian in the $(x, p_x, y, p_y, z, \delta)$ phase space recovers the map (4.316).

The off-momentum fixed point in the (x, p_x, y, p_y) space is, according to Eq. (4.352), given by the first four components of the vector $\mathbf{A_2}U_0$. With $\mathbf{A_2}$ given by Eq. (4.319) and U_0 by Eq. (4.351), it follows that the fixed point is simply

$$\begin{bmatrix} x \\ p_x \\ y \\ p_y \end{bmatrix} = \delta_0 \begin{bmatrix} \eta \\ \eta' \\ 0 \\ 0 \end{bmatrix},$$

for a particle whose relative momentum error is $\delta = \delta_0$.

Consideration of fixed point does not have to be limited to off-momentum particles. Homework 4.104 applies the fixed point consideration to the case of a closed-orbit for on-momentum particles kicked by a localized dipole magnet. Fixed point of off-momentum particles are related to the dispersion functions. Fixed point of on-momentum particles are related to the closed-orbits. Both considerations can be extended to nonlinear cases, yielding nonlinear dispersion functions and nonlinear closed-orbits. Homework 4.105 generalizes the analysis of the off-momentum fixed point to the case of a nonlinear kick due to a localized sextupole.

Homework 4.104 Consider the uncoupled linear system of Homework 4.100. Let there be a static angular kick $\Delta x' = \theta$ at position $s = 0$ of this accelerator due to a dipole magnet. Apply the fixed point analysis to find the steady state motion of the particles. This fixed point defines the closed-orbit distortion of the stored beam in the accelerator. Note this homework asks you to find a 6-D closed-orbit in the phase space, not the conventional closed-orbit displacement in the kicking plane.

Solution The one-turn map around $s = 0$ is given by the matrix M of Eq. (4.316), with $\alpha, \beta, \gamma, \eta,$ and η' evaluated at the position $s = 0$. The steady state $(x, x', y, y', z, \delta)$ observed at the exit of the angular kick satisfies the condition

$$
M \begin{bmatrix} x \\ x' \\ y \\ y' \\ z \\ \delta \end{bmatrix} + \begin{bmatrix} 0 \\ \theta \\ 0 \\ 0 \\ 0 \\ 0 \end{bmatrix} = \begin{bmatrix} x \\ x' \\ y \\ y' \\ z_1 \\ \delta \end{bmatrix} . \tag{4.354}
$$

The quantity z_1 on the right-hand-side is not the same as z. This is because z does not have a steady state. We need to analyze the motion described by Eq. (4.354).

Obviously we have $y = 0, y' = 0$. The remaining of Eq. (4.354) can be written as

$$
\begin{aligned}
R_{11}(x - \eta\delta) + R_{12}(x' - \eta'\delta) &= (x - \eta\delta) , \\
R_{21}(x - \eta\delta) + R_{22}(x' - \eta'\delta) + \theta &= (x' - \eta'\delta) , \\
(\eta' - \eta' R_{11} + \eta R_{21})x + (-\eta + \eta R_{22} - \eta' R_{12})x' + z & \\
+ (-\alpha_c C + A_\eta \sin \mu_x)\delta &= z_1 ,
\end{aligned} \tag{4.355}
$$

where the definitions of the various quantities can be found with Eq. (4.316).

The first two of Eq. (4.355) can be solved for x and x',

$$
\begin{aligned}
x &= \frac{\theta}{2}\beta_x \cot \frac{\mu_x}{2} + \eta\delta , \\
x' &= \frac{\theta}{2}(1 - \alpha_x \cot \frac{\mu_x}{2}) + \eta'\delta .
\end{aligned} \tag{4.356}
$$

Equation (4.356) gives the closed-orbit displacement of the beam particles in the accelerator observed at the exit of the angular kick. The third equation in (4.355) gives the path length slippage per turn in the "steady state". When x and x' are substituted using Eq. (4.356), one obtains

$$
z_1 - z = -\alpha_c C\delta - \eta\theta .
$$

The path length slips by $-C\alpha_c\delta$ due to a momentum error, and by $-\eta\theta$ due to an angular kick θ, and this slippage occurs every turn.

Due to the term $-\eta\theta$ in the path length slippage, a horizontal kick in x' is intrinsically coupled to the longitudinal motion. This intrinsic link is a necessary consequence of the symplecticity of the system. This coupling does not occur in the y-dimension because there is no dispersion in the y-dimension in the ideal case being studied.

Homework 4.105 Replace the horizontal angular kick of Homework 4.104 by a thin-lens sextupole of strength λ. Find the nonlinear dispersion functions at the exit of the sextupole to order δ^2. In particular, we know that in this accelerator there is no linear y-dispersion, but does the sextupole generate a nonlinear y-dispersion?

4.9.3   Tune shift and chromaticity

In the earlier sections, we developed a normal form scheme away from resonances. In terms of the canonical coordinates $U = (\phi_x, A_x, \phi_y, A_y, z, \delta)$, we showed that, away from resonances, a nonlinear one-turn map can be transformed into a simple form

$$N = e^{-:H(A_x, A_y, \delta):} ,$$

where H is the effective Hamiltonian. The Hamilton equations of motion read

$$
\begin{aligned}
\frac{d\phi_x}{dn} &= \frac{\partial H}{\partial A_x}, & \frac{dA_x}{dn} &= -\frac{\partial H}{\partial \phi_x}, \\
\frac{d\phi_y}{dn} &= \frac{\partial H}{\partial A_y}, & \frac{dA_y}{dn} &= -\frac{\partial H}{\partial \phi_y}, \\
\frac{dz}{dn} &= \frac{\partial H}{\partial \delta}, & \frac{d\delta}{dn} &= -\frac{\partial H}{\partial z},
\end{aligned}
\tag{4.357}
$$

where n is the turn index which runs from 0 to 1 for one turn.

Since H is independent of ϕ_x, ϕ_y, and z, it follows from Eq. (4.357) that A_x, A_y, and δ are constants of the motion. With $H = H(A_x, A_y, \delta)$, independent of n, the effective Hamiltonian H is also a constant of the motion. It then follows that $\frac{d\phi_x}{dn}, \frac{d\phi_y}{dn}$, and $\frac{dz}{dn}$ are constants independent of n. Integrations over one turn then give the one-turn map which reads

$$
\begin{aligned}
\Delta\phi_x &= \frac{\partial H}{\partial A_x}, & \Delta A_x &= 0, \\
\Delta\phi_y &= \frac{\partial H}{\partial A_y}, & \Delta A_y &= 0, \\
\Delta z &= \frac{\partial H}{\partial \delta}, & \Delta\delta &= 0,
\end{aligned}
\tag{4.358}
$$

where Δ of a quantity means the change of this quantity over one turn. In particular, the phases $\phi_{x,y}$ and z advance by an amount specified by Eq. (4.358) per iteration of the map. These advances depend on the values of $A_{x,y}$ and δ, which are themselves invariant under the application of the map.

As the map being considered is a one-turn map around a circular accelerator, one can identify the *tunes* to be

$$
\begin{aligned}
\nu_x(A_x, A_y, \delta) &= \frac{\Delta\phi_x}{2\pi} = \frac{1}{2\pi}\frac{\partial H}{\partial A_x}, \\
\nu_y(A_x, A_y, \delta) &= \frac{\Delta\phi_y}{2\pi} = \frac{1}{2\pi}\frac{\partial H}{\partial A_y}.
\end{aligned}
\tag{4.359}
$$

Different particles with different $A_{x,y}$ and δ have different tunes. The dependencies of the tunes on $A_{x,y}$ are called the *detuning* effects. The dependencies on δ are called the *chromaticities*.

We showed earlier that, to third order in U, the effective Hamiltonian, away from resonances, can be written as [see Eqs. (4.336) and (4.346–4.347)]

$$
\begin{aligned}
H &= -f_2 - h_3 - h_4, \tag{4.360}\\
f_2 &= -\mu_x A_x - \mu_y A_y - \frac{1}{2}\alpha_c\delta^2, \\
h_3 &= -C_{x1}A_x\delta - C_{y1}A_y\delta - C_3\delta^3, \\
h_4 &= -C_{xx}A_x^2 - C_{xy}A_x A_y - C_{yy}A_y^2 - C_{x2}A_x\delta^2 - C_{y2}A_y\delta^2 - C_4\delta^4,
\end{aligned}
$$

where $\frac{\mu_{x,y}}{2\pi}$ are the *unperturbed* tunes of the linear system.[40] This gives, using Eq. (4.359),

$$
\begin{aligned}
\nu_x(A_x, A_y, \delta) &= \frac{1}{2\pi}\left(\mu_x + C_{x1}\delta + 2C_{xx}A_x + C_{xy}A_y + C_{x2}\delta^2\right), \\
\nu_y(A_x, A_y, \delta) &= \frac{1}{2\pi}\left(\mu_y + C_{y1}\delta + C_{xy}A_x + 2C_{yy}A_y + C_{y2}\delta^2\right).
\end{aligned}
\tag{4.361}
$$

In particular, the chromatic effects are seen by setting $A_x = A_y = 0$,

$$
\begin{aligned}
\nu_x(\delta) &= \frac{1}{2\pi}(\mu_x + C_{x1}\delta + C_{x2}\delta^2), \\
\nu_y(\delta) &= \frac{1}{2\pi}(\mu_y + C_{y1}\delta + C_{y2}\delta^2),
\end{aligned}
\tag{4.362}
$$

and the detuning effects are seen by setting $\delta = 0$,

$$
\begin{aligned}
\nu_x(A_x, A_y) &= \frac{1}{2\pi}\left(\mu_x + 2C_{xx}A_x + C_{xy}A_y\right), \\
\nu_y(A_x, A_y) &= \frac{1}{2\pi}\left(\mu_y + C_{xy}A_x + 2C_{yy}A_y\right).
\end{aligned}
\tag{4.363}
$$

When $A_x = A_y = 0$ and $\delta = 0$, the tunes are of course just given by $\frac{\mu_{x,y}}{2\pi}$.

Note that all terms contained in Eq. (4.361) appear in either (4.362) or (4.363). This however is not true to higher orders because then there will

---

[40] Unperturbed by nonlinearities, but perturbed by the linear coupling effects. See footnote 39.

be synchrobetatron coupling terms that involve both $A_{x,y}$ and δ in the tune expressions. In other words, all nonlinear beam dynamics effects up to the third order are covered by the detunings and chromaticities away from resonances, but this is no longer true to higher orders.

Note also from Eq. (4.363) that the A_x-dependence of ν_y is the same as the A_y-dependence of ν_x — both are described by the coefficient C_{xy},

$$\frac{\partial \nu_y}{\partial A_x} = \frac{\partial \nu_x}{\partial A_y}. \tag{4.364}$$

The Hamiltonian (4.360) also describes the behavior of the path length according to

$$\Delta z = \alpha_c \delta + 3C_3 \delta^2 + 4C_4 \delta^3 + C_{x1} A_x + C_{y1} A_y + 2C_{x2} A_x \delta + 2C_{y2} A_y \delta. \tag{4.365}$$

The path length change per revolution depends on the betatron amplitudes $A_{x,y}$ and δ. In particular, the A_x-dependence of Δz is the same as the δ-dependence of ν_x and the A_y-dependence of Δz is the same as the δ-dependence of ν_y,

$$\frac{\partial \Delta z}{\partial A_x} = 2\pi \frac{\partial \nu_x}{\partial \delta}, \qquad \frac{\partial \Delta z}{\partial A_y} = 2\pi \frac{\partial \nu_y}{\partial \delta}. \tag{4.366}$$

An intriguing observation follows: chromaticities, quantities seemingly relevant only to the transverse dynamics, are intimately related to the longitudinal dynamics involving the path length.

The two important beam dynamical theorems (4.364) and (4.366), otherwise obscure, follow naturally from the fact that the system is Hamiltonian. We will extend this observation further by a discussion following Eq. (4.405).

4.9.4 Smooth approximation

In this section, we will introduce a tune shift treatment, sometimes called the *smooth approximation*, which does not use the Lie language.[41] Applications of the results in this section can be found later when we elaborate more on single sextupole and the beam-beam interaction. A comparison will then be made with what is obtained using the Lie language. For the tune shift consideration here, we will ignore the resonance effects, even though the smooth approximation can be applied to the case of isolated resonances as well.

Consider a 1-D motion of a particle described by the equation

$$x'' + K(s)x = f(x, s), \tag{4.367}$$

where $K(s)$ is the focusing function of the linear, unperturbed accelerator environment, $f(x, s)$ is a perturbation that depends on the instantaneous displacement of the particle x and the position coordinate s. The perturbation $f(x, s)$

---

[41]The smooth approximation is an analysis old-fashioned by today's standard. However, it still provides important basis of the more modern analyses. See A. Schoch, CERN report 57-23 (1958); P.A. Sturrock, Ann. Phys. 3, 113 (1958); G. Guignard, CERN report 76-06 (1976); R. Bartolini, et al., Phys. Rev. Special Topics — Accel. & Beams 11, 104002, (2008); A.W. Chao, Lectures on Accelerator Physics, World Scientific (2020).

is considered to be periodic in s with period $2\pi R$, the circumference of the accelerator. The system can be described by the Hamiltonian[42]

$$H(x, p_x, s) \;=\; \frac{1}{2}p_x^2 + \frac{1}{2}K(s)x^2 - \int_0^x dx'\, f(x', s)\,. \tag{4.368}$$

Let us start with the equation of motion (4.367). We learned from the Courant–Snyder treatment that the problem simplifies if we introduce the transformation

$$x \;=\; \sqrt{\beta(s)}\, u\,,$$
$$\theta \;=\; \frac{\psi(s)}{\nu} \;=\; \frac{1}{\nu}\int_0^s \frac{ds'}{\beta(s')}\,, \tag{4.369}$$

where $\beta(s)$ is the unperturbed β-function found in the absence of $f(x, s)$, and ν is the unperturbed tune. With Eq. (4.369), the dynamic variable x is replaced by u, and the independent time-variable s becomes θ. The functions $\beta(s), K(s)$ and $f(x, s)$ become $\beta(\theta), K(\theta)$ and $f(u, \theta)$, and are periodic in θ with period 2π. Equation (4.367) in the new variables reads

$$\frac{d^2 u}{d\theta^2} + \nu^2 u \;=\; \nu^2 \beta^{3/2}(\theta) f(u, \theta) \;\equiv\; F(u, \theta)\,. \tag{4.370}$$

In the absence of perturbation, u is described by a simple harmonic motion in θ. This means we can write

$$u \;=\; \sqrt{2A}\sin\phi\,, \tag{4.371}$$

and

$$\frac{du}{d\theta} \;=\; \nu\sqrt{2A}\cos\phi\,, \tag{4.372}$$

where A and $\frac{d\phi}{d\theta} = \nu$ are constants.

With a perturbation, we could insist on the action-angle transformation of the form (4.371–4.372) except that A and $\frac{d\phi}{d\theta}$ now depend on θ and we need to solve for them. To do so, first note that by substituting (4.371) into (4.372), one obtains a self-consistency condition

$$\frac{1}{2}\frac{dA}{d\theta}\sin\phi + \left(\frac{d\phi}{d\theta} - \nu\right) A\cos\phi \;=\; 0\,. \tag{4.373}$$

Secondly, substituting Eqs. (4.371–4.372) into Eq. (4.370) gives another condition

$$\frac{1}{2}\frac{dA}{d\theta}\cos\phi - \left(\frac{d\phi}{d\theta} - \nu\right) A\sin\phi \;=\; \frac{\sqrt{A}}{\sqrt{2\nu}}F(\sqrt{2A}\sin\phi, \theta)\,. \tag{4.374}$$

---

[42]Do not write (4.368) with p_x replaced by x'. The fact that $p_x = x'$ is a *consequence* of the Hamiltonian (4.368). Replacing p_x by x' at the Hamiltonian level is a misuse of the Hamiltonian.

Equations (4.373–4.374) can be combined to give

$$\frac{dA}{d\theta} = \frac{\sqrt{2A}}{\nu}\cos\phi\, F(\sqrt{2A}\sin\phi,\theta)\,,$$

$$\frac{d\phi}{d\theta} = \nu - \frac{1}{\nu\sqrt{2A}}\sin\phi\, F(\sqrt{2A}\sin\phi,\theta)\,. \tag{4.375}$$

Effectively we have decomposed a second order differential equation (4.370) of one variable u into two first order differential equations (4.375) of two variables A and ϕ. So far no approximations have been made; Eq. (4.375) is exact.

If the perturbation is sufficiently weak, the quantities A and $\frac{d\phi}{d\theta}$ are approximately constants of the motion, i.e. they change slowly with time variable θ. We can approximate the expressions (4.375) by keeping only the slowly varying terms and dropping all fast oscillating terms on the right-hand-sides. In other words, we make the "smooth approximation",

$$\frac{dA}{d\theta} \approx \frac{\sqrt{2A}}{\nu}\langle\cos\phi\, F(\sqrt{2A}\sin\phi,\theta)\rangle\,,$$

$$\frac{d\phi}{d\theta} \approx \nu - \frac{1}{\nu\sqrt{2A}}\langle\sin\phi\, F(\sqrt{2A}\sin\phi,\theta)\rangle\,. \tag{4.376}$$

where the angular brackets mean the smoothing procedure. Fast and slow refer to a comparison with the revolutions. The angular variables θ and ϕ are considered to be fast variables, while the changes in A and $\frac{d\phi}{d\theta}$ are considered to be slow for sufficiently weak perturbations.

Consider the case when the tune is away from resonances.[43] The smoothing in this case is obtained by a straightforward averaging over the fast angular variables θ and ϕ, i.e.

$$\langle\ \rangle = \frac{1}{4\pi^2}\int_0^{2\pi} d\phi \int_0^{2\pi} d\theta\,. \tag{4.377}$$

After averaging, Eq. (4.376) becomes

$$\frac{dA}{d\theta} \approx 0\,,$$

$$\frac{d\phi}{d\theta} \approx \nu - \frac{1}{4\pi^2\nu\sqrt{2A}}\int_0^{2\pi} d\phi \int_0^{2\pi} d\theta \sin\phi\, F(\sqrt{2A}\sin\phi,\theta)\,. \tag{4.378}$$

The reason $\frac{dA}{d\theta}$ vanishes is that $F(\sqrt{2A}\sin\phi,\theta)$ is an even function in $\phi' \equiv \phi - \frac{\pi}{2}$; the product $\cos\phi\, F(\sqrt{2A}\sin\phi,\theta)$ is an odd function of ϕ' which averages to zero. The fact that we are averaging θ and ϕ independently of each other assumes that they are not correlated in some way. Such correlations would be a consequence of resonances, which we ignore in this section but will be discussed in Sec. 4.10.3.

---

[43] For a more detailed discussion on smooth approximation near an isolated resonance, see A.W. Chao, Lectures on Accelerator Physics, World Scientific (2020).

Equation (4.378) says that the action variable A is a constant of the motion even in the presence of perturbations (valid to 1st order in the perturbation strengths) provided the unperturbed tune is away from resonances. This is already an important result of the smooth approximation. Note that A still changes rapidly within each revolution, Eq. (4.378) only says that these rapid changes do not accumulate from turn to turn and as a result, A appears to be constant.

The quantity $\frac{d\phi}{d\theta}$ in Eq. (4.378), which assumes the physical meaning of the perturbed tune, depends on A. The motion of particles in the polar $(\sqrt{2A},\phi)$ space is such that a particle move along circles of $\sqrt{2A} = $ constant with a constant angular speed. This constant angular speed, however, is different for different particles, creating a sheering effect on the distribution of particles in the phase space. The dependence of the perturbed tune on A specifies the detuning. The tune shift for a particle with amplitude A is given by

$$\Delta\nu(A) \;=\; -\frac{1}{4\pi^2\nu\sqrt{2A}}\int_0^{2\pi} d\phi \int_0^{2\pi} d\theta \sin\phi \, F(\sqrt{2A}\sin\phi,\theta)\,. \tag{4.379}$$

From Eq. (4.379), one notes that if the perturbation $f(x,s)$ is an even function of x, then to 1st order of the perturbation, the tune shift vanishes. Since the perturbation due to sextupoles, decapoles, etc. are even in x, these multipoles do not contribute to tune shifts for on-momentum particles to 1st order of their strengths. Quadrupoles, octupoles, and the beam-beam force, on the other hand, do contribute to 1st order tune shifts.

Noting the fact that the action and angle are canonical variables, one can cast Eq. (4.378) in a Hamiltonian form. Indeed, with the Hamiltonian

$$H(\phi, A, \theta) \;=\; \nu A - \frac{1}{4\pi^2\nu}\int_0^{2\pi} d\phi' \int_0^{2\pi} d\theta \sin\phi' \int_0^{A} \frac{dA'}{\sqrt{2A'}} F(\sqrt{2A'}\sin\phi',\theta)\,,$$

$$\tag{4.380}$$

Eq. (4.379) is recovered by the Hamilton equations

$$\frac{d\phi}{d\theta} \;=\; \frac{\partial H}{\partial A}\,, \qquad \text{and} \qquad \frac{dA}{d\theta} \;=\; -\frac{\partial H}{\partial\phi} \;=\; 0\,. \tag{4.381}$$

Since the Hamiltonian H is independent of the time-variable θ, it is a constant of the motion. The system is solvable, or integrable, away from resonances.[44]

Error quadrupole   Let us consider a few examples. Consider first a thin-lens error quadrupole with focal length f located as the position $s = 0$ in the circular accelerator. In this case, we have

$$f(x,s) \;=\; -\frac{x}{f}\,\delta_p(s)\,,$$

---

[44]The fact that H is also independent of ϕ gives the additional nice feature $A = $ constant, but this property is not a necessary condition for the system to be called integrable.

where $\delta_p(s)$ is the periodic δ-function with period $2\pi R$. The kick a particle receives as it traverses the error quadrupole is $\Delta x' = -\frac{x}{f}$. This means

$$F(u,\theta) = \nu^2 \beta^{3/2}(0) f(u,\theta) = -\nu \beta(0) \frac{u}{f} \delta_p(\theta).$$

Substitution into Eq. (4.379) yields

$$\Delta \nu \approx \frac{\beta(0)}{4\pi f}, \tag{4.382}$$

which is a familiar result of accelerator optics. In particular, in this case, $\Delta \nu$ is independent of A. All particles receive the same tune shift. There is no detuning.

Error octupole   Now consider a more complex problem, namely the case of a thin-lens error octupole with

$$f(x,s) = \epsilon x^3 \delta_p(s). \tag{4.383}$$

Substituting into Eq. (4.379) gives

$$\Delta \nu \approx -\frac{1}{2\pi^2} \epsilon \beta^2(0) A \int_0^{2\pi} d\phi \cos^4 \phi = -\frac{3}{8\pi} \epsilon \beta^2(0) A. \tag{4.384}$$

This tune shift is proportional to the action A.

What we smooth is the Hamiltonian   The Hamiltonian (4.380) is in fact simply the Hamiltonian (4.368) averaged over θ and ϕ. Contained in the Hamiltonian is the potential of the force,

$$V(x,s) = -\int_0^x dx' f(x',s).$$

One then recognizes that what we observed in the smooth approximation developed from Eq. (4.367) to Eq. (4.381), what is being smoothed over is the Hamiltonian, and in particular, it potential term.

Take the octupole case again for example. The averaged potential is given by

$$\langle V \rangle = -\epsilon \beta^2(0) A^2 \langle \sin^4 \phi \rangle \delta_p(s) = -\frac{3}{8} \epsilon \beta^2(0) A^2 \delta_p(s). \tag{4.385}$$

In terms of the potential V, the tune shift (4.379) can be written as

$$\Delta \nu = \frac{1}{2\pi} \int ds \frac{\partial \langle V \rangle}{\partial A}. \tag{4.386}$$

Substituting Eq. (4.385) into Eq. (4.386), we recover the tune shift (4.384).

In retrospect, the step of smoothing the equation of motion as we did leading to Eq. (4.378) is a bit risky. Although they look perfectly justified, there is no guarantee that such an approximation has not led to nonsymplecticity. A much safer procedure is to resort to the Hamiltonian. When approximations or truncations are to be made, they should be made directly on the Hamiltonians, not on the equations of motion. Here, fortunately, the smooth approximation on the equation of motion is indeed symplectic.

Homework 4.106 Consider an accelerator with uniform focusing and a uniform distribution of octupole error. The equation of motion is

$$\frac{d^2x}{d\theta^2} + \nu^2 x = \epsilon x^3 . \tag{4.387}$$

The tune shift can be obtained from Eq. (4.379),

$$\Delta\nu = -\frac{\epsilon A}{\pi\nu} \int_0^{2\pi} d\phi \sin^4\phi = -\frac{3\epsilon A}{4\nu} . \tag{4.388}$$

But Eq. (4.387) — when the perturbation is uniformly distributed — can be solved exactly. Show that the tune shift (4.388) agrees with this exact solution to 1st order in ϵ.

Solution Equation (4.387) has a constant of the motion, which we designate as $\nu^2 A$, namely,

$$\frac{1}{2}\left(\frac{dx}{d\theta}\right)^2 + \frac{\nu^2}{2}x^2 - \frac{\epsilon}{4}x^4 = \text{constant} = \nu^2 A ,$$

which in turn gives

$$\frac{dx}{d\theta} = \pm\sqrt{2\nu^2 A - \nu^2 x^2 + \frac{\epsilon}{2}x^4} . \tag{4.389}$$

Integrating the periodic motion (4.389) gives a period

$$\Theta = 4 \int_0^{\hat{x}} \frac{dx}{\sqrt{2\nu^2 A - \nu^2 x^2 + \frac{\epsilon}{2}x^4}} , \tag{4.390}$$

where $\hat{x}$ is the peak amplitude of the motion with

$$2\nu^2 A - \nu^2 \hat{x}^2 + \frac{\epsilon}{2}\hat{x}^4 = 0 .$$

Equation (4.390) gives an exact expression of the tune shift $\Delta\nu$. When $\epsilon = 0$, the unperturbed case has

$$\hat{x}_0 = \sqrt{2A} , \qquad \text{and} \qquad \Theta_0 = \frac{2\pi}{\nu} .$$

When $\epsilon \neq 0$, the perturbed tune is related to Θ by

$$\nu + \Delta\nu = \frac{2\pi}{\Theta} . \tag{4.391}$$

To compute $\Delta\nu$, we need to compute Θ of Eq. (4.390). We will do this to 1st order in ϵ. By a change of variable from x to u, where

$$u = x^2 - \frac{\epsilon}{2\nu^2}x^4$$

$$\implies \quad dx = \frac{du}{2\sqrt{u}}\left(1 + \frac{3\epsilon}{4\nu^2}u + \mathcal{O}(\epsilon^2)\right),$$

we find

$$\Theta = \frac{2}{\nu}\int_0^{2A} du\,\frac{1 + \frac{3\epsilon}{4\nu^2}u}{\sqrt{u(2A-u)}}$$

$$= \frac{2\pi}{\nu}\left(1 + \frac{3\epsilon}{4\nu^2}A\right).$$

Substituting into Eq. (4.391) gives a tune shift which agrees with Eq. (4.388).

4.9.5   Tune shift using Lie algebra

Now let us try to connect these results based on the smooth approximation to Lie algebra. Consider a localized perturbation,

$$f(x,s) = \epsilon(x)\,\delta_p(s).$$

The tune shift with amplitude, the detuning, Eq. (4.379) or Eq. (4.386), reads

$$\Delta\nu(A) = -\frac{\sqrt{\beta(0)}}{4\pi^2\sqrt{2A}}\int_0^{2\pi} d\phi\,\sin\phi\,\epsilon\left(\sqrt{2A\beta(0)}\sin\phi\right). \tag{4.392}$$

The Hamiltonian (4.380) reads

$$H = \nu A - \frac{1}{4\pi^2}\int_0^{2\pi}d\phi\int_0^{\sqrt{2A\beta(0)}\sin\phi} dx'\epsilon(x'). \tag{4.393}$$

This system can also be described in the Lie language. If one observes the particle motion at the exit of the perturbation, the one-turn map, in the (ϕ, A) phase space, is

$$e^{:-2\pi\nu A:}\exp\left[:\int_0^{\sqrt{2A\beta(0)}\sin\phi} dx'\epsilon(x'):\right].$$

To 1st order in the perturbation strength, this map can be concatenated using the BCH formula to yield

$$\exp\left[:-2\pi\nu A + \left(\frac{:-2\pi\nu A:}{1 - e^{:2\pi\nu A:}}\right)\int_0^{\sqrt{2A\beta(0)}\sin\phi} dx'\epsilon(x'):\right] \tag{4.394}$$

Let us decompose

$$\int_0^{\sqrt{2A\beta(0)}\sin\phi} dx'\epsilon(x') = \sum_{k=-\infty}^{\infty} C_k(A)e^{ik\phi}, \tag{4.395}$$

then we have

$$\left(\frac{: - 2\pi\nu A:}{1 - e^{:2\pi\nu A:}}\right) \int_0^{\sqrt{2A\beta(0)}\,\sin\phi} dx'\epsilon(x') \;=\; \sum_{k=-\infty}^{\infty} C_k(A) \left(\frac{2\pi i\,k\nu}{1 - e^{-2\pi i\,k\nu}}\right) e^{ik\phi}.$$

Away from resonances, terms with $k \neq 0$ can be transformed away by a normal form transformation. After the transformation, the map (4.394) becomes

$$N \;=\; \exp[: - 2\pi\nu A + C_0(A):]\,,$$

where

$$C_0(A) \;=\; \frac{1}{2\pi} \int_0^{2\pi} d\phi \int_0^{\sqrt{2A\beta(0)}\,\sin\phi} dx'\epsilon(x')\,.$$

It is easy to see that this map is described by the effective Hamiltonian (4.393), and it has the tune shift (4.392). The actual normal form transformations are short circuited here. This connects the Lie analysis and the smooth approximation at least away from resonances. The normal form transformation (what has been short-circuited just now) and the step of smoothing in the smooth approximation are actually the same procedure — both involve observing the system in a stroboscopic slow frame. In a sense, to first order of perturbation,

$$\text{Smooth approximation in Hamiltonian dynamics}$$
$$=\; \text{Normal form transformation in Lie algebra}\,.$$

Note that it is the zeroth Fourier harmonic of the potential (4.395) that contributes to the tune shift away from resonances. The other harmonics will reemerge when resonances appear, as to be discussed next.

Homework 4.107 As a somewhat basic exercise of a tune shift effect, consider a simple harmonic oscillator. The map that brings the canonical coordinates (ϕ, A) from time $t = 0$ to time t is

$$M \;=\; e^{:-\omega At:}\,, \tag{4.396}$$

where ω is the angular frequency of the oscillator. One has

$$M\phi \;=\; \phi + \omega t\,, \qquad \text{and} \qquad MA \;=\; A\,,$$

which means the map can be written as

$$\phi(t) \;=\; \phi(0) + \omega t\,, \qquad \text{and} \qquad A(t) \;=\; A(0)\,,$$

as expected.

One can also describe the system in terms of the Hamiltonian language with the Hamiltonian

$$H(\phi, A, t) \;=\; \omega A\,.$$

Indeed, the Hamilton equations read

$$\dot{\phi} = \frac{\partial H}{\partial A} = \omega, \qquad \text{and} \qquad \dot{A} = -\frac{\partial H}{\partial \phi} = 0.$$

Now describe the system in a rotating frame that has a rotation frequency Ω so that the oscillator has an angular frequency of $\omega - \Omega$. Describe it in both the Lie and the Hamiltonian languages.

Solution In the Hamiltonian language, we introduce a generating function G which transforms the coordinates from (ϕ, A) to (ϕ', A'), where

$$G(\phi, A', t) = (\phi - \Omega t)A'.$$

The new coordinates are then related to the old ones by

$$\phi' = \frac{\partial G}{\partial A'} = \phi - \Omega t, \qquad \text{and} \qquad A = \frac{\partial G}{\partial \phi} = A'.$$

The new Hamiltonian is

$$H' = H + \frac{\partial G}{\partial t} = (\omega - \Omega)A. \tag{4.397}$$

The new equations of motion are

$$\dot{\phi}' = \omega - \Omega, \qquad \text{and} \qquad \dot{A}' = 0.$$

In the Lie language, the old map is given by Eq. (4.396). The coordinate transformation can be described by the Lie map (see the coordinate shift map of Table 4.3),

$$R(t) = e^{:\Omega A t:}$$
$$\implies \quad R\phi = \phi - \Omega t = \phi', \quad \text{and} \quad RA = A = A'. \tag{4.398}$$

The new Lie map must involve a transformation from the new coordinates to the old ones at time $t = 0$ and from the old coordinates back to the new ones at time t, i.e.,

$$M' = R(t)MR^{-1}(0). \tag{4.399}$$

Note that because the coordinate transformation depends on time, the map is not transformed according to a similarity transformation as prescribed, e.g., by Eq. (4.46).

Substituting R from Eq. (4.398) and M from Eq. (4.396) into Eq. (4.399) gives

$$M' = e^{:\Omega A t:}e^{:-\omega A t:} = e^{:-(\omega - \Omega)A t:}, \tag{4.400}$$

which is consistent with the new Hamiltonian (4.397). This exercise will be used later when we discuss the resonances.

Homework 4.108 Extend the 1-D Lie algebra analysis of this section to 2-D for both x- and y-motions. Find the tune shifts $\Delta\nu_x(A_x, A_y)$ and $\Delta\nu_y(A_x, A_y)$ in the presence of a localized weak single thin-lens octupole. It is not as difficult as it might sound.

Homework 4.109 Refer to Eq. (4.207) and the effective Hamiltonian it yields. Find the dependence of the synchrotron tune on the synchrotron oscillation amplitude.

4.10   Isolated resonance

The normal form analysis so far applies when the betatron frequencies are away from resonances. When a resonance condition [see also Eq. (4.333)]

$$m\nu_x + n\nu_y = p \tag{4.401}$$

is valid or approximately valid, we have to modify our analysis. Be reminded that ν_x and ν_y are the tunes perturbed by linear coupling optics but unperturbed by nonlinear effects. We will consider the idealized situation when there is one and only one isolated resonance nearby that dominates the dynamics. We will not discuss the case with two or more interplaying resonances. This idealization excludes the study of chaos.

A side comment is obligatory here. Alert readers have noted that all along we have been *assuming* the problem is integrable. This assumption is hidden in that we assumed our expressions of the effective Hamiltonian and the invariants converge in whichever perturbation parameter we have chosen. Chaos is actually a result when the system is nonintegrable. One way to exhibit the convergence problem is to look at the map of resonance lines in the (ν_x, ν_y) plane. As the order of resonances (which is defined as $|m| + |n|$) is raised, we have an ever increasingly dense web of resonance lines. Obviously the number of neighboring resonances does not converge as one increases the resonance order. In fact, one might summarize the situation as having

"an infinite number of infinitely weak resonances infinitely close by",

and we are asked whether the particle motion would be stable under the circumstance. On the other hand, the hope is that the strengths (or the width in the tune space) of resonances decrease sufficiently rapidly with the resonance order, and a finite detuning effect plays the role to suppress these ever weakening resonances, so that the effect on beam dynamics actually converges. As a result, hopefully only one of the relatively low order resonances plays a dominating role, and our analysis of this section approximately applies. In particular, the system then becomes integrable.

4.10.1   Normal form near isolated resonance

Now assuming integrability, consider a second order map (4.322). We have described a normal form analysis of this map, and shown that it can be transformed into the form of Eq. (4.324). We next decompose f_3 according to Eq. (4.330). Away from resonances, we choose F_3 according to Eq. (4.335). When a resonance is nearby, however, the normal form analysis as described so far must be modified. The basic idea remains to seek a transformation to make the beam dynamics "as simple as possible". However, in this case, the simplest possible form of the map no longer gives an effective Hamiltonian that depends only on A_x, A_y, and δ. Instead, it is now more complicated.

Near a resonance, we now choose F_3 similarly to Eq. (4.335) except that terms excluded from the summation are not only those with ($a = b$ and $c = d$), but also those resonant terms satisfying ($a - b = km$ and $c - d = kn$) for some integer k [see Eq. (4.334)]. We then find that the best choice of F_3 will be such that

$$(e^{-:f_2:} - 1)F_3 + f_3 \;=\; h_3 + h_3^{(r)} \,,$$

where h_3 has been defined in Eq. (4.336) and is a function of A_x, A_y and δ only. The function $h_3^{(r)}$ is given by

$$h_3^{(r)} \;=\; \sum_k \sum_{a,c,e} C^{(3)}_{a,a-km,c,c-kn,e} A_x^{a-\frac{1}{2}km} A_y^{c-\frac{1}{2}kn} e^{ik(m\phi_x+n\phi_y)} \delta^e \,. \qquad (4.402)$$

The summation in Eq. (4.402) contains terms that satisfy the conditions

$$\begin{aligned}
k &\neq 0 \,, \\
2a + 2c - k(m+n) + e &= 3 \,, \\
0 \leq a, a - km, \; c, c - kn, e &\leq 3 \,. \qquad (4.403)
\end{aligned}$$

The reason the $k = 0$ terms are excluded from the summation is because they have already been included in h_3.

We see now that the "simplest" form depends not only on A_x, A_y and δ, but also on ϕ_x and ϕ_y. However, it is important to note that it is only the combined variable $m\phi_x + n\phi_y$ that appears in the normal form.

Take the resonance $3\nu_x = p$ for example. We have $m = 3, n = 0$. There are only two terms that satisfy the conditions (4.403) and therefore contribute to $h_3^{(r)}$. The result is

$$h_3^{(r)} \;=\; A_x^{3/2} \left(C^{(3)}_{3000,0} e^{i3\phi_x} + C^{(3)}_{0300,0} e^{-i3\phi_x} \right) . \qquad (4.404)$$

Similarly we can work out for all the other resonances excited by this second order map. The results are

resonance $2\nu_x + \nu_y = p$

$$h_3^{(r)} \;=\; A_x A_y^{1/2} \left(C^{(3)}_{2010,0} e^{i2\phi_x+i\phi_y} + C^{(3)}_{0201,0} e^{-i2\phi_x-i\phi_y} \right) ,$$

resonance $2\nu_x - \nu_y = p$

$$h_3^{(r)} \;=\; A_x A_y^{1/2} \left(C^{(3)}_{2001,0} e^{i2\phi_x-i\phi_y} + C^{(3)}_{0210,0} e^{-i2\phi_x+i\phi_y} \right) ,$$

resonance $\nu_x + 2\nu_y = p$

$$h_3^{(r)} \;=\; A_x^{1/2} A_y \left(C^{(3)}_{1020,0} e^{i\phi_x+i2\phi_y} + C^{(3)}_{0102,0} e^{-i\phi_x-i2\phi_y} \right) ,$$

resonance $\nu_x - 2\nu_y = p$

$$h_3^{(r)} = A_x^{1/2} A_y \left(C_{1002,0}^{(3)} e^{i\phi_x - i2\phi_y} + C_{0120,0}^{(3)} e^{-i\phi_x + i2\phi_y} \right),$$

resonance $3\nu_y = p$

$$h_3^{(r)} = A_y^{3/2} \left(C_{0030,0}^{(3)} e^{i3\phi_y} + C_{0003,0}^{(3)} e^{-i3\phi_y} \right),$$

resonance $2\nu_x = p$

$$h_3^{(r)} = A_x \delta \left(C_{2000,1}^{(3)} e^{i2\phi_x} + C_{0200,1}^{(3)} e^{-i2\phi_x} \right),$$

resonance $\nu_x + \nu_y = p$

$$h_3^{(r)} = A_x^{1/2} A_y^{1/2} \delta \left(C_{1010,1}^{(3)} e^{i\phi_x + i\phi_y} + C_{0101,1}^{(3)} e^{-i\phi_x - i\phi_y} \right),$$

resonance $\nu_x - \nu_y = p$

$$h_3^{(r)} = A_x^{1/2} A_y^{1/2} \delta \left(C_{1001,1}^{(3)} e^{i\phi_x - i\phi_y} + C_{0110,1}^{(3)} e^{-i\phi_x + i\phi_y} \right),$$

resonance $2\nu_y = p$

$$h_3^{(r)} = A_y \delta \left(C_{0020,1}^{(3)} e^{i2\phi_y} + C_{0002,1}^{(3)} e^{-i2\phi_y} \right),$$

resonance $\nu_x = p$

$$
\begin{aligned}
h_3^{(r)} = {} & A_x^{1/2} \delta^2 \left(C_{1000,2}^{(3)} e^{i\phi_x} + C_{0100,2}^{(3)} e^{-i\phi_x} \right) \\
& + A_x \delta \left(C_{2000,1}^{(3)} e^{i2\phi_x} + C_{0200,1}^{(3)} e^{-i2\phi_x} \right) \\
& + A_x^{1/2} A_y \left(C_{1011,0}^{(3)} e^{i\phi_x} + C_{0111,0}^{(3)} e^{-i\phi_x} \right) \\
& + A_x^{3/2} \left(C_{3000,0}^{(3)} e^{i3\phi_x} + C_{0300,0}^{(3)} e^{-i3\phi_x} \right) \\
& + A_x^{3/2} \left(C_{2100,0}^{(3)} e^{i\phi_x} + C_{1200,0}^{(3)} e^{-i\phi_x} \right),
\end{aligned}
$$

resonance $\nu_y = p$

$$
\begin{aligned}
h_3^{(r)} = {} & A_y^{1/2} \delta^2 \left(C_{0010,2}^{(3)} e^{i\phi_y} + C_{0001,2}^{(3)} e^{-i\phi_y} \right) \\
& + A_y \delta \left(C_{0020,1}^{(3)} e^{i2\phi_y} + C_{0002,1}^{(3)} e^{-i2\phi_y} \right) \\
& + A_x A_y^{1/2} \left(C_{1110,0}^{(3)} e^{i\phi_y} + C_{1101,0}^{(3)} e^{-i\phi_y} \right) \\
& + A_y^{3/2} \left(C_{0030,0}^{(3)} e^{i3\phi_y} + C_{0003,0}^{(3)} e^{-i3\phi_y} \right) \\
& + A_y^{3/2} \left(C_{0021,0}^{(3)} e^{i\phi_y} + C_{0012,0}^{(3)} e^{-i\phi_y} \right).
\end{aligned}
\tag{4.405}
$$

The term h_3 is the same for all resonances.

There are a total of 35 C-coefficients of third order. The resonances mentioned above involve 32 C-coefficients (8 of them involved twice). They are the resonance driving terms. The remaining 3 appeared in the function h_3.

A comment is to be made here. We emphasized before that in the perturbative approach, such as in the Lie algebra approach, we need first to identify the small perturbative parameters. In this section, we have assumed the small parameters to be the components of X, and as a result, our Taylor expansions are in terms of powers of $A_x^{1/2}, A_y^{1/2}, \delta$. Take Eq. (4.402) for example. We chose to expand it in terms of the eigenmodes proportional to integral powers of $A_x^{1/2}, A_y^{1/2}, \delta$ for this reason. In case we choose a different parameter as the expansion small parameter, the Taylor expansion will look different, and the choice of the eigenmodes will need to be changed. We will see one example later in Sec. 4.10.6 for the case of the beam-beam interaction, when we choose the beam-beam kick strength as the expansion parameter while there is no restriction on X. The idea of expanding in terms of eigenmodes remains, but the choice of eigenmodes will be different.

Normal form theorems In Sec. 4.9.3, we mentioned a few intriguing properties of nonlinear dynamics such as Eqs. (4.364) and (4.366). We emphasized at the time that these properties follow directly from the Hamiltonian nature of the system. They are not at all obvious results if viewed from the equations of motion. Here near an isolated resonance, we find there are eight coefficients involved twice in the effective Hamiltonian of the system. It follows that there are eight additional properties like Eqs. (4.364) and (4.366) linking up the inner working of the nonlinear dynamics of the accelerator, and to be viewed as direct consequences of the system's symplecticity. We are not listing these eight dynamical properties explicitly, suggesting the reader only be aware of the existence of these hidden theorems, each as intriguing as Eqs. (4.364) and (4.366).

4.10.2 1-D nonlinear resonance

Near the resonance (4.401), we now have

$$N = e^{:f_2:} e^{:h_3 + h_3^{(r)}:}, \tag{4.406}$$

where h_3 is given by Eq. (4.336) and $h_3^{(r)}$ is given by Eqs. (4.404–4.405).

Given Eq. (4.406), we next try to obtain the effective Hamiltonian of the system. To do so, it seems that we need to concatenate the two factor maps on the right-hand-side, which gives, to 1st order in $h_3^{(r)}$,

$$
\begin{aligned}
N &= \exp\left[:f_2 + \left(\frac{:f_2:}{1 - e^{-:f_2:}}\right)(h_3 + h_3^{(r)}):\right] \\
&= \exp\left[:f_2 + h_3 + \left(\frac{:f_2:}{1 - e^{-:f_2:}}\right) h_3^{(r)}:\right].
\end{aligned} \tag{4.407}
$$

However, expression (4.407) contains the term

$$
\left(\frac{:f_2:}{1 - e^{-:f_2:}}\right) h_3^{(r)} = \sum_{k\neq 0}\sum_{ace} C^{(3)}_{a,a-km,c,c-kn,e}
$$
$$
\times \frac{ik(m\mu_x + n\mu_y)}{1 - e^{-ik(m\mu_x + n\mu_y)}} A_x^{a - \frac{1}{2}km} A_y^{c - \frac{1}{2}kn} e^{ik(m\phi_x + n\phi_y)} \delta^e,
$$

which is problematic because it diverges near the resonance.

Rotating frame　To get around this problem, we will consider observing the particle motion in a strobe frame which rotates in the phase space. Take the 1-D resonance $3\nu_x = p$ for example. We consider a transformation of coordinates from $U = (\phi_x, A_x, \phi_y, A_y, z, \delta)$ to $U' = (\phi'_x, A_x, \phi_y, A_y, z, \delta)$, where

$$\phi'_x = \phi_x - \frac{2\pi}{3} pk. \tag{4.408}$$

In Eq. (4.408), k is the turn index which increases by one unit per turn and assumes the role of the time variable. The new variable ϕ'_x, the angle variable in the rotating frame, is a slow variable because it increases by $\frac{2\pi d}{3}$ per turn, where

$$d = 3\nu_x - p$$

is small, $|d| \ll 1$ [see also Eq. (4.232)].

To facilitate this change of coordinates, we consider the coordinate shift map (see Homework 4.107)

$$R(k) = e^{:\frac{2\pi}{3} pk A_x:}$$
$$\implies R\phi_x = \phi_x - \frac{2\pi}{3} pk = \phi'_x.$$

In the new coordinate system, the map for the k-th turn (which takes the particle motion as the turn index increases from k to $k+1$) can be written as

$$\begin{aligned}
N' &= e^{:\frac{2\pi}{3} p(k+1) A_x:} N e^{-:\frac{2\pi}{3} pk A_x:} \\
&= \left[e^{:\frac{2\pi}{3} p(k+1) A_x:} e^{:f_2:} e^{-:\frac{2\pi}{3} pk A_x:} \right] \left[e^{:\frac{2\pi}{3} pk A_x:} e^{:h_3 + h_3^{(r)}:} e^{-:\frac{2\pi}{3} pk A_x:} \right].
\end{aligned} \tag{4.409}$$

The two factor maps in the square brackets in Eq. (4.409) can be rewritten in more convenient forms. The first factor map reads

$$e^{:\frac{2\pi}{3} p(k+1) A_x:} e^{:f_2:} e^{-:\frac{2\pi}{3} pk A_x:} = e^{:f'_2:}, \tag{4.410}$$

where [see also Eq. (4.315)]

$$f'_2 = -\frac{2\pi}{3} d A_x - \mu_y A_y - \frac{1}{2} \alpha_c C \delta^2, \tag{4.411}$$

where α_c is the momentum compaction factor, C is the storage ring circumference. Equations (4.410–4.411) follow because A_x commutes with f_2. The other factor map in Eq. (4.409) can be written as

$$\begin{aligned}
e^{:\frac{2\pi}{3} pk A_x:} & e^{:h_3 + h_3^{(r)}:} e^{-:\frac{2\pi}{3} pk A_x:} \\
&= \exp\left[:e^{:\frac{2\pi}{3} pk A_x:} \left(h_3 + h_3^{(r)} \right): \right] \\
&= \exp\left[:h_3 + e^{:\frac{2\pi}{3} pk A_x:} h_3^{(r)}: \right] \\
&= \exp\left[:h_3 + 2 A_x^{3/2} \operatorname{Re}\left(C_{3000,0}^{(3)} e^{i3\phi'_x} \right): \right],
\end{aligned} \tag{4.412}$$

where we have substituted expression (4.404) for $h_3^{(r)}$, and $\text{Re}\,(\cdots)$ means taking the real part of the quantity $(\cdots)$. We have also applied Eq. (4.329) and used

$$e^{:\frac{2\pi}{3}pkA_x:}e^{i3\phi_x} \;=\; e^{i3\left(\phi_x-\frac{2\pi}{3}pk\right)} \;=\; e^{i3\phi'_x}\,.$$

In the above procedure, note in Eq. (4.411) that the quantity μ_x has been effectively replaced by the small parameter $\frac{2\pi d}{3}$, and in Eq. (4.412), ϕ_x has been replaced by the slow phase ϕ'_x. In the new rotating system, we are now ready to concatenate the map (4.409) without introducing a small denominator, to 1st order in the nonlinearity strength,

$$
\begin{aligned}
N' \;&=\; e^{:f'_2:}\exp\left[:h_3 + 2A_x^{3/2}\,\text{Re}\left(C^{(3)}_{3000,0}e^{i3\phi'_x}\right):\right]\\[2mm]
&=\; \exp\left[:f'_2 + h_3 + \left(\frac{:f'_2:}{1-e^{-:f'_2:}}\right)2A_x^{3/2}\,\text{Re}\left(C^{(3)}_{3000,0}e^{i3\phi'_x}\right):\right]\\[2mm]
&=\; \exp\left\{:f'_2 + h_3 + 2A_x^{3/2}\,\text{Re}\left[C^{(3)}_{3000,0}\left(\frac{i2\pi d}{1-e^{-i2\pi d}}\right)e^{i3\phi'_x}\right]:\right\},
\end{aligned}
\tag{4.413}
$$

where we have used

$$:f'_2:e^{i3\phi'_x} \;=\; i2\pi d\,e^{i3\phi'_x}\,.$$

The expression in Eq. (4.413) is now well-behaved when $d \to 0$. The effective Hamiltonian for the one-turn map is then, in the rotating frame,

$$
\begin{aligned}
H(\phi'_x, A_x, \phi_y, A_y, z, \delta, k) \;=\;& \frac{2\pi d}{3}A_x + \mu_y A_y + \frac{1}{2}\alpha_c C\delta^2 - C^{(3)}_{0000,3}\delta^3\\[2mm]
& -C^{(3)}_{1100,1}A_x\delta - C^{(3)}_{0011,1}A_y\delta - 2A_x^{3/2}\,\text{Re}\left(C^{(3)}_{3000,0}\frac{i2\pi d}{1-e^{-i2\pi d}}\,e^{i3\phi'_x}\right).
\end{aligned}
\tag{4.414}
$$

Stroboscopic frame   The rotating frame thus avoids the small denominator problem. Alternatively to a rotating frame, one could apply a trick by raising the map to some integer power. This amounts to observing the particle motion not every turn, but every few turns. For a third order resonance, for example, one would observe the system once every three turns. By "strobing" the system at three-turn intervals, one expects that the dynamics of the resonance to become more visible.

We thus consider the map $M^3 = (A^{-1}NA)^3 = A^{-1}N^3A$. The dynamics are contained in the map N^3 which reads

$$
\begin{aligned}
N^3 \;&=\; \left(e^{:f_2:}e^{:h_3+h_3^{(r)}:}\right)^3\\[2mm]
&=\; e^{3:f_2:}\left(e^{-2:f_2:}e^{:h_3+h_3^{(r)}:}e^{2:f_2:}\right)\left(e^{-:f_2:}e^{:h_3+h_3^{(r)}:}e^{:f:}\right)e^{:h_3+h_3^{(r)}:}\\[2mm]
&=\; e^{3:f_2:}\exp\left[:e^{-2:f_2:}(h_3+h_3^{(r)}):\right]\exp\left[:e^{-:f_2:}(h_3+h_3^{(r)}):\right]e^{:h_3+h_3^{(r)}:}.
\end{aligned}
\tag{4.415}
$$

To 1st order in h_3 and $h_3^{(r)}$, this gives

$$N^3 \;=\; e^{3:f_2:}\exp\left[:\sum_{k=0}^{2}e^{-k:f_2:}(h_3+h_3^{(r)}):\right].\tag{4.416}$$

Since f_2 and h_3 both depend only on A_x, A_y, and δ, we have

$$\sum_{k=0}^{2} e^{-k:f_2:} h_3 \;=\; 3h_3 \,.$$

The other quantity in Eq. (4.416) reads

$$\sum_{k=0}^{2} e^{-k:f_2:} h_3^{(r)} \;=\; 2A_x^{3/2}\,\mathrm{Re}\left[C_{3000,0}^{(3)} e^{i3\phi_x} \left(\sum_{k=0}^{2} e^{-i3k\mu_x} \right) \right]$$

$$\;=\; 2A_x^{3/2}\,\mathrm{Re}\left[C_{3000,0}^{(3)} e^{i3\phi_x}\, \frac{1 - e^{-i9\mu_x}}{1 - e^{-i3\mu_x}} \right]$$

$$\;=\; 2A_x^{3/2}\,\mathrm{Re}\left[C_{3000,0}^{(3)} e^{i3\phi_x}\, \frac{1 - e^{-i6\pi d}}{1 - e^{-i2\pi d}} \right]\,.$$

We next note that the factor $e^{:3f_2:}$ in Eq. (4.416) can be written as [see Eq. (4.234)]

$$e^{:3f_2:} \;=\; e^{:3f_2':}\,.$$

Combining the results gives, to 1st order in h_3 and $h_3^{(r)}$,

$$N^3 \;=\; e^{:3f_2':} \exp\left[:3h_3 + 2A_x^{3/2}\,\mathrm{Re}\left(C_{3000,0}^{(3)} e^{i3\phi_x}\, \frac{1 - e^{-i6\pi d}}{1 - e^{-i2\pi d}} \right): \right]$$

$$\;=\; \exp\left\{ :3f_2' + 3h_3 + 2A_x^{3/2}\,\mathrm{Re}\left[C_{3000,0}^{(3)} \frac{1 - e^{-i6\pi d}}{1 - e^{-i2\pi d}} \left(\frac{3:f_2':}{1 - e^{-3:f_2':}} \right) e^{i3\phi_x} \right]: \right\}$$

$$\;=\; \exp\left\{ :3f_2' + 3h_3 + 2A_x^{3/2}\,\mathrm{Re}\left[C_{3000,0}^{(3)} \frac{i6\pi d}{1 - e^{-i2\pi d}} e^{i3\phi_x} \right]: \right\}\,. \qquad (4.417)$$

Equation (4.417) obtained by strobing agrees with Eq. (4.413) obtained by rotating the phase space. In particular, the small denominator problem is avoided. Note that at three-turn intervals, the difference between ϕ_x and ϕ_x' disappears.

Hamilton equation   Let k be the turn number index. The one-turn Hamilton equations of motion are, using the effective Hamiltonian (4.414),

$$\frac{d\phi_x'}{dk} = \frac{\partial H}{\partial A_x} = \frac{2\pi}{3} d - C_{1100,1}^{(3)} \delta - 3A_x^{1/2}\,\mathrm{Re}\left[C_{3000,0}^{(3)} \frac{i2\pi d}{1 - e^{-i2\pi d}} e^{i3\phi_x'} \right],$$

$$\frac{dA_x}{dk} = -\frac{\partial H}{\partial \phi_x'} = -2A_x^{3/2}\,\mathrm{Re}\left[C_{3000,0}^{(3)} \frac{6\pi d}{1 - e^{-i2\pi d}} e^{i3\phi_x'} \right],$$

$$\frac{d\phi_y}{dk} = \frac{\partial H}{\partial A_y} = \mu_y - C_{0011,1}^{(3)} \delta \,,$$

$$\frac{dA_y}{dk} = -\frac{\partial H}{\partial \phi_y} = 0\,,$$

$$\frac{dz}{dk} = \frac{\partial H}{\partial \delta} = \alpha_c C\delta - 3C^{(3)}_{0000,3}\delta^2 - C^{(3)}_{1100,1}A_x - C^{(3)}_{0011,1}A_y,$$

$$\frac{d\delta}{dk} = -\frac{\partial H}{\partial z} = 0. \tag{4.418}$$

Take the first member of Eq. (4.418) for example. There are three terms on the right-hand-side. The first term describes the unperturbed betatron phase advance per turn. The second term gives the nonlinear effect which is not driven by resonance, i.e. the chromaticity. The third term is the nonlinear resonance driven term.

There are three constants of the motion: δ, A_y, and

$$h_x = \frac{2\pi d}{3}A_x - C^{(3)}_{1100,0}A_x\delta - 2A_x^{3/2}\,\mathrm{Re}\left[C^{(3)}_{3000,0}\frac{i2\pi d}{1 - e^{-i2\pi d}}e^{i3\phi'_x}\right].$$

The quantity h_x can be regarded as the effective Hamiltonian in the x phase space. For small d, and consider on-momentum particles with $\delta = 0$, we have

$$h_x \approx \frac{2\pi d}{3}A_x - 2A_x^{3/2}\,\mathrm{Re}(C^{(3)}_{3000,0}e^{i3\phi'_x}). \tag{4.419}$$

The reader is suggested to pay attention to the structure of the second term, the resonance driving term. in particular, note its dependence on A_x and associate a physical meaning to the resonance driving coefficient $C^{(3)}_{3000,0}$.

In the next section [see Eq. (4.430)], we will show that for the case of a single thin-lens sextupole and observing the particle motion at the exit point of the sextupole, $C^{(3)}_{3000,0} = \frac{i}{8}\lambda(2\beta_x)^{3/2}$, where λ is the sextupole strength. For that case,

$$h_x = \frac{2\pi d}{3}A_x + \frac{\lambda}{\sqrt{2}}(\beta_x A_x)^{3/2}\sin 3\phi'_x,$$

which agrees with Eq. (4.238).

Resonance tune footprint From the first member of Eq. (4.418), it follows that, as ϕ'_x varies with time, $\frac{1}{2\pi}\frac{d\phi'_x}{dk}$ oscillates around the value $\frac{d}{3}$ with an amplitude $\frac{\Delta}{3}$, where, for small d and on-momentum,

$$\Delta = \frac{9}{2\pi}A_x^{1/2}\left|C^{(3)}_{3000,0}\right|. \tag{4.420}$$

This quantity Δ can be loosely associated with a tune resonance footprint in the sense that the "instantaneous tune" of the particle wobbles in time slowly with a half-width of $\frac{\Delta}{3}$.

Whether the proximity to the resonance plays a significant role in the dynamics depends on which of the two terms in Eq. (4.419) dominates. If a particle has $\Delta \gtrsim |d|$, which happens when the particle has an unperturbed tune close to $\frac{1}{3}$, and its amplitude A_x is sufficiently large, its motion will exhibit a pronounced resonance response. When $\Delta \lesssim |d|$, the resonance behavior would be weak. In this sense, this tune footprint width Δ is loosely — but only loosely because resonance width will still require a more rigorous definition — associated with the width of the resonance in tune units.

4.10.3   Coupled nonlinear resonance

So far we have been considering the resonance $3\nu_x \approx p$ which involves only the 1-D x-motion. The analysis can be extended to the coupling resonances such as the $2\nu_x + \nu_y \approx p$ resonance. The corresponding $h_3^{(r)}$ is given by Eq. (4.405).

Let us define two reference tunes ν_{x0} and ν_{y0} in such a way that $\nu_x \approx \nu_{x0}$ and $\nu_y \approx \nu_{y0}$, where $\nu_{x0,y0}$ satisfy the resonance condition $2\nu_{x0} + \nu_{y0} = p$ exactly. (This definition does not give $\nu_{x0,y0}$ uniquely, but this ambiguity does not matter.) Let $2\nu_x + \nu_y = p + d$ with a small d, $|d| \ll 1$. Consider then a frame that "rotates" with frequency ν_{x0} in the x phase space and frequency ν_{y0} in the y phase space. The change of coordinates from $(\phi_x, A_x, \phi_y, A_y, z, \delta)$ to $(\phi'_x, A_x, \phi'_y, A_y, z, \delta)$, where

$$\phi'_x \;=\; \phi_x - 2\pi\nu_{x0}k, \qquad \text{and} \qquad \phi'_y \;=\; \phi_y - 2\pi\nu_{y0}k \tag{4.421}$$

can be facilitated by the map

$$R(k) \;=\; e^{:2\pi(\nu_{x0}A_x + \nu_{y0}A_y)k:} \,.$$

The new map is then

$$\begin{aligned}
N' \;&=\; e^{:2\pi(\nu_{x0}A_x + \nu_{y0}A_y)(k+1):}\, N\, e^{:-2\pi(\nu_{x0}A_x + \nu_{y0}A_y)k:}\\[4pt]
&=\; e^{:f'_2:}\exp\left[:h_3 + e^{:2\pi(\nu_{x0}A_x + \nu_{y0}A_y)k:}h_3^{(r)}:\right]\\[4pt]
&=\; e^{:f'_2:}\exp\left[:h_3 + 2A_x A_y^{1/2}\,\mathrm{Re}(C_{2010,0}^{(3)}e^{i2\phi'_x + i\phi'_y}):\right],
\end{aligned} \tag{4.422}$$

where

$$f'_2 \;=\; -2\pi(\nu_x - \nu_{x0})A_x - 2\pi(\nu_y - \nu_{y0})A_y - \frac{1}{2}\alpha_c C\delta^2 \,.$$

Concatenating the two factor maps in Eq. (4.422) gives

$$N' \;=\; e^{:-H:},$$

where H is the effective Hamiltonian,

$$\begin{aligned}
H(\phi'_x, A_x, \phi'_y, A_y, z, \delta, k) \;=\;& 2\pi(\nu_x - \nu_{x0})A_x + 2\pi(\nu_y - \nu_{y0})A_y\\[4pt]
&+ \frac{1}{2}\alpha_c C\delta^2 - C_{0000,3}^{(3)}\delta^3 - C_{1100,1}^{(3)}A_x\delta - C_{0011,1}^{(3)}A_y\delta\\[4pt]
&- 2A_x A_y^{1/2}\,\mathrm{Re}\left[C_{2010,0}^{(3)}\frac{i2\pi d}{1 - e^{-i2\pi d}}e^{i2\phi'_x + i\phi'_y}\right].
\end{aligned}$$

The Hamilton equations are

$$\begin{aligned}
\frac{d\phi'_x}{dk} \;=\; \frac{\partial H}{\partial A_x} \;=\;& 2\pi(\nu_x - \nu_{x0}) - C_{1100,1}^{(3)}\delta\\[4pt]
&- 2A_y^{1/2}\,\mathrm{Re}\left[C_{2010,0}^{(3)}\frac{i2\pi d}{1 - e^{-i2\pi d}}e^{i2\phi'_x + i\phi'_y}\right],
\end{aligned}$$

$$\frac{dA_x}{dk} = -\frac{\partial H}{\partial \phi'_x} = -2A_x A_y^{1/2} \operatorname{Re}\left[C^{(3)}_{2010,0}\frac{4\pi d}{1 - e^{-i2\pi d}}e^{i2\phi'_x + i\phi'_y}\right],$$

$$\frac{d\phi'_y}{dk} = \frac{\partial H}{\partial A_y} = 2\pi(\nu_y - \nu_{y0}) - C^{(3)}_{0011,1}\delta$$

$$-A_x A_y^{-1/2} \operatorname{Re}\left[C^{(3)}_{2010,0}\frac{i2\pi d}{1 - e^{-i2\pi d}}e^{i2\phi'_x + i\phi'_y}\right],$$

$$\frac{dA_y}{dk} = -\frac{\partial H}{\partial \phi_y} = -2A_x A_y^{1/2} \operatorname{Re}\left[C^{(3)}_{2010,0}\frac{2\pi d}{1 - e^{-i2\pi d}}e^{i2\phi'_x + i\phi'_y}\right]. \quad (4.423)$$

The equations for $\frac{dz}{dk}$ and $\frac{d\delta}{dk}$ are the same as those in Eq. (4.418).

As can be seen from Eq. (4.423), the resonance affects both x and y motions. There is an apparent divergence of $\frac{d\phi'_y}{dk}$ when $A_y \to 0$, but this does not cause problem in actual beam dynamics — it is a kinematic factor due to the polar coordinates chosen.

There are three constants of the motion. The first two are δ and H. An inspection of Eq. (4.423) gives a third constant, i.e.,

$$\frac{d}{dk}(A_x - 2A_y) = 0 \quad \Longrightarrow \quad A_x - 2A_y = \text{constant}. \quad (4.424)$$

In general, for resonances of third order and lower, the one-turn effective Hamiltonian can be written as

$$H = 2\pi(\nu_x - \nu_{x0})A_x + 2\pi(\nu_y - \nu_{y0})A_y + \frac{1}{2}\alpha_c C\delta^2$$
$$-C^{(3)}_{0000,3}\delta^3 - C^{(3)}_{1100,1}A_x\delta - C^{(3)}_{0011,1}A_y\delta + H',$$

where H' has the same expression as $-h_3^{(r)}$ in Eqs. (4.404–4.405) for the corresponding resonance, except that $\phi_{x,y}$ are replaced by $\phi'_{x,y}$.

Sum and difference resonances One can repeat the above analysis for a general resonance described as $\nu_x \approx \nu_{x0}, \nu_y \approx \nu_{y0}$ with $m\nu_{x0} + n\nu_{y0} = p$. The one-turn effective Hamiltonian, observed in the rotating frame defined by Eq. (4.421), would have the general form

$$H = 2\pi(\nu_x - \nu_{x0})A_x + 2\pi(\nu_y - \nu_{y0})A_y + \frac{1}{2}\alpha_c C\delta^2$$
$$-h(A_x, A_y, \delta) - A_x^{|m|/2}A_y^{|n|/2}\operatorname{Re}(\epsilon e^{im\phi'_x + in\phi'_y}), \quad (4.425)$$

where h is the nonresonant normal form contribution and depends on A_x, A_y, and δ only, ϵ is some complex coefficient related to the strength of the resonance-driving nonlinearity. For resonances of third order, for example, we have

$$
|\epsilon| \;=\;
\begin{cases}
\left| C^{(3)}_{3000,0} \right|, & 3\nu_x = p, \\[2mm]
\left| C^{(3)}_{2010,0} \right|, & 2\nu_x + \nu_y = p, \\[2mm]
\left| C^{(3)}_{2001,0} \right|, & 2\nu_x - \nu_y = p, \\[2mm]
\left| C^{(3)}_{1020,0} \right|, & \nu_x + 2\nu_y = p, \\[2mm]
\left| C^{(3)}_{1002,0} \right|, & \nu_x - 2\nu_y = p, \\[2mm]
\left| C^{(3)}_{0030,0} \right|, & 3\nu_y = p.
\end{cases}
$$

It follows from the Hamilton equation with the Hamiltonian (4.425) that

$$
\frac{d}{dk}(nA_x - mA_y) \;=\; -\left(n\frac{\partial H}{\partial \phi'_x} - m\frac{\partial H}{\partial \phi'_y} \right) \;=\; 0, \qquad (4.426)
$$

which means the quantity $nA_x - mA_y$ is a constant of the motion. The quantity (4.424) is just its special case. In this system, the three constants of the motion are δ, H, and $nA_x - mA_y$. The system has been fully integrated. Property (4.426) follows because ϕ'_x and ϕ'_y appear in H as a combined quantity $m\phi'_x + n\phi'_y$.

It may worth a comment here about the observation that there is a conservation law each time there is a missing dependance in the Hamiltonian. In the present case, the missing dependance of the Hamiltonian on the time variable k gives the conservation of H; missing of the z-dependance gives the conservation of δ; and the combined $m\phi'_x + n\phi'_y$ dependence gives the conservation of $nA_x - mA_y$.

The invariance of $nA_x - mA_y$, valid to 1st order in the nonlinearity strength, is an important observation, even if it is perhaps a rather nontrivial result of the Hamiltonian analysis. When m and n have opposite signs, i.e. the resonance is a "difference resonance", the invariance of $nA_x - mA_y$ imposes a constraint on both of the oscillation amplitudes A_x and A_y. The particle motion is then necessarily bounded and is therefore necessarily stable. For the "sum resonances", on the other hand, the invariance of $nA_x - mA_y$ does not impose a constraint on either A_x or A_y. The particle motion is potentially unbounded. Furthermore, it predicts that if one is unbounded, so must also be the other.

Resonance tune footprint   We defined a tune footprint width in Eq. (4.420) for the resonance $3\nu_x \approx p$. Similarly, for the resonance $2\nu_x + \nu_y \approx p$, one can obtain a feel of the resonance strength by associating it to the amplitude of variation of the quantity $\frac{1}{2\pi}\frac{d}{dk}(2\phi'_x + \phi'_y)$, which according to Eq. (4.423), for small d and $\delta = 0$, is given by

$$
\frac{1}{2\pi}\frac{d}{dk}(2\phi'_x + \phi'_y) \;=\; d - \frac{1}{2\pi}\left(4A_y^{1/2} + A_x A_y^{-1/2} \right) \mathrm{Re}\!\left(C^{(3)}_{2010,0}\, e^{i2\phi'_x + i\phi'_y} \right).
$$

In analogy to the tune footprint width defined in Eq. (4.420), one may be tempted to introduce

$$\Delta = (4A_y^{1/2} + A_x A_y^{-1/2})\left|C^{(3)}_{2010,0}\right| .$$

This definition of resonance width, however, has the drawback of an apparent divergence as $A_y \to 0$.

A general expression for the tune footprint width is, for the system with effective Hamiltonian (4.425),

$$\Delta = \frac{1}{2}\left(\frac{m|m|}{A_x} + \frac{n|n|}{A_y}\right) A_x^{|m|/2} A_y^{|n|/2} |\epsilon| ,$$

where we have set $d = m\nu_x + n\nu_y - p$ and $\delta = 0$.

If one would like a geometric definition of a width in the ν_x-ν_y plane around the resonance line $m\nu_x + n\nu_y = p$, it may be more appropriate to call $\frac{\Delta}{\sqrt{m^2+n^2}}$ the resonance width as illustrated in Fig. 4.8.

It should be mentioned here that this definition of resonance tune footprint widths does not take into account the effect of resonance stopbands. A more careful definition of resonance width according to its resonance stopband will be more involved.

Figure 4.8: A resonance tune footprint width in the tune plot.

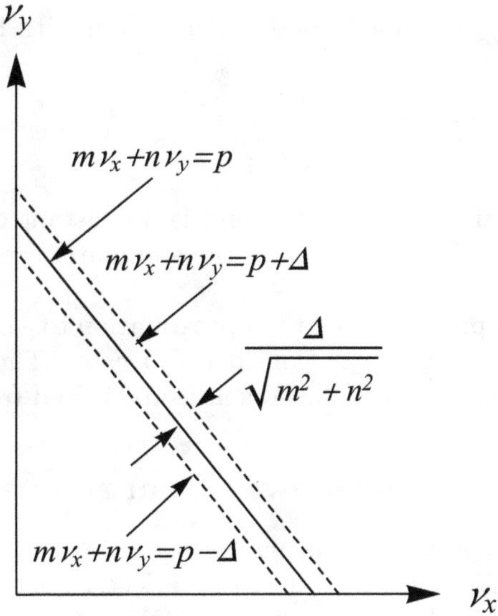

Smooth approximation In Sec. 4.9.4, we introduced a smooth approximation for a 1-D system, under which the equation of motion of a particle can be approximately written as Eq. (4.376), where the brackets mean keeping the slowly varying terms. Away from resonances, the slowly varying terms are extracted by (4.377). When the tune is close to a rational number $\frac{p}{n}$, there are additional slow terms to be included in the smoothing process. To see this, let us first change variable from ϕ to

$$\psi = \phi - \frac{p}{n}\theta.$$

Since ϕ advances by $2\pi\nu$ per turn, and θ advances by 2π per turn, and ν is close to $\frac{p}{n}$, it is easy to see that ψ is slowly varying on a turn-by-turn basis.

We further Fourier decompose[45]

$$F\left(\sqrt{2A}\sin\phi,\theta\right) = \sum_{m=-\infty}^{\infty}\sum_{k=-\infty}^{\infty} f_{mk}(A)e^{ik\phi+im\theta},$$

then

$$\langle\cos\phi\, F\rangle = \frac{1}{2}\sum_{q=-\infty}^{\infty} e^{iqn\psi}[f_{(-pq)(qn-1)} + f_{(-pq)(qn+1)}],$$

$$\langle\sin\phi\, F\rangle = \frac{1}{2i}\sum_{q=-\infty}^{\infty} e^{iqn\psi}[f_{(-pq)(qn-1)} - f_{(-pq)(qn+1)}]. \qquad (4.427)$$

Equations (4.376) and (4.427) describe a system with Hamiltonian

$$H(\psi,A,\theta) = \left(\nu-\frac{p}{n}\right)A - \frac{\sqrt{2A}}{2\nu}\sum_{q=-\infty}^{\infty}\frac{e^{iqn\psi}}{iqn}[f_{(-pq)(qn-1)} + f_{(-pq)(qn+1)}].$$

This Hamiltonian is independent of s, and is a constant of the motion. Away from resonances, only the $q = 0$ term in the summation remains.

Homework 4.110 Apply the smooth approximation to a single sextupole problem near the resonance $\nu_x = \frac{p}{3}$. Consider 1-D case. The result should agree with Eq. (4.414) when the C-coefficients are substituted from Eq. (4.430) later.

4.10.4   Single sextupole, away from resonance

In this section, we illustrate our normal form results by applying it to a simple nonlinear system. Consider an otherwise-perfectly-linear circular accelerator which contains a single thin-lens sextupole. Observed at the exit point of the sextupole, the one-turn Lie map is

$$M = e^{:f_2(X):}e^{:f_3(X):},$$

---

[45]This is possible because F is periodic in both θ and ϕ with period 2π.

where $f_2(X)$, given by Eq. (4.317) with $\alpha_{x,y}, \beta_{x,y}, \gamma_{x,y}, \eta$, and η' the unperturbed lattice functions evaluated at the exit of the sextupole, describes the linear transformation of the accelerator, and

$$f_3(X) \;=\; \lambda(x^3 - 3xy^2) \tag{4.428}$$

is due to the thin-lens sextupole whose strength is given by Eq. (4.214).

Consider the case away from resonances. We first carry out a third order normal form transformation according to Eq. (4.323), i.e.

$$A \;=\; e^{:F_3(U):} A_2 \,,$$

where A_2 is the linear map whose matrix representation is given by Eq. (4.319). The transformed map reads

$$N \;=\; AMA^{-1} \;=\; e^{:f_2 + h_3:} \,,$$

where f_2 here is given by Eq. (4.318).

To find expressions for F_3 and h_3, we follow the normal form procedure. First, we need to express f_3 as

$$\begin{aligned}
f_3(A_2 U) \;&=\; \lambda(x^3 - 3xy^2)\big|_{x=\sqrt{\beta_x}\bar{x}+\eta\delta,\; y=\sqrt{\beta_y}\bar{y}} \\
&=\; \lambda(\sqrt{\beta_x}\bar{x} + \eta\delta)^3 - 3\lambda(\sqrt{\beta_x}\bar{x} + \eta\delta)(\sqrt{\beta_y}\bar{y})^2 \\
&=\; \lambda(\sqrt{2\beta_x A_x}\sin\phi_x + \eta\delta)^3 \\
&\quad - 3\lambda(\sqrt{2\beta_x A_x}\sin\phi_x + \eta\delta)(\sqrt{2\beta_y A_y}\sin\phi_y)^2 \,.
\end{aligned} \tag{4.429}$$

We need to express the above in terms of the eigenmode functions $|abcd, e\rangle$ defined by Eq. (4.327). The result is given by Eq. (4.330) with coefficients

$$\begin{aligned}
C^{(3)}_{0000,3} \;&=\; \lambda\eta^3 \,, \\[4pt]
C^{(3)}_{1000,2} = -C^{(3)}_{0100,2} \;&=\; -\frac{3i}{2}\lambda\sqrt{2\beta_x}\,\eta^2 \,, \\[4pt]
C^{(3)}_{2000,1} = C^{(3)}_{0200,1} \;&=\; -\frac{3}{2}\lambda\beta_x\eta \,, \\[4pt]
C^{(3)}_{1100,1} \;&=\; 3\lambda\beta_x\eta \,, \\[4pt]
C^{(3)}_{0020,1} = C^{(3)}_{0002,1} \;&=\; \frac{3}{2}\lambda\beta_y\eta \,, \\[4pt]
C^{(3)}_{0011,1} \;&=\; -3\lambda\beta_y\eta \,, \\[4pt]
C^{(3)}_{3000,0} = -C^{(3)}_{0300,0} \;&=\; \frac{i}{8}\lambda(2\beta_x)^{3/2} \,, \\[4pt]
C^{(3)}_{2100,0} = -C^{(3)}_{1200,0} \;&=\; -\frac{3i}{8}\lambda(2\beta_x)^{3/2} \,, \\[4pt]
C^{(3)}_{1020,0} = -C^{(3)}_{0102,0} = C^{(3)}_{1002,0} = -C^{(3)}_{0120,0} \;&=\; -\frac{3i}{4}\lambda\sqrt{2\beta_x}\,\beta_y \,, \\[4pt]
C^{(3)}_{1011,0} = -C^{(3)}_{0111,0} \;&=\; \frac{3i}{2}\lambda\sqrt{2\beta_x}\,\beta_y \,.
\end{aligned} \tag{4.430}$$

The unlisted coefficients vanish.

Effective Hamiltonian　Away from resonances, we have then from Eq. (4.336),

$$
\begin{aligned}
h_3 &= C^{(3)}_{0000,3}\delta^3 + C^{(3)}_{1100,1}A_x\delta + C^{(3)}_{0011,1}A_y\delta \\
&= \lambda\eta^3\delta^3 + 3\lambda\beta_x A_x\eta\delta - 3\lambda\beta_y A_y\eta\delta\,.
\end{aligned}
\tag{4.431}
$$

The effective Hamiltonian, to 1st order in λ and away from resonances, is

$$
\begin{aligned}
H(\phi_x, A_x, \phi_y, A_y, z, \delta, n) &= -f_2 - h_3 \\
&= \mu_x A_x + \mu_y A_y + \frac{1}{2}\alpha_c C\delta^2 - \lambda\eta^3\delta^3 - 3\lambda\beta_x A_x\eta\delta + 3\lambda\beta_y A_y\eta\delta\,.
\end{aligned}
\tag{4.432}
$$

In this Hamiltonian, the time variable is the turn number n. This gives the betatron tunes as

$$
\begin{aligned}
\nu_x(A_x, A_y, \delta) &= \frac{1}{2\pi}\frac{\partial H}{\partial A_x} = \frac{1}{2\pi}(\mu_x - 3\lambda\beta_x\eta\delta)\,, \\
\nu_y(A_x, A_y, \delta) &= \frac{1}{2\pi}\frac{\partial H}{\partial A_y} = \frac{1}{2\pi}(\mu_y + 3\lambda\beta_y\eta\delta)\,.
\end{aligned}
\tag{4.433}
$$

The terms proportional to δ are contributions of the sextupole to the linear chromaticities.

The effective Hamiltonian also gives an expression of the path length change per turn as

$$
\Delta z(A_x, A_y, \delta) = \frac{\partial H}{\partial \delta} = \alpha_c C\delta - 3\lambda\eta^3\delta^2 - 3\lambda\eta(\beta_x A_x - \beta_y A_y)\,.
\tag{4.434}
$$

As pointed out after Eqs. (4.365) and (4.366), the linear chromaticities also appear in the path length dependencies on the betatron amplitudes.

Equation (4.433) is a well-known formula. As a particle passes through the sextupole with transverse displacements x and y, it receives angular kicks

$$
\begin{aligned}
\Delta x' &= -\frac{B_y L}{B\rho} = 3\lambda(x^2 - y^2)\,, \\
\Delta y' &= \frac{B_x L}{B\rho} = -6\lambda xy\,.
\end{aligned}
\tag{4.435}
$$

For an off-momentum particle, the particle executes betatron oscillation relative to a displaced position $x = \eta\delta$. This makes the sextupole behave as if it is a quadrupole with strength $\frac{L}{B\rho}\frac{\partial B_y}{\partial x} = -6\lambda\eta\delta$. This leads to the chromatic part of the tune shifts as described in Eq. (4.433).

To understand Eq. (4.434), substitute

$$
x = \sqrt{2\beta_x A_x}\sin\phi_x + \eta\delta, \qquad y = \sqrt{2\beta_y A_y}\sin\phi_y
$$

into Eq. (4.435), and recognize the fact that $\sin\phi_{x,y}$ are rapidly oscillating from turn to turn, which means we can take averages over ϕ_x and ϕ_y to obtain

$$
\begin{aligned}
\Delta x' &= 3\lambda(\eta^2\delta^2 + \beta_x A_x - \beta_y A_y)\,, \\
\Delta y' &= 0\,.
\end{aligned}
$$

Equation (4.434) then follows by noting that the path length change Δz is related to $\Delta x'$ by $\Delta z = -\eta\Delta x'$ (see Homework 4.104).

Normal form transformation　We also need to examine what transformation we made to bring the system from the physical coordinates to the normal form coordinates with the effective Hamiltonian (4.432). This is done by working out the expression (4.335),

$$
\begin{aligned}
F_3 \;=\; & -\frac{3}{2}\lambda\sqrt{2\beta_x A_x}\,\eta^2\delta^2\,\frac{\cos(\phi_x+\frac{\mu_x}{2})}{\sin\frac{\mu_x}{2}} \\[6pt]
& -\frac{3}{2}\lambda\eta\delta\left[A_x\beta_x\frac{\sin(2\phi_x+\mu_x)}{\sin\mu_x}-A_y\beta_y\frac{\sin(2\phi_y+\mu_y)}{\sin\mu_y}\right] \\[6pt]
& +\frac{1}{8}\lambda(2\beta_x A_x)^{3/2}\left[\frac{\cos(3\phi_x+\frac{3}{2}\mu_x)}{\sin\frac{3\mu_x}{2}}-3\frac{\cos(\phi_x+\frac{\mu_x}{2})}{\sin\frac{\mu_x}{2}}\right] \\[6pt]
& -\frac{3}{4}\lambda\sqrt{2\beta_x A_x}\,\beta_y A_y\left[\frac{\cos(\phi_x+2\phi_y+\frac{\mu_x}{2}+\mu_y)}{\sin(\frac{\mu_x}{2}+\mu_y)}\right. \\[6pt]
& \left.+\frac{\cos(\phi_x-2\phi_y+\frac{\mu_x}{2}-\mu_y)}{\sin(\frac{\mu_x}{2}-\mu_y)}-2\frac{\cos(\phi_x+\frac{\mu_x}{2})}{\sin\frac{\mu_x}{2}}\right]. \quad (4.436)
\end{aligned}
$$

Much useful information are hidden in the transformation $e^{:F_3:}$. Note that although we are discussing the case away from resonances, this transformation contains resonance denominators, as evidenced by an inspection of Eq. (4.436). The effective Hamiltonian, on the other hand, does not contain terms with small denominators because, away from resonances, they have all been absorbed into the normal form transformation.

The reader is reminded that the effective Hamiltonian contains all the information for the multiturn dynamics. The kinematics within each revolution are all absorbed in the normal form transformation. This situation is analogous to the linear case. In the linear case, the effective Hamiltonian contains the tune; the normal form transformation contains the Courant–Snyder functions α,β,γ. The tune is independent of s; the Courant–Snyder functions are periodic functions of s.

How are the normalized coordinates related to the physical coordinates? Equation (4.436) gives an expression of F_3 in terms of ϕ_x, A_x, ϕ_y, A_y, and δ. One may also express it in terms of $\bar{x},\bar{p}_x,\bar{y},\bar{p}_y$, and δ [see Eq. (4.314)],

$$
\begin{aligned}
F_3 \;=\; & \frac{3}{2}\lambda\sqrt{\beta_x}\left(\bar{x}-\cot\frac{\mu_x}{2}\bar{p}_x\right)\eta^2\delta^2 \\[6pt]
& +\frac{3}{4}\lambda\left[\beta_x(\bar{x}^2-2\cot\mu_x\,\bar{x}\bar{p}_x-\bar{p}_x^2)-\beta_y(\bar{y}^2-2\cot\mu_y\,\bar{y}\bar{p}_y-\bar{p}_y^2)\right]\eta\delta \\[6pt]
& +\frac{3}{8}\lambda\sqrt{\beta_x}\beta_y\left[-4\bar{x}\bar{y}^2+2\bar{p}_x(\bar{y}^2+\bar{p}_y^2)\cot\frac{\mu_x}{2}\right. \\[6pt]
& \qquad -(2\bar{x}\bar{y}\bar{p}_y+\bar{p}_x\bar{p}_y^2-\bar{p}_x\bar{y}^2)\cot\left(\frac{\mu_x}{2}-\mu_y\right) \\[6pt]
& \qquad \left.+(2\bar{x}\bar{y}\bar{p}_y-\bar{p}_x\bar{p}_y^2+\bar{p}_x\bar{y}^2)\cot\left(\frac{\mu_x}{2}+\mu_y\right)\right] \\[6pt]
& +\frac{1}{8}\lambda\beta_x^{3/2}\left[4\bar{x}^3-3\bar{p}_x(\bar{x}^2+\bar{p}_x^2)\cot\frac{\mu_x}{2}-\bar{p}_x(3\bar{x}^2-\bar{p}_x^2)\cot\frac{3\mu_x}{2}\right]. \quad (4.437)
\end{aligned}
$$

The normalized coordinates U are given by

$$
\begin{aligned}
U \;=\;& A^{-1}X \;=\; e^{-:F_3:}A_2 X \\[2mm]
=\;& e^{-:F_3:}
\begin{bmatrix} \bar{x} \\ \bar{p}_x \\ \bar{y} \\ \bar{p}_y \\ \bar{z} \\ \delta \end{bmatrix}
=
\begin{bmatrix} \bar{x} \\ \bar{p}_x \\ \bar{y} \\ \bar{p}_y \\ \bar{z} \\ \delta \end{bmatrix}
- :F_3:
\begin{bmatrix} \bar{x} \\ \bar{p}_x \\ \bar{y} \\ \bar{p}_y \\ \bar{z} \\ \delta \end{bmatrix}
+ \mathcal{O}(\lambda^2 X^3).
\end{aligned}
\tag{4.438}
$$

Substituting Eq. (4.437) into Eq. (4.438) gives explicit expressions of the U-coordinates, i.e. the normal form coordinates, in terms of the coordinates $(\bar{x}, \bar{p}_x, \bar{y}, \bar{p}_y, \bar{z}, \delta)$.

By relabeling $(\bar{x}, \bar{p}_x, \bar{y}, \bar{p}_y, \bar{z}, \delta)$ as $(\bar{x}_1, \bar{p}_{x1}, \bar{y}_1, \bar{p}_{y1}, \bar{z}_1, \delta)$ and substituting for $(\bar{x}_1, \bar{p}_{x1}, \bar{y}_1, \bar{p}_{y1}, \bar{z}_1, \delta)$ by Eq. (4.318), and reserving the symbols $(\bar{x}, \bar{p}_x, \bar{y}, \bar{p}_y, \bar{z}, \delta)$ for the components of U, we obtain the relation between the final normal form coordinates in terms of the original physical coordinates $(x, p_x, y, p_y, z, \delta)$,

$$
\begin{aligned}
\bar{x} \;=\;& \bar{x}_1 - \frac{3}{2}\lambda\sqrt{\beta_x}\,\eta^2\delta^2 \cot\frac{\mu_x}{2} - \frac{3}{2}\lambda\beta_x\eta\delta(\bar{x}_1\cot\mu_x + \bar{p}_{x1}) \\[1mm]
& + \frac{3}{8}\lambda\sqrt{\beta_x}\,\beta_y \left\{ 2(\bar{y}_1^2 + \bar{p}_{y1}^2)\cot\frac{\mu_x}{2} \right. \\[1mm]
& \qquad\qquad \left. + (\bar{y}_1^2 - \bar{p}_{y1}^2)\left[\cot\left(\frac{\mu_x}{2}+\mu_y\right) + \cot\left(\frac{\mu_x}{2}-\mu_y\right)\right] \right\} \\[1mm]
& - \frac{3}{8}\lambda\beta_x^{3/2}\left[(\bar{x}_1^2 + 3\bar{p}_{x1}^2)\cot\frac{\mu_x}{2} + (\bar{x}_1^2 - \bar{p}_{x1}^2)\cot\frac{3\mu_x}{2}\right] + \mathcal{O}(\lambda^2 X_1^3) \\[2mm]
=\;& \frac{x - \eta\delta}{\sqrt{\beta_x}} - \frac{3}{2}\lambda\sqrt{\beta_x}\,\eta^2\delta^2\cot\frac{\mu_x}{2} \\[1mm]
& - \frac{3}{2}\lambda\sqrt{\beta_x}\,\eta\delta\left[(x-\eta\delta)\cot\mu_x + \alpha_x x + \beta_x x' - \alpha_x\eta\delta - \beta_x\eta'\delta\right] \\[1mm]
& + \frac{3}{8}\lambda\sqrt{\beta_x}\left\{2[y^2 + (\alpha_y y + \beta_y y')^2]\cot\frac{\mu_x}{2}\right. \\[1mm]
& \qquad\qquad \left. + [y^2 - (\alpha_y y + \beta_y y')^2]\left[\cot\left(\frac{\mu_x}{2}+\mu_y\right) + \cot\left(\frac{\mu_x}{2}-\mu_y\right)\right]\right\} \\[1mm]
& - \frac{3}{8}\lambda\sqrt{\beta_x}\left\{\left[(x-\eta\delta)^2 + 3(\alpha_x x + \beta_x x' - \alpha_x\eta\delta - \beta_x\eta'\delta)^2\right]\cot\frac{\mu_x}{2}\right. \\[1mm]
& \qquad\qquad \left. + \left[(x-\eta\delta)^2 - (\alpha_x x + \beta_x x' - \alpha_x\eta\delta - \beta_x\eta'\delta)^2\right]\cot\frac{3\mu_x}{2}\right\} + \mathcal{O}(\lambda^2 X^3),
\end{aligned}
$$

$$
\begin{aligned}
\bar{p}_x \;=\;& \bar{p}_{x1} - \frac{3}{2}\lambda\sqrt{\beta_x}\,\eta^2\delta^2 - \frac{3}{2}\lambda\beta_x\eta\delta(\bar{x}_1 - \bar{p}_{x1}\cot\mu_x) \\[1mm]
& + \frac{3}{4}\lambda\sqrt{\beta_x}\,\beta_y\left\{2\bar{y}_1^2 + \bar{y}_1\bar{p}_{y1}\left[\cot\left(\frac{\mu_x}{2}-\mu_y\right) - \cot\left(\frac{\mu_x}{2}+\mu_y\right)\right]\right\} \\[1mm]
& - \frac{3}{4}\lambda\beta_x^{3/2}\bar{x}_1\left[2\bar{x}_1 - \bar{p}_{x1}\left(\cot\frac{\mu_x}{2} + \cot\frac{3\mu_x}{2}\right)\right] + \mathcal{O}(\lambda^2 X_1^3) \\[2mm]
=\;& \frac{\alpha_x(x-\eta\delta) + \beta_x(x' - \eta'\delta)}{\sqrt{\beta_x}} - \frac{3}{2}\lambda\sqrt{\beta_x}\,\eta^2\delta^2
\end{aligned}
$$

$$-\frac{3}{2}\lambda\sqrt{\beta_x}\,\eta\delta\left[(x-\eta\delta)-(\alpha_x x+\beta_x x'-\alpha_x\eta\delta-\beta_x\eta'\delta)\cot\mu_x\right]$$

$$+\frac{3}{4}\lambda\sqrt{\beta_x}\left\{2y^2+y(\alpha_y+\beta_y y')\left[\cot\left(\frac{\mu_x}{2}-\mu_y\right)-\cot\left(\frac{\mu_x}{2}+\mu_y\right)\right]\right\}$$

$$-\frac{3}{4}\lambda\sqrt{\beta_x}(x-\eta\delta)\Big[2(x-\eta\delta)$$

$$-(\alpha_x x+\beta_x x'-\alpha_x\eta\delta-\beta_x\eta'\delta)\left(\cot\frac{\mu_x}{2}+\cot\frac{3\mu_x}{2}\right)\Big]+\mathcal{O}(\lambda^2 X^3),$$

$$\bar{y}=\bar{y}_1+\frac{3}{2}\lambda\beta_y\eta\delta(\bar{y}_1\cot\mu_y+\bar{p}_{y1})$$

$$-\frac{3}{4}\lambda\sqrt{\beta_x}\beta_y\left[-2\bar{p}_{x1}\bar{p}_{y1}\cot\frac{\mu_x}{2}+(\bar{x}_1\bar{y}_1+\bar{p}_{x1}\bar{p}_{y1})\cot\left(\frac{\mu_x}{2}-\mu_y\right)\right.$$

$$\left.-(\bar{x}_1\bar{y}_1-\bar{p}_{x1}\bar{p}_{y1})\cot\left(\frac{\mu_x}{2}+\mu_y\right)\right]+\mathcal{O}(\lambda^2 X_1^3)$$

$$=\frac{y}{\sqrt{\beta_y}}+\frac{3}{2}\lambda\sqrt{\beta_y}\,\eta\delta(y\cot\mu_y+\alpha_y y+\beta_y y')$$

$$-\frac{3}{4}\lambda\sqrt{\beta_y}\left\{-2(\alpha_x x+\beta_x x'-\alpha_x\eta\delta-\beta_x\eta'\delta)(\alpha_y y+\beta_y y')\cot\frac{\mu_x}{2}\right.$$

$$+[(x-\eta\delta)y+(\alpha_x x+\beta_x x'-\alpha_x\eta\delta-\beta_x\eta'\delta)(\alpha_y y+\beta_y y')]\cot\left(\frac{\mu_x}{2}-\mu_y\right)$$

$$\left.-[(x-\eta\delta)y-(\alpha_x x+\beta_x x'-\alpha_x\eta\delta-\beta_x\eta'\delta)(\alpha_y y+\beta_y y')]\cot\left(\frac{\mu_x}{2}+\mu_y\right)\right\}$$

$$+\mathcal{O}(\lambda^2 X^3),$$

$$\bar{p}_y=\bar{p}_{y1}+\frac{3}{2}\lambda\beta_y\eta\delta(\bar{y}_1-\bar{p}_{y1}\cot\mu_y])+\frac{3}{4}\lambda\sqrt{\beta_x}\beta_y\left[4\bar{x}_1\bar{y}_1-2\bar{p}_{x1}\bar{y}_1\cot\frac{\mu_x}{2}\right.$$

$$\left.+(\bar{x}_1\bar{p}_{y1}-\bar{p}_{x1}\bar{y}_1)\cot\left(\frac{\mu_x}{2}-\mu_y\right)-(\bar{x}_1\bar{p}_{y1}+\bar{p}_{x1}\bar{y}_1)\cot\left(\frac{\mu_x}{2}+\mu_y\right)\right]$$

$$+\mathcal{O}(\lambda^2 X_1^3)$$

$$=\frac{\alpha_y y+\beta_y y'}{\sqrt{\beta_y}}+\frac{3}{2}\lambda\sqrt{\beta_y}\,\eta\delta[y-(\alpha_y y+\beta_y y')\cot\mu_y]$$

$$+\frac{3}{4}\lambda\sqrt{\beta_y}\left\{4(x-\eta\delta)y-2(\alpha_x x+\beta_x x'-\alpha_x\eta\delta-\beta_x\eta'\delta)y\cot\frac{\mu_x}{2}\right.$$

$$+[(x-\eta\delta)(\alpha_y y+\beta_y y')-(\alpha_x x+\beta_x x'-\alpha_x\eta\delta-\beta_x\eta'\delta)y]\cot\left(\frac{\mu_x}{2}-\mu_y\right)$$

$$\left.-[(x-\eta\delta)(\alpha_y y+\beta_y y')+(\alpha_x x+\beta_x x'-\alpha_x\eta\delta-\beta_x\eta'\delta)y]\cot\left(\frac{\mu_x}{2}+\mu_y\right)\right\}$$

$$+\mathcal{O}(\lambda^2 X^3)$$

$$\bar{z}=\bar{z}_1+3\lambda\eta^2\delta\sqrt{\beta_x}\left(\bar{x}_1-\bar{p}_{x1}\cot\frac{\mu_x}{2}\right)+\frac{3}{4}\lambda\beta_x\eta(\bar{x}_1^2-2\cot\mu_x\bar{x}_1\bar{p}_{x1}-\bar{p}_{x1}^2)$$

$$-\frac{3}{4}\lambda\beta_y\eta(\bar{y}_1^2-2\cot\mu_y\bar{y}_1\bar{p}_{y1}+\bar{p}_{y1}^2)+\mathcal{O}(\lambda^2 X_1^3)$$

$$=z+\eta'x-\eta x'+3\lambda\eta^2\delta\left[x-\eta\delta-(\alpha_x x+\beta_x x'-\alpha_x\eta\delta-\beta_x\eta'\delta)\cot\frac{\mu_x}{2}\right]$$

$$+\frac{3}{4}\lambda\eta\Big[(x-\eta\delta)^2 - 2\cot\mu_x(x-\eta\delta)(\alpha_x x + \beta_x x' - \alpha_x\eta\delta - \beta_x\eta'\delta)$$
$$-(\alpha_x x + \beta_x x' - \alpha_x\eta\delta - \beta_x\eta'\delta)^2$$
$$-y^2 + 2\cot\mu_y y(\alpha_y y + \beta_y y') - (\alpha_y y + \beta_y y')^2\Big] + \mathcal{O}(\lambda^2 X^3)\,.$$

$$(4.439)$$

Invariant of the motion   Having obtained Eq. (4.436), one can calculate the two betatron invariants $W_{x,y}$ in terms of the linear amplitudes $A_{x,y}$ and the phases $\phi_{x,y}$ as follows,

$$W_{x,y} = A^{-1}A_{x,y} = e^{-:F_3:}A_{x,y} = A_{x,y} - :F_3:A_{x,y} + \mathcal{O}(\lambda^2 X^4)\,.$$

Noting that

$$:F_3:A_{x,y} \;=\; [F_3, A_{x,y}] \;=\; \frac{\partial F_3}{\partial\phi_{x,y}}\,,$$

we obtain the results

$$W_x \;=\; A_x - \frac{3}{2}\lambda\sqrt{2\beta_x A_x}\,\eta^2\delta^2\frac{\sin(\phi_x + \frac{\mu_x}{2})}{\sin\frac{\mu_x}{2}} + 3\lambda\beta_x A_x\eta\delta\frac{\cos(2\phi_x + \mu_x)}{\sin\mu_x}$$
$$-\frac{3}{4}\lambda\sqrt{2\beta_x A_x}\,\beta_y A_y\left[\frac{\sin(\phi_x + 2\phi_y + \frac{\mu_x}{2} + \mu_y)}{\sin(\frac{\mu_x}{2} + \mu_y)}\right.$$
$$\left.+\frac{\sin(\phi_x - 2\phi_y + \frac{\mu_x}{2} - \mu_y)}{\sin(\frac{\mu_x}{2} - \mu_y)} - 2\frac{\sin(\phi_x + \frac{\mu_x}{2})}{\sin\frac{\mu_x}{2}}\right]$$
$$-\frac{3}{8}\lambda(2\beta_x A_x)^{3/2}\left[\frac{\sin(\phi_x + \frac{\mu_x}{2})}{\sin\frac{\mu_x}{2}} - \frac{\sin(3\phi_x + \frac{3\mu_x}{2})}{\sin\frac{3\mu_x}{2}}\right] + \mathcal{O}(\lambda^2 X^4)\,,$$

$$W_y \;=\; A_y - 3\lambda\beta_y A_y\eta\delta\frac{\cos(2\phi_y + \mu_y)}{\sin\mu_y}$$
$$-\frac{3}{2}\lambda\sqrt{2\beta_x A_x}\,\beta_y A_y\left[\frac{\sin(\phi_x + 2\phi_y + \frac{\mu_x}{2} + \mu_y)}{\sin(\frac{\mu_x}{2} + \mu_y)} - \frac{\sin(\phi_x - 2\phi_y + \frac{\mu_x}{2} - \mu_y)}{\sin(\frac{\mu_x}{2} - \mu_y)}\right]$$
$$+\mathcal{O}(\lambda^2 X^4)\,.$$

$$(4.440)$$

These expressions are identical to Eq. (4.246) if we drop the terms proportional to δ.

One can calculate the invariants W_x and W_y to order $\mathcal{O}(\lambda^2 X^4)$ by substituting the expressions of $\bar{x}$ and $\bar{p}_x$ in terms of $(\bar{x}_1, \bar{p}_{x1}, \bar{y}, \bar{p}_{y1}, \bar{z}_1)$ in Eq. (4.439) into $W_x = \frac{1}{2}(\bar{x}^2 + \bar{p}_x^2)$ and $W_y = \frac{1}{2}(\bar{y}^2 + \bar{p}_y^2)$. This gives

$$W_x \;=\; \frac{1}{2}(\bar{x}_1^2 + \bar{p}_{x1}^2) - \frac{3}{2}\lambda\sqrt{\beta_x}\,\eta^2\delta^2\Big(\bar{x}_1\cot\frac{\mu_x}{2} + \bar{p}_{x1}\Big)$$
$$-\frac{3}{2}\lambda\beta_x\eta\delta\left[(\bar{x}_1^2 - \bar{p}_{x1}^2)\cot\mu_x + 2\bar{x}_1\bar{p}_{x1}\right]$$
$$+\frac{3}{8}\lambda\sqrt{\beta_x}\,\beta_y\left\{4\bar{p}_{x1}\bar{y}_1^2 + 2\bar{x}_1(\bar{y}_1^2 + \bar{p}_{y1}^2)\cot\frac{\mu_x}{2}\right.$$

$$+\bar{x}_1(\bar{y}_1^2 - \bar{p}_{y1}^2)\left[\cot\left(\frac{\mu_x}{2} + \mu_y\right) + \cot\left(\frac{\mu_x}{2} - \mu_y\right)\right]$$

$$-2\bar{p}_{x1}\bar{y}_1\bar{p}_{y1}\left[\cot\left(\frac{\mu_x}{2} + \mu_y\right) - \cot\left(\frac{\mu_x}{2} - \mu_y\right)\right]\Bigg\}$$

$$-\frac{3}{8}\lambda\beta_x^{3/2}\left\{\bar{x}_1(\bar{x}_1^2 + 3\bar{p}_{x1}^2)\cot\frac{\mu_x}{2} + \bar{x}_1(\bar{x}_1^2 - \bar{p}_{x1}^2)\cot\frac{3\mu_x}{2}\right.$$

$$\left. +2\bar{x}_1\bar{p}_{x1}\left[2\bar{x}_1 - \left(\cot\frac{\mu_x}{2} + \cot\frac{3\mu_x}{2}\right)\bar{p}_{x1}\right]\right\} + \mathcal{O}(\lambda^2 X_1^4),$$

$$\begin{aligned}
W_y &= \frac{1}{2}(\bar{y}_1^2 + \bar{p}_{y1}^2) + \frac{3}{2}\lambda\beta_y\eta\delta[(\bar{y}_1^2 - \bar{p}_{y1}^2)\cot\mu_y + 2\bar{y}_1\bar{p}_{y1}] \\
&\quad + \frac{3}{4}\lambda\sqrt{\beta_x}\beta_y\left\{4\bar{x}_1\bar{y}_1\bar{p}_{y1}\right. \\
&\quad -[\bar{y}_1(\bar{x}_1\bar{y}_1 + \bar{p}_{x1}\bar{p}_{y1}) - \bar{p}_{y1}(\bar{x}_1\bar{p}_{y1} - \bar{p}_{x1}\bar{y}_1)]\cot\left(\frac{\mu_x}{2} - \mu_y\right) \\
&\quad \left. +[\bar{y}_1(\bar{x}_1\bar{y}_1 - \bar{p}_{x1}\bar{p}_{y1}) - \bar{p}_{y1}(\bar{x}_1\bar{p}_{y1} + \bar{p}_{x1}\bar{y}_1)]\cot\left(\frac{\mu_x}{2} + \mu_y\right)\right\} \\
&\quad + \mathcal{O}(\lambda^2 X_1^4).
\end{aligned} \tag{4.441}$$

If we substitute $\bar{x}_1, \bar{p}_{x1}, \bar{y}, \bar{p}_{y1}$ in Eq. (4.441) respectively by $\sqrt{2A_x}\sin\phi_x$, $\sqrt{2A_x}\cos\phi_x$, $\sqrt{2A_y}\sin\phi_y$, and $\sqrt{2A_y}\cos\phi_y$, one recovers Eq. (4.440). As was mentioned there, these results also agree with those obtained in Eq. (4.246) by manipulating the BCH formula. Note also that one can obtain the connections between the invariants and the physical coordinates $(x, x', y, y', z, \delta)$ by substituting $\bar{x}_1, \bar{p}_{x1}, \bar{y}, \bar{p}_{y1}$ in Eq. (4.441) using Eq. (4.318).

Higher order So far we have looked at the second order effects. One can also explore the third order effects of this accelerator with a single sextupole. This accelerator is modeled by Eq. (4.339) with f_2 given by Eq. (4.317), f_3 given by Eq. (4.428), and $f_4 = 0$. Substituting Eq. (4.429) for f_3, (4.436) for F_3 and (4.431) for h_3 into Eq. (4.344), and after performing some algebra, one obtains, using Eq. (4.347),

$$\begin{aligned}
h_4 &= \frac{9}{16}\lambda^2\beta_x\beta_y^2 A_y^2\left[4\cot\frac{\mu_x}{2} + \cot\left(\frac{\mu_x}{2} + \mu_y\right) + \cot\left(\frac{\mu_x}{2} - \mu_y\right)\right] \\
&\quad -\frac{9}{4}\lambda^2\beta_x\beta_y A_x A_y\left[2\beta_x\cot\frac{\mu_x}{2} + \beta_y\cot\left(\frac{\mu_x}{2} - \mu_y\right) - \beta_y\cot\left(\frac{\mu_x}{2} + \mu_y\right)\right] \\
&\quad +\frac{9}{16}\lambda^2\beta_x^3 A_x^2\left(3\cot\frac{\mu_x}{2} + \cot\frac{3\mu_x}{2}\right) \\
&\quad -\frac{9}{2}\lambda^2\beta_y A_y\eta^2\delta^2\left(\beta_x\cot\frac{\mu_x}{2} - \beta_y\cot\mu_y\right) \\
&\quad +\frac{9}{2}\lambda^2\beta_x^2 A_x\eta^2\delta^2\left(\cot\frac{\mu_x}{2} + \cot\mu_x\right) + \frac{9}{4}\lambda^2\beta_x\eta^4\delta^4\cot\frac{\mu_x}{2}.
\end{aligned}$$

The effective Hamiltonian is $-f_2 - h_3 - h_4$. The betatron tunes can now be carried to include terms higher order than Eq. (4.433). This gives

$$
\begin{aligned}
\nu_x(A_x, A_y, \delta) &= \frac{1}{2\pi}\mu_x - \frac{3}{2\pi}\lambda\beta_x\eta\delta \\
&+ \frac{9}{8\pi}\lambda^2\beta_x\beta_y A_y \left[2\beta_x \cot\frac{\mu_x}{2} + \beta_y \cot\left(\frac{\mu_x}{2} - \mu_y\right) - \beta_y \cot\left(\frac{\mu_x}{2} + \mu_y\right) \right] \\
&- \frac{9}{16\pi}\lambda^2\beta_x^3 A_x \left(3\cot\frac{\mu_x}{2} + \cot\frac{3\mu_x}{2} \right) \\
&- \frac{9}{4\pi}\lambda^2\beta_x^2\eta^2\delta^2 \left(\cot\frac{\mu_x}{2} + \cot\mu_x \right) + \mathcal{O}(\lambda^3 X^3) , \\
\nu_y(A_x, A_y, \delta) &= \frac{1}{2\pi}\mu_y + \frac{3}{2\pi}\lambda\beta_y\eta\delta \\
&- \frac{9}{16\pi}\lambda^2\beta_x\beta_y^2 A_y \left[4\cot\frac{\mu_x}{2} + \cot\left(\frac{\mu_x}{2} + \mu_y\right) + \cot\left(\frac{\mu_x}{2} - \mu_y\right) \right] \\
&+ \frac{9}{8\pi}\lambda^2\beta_x\beta_y A_x \left[2\beta_x \cot\frac{\mu_x}{2} + \beta_y \cot\left(\frac{\mu_x}{2} - \mu_y\right) - \beta_y \cot\left(\frac{\mu_x}{2} + \mu_y\right) \right] \\
&+ \frac{9}{4\pi}\lambda^2\beta_y\eta^2\delta^2 \left(\beta_x \cot\frac{\mu_x}{2} - \beta_y \cot\mu_y \right) + \mathcal{O}(\lambda^3 X^3) .
\end{aligned}
$$

As mentioned before, the leading terms in the shifts of the betatron tunes from the ideal values are proportional to δ, and these are the linear chromaticity terms. In particular, on-momentum particles do not experience betatron tune shifts due to sextupoles to 1st order in the sextupole strength. The higher order terms, on the other hand, contains terms that do not vanish for $\delta = 0$. Also as pointed out in Eq. (4.364), the ν_x dependence on A_y is the same as the ν_y dependence on A_x.

One also obtains a higher order counterpart of Eq. (4.434) for the path length,

$$
\begin{aligned}
\Delta z(A_x, A_y, \delta) &= \alpha_c\delta - 3\lambda\eta^3\delta^3 - 3\lambda\eta(\beta_x A_x - \beta_y A_y) \\
&+ 9\lambda^2\beta_y A_y\eta^2\delta \left(\beta_x \cot\frac{\mu_x}{2} - \beta_y \cot\mu_y \right) \\
&- 9\lambda^2\beta_x^2 A_x\eta^2\delta \left(\cot\frac{\mu_x}{2} + \cot\mu_y \right) \\
&- 9\lambda^2\beta_x\eta^4\delta^3 \cot\frac{\mu_x}{2} + \mathcal{O}(\lambda^3 X^4) .
\end{aligned}
$$

Smear and distortion function   Equation (4.440) can be used to extract the phenomenological quantities called the "smear" and the "distortion function". The x-smear S_x is the relative variation of the Courant–Snyder invariant A_x (which is a true invariant only in the linear case without the sextupole perturbation) as a function of time for an on-momentum particle. Let the peak-to-peak variation of A_x be $2\Delta A_x$. Similarly we define the y-smear S_y. To 1st order in λ, since W_x are W_y are true invariants, the smears are given by

$$
S_x = \frac{\Delta A_x}{A_x} = \frac{3}{4}\lambda\sqrt{\frac{2\beta_x}{A_x}}\beta_y A_y \left(\frac{1}{|\sin(\frac{\mu_x}{2} + \mu_y)|} + \frac{1}{|\sin(\frac{\mu_x}{2} - \mu_y)|} + \frac{2}{|\sin\frac{\mu_x}{2}|} \right)
$$

$$+\frac{3}{8}\lambda(2\beta_x)^{3/2}A_x^{1/2}\left(\frac{1}{|\sin\frac{\mu_x}{2}|}+\frac{1}{|\sin\frac{3\mu_x}{2}|}\right),$$

$$S_y=\frac{\Delta A_y}{A_y}=\frac{3}{2}\lambda\sqrt{2\beta_x A_x}\beta_y\left(\frac{1}{|\sin(\frac{\mu_x}{2}+\mu_y)|}+\frac{1}{|\sin(\frac{\mu_x}{2}-\mu_y)|}\right).$$

The smears also contain the resonance denominators. Near resonances, the smears become large. The smears are sometimes used as indicators of how the system deviates from a perfectly linear system.

Homework 4.111 Although away from resonances, the normal form transformation contains terms that exhibit resonant small denominators. It is instructive exercise to go through the steps to derive Eq. (4.436). You need to establish Eq. (4.430) first and use Eq. (4.335). After deriving Eq. (4.436), go one more step to derive Eqs. (4.437) and (4.439). Pay attention to which quantities contain small denominators and which ones do not.

Homework 4.112 One may repeat the analysis in the text for the case of a single octupole. Consider 1-D motion for simplicity. The one-turn map around the exit of the octupole is

$$e^{-:\mu A:}e^{:\lambda x^4:},$$

where λ is the octupole strength, and $A=\frac{1}{2}(\gamma x^2+2\alpha xp+\beta p^2)$. Use a computer analytical tool to do the algebra if available.

(a) Find the invariant of the motion to 1st order in λ. Express it in terms of the action-angle coordinates (ϕ, A). Assume there are no resonances nearby.

(b) Find the tune shift with amplitude A to 1st order in λ.

(c) Find the invariant when the tune is close to a resonance $\nu=\frac{p}{4}$.

This somewhat extensive homework actually is a rather practical one.

Homework 4.113 Consider a circular accelerator, which has a perfectly linear optics in its x and y motions, but is perturbed by a thin, weak quadrupole of strength $k=\frac{\ell}{B\rho}\frac{\partial B_y}{\partial x}$.

(a) Perform a normal form analysis to obtain the x- and y-tune shifts to 1st order in k. Also calculate the perturbation on the Courant–Snyder β- and α-functions to 1st order in k at the exit location of the perturbing quadrupole.

(b) Extend the analysis to 2nd order in k.

Compare the results with an exact calculation using matrices. Note that the small parameter of this problem is k.

Homework 4.114 Consider a weak thin-lens skew quadrupole with strength $k=\frac{\ell}{B\rho}\frac{\partial B_y}{\partial y}$. Compute the tune shifts to order $\mathcal{O}(k^2)$,

(a) away from resonances,

(b) near a $\nu_x\pm\nu_y=p$ resonance.

4.10.5 Sextupole pairing and achromat

Accelerator lattice is often constructed from building blocks. One very useful building block is a section of magnets arranged in such a way that it is transparent to particle motion, i.e. the section is there only to physically transport the beam but otherwise leaving all beam dynamics intact. In that case, this building block beamline section can be a "unit transformation".

However there is a question of how accurately the section acts as a unit transformation. An array of quadrupoles can easily make a unit transformation in the sense that its 6×6 map is the unit matrix I, but this is done only to 1st order in $X = (x, x', y, y', z, \delta)$. Particularly, the higher order δ-dependence of the map often contributes to chromatic aberrations that must be compensated for. Adding sextupoles to this array of quadrupoles can make the section act as a unit transformation to 2nd order in X, thus providing a more accurate unit transformation section, which we will call an achromat — in this case, a second order achromat. Similarly, if the array includes octupoles, the unit transformation can be made into a third order achromat, etc. There are no perfect achromats. There are only increasingly higher ordered ones.

Sextupoles are needed to correct for the first order chromatic aberrations when $\delta \neq 0$. We want the sextupoles to act according to the particle's energy deviation δ. In doing so, however, we must make sure that they do not affect too much for particles with $\delta = 0$. In this section, we consider the on-momentum particles, and try to arrange the sextupoles so as not to disturb them too much. The trick is sextupole pairing. The analysis will apply the Lie algebra techniques we have learned.

First consider two thin sextupoles of strengths λ_1 and λ_2 at locations 1 and 2 in an accelerator. Let us consider only the 1-D x-motion for now. Let ψ be the betatron phase advance between these two locations. Let $\alpha_{1,2}, \beta_{1,2}$, and $\gamma_{1,2}$ be the Courant–Snyder parameters at these two locations. See Fig. 4.9(a). The map from the entrance of sextupole 1 to the exit of the sextupole 2 is

$$M = e^{:\lambda_1 x^3:} e^{:f_2:} e^{:\lambda_2 x^3:},$$

where f_2 is the linear transformation from location 1 to location 2. We have

$$M = e^{:f_2:} \exp\left(:e^{-:f_2:}\lambda_1 x^3:\right) e^{:\lambda_2 x^3:}.$$

Using Eq. (4.80), we have

$$e^{-:f_2:}x = x\sqrt{\frac{\beta_1}{\beta_2}}(\cos\psi - \alpha_2\sin\psi) - x'\sqrt{\beta_1\beta_2}\sin\psi.$$

This leads to

$$M = e^{:f_2:}\exp\left\{:\lambda_1\left[x\sqrt{\frac{\beta_1}{\beta_2}}(\cos\psi - \alpha_2\sin\psi) - x'\sqrt{\beta_1\beta_2}\sin\psi\right]^3:\right\}e^{:\lambda_2 x^3:}.$$

$$(4.442)$$

Figure 4.9: Optimal arrangement of two sextupoles. (a) thin sextupoles, (b) thick sextupoles.

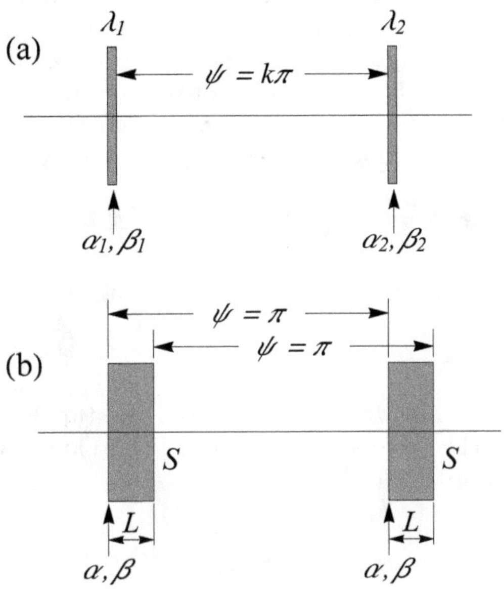

We now observe that the two factor exponential maps on the right side of Eq. (4.442) — those involving the sextupoles — commute if $\sin \psi = 0$. In other words, if $\psi = k\pi$ for some integer k, the total map of this section reads

$$M \; = \; e^{:f_2:} \exp\left[:\lambda_1 x^3 \left(\frac{\beta_1}{\beta_2}\right)^{3/2} (-1)^{3k} + \lambda_2 x^3:\right].$$

If we further choose the sextupole strengths so that

$$\lambda_1 \left(\frac{\beta_1}{\beta_2}\right)^{3/2} (-1)^{3k} + \lambda_2 \; = \; 0, \tag{4.443}$$

then the map of the section is $e^{:f_2:}$ just as if the sextupoles were not present. For on-momentum particles, the effects of the two sextupoles have canceled each other, and the section is made optically transparent when observed outside the section.

Note that the cancellation is exact. Equation (4.442) is not a perturbation treatment. Note also that the cancellation does not apply if observed inside the section — we have not applied the BCH formula in its derivation.

In the special case when $\beta_1 = \beta_2$, the cancellation condition reads

$$\begin{cases} \lambda_1 = \lambda_2 & \text{if } \psi = \text{odd multipole of } \pi, \\ \lambda_1 = -\lambda_2 & \text{if } \psi = \text{even multipole of } \pi. \end{cases}$$

Now let us look into what effects would affect the exact cancellation mentioned above. One situation occurs when the condition $\sin\psi$ is not exactly 0. This will be studied in Homework 4.115. Another situation occurs when the sextupoles are thick, as shown in Fig. 4.9(b). To be specific, let ψ — measured from the entrance of sextupole 1 to the entrance of sextupole 2 (or measured between the two exits) — be exactly equal to π, and let $\beta_1 = \beta_2$ and $\alpha_1 = \alpha_2$. Let the two sextupoles both have strength S and length L. Let us still consider 1-D motion. Inside the sextupole, the equation of motion is

$$x'' = Sx^2, \qquad \lambda = \frac{SL}{3}.$$

The map from the entrance of sextupole 1 to the exit of sextupole 2 is

$$M = e^{:-\frac{L}{2}x'^2+\frac{1}{3}SLx^3:}\,e^{:\frac{L}{2}x'^2:}\,e^{:f_2:}\,e^{:-\frac{L}{2}x'^2+\frac{1}{3}SLx^3:}\,,$$

where $e^{:f_2:}$ is the linear map from the entrance of sextupole 1 to the entrance of sextupole 2 without the sextupoles. We would like to know what is the residual error of the map, to order $\mathcal{O}(L^2)$.

We have

$$M = e^{:f_2:}\left(e^{-:f_2:}e^{:\frac{L}{2}x'^2:}e^{:f_2:}\right)\left(e^{-:f_2:}e^{:-\frac{L}{2}x'^2:}e^{:f_2:}\right)$$
$$\times\left(e^{-:f_2:}e^{:-\frac{L}{2}x'^2+\frac{1}{3}SLx^3:}e^{:f_2:}\right)\left(e^{-:f_2:}e^{:\frac{L}{2}x'^2:}e^{:f_2:}\right)e^{:-\frac{L}{2}x'^2+\frac{1}{3}SLx^3:}\,.$$

Using the fact that

$$e^{-:f_2:}x = -x, \qquad e^{-:f_2:}x' = -x',$$

it follows

$$\begin{aligned}
M &= e^{:f_2:}e^{:\frac{L}{2}x'^2:}e^{:-\frac{L}{2}x'^2:}e^{:-\frac{L}{2}x'^2-\frac{1}{3}SLx^3:}e^{:\frac{L}{2}x'^2:}e^{:-\frac{L}{2}x'^2+\frac{1}{3}SLx^3:}\\
&= e^{:f_2:}e^{:-\frac{L}{2}x'^2:}\left(e^{:\frac{L}{2}x'^2:}e^{:-\frac{L}{2}x'^2-\frac{1}{3}SLx^3:}\right)\left(e^{:\frac{L}{2}x'^2:}e^{:-\frac{L}{2}x'^2+\frac{1}{3}SLx^3:}\right).
\end{aligned}$$

The left-most two exponential factor maps $e^{:f_2:}e^{:-\frac{L}{2}x'^2:}$ give the intended map if the scheme has worked perfectly. The "residual error map" due to the fact that the sextupoles are thick is therefore given by

$$E = \left(e^{:\frac{L}{2}x'^2:}e^{:-\frac{L}{2}x'^2-\frac{1}{3}SLx^3:}\right)\left(e^{:\frac{L}{2}x'^2:}e^{:-\frac{L}{2}x'^2+\frac{1}{3}SLx^3:}\right), \qquad (4.444)$$

and we need to compute E to order $\mathcal{O}(L^3)$. To do so, one writes

$$e^{:\frac{L}{2}x'^2:}e^{:-\frac{L}{2}x'^2+\frac{1}{3}SLx^3:} = e^{:h:}\,,$$

where, h can be found by applying the BCH formula Eq. (4.137) as

$$h = \frac{1}{3}SLx^3 - \frac{1}{2}SL^2x'x^2 + \frac{1}{3}SL^3x'^2x + \frac{1}{12}S^2L^3x^4 + \mathcal{O}(L^4).$$

Inserting this into Eq. (4.444) then yields the 1-D on-momentum error map,

$$
\begin{aligned}
E &= e^{:-\frac{1}{3}SLx^3 + \frac{1}{2}SL^2 x'x^2 - \frac{1}{3}SL^3 x'^2 x + \frac{1}{12}S^2 L^3 x^4:} \\
&\quad \times\ e^{:\frac{1}{3}SLx^3 - \frac{1}{2}SL^2 x'x^2 + \frac{1}{3}SL^3 x'^2 x + \frac{1}{12}S^2 L^3 x^4:} \\
&= e^{:\frac{1}{6}S^2 L^3 x^4 + \mathcal{O}(L^4):}\,.
\end{aligned}
\tag{4.445}
$$

The leading term in the error map is of the order $\mathcal{O}(L^3)$ and it behaves like a thin octupole.

In case a sextupole pair is used in an achromat design and if the sextupoles are strong, it is possible to compensate their residual nonlinear effect by inserting at the beginning or at the end of the sextupole pair section a thin octupole whose kick is (see Homework 4.117)

$$
\Delta x' = -\frac{2}{3}S^2 L^3 x^3\,.
$$

Homework 4.115 Consider the 1-D motion for the system shown in Fig. 4.9(a) with thin sextupoles. What happens if the conditions $\psi = k\pi$ or (4.443) are not satisfied exactly? Consider the case when $k = 1$, (a) what if $(\beta_1/\beta_2)^{3/2}\lambda_1 = \lambda_2 + \Delta\lambda$? (b) what if $\psi = \pi + \Delta\psi$? Give the error map to 1st order in $\Delta\lambda$ and $\Delta\psi$.

Solution

(a) The error map is

$$
E = e^{:-\Delta\lambda x^3:}
\tag{4.446}
$$

The net effect is as if there is a thin sextupole of strength $-\Delta\lambda$ at position 2.

(b) Here we assume condition (4.443) is satisfied. to 1st order in $\Delta\psi$,

$$
\begin{aligned}
M &= e^{:f_2:} \exp\left[: -\lambda_2 x^3 + 3\Delta\psi\lambda_2 x^2(\alpha_2 x + \beta_2 x'):\right] e^{:\lambda_2 x^3:} \\
&= e^{:f_2:} \exp\left[:3\Delta\psi\lambda_2 x^2(\alpha_2 x + \beta_2 x') + \mathcal{O}(\Delta\psi^2):\right].
\end{aligned}
$$

At the exit end of sextupole 2, the trajectory of a particle gets an additional kick, to order $\mathcal{O}(\Delta\psi)$, of

$$
\begin{aligned}
\Delta x &= -3\beta_2\Delta\psi\lambda_2 x^2 + \mathcal{O}(\Delta\psi^2)\,, \\
\Delta x' &= 3\Delta\psi\lambda_2(3\alpha_2 x^2 + 2\beta_2 xx') + \mathcal{O}(\Delta\psi^2)\,.
\end{aligned}
$$

This map however is nonsymplectic if truncated. The symplectic error map requires calculating

$$
\exp\left[:3\Delta\psi\lambda_2 x^2(\alpha_2 x + \beta_2 x'):\right] \begin{bmatrix} x \\ x' \end{bmatrix}
$$

to all orders in $\Delta\psi$. This can be worked out as a continuation of this homework.

Homework 4.116 The text showed that a sextupole pair as shown in Fig. 4.9(a) has a net map that the two sextupoles cancel each other perfectly. What happens if the sextupoles are not too thin and we have the situation in Fig. 4.9(b), and if we make thin-lens approximation of the sextupoles (a drift of $\frac{L}{2}$ followed by a lumped kick followed by another drift of $\frac{L}{2}$)? The map of single sextupole has been given by Eq. (2.38) of Chapter 2, which is symplectic. Do these two sextupoles cancel each other when modeled this way?

Solution It is suggested that you guess at the answer first. Follow the motion through the sextupole pair. Show explicitly that the next result is

$$x_4 = -x_0 - x_0'L, \qquad x_4' = -x_0'.$$

The two sextupoles also cancel each other perfectly in this model.

Homework 4.117 In Homework 4.116, we modeled the sextupoles as thin-lenses. We might also consider the Runge–Kutta approximation, which is more numerically accurate but is nonsymplectic.

Tracing a particle with initial conditions (x_0, x_0') through the thick-sextupole pair system of Fig. 4.9(b), the particle coordinate at the exit of the system, to order $\mathcal{O}(L^3)$, is determined by

$$\begin{cases} x_1 = x_0 + x_0'L + \frac{1}{2}Sx_0^2L^2 + \frac{1}{3}Sx_0x_0'L^3 + \mathcal{O}(L^4), \\ x_1' = x_0' + Sx_0^2L + Sx_0x_0'L^2 + \frac{1}{3}S(x_0'^2 + Sx_0^3)L^3 + \mathcal{O}(L^4), \end{cases}$$

$$\begin{cases} x_2 = -x_1, \\ x_2' = -x_1', \end{cases}$$

$$\begin{cases} x_3 = x_2 - Lx_2', \\ x_3' = x_2', \end{cases}$$

$$\begin{cases} x_4 = x_3 + x_3'L + \frac{1}{2}Sx_3^2L^2 + \frac{1}{3}Sx_3x_3'L^3 + \mathcal{O}(L^4), \\ x_4' = x_3' + Sx_3^2L + Sx_3x_3'L^2 + \frac{1}{3}S(x_3'^2 + Sx_3^3)L^3 + \mathcal{O}(L^4), \end{cases} \qquad (4.447)$$

where $(x_1, x_1'), (x_2, x_2'), (x_3, x_3')$, and (x_4, x_4') are the coordinates of the particle at different stages in passing through the system with (x_4, x_4') the coordinates at the exit of sextupole 2. The connection between (x_1, x_1') and (x_0, x_0') and the connection between (x_4, x_4') and (x_3, x_3') are according to Eq. (2.44) of Chapter 2.

(a) Do the two sextupoles cancel each other in this approximation?

(b) Analyze the problem in Lie language making use of the result Eq. (4.445) in the text. This is legitimate because both the Ronge-Kutta and the error map (4.445) are approximations valid to $\mathcal{O}(L^3)$.

Solution

(a)

$$\begin{cases} x_4 = -x_0 - x_0'L + \mathcal{O}(L^4), \\ x_4' = -x_0' - \frac{2}{3}S^2L^3x_0^3 + \mathcal{O}(L^4). \end{cases} \qquad (4.448)$$

(b) In the Lie language, the map is given by

$$M = e^{:f_2:}e^{-\frac{1}{2}Lx'^2:}E,$$

where E is the error map (4.445). We need to find X_{final} where

$$X_{\text{final}} = MX\Big|_{x=x_0, x'=x'_0}.$$

Note the ordering of factor maps of M is such that earlier maps occur to the left, but when performing the computations, operators to the right are applied first. After computation is completed, x and x' are set to x_0 and x'_0.

We first find

$$E\begin{bmatrix} x \\ x' \end{bmatrix} = \begin{bmatrix} x \\ x' + \tfrac{2}{3}S^2 L^3 x^3 \end{bmatrix} + \mathcal{O}(L^4).$$

This then leads to

$$X_{\text{final}} = M\begin{bmatrix} x \\ x' \end{bmatrix} = \begin{bmatrix} -x_0 - Lx'_0 \\ -x'_0 - \tfrac{2}{3}S^2 L^3 x_0^3 \end{bmatrix},$$

which agrees with Eq. (4.448). Did you expect them to agree?

4.10.6 Beam-beam interaction

One of the sources of nonlinear perturbation in the storage ring collider accelerators is the beam-beam interaction. Particles in one beam, as they circulate around the storage ring, encounter the electromagnetic fields generated by the on-coming beam every time it passes through the point of collision $s = 0$. These fields perturb the motion of the particles being considered. This perturbation is very localized at the point of collision and is very nonlinear. The beam-beam interaction perturbation is one of the main limitations on the beam intensity in colliders. In this section, we use the beam-beam interaction as an illustration of the Lie algebra technique. A more elaborate discussion of the beam-beam problem using non-Lie algebra analyses is to be given in Chapter 8.

Perhaps here a comment can be made in passing about the beam-beam perturbation. How strongly does a nonlinear perturbation influence the beam dynamics in a storage ring basically depends on three ingredients, the strength of the perturbation, how nonlinear it is, and how localized it is. The beam-beam perturbation in a collider is made as strong as possible to maximize the luminosity; it is extremely nonlinear by its nature, and it is sharply localized at the collision point. All three ingredients point to the beam-beam interaction to have a strong impact on beam dynamics.

As should be familiar by now, Lie analysis basically follow three steps.

1. Form the element Lie maps.

2. Concatenate the element maps into a one-turn Lie map using BCH formula. As emphasized, the key step here is to identify the appropriate small parameter to expand the BCH formula with.

3. Normal form analyze the one-turn map.

Consider the 1-D motion only. In the absence of the beam-beam perturbation, the one-turn map of the accelerator around $s = 0$ can be written as

$$\begin{aligned}
x &= x_0 \cos \mu + \beta p_0 \sin \mu \,, \\
p &= -\frac{x_0}{\beta} \sin \mu + p_0 \cos \mu \,,
\end{aligned} \tag{4.449}$$

where β is the β-function at $s = 0$, $\nu = \frac{\mu}{2\pi}$ is the betatron tune, and we have assumed the parameter $\alpha = 0$ at $s = 0$. Both β and μ are the nominal values unperturbed by the beam-beam interaction. In matrix form, we can represent (4.449) by the Courant–Snyder map

$$\begin{bmatrix} \cos \mu & \beta \sin \mu \\ -\frac{1}{\beta} \sin \mu & \cos \mu \end{bmatrix} . \tag{4.450}$$

In Lie algebraic form, we can write the map as

$$e^{:f_2:}, \qquad \text{where} \qquad f_2 = -\frac{\mu}{2}\left(\frac{x^2}{\beta} + \beta p^2\right), \tag{4.451}$$

i.e.,

$$x = e^{:f_2:} x \Big|_{x=x_0, p=p_0}, \qquad \text{and} \qquad p = e^{:f_2:} p \Big|_{x=x_0, p=p_0}.$$

We now introduce the beam-beam interaction. Consider the case when there is only one collision point around the circumference. The beam-beam perturbation is represented as a δ-function kick at $s = 0$,

$$x = x_0 \qquad \text{and} \qquad p = p_0 + f(x_0) \,, \tag{4.452}$$

where $f(x)$ is the nonlinear beam-beam kick. The form of $f(x)$ depends on how particles are distributed in the on-coming beam at the collision point. For example,

$$f(x) = \frac{Nr_0}{\gamma} \begin{cases} \frac{2}{x}\left(1 - e^{-x^2/2\sigma^2}\right), & \text{round Gaussian of rms radius } \sigma, \\ \frac{\sqrt{2\pi}}{W} \int_{-x}^{x} \frac{dx'}{\sigma} e^{-x'^2/2\sigma^2}, & \text{flat Gaussian of width } W, \\ \frac{2x}{a^2} \text{ if } x < a, \quad \frac{2}{x} \text{ if } x > a, & \text{uniform disk of radius } a, \end{cases} \tag{4.453}$$

where N is the number of particles in the on-coming beam, r_0 is the classical radius of the particle, γ is the relativistic Lorentz factor of the particle under consideration (not necessarily equal to that of the on-coming beam). We have assumed the two beam particles have the same sign of charge so that the beam-beam kick is repulsive.

The beam-beam map (4.452) can be written in Lie form as

$$e^{:F:} \qquad \text{where} \qquad F = \int_0^x dx' f(x') \,.$$

The quantity $-F$ is the potential due to the beam-beam force.

Observe the particle motion at the exit of the collision point, the one-turn map of the particle motion, in the Lie formulation, reads

$$e^{:f_2:}e^{:F:}\,.$$

$$(4.454)$$

We now need to identify the small parameter to carry out our analysis. In our previous cases of sextupoles, we have chosen the coordinates of X to be the small parameters — our Taylor expansion are made in terms of $f_2, f_3, f_4, \ldots$, etc. In the treatment of the beam-beam interaction, we will choose the beam-beam kick strength to be the small parameter. The coordinates of X are not considered small so that it is possible to examine the motion of particles in the tails of the beam distribution.

If the beam-beam perturbation is weak, we can apply the BCH formula (4.140) to concatenate the map (4.454) to obtain the effective Hamiltonian of the system. To proceed, we introduce the action-angle variables (ϕ, A) according to

$$x \;=\; \sqrt{2A\beta}\,\sin\phi, \qquad p \;=\; \sqrt{\frac{2A}{\beta}}\,\cos\phi\,,$$

and decompose $F(x)$ as a Fourier series in ϕ as

$$F(x) \;=\; \sum_{n=-\infty}^{\infty} c_n(A)e^{in\phi}\,,$$

with $c_{-n} = c_n^*$.

In previous studies we have $c_n(A)$ always expressed as A to some power and higher powers in A are truncated. This does not have to be the case. For the beam-beam problem, for example, as we will soon see, $c_n(A)$ are more related to the Bessel functions. As just mentioned, we do not consider A to be small here. See also the side comment on page 297.

The function f_2 becomes $f_2 = -\mu A$, and we have the properties

$$:f_2:g(A) \;=\; 0, \qquad :f_2:e^{in\phi} \;=\; in\mu e^{in\phi}\,,$$

for arbitrary function $g(A)$. We then have, to 1st order in the beam-beam perturbation strength,

$$\begin{aligned}
h \;&=\; f_2 + \left(\frac{:f_2:}{1 - e^{-:f_2:}}\right) F + \mathcal{O}(F^2)\\[2mm]
&=\; -\mu A + \sum_n c_n(A)\left(\frac{in\mu}{1 - e^{-in\mu}}\right)e^{in\phi}\\[2mm]
&=\; -\mu A + \sum_n c_n(A)\frac{n\mu}{2\sin\frac{n\mu}{2}}e^{in\phi + i\frac{n\mu}{2}}\,,
\end{aligned}$$

$$(4.455)$$

where the one-turn map (4.454) has been concatenated into the form $e^{:h:}$.

Equation (4.455) is the expression of the beam-beam perturbed invariant ($-h$ is the effective Hamiltonian), to first order of perturbation. Resonances appear at

$$\nu = \frac{\mu}{2\pi} = \frac{p}{n}, \tag{4.456}$$

for all integers p and n, as long as $c_n \neq 0$. Near resonances, the expansion (4.455) diverges due to the small denominators of Eq. (4.455).

If we take the round Gaussian beam distribution in Eq. (4.453), we have

$$F(x) = \frac{Nr_0}{\gamma} \int_0^{A\beta/2\sigma^2} \frac{d\alpha}{\alpha} \left(1 - e^{-2\alpha \sin^2 \phi}\right).$$

The decomposition coefficients c_n are

$$
\begin{aligned}
c_n(A) &= \frac{1}{2\pi} \int_0^{2\pi} d\phi \, e^{-in\phi} F(x) \\
&= \frac{Nr_0}{\gamma} \int_0^{A\beta/2\sigma^2} \frac{d\alpha}{\alpha} \frac{1}{2\pi} \int_0^{2\pi} d\phi \, e^{-in\phi} \left(1 - e^{-2\alpha \sin^2 \phi}\right) \\
&= \frac{Nr_0}{\gamma} \int_0^{A\beta/2\sigma^2} \frac{d\alpha}{\alpha} \frac{1}{2\pi} \int_0^{2\pi} d\phi \, e^{-in\phi} \left[1 - e^{-\alpha} \sum_k I_k(\alpha) e^{2ik\phi}\right] \\
&= \frac{Nr_0}{\gamma} \int_0^{A\beta/2\sigma^2} \frac{d\alpha}{\alpha} \begin{cases} 1 - e^{-\alpha} I_0(\alpha), & \text{if } n = 0, \\ -e^{-\alpha} I_{n/2}(\alpha), & \text{if } n = \text{even} \neq 0, \\ 0, & \text{otherwise.} \end{cases}
\end{aligned}
\tag{4.457}
$$

In the derivation of (4.457), we have used

$$e^{x \cos y} = \sum_{k=-\infty}^{\infty} I_k(x) e^{iky},$$

where $I_k(x)$ is the Bessel function.

One observes that due to the symmetry of the beam-beam force, only even order resonances with even n are excited. Figure 4.10 shows the behavior of c_n as functions of A.

Beam-beam tune shift parameter Away from resonances, one can make a normal form transformation that removes the oscillating terms in Eq. (4.455). The only term left will be the zero-th Fourier harmonic term, and the effective Hamiltonian becomes

$$-h(\phi, A, k) = \mu A - c_0(A),$$

where k is the revolution number serving as the time variable.

The tune shift due to the beam-beam perturbation is given by

$$\Delta\nu = -\frac{1}{2\pi} \frac{dc_0(A)}{dA}. \tag{4.458}$$

Figure 4.10: Beam-beam detuning and resonance strength functions. The vertical scale is $\pm c_n / \frac{Nr_0}{\gamma}$. The horizontal scale is $\frac{A\beta}{2\sigma^2}$.

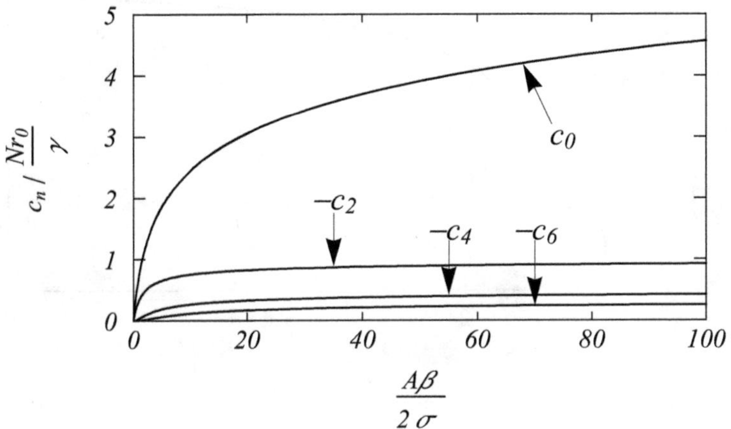

This tune shift is a function of the amplitude A, and A is a constant of the motion away from resonances.

The tune shift (4.458) reads, for a round Gaussian beam,

$$\Delta\nu(A) = -\frac{1}{2\pi}\frac{Nr_0}{\gamma}\frac{d}{dA}\left[\int_0^{A\beta/2\sigma^2}\frac{d\alpha}{\alpha}\left(1 - e^{-\alpha}I_0(\alpha)\right)\right]$$

$$= -\frac{1}{2\pi}\frac{Nr_0}{\gamma A}\left[1 - e^{-A\beta/2\sigma^2}I_0\left(\frac{A\beta}{2\sigma^2}\right)\right].$$

Plotted in Fig. 4.11 is the function

$$\Delta\nu\Big/\frac{Nr_0\beta}{4\pi\gamma\sigma^2} = \frac{1}{x}\left[e^{-x}I_0(x) - 1\right],$$

where $x = \frac{A\beta}{2\sigma^2}$. For small betatron amplitudes, the tune shift is given by

$$\Delta\nu(A=0) = -\frac{Nr_0\beta}{4\pi\gamma\sigma^2}.$$

The overall minus sign comes from the assumption the two colliding beams have the same sign of charge. This parameter is called the beam-beam tune shift parameter, although it refers only to the tune shift of small-amplitude particles. As can be seen in Fig. 4.11, particles with large amplitudes do not experience much tune shift. This is because, unlike forces due to magnet multipoles, the beam-beam force decreases rapidly with amplitude. An inspection of Fig. 4.11 indicates that the beam-beam tune shift also has the physical meaning of the beam-beam induced tune spread of the beam.

Figure 4.11: Beam-beam tune shift.

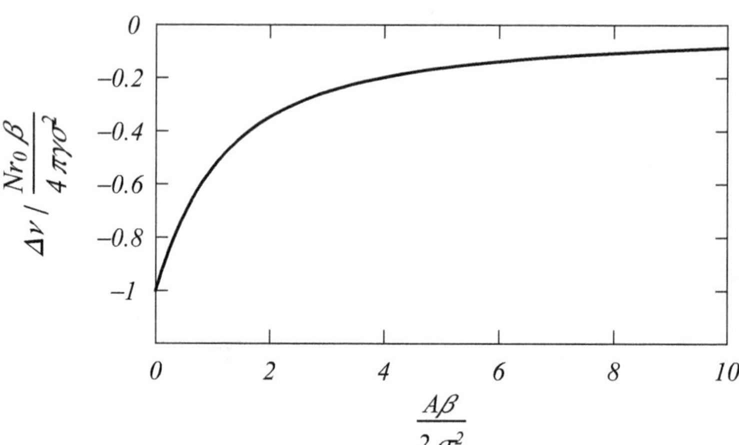

Beam-beam resonance   In case the tune is close to the resonance (4.456) for some integers n and p, let

$$\nu = \frac{p}{n} + d, \quad |d| \ll 1.$$

As mentioned in Sec. 4.8, the expansion (4.455) diverges. One way to proceed is to consider a rotating observer; another way is to consider an n-turn stroboscopic map. Following the development similar to Eqs. (4.415–4.417), the n-turn map can be written as

$$e^{:n\bar{h}:},$$

where $\bar{h}$ (the n-turn effective Hamiltonian is $-n\bar{h}$) is given by

$$\bar{h} = \frac{d}{\nu}f_2 + \frac{d}{\nu}\left(\frac{:f_2:}{1 - e^{-nd:f_2:/\nu}}\right)\left[1 + e^{-:f_2:} + \cdots e^{-(n-1):f_2:}\right]F + \mathcal{O}(F^2).$$

After some algebraic manipulations, we arrive at

$$\bar{h} = -2\pi dA + c_0(A) + \sum_{k=1}^{\infty} c_{kn}(A)\,\frac{2\pi knd}{\sin(\pi knd)}\cos(kn\phi + \pi knd). \qquad (4.459)$$

The expression (4.459) is well-behaved at the resonance, i.e. there is no small-denominator problem when $d \to 0$. As pointed out before, it can be obtained by considering the expression (4.455) and taking the limit approaching the resonance. These results are to be compared with those obtained by the smooth approximation. Once the effective Hamiltonian is obtained, particle motion in the phase space can be explored.

Homework 4.118 This homework serves as a good review of the subject.

(a) Follow the text to fill in the derivation leading to Eq. (4.459).

(b) Assume a reasonable value of the beam-beam tune shift parameter, assume proximity to a 4th order resonance, explore the particle motion in more detail. In particular plot the constant-Hamiltonian contours in the (ϕ, A) phase space.

Solution (b) An example case of the phase space contour plot is shown in Fig. 4.12. Tune shift parameter $\frac{N r_0 \beta}{4\pi\gamma\sigma^2} = 0.01$, and the polar space $(\phi, \frac{\beta A}{2\sigma^2})$ is used for the plot. Cases (a) to (d) are for $d = -0.02, -0.01, 0, 0.01$, respectively. Note that the case when the linear term in the Hamiltonian vanishes is not the case when $d = 0$ but the case when $d = -\frac{N r_0 \beta}{4\pi\gamma\sigma^2}$.

Figure 4.12: An example constant-Hamiltonian contour plot for the case of a beam-beam interaction near a 4th order resonance. Cases (a) to (d) are for $d = -0.02, -0.01, 0, 0.01$, respectively. Figures are plotted in the polar space $(\phi, \frac{\beta A}{2\sigma^2})$.

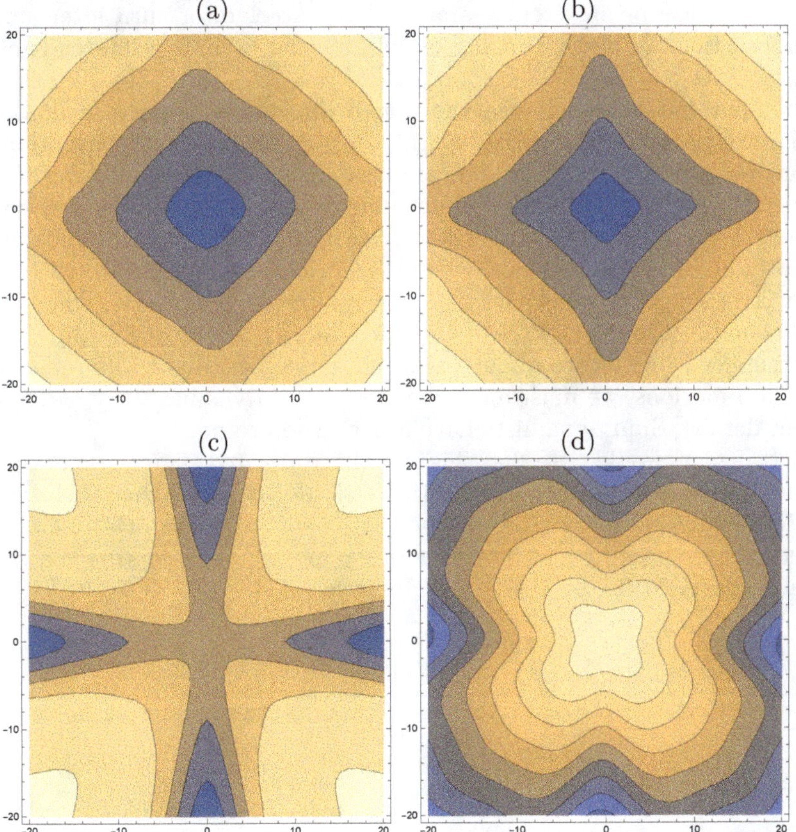

Chapter 5

Spin Dynamics of Proton

To study the spin effects in particle and nuclear physics, it often requires beams with high degree of spin polarization. For example, it is desired to have all beam particles in a circular collider align their spins, and furthermore keep this alignment as the beam circulates around the accelerator turn after turn. To accomplish this, an understanding of spin motion in an accelerator is of great interest.[1]

The spin can be considered the fourth dimension of particle motion. In addition to the six orbital variables $(x, x', y, y', z, \delta)$ in a more conventional 3-D motion, we add two more dynamical variables for the spin, making the total phase space 8-dimensional. The only difference is that a particle's dynamics of the spin variables depend sensitively on its orbital variables, but dynamics of its orbital variables does not depend (ignoring terms proportional to $\hbar$ such as the Stern–Gerlach effect) on its spin variables. As a result, when we care only about the orbital motion of the particles, we ignore their spins. On the other hand, when we care about their spin dynamics, due to its sensitivity to the orbital motions, we must analyze their orbital dynamics very carefully and in great details. Spin dynamists have a harder job to do.

Alert readers would note that this must have violated symplecticity condition of Hamiltonian systems — a simplistic version of it states that if A affects B, then B must also affect A. Indeed, the part of spin affecting orbital motion will be ignored here because it is numerically too minuscule, accepting the violation of symplecticity in the process.[2] See Homework 5.4.

---

[1] For a review of beam polarization, see for example Sec. 2.6, Handbook Accel. Phys. & Eng., 2nd ed., World Scientific (2013).

[2] One way to recover the symplecticity using Lie algebra is discussed in K. Yokoya, Nucl. Instr. Meth. A258, 149 (1987).

5.1 Thomas–BMT equation

Spin of a particle interacts with an electromagnetic field through the magnetic moment associated with its spin. Let $\hbar\vec{S}$ be the spin of a particle at rest represented as a 3-D vector. The associated magnetic moment is

$$\vec{\mu} = \frac{ge}{2m}\hbar\vec{S}, \tag{5.1}$$

where g, a fundamental physical dimensionless constant, is the gyromagnetic ratio of the particle under consideration; $\frac{e}{m}$ is the charge-to-mass ratio of the particle under consideration. We use the SI unit system.

Underlying Eq. (5.1) is the statement that the magnetic moment of a particle is fixed on top of its spin; $\vec{\mu}$ and $\vec{S}$ are always aligned with each other and they are always connected as Eq. (5.1) dictates under all circumstances.

For a Dirac particle, g is nominally equal to 2. The deviation of g from 2, attributed to an "anomalous" magnetic moment of the particle, is specified by the parameter

$$a = \frac{g-2}{2}.$$

For electrons and muons, according to first order perturbative quantum electrodynamics, the value of a is approximately equal to the fine structure constant $\alpha_F \approx \frac{1}{137.036}$ divided by 2π. But this approximation breaks down badly for hadronic particles such as protons and deuterons when the electromagnetic force is overwhelmed by the strong force. More accurately, the measured values are

$$a = \begin{cases} 0.00115965219, & \text{electron, positron,} \\ 0.001165923, & \text{muon,} \\ 1.79284739, & \text{proton, antiproton,} \\ -0.1429878, & \text{deuteron.} \end{cases}$$

It should be mentioned in passing that in Eq. (5.1), the natural unit of the magnetic moment is $\frac{e\hbar}{2m}$, the Bohr magneton. This means the low-massed electrons have a much larger natural magnetic moment than the high-massed protons, in spite of the facts that their g factors are of the same order and that they have exactly the same spin $\frac{\hbar}{2}$.

Consider a particle at rest in a magnetic field $\vec{B}$. In this frame, the spin of the particle simply precesses around the magnetic field $\vec{B}$. What happens is that the particle's angular momentum $\hbar\vec{S}$, its spin, is experiencing a torque $\vec{\mu} \times \vec{B}$ in the magnetic field, i.e.

$$\frac{d}{dt}\left(\hbar\vec{S}\right) = \vec{\mu} \times \vec{B}.$$

As the spin is locked to the magnetic moment, the torque due to the magnetic moment causes the spin to precess. The precession equation of motion for the spin then follows,

$$\frac{d\vec{S}}{dt} = \vec{\Omega} \times \vec{S}, \tag{5.2}$$

with the precession angular velocity

$$\vec{\Omega} = -\frac{ge}{2m}\vec{B}\,. \tag{5.3}$$

In the rest frame, spin of a particle precesses according to magnetic fields; it pays no attention to electric fields if any.

Equations (5.2–5.3) describe the spin precession for a stationary particle, but we need an equation for a relativistic particle moving in an electromagnetic field $\vec{E}$ and $\vec{B}$ in an accelerator. Let $c\vec{\beta}$ be the instantaneous velocity of the particle; it is obvious that we need to make a Lorentz transformation to the particle's rest frame. When doing so, the form of the spin precession equation remains to be (5.2); only $\vec{\Omega}$ needs to be transformed.

An important point is to be made here. Note that we are not Lorentz transforming $\vec{S}$, so in the final equation, $\vec{S}$ will be a quantity in the particle's rest frame, while all other quantities, $t, \vec{E}, \vec{B}, \vec{\beta}$ refer to the laboratory frame. The equation we shall derive momentarily therefore mixes quantities in two different reference frames![3] This is a unique peculiarity of the Thomas–BMT equation;[4] it is not formally Lorentz covariant. A covariant description of spin motion of course does exist, but we don't need it here.

The magnetic field in the rest frame is given by a Lorentz transformation from the laboratory frame,

$$\vec{B}_R = \gamma\vec{B}_\perp + \vec{B}_\| - \frac{\gamma\vec{\beta}}{c} \times \vec{E}\,, \tag{5.4}$$

with $\vec{B}_\perp$ and $\vec{B}_\|$ the components of $\vec{B}$ perpendicular and parallel to $\vec{\beta}$, respectively in the laboratory frame. Note that the electric field $\vec{E}$ in the laboratory frame contributes to the magnetic field $\vec{B}_R$ in the rest frame. There is also an electric field $\vec{E}_R$ in the rest frame, but as mentioned, we do not care about it.[5] Also note that the transverse components of $\vec{B}$ and $\vec{E}$ have magnified effect by a factor γ, while $\vec{B}_\|$ does not, and $\vec{E}_\|$ has no effect at all.

The angular velocity $\vec{\Omega}$ consists of two terms. The first term follows from Eq. (5.3),

$$-\frac{1}{\gamma}\frac{ge}{2m}\vec{B}_R\,, \tag{5.5}$$

where $\vec{B}_R$ is given by Eq. (5.4), and we have included an extra factor of $\frac{1}{\gamma}$ to take care of the time dilation.

---

[3] A deeper reason allowing this to be tolerated — or in fact even preferred — is due to the fact that we actually only care to know $\vec{S}$ in the particle's rest frame — it is the spin in the rest frame that plays a role in a particle collision experiment.

[4] L.H. Thomas, Phil. Mag. 3,1 (1927); L.H. Thomas, Conf. High Energy Spin Physics, AIP Proc. 95, p. 4 (1982); V. Bargmann, L. Michel, and V.L. Telegdi, Phys. Rev. Lett. 2, 435 (1959); S.Y. Lee, Spin Dynamics and Snakes in Synchrotrons, World Scientific (1997); J.D. Jackson, Classical Electrodynamics, 3rd ed., Wiley, N.Y. (1998).

[5] Observe that $\vec{E}_R$ acts on electric monopole but ignores magnetic dipole; while $\vec{B}_R$ acts on magnetic dipole but ignores electric monopole.

But there is also a second more subtle term; it is due to Thomas precession which contributes an additional term to angular velocity when the particle is accelerated *sideways,*

$$-\frac{\gamma-1}{\beta^2}\vec{\beta}\times\dot{\vec{\beta}} = -\frac{e\gamma}{m(\gamma+1)}\left(\frac{\vec{\beta}}{c}\times\vec{E}-\beta^2\vec{B}_\perp\right), \tag{5.6}$$

where the last step is obtained after considering the Lorentz force equation

$$\frac{d}{dt}(m\gamma c\vec{\beta}) = e(\vec{E}+c\vec{\beta}\times\vec{B}).$$

Adding the two contributions (5.5) and (5.6), we obtain

$$\vec{\Omega} = -\frac{e}{m}\left[\left(a+\frac{1}{\gamma}\right)\vec{B}-\frac{a\gamma}{\gamma+1}\vec{\beta}(\vec{\beta}\cdot\vec{B})-\left(a+\frac{1}{\gamma+1}\right)\frac{\vec{\beta}}{c}\times\vec{E}\right]. \tag{5.7}$$

which, when substituted into (5.2), is called the Thomas–BMT equation, where BMT stands for Bargmann, Michel, and Telegdi (1959).

Thomas precession Where does this extra Thomas precession (5.6) come from? The origin of Thomas precession is relativistic kinematics — pure kinematics, no dynamics. Two successive Lorentz transformations along $\vec{\beta}_1$ and $\vec{\beta}_2$ can be combined into one single Lorentz transformation only if $\vec{\beta}_1 \parallel \vec{\beta}_2$. Otherwise, they can be combined into a Lorentz transformation plus a rotation. This additional rotation needed here is along the axis $\vec{\beta}_1 \times \vec{\beta}_2$ and is the origin of the Thomas precession.

Perhaps a comment can help here. As we know, Lorentz transformation along $\vec{\beta}$ transforms longitudinal and transverse components of a vector differently. (Longitudinal and transverse refer to the direction of $\vec{\beta}$.) If $\vec{\beta}_1 \parallel \vec{\beta}_2$, there is no mixing among the transverse and longitudinal components and the combined transformation remains Lorentzian. If $\vec{\beta}_1$ and $\vec{\beta}_2$ are not in the same direction, however, there must be some unpleasant mixing among these longitudinal and the transverse components in the two successive Lorentz transformations. You can imagine that this mixing is rather complicated and will require pages of algebra to sort it out. The beauty of Thomas precession in fact is that this complicated mixing can be nicely summarized by a coordinate rotation.

The reason Thomas precession has the form $\vec{\Omega} \propto \vec{\beta} \times \dot{\vec{\beta}}$ should now be clear. Consider two reference frames at a time t and a slightly later time $t + \Delta t$. The later frame has rotated relative to the earlier one by an angle $\propto \vec{\beta} \times (\vec{\beta} + \dot{\vec{\beta}}\,\Delta t) = \vec{\beta} \times \dot{\vec{\beta}}\,\Delta t$. The reference frame's rotation angular velocity therefore is $\propto \vec{\beta} \times \dot{\vec{\beta}}$. See also Homework 5.3.

Homework 5.1 Familiarize with spin motion considering this homework.

(a) Referring to Eq. (5.2) and Fig. 5.1, show that the precession of $\vec{S}$ around $\vec{\Omega}$ obeys the right-hand rule.

Figure 5.1: Spin precession of a particle in its rest frame.

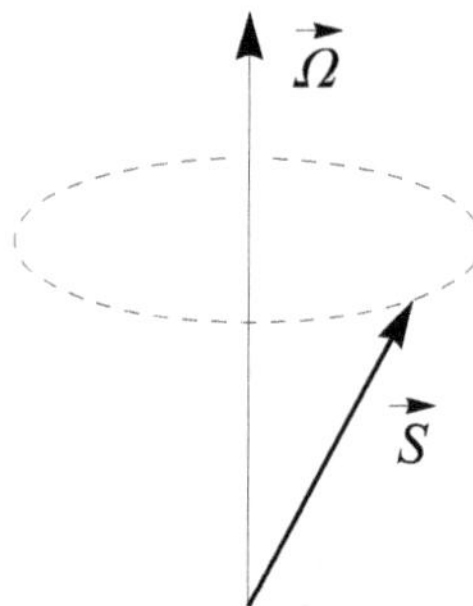

(b) Consider an electron, a positron, a muon, a proton, and a deuteron, all embedded at rest in a magnetic field $\vec{B}$. Do all particles precess right-handedly and/or left-handedly around $\vec{B}$ regardless of the signs of their electric charge? Which particle precesses most rapidly, and which most slowly?

(c) A curiosity question: The deuteron has $a < 0$. Can $a < -1$ for some particles? In those cases — most likely a complex nuclei, the magnetic moment of the particle has a curious opposite sign to its spin, thus violating the right-hand rule of (a).

Homework 5.2 It would be instructive to follow the steps in the text and derive the Thomas–BMT equation. Learn some subtleties in the process, e.g. just to name one, learn that the longitudinal magnetic field $\vec{B}_{\parallel}$ contributes to spin precession but not to Thomas precession. Verify that the left-hand-side can indeed be expressed by the right-hand-side.

Homework 5.3 It was mentioned in the text that two successive Lorentz transformations can be combined into a single Lorentz transformation together with a spatial rotation, and this additionally required rotation was the source of Thomas precession. It was also pointed out that this additional rotation is not needed when the reference frames move collinearly. Consider three reference frames K_0, K_1, K_2, where K_1 moves with velocity $\vec{v}_a = \beta_a c \hat{x}$ relative to K_0, and K_2 moves with a perpendicular velocity $\vec{v}_b = \beta_b c \hat{y}$ relative to K_1.

(a) Find the Lorentz transformations explicitly in terms of the coordinate transformations from (x_0, y_0, z_0, t_0) to (x_1, y_1, z_1, t_1) and from (x_1, y_1, z_1, t_1) to (x_2, y_2, z_2, t_2).

(b) Find the net transformation from (x_0, y_0, z_0, t_0) to (x_2, y_2, z_2, t_2).

(c) Find the velocity of K_2 as observed relative to K_0, and find the corresponding Lorentz transformation from K_0 to K_2.

(d) Show that the transformation obtained in (b) is not the same as that obtained in (c). But there is an additional spatial rotation required to match them. Find the rotation angle required.

(e) Link the result obtained in (d) to the Thomas precession angular velocity Eq. (5.6).

Solution Let $\gamma_a = \frac{1}{\sqrt{1-\beta_a^2}}, \gamma_b = \frac{1}{\sqrt{1-\beta_b^2}}$.

(a)

$$\begin{bmatrix} x_1 \\ y_1 \\ z_1 \\ ct_1 \end{bmatrix} = \begin{bmatrix} \gamma_a(x_0 - \beta_a ct_0) \\ y_0 \\ z_0 \\ \gamma_a(ct_0 - \beta_a x_0) \end{bmatrix}, \qquad \begin{bmatrix} x_2 \\ y_2 \\ z_2 \\ ct_2 \end{bmatrix} = \begin{bmatrix} x_1 \\ \gamma_b(y_1 - \beta_b ct_1) \\ z_1 \\ \gamma_b(ct_1 - \beta_b y_1) \end{bmatrix}.$$

(b)

$$\begin{bmatrix} x_2 \\ y_2 \\ z_2 \\ ct_2 \end{bmatrix} = \begin{bmatrix} \gamma_a(x_0 - \beta_a ct_0) \\ \gamma_b y_0 + \beta_b \gamma_a \gamma_b(\beta_a x_0 - ct_0) \\ z_0 \\ \gamma_a \gamma_b(ct_0 - \beta_a x_0) - \beta_b \gamma_b y_0 \end{bmatrix}.$$

Needless to say, the transformation can be written in a matrix form $X_2 = R X_0$ with $\det R = 1$.

(c) The velocity of K_2 as observed in K_0 is $\vec{v}_c = \beta_a c \hat{x} + \frac{\beta_b c}{\gamma_a} \hat{y}$.

(d) The required additional rotation is

$$\begin{bmatrix} \frac{\gamma_a + \gamma_b}{1 + \gamma_a \gamma_b} & -\frac{\beta_a \gamma_a \beta_b \gamma_b}{1 + \gamma_a \gamma_b} & 0 & 0 \\ \frac{\beta_a \gamma_a \beta_b \gamma_b}{1 + \gamma_a \gamma_b} & \frac{\gamma_a + \gamma_b}{1 + \gamma_a \gamma_b} & 0 & 0 \\ 0 & 0 & 1 & 0 \\ 0 & 0 & 0 & 1 \end{bmatrix}.$$

(e) Consider a particle moving in the x-direction and accelerated sideways in the y-direction for an infinitesimal time period Δt. The rotation angle as observed in the rest frame of the particle can be found to be

$$\Delta \theta = -\frac{\beta_a \gamma_a}{1 + \gamma_a} \dot{\beta} \Delta t.$$

The precession angular frequency is then $\frac{1}{\gamma_a} \frac{\Delta \theta}{\Delta t}$, where the factor γ_a is time dilation back to the laboratory frame.

5.2 Spin motion in an accelerator

To describe the spin motion in a circular accelerator, it is more convenient to change the time variable t to the distance variable $s = \beta ct$ of a reference particle in the beam. In an accelerator, the electric and magnetic fields seen by a particle at a given position s depend on the particle's orbital coordinates (x, y, z), all measured relative to the reference particle. For example, a particle with a transverse displacement will see a focusing or defocusing magnetic field in a quadrupole magnet. A particle on the reference orbit, however, sees no fields in a quadrupole. It sees only the bending magnetic field in the dipole magnets

Figure 5.2: Spin rotation going through a single bending magnet.

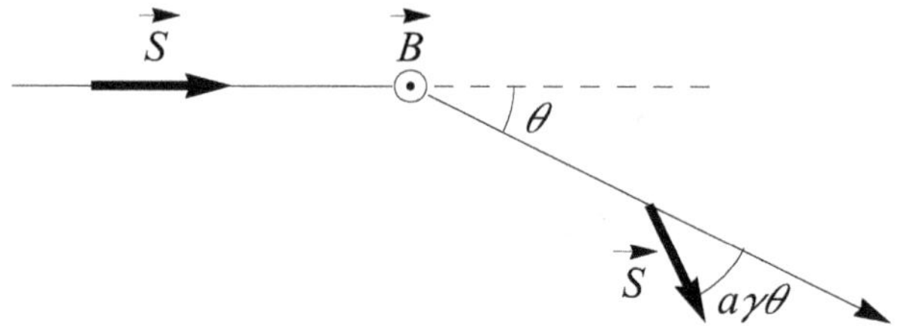

and the accelerating electric field in the RF cavities. The accelerating electric field does not contribute to spin precession for the reference particle, according to Eq. (5.7), because it is parallel to the velocity $\vec{\beta}$. The bending magnetic field

$$\vec{B} = B_0(s)\hat{y}, \qquad \text{with} \qquad B_0(s + 2\pi R) = B_0(s),$$

on the other hand, does give rise to a precession, according to Eq. (5.7),

$$\frac{d\vec{S}}{ds} = -\frac{eB_0(s)}{P_0}(1 + a\gamma)\hat{y} \times \vec{S}, \tag{5.8}$$

where $P_0 = mc\beta\gamma$. In a planar storage ring, all bending magnetic fields are aligned in the $\hat{y}$ direction. With the precession axis always along $\hat{y}$, the y-component of the spin S_y is preserved.

The spin rotation going through a single bending magnet is sketched in Fig. 5.2. The orbital trajectory of the particle is bent by an angle θ around $\vec{B}$ due to the bending action. In the mean time, the spin precesses also around $\vec{B}$ by an angle $(1 + a\gamma)\theta$, according to Eq. (5.8). Relative to the bent trajectory, the spin has precessed by an angle $a\gamma\theta$. Note that the quantity we care most about is not the total spin precession angle but the precession angle relative to the instantaneous design trajectory.

If we adopt the coordinate system $(\hat{x}, \hat{y}, \hat{z})$ that rotates with the design trajectory of the reference particle, then in this rotating frame, the spin components S_x and S_z rotate with the angular speed $a\gamma\frac{eB_0c}{P_0}$ which is $a\gamma$ times the speed at which the coordinate system rotates. As the particle completes one turn, the coordinate system rotates by 2π and the spin of the reference particle precesses around $\hat{y}$ by an angle $2\pi a\gamma$ relative to the coordinate system. In analogy to the definition of orbital horizontal betatron, vertical betatron and synchrotron tunes ν_x, ν_y, ν_s, we define a spin tune,

$$\nu_{\text{spin}} = a\gamma = \begin{cases} E_0/(0.440648 \text{ GeV}), & \text{electron, positron,} \\ E_0/(90.6221 \text{ GeV}), & \text{muon,} \\ E_0/(0.523342 \text{ GeV}), & \text{proton, antiproton,} \\ -E_0/(13.117 \text{ GeV}), & \text{deuteron,} \end{cases} \tag{5.9}$$

where E_0 means the total energy (not the kinetic energy) of the particle.

According to Eq. (5.8), a pure Dirac particle will not spin precess in a circular accelerator; its direction is frozen with its orbital trajectory. The spin precession comes about only if the particle's anomalous magnetic moment $a \neq 0$.

We so far considered the spin motion of the reference particle following the design trajectory. We next consider a nonideal particle executing some transverse motion in an ideal planar accelerator. This particle sees a magnetic field in the laboratory frame,

$$\vec{B} = G(s)(y\hat{x} - x\hat{y}) + B_0(s)\hat{y},$$

where $G(s)$ is the quadrupole strength, and an electric field

$$\vec{E} = E_z(s)\hat{z}.$$

Substituting into Eq. (5.7) yields

$$\vec{\Omega} = -\frac{e}{m}\left[\left(a+\frac{1}{\gamma}\right)B_0(s)\hat{y} + \left(a+\frac{1}{\gamma}\right)G(s)(y\hat{x} - x\hat{y}) - a\left(1-\frac{1}{\gamma}\right)y'B_0(s)\hat{z}\right.$$
$$\left. + \left(a+\frac{1}{\gamma+1}\right)\frac{\beta}{c}E_z(s)(x'\hat{y} - y'\hat{x})\right],$$

where we have used $\vec{\beta} \approx \beta(\hat{z} + x'\hat{x} + y'\hat{y})$ and kept terms only up to linear order in x, x', y, y'. We see that because the orbital motion now acquires angles x' and y', the electric field also contributes to spin precession.

We now change the "time" variable from s to

$$\theta \equiv \int_0^s \frac{ds'}{\rho(s')} = \text{accumulated bending angle},$$

(θ advances by 2π per revolution) and obtain

$$\frac{d}{d\theta}\vec{S} = \vec{k} \times \vec{S}, \tag{5.10}$$

where

$$\vec{k} = -\left[(a\gamma+1)\,\hat{y} + (a\gamma+1)\left(\frac{G}{B\rho}\right)\rho(y\hat{x} - x\hat{y}) - a\,(\gamma-1)\,y'\hat{z}\right.$$
$$\left. + \left(a\gamma + \frac{\gamma}{\gamma+1}\right)\frac{\beta}{c}\left(\frac{E_z}{B\rho}\right)\rho(x'\hat{y} - y'\hat{x})\right], \tag{5.11}$$

with $B\rho = \frac{p_0}{e} = $ magnetic rigidity of the design particle.

We ignore the last term, the term involving electric field, in Eq. (5.11) in the following unless otherwise noted because its contribution is typically small. To arrive at this conclusion, we compare the two relevant terms $(a\gamma+1)\left(\frac{G}{B\rho}\right)\rho y$ and $\left(a\gamma + \frac{\gamma}{\gamma+1}\right)\frac{\beta}{c}\left(\frac{E_z}{B\rho}\right)\rho y'$, and then observe that typically, $y' \sim \frac{y}{\beta_y}$ (β_y is the typical vertical beta-function) and

$$G\beta_y \gg \frac{E_z}{c}.$$

Equations (5.10) and (5.11) describe the spin motion relative to a fixed coordinate system $\hat{x}, \hat{y}, \hat{z}$. But we want to describe it in a rotating frame with

$$\frac{d\hat{x}}{ds} = \frac{1}{\rho}\hat{z}, \qquad \frac{d\hat{y}}{ds} = 0, \qquad \frac{d\hat{z}}{ds} = -\frac{1}{\rho}\hat{x}.$$

In this rotating frame, the equation of spin precession motion is given by

$$\frac{d}{d\theta}\vec{S} = \vec{h} \times \vec{S}, \tag{5.12}$$

where the angular velocity is

$$
\begin{aligned}
\vec{h} &= \vec{k} + \hat{y} = h_x\hat{x} + h_y\hat{y} + h_z\hat{z}, \\
h_x &= -(a\gamma + 1)\left(\frac{G}{B\rho}\right)\rho y, \\
h_y &= -a\gamma + (a\gamma + 1)\left(\frac{G}{B\rho}\right)\rho x, \\
h_z &= a(\gamma - 1)y'.
\end{aligned}
\tag{5.13}
$$

Among all the terms in $h_{x,y,z}$ of Eq. (5.13), the leading term is the $-a\gamma$ term in h_y. It is the spin precession seen by the ideal reference particle. All other terms contain x or y, and are seen only by particles deviating from the ideal trajectory.

The spin dynamics described by Eqs. (5.12) and (5.13) assumes the orbital motion x, x', y, y' are given. These orbital coordinates are supposed to be calculated by Lorentz force equation without taking into account of the spin. After the orbital motion is solved, we apply the result to the spin motion using the Thomas–BMT equation, in a perturbative approach,

$$
\begin{aligned}
\text{orbital motion:} &\quad \text{Lorentz equation,} \\
\text{spin motion:} &\quad \text{Thomas–BMT equation.}
\end{aligned}
$$

In other words, we have made an approximation that orbital motion affects the spin while the spin does not affect orbital motion. As mentioned earlier, this violates symplecticity. The justification of this approximation lies in the extreme weakness of the spin influence, for example the Stern–Gerlach effect,[6] on the orbital motion.

Electrostatic storage ring We have been ignoring $\vec{E}$ because most likely $\vec{E}$ is in the longitudinal direction, and is approximately parallel to $\vec{\beta}$, which in turn means it does not contribute much to spin precession. However, if we use a transverse $\vec{E}$, e.g. in electrostatic beam separators or electrostatic lenses, then we must take into account of its effects. Transverse electric fields also appear when we consider the space charge, beam-beam or collective wakefield effects.

---

[6]W. Gerlach, O. Stern, Zeitschrift für Physik, 9 (1), 349 (1922).

In the presence of both $\vec{B}$ and $\vec{E}$, we first show that the orbital motion contains a rotational motion described by (Homework 5.5)

$$\dot{\hat{\beta}} = \vec{\Omega}_{\text{orb}} \times \hat{\beta}, \qquad (5.14)$$

with $\hat{\beta} = \frac{\vec{\beta}}{\beta}$, and

$$\vec{\Omega}_{\text{orb}} = -\frac{e}{m}\left(\frac{\vec{B}}{\gamma} - \frac{1}{\gamma\beta c}\hat{\beta}\times\vec{E}\right). \qquad (5.15)$$

Note that in equations like (5.12) and (5.14) that describe precession or rotation, they are physically meaningful only when the vector being rotated has a constant magnitude. In (5.12), $|\vec{S}|$ has a constant magnitude. In (5.14), $|\hat{\beta}| = 1$. For instance, it would not make sense if Eq. (5.14) reads like $\dot{\vec{\beta}} = \vec{\Omega}_{\text{orb}} \times \vec{\beta}$.

Consider a special storage ring whose bending and focusing are provided purely by static transverse electric fields — transverse bending electroplates and electrostatic focusing lenses. The spin precession is given by, according to Eq. (5.7),

$$\vec{\Omega} = \frac{e}{m}\left(a + \frac{1}{\gamma+1}\right)\frac{\beta}{c}\hat{\beta}\times\vec{E},$$

while the orbital rotation is given by

$$\vec{\Omega}_{\text{orb}} = \frac{e}{m}\frac{1}{\gamma\beta c}\hat{\beta}\times\vec{E}.$$

It follows a clever idea that there is a magic energy when the spin precession coincides with the orbital rotation, i.e. the spin is frozen relative to the particle's orbital motion.[7] This magic energy occurs when

$$\left(a + \frac{1}{\gamma+1}\right)\beta = \frac{1}{\gamma\beta},$$

or

$$\gamma = \sqrt{1 + \frac{1}{a}}. \qquad (5.16)$$

The magic energy for different particles are

$$mc^2\gamma = \begin{cases} 15.0143926 \text{ MeV}, & \text{electron}, \\ 1171.0642 \text{ MeV}, & \text{proton}, \\ 3096.150 \text{ MeV}, & \text{muon}. \end{cases}$$

For muons, for example, this magic energy is about 3.1 GeV. It follows that if one uses electrostatic bends instead of dipole magnets and electrostatic lenses

---

[7]See for example F. Farley, Progress in Particle & Nuclear Physics, 52 (1): 1 (2004). For further reading on the application to detect electric dipole moments using electrostatic storage rings, see e.g. R. Talman, Rev. Accel. Sci. Tech., Vol. 10, 267 (2019).

instead of quadrupole magnets for such a muon storage ring at this magic energy, one can attempt to make precise measurement of the muon's $g{-}2$ value.

An incidental comment is in order. We mentioned earlier that in a purely magnetic ring, the spin rotation coincides with the coordinate rotation if the anomalous gyromagnetic ratio $a = 0$. That case is not to be confused with the present case which applies to an all-electric ring which works only when $a \neq 0$.

Homework 5.4 Consider a proton of kinetic energy of 1 GeV traversing a quadrupole magnet whose pole tip field is 1 Tesla at pole tip radius of 5 cm. Compute the Lorentz force the proton experiences as it traverses the magnet at a radius of 1 cm. Then compute the Stern–Gerlach force,

$$ F \;=\; \frac{e\hbar g_p}{4m_p}\, G\,, \tag{5.17} $$

in the same quadrupole magnet, where $g_p = 5.5856$ is the proton gyromagnetic ratio and G is the quadrupole gradient. Compare their magnitudes.

Solution You should first try to derive Eq. (5.17) from first principles. The two forces are about 13 orders of magnitude apart. In order for the spin effect to become significant, its effect needs to be accumulated coherently over at least $\sim 10^{13}$ revolutions, and here lies the difficult challenge of measuring $g - 2$ of elementary particles using electrostatic storage rings, as mentioned in the text.

Homework 5.5
    (a) Derive Eqs. (5.14) and (5.15).
    (b) Derive Eq. (5.16).

Solution (a) First note that $\hat{\beta} = \frac{\vec{\beta}}{\beta} = \frac{\vec{p}}{p}$. Show that $\dot{p} = \hat{\beta} \cdot e\vec{E}$. Then, it follows

$$
\begin{aligned}
\dot{\hat{\beta}} &= \frac{1}{p}\dot{\vec{p}} - \frac{\dot{p}}{p^2}\vec{p} \;=\; \frac{1}{p}\dot{\vec{p}} - \frac{\dot{p}}{p}\hat{\beta} \\[4pt]
&= \frac{1}{p}\left(e\vec{E} + ec\vec{\beta} \times \vec{B}\right) - \frac{e}{p}(\hat{\beta} \cdot \vec{E})\hat{\beta} \\[4pt]
&= \frac{ec}{p}\vec{\beta} \times \vec{B} + \frac{e}{p}\vec{E}_\perp\,,
\end{aligned}
$$

which in turn yields Eqs. (5.14) and (5.15).

Homework 5.6 We have not considered solenoids. Show that the spin precession going through a solenoid $\vec{B} = B_s\hat{z}$ is a rotation around $\hat{z}$ by an angle

$$ -\frac{eB_s\ell}{P_0}(1+a) \;=\; -\frac{B_s\ell}{(B\rho)}(1+a) \;=\; -\frac{B_s\ell}{(B\rho)}\frac{g}{2}\,. $$

Observe and explain the minus sign.

Homework 5.7 It may be useful to have a device which rotates the spin without affecting the particle trajectory. One such device is solenoid $\vec{B} = B_s\hat{z}$. Another device, called the Wien filter, consists of transverse $\vec{E}$ and $\vec{B}$ with $\vec{E} = -\beta c\hat{z} \times \vec{B}$, where we have assumed the nominal direction of beam motion is along $\hat{z}$. Show that the beam trajectory is unperturbed, while its spin precession is given by

$$\vec{\Omega} = -\frac{eg}{2m\gamma^2}\vec{B}.$$

The factor of $\frac{1}{\gamma^2}$ indicates this device works mainly for nonrelativistic particles. Still more examples of using solenoids for spin manipulation will be discussed when we talk about Siberian snakes.

5.3 Spinor algebra

5.3.1 The spinor

To describe a spin, one can consider it as a 3-D vector, and use Eqs. (5.10) or (5.12) to describe the evolution of the three vector components $S_{x,y,z}$. Alternatively, one can also use a spinor algebra language borrowed from quantum mechanics to describe it. To proceed, one first introduces the Pauli matrices,

$$\sigma_x = \begin{bmatrix} 0 & -i \\ i & 0 \end{bmatrix}, \qquad \sigma_y = \begin{bmatrix} 1 & 0 \\ 0 & -1 \end{bmatrix}, \qquad \sigma_z = \begin{bmatrix} 0 & 1 \\ 1 & 0 \end{bmatrix}. \tag{5.18}$$

We will discuss how to apply spinor algebra for spin dynamics in the next section 5.3.2. In this section, we only want to introduce the spinors and their algebra for later applications.

We make a few comments before proceeding.

- We have made a cyclic permutation of the usual convention of the Pauli matrices. This permuted definition is more natural for us because we have $\hat{y}$, instead of the conventional $\hat{z}$, as the rotation axis, at least for the ideal reference particle. Such a cyclic permutation only means a change of representation model and does not affect any physical results.

- Both the 3-D vector and the 2-D complex spinor are representations of a spin. There is one and the same spin, and we have two ways to describe it. These two descriptions are equivalent yielding identical physical contents.

- The Pauli matrices are a way to represent taking square roots. For a particle with spin $\frac{\hbar}{2}$, spin is a quantum number it carries to represent its double-valued phase space (spin up and spin down). In a sense, one is required to take the square root of the unit matrix I, and there are four solutions: $I, \sigma_x, \sigma_y, \sigma_z$ — note that we are talking about square roots of I, not square roots of $-I$. The Pauli matrices are a mathematical representation to complete these four solutions.

The algebra　It is straightforward to show that

$$\sigma_i^\dagger = \sigma_i, \qquad \text{i.e. } \sigma_i \text{ is Hermitian,}$$

$$\sigma_i^\dagger \sigma_i = \sigma_i^2 = I, \qquad \text{i.e. } \sigma_i \text{ is unitary,}$$

$$\vec{\sigma}^\dagger \cdot \vec{\sigma} = 3I,$$

$$\sigma_x \sigma_y = -\sigma_y \sigma_x = i\sigma_z, \quad \sigma_y \sigma_z = -\sigma_z \sigma_y = i\sigma_x, \quad \sigma_z \sigma_x = -\sigma_x \sigma_z = i\sigma_y,$$

$$\sigma_i \sigma_j = \delta_{ij} I + i \sum_k \epsilon_{ijk} \sigma_k,$$

$$\left[\frac{\sigma_i}{2}, \frac{\sigma_j}{2} \right] = i \sum_k \epsilon_{ijk} \frac{\sigma_k}{2}$$

$$\text{(commutator of angular momentum components)},$$

$$\vec{\sigma} \times \vec{\sigma} = 2i\vec{\sigma},$$

$$\det(\sigma_i) = -1,$$

$$\text{tr}(\sigma_i) = 0,$$

$$\det(\vec{\sigma} \cdot \vec{a}) = -|\vec{a}|^2,$$

$$(\vec{\sigma} \cdot \vec{a})^2 = |\vec{a}|^2 I,$$

$$(\vec{\sigma} \cdot \vec{a})(\vec{\sigma} \cdot \vec{b}) = (\vec{a} \cdot \vec{b}) I + i\vec{\sigma} \cdot (\vec{a} \times \vec{b}),$$

$$(\vec{\sigma} \cdot \vec{a}) \text{ and } (\vec{\sigma} \cdot \vec{b}) \quad \begin{cases} \text{commute,} & \text{if } \vec{a} \parallel \vec{b}, \\ \text{anticommute,} & \text{if } \vec{a} \perp \vec{b}, \end{cases}$$

$$\text{tr}(\vec{\sigma}(\vec{\sigma} \cdot \vec{a})) = \text{tr}((\vec{\sigma} \cdot \vec{a})\vec{\sigma}) = 2\vec{a},$$

$$\vec{\sigma}(\vec{\sigma} \cdot \vec{a}) = \vec{a} I - i\vec{\sigma} \times \vec{a},$$

$$(\vec{\sigma} \cdot \vec{a})\vec{\sigma} - \vec{\sigma}(\vec{\sigma} \cdot \vec{a}) = 2i\vec{\sigma} \times \vec{a},$$

$$(\vec{\sigma} \cdot \vec{a})(\vec{\sigma} \cdot \vec{b}) - (\vec{\sigma} \cdot \vec{b})(\vec{\sigma} \cdot \vec{a}) = 2i\vec{\sigma} \cdot (\vec{a} \times \vec{b}), \tag{5.19}$$

where a dagger means taking transpose and complex conjugate of a matrix, i.e. $M^\dagger = \tilde{M}^*$, and δ_{ij} is the Kronecker delta index, ϵ_{ijk} is the permutation index.

In addition, in preparation for the next section, we mention an elegant and important formula,

$$e^{-\frac{i}{2}\vec{\sigma} \cdot \vec{\phi}} = I \cos\frac{\phi}{2} - i(\vec{\sigma} \cdot \hat{\phi}) \sin\frac{\phi}{2}. \tag{5.20}$$

It is easy to see that this matrix is not Hermitian, but is unitary.

The left-hand-side of Eq. (5.20) is a spinor representation (a spinor map) of a rotation around $\hat{\phi}$ with an angle ϕ, i.e. it is a 2×2 representation of a 3-D rotation. The map appears as an exponential form, but it can be expanded by using Eq. (5.20). Note that on the left-hand-side, it is a matrix that sits inside the exponent. How these spinor maps are to be used to describe spin motion is the subject of the next section. But here we just mention that, for example, a spin precession of angle $2\pi a\gamma$ around the $\hat{y}$-direction would be described by the map

$$e^{-i\pi a\gamma \sigma_y},$$

which, according to Eq. (5.20), can be expanded as $I \cos(\pi a\gamma) - i\sigma_y \sin(\pi a\gamma)$.

Normally one would expect to represent a 3-D rotation using 3×3 matrices. The reason we can do the same using much simpler 2×2 matrices in the spinor language is because we are now using complex numbers, while the elements in the 3×3 matrices are real. We are just trading 3×3 real orthogonal matrices by 2×2 complex unitary matrices.

We should mention one peculiarity of Eq. (5.20). Consider a rotation angle $\phi = 2\pi$ around some axis $\hat{\phi}$. Such a rotation ought to give a complete turn of the spin and no net change should result from it. Equation (5.20), however, gives a spinor map $e^{-\vec{\sigma} \cdot \vec{\phi}/2} = -I$, not I. A complete $360°$ rotation in spinor language gives rise to a sign change. As will be seen later, however, all physical quantities involves quadratic forms of the spinor state vectors. Such sign changes do not affect any physical results, including the expectation values of particle's spins. The double-valuedness is appropriate to represent spin-$\frac{1}{2}$ particles (spin up and spin down).[8] The spinor algebra we are going to use applies only to the spin dynamics of spin-$\frac{1}{2}$ particles.

Another way to view Eq. (5.20) is to note that when the left-hand-side is Taylor expanded, it contains powers of the matrix $\vec{\sigma} \cdot \vec{\phi}$. By the properties of the Pauli matrices, however, all these terms condense into two terms, one independent of $\vec{\sigma}$, the other linear in $\vec{\sigma}$, as given by the right-hand-side. Equation (5.20) is quite elegant. [See also the discussion following Eq. (4.62) when we introduced the Hamilton–Caley theorem. See Homework 5.9(d).]

Equation (5.20) gives the spinor rotation map once the rotation angle and rotation axis are given. Sometimes one needs to do the reverse, i.e. one has to find the rotation angle and rotation axis once its spinor map — let it be M — is given. Equation (5.20) can be inverted for that purpose, and the answer is

$$\cos\frac{\phi}{2} = \frac{1}{2}\,\mathrm{tr}(M),$$

$$\hat{\phi} = \frac{i}{2\sin\frac{\phi}{2}}\,\mathrm{tr}(M\vec{\sigma}). \tag{5.21}$$

Note that in performing these calculations, the matrix M must satisfy some conditions. In particular, $\mathrm{tr}M$ must be between -2 and 2, and the vector $\mathrm{tr}(M\vec{\sigma})$ must be purely imaginary. Otherwise something is wrong with the input matrix M. The fact that $\hat{\phi}$ obtained this way has a unit magnitude is due to the unitarity of M.

Homework 5.8 The Pauli matrices $\vec{\sigma}$ is a peculiar object because it is simultaneously a vector and a matrix, and it is complex.

(a) Show Eq. (5.19) as a good entry practice.

(b) As stated in the text, there are three 2×2 Pauli matrices $\sigma_{x,y,z}$. Show that there are no more Pauli matrices that satisfy the conditions being Hermitian and anticommute with $\sigma_{x,y,z}$.

---

[8]Beware of something we are not saying. We are not saying that a spin rotation by 2π transforms a spin-up state into a spin-down state and vice versa. After rotating by 2π, spin-up remains spin-up; spin-down remains spin-down.

Homework 5.9
(a) Prove Eq. (5.20).
(b) Invert Eq. (5.20) to prove Eq. (5.21).
(c) Show that the vector $\hat{\phi}$ obtained by Eq. (5.21) has a unity magnitude.
(d) Prove Eq. (5.20) using the Hamilton–Caley theorem (4.61–4.62).

Solution (a) Use the identity $(\vec{\sigma}\cdot\vec{a})^2 = |\vec{a}|^2 I$ of Eq. (5.19) and Taylor expansions to show

$$
\begin{aligned}
\cos(\vec{\sigma}\cdot\vec{\phi}) &= I\cos\phi\,,\\
\sin(\vec{\sigma}\cdot\vec{\phi}) &= (\vec{\sigma}\cdot\hat{\phi})\sin\phi\,.
\end{aligned}
$$

Homework 5.10 Show that

$$
\begin{aligned}
e^{-\frac{i}{2}\vec{\sigma}\cdot\vec{\phi}}(\vec{\sigma}\cdot\vec{a}) &= (\vec{\sigma}\cdot\vec{a}_\perp)e^{\frac{i}{2}\vec{\sigma}\cdot\vec{\phi}} + (\vec{\sigma}\cdot\vec{a}_\parallel)e^{-\frac{i}{2}\vec{\sigma}\cdot\vec{\phi}}\,,\\
(\vec{\sigma}\cdot\vec{a})e^{-\frac{i}{2}\vec{\sigma}\cdot\vec{\phi}} &= e^{\frac{i}{2}\vec{\sigma}\cdot\vec{\phi}}(\vec{\sigma}\cdot\vec{a}_\perp) + e^{-\frac{i}{2}\vec{\sigma}\cdot\vec{\phi}}(\vec{\sigma}\cdot\vec{a}_\parallel)\,,
\end{aligned}
\qquad (5.22)
$$

where $\vec{a} = \vec{a}_\parallel + \vec{a}_\perp$ with $\vec{a}_\parallel = (\vec{a}\cdot\hat{\phi})\hat{\phi}$ and $\vec{a}_\perp = (\hat{\phi}\times\vec{a})\times\hat{\phi}$. These complicated-looking expressions actually become handy later when performing more sophisticated spinor algebra manipulations.

Homework 5.11 Use the result of Homework 5.10 to expand expressions for

$$
e^{-\frac{i}{2}\vec{\sigma}\cdot\vec{\phi}}(\vec{\sigma}\cdot\vec{a})e^{-\frac{i}{2}\vec{\sigma}\cdot\vec{\phi}}\,,
$$

and

$$
e^{-\frac{i}{2}\vec{\sigma}\cdot\vec{\phi}}(\vec{\sigma}\cdot\vec{a})e^{\frac{i}{2}\vec{\sigma}\cdot\vec{\phi}}\,.
$$

Homework 5.12 By an inspection of Eq. (5.20), answer the question can I be a rotation map? Similarly, can $-I, \sigma_x, \sigma_y, \sigma_z$ be rotation maps? How about $\pm i\sigma_{x,y,z}$? Consult the discussion following Eq. (5.21).

Homework 5.13 A rigid body is being rotated by an angle $\frac{\pi}{2}$ around the $\hat{x}$-axis. It is then followed by a rotation of $-\frac{\pi}{2}$ around $\hat{y}$ and then again by $\frac{\pi}{2}$ around $\hat{z}$. The net result can be described as a single rotation. Use spinor algebra to find the rotation angle ϕ and the rotation axis $\hat{\phi}$ of the net rotation.

Solution The net rotation is given by

$$
\begin{aligned}
M_{\text{tot}} &= -\frac{i}{\sqrt{2}}(\sigma_x + \sigma_z)\,,\\
\phi &= \pi\,,\\
\hat{\phi} &= \frac{1}{\sqrt{2}}(\hat{x} + \hat{z})\,.
\end{aligned}
$$

Homework 5.14 Show that any complex 2×2 matrix can be expressed as a linear combination of $I, \sigma_x, \sigma_y, \sigma_z$. Show the identity

$$
M = \frac{1}{2}\{(\mathrm{tr}M)I + [\mathrm{tr}(M\vec{\sigma})]\cdot\vec{\sigma}\}\,,
$$

for arbitrary complex 2×2 matrix M.

5.3.2 Spin dynamics using spinor

In the 3-D vector language, a spin is described by $\vec{S}$. In the spinor algebra language,[9] it is represented by a complex 2-component column vector ψ, which is connected to $\vec{S}$ by

$$\vec{S} = \psi^\dagger \vec{\sigma} \psi. \tag{5.23}$$

In the 3-D vector language, spin motion is given by Eq. (5.12) in an accelerator. In the spinor language, it is described by the time evolution of ψ,

$$\frac{d\psi}{d\theta} = -\frac{i}{2}(\vec{h} \cdot \vec{\sigma})\psi. \tag{5.24}$$

Indeed, one can show using the properties (5.19) that Eq. (5.12) follows from Eqs. (5.23–5.24). In the spinor language, Eq. (5.24) replaces Eq. (5.12). Once ψ is determined from Eq. (5.24), the spin vector $\vec{S}$ can be calculated from Eq. (5.23) if one so desires.

Before we accept Eqs. (5.23) and (5.24), there are a couple of self-consistency conditions for the formalism to make sense, and we need to establish them first (Homework 5.17),

$$\begin{aligned} \vec{S} &= \text{real}, \\ \psi^\dagger \psi &= \text{constant}, \\ (\psi^\dagger \vec{\sigma} \psi) \cdot (\psi^\dagger \vec{\sigma} \psi) &= \text{constant}. \end{aligned} \tag{5.25}$$

The reason for these conditions should be obvious.

For a noncompelling, historical reason, perhaps more often the 3-D vector language has been adopted by electron spin accelerator physicists, while the spinor language more often by proton spin accelerator physicists.

If $\vec{h}$ is a constant, i.e. independent of θ, Eq. (5.24) has the solution

$$\psi(\theta) = e^{-\frac{i}{2}\vec{h}\cdot\vec{\sigma}\theta}\psi(0). \tag{5.26}$$

Since $\vec{h}$ is the angular velocity, the map (5.26) is the spinor representation of a spin rotation by an angle of $\vec{\phi} = \vec{h}\theta$.

As mentioned earlier, when $\phi = 2\pi$, meaning a rotation for a full turn, the rotation map is not equal to I, but equal to $-I$. This is not a problem because the spin, given by Eq. (5.23), is quadratic in ψ. In particular, it does not mean the spin of a particle flips every turn of revolution.

In general, $\vec{h}$ is not a constant but depends on θ. To study the spin dynamics in this general case, we need to examine the quantity

$$\vec{h} \cdot \vec{\sigma} = \begin{bmatrix} h_y & h_z - ih_x \\ h_z + ih_x & -h_y \end{bmatrix} \tag{5.27}$$

$$= \begin{bmatrix} -a\gamma + (a\gamma + 1)\left(\frac{G}{B\rho}\right)\rho x & a(\gamma - 1)y' + i(a\gamma + 1)\left(\frac{G}{B\rho}\right)\rho y \\ a(\gamma - 1)y' - i(a\gamma + 1)\left(\frac{G}{B\rho}\right)\rho y & a\gamma - (a\gamma + 1)\left(\frac{G}{B\rho}\right)\rho x \end{bmatrix},$$

---

[9]See for example S.Y. Lee, Spin Dynamics and Snakes in Synchrotrons, World Scientific (1997); B.W. Montague, Phys. Reports, 113, 1 (1984).

which appears prominently in Eq. (5.24). This examination is the subject of the next section.

Incidentally, the time variable θ here refers to the bending angle of the beam trajectory. The factor ρ appearing in Eq. (5.27) does not impose a problem in a straight section with no magnetic fields and $\rho \to \infty$ because $\rho d\theta = ds$.

Rotating frame   The spin precession rotation has an angular velocity $\vec{h}$. If we observe the spin motion in a rotating frame that has a angular velocity $\vec{h}_0$, the spin state in the rotating frame is given by ψ_1, where

$$\psi = e^{-\frac{i}{2}\vec{h}_0 \cdot \vec{\sigma}\theta} \psi_1 . \tag{5.28}$$

The equation of motion for ψ_1 is (Homework 5.18) given by

$$\frac{d\psi_1}{d\theta} = -\frac{i}{2} e^{\frac{i}{2}\vec{h}_0 \cdot \vec{\sigma}\theta} \left[(\vec{h} - \vec{h}_0) \cdot \vec{\sigma} \right] e^{-\frac{i}{2}\vec{h}_0 \cdot \vec{\sigma}\theta} \psi_1 . \tag{5.29}$$

Physically, Eq. (5.29) asks ψ_1 first be rotated by $e^{-\frac{i}{2}\vec{h}_0 \cdot \vec{\sigma}\theta}$ to become ψ, then advance ψ by a precession of $(\vec{h} - \vec{h}_0)\theta$, and then rotate back to the ψ_1 spinor by $e^{\frac{i}{2}\vec{h}_0 \cdot \vec{\sigma}\theta}$. Incidentally, the connection between ψ and ψ_1 is given by Eq. (5.28). It would be a mess if they get connected by, e.g.

$$\psi = \psi_1 e^{-\frac{i}{2}\vec{h}_0 \cdot \vec{\sigma}\theta} .$$

Indeed if this definition is adopted, the above physical meaning would not apply.

Homework 5.15 The 3-D spin vector $\vec{S}$ is related to the spinor state ψ by Eq. (5.23).
    (a) Find the spinor states that represent $\vec{S} = \hat{x}, -\hat{x}, \hat{y}, -\hat{y}, \hat{z}, -\hat{z}$.
    (b) Find a spinor state that represents $\vec{S} = \hat{x}\sin\theta\sin\phi + \hat{y}\cos\theta + \hat{z}\sin\theta\cos\phi$.

Solution (a) They are respectively,

$$\begin{bmatrix} \frac{1}{\sqrt{2}} \\ \frac{i}{\sqrt{2}} \end{bmatrix}, \quad \begin{bmatrix} \frac{1}{\sqrt{2}} \\ -\frac{i}{\sqrt{2}} \end{bmatrix}, \quad \begin{bmatrix} 1 \\ 0 \end{bmatrix}, \quad \begin{bmatrix} 0 \\ 1 \end{bmatrix}, \quad \begin{bmatrix} \frac{1}{\sqrt{2}} \\ \frac{1}{\sqrt{2}} \end{bmatrix}, \quad \begin{bmatrix} \frac{1}{\sqrt{2}} \\ -\frac{1}{\sqrt{2}} \end{bmatrix} .$$

Each answer is not unique, arbitrary up to a phase factor.
    (b) One solution is

$$\begin{bmatrix} \sqrt{\dfrac{1+\cos\theta}{2}} \\[2mm] \dfrac{e^{i\phi}}{\sqrt{2(1+\cos\theta)}} \end{bmatrix} .$$

Homework 5.16 Verify the statement in the text that Eq. (5.24) in the spinor language replaces Eq. (5.12) in the vector language as long as the connection is made with Eq. (5.23).

Homework 5.17 Prove Eq. (5.25) using the spinor algebra.

Homework 5.18

(a) Derive Eq. (5.29).

(b) If we decompose $\vec{\Delta} = \vec{h} - \vec{h}_0$ into $\vec{\Delta}_\parallel + \vec{\Delta}_\perp$ where $\parallel$ and $\perp$ refer to parallel and perpendicular to $\vec{h}_0$, use Eq. (5.29) and the result of Homework 5.10 to show that

$$\frac{d\psi_1}{d\theta} = -\frac{i}{2}\left[e^{i\vec{h}_0 \cdot \vec{\sigma}\theta}(\vec{\Delta}_\perp \cdot \vec{\sigma}) + \vec{\Delta}_\parallel \cdot \vec{\sigma}\right]\psi_1 . \tag{5.30}$$

The parallel and perpendicular components of $\vec{\Delta}$ induce very different spin dynamics. We will use this equation later in Eq. (5.37) when we analyze the spin motion near a depolarization resonance, when $\vec{\Delta}_\perp$ will represent the resonance strength parameter and $\vec{\Delta}_\parallel$ will represent the spacing of spin tune to the exact resonant value.

5.4   Depolarization resonance

5.4.1   Isolated resonance

Our job is to solve Eqs. (5.24) and (5.27) for $\psi(\theta)$. To do so, we will make a few approximations first.

We note that the leading terms in (5.27) are the $\pm a\gamma$ terms in the diagonal elements. These are constants, independent of θ. The dominating spin motion is therefore a constant precession around $\hat{y}$ with spin tune of $a\gamma$. It is seen by the ideal particle moving along the design trajectory.

Here it reveals the reason why we choose the specific time variable θ in our analysis. Only when choosing the bending angle θ as the time variable would we see a constant precession rate.

The remaining terms depend linearly on x and y and are seen only by non-ideal particles with transverse displacements. This means they have a θ dependence of

$$\begin{cases} e^{\pm i\nu_{x,y}\theta + iK\theta}, & \text{if } x, y \text{ come from betatron oscillations,} \\ e^{iK\theta}, & \text{if } x, y \text{ come from closed orbit distortions,} \end{cases} \tag{5.31}$$

where K is any integer (positive or negative). These terms, although small, perturb the spin motion, and their perturbation tunes are $K \pm \nu_{x,y}$ and K, depending on the origin of x and y, according to Eq. (5.31). One expects a resonance to occur when one of these spin-perturbation tunes hits the unperturbed spin tune, i.e. when

$$a\gamma = \begin{cases} K \pm \nu_{x,y}, & \text{if } x, y \text{ come from betatron oscillations,} \\ K, & \text{if } x, y \text{ come from closed orbit distortions.} \end{cases} \tag{5.32}$$

Note the $\pm$ sign. Unlike the orbital resonances where only the sum resonances are unstable and difference resonances are stable, here both the sum and difference resonances depolarize. Only their strengths differ according to the strengths of the corresponding driving harmonics.

Why do we care about these resonances? What happens when a resonance condition is fulfilled or approximately fulfilled? When a beam is injected into an accelerator, we carefully align its polarization along $\hat{y}$. The ideal particle, whose spin motion is simple precession around $\hat{y}$, would keep its spin aligned to $\hat{y}$ turn after turn. The hope is that the spins of the nonideal particles would not deviate far from $\hat{y}$ so that the net polarization of the whole beam does not suffer. In general, this is not a problem because the perturbations are weak. However, when a resonance occurs, these weak perturbations accumulate turn by turn to cause large deviations of spin from $\hat{y}$, at least for the nonideal particles with large enough orbital deviations, and the beam polarization would be compromised or lost. For this reason, resonances (5.32) are called *depolarization resonances*. These resonances do not affect particle's orbital x-, y- and z-motions — ignoring Stern–Gerlach effects.

We now start to trim Eq. (5.27) by making some approximations. A closer examination shows that the perturbation due to x-motion is not important. An x-motion contributes only to diagonal elements of (5.27) and thus perturbs only rotations around $\hat{y}$, which do not depolarize. Such perturbations do not worry us; we only worry about rotations perpendicular to $\hat{y}$. This means we can drop the x-perturbation terms from Eq. (5.27).

Next we make an approximation that there is one and only one resonance causing depolarization. This means all resonances are sufficiently far apart so that we can deal with them one by one isolated from others. Let the depolarization condition (5.32) be written as

$$a\gamma = \kappa.$$

Then, under this isolated-resonance assumption, one can approximate the off-diagonal elements of (5.27) by filtering out and keeping only their κ-th Fourier components.

One word of caution on the isolated-resonance assumption is in order. This is a very useful and fruitful approximation as we will soon see; however, it is also a subtle one. Making this approximation, for example, would lose some important and interesting spin dynamics such as interference effects between nearby resonances and another interesting phenomenon called spin echo.[10] When nonlinearities are added, there will also be spin chaos. We will adopt the isolated resonance approximation in the following. Spin echo will be mentioned in Secs. 5.5 and 7.5.

Let us make one more comment concerning Eq. (5.32). In general, the right-hand-side of Eq. (5.32) contains the frequency content of the 3-D orbital motion. When orbital nonlinearities are included, the right-hand-side will become

$$K + n_x \nu_x + n_y \nu_y + n_s \nu_s,$$

where $n_{x,y,s}$ specifies the order of the orbital nonlinearities involved, and higher order depolarization resonances can occur accordingly at

$$a\gamma = K + n_x \nu_x + n_y \nu_y + n_s \nu_s. \tag{5.33}$$

<hr>

[10]A.W. Chao and E.D. Courant, Phys. Rev. ST Accel. Beams 10, 014001 (2007).

These nonlinear depolarization resonances, however, comes only from orbital motion. The spin motion remains linear as the left-hand-side remains to be the natural spin tune $a\gamma$. Nonlinear depolarization resonances do not involve high order spin motion. The left-hand-side of Eq. (5.33) will not read $n_{\text{spin}} \times a\gamma$ until Stern–Gerlach effect becomes important.

After making these approximations, we obtain[11]

$$\vec{h} \cdot \vec{\sigma} \approx \begin{bmatrix} -a\gamma & \epsilon e^{i\kappa\theta} \\ \epsilon^* e^{-i\kappa\theta} & a\gamma \end{bmatrix}, \tag{5.34}$$

with

$$\begin{aligned} \epsilon &= \frac{1}{2\pi} \int_0^{2\pi} d\theta \; e^{-i\kappa\theta} (h_z - i h_x) \\ &= \frac{1}{2\pi} \int_0^{2\pi} d\theta \; e^{-i\kappa\theta} \left[a(\gamma - 1) y' + i(a\gamma + 1)\left(\frac{G}{B\rho}\right) \rho y \right], \end{aligned} \tag{5.35}$$

where y and y' refer to the betatron contribution if $\kappa = K \pm \nu_y$, and refer to the closed orbit contribution if $\kappa = K$. There is no driving of the $\kappa = K \pm \nu_x$ resonances because the x-contributions have been dropped. The integrand of Eq. (5.35) is periodic in θ with period 2π — otherwise the Fourier analysis would be meaningless. Keep in mind that the time variable θ has to be the accumulated bending angle of the ideal reference particle.

The important quantity ϵ is independent of θ, and is the complex *depolarization resonance strength*. It has the same unit and the same dimensionality of the spin tune. Contained in Eq. (5.35) is the statement that depolarization action occurs from having vertical excursion. The first term comes from dipoles; the second term from quadrupoles. The second term typically dominates.

The expression (5.35) for the resonance strength is derived when the resonance is observed at $\theta = 0$, a specific location of the ring or the synchrotron. If the spin is observed at another position with $\theta = \theta^*$, the resonance strength acquires an additional phase factor $e^{-iK\theta^*}$. The magnitude $|\epsilon|$ is invariant around the accelerator, but its phase varies.

5.4.2 Resonance strength ϵ

The $a\gamma = K$ resonances are called *imperfection* resonances. They come from y-closed-orbit distortion which results from imperfections. By a good orbit correction — a good small rms orbit distortion, and if necessary, supplemented by an additional clever harmonic matching algorithm to be mentioned later (Sec. 5.4.5) — effect of these resonances can be alleviated.

The $a\gamma = K \pm \nu_y$ resonances are called *intrinsic* resonances. They come from vertical betatron oscillations of the particles. Since the beam intrinsically

---

[11]E.D. Courant and R.D. Ruth, Brookhaven Nat. Lab. Report, BNL-51270 (1980); L. Teng, Conf. High Energy Physics with Polarized Beams and Polarized Targets, AIP Proc. No. 51, Argonne, p. 248 (1978); S.Y. Lee, Spin Dynamics and Snakes in Synchrotrons, World Scientific (1997).

has a finite emittance, these resonances cannot be avoided by some closed-orbit correction scheme. Note that all particles have the same strength of imperfection resonances because they all have the same y-closed-orbit, but the strengths of their intrinsic resonances differ according to their vertical betatron amplitudes. The ideal on-orbit particle, for example, would have zero intrinsic resonance strength.

In case the circular accelerator or synchrotron has a superperiod P_{sup}, then the intrinsic resonances occur only when $K = mP_{\text{sup}}$ is an integral multiple of P_{sup}. The driving terms of the other resonances vanish due to destructive interference among the superperiods. Occurrence of intrinsic resonances therefore are P_{sup} times more sparse than the imperfection resonances. On the other hand, the ones with $K = mP_{\text{sup}}$ add constructively so they become P_{sup} times stronger.

To calculate the resonance strength ϵ, we set y to be the vertical closed orbit for the imperfection resonances, and set it to be the vertical betatron displacement for the intrinsic resonances. In a computer code, to calculate ϵ for the imperfection resonances, the closed orbit is generated by a random error orbit kicks with some orbit correction algorithm applied. To calculate for the intrinsic resonances, the betatron oscillation comes from an intrinsic betatron emittance and no random error generation is invoked.

For intrinsic resonances, as mentioned earlier, ϵ depends on the betatron oscillation of the particle under consideration. In addition, the value of ϵ calculated contains an overall factor of the particle's initial betatron oscillation phase factor. For this reason, when calculating ϵ of an intrinsic resonance, we often consider a particle at 1-σ betatron amplitude and we often only quote the value of $|\epsilon|$.

In the case of a synchrotron, as the beam energy increases in an acceleration process, the spin tune $a\gamma$ increases proportionally. As a result, there is one imperfection resonance to cross every 0.5233 GeV for protons and 0.4406 GeV for electrons. To accelerate to high energies, a large number of imperfection resonances are to be crossed. The number of intrinsic resonances to cross will be a factor of $\approx \frac{P_{\text{sup}}}{2}$ less — the factor of P_{sup} is because the intrinsic resonances occur only at intervals of $P_{\text{sup}}\nu_y$; the extra factor of 2 is to take into account there are both sum and difference resonances. For example, to accelerate a proton beam to 30 GeV in the Brookhaven AGS synchrotron, there are 56 imperfection and 9 intrinsic resonances to cross. To accelerate from 8 to 400 GeV in the Fermilab Main Ring, there are 749 imperfection and 250 intrinsic resonances to cross.

So the higher the beam energy, the more resonances to be crossed. Furthermore, the higher the beam energy, also the stronger the resonances. This is because the spin precession angle is $a\gamma + 1$ times larger than the orbital perturbation angles, and $a\gamma$ becomes larger at higher energies. As an illustration of this tendency, Fig. 5.3 shows a calculation of the intrinsic resonance strength ϵ

Figure 5.3: Calculated strengths $|\epsilon|$ for the intrinsic resonances of the AGS synchrotron (see footnote 12). The superperiodicity of AGS is $P_{\mathrm{sup}} = 12$.

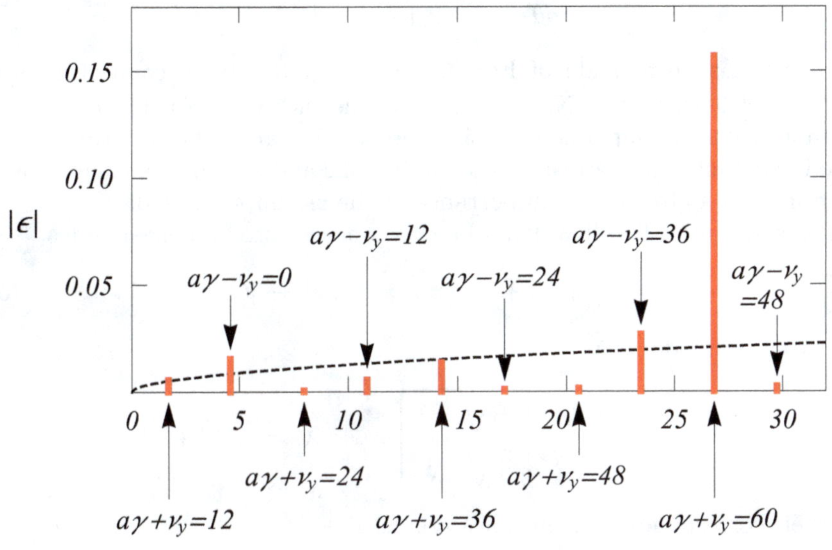

for the AGS.[12] The dashed curve in Fig. 5.3 shows the rough numerical scaling (5.52) for the intrinsic resonances, as will be mentioned in Sec. 5.4.5.

Homework 5.19 Show that an estimate expression for the rms value of imperfection resonance strength ϵ for a proton synchrotron with energy E_0, rms vertical orbit y_{rms}, typical quadrupole focal length f_{quad}, and total number of quadrupoles N_{quad} can be written as

$$\epsilon_{\mathrm{rms}} \approx \frac{a\gamma + 1}{2\pi} \sqrt{\frac{N_{\mathrm{quad}}}{2}} \frac{y_{\mathrm{rms}}}{f_{\mathrm{quad}}}.$$

Take for example $E_0 = 200$ GeV, $N_{\mathrm{quad}} = 100$, $f_{\mathrm{quad}} = 15$ m, $y_{\mathrm{rms}} = 1$ mm, give an estimate of ϵ_{rms}.

5.4.3 Spin motion near resonance

How does the spin move near an isolated resonance $a\gamma = \kappa$? Let $\delta = a\gamma - \kappa$ be the separation of the spin tune from the resonance, and observe the spin motion in a rotating frame as [see Eqs. (5.28–5.29)],

$$\psi_1 = e^{-\frac{i}{2}\kappa\theta\sigma_y}\psi. \tag{5.36}$$

---

[12]L. Teng, AIP Proc. No. 51, Argonne, p. 248 (1978); E.D. Courant and R.D. Ruth, Brookhaven Laboratory Report BNL-51270 (1980).

Then the evolution of ψ_1 involves only a constant map independent of θ,

$$\frac{d\psi_1}{d\theta} = -\frac{i}{2}\begin{bmatrix} -\delta & \epsilon \\ \epsilon^* & \delta \end{bmatrix}\psi_1 , \qquad (5.37)$$

where use has been made of Eq. (5.22) or Eq. (5.30) by connecting ϵ with $e^{i\kappa\theta}\vec{\Delta}_\perp$ and δ with $\vec{\Delta}_\parallel$. Needless to say, the fact that Eq. (5.37) involves a constant evolution map assumes using the bending angle θ as the time variable. More importantly, it is a result of assuming one and only one resonance nearby, which in turn addresses the importance of the assumption made.

The solution to Eq. (5.37) can be decomposed into two eigenmodes,

$$\psi_1(\theta) = C_+\psi_+ + C_-\psi_-, \qquad |C_+|^2 + |C_-|^2 = 1 . \qquad (5.38)$$

The two eigenmodes are

$$\psi_\pm = e^{\pm i\lambda\theta/2}\begin{bmatrix} \frac{\epsilon}{|\epsilon|}\sqrt{\frac{\lambda\pm\delta}{2\lambda}} \\ \mp\sqrt{\frac{\lambda\mp\delta}{2\lambda}} \end{bmatrix} , \qquad \lambda = \sqrt{\delta^2 + |\epsilon|^2} , \qquad (5.39)$$

and their corresponding eigenvalues are $\pm\lambda$, i.e.

$$\frac{d\psi_\pm}{d\theta} = \frac{i}{2}(\pm\lambda)\psi_\pm .$$

The orthonormality condition is obeyed according to

$$\psi_+^\dagger\psi_- = \psi_-^\dagger\psi_+ = 0, \qquad \psi_+^\dagger\psi_+ = \psi_-^\dagger\psi_- = 1 .$$

It is easy to show that $\psi_1^\dagger(\theta)\psi_1(\theta)$ is conserved. We have normalized it to 1.

The spin projection onto the y-axis is

$$\begin{aligned} S_y &= \psi^\dagger\sigma_y\psi = \psi_1^\dagger\sigma_y\psi_1 \\ &= \frac{\delta}{\lambda}(|C_+|^2 - |C_-|^2) + \frac{2|\epsilon|}{\lambda}\,\mathrm{Re}[C_+C_-^*e^{i\lambda\theta}] . \end{aligned} \qquad (5.40)$$

If we consider a particle initially fully polarized along $\hat{y}$ with

$$\psi(\theta = 0) = \begin{bmatrix} 1 \\ 0 \end{bmatrix} ,$$

then we have

$$C_+ = \frac{|\epsilon|}{\epsilon}\sqrt{\frac{\lambda + \delta}{2\lambda}} , \qquad C_- = \frac{|\epsilon|}{\epsilon}\sqrt{\frac{\lambda - \delta}{2\lambda}} .$$

Substituting into Eq. (5.40) yields

$$S_y = \frac{\delta^2 + |\epsilon|^2\cos\lambda\theta}{\delta^2 + |\epsilon|^2} . \qquad (5.41)$$

Figure 5.4: Time evolution of the y-component of the spin (a) exactly on resonance $\frac{\delta}{|\epsilon|} = 0$, (b) near resonance $\frac{\delta}{|\epsilon|} = 1$, (c) away from resonance $\frac{\delta}{|\epsilon|} = 3$.

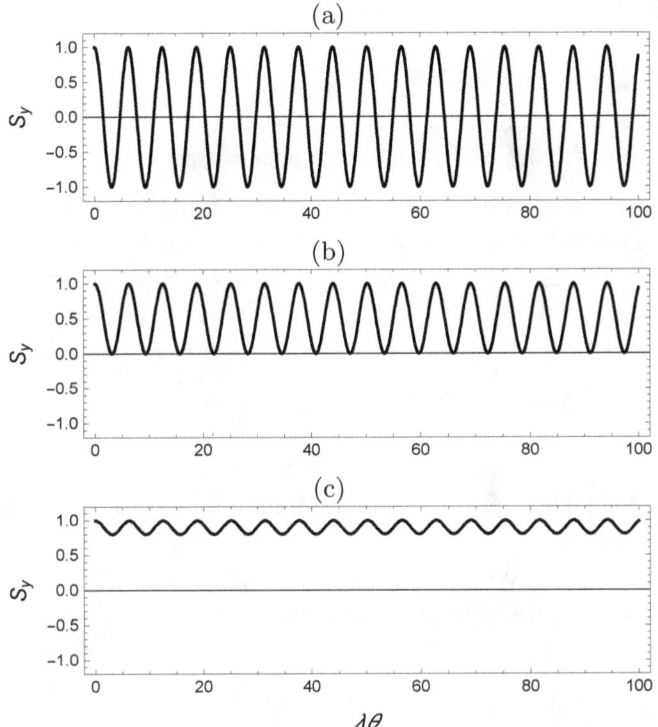

In particular, one finds that S_y oscillates between 1 and $\frac{\delta^2 - |\epsilon|^2}{\delta^2 + |\epsilon|^2}$. The oscillation tune is equal to λ. Figure 5.4 shows the time evolution of S_y.

The oscillation of S_y is slow. This is because the oscillation tune $\lambda = \sqrt{\delta^2 + |\epsilon|^2}$ and both δ and $|\epsilon|$ are much less than 1. This means S_y changes very little per revolution of the particle. In contrast, the spin components S_x and S_z oscillate rapidly from turn to turn — unless they are observed in a rotating frame. See Homework 5.21.

If we ask what is the time averaged polarization of this particle, projected onto the $\hat{y}$-axis, we obtain

$$\langle S_y \rangle = \frac{\delta^2}{\delta^2 + |\epsilon|^2}, \tag{5.42}$$

which is reduced from the initial polarization of $S_y = 1$ due to the proximity to the resonance. If we have a beam of particles with random spinor phases, polarization of the entire beam would have the value $\langle S_y \rangle$ after this initially fully polarized beam resettles.

Away from the resonance, $\delta \to \pm\infty$, we find $\langle S_y \rangle = 1$. The particle (or the beam) remains fully polarized. Closer to the resonance, the polarization decreases. Right on the resonance, we have $\langle S_y \rangle = 0$. The width of the resonance is therefore given by $|a\gamma - \kappa| = |\epsilon|$. For this reason, $|\epsilon|$ (not ϵ — keep in mind that ϵ is a complex quantity) has the physical meaning of resonance width in spin tune units.

Homework 5.20

(a) Prove Eqs. (5.37).

(b) Prove Eqs. (5.39) and (5.40) for a general spinor (5.38).

Solution (b) There are two ways to prove (5.39). One way is to back substitute (5.38) into (5.37) and verify it. This solves the homework problem, but you don't really learn anything. The other way is to diagonalize

$$\begin{bmatrix} -\delta & \epsilon \\ \epsilon^* & \delta \end{bmatrix} = T^{-1} \begin{bmatrix} \lambda_+ & 0 \\ 0 & \lambda_- \end{bmatrix} T,$$

and solve for $T\psi_1$.

Homework 5.21 For a particle initially fully polarized in the $\hat{y}$-direction,

(a) Prove Eq. (5.41).

(b) Give the spinor vector $\psi_1(\theta)$ for arbitrary time θ.

(c) Calculate $S_x(\theta)$ and $S_z(\theta)$. Make sure that your result gives $S_x^2 + S_y^2 + S_z^2 = 1$.

Solution

$$\psi_1(\theta) = \begin{bmatrix} \cos \frac{\lambda\theta}{2} + i\frac{\delta}{\lambda} \sin \frac{\lambda\theta}{2} \\ -i\frac{\epsilon^*}{\lambda} \sin \frac{\lambda\theta}{2} \end{bmatrix},$$

$$S_x(\theta) = -\frac{\mathrm{Re}(\epsilon)}{\lambda} \sin \lambda\theta + \frac{2\delta \, \mathrm{Im}(\epsilon)}{\lambda^2} \sin^2 \frac{\lambda\theta}{2},$$

$$S_z(\theta) = -\frac{\mathrm{Im}(\epsilon)}{\lambda} \sin \lambda\theta - \frac{2\delta \, \mathrm{Re}(\epsilon)}{\lambda^2} \sin^2 \frac{\lambda\theta}{2}.$$

Note that $S_x(\theta)$ and $S_z(\theta)$ are components in the rotating frame with angular velocity $\kappa\hat{y}$. In this rotating frame, the spin $\vec{S}$ precesses slowly near the resonance with precession frequency λ, not the fast spin tune $a\gamma$.

Homework 5.22 Consider a particle that enters a region with dipole magnetic field $\vec{B} = B_0\hat{y}$ with an initial spin $\vec{S}(0)$. Compute the spin at the exit of the magnet $\vec{S}(L)$ using spinor algebra.

Homework 5.23 Repeat Homework 5.22 for a quadrupole magnet with length ℓ and $\vec{B} = G(y\hat{x} + x\hat{y})$. Consider the case when the particle enters the quadrupole with x_0 (let $y_0 = 0$, $x_0' = y_0' = 0$). There is not a closed-form solution for this problem.

5.4.4 Froissart–Stora formula

We have assumed the particle energy, and therefore the spin tune $a\gamma$, is fixed in the above analysis. In a synchrotron, the beam energy is being changed during acceleration. As a result, the spin tune $a\gamma$ increases with time, and we must therefore consider a time-dependent spin tune, and in particular the problem of crossing depolarization resonances.

Let us again consider one and only one resonance, and we are crossing it with

$$a\gamma = \kappa + \Gamma\theta. \tag{5.43}$$

The resonance is crossed at time $\theta = 0$, and Γ is the crossing speed,

$$\Gamma = \frac{1}{2\pi}\frac{d}{dn}(a\gamma - \kappa),$$

where $\frac{d}{dn}$ indicates the rate of change per turn. For imperfection resonances, $\kappa = K$ is fixed, γ is determined by the acceleration gradient, we have

$$\Gamma_{\text{imp}} = \frac{a}{2\pi}\frac{d\gamma}{dn}. \tag{5.44}$$

For the intrinsic resonances, in addition to the acceleration rate, the crossing speed can also be controlled by introducing a pulsed jump of the tune ν_y at the moment of crossing. In fact, the tune jump is usually made so that it dominates over the acceleration rate, so that

$$\Gamma_{\text{int}} = \pm\frac{1}{2\pi}\frac{d\nu_y}{dn}, \tag{5.45}$$

where the $\pm$ sign depending on whether it is sum or difference resonance that is being crossed.

We need now to solve Eqs. (5.24), (5.34) and (5.43). We make a transformation

$$\psi_1 = e^{-\frac{i}{2}(\kappa\theta + \frac{1}{2}\Gamma\theta^2)\sigma_y}\psi.$$

Then the evolution of ψ_1 is described by

$$\frac{d\psi_1}{d\theta} = -\frac{i}{2}\begin{bmatrix} 0 & \epsilon e^{-\frac{i}{2}\Gamma\theta^2} \\ \epsilon^* e^{\frac{i}{2}\Gamma\theta^2} & 0 \end{bmatrix}\psi_1, \tag{5.46}$$

where use has been made of Eq. (5.22). This time the evolution map for ψ_1 is time dependent.

Equation (5.46) looks simple, but its solution is a bit involved. Let

$$\psi_1(\theta) = \begin{bmatrix} f(\theta) \\ g(\theta) \end{bmatrix},$$

with $|f|^2 + |g|^2 = 1$, it follows that

$$\frac{d^2 f}{d\theta^2} + i\Gamma\theta\frac{df}{d\theta} + \frac{|\epsilon|^2}{4}f = 0. \tag{5.47}$$

Define $x = \sqrt{\Gamma}\theta$, assuming $\Gamma > 0$, and make a transformation

$$f = e^{-\frac{i}{4}x^2} F,$$

then

$$\frac{d^2 F}{dx^2} + \left(\frac{x^2}{4} - a\right) F = 0, \qquad \text{with} \quad a = \frac{i}{2} - \frac{|\epsilon|^2}{4\Gamma}. \tag{5.48}$$

The solutions to Eq. (5.48) are the parabolic cylindrical functions $E(a,x)$ and $E^*(a,x)$.[13] With the initial condition $|f(-\infty)| = 1$, the relevant solution is $E(a,x)$. Substituting a from Eq. (5.48), we obtain

$$f(\theta) = \frac{1}{\sqrt{2}}\, e^{-\pi|\epsilon|^2/4\Gamma} E\left(\frac{i}{2} - \frac{|\epsilon|^2}{4\Gamma}, \sqrt{\Gamma}\theta\right).$$

We are interested in the asymptotic behavior of the function $E(a,x)$,

$$\lim_{x\to\infty} E(a,x) = \sqrt{\frac{2}{x}} \exp\left(\frac{i}{4}x^2 - ia\ln x + i\Phi\right),$$

$$\lim_{x\to-\infty} E(a,x) = \sqrt{\frac{2}{|x|}} \exp\left(\frac{i}{4}x^2 - ia\ln|x| - a\pi + i\Phi\right),$$

$$\Phi = \frac{\pi}{4} + \frac{1}{2}\arg\Gamma\left(\frac{1}{2} + ia\right). \tag{5.49}$$

The polarization is given by

$$S_y = |f|^2 - |g|^2 = 2|f|^2 - 1.$$

For $\theta \to -\infty$, we have $S_y(-\infty) = 1$. Long after the beam is accelerated across the resonance, the polarization is reduced to

$$S_y(\infty) = 2|f(\infty)|^2 - 1$$
$$= 2e^{-\pi|\epsilon|^2/2\Gamma} - 1. \tag{5.50}$$

---

[13] Parabolic cylindrical function,

$$E(a,x) = 2^{\frac{1}{4}+\frac{a}{2}} x^{-\frac{1}{2}} W_{\frac{1}{4}+\frac{a}{2},-\frac{1}{4}}\left(\frac{x^2}{2}\right)$$

$$= 2^{\frac{1}{4}+\frac{a}{2}} x^{-\frac{1}{2}} e^{-\frac{z}{2}} z^{\frac{1}{2}+\mu}\, {}_1F_1\left(\frac{1}{2}+\mu-k; 1+2\mu; z\right)$$

$$= 2^{\frac{1}{4}+\frac{a}{2}} x^{-\frac{1}{2}} e^{-\frac{1}{4}x^2} \left(\frac{x^2}{2}\right)^{\frac{1}{4}}\, {}_1F_1\left(-\frac{a}{2}; \frac{1}{2}; \frac{x^2}{2}\right)$$

$$= 2^{\frac{a}{2}} e^{-\frac{1}{4}x^2}\, {}_1F_1\left(-\frac{a}{2}; \frac{1}{2}; \frac{x^2}{2}\right),$$

where $k = \frac{1}{4} + \frac{a}{2}, \mu = -\frac{1}{4}, z = \frac{1}{2}x^2$. See M. Abramowitz and I. Stegun, Handbook of Mathematical Functions.

Figure 5.5: Froissart–Stora formula. There is little loss of polarization if $\frac{|\epsilon|}{\sqrt{\Gamma}} \ll 1$ or when $\frac{|\epsilon|}{\sqrt{\Gamma}} \gg 1$ (polarization flip).

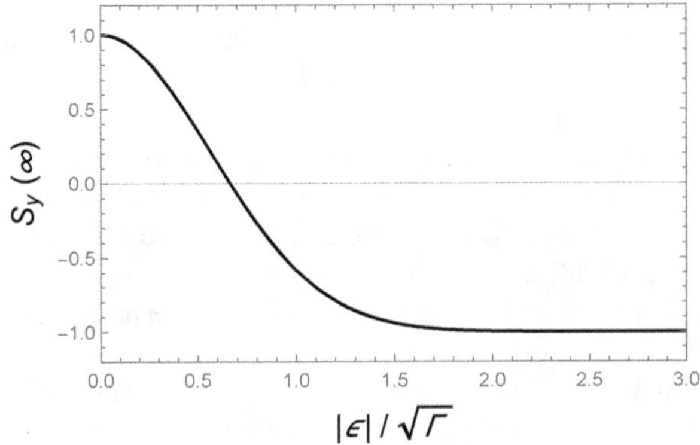

This is a surprisingly simple final result given the complicated analysis. See also Homework 5.25. We will offer another derivation of the Froissart–Stora formula when we come to a matrix formalism in Sec. 5.5.3.

Equation (5.50) was first derived by Froissart and Stora.[14] It says that if a resonance is weak and/or it is crossed quickly ($\frac{|\epsilon|^2}{\Gamma} \ll 1$), there is little loss of polarization. When it is strong and/or it is crossed slowly ($\frac{|\epsilon|^2}{\Gamma} \gg 1$), the polarization is flipped, and there is, somewhat surprisingly, also little loss of polarization. Loss of polarization occurs when the crossing speed Γ is comparable to $|\epsilon|^2$, and that is to be avoided. Figure 5.5 shows the dependence of $S_y(\infty)$ on $\frac{|\epsilon|}{\sqrt{\Gamma}}$.

It is true that, according to the Froissart–Stora formula, both fast crossing and slow crossing of a resonance would preserve polarization of a beam. There is a caveat, however, when one tries to apply slow-crossing to a strong intrinsic resonance. This is because to preserve polarization in slow-crossing, it is necessary that $\frac{|\epsilon|^2}{\Gamma} \ll 1$. But for intrinsic resonances, the resonance strength of a particle depends on the particle's betatron oscillation amplitude. As a result, particles at the core of the beam will not satisfy the slow-crossing criterion. Not all particles get spin flipped, and there will be a net loss of polarization in these cases. See Homework 5.27.

For the case of fast crossing of weak resonances, the Froissart–Stora formula predicts a loss of polarization by

$$\frac{\Delta P}{P} = -\frac{\pi |\epsilon|^2}{\Gamma}.$$

(5.51)

---

[14]M. Froissart and R. Stora, Nucl. Instr. Meth. 7, 297 (1960).

Homework 5.24
   (a) Verify Eq. (5.46).
   (b) Verify Eq. (5.48).

Homework 5.25 It may be instructive to try investigating a bit the Froissart–Stora equation without delving into the very special parabolic cylindrical functions by a homework like this one.

   (a) Some peripheral investigation on the asymptotic limit (5.49) can be obtained without resorting to special functions as in the text. Start with Eq. (5.48), and assume an ansatz for the asymptotic form for the function $F(x) \approx A x^\beta e^{i\gamma x^2}$. Substitute into Eq. (5.48) to obtain $\gamma = \pm\frac{1}{4}$ and $\beta = -\frac{1}{2} \mp ia$.

   (b) Use Eq. (5.49) to verify the Froissart–Stora formula (5.50).

Homework 5.26 Write a numerical program to solve the differential equation for $f(\theta)$ [or $F(\theta)$] as a function of θ and observe its behavior as its precession tune sweeps through a resonance. Pay attention to the behavior when the spin tune is far away from the resonance, when it gets close to the resonance, and then when it emerges from the resonance.

Solution Be careful with the $\frac{x^2}{4}$ term in Eq. (5.48) or the $i\Gamma\theta\,\frac{df}{d\theta}$ term in Eq. (5.47). They give troubles when $|x| \to \infty$ or $|\theta| \to \infty$. In fact, these are the unavoidable subtleties of the parabolic cylindrical functions.

Homework 5.27 Consider a fully vertically polarized beam being brought to cross an intrinsic resonance. The resonance strength $|\epsilon|$ in this case depends on the vertical betatron amplitude a_y of the individual particles. Let the beam have a Gaussian distribution of betatron amplitudes as

$$\psi(a_y) \;=\; \frac{1}{\sigma_y^2}\, e^{-a_y^2/2\sigma_y^2}\,,$$

with normalization $\int_0^\infty \psi(a_y) a_y da_y = 1$ and $\int_0^\infty \psi(a_y) a_y^2\, a_y da_y = 2\sigma_y^2$. Let the resonance strength be given by $|\epsilon|^2 = \beta a_y^2$ with some known coefficient β. The beam core will maintain its initial polarization $+1$. The beam tail will flip its spin to -1. Show that the net beam polarization after the beam completes the resonance crossing is given by

$$S_y(\infty) \;=\; \frac{\Gamma - \pi\beta\sigma_y^2}{\Gamma + \pi\beta\sigma_y^2}\,.$$

5.4.5   Harmonic matching

Numerical scaling of ϵ   An inspection of Fig. 5.3 shows a large fluctuation of strengths among the different resonances — the $a\gamma + \nu_y = 60$ resonance stands out particularly strong. However, the fact that the strengths increase with beam energy is a general trend. A rough general empirical universal scaling covering all existing proton synchrotrons has been suggested,[15]

$$|\epsilon_{\text{imp}}| \approx 3 \times 10^{-4} \left(\frac{E}{25\,\text{GeV}} \right),$$

$$|\epsilon_{\text{int}}| \approx 0.02 \left(\frac{E}{25\,\text{GeV}} \right)^{1/2}. \tag{5.52}$$

The imperfection and the intrinsic resonances scale differently with beam energy E. The scaling for $|\epsilon_{\text{int}}|$ was shown as dashed curve in Fig. 5.3 for the AGS. The case of imperfection resonances is meant to be that after an orbit correction has been applied.

Acceleration to high energy　If we then take the Froissart–Stora formula (5.51), we expect, after accelerating to the top energy crossing all the resonances rapidly, a polarization loss

$$\frac{\Delta P}{P} = -\sum_{i \in \text{imp}} \frac{\pi |\epsilon_i|^2}{\Gamma_{\text{imp}}} - \sum_{i \in \text{int}} \frac{\pi |\epsilon_i|^2}{\Gamma_{\text{int}}}.$$

Assuming all intrinsic resonances are crossed by tune jumps and using Eqs. (5.44) and (5.45), we have

$$\frac{\Delta P}{P} = -2\pi^2 \left(\sum_{i \in \text{imp}} \frac{|\epsilon_i|^2}{a\gamma'} + \sum_{i \in \text{int}} \frac{|\epsilon_i|^2}{\nu_y'} \right),$$

where γ' and ν_y' refer to the change rate per turn. Assuming the tune jump is made approximately 0.2 units in 1 μs, and acceleration rate is approximately 25 GeV in 0.2 s, we might take

$$\Gamma_{\text{imp}} = 1 \times 10^{-4} \left(\frac{\hat{E}}{25\,\text{GeV}} \right),$$

$$\Gamma_{\text{int}} = 0.085 \left(\frac{\hat{E}}{25\,\text{GeV}} \right),$$

where $\hat{E}$ is the maximum energy of the synchrotron; it is included so that the result can scale universally to all proton synchrotrons assuming the synchrotron circumference is proportional to $\hat{E}$.

To accelerate to the top energy $\hat{E}$, the beam crosses approximately $\frac{\hat{E}}{0.523\,\text{GeV}}$ imperfection resonances and $\frac{2}{P_{\text{sup}}} \frac{\hat{E}}{0.523\,\text{GeV}}$ intrinsic resonances with P_{sup} the accelerator superperiodicity. Assuming the rough scaling (5.52), and summed

---

[15]R.D. Ruth, 12th Int. Conf. High Energy Accel., Chicago (1983).

over all resonances, we obtain a rough estimate of polarization loss for a proton beam to be accelerated to a top energy $\hat{E}$,[16]

$$\frac{\Delta P}{P} \approx -0.7 \times 10^{-4}(\hat{E}[\text{GeV}])^2 - \frac{0.028}{P_\text{sup}}\,\hat{E}[\text{GeV}]\,,$$

where the first term is from crossing the imperfection resonances, the second term due to the intrinsic resonances.

The loss of polarization becomes increasingly severe as the peak energy increases, i.e. for larger synchrotrons. For example, this estimate gives $\frac{\Delta P}{P} \approx$ -10% for the AGS 25 GeV with $P_\text{sup} = 12$, while it predicts $\frac{\Delta P}{P} \approx -13$ for the Fermilab Main Ring 400 GeV with $P_\text{sup} = 6$. The polarization would most likely be destroyed if the top beam energy of the synchrotron $\hat{E} \gtrsim 100$ GeV.

Harmonic matching   We now conclude that to accelerate to energies $\gtrsim$ 100 GeV, one needs to do better than a straight acceleration even with fast tune jumps. One needs to do better than the condition when Eq. (5.52) applies. One way to proceed, applied successfully to several synchrotrons, is to implement a harmonic correction of the resonance strength.[17] The idea is to excite a special correction synchronized so that $\epsilon = 0$ at the moment when a strong resonance is to be crossed. By correcting each of the strong resonances individually during the energy ramping, admittedly a tedious process, a higher beam polarization can be reached.

To harmonic correct an imperfection resonance, for example, two sets of vertical closed orbit knobs are powered at the moment of crossing to contribute an artificial ϵ,

$$\epsilon = \frac{i(\kappa + 1)}{2\pi} \int_0^{2\pi} d\theta\, e^{-i\kappa\theta}\, \frac{G(\theta)}{B\rho}\,\rho\Delta y_\text{COD}\,, \tag{5.53}$$

that exactly compensates the contribution from the error fields, where Δy_COD is the vertical closed orbit contribution due to the harmonic orbit kickers. They are turned on adiabatically before the resonance crossing and turned off adiabatically after crossing, while the acceleration continues throughout the process.

Equation (5.53) follows directly from Eq. (5.35). The reason of needing two sets of orbit kickers is that ϵ is a complex quantity. With a good orbit correction performed ahead of time, the required orbit manipulation for harmonic correction is typically very small. The setting of these two orbit correctors is obtained empirically. Each strong resonance requires its own specific harmonic correctors.

To harmonic match the intrinsic resonances, two sets of quadrupoles are

---

[16]Strictly speaking, we need to demand $\frac{|\epsilon|^2}{\Gamma} \ll 1$ for all resonances crossed. This puts a limit on $E \lesssim 225$ GeV.

[17]H.D. Bremer, et al., Conf. High Energy Spin Physics, AIP Proc. No. 95, Brookhaven, p. 407 (1982); A.W. Chao, AIP Proc. No. 95, Brookhaven, p. 458 (1982).

powered to generate an artificial ϵ, again using Eq. (5.35),

$$\frac{\epsilon}{\sqrt{J_y}} = \frac{i(\kappa+1)}{2\pi} \int_0^{2\pi} d\theta \; e^{-i\kappa\theta+i\psi_y(\theta)} \frac{\Delta G(\theta)}{B\rho} \rho\sqrt{\beta_y(\theta)},$$

to compensate for the resonance to be crossed, where ΔG specifies the harmonic quadrupole correctors.

To minimize the effect on the betatron motion, the quadrupoles need to be changed in such a way that the two betatron tunes do not change — doing so may require more than two quadrupoles. These corrections are definitive and can be calculated ahead of time and not determined empirically on line. We loosely refer to the imperfection resonance compensation as harmonic correction, and refer to the intrinsic resonance compensation as harmonic matching.

For higher energies beyond the reach of harmonic correction and harmonic matching,[18] we will then consider the idea of Siberian snakes, a subject to be introduced in Sec. 5.6. Later on page 477 we will also discuss an analogous harmonic matching approach for electron storage rings.

Homework 5.28 A polarized beam is to be accelerated through the imperfection resonance $a\gamma = 4$. The synchrotron is known to have a complex resonance strength $\epsilon = \epsilon_1 + i\epsilon_2$ when observed at the position $\theta = 0$. Two localized vertical orbit bumps are to be used to compensate for the resonance strength. Assume there is only one quadrupole with $\Delta y_{\mathrm{COD}} \neq 0$ in each of the orbit bumps and the quadrupoles $\left(\frac{G\ell}{B\rho}\right)_{1,2}$ are at positions $\theta_{1,2}$, respectively. Find the required orbit bumps $\Delta y_{\mathrm{COD},1}$ and $\Delta y_{\mathrm{COD},2}$.

5.5   Spinor matrix formalism

The Froissart–Stora formula applies to crossing the resonance from $\theta = -\infty$ all the way to $\theta = \infty$ linearly in time. The spin dynamics during the resonance passage is found to be rather complex, reflected by the need of complex special functions in the analysis. Only the asymptotic behavior becomes simple again, leading to the Froissart–Stora formula. On the other hand, question arises as to what happens to the spin motion when the resonance crossing has a different crossing pattern than linear. Subsequent research of spin dynamics begins to study the transient spin behavior, and the resulting intricate spin interference and echo behaviors have been investigated theoretically and experimentally.

Several different resonance crossing patterns are of interest. One example is to start with a spin tune at a certain finite time (instead of $-\infty$) below the resonance, cross the resonance linearly with time, and then park the spin tune a finite time (instead of $+\infty$) above the resonance after crossing. Another example could be crossing the resonance back and forth multiple times linearly in time, and park the spin tune at some location after the process. Still another

---

[18]For one thing, harmonic correcting thousands of resonances one by one is not practical.

possibility is to cross the resonance by a sudden jump of spin tune. To study these patterns will require an extension of the Froissart–Stora analysis.

A matrix formalism has been developed for the purpose.[19] The spin motion can be expressed in terms of hypergeometric functions if the acceleration pattern consists of piecewise linear segments, including the possibility of crossing the resonance multiple number of times, thus allowing the study of interference effects of resonance crossings. Such intricate spin behavior was also experimentally tested at the COSY synchrotron with high accuracy.[20]

5.5.1   Equation of motion

We assume the spin dynamics is determined by one and only one depolarization resonance of strength ϵ at $a\gamma = \kappa$. The resonance crossing is prescribed according to a pattern

$$a\gamma(\theta) \; = \; \kappa + \alpha(\theta)\,,$$

with a time-dependent $\alpha(\theta)$.

The spinor equation of motion is, from Eqs. (5.24) and (5.34),

$$\frac{d\psi}{d\theta} \; = \; -\frac{i}{2}\begin{bmatrix} -a\gamma & \epsilon e^{i\kappa\theta} \\ \epsilon^* e^{-i\kappa\theta} & a\gamma \end{bmatrix}\psi\,. \tag{5.54}$$

We will also allow ϵ to depend on time θ. This spinor equation has implicitly assumed that $\alpha(\theta)$ and $\epsilon(\theta)$ as functions of time vary slowly compared to spin precession. This is a good assumption because spin precession is fast.

We define

$$\beta(\theta) \; = \; \int_{\theta_0}^{\theta} d\theta'\,\alpha(\theta')\,,$$

where θ_0 refers to the initial time when the spin dynamics is launched, and make the transformation

$$\psi_1 \; = \; e^{-\frac{i}{2}[\kappa\theta+\beta(\theta)]\sigma_y}\,\psi\,.$$

It follows that

$$\frac{d\psi_1}{d\theta} \; = \; -\frac{i}{2}\begin{bmatrix} 0 & \epsilon(\theta)e^{-i\beta(\theta)} \\ \epsilon^*(\theta)e^{i\beta(\theta)} & 0 \end{bmatrix}\psi_1\,.$$

We then let

$$\psi_1(\theta) \; = \; \begin{bmatrix} f(\theta) \\ g(\theta)e^{i\beta(\theta)} \end{bmatrix}\,, \tag{5.55}$$

[19] A.W. Chao, Phys. Rev. Special Topics - Accel. & Beams 8, 104001 (2005).
[20] V.S. Morozov, et al., Phys. Rev. Lett. 100, 054801 (2008).

to obtain

$$f' = -\frac{i\epsilon}{2}g,$$
$$g' = -i\alpha g - \frac{i\epsilon^*}{2}f, \tag{5.56}$$

where a prime designates $\frac{d}{d\theta}$.

It follows from Eq. (5.56) that $|f|^2 + |g|^2$ is a constant of the motion. We choose the normalization

$$|f(\theta)|^2 + |g(\theta)|^2 = 1 \qquad \text{for all } \theta.$$

This condition holds even when α and ϵ are functions of θ.

We will primarily be interested in P_y, the y-component of polarization, in this planar accelerator. This y-component at any time θ is given by

$$\begin{aligned}
P_y(\theta) &= \psi^\dagger \sigma_y \psi = \psi_1^\dagger \sigma_y \psi_1 \\
&= |f(\theta)|^2 - |g(\theta)|^2 = 2|f(\theta)|^2 - 1.
\end{aligned}$$

Eliminating g from Eq. (5.56) yields

$$f''(\theta) + \left[i\alpha(\theta) - \frac{\epsilon'(\theta)}{\epsilon(\theta)}\right] f'(\theta) + \frac{|\epsilon|^2}{4} f(\theta) = 0. \tag{5.57}$$

For the case when ϵ is constant in time, it is $|\epsilon|^2$ that plays the role in determining the polarization; the phase of ϵ does not matter.

We need to specify the initial conditions of the spin at time $\theta = \theta_0$. Let us designate $\alpha_0 = \alpha(\theta_0)$ and $\epsilon_0 = \epsilon(\theta_0)$. We assume that these parameters have been held at these values, and that the spin has been in an eigenstate, from $\theta = -\infty$ till $\theta = \theta_0$. The initial spin is given by the eigenstate,

$$\begin{bmatrix} f \\ g \end{bmatrix}_{\theta_0} = \sqrt{\frac{\Omega + |\alpha_0|}{2\Omega}} \begin{bmatrix} 1 \\ -\frac{\mathrm{sgn}(\alpha_0)}{\epsilon_0}(\Omega - |\alpha_0|) \end{bmatrix}, \tag{5.58}$$

where

$$\Omega = \sqrt{\alpha_0^2 + |\epsilon_0|^2}. \tag{5.59}$$

The initial polarization of this eigenstate is

$$P_y(\theta_0) = |f(\theta_0)|^2 - |g(\theta_0)|^2 = \frac{|\alpha_0|}{\Omega},$$

which approaches 100% when $|\alpha_0| \gg |\epsilon_0|$ as one would expect because the spin tune is far from the resonance.

Note that being in a pure eigenstate, the spin is initially 100% polarized even though its y-component $|P_y(\theta_0)| < 1$. Indeed, one may compute the x- and z-components of the polarization to obtain

$$P_x(\theta_0) = \frac{\mathrm{sgn}(\alpha_0)}{\Omega} \mathrm{Im}(\epsilon_0 e^{i\kappa\theta_0}),$$
$$P_z(\theta_0) = -\frac{\mathrm{sgn}(\alpha_0)}{\Omega} \mathrm{Re}(\epsilon_0 e^{i\kappa\theta_0}),$$

and $P_x^2(\theta_0) + P_z^2(\theta_0) + P_y^2(\theta_0) = 1$. But as mentioned, we will be interested only in the behavior of $P_y(\theta)$.

5.5.2   Piecewise constant α and ϵ

We are now ready to treat some specific cases. Consider the case when α and ϵ stay at constant values from time θ_1 to θ_2, i.e. when

$$\alpha(\theta) \;=\; \alpha_0, \qquad \epsilon(\theta) \;=\; \epsilon_0, \qquad \theta_2 > \theta > \theta_1 \,.$$

The spin motion in this period can be written in a matrix form,

$$\begin{bmatrix} f \\ g \end{bmatrix}_{\theta_2 > \theta > \theta_1} \;=\; T_{\alpha_0,\epsilon_0}(\theta,\theta_1) \begin{bmatrix} f \\ g \end{bmatrix}_{\theta_1} ,$$

where the map that brings the spin state from θ_1 to θ is designated by a matrix $T_{\alpha_0,\epsilon_0}(\theta,\theta_1)$. In this simple case, it is given by solving Eq. (5.56). The result is

$$T_{\alpha_0,\epsilon_0}(\theta,\theta_1) \;=\; e^{-\frac{i}{2}\alpha_0(\theta-\theta_1)} \begin{bmatrix} 1 & 0 \\ \frac{\alpha_0}{\epsilon_0} & \frac{i\Omega}{\epsilon_0} \end{bmatrix} \begin{bmatrix} \cos\Theta & \sin\Theta \\ -\sin\Theta & \cos\Theta \end{bmatrix} \begin{bmatrix} 1 & 0 \\ \frac{i\alpha_0}{\Omega} & -\frac{i\epsilon_0}{\Omega} \end{bmatrix} ,$$

$$\tag{5.60}$$

where

$$\Theta \;=\; \frac{\Omega}{2}\,(\theta-\theta_1)\,.$$

One can check that the normalization is preserved with this map, i.e., $|f(\theta)|^2 + |g(\theta)|^2 = |f(\theta_1)|^2 + |g(\theta_1)|^2$.

The special case when $\epsilon_0 = 0$ — the case without resonance — gives

$$T_{\alpha_0,0}(\theta,\theta_1) \;=\; \begin{bmatrix} 1 & 0 \\ 0 & e^{-i\alpha_0(\theta-\theta_1)} \end{bmatrix} .$$

Another special case occurs when $\Theta = m\pi$ where m is an integer, i.e. when the accumulated spin precession angle $\Omega(\theta-\theta_1)$ is an integral multiple of 2π. Then

$$T_{\alpha_0,\epsilon_0}(\theta,\theta_1) \;=\; (-1)^m e^{-i\frac{\alpha_0}{\Omega}m\pi} \begin{bmatrix} 1 & 0 \\ 0 & 1 \end{bmatrix} .$$

Other than an overall phase factor, the spin motion in this period is a unit transformation. This period from θ_1 to θ is therefore spin transparent, and has no net effect on the spin motion.

Still another special case is when $\Theta = \left(m+\frac{1}{2}\right)\pi$, in which case we have

$$T_{\alpha_0,\epsilon_0}(\theta,\theta_1) \;=\; (-1)^m e^{-i\frac{\alpha_0}{2\Omega}(2m+1)\pi} \frac{1}{\Omega} \begin{bmatrix} i\alpha_0 & -i\epsilon_0 \\ -i\epsilon_0^* & -i\alpha_0 \end{bmatrix} .$$

Application 1 As a first application of this case of constant α and constant ϵ, consider a spin launched at time θ_0 with initial conditions (5.58), and let α and ϵ be kept at their initial values α_0 and ϵ_0. The spin will stay in the eigenstate, with

$$
\begin{bmatrix} f \\ g \end{bmatrix}_{\theta > \theta_0} = T_{\alpha_0, \epsilon_0}(\theta, \theta_0) \begin{bmatrix} f \\ g \end{bmatrix}_{\theta_0}
$$

$$
= e^{\frac{i}{2}(\theta - \theta_0)[\mathrm{sgn}(\alpha_0)\Omega - \alpha_0]} \begin{bmatrix} f \\ g \end{bmatrix}_{\theta_0}. \tag{5.61}
$$

In this case, polarization in this time period is preserved,

$$
P_y(\theta) = |f(\theta)|^2 - |g(\theta)|^2 = |f(\theta_0)|^2 - |g(\theta_0)|^2 = \frac{|\alpha_0|}{\Omega}.
$$

Application 2 Our formulation allows the case of a spin tune jump across the resonance. As a second application, consider the case when the resonance strength ϵ stays constant at the value ϵ_0, while the spin tune crosses the resonance at time $\theta = 0$ according to a tune-jump pattern,

$$
\alpha(\theta) = \begin{cases} -A, & \text{if } \theta < 0, \\ A, & \text{if } \theta > 0, \end{cases}
$$

as illustrated in Fig. 5.6(a). The sign is such that the resonance is crossed from below when $A > 0$ and from above if $A < 0$.

In order for the single-resonance approximation, i.e. Eq. (5.54), to hold, the sudden change of α is assumed to take place over at least several spin precession periods. At the same time, it should be fast enough to qualify for a sudden jump. These conditions together require that the time during which the jump takes place, measured in number of turns n_{jump}, should satisfy

$$
\frac{1}{a\gamma} \ll n_{\mathrm{jump}} \ll \frac{1}{\Omega},
$$

where Ω is given by Eq. (5.59), $\Omega = \sqrt{A^2 + |\epsilon_0|^2}$.

The spin state after crossing is given by

$$
\begin{bmatrix} f \\ g \end{bmatrix}_{\theta > 0} = T_{A, \epsilon_0}(\theta, 0)\, T_{-A, \epsilon_0}(0, \theta_0) \begin{bmatrix} f \\ g \end{bmatrix}_{\theta_0}.
$$

When the initial conditions are specified by (5.58), then

$$
\begin{bmatrix} f \\ g \end{bmatrix}_{\theta > 0} = e^{-\frac{i}{2}A(\theta + \theta_0)} e^{-\frac{i}{2}\mathrm{sgn}(A)\Omega(\theta - \theta_0)} \sqrt{\frac{\Omega + |A|}{2\Omega}}
$$

$$
\times \begin{bmatrix} 1 + 2i\frac{A}{\Omega} \sin\frac{\Omega\theta}{2} e^{\frac{i}{2}\mathrm{sgn}(A)\Omega\theta} \\ \frac{\mathrm{sgn}(A)}{\epsilon_0}(\Omega - |A|)\left(1 - 2i\frac{A}{\Omega} \sin\frac{\Omega\theta}{2} e^{\frac{i}{2}\mathrm{sgn}(A)\Omega\theta}\right) \end{bmatrix}.
$$

Figure 5.6: A one-step spin tune jump. (a) Spin tune distance to resonance, α, as a function of time θ jumping across the resonance once. (b) Transient behavior of the spin polarization component P_y for three cases $\frac{|\epsilon|}{A} = 0.3, 1, 3$, respectively (red, blue, green). The horizontal axis of (a) is θ. That of (b) is $\Omega\theta$.

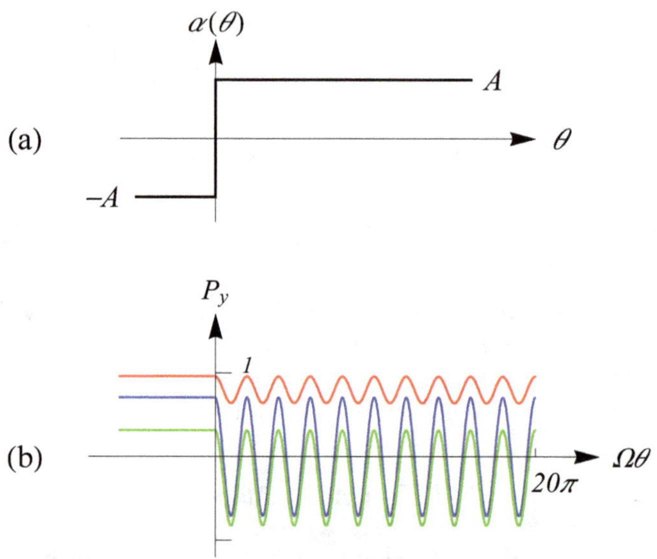

Again one can verify that the normalization $|f(\theta)|^2 + |g(\theta)|^2 = 1$ holds. The polarization after the spin tune jump is found to be

$$
\begin{aligned}
P_y(\theta > 0) &= 2|f(\theta)|^2 - 1 \\
&= \frac{|A|}{(A^2 + |\epsilon_0|^2)^{3/2}} \left[(A^2 - |\epsilon_0|^2) + 2|\epsilon_0|^2 \cos\Omega\theta \right].
\end{aligned}
\tag{5.62}
$$

This polarization P_y is plotted in Fig. 5.6(b) for three cases for illustration. It oscillates in time θ around the mean value $\frac{|A|\,(A^2 - |\epsilon_0|^2)}{(A^2 + |\epsilon_0|^2)^{3/2}}$ with frequency Ω, and it oscillates between the maximum value $\frac{|A|}{(A^2 + |\epsilon_0|^2)^{1/2}}$ and the minimum value $\frac{|A|(A^2 - 3|\epsilon_0|^2)}{(A^2 + |\epsilon_0|^2)^{3/2}}$, where the maximum value is actually equal to the initial polarization at time θ_0. Although the final polarization depends on the observation time θ, it does not depend on the initial time θ_0 because the spin was launched in its eigenstate.

Application 3   As a third application, consider an experiment that turns off the resonance at time $\theta = \theta_1 > 0$ after a tune jump at time $\theta = 0$,

$$
\alpha(\theta) = \begin{cases} -A, & \text{if } \theta < 0, \\ A, & \text{if } \theta > 0, \end{cases}
$$

$$\epsilon(\theta) \;=\; \begin{cases} \epsilon_0, & \text{if } \theta < \theta_1, \\ 0, & \text{if } \theta > \theta_1. \end{cases}$$

The spin state after switching off the resonance is

$$\begin{bmatrix} f \\ g \end{bmatrix}_{\theta>\theta_1} = T_{A,0}(\theta,\theta_1) T_{A,\epsilon_0}(\theta_1,0) T_{-A,\epsilon_0}(0,\theta_0) \begin{bmatrix} f \\ g \end{bmatrix}_{\theta_0}. \tag{5.63}$$

One can then calculate the polarization $P_y(\theta > \theta_1)$ and find that it does not depend on θ after the resonance is switched off, and is given by Eq. (5.62) with θ set to θ_1, as one might have expected.

Application 4 In this application, we keep the resonance strength ϵ_0 fixed, but make a spin tune jump across the resonance at time $\theta = 0$, and double-jump across the resonance back again at time $\theta_1 > 0$, i.e.

$$\alpha(\theta) \;=\; \begin{cases} -A, & \text{if } \theta < 0, \\ A, & \text{if } 0 < \theta < \theta_1, \\ -A, & \text{if } \theta_1 < \theta, \end{cases} \tag{5.64}$$

as seen in Fig. 5.7.

Figure 5.7: A two-step spin tune jump. Spin tune distance to resonance, α, as a function of time θ jumping across the resonance twice.

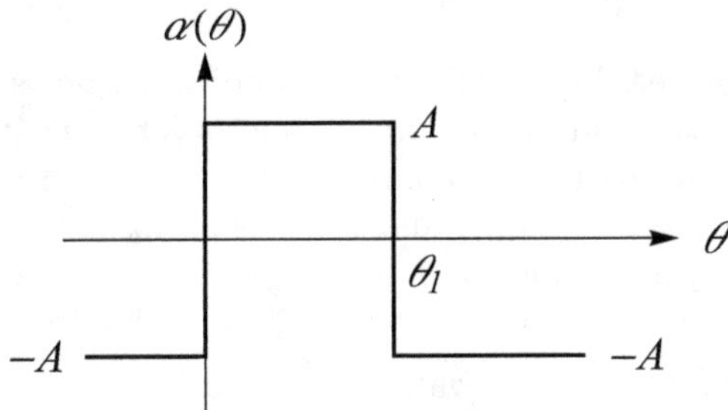

This example is of interest to study the interference effect of two resonance crossings. In general, one may consider crossing two nearby resonances in one tune ramping process. Here we try to gain some insight by considering crossing a single resonance twice. In this example, the crossings are performed by sudden tune jumps forward and backward, and the two crossings are separated in time by θ_1.

The polarization during $0 < \theta < \theta_1$ is already found in Eq. (5.62). The spin state after the second crossing, $\theta > \theta_1$, is given by

$$\begin{bmatrix} f \\ g \end{bmatrix}_{\theta>\theta_1} = T_{-A,\epsilon_0}(\theta,\theta_1) T_{A,\epsilon_0}(\theta_1,0) T_{-A,\epsilon_0}(0,\theta_0) \begin{bmatrix} f \\ g \end{bmatrix}_{\theta_0}.$$

Polarization is then found to be, after some algebra,

$$P_y(\theta) = \frac{|A|}{\Omega^5}\left\{\left[(\Omega^2 - 2A^2)^2 + 4A^2(\Omega^2 - A^2)\cos\Omega\theta_1\right]\right.$$
$$+\ 2(\Omega^2 - A^2)(\Omega^2 + 2A^2\cos\Omega\theta_1)\cos\Omega\theta(1 - \cos\Omega\theta_1)$$
$$\left.-\ 2(\Omega^2 - A^2)\sin\Omega\theta\sin\Omega\theta_1\left[\Omega^2 - 2A^2(1 - \cos\Omega\theta_1)\right]\right\}. \qquad (5.65)$$

There are three terms in Eq. (5.65). The second and third terms combine to give an oscillatory contribution P_{osc} of the polarization after the double spin tune jumps. The oscillation frequency is Ω. The average polarization P_{av} is given by the first term of Eq. (5.65).

We see that P_{osc} and P_{av} depend on θ_1, the time separation between the two jumps. As promised, Eq. (5.65) contains information concerning the interference between the two jump crossings and suggests an echo phenomenon to be explored. Before continuing to other applications of the matrix formalism, let us pause to consider these effects.

Resonance interference   According to Eq. (5.65), a destructive interference between the two crossings will yield a maximum average polarization, which occurs when

$$\Omega\theta_1 = \text{even multiple of } \pi\,, \qquad (5.66)$$
$$P_{\text{av}} = \frac{|A|}{\Omega}\,,$$
$$P_{\text{osc}} = 0\,.$$

As mentioned earlier, Eq. (5.66) is the condition when the spin motion between the two spin tune jumps is transparent. When that happens, the two equal and opposite jumps have a complete destructive interference. The final polarization keeps its initial value, and there is no loss of polarization.

If the spin tune jumps not to A but to some other value A' for the time period between 0 and θ_1, the condition for destructive interference still occurs at $\sqrt{A'^2 + |\epsilon|^2}\,\theta_1 = 2m\pi$. This is valid even if A' is negative, i.e. the resonance is not actually crossed.

A constructive interference yields a minimum average polarization, which occurs with

$$\Omega\theta_1 = \text{odd multiple of } \pi\,,$$
$$P_{\text{av}} = \frac{|A|}{\Omega^5}(\Omega^4 - 8A^2\Omega^2 + 8A^4)\,,$$
$$P_{\text{osc}} = \frac{4|A|}{\Omega^5}(\Omega^2 - A^2)(\Omega^2 - 2A^2)\cos\Omega\theta\,.$$

Figure 5.8 shows the polarization of a particle going through the double resonance crossing destructively and constructively. The parameters used are $|\epsilon_0| = 0.08, a\gamma = 4, A = 0.1$. The destructive case (a) has $\Omega\theta_1 = 16\pi$; the constructive case (b) has $\Omega\theta_1 = 15\pi$.

Figure 5.8: Behavior of particle polarization successively crossing a resonance twice, (a) destructively and (b) constructively.

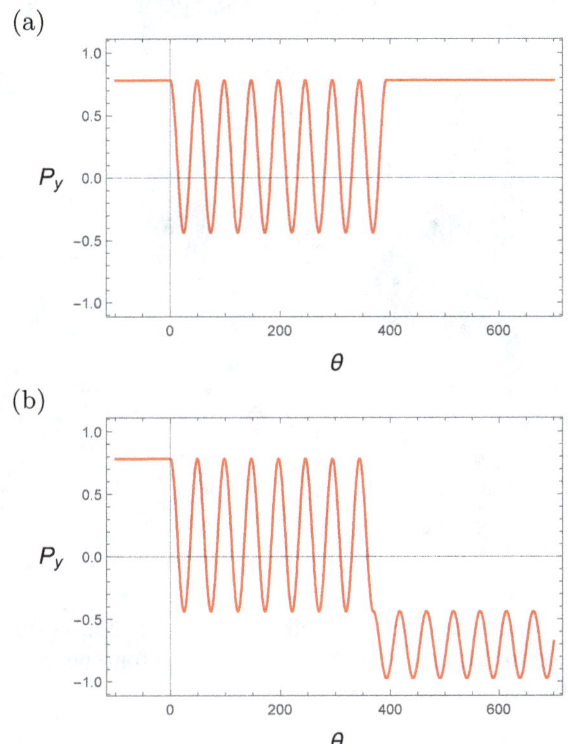

It should be pointed out that this interference behavior is independent of whether $|A|$ is small or large compared with $|\epsilon|$. It also does not matter whether θ_1 is small or large as long as $\Omega\theta_1$ is an even or odd multiple of π. In other words, at least in principle, after crossing a resonance, the potential of interfering strongly with a next resonance crossing lasts indefinitely in time, and occurs even when the resonance is crossed "cleanly" in each crossing (meaning $|A| \gg |\epsilon|$). In an acceleration process, for example, it is not how long ago a resonance was crossed or how far the spin tune has passed the resonance that determines whether this resonance crossing can be considered an isolated event. The effect of crossing a resonance carries with it an infinitely long memory.

But this intriguing observation has a catch. It applies only to the case of a single particle. For a beam of particles with a finite energy spread among the particles, an averaging on our results over the beam's energy distribution will have to be carried out. It is then expected that some of the interference features we observe here, particularly its infinitely long memory, will be diluted.

Figure 5.9: Response of a polarized beam to a double-jump across a resonance. Three curves correspond to three different energy spreads of the beam. The red, green, blue curves are for a beam with energy spread $\sigma_\delta = 0, 0.1\%$ and 0.3%, respectively.

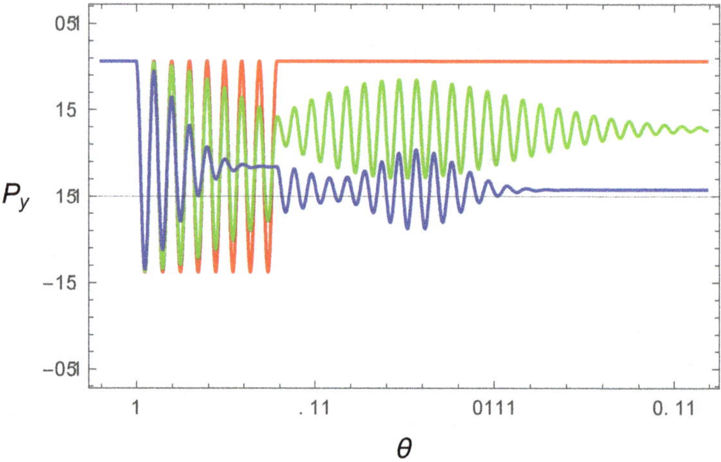

Effect of energy spread More specifically for the double-jump application here, the polarization for an off-energy particle after the first jump will be determined by

$$
\begin{bmatrix} f \\ g \end{bmatrix}_{\theta_1 > \theta > 0} = T_{A+a\gamma\delta,\epsilon_0}(\theta,0)\, T_{-A+a\gamma\delta,\epsilon_0}(0,\theta_0) \begin{bmatrix} f \\ g \end{bmatrix}_{\theta_0},
$$

where $\delta = \frac{\Delta\gamma}{\gamma}$ is the relative energy deviation of the particle, while after the second jump, we have

$$
\begin{bmatrix} f \\ g \end{bmatrix}_{\theta > \theta_1} = T_{-A+a\gamma\delta,\epsilon_0}(\theta,\theta_1)\, T_{A+a\gamma\delta,\epsilon_0}(\theta_1,0)\, T_{-A+a\gamma\delta,\epsilon_0}(0,\theta_0) \begin{bmatrix} f \\ g \end{bmatrix}_{\theta_0}. \quad (5.67)
$$

The net beam polarization will then be obtained by integrating the single-particle polarization over the δ-distribution of the beam — more detail of this derivation will be given in Sec. 7.5.2.

Figure 5.9 is an illustration of the transition from a single particle to a beam. Parameters used are the same as that of the destructive case of Fig. 5.8. The red curve shows the polarization P_y as a function of θ for a single on-momentum particle, or a beam with no energy spread, and it reproduces Fig. 5.8(a). The green and blue curves are what happens to a beam of Gaussian energy distribution with rms $\sigma_\delta = 0.1\%$ and 0.3%, respectively.

For the blue curve with a large energy spread, one sees that the polarization after each jump settles down to some equilibrium value, and each jump reduces the polarization level by a factor $\frac{A^2 - |\epsilon_0|^2}{A^2 + |\epsilon_0|^2} = 0.22$, which is what is expected if each jump is to be regarded as an isolated event and if the oscillatory term of polarization after a jump is ignored [keeping only the first term in Eq. (5.62)]. After each jump, the memory of the jump is kept for a time

$$\theta_{\text{memory}} \sim \frac{\pi\Omega}{Aa\gamma\sigma_\delta},$$

which for the parameters of the blue curve gives $\theta_{\text{memory}} \sim 300$. After a time of θ_{memory}, the jump can be considered an isolated event. When that happens, it does not matter whether the two crossings are destructive or constructive; they both yield the same final beam polarization.

As a result, a single particle remembers resonances forever, while a beam (with enough energy spread) treats each resonance crossing as isolated independent events. What is happening is quite amazing. In this sense, it should be kept in mind that, strictly speaking, the Froissart–Stora formula applies only to a polarized beam; it does not apply to single particles.

Spin echo One should keep in mind that the loss of beam polarization is only due to the beam's energy spread; individual particles remember the jumps indefinitely. This effect provides the possibility of a spin echo phenomena to be explored in a circular accelerator.

The general fact that a single particle keeps indefinite memory of its detailed history while a beam smears these memories out provides the circumstance of an echo effect will be the subject of Chapter 7. As we will mention then, echoes are very wide-spread and common phenomena in accelerators.

As a prelude to spin echo, let us inspect Fig. 5.9 a bit closer. In a double-jump example, even if the beam has a large energy spread so that each of the resonance crossing can be regarded as independent event as far as interference effects are concerned, we observe that an echo of the spin polarization still occurs unexpectedly as a sudden recoherence of spin motion at the time $\theta = 2\theta_1$.[21] In the above example of Fig. 5.9, the echo already shows up curiously in both the green and the blue curves as secondary excitations around $\theta = 2\theta_1 \approx 785$. There is an echo signal even when θ_1 is much larger than θ_{memory}.

We will return to the subject of spin echo in Sec. 7.5 with a more detailed derivation of the echo signal. To wrap up discussion in this section, however, we show here another more exaggerated case in Fig. 5.10 when the energy spread is increased to $\sigma_\delta = 0.8\%$, θ_1 is extended to $\Omega\theta_1 = 41\pi$ and A increased to 0.15.[22] The three vertical lines indicate the locations of $\theta = 0, \theta_1, 2\theta_1$. One sees that, although the large energy spread of the beam has apparently completely suppressed all transient behavior, there remains a prominent echo of polarization near time $2\theta_1$.

---

[21]A.W. Chao and E.D. Courant, Phys. Rev. ST-Accel. & Beams, 10, 014001 (2007); M. Bai, A. Chao, and T. Roser, AIP Conf. Proc. 1149, 785 (2009).

[22]We could have taken $\Omega\theta_1 = 40\pi$. It does not matter whether the interference is nominally constructive or destructive, the echo occurs in either case.

Figure 5.10: An echo effect occurs long time after resonance crossings even when the beam energy spread is large.

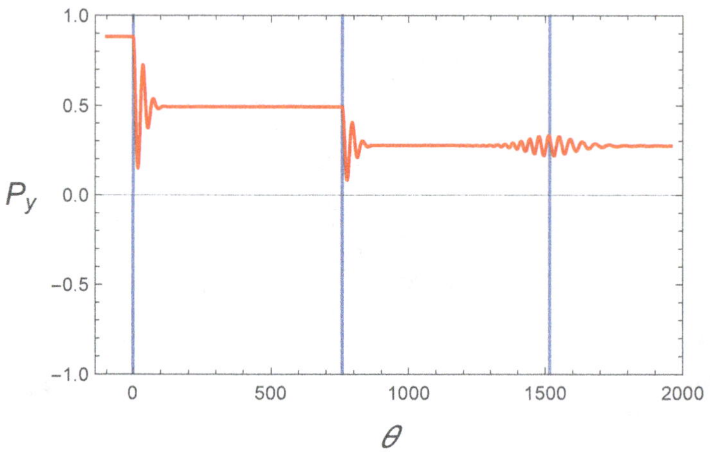

Homework 5.29
  (a) Verify Eq. (5.62).
  (b) Verify Eq. (5.65).

Homework 5.30 The text gave an application (Application 2) of the matrix formalism when the resonance strength ϵ is held fixed while the spin tune makes a sudden jump. Follow the analysis to study the case when the spin tune is held fixed while the resonance strength makes a sudden jump from ϵ_0 to ϵ_1. Give the special cases when
  (a) $\epsilon_1 = -\epsilon_0$.
  (b) $\epsilon_1 = \epsilon_0^*$.
  (c) $\epsilon_1 = 2\epsilon_0$.
Change of resonance strength can be made by a change of the vertical closed orbit (imperfection resonance) or a kick of the vertical betatron oscillation (intrinsic resonance) in a synchrotron. In this case, the beam is not accelerated and an experiment should be easier to perform.

5.5.3   Constant ϵ, piecewise-linear $\alpha(\theta)$

So far we have considered several applications of the matrix formalism when the resonance is crossed by sudden jumps. The formalism also applies when the resonance is crossed by varying the spin tune piecewise-linearly in time.

In this case, we consider a resonance strength ϵ that is constant in time, while the spin tune has a linear dependence,

$$\alpha(\theta) = (\theta - \theta_c)\Gamma, \tag{5.68}$$

with Γ specifying the rate of change of the spin tune (Γ can have either sign) and θ_c specifying the time when spin tune crosses the resonance (by linear extrapolation if resonance is not actually crossed).

Equation of motion We consider the time range over which Eq. (5.68) applies to be from $\theta = \theta_1$ to θ_2. Between θ_1 and θ_2, we have from Eq. (5.57),

$$f''(\theta) + i\Gamma(\theta - \theta_c)f'(\theta) + \frac{|\epsilon|^2}{4} f(\theta) \; = \; 0 \,.$$

We now follow steps similar to the Froissart–Stora case and make a transformation

$$f(\theta) \; = \; F(\theta)\, e^{-i\frac{\Gamma}{4}(\theta - \theta_c)^2} \,,$$

and change variable from θ to

$$x \; = \; \sqrt{|\Gamma|}\,(\theta - \theta_c)\,.$$

We then obtain [see also Eq. (5.48)]

$$\frac{d^2 F}{dx^2} + \left(\frac{x^2}{4} - a \right) F \; = \; 0 \,,$$

where

$$a \; = \; \frac{i}{2}\mathrm{sgn}(\Gamma) - \frac{|\epsilon|^2}{4|\Gamma|} \,.$$

Note that in this case, only the absolute value $|\epsilon|$ enters. The phase of ϵ matters only as an overall phase in determining g of Eq. (5.55), which does not have physical consequences on the beam polarization along the $\hat{y}$-direction. Given $F(\theta)$, polarization is given by $P_y(\theta) = 2|F(\theta)|^2 - 1$.

Two independent solutions for F can be expressed in terms of the degenerate hypergeometric function $_1F_1$,[23]

$$y_1(x) \; = \; e^{-i\frac{x^2}{4}}\,{}_1F_1\left(\frac{1}{4} - i\frac{a}{2}; \frac{1}{2}; i\frac{x^2}{2} \right) \,,$$

$$y_2(x) \; = \; x e^{i\frac{x^2}{4}}\,{}_1F_1\left(\frac{3}{4} + i\frac{a}{2}; \frac{3}{2}; -i\frac{x^2}{2} \right) \,.$$

It is useful here to give explicit expressions also for their derivatives,

$$\frac{dy_1}{dx} \; = \; \frac{x}{2}e^{-i\frac{x^2}{4}}\left[-i\,{}_1F_1\left(\frac{1}{4} - i\frac{a}{2}; \frac{1}{2}; i\frac{x^2}{2} \right) + (i+2a)\,{}_1F_1\left(\frac{5}{4} - i\frac{a}{2}; \frac{3}{2}; i\frac{x^2}{2} \right) \right] \,,$$

$$\frac{dy_2}{dx} \; = \; \frac{e^{i\frac{x^2}{4}}}{6}\left[3(2+ix^2)\,{}_1F_1\left(\frac{3}{4} + i\frac{a}{2}; \frac{3}{2}; -i\frac{x^2}{2} \right) - (3i-2a)x^2\,{}_1F_1\left(\frac{7}{4} + i\frac{a}{2}; \frac{5}{2}; -i\frac{x^2}{2} \right) \right] \,.$$

---

[23]P.M. Morse and H. Feshbach, Methods of Theoretical Physics, McGraw-Hill (1978). The notation there is $_1F_1(\alpha; \gamma; z) = F(\alpha|\gamma|z)$.

It is easy to show that the Wronskian

$$W(x) \;=\; y_1(x)\frac{dy_2}{dx}(x) - \frac{dy_1}{dx}(x)y_2(x)$$

is a constant in x, and is equal to 1.

The general solution for F is

$$F(x) \;=\; C_1 y_1(x) + C_2 y_2(x),$$

where $C_{1,2}$ are determined by the initial conditions. Given $f(\theta_1)$ and $g(\theta_1)$ at initial time θ_1, we have the initial conditions,

$$F(x_1) \;=\; f(x_1)e^{\frac{i}{4}\mathrm{sgn}(\Gamma)x_1^2},$$

$$\frac{dF}{dx}(x_1) \;=\; \left[-\frac{i\epsilon}{2\sqrt{|\Gamma|}}g(x_1) + \frac{i}{2}\mathrm{sgn}(\Gamma)x_1 f(x_1)\right]e^{\frac{i}{4}\mathrm{sgn}(\Gamma)x_1^2},$$

where $x_1 = \sqrt{|\Gamma|}(\theta_1 - \theta_c)$. We then have, now expressed in matrix form,

$$\begin{bmatrix} C_1 \\ C_2 \end{bmatrix} = \begin{bmatrix} \frac{dy_2}{dx} & -y_2 \\ -\frac{dy_1}{dx} & y_1 \end{bmatrix}_{x_1} e^{\frac{i}{4}\mathrm{sgn}(\Gamma)x_1^2} \begin{bmatrix} 1 & 0 \\ \frac{ix_1}{2}\mathrm{sgn}(\Gamma) & -\frac{i\epsilon}{2\sqrt{|\Gamma|}} \end{bmatrix} \begin{bmatrix} f \\ g \end{bmatrix}_{x_1}.$$

For time $\theta_2 > \theta > \theta_1$, we have

$$\begin{bmatrix} F \\ \frac{dF}{dx} \end{bmatrix}_x = \begin{bmatrix} y_1 & y_2 \\ \frac{dy_1}{dx} & \frac{dy_2}{dx} \end{bmatrix}_x \begin{bmatrix} C_1 \\ C_2 \end{bmatrix},$$

which then finally yields

$$\begin{bmatrix} f \\ g \end{bmatrix}_\theta \;=\; U_{\Gamma,\theta_c,\epsilon}(\theta,\theta_1)\begin{bmatrix} f \\ g \end{bmatrix}_{\theta_1},$$

where a map has been defined,

$$U_{\Gamma,\theta_c,\epsilon}(\theta,\theta_1) \;=\; e^{\frac{i}{4}\mathrm{sgn}(\Gamma)(x_1^2 - x^2)} \begin{bmatrix} 1 & 0 \\ \frac{\mathrm{sgn}(\Gamma)\sqrt{|\Gamma|}}{\epsilon}x & 2i\frac{\sqrt{|\Gamma|}}{\epsilon} \end{bmatrix} \begin{bmatrix} y_1 & y_2 \\ \frac{dy_1}{dx} & \frac{dy_2}{dx} \end{bmatrix}_x$$

$$\times \begin{bmatrix} \frac{dy_2}{dx} & -y_2 \\ -\frac{dy_1}{dx} & y_1 \end{bmatrix}_{x_1} \begin{bmatrix} 1 & 0 \\ \frac{ix_1}{2}\mathrm{sgn}(\Gamma) & -\frac{i\epsilon}{2\sqrt{|\Gamma|}} \end{bmatrix}. \qquad (5.69)$$

Other than a phase factor, the outer two matrices of the above product bring the f,g representation to $F, \frac{dF}{dx}$, while the middle two matrices describe the dynamics of $F, \frac{dF}{dx}$. When $\theta = \theta_1$, we have $U = I$.

Application 5

As an application of the case with linear ramping in spin tune, let us start with initial condition (5.58) at time θ_0, and cross the resonance in a sloped-step pattern,

$$\alpha(\theta) = \begin{cases} -A, & \text{if } \theta < -\frac{A}{\Gamma}, \\ \Gamma\theta, & \text{if } \frac{A}{\Gamma} > \theta > -\frac{A}{\Gamma}, \\ A, & \text{if } \theta > \frac{A}{\Gamma}, \end{cases}$$

as shown in Fig. 5.11.

Figure 5.11: Crossing a resonance in a sloped-step pattern.

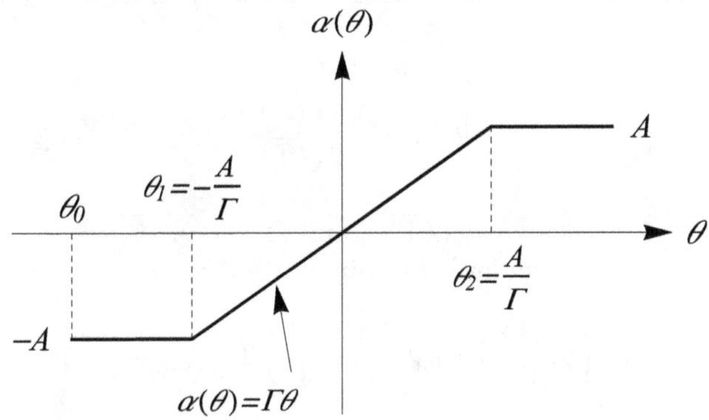

The crossing speed is specified by the slope Γ. The spin tune swings from $-A$ to A linearly. We will consider Γ and A to have the same sign. When $A > 0, \Gamma > 0$, the resonance is crossed from below as shown in Fig. 5.11. When $A < 0, \Gamma < 0$, the resonance is crossed from above. The following analysis applies to both cases. The resonance strength has a constant value ϵ throughout the process.

For time $\theta_1 > \theta > \theta_0$ before starting the crossing, spin state is mapped by

$$\begin{bmatrix} f \\ g \end{bmatrix}_\theta = T_{-A,\epsilon}(\theta, \theta_0) \begin{bmatrix} f \\ g \end{bmatrix}_{\theta_0},$$

where $f(\theta_0)$ and $g(\theta_0)$ are given by Eq. (5.58) with α_0 and ϵ_0 replaced by $-A$ and ϵ, respectively. The result has been given by Eq. (5.61). The polarization is constant,

$$P_y(\theta) = \frac{|A|}{\sqrt{A^2 + |\epsilon|^2}}. \tag{5.70}$$

During the crossing, $\theta_2 > \theta > \theta_1$, the spin state is given by

$$\begin{bmatrix} f \\ g \end{bmatrix}_\theta = U_{\Gamma,0,\epsilon}\left(\theta, -\frac{A}{\Gamma}\right) T_{-A,\epsilon}\left(-\frac{A}{\Gamma}, \theta_0\right) \begin{bmatrix} f \\ g \end{bmatrix}_{\theta_0}. \tag{5.71}$$

We now define four quantities

$$
Y_k = y_k(x_2) = y_k\left(\frac{|A|}{\sqrt{|\Gamma|}}\right),
$$

$$
Y_k' = \frac{dy_k}{dx}(x_2) = \frac{dy_k}{dx}\left(\frac{|A|}{\sqrt{|\Gamma|}}\right), \quad k = 1, 2. \tag{5.72}
$$

Since $y_1(x)$ and $\frac{dy_2}{dx}(x)$ are even functions of x, $y_2(x)$ and $\frac{dy_1}{dx}(x)$ are odd functions of x, and $x_1 = -x_2$, we have $y_1(x_1) = Y_1$, $\frac{dy_1}{dx}(x_1) = -Y_1'$, $y_2(x_1) = -Y_2$, $\frac{dy_2}{dx}(x_1) = Y_2'$.

After some algebra, we find by using Eq. (5.71) that, for $\theta_2 > \theta > \theta_1$, the polarization is given by

$$
P_y(\theta) = \frac{\Omega + |A|}{\Omega} \left| y_1 Y_2' + y_2 Y_1' - \frac{i\,\mathrm{sgn}(A)\Omega}{2\sqrt{|\Gamma|}}(y_1 Y_2 + y_2 Y_1) \right|^2 - 1, \tag{5.73}
$$

where $y_{1,2}$ are evaluated at $x = \sqrt{|\Gamma|}\,\theta$, and $\Omega = \sqrt{A^2 + |\epsilon|^2}$.

We then proceed to calculate the spin state after crossing, $\theta > \theta_2$,

$$
\begin{bmatrix} f \\ g \end{bmatrix}_\theta = T_{A,\epsilon}\left(\theta, \frac{A}{\Gamma}\right) U_{\Gamma,0,\epsilon}\left(\frac{A}{\Gamma}, -\frac{A}{\Gamma}\right) T_{-A,\epsilon}\left(-\frac{A}{\Gamma}, \theta_0\right) \begin{bmatrix} f \\ g \end{bmatrix}_{\theta_0}.
$$

After some algebra, we find

$$
\begin{aligned}
P_y(\theta) = \frac{\Omega + |A|}{\Omega} \Bigg| &(Y_1 Y_2' + Y_2 Y_1')e^{-i\,\mathrm{sgn}(A)\Theta} \\
&- \frac{i\,\mathrm{sgn}(A)\Omega}{\sqrt{|\Gamma|}} Y_1 Y_2 \cos\Theta + \frac{4\sqrt{|\Gamma|}}{\Omega} Y_1' Y_2' \sin\Theta \Bigg|^2 - 1,
\end{aligned} \tag{5.74}
$$

where $\Theta = \frac{\Omega}{2}\left(\theta - \frac{A}{\Gamma}\right)$. This polarization oscillates in time θ with frequency Ω. Equation (5.74) reduces to (5.62) in the limit $|\Gamma| \to \infty$, the case of a spin tune jump, as it should.

Equations (5.70, 5.73, 5.74) combine to describe the polarization as a function of time during the crossing process. Figure 5.12 shows $P_y(\theta)$ as a function of $\sqrt{\Gamma}\theta$ for three cases (with $A > 0, \Gamma > 0$),

$$
\frac{A}{\sqrt{\Gamma}} = \pi, \quad \frac{|\epsilon|}{\sqrt{\Gamma}} = 1, \qquad \text{red},
$$

$$
\frac{A}{\sqrt{\Gamma}} = \pi, \quad \frac{|\epsilon|}{\sqrt{\Gamma}} = 4, \qquad \text{green},
$$

$$
\frac{A}{\sqrt{\Gamma}} = 1, \quad \frac{|\epsilon|}{\sqrt{\Gamma}} = 1, \qquad \text{blue}.
$$

Figure 5.12: Single-particle polarization as function of time during a sloped-step resonance crossing for three example cases.

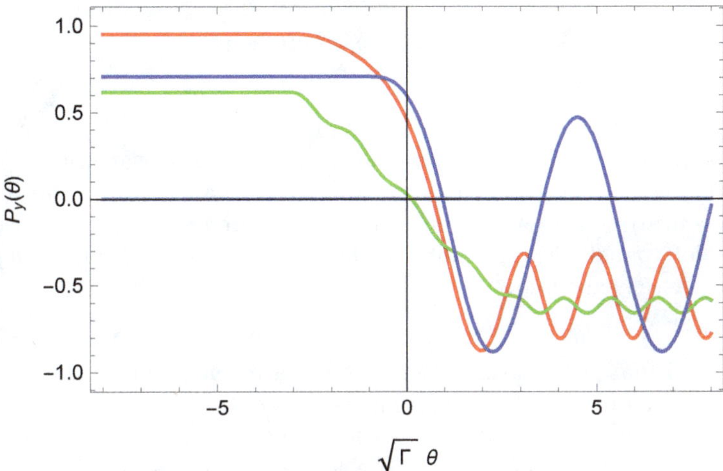

Comparison with Froissart–Stora formula   Equation (5.74) may be considered an extension of the Froissart–Stora formula, which was derived assuming the resonance was crossed linearly with an infinite step-size from $\theta = -\infty$ to $\theta = +\infty$. In an experiment with finite step-size, we will consider Eq. (5.74) to be the "final polarization" after crossing. Unlike the Froissart–Stora formula, this final polarization depends on the time of observation, and it oscillates sinusoidally with a frequency $\Omega = \sqrt{A^2 + |\epsilon|^2}$ and an amplitude $\pm P_{\text{osc}} = \pm |D_1^2 + D_2^2|$, centered around $P_{\text{av}} = |D_1|^2 + |D_2|^2 - 1$, i.e.

$$P_{\text{final}} = (|D_1|^2 + |D_2|^2 - 1) \pm |D_1^2 + D_2^2|, \tag{5.75}$$

with

$$D_1 = \sqrt{\frac{\Omega + |A|}{2\Omega}} \left[Y_1 Y_2' + Y_1' Y_2 - i \frac{\text{sgn}(A)\Omega}{\sqrt{|\Gamma|}} Y_1 Y_2 \right],$$

$$D_2 = \sqrt{\frac{\Omega + |A|}{2\Omega}} \left[\frac{4\sqrt{|\Gamma|}}{\Omega} Y_1' Y_2' - i\, \text{sgn}(A)(Y_1 Y_2' + Y_1' Y_2) \right].$$

This final polarization depends only on two scaled variables

$$\bar{A} = \frac{A}{2\sqrt{|\Gamma|}}, \qquad \bar{\epsilon} = \frac{\epsilon}{2\sqrt{|\Gamma|}}, \tag{5.76}$$

where $\bar{A}$ is the normalized step size, $\bar{\epsilon}$ is the normalized resonance width.

The red curves in Fig. 5.13 are the mean, the upper bound, and the lower bound values of the final polarization as functions of $|\bar{\epsilon}|$ for four cases $\bar{A} = 0.5$

(upper left), 1 (upper right), 2 (lower left), and 10 (lower right). The blue curves
are the Froissart–Stora formula

$$P_{\text{final,FS}} \;=\; 2e^{-2\pi|\bar\epsilon|^2} - 1 \,.$$

The Froissart–Stora formula of course does not depend on $\bar A$. The blue curves in
all four panels are therefore the same. Naturally, the Froissart–Stora prediction
is accurate when $|\bar A| \gg |\bar\epsilon|$ such as the lower right panel shows, and it becomes
rather inaccurate when $|\bar A|$ becomes comparable or less than $|\bar\epsilon|$ as seen e.g. in
the upper left panel. Applicability of the Froissart–Stora formula depends on
the value of $\bar A$ as expected.

Figure 5.13: Final polarization after crossing a resonance in a sloped-step pat-
tern as function of $|\bar\epsilon|$ for four different values of $\bar A$. Red curves give the mean,
upper bound, and lower bound of the oscillating final polarization. Blue curves
are the Froissart–Stora prediction. The four cases are $\bar A = 0.5$ (upper left), 1
(upper right), 2 (lower left), and 10 (lower right).

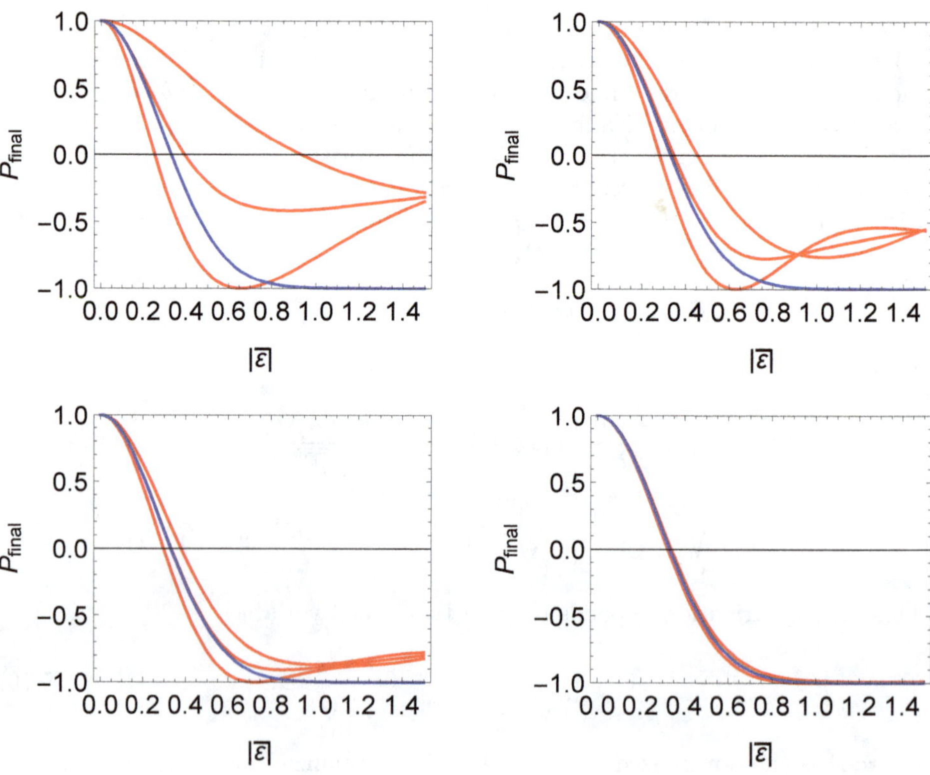

Application 6

The previous application 5 is somewhat sophisticated. The matrix formalism can be extended to more sophisticated applications. For example, a double-crossing pattern as shown in Fig. 5.14,

$$
\alpha(\theta) \;=\;
\begin{cases}
-A, & \text{if } \theta < -\frac{2A}{\Gamma}, \\
\Gamma\theta + A, & \text{if } 0 > \theta > -\frac{2A}{\Gamma}, \\
-\Gamma\theta + A, & \text{if } \frac{2A}{\Gamma} > \theta > 0, \\
-A, & \text{if } \theta > \frac{2A}{\Gamma},
\end{cases}
$$

would be useful to investigate the interference behavior between the two crossings. We consider the case $A > 0, \Gamma > 0$.

Figure 5.14: Double crossing of a resonance in a sloped-step pattern.

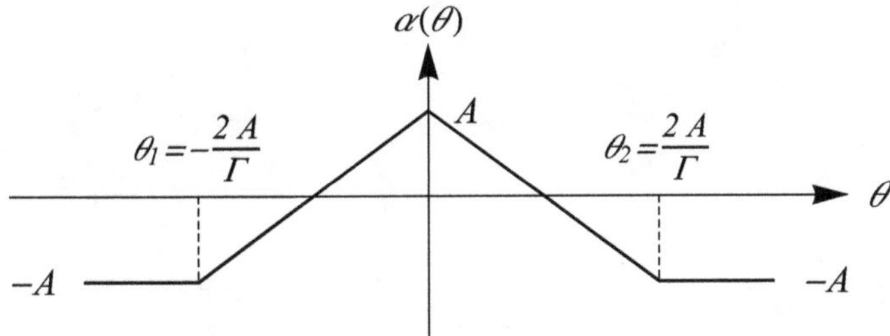

Spin state after crossing $(\theta > \frac{2A}{\Gamma})$ is given by

$$
\begin{bmatrix} f \\ g \end{bmatrix}_{\theta}
= T_{-A,\epsilon}\!\left(\theta, \frac{2A}{\Gamma}\right)
U_{-\Gamma,\frac{A}{\Gamma},\epsilon}\!\left(\frac{2A}{\Gamma}, 0\right)
U_{\Gamma,-\frac{A}{\Gamma},\epsilon}\!\left(0, -\frac{2A}{\Gamma}\right)
T_{-A,\epsilon}\!\left(-\frac{2A}{\Gamma}, \theta_0\right)
\begin{bmatrix} f \\ g \end{bmatrix}_{\theta_0}.
$$

Like before, polarization after the two resonance crossings, given by $2|f|^2 - 1$, oscillates in time θ with frequency $\Omega = \sqrt{A^2 + |\epsilon|^2}$ around a mean value. Figure 5.15 shows the mean, the upper bound, and the lower bound values of the final polarization as a function of $|\bar{\epsilon}|$, again for four cases $\bar{A} = 0.5$ (upper left), 1 (upper right), 2 (lower left), 10 (lower right).

Figure 5.15 exhibits a complex dependence of the polarization on the resonance and spin tune parameters. Of particular interest is the interference pattern of the two crossings. In particular, one notices the quasiperiodic behavior of the polarization as a function of $|\bar{\epsilon}|$. To explore the condition for the interference pattern, consider the spin precession angle $\Delta\Psi$ accumulated between the two resonance crossings — from crossing point to crossing point. Exact value of $\Delta\Psi$ has to be found by the hypergeometric functions, but as an approximation, we write

$$
\Delta\Psi \approx \int_{-A/\Gamma}^{A/\Gamma} d\theta \, \sqrt{\alpha^2(\theta) + |\epsilon|^2}
= 8 \int_{0}^{\bar{A}} dx \, \sqrt{x^2 + |\bar{\epsilon}|^2}.
$$

Figure 5.15: Final polarization after twice crossing a resonance in a sloped-step pattern as function of $|\bar{\epsilon}|$ for four different values of $\bar{A}$. Red curves give the mean, upper bound, and lower bound of the oscillating final polarization. The four cases are $\bar{A} = 0.5$ (upper left), 1 (upper right), 2 (lower left), and 10 (lower right).

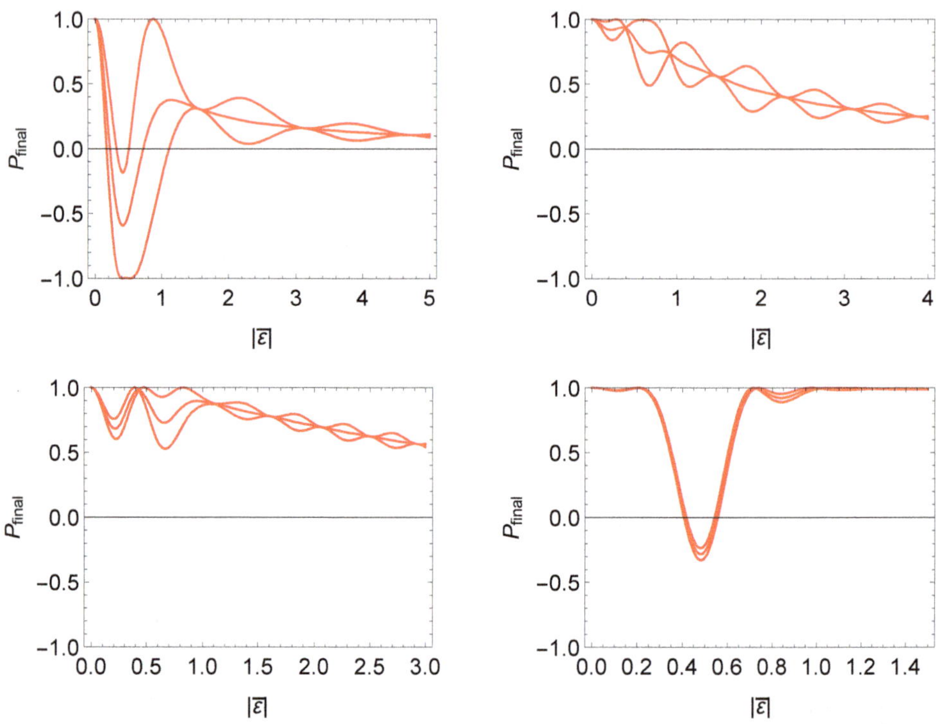

Following the discussion in Eq. (5.66), one anticipates a destructive interference to occur when $\Delta\Psi$ is equal to an even multiple of π. When $\bar{A} \ll |\bar{\epsilon}|$, $\Delta\Psi \approx 8|\bar{\epsilon}|\bar{A}$, this means destructive interferences occur when

$$|\bar{\epsilon}| \approx \frac{m\pi}{4\bar{A}}, \qquad m = \text{integer}.$$

It is further noted that when a destructive interference occurs, not only the final polarization ceases to oscillate, but also it is equal to the initial eigenstate polarization $P_y = \frac{A}{\Omega}$. There is therefore no net depolarization in the double resonance crossing. These conclusions agree with Fig. 5.15. See Homework 5.31.

COSY experiment　　Once a ramping procedure is given consisting of piecewise constant, piecewise linear, and sudden jump segments, a sequential multiplication of matrices, either in the form of $T_{\alpha,\epsilon}(\theta_1, \theta_2)$ of Eq. (5.60) or in the form of $U_{\Gamma,\theta_c,\epsilon}(\theta_1, \theta_2)$ in Eq. (5.69), will give the spin state at any time θ.

This matrix formalism can then be applied to various experimental conditions such as to extend the Froissart–Stora analysis to that with a finite spin tune ramping and to study interference effects of repeated resonance crossings. Unlike the Froissart–Stora formula that focuses on the final beam polarization, the beam polarization during the process can also be predicted in detail.

An experiment was carried out at the COoler SYnchrotron (COSY), Jülich with a 1.85-GeV/c polarized deuteron beam to test these predictions (see footnote 20). The deuteron beam was first cooled by electron beam cooling. An artificial depolarization resonance was introduced by an RF-solenoid whose RF frequency f_{RFS} is varied. The depolarization resonance occurs when the spin tune $a\gamma$ hits the tune of the solenoid's oscillation tune (modulus integer). The resonance crossing pattern was then controlled by varying f_{RFS}. Beam energy was not ramped.

Figure 5.16 shows some of the results. It shows a remarkable agreement with the predicted spin behavior for various crossing patterns. The curves are predictions using the matrix formalism. In the analysis, effect of the beam's energy spread needs to be included and appears as a fitting parameter to the data — the only fitting parameter in the experiment.

Figure 5.16: Experimental test of the spin crossing matrix formalism carried out at the COSY synchrotron (see footnote 20). Various crossing patterns were tested and compared with predictions using the matrix formalism. The dots are experimental data and the curves are the predictions using the matrix formalism.

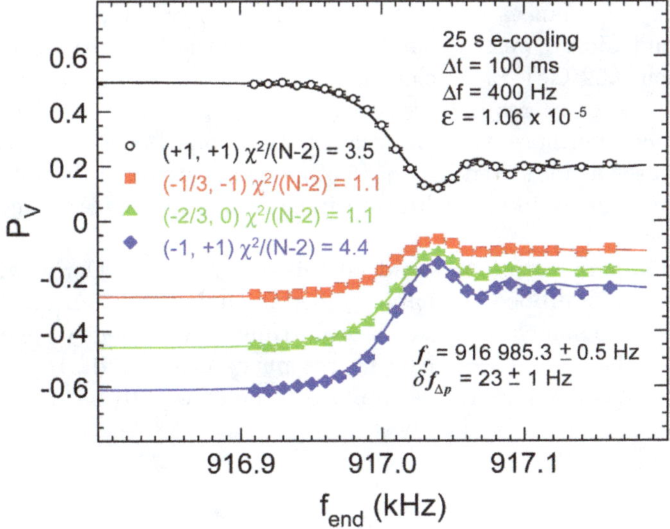

Homework 5.31 In Fig. 5.15, we showed the spin final state for the case of double sloped crossing, and it exhibits a complicated behavior. Of special

interest is the interference effect between the two crossings. Follow the text to examine this effect. In particular, by examining Fig. 5.15, convince yourself that

(a) the destructive interference occurs when $\Delta\Psi \approx 2m\pi$, and when it occurs, the final polarization does not oscillate, and the average polarization is given by $\frac{A}{\Omega}$, same as the initial polarization before the double crossing started.

(b) the constructive interference occurs when $\Delta\Psi \approx (2m + 1)\pi$, and when it occurs, the final polarization is approximately given by the Froissart–Stora formula with an effective resonance strength of 2ϵ.

Solution For completeness, first show that

$$\Delta\Psi \;=\; 4\bar{A}\sqrt{\bar{A}^2 + |\bar{\epsilon}|^2} + 4|\bar{\epsilon}|^2 \ln\left(\frac{\bar{A} + \sqrt{\bar{A}^2 + |\bar{\epsilon}|^2}}{|\bar{\epsilon}|}\right).$$

Homework 5.32 Perhaps a bit ambitious, a homework can be considered to apply the matrix analysis to the case when the spin tune is held fixed in time, while the resonance strength ϵ is programmed to make a linear excursion of either a sloped-step or a sloped double-crossing pattern.

5.6   Siberian snake

As discussed so far, it evidently takes some effort and some loss of polarization to cross a depolarization resonance. For a high energy synchrotron, there may be many resonances to cross during the energy ramping process. From Eqs. (5.9) and (5.32), we see that there are potentially multiple resonances to cross for each 0.52 GeV of acceleration (for a proton beam). A 1-TeV proton synchrotron therefore requires crossing about a few thousand depolarization resonances. Furthermore, the strengths of the resonances increase as the beam energy increases as indicated by Eq. (5.52). Trying to maintain polarization for a 1-TeV proton beam in a synchrotron in such a scenario would be impractical to say the least.

An ingenious invention to avoid all this was made by Derbenev and Kondratenko,[24] and is dubbed the name "Siberian snake" by Courant.

The idea is to somehow make the spin tune independent of energy. If this can be done, then there would be no resonance crossing during acceleration. Furthermore, this fixed spin tune would best be chosen to be as far away from the resonances as possible. The best choice is $\nu_{\text{spin}} = K + \frac{1}{2}$. Siberian snake is a device to do just that.

5.6.1   Type-1 snake

A type-1 snake is a magnet or a set of magnets whose net effect on the spin of a particle is to rotate it by π around the $\hat{z}$-axis as it passes through the device.

---

[24]Ya.S. Derbenev and A.M. Kondratenko, Sov. Phys. Doklady **20**, 562 (1976); Ya.S. Derbenev, et al., Part. Accel. **8**, 115 (1978).

Consider an ideal planar storage ring that has no depolarization fields. We now insert a type-1 snake in this ring. The spin going through the snake magnet makes a rotation by an angle of π around the $\hat{z}$-axis. The rotation in the rest of the accelerator is a rotation by an angle $2\pi a\gamma$ around $\hat{y}$. The total spin rotation for one turn observed at the entrance of the snake is therefore

$$
\begin{aligned}
M_{\text{tot}} &= e^{-\frac{i}{2}\vec{\sigma}\cdot 2\pi a\gamma\hat{y}}\, e^{-\frac{i}{2}\vec{\sigma}\cdot\pi\hat{z}} \\
&= [I\cos(\pi a\gamma) - i\sigma_y\sin(\pi a\gamma)](-i\sigma_z) \\
&= -i[\sigma_z\cos(\pi a\gamma) + \sigma_x\sin(\pi a\gamma)]\,.
\end{aligned}
$$

Equation (5.21) is then used to obtain the net spin precession angle. This angle by definition is $2\pi\nu_{\text{spin}}$. Thus

$$
\cos(\pi\nu_{\text{spin}}) = \frac{1}{2}\operatorname{tr}(M_{\text{tot}}) = 0\,,
$$
$$
\implies \quad \nu_{\text{spin}} = K + \frac{1}{2}\,.
$$

The spin tune is thus made energy independent and is a half-integer by this simple device!

One can also calculate the spin rotation axis by using Eq. (5.21). It is found to be (changing notation from $\hat{\phi}$ to $\hat{n}$)

$$
\hat{n} = \frac{i}{2}\operatorname{tr}(M_{\text{tot}}\vec{\sigma}) = \hat{z}\cos(\pi a\gamma) - \hat{x}\sin(\pi a\gamma)\,. \tag{5.77}
$$

The rotation axis depends on the location of observation. The axis $\hat{n}$ in Eq. (5.77) refers to the entrance of the snake where M_{tot} was calculated.

One can repeat the calculation for another observation position around the ring. A similarity transformation is then made on M_{tot}. It can then be shown that the spin tune remains $K + \frac{1}{2}$ while the rotation axis lies in the x-z plane, perpendicular to $\hat{y}$ and precesses around $\hat{y}$ as the particle circles around the storage ring.

Figure 5.17 shows the spin dynamics of a Siberian snake of type 1.[25] A particle is launched from the location opposite to the snake with three initial spin orientations, as shown in Figs. 5.17(a), (b) and (c). Figures 5.17(a) and (b) demonstrate why the spin tune is $\frac{1}{2}$. It turns out that the spin direction shown in Fig. 5.17(c) is also the spin precession direction $\hat{n}$ at any location in the accelerator. The reason we know that Fig. 5.17(c) has this distinction, and Figs. 5.17(a) and (b) do not, is that the spin in Fig. 5.17(c) repeats itself turn after turn. By examining Fig. 5.17, therefore, one sees that a spin in general will precess around $\hat{n}$ turn after turn no matter where the observation point is.

The precession axis $\hat{n}$ also has the significance that if the stored beam has a net polarization, it must be along $\hat{n}$ or $-\hat{n}$ because any initial polarization perpendicular to $\hat{n}$ will smear out to zero as particles precess around $\hat{n}$ independently of one another. If a beam with initial polarization $P_i\hat{n}_i$ is injected

---

[25] B.W. Montague, Phys. Reports, 113, 1 (1984).

Figure 5.17: Spin motion in a storage ring with Siberian snake of type-1. (a),
(b) are motion of spin perpendicular to the equilibrium direction, demonstrating
the spin tune is $\frac{1}{2}$. (c) shows the equilibrium spin direction. The angle $\phi = \frac{a\gamma}{2}$
lies in the x-z plane. The shaded arrows represent the spin of the particle under
consideration.

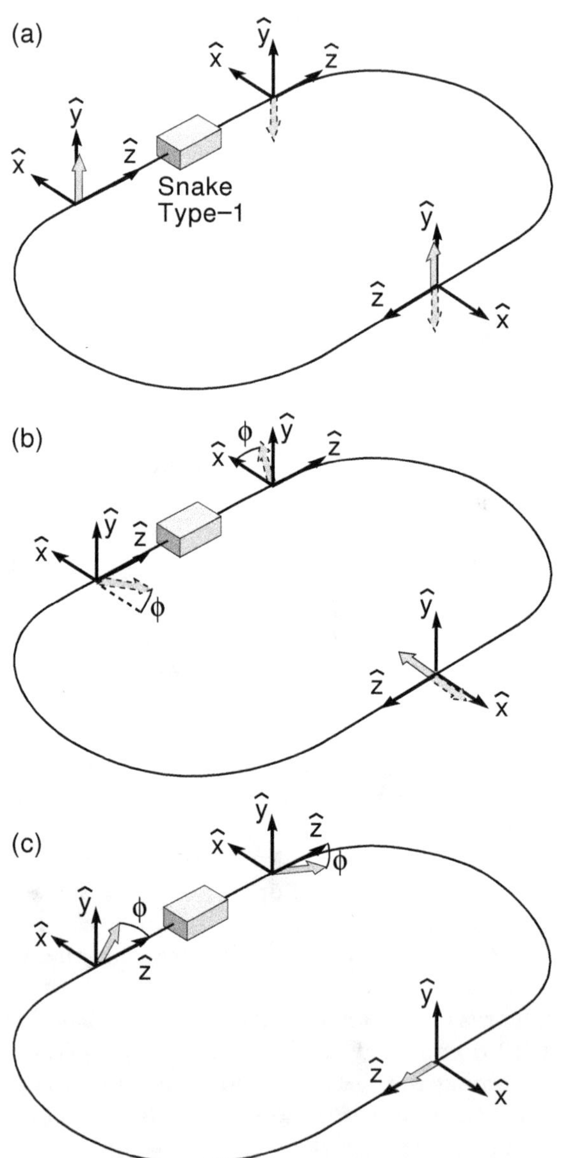

into this synchrotron, then after injection transient, the beam polarization will suffer a loss by a cosine factor $\hat{n}_i \cdot \hat{n}$ and settle down to the value $P_i(\hat{n}_i \cdot \hat{n})\hat{n}$.

The spin direction $\hat{n}$, of which Fig. 5.17(c) gives an example, has therefore simultaneously three physical meanings,

- spin closed-orbit — a spin motion that closes on itself turn after turn;
- axis of spin precession — all spins not exactly aligned with it will precess around it;
- direction of equilibrium polarization for the whole beam.

For a Siberian snake of type-1, Fig. 5.17(c) shows that $\hat{n}$ at the location exactly opposite to the snake must be longitudinal, $\hat{n} = \hat{z}$. At the entrance to the snake, Fig. 5.17(c) shows $\hat{n}$ is indeed given by Eq. (5.77).

Since high energy experiments often require longitudinal beam polarization, a type-1 Siberian snake offers a convenient location to install the collision point and the collider detector at the opposite point of the snake.

Spin rotator is not Siberian snake   Incidentally, there is another device called spin rotator (see page 468 for its discussion in more detail), which are not to be confused with Siberian snakes.

- A Siberian snake rotates the spins of all particles in the beam by a specific angle around a specific axis. The rotation applies to all spins, including those deviating from the nominal polarization direction.

- A spin rotator rotates the nominal polarization — the beam's equilibrium polarization — from one specific incoming direction to a specific outgoing direction. This rotation is important only for the nominal polarization direction. How does a nonnominal spin rotate by the action of the device does not matter.

A Siberian snake is thus a much more intricate device than a spin rotator. In our effort to demonstrate the mechanism of a type-1 snake in Fig. 5.17, for example, we needed to examine spin motion for all three possible spin orientations. For a spin rotator, we only have to check if the nominal spin orientation is rotated as desired while the other two orientations do not matter. A spin rotator that gives a π rotation along the $\hat{z}$-direction, for example, will generally not give a spin tune of $K + \frac{1}{2}$.

Solenoid snake   The simplest type-1 snake is a solenoid. The required strength is (see Homework 5.6)

$$\frac{B_s \ell}{B\rho} = \frac{\pi}{1+a} = \frac{2\pi}{g}.$$
(5.78)

This means $B_s \ell$ increases as beam energy increases. A solenoid snake therefore works only for low energy beams. For protons higher than several GeV, solenoid snakes become unpractical.

Although conceptually the simplest, a solenoid serving as a type-1 snake is a hefty strong device. If installed without compensation into a storage ring, it affects the stability of the storage ring lattice significantly. Homework 5.34 illustrates this point.

Dipole snake Another type of snake is to use multiple dipole magnets, whose fields are either along $\hat{x}$ or along $\hat{y}$ (or at least in the x-y plane), in such a way that their net effect on spin is a rotation around $\hat{z}$ by an angle π. Compared with a solenoid snake, these snakes have the disadvantage that they distort the closed orbit of the beam. The advantage, however, is that their strengths do not increase with beam energy.

There are many practical designs of this type of snakes. We will mention a couple of designs later in Sec. 5.6.8. Suffice it to say here that this dipole snake design works at high energies because the needed strengths $B\ell$ are independent of beam energy. On the other hand, they do not work at low energies because the orbit distortion associated with these transverse bending magnets becomes excessive. For low energy applications, solenoids are more practical.

Homework 5.33

(a) Show that a type-1 solenoid snake would have a strength given by Eq. (5.78).

(b) For a proton beam of 10 GeV/c, calculate the required solenoid field strength B_s if the solenoid has a 10 m length.

Homework 5.34 When a type-1 solenoid snake is inserted into an otherwise perfect storage ring, particles' orbital motions are seriously perturbed — after all, this solenoid would be very strong. Use a 4×4 transport matrix analysis to evaluate the orbital stability conditions of this storage ring.

Solution Let ℓ be the solenoid's length, and let $K = \frac{B_s}{B\rho}$ with B_s the solenoid's magnetic field. The 4×4 transport matrix through the solenoid is

$$
M_{\text{sol}} =
\begin{bmatrix}
\frac{1}{2}(1+\cos K\ell) & \frac{1}{K}\sin K\ell & \frac{1}{2}\sin K\ell & \frac{1}{K}(1-\cos K\ell) \\
-\frac{K}{4}\sin K\ell & \frac{1}{2}(1+\cos K\ell) & -\frac{K}{4}(1-\cos K\ell) & \frac{1}{2}\sin K\ell \\
-\frac{1}{2}\sin K\ell & -\frac{1}{K}(1-\cos K\ell) & \frac{1}{2}(1+\cos K\ell) & \frac{1}{K}\sin K\ell \\
\frac{K}{4}(1-\cos K\ell) & -\frac{1}{2}\sin K\ell & -\frac{K}{4}\sin K\ell & \frac{1}{2}(1+\cos K\ell)
\end{bmatrix}.
$$

For a type-1 snake, $K\ell = \frac{2\pi}{g}$. For protons, $g = 5.5857$.

Let the total transformation for one turn around the middle of the solenoid, in the absence of the solenoid, be

$$
M_0 =
\begin{bmatrix}
\cos\mu_x & \beta_x \sin\mu_x & 0 & 0 \\
-\frac{1}{\beta_x}\sin\mu_x & \cos\mu_x & 0 & 0 \\
0 & 0 & \cos\mu_y & \beta_y \sin\mu_y \\
0 & 0 & -\frac{1}{\beta_y}\sin\mu_y & \cos\mu_y
\end{bmatrix},
$$

where $\mu_{x,y} = 2\pi\nu_{x,y}$, and $\beta_{x,y}, \alpha_{x,y}$ are the Courant–Snyder functions with $\alpha_x = \alpha_y = 0$ assumed for simplicity. The one-turn transport matrix including the solenoid (observed at the exit of the solenoid) is therefore

$$M_{\text{tot}} = M_{\text{sol}} M_{\ell/2}^{-1} M_0 M_{\ell/2}^{-1},$$

where $M_{\ell/2}$ is the drift matrix for half the solenoid length,

$$M_{\ell/2} = \begin{bmatrix} 1 & \frac{\ell}{2} & 0 & 0 \\ 0 & 1 & 0 & 0 \\ 0 & 0 & 1 & \frac{\ell}{2} \\ 0 & 0 & 0 & 1 \end{bmatrix}.$$

To find the limitation on the choice of the working point in the (ν_x, ν_y) plane, we calculate the eigenvalues of M_{tot}. Stability requires all four eigenvalues to have absolute value of unity. Figure 5.18 shows the stability region in the (ν_x, ν_y) space by two examples.

Figure 5.18: Stability region of a full solenoid snake in the working point space (ν_x, ν_y). Darkened regions are unstable. Figure on the left is when $\beta_x = 2.8$ m, $\beta_y = 1$ m, and $\ell = 5$ m. The one on the right has changed β_y to 0.4 m. The unstable regions are periodic in ν_x and ν_y with period of 1.

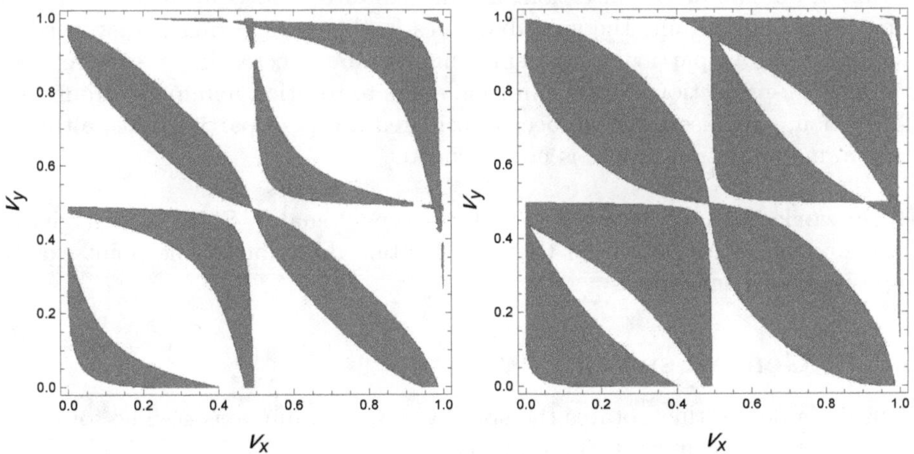

Homework 5.35 Homework 5.34 is for the case of a proton beam. For electrons, there is a simplification because $a \approx 0$, and $K\ell = \pi$. Show that in that case, the stability criterion can be obtained analytically,

$$|S| \leq 2,$$

where

$$S = \sin \mu_x \sin \mu_y \left[\frac{\pi^2 \beta_x \beta_y}{4\ell^2} - \frac{\pi^2}{16}\left(\frac{\beta_x}{\beta_y} + \frac{\beta_y}{\beta_x}\right) + \frac{\pi^2 \ell^2}{64\beta_x\beta_y} - \frac{\ell^2}{2\beta_x\beta_y} + \frac{4\ell^2}{\pi^2\beta_x\beta_y} \right]$$

$$+ \sin \mu_x \cos \mu_y \left(-\frac{\pi^2 \beta_x}{4\ell} + \frac{\pi^2 \ell}{16\beta_x} - \frac{\ell}{\beta_x} \right)$$

$$+ \cos \mu_x \sin \mu_y \left(-\frac{\pi^2 \beta_y}{4\ell} + \frac{\pi^2 \ell}{16\beta_y} - \frac{\ell}{\beta_y} \right) + \cos \mu_x \cos \mu_y \left(\frac{\pi^2}{4} - 2 \right).$$

5.6.2   Type-2 snake

If the device rotates the spin by an angle π around the $\hat{x}$-axis, it is called Siberian snake of type 2. The total spin rotation observed at the entrance of the snake is

$$M_{\text{tot}} = e^{-\frac{i}{2}\vec{\sigma}\cdot 2\pi a\gamma \hat{y}} e^{-\frac{i}{2}\vec{\sigma}\cdot \pi \hat{x}} = -i[\sigma_x \cos(\pi a\gamma) - \sigma_z \sin(\pi a\gamma)]. \tag{5.79}$$

Again one finds a half-integral spin tune $\nu_{\text{spin}} = K + \frac{1}{2}$. The rotation direction at the point exactly opposite to the snake is $\hat{n} = \hat{x}$.

In passing, it should be mentioned that there are circumstances when the synchrotron contains special magnet insertions that include horizontal bends, vertical bends, and solenoids. Even when the insertion does not intend to serve as Siberian snake, and even if the insertion does not generate a net closed orbit deviation outside of the insertion, it can facilitate a net spin rotation around some axis. As a result, the spin dynamics in the storage ring is disturbed by this insertion. In particular, the spin tune can be affected if the spin rotation contains a $\hat{y}$-projection. Since spin tune refers to rotation nominally around the $\hat{y}$-direction, this insertion has been considered a type-3 partial snake although its spin manipulation aspect is not intended.[26]

Homework 5.36 Verify Eq. (5.79) for a type-2 snake. Show that it gives a spin tune $\nu_{\text{spin}} = K + \frac{1}{2}$, and that the rotation direction at the point exactly opposite to the snake is $\hat{n} = \hat{x}$.

5.6.3   General snake

Consider a device that rotates the spin by π around any axis $\hat{\phi}$. The total spin rotation at the entrance to the snake is

$$M_{\text{tot}} = -i(\vec{\sigma}\cdot\hat{\phi})\cos(\pi a\gamma) - (\hat{y}\cdot\hat{\phi})I\sin(\pi a\gamma) - i\vec{\sigma}\cdot(\hat{y}\times\hat{\phi})\sin(\pi a\gamma). \tag{5.80}$$

The spin precession axis at the snake entrance is

$$\hat{n} = \cos(\pi a\gamma)\hat{\phi} + \sin(\pi a\gamma)(\hat{y}\times\hat{\phi}). \tag{5.81}$$

This device constitutes a Siberian snake with $\nu_{\text{spin}} = K + \frac{1}{2}$ if and only if $\hat{\phi}\perp\hat{y}$. In particular, when $\hat{\phi} = \hat{z}$, it is a type-1 snake, and when $\hat{\phi} = \hat{x}$, it is a type-2 snake.

---

[26] M.G. Minty, et al., Phys. Rev. D, 44, R1361 (1991).

Homework 5.37

(a) Verify Eqs. (5.80) and (5.81).

(b) Find the general expression of the spin tune. Show that it gives a half-integral spin tune if and only if $\hat{\phi} \perp \hat{y}$.

(c) Find the net polarization direction for the general snake at the point exactly opposite to the snake.

(d) Repeat (c) for the location exactly in the middle of the snake.

5.6.4 Helical dipole snake

As mentioned earlier, a solenoid snake is used mainly for low energy beams and dipole snakes are more applicable at higher energies. The reason dipole snakes are not useful at low energies is because the required integrated magnetic field strength $B\ell$ is fixed, independent of energy, and at low energies, the beam trajectory will be distorted too excessively. To minimize this trajectory distortion effect, a compact snake of a helical type was invented.[27]

Consider a helical dipole with magnetic field,

$$\vec{B}(s) = B_0 \left(\hat{x} \cos \frac{2\pi s}{\lambda} + \hat{y} \sin \frac{2\pi s}{\lambda} \right).$$

This magnet constitutes a full twist helical magnet if its length covers from $s = 0$ to $s = \lambda$.

The spin map for a full passage of this magnet is found to be[28]

$$M = \cos \frac{\phi}{2} - \frac{i \sin \frac{\phi}{2}}{\sqrt{1 + \xi^2}} (\sigma_z + \xi \sigma_x),$$

where

$$\xi = (a\gamma + 1) \frac{B_0 \lambda}{2\pi B \rho}, \qquad \phi = 2\pi(\sqrt{1 + \xi^2} - 1).$$

The spin precesses by an angle ϕ around the axis $\frac{\hat{z} + \xi \hat{x}}{\sqrt{1 + \xi^2}}$, which lies in the horizontal plane as required for a general snake. To provide a Siberian snake of spin tune $\frac{1}{2}$, ϕ is chosen to be π.

5.6.5 Double snake

With a single Siberian snake in the accelerator, the equilibrium beam polarization along the direction $\hat{n}$ is perpendicular to $\hat{y}$ and executes rapid precession going through the arcs. This is an undesirable arrangement. For one reason, the polarization is going to be sensitive to the energy spread of the beam. As snake designs get increasingly sophisticated, it is possible to avoid this by installing

---

[27]E. Ludmirsky, Proc. Part. Accel. Conf., Dallas, USA, 793 (1993); T. Roser, Handbook Accel. Phys. & Eng., 2nd ed., Sec. 2.6.3, World Scientific (2013).

[28]E.D. Courant, BNL report BNL-61920, AD/RHIC-133 (1995); S.Y. Lee, Spin Dynamics and Snakes in Synchrotrons, World Scientific (1997).

two snakes at opposite locations in the accelerator. For example, if we have one type-1 snake and one type-2 snake, as sketched in Fig. 5.19, the equilibrium polarization direction would be $\hat{y}$ in one half of the accelerator and $-\hat{y}$ in the other half. The spin rotation of this accelerator, observed at the entrance to the type-2 snake, is

$$M_{\mathrm{tot}} = e^{-\frac{i}{2}\vec{\sigma}\cdot\pi a\gamma\hat{y}}e^{-\frac{i}{2}\vec{\sigma}\cdot\pi\hat{z}}e^{-\frac{i}{2}\vec{\sigma}\cdot\pi a\gamma\hat{y}}e^{-\frac{i}{2}\vec{\sigma}\cdot\pi\hat{x}} = -i\sigma_y .$$

The spin tune is still $K + \frac{1}{2}$.

Incidentally, Fig. 5.19 serves as another illustration of the difference between a Siberian snake and a spin rotator. The two snakes in Fig. 5.19 are distinctly different as far as Siberian snakes go, but if they were just to serve as spin rotators, they are the same and can both be of type 1 or type 2 and no one would care.

Figure 5.19: A storage ring with double Siberian snakes.

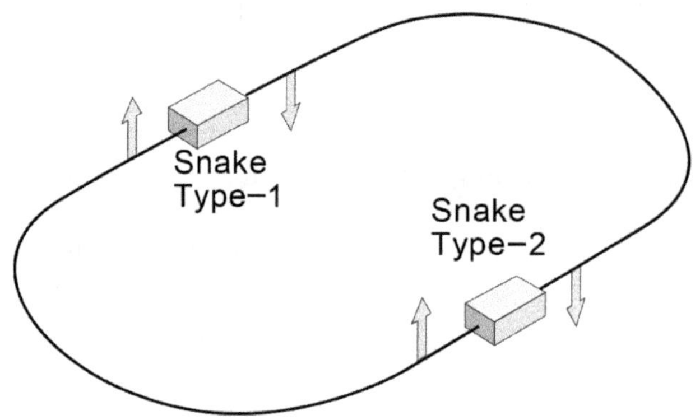

Homework 5.38 Find the spin tune of a double-snake scheme when the type-1 and the type-2 snakes are not located exactly opposite to each other, but by an arc with $2\pi\alpha$ bending and another arc with $2\pi(1-\alpha)$ bending. If α represents an error in installation of the two snakes, how large an α can be tolerated?

Homework 5.39 What happens if both snakes in the double snake design are of type-1? or type-2?

5.6.6 Partial snake

Sometimes there may not be sufficient space (or budget) to install a full Siberian snake. In this case, one may consider the idea of a partial snake.[29] A partial snake rotates the spin by an angle $\phi < \pi$. Let the rotation axis be $\hat{\phi}$ and consider a single snake in the circular accelerator.

---

[29]T. Roser, Proc. Workshop on Siberian snakes and depolarizing techniques, Minneapolis, (1989).

The one-turn spin rotation map at the entrance to the snake reads

$$
\begin{aligned}
M_{\text{tot}} \;=\;& I \cos(\pi a\gamma)\cos\frac{\phi}{2} - \sin(\pi a\gamma)\sin\frac{\phi}{2}\,(\hat{y}\cdot\hat{\phi}) - i\sin(\pi a\gamma)\cos\frac{\phi}{2}\,\sigma_y \\
& -i\cos(\pi a\gamma)\sin\frac{\phi}{2}\,(\vec{\sigma}\cdot\hat{\phi}) - i\sin(\pi a\gamma)\sin\frac{\phi}{2}\,\vec{\sigma}\cdot(\hat{y}\times\hat{\phi})\,.
\end{aligned}
$$

When $\phi = \pi$, this reduces to the case of the general snake, Eq. (5.80).

The spin tune is determined by

$$
\cos\pi\nu_{\text{spin}} \;=\; \cos(\pi a\gamma)\cos\frac{\phi}{2} - (\hat{y}\cdot\hat{\phi})\sin(\pi a\gamma)\sin\frac{\phi}{2}\,. \tag{5.82}
$$

If the snake rotation axis $\hat{\phi}\perp\hat{y}$, we have

$$
\cos\pi\nu_{\text{spin}} \;=\; \cos(\pi a\gamma)\cos\frac{\phi}{2}\,.
$$

Figure 5.20 shows the spin tune as a function of $a\gamma$ for this case. When $\phi = 0$, we have $\nu_{\text{spin}} = a\gamma$. When $\phi = \pi$, we have a full snake with $\nu_{\text{spin}} = K + \frac{1}{2}$. When $\phi < \pi$, we have a partial snake. The spin tune has stopbands around each integer with a stopband width of

$$
\Delta\nu_{\text{spin}} \;=\; \pm\frac{\phi}{2\pi}\,.
$$

It is clear from Fig. 5.20 that a partial snake avoids integer (imperfection) resonances even with a rather modest value of ϕ because the spin tune will now never be equal to an integer. A 10% snake for example can help greatly in avoiding the imperfection resonances. If ϕ is sufficiently large, the intrinsic resonances can be avoided as well. A 20% snake for example can be used to avoid intrinsic resonances for ν_y (fractional part) up to ± 0.1. Solenoids become a practical partial snake because the needed strength becomes accessible up to modestly high energy proton synchrotrons such as the AGS, and no orbit distortion issue needs to be considered.

Homework 5.340

(a) Prove Eq. (5.82).

(b) Show that the polarization direction with a partial snake is given by

$$
\begin{aligned}
\hat{n} \;=\;& \frac{1}{\sin(\pi\nu_{\text{spin}})}\left[\hat{y}\left(\sin(\pi a\gamma)\cos\frac{\phi}{2} - 2\sin\frac{\phi}{2}\sin\frac{a\gamma\theta}{2}\sin\frac{a\gamma(2\pi-\theta)}{2}(\hat{y}\cdot\hat{\phi})\right)\right. \\
& \left. + \hat{\phi}\cos[(\pi-\theta)a\gamma]\sin\frac{\phi}{2} + (\hat{y}\times\hat{\phi})\sin[(\pi-\theta)a\gamma]\sin\frac{\phi}{2}\right]\,,
\end{aligned}
$$

where $2\pi > \theta > 0$ specifies the location of observation point with $\theta = 0$ at the entrance of the snake and $\theta = 2\pi$ at the exit.

(c) Confirm that this result of $\hat{n}$ has a unit magnitude.

Figure 5.20: Spin tune ν_{spin} as a function of $a\gamma$ for a partial Siberian snake. The cases $\phi = 0, \frac{\pi}{3}, \frac{2\pi}{3}$, and π correspond to no snake, $\frac{1}{3}$ snake, $\frac{2}{3}$ snake, and full snake, respectively.

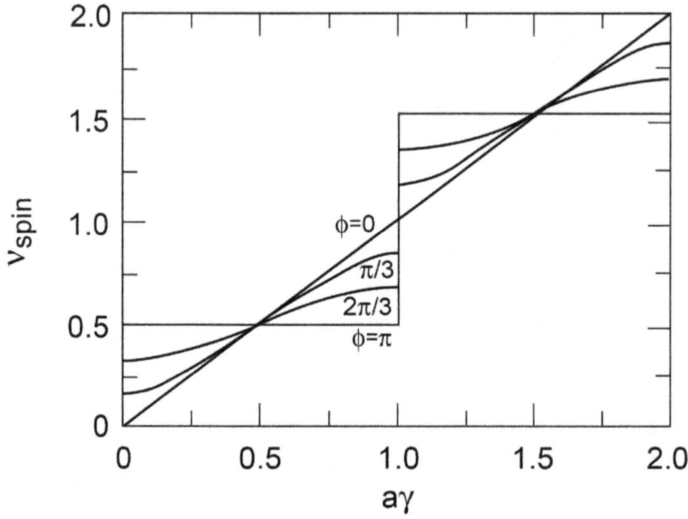

Homework 5.41 Following Eq. (5.82), we choose $\hat{\phi}\perp\hat{y}$ for the partial snake. Would relaxing this constraint allow more flexibility to be gained?

Solution Not much to gain if spin tune is being considered as we do here. The condition $\hat{\phi}\perp\hat{y}$ controls the spin tune. The condition $\hat{\phi} \parallel \hat{y}$ affects the spin precession axis. If spin axis is of more concern to you for some reason, you might consider $\hat{\phi}$ to have a component along $\hat{y}$.

5.6.7   Depolarization due to snake

It might now be expected that since the spin tune is made $K + \frac{1}{2}$, all depolarization effects must be minuscule because we are so far away from all resonances. This is not the case. The perturbation of a snake on spin is so huge — rotation by as much as π! — that even though the spin tune is far away from resonances, one still feels the remote effect of the integer resonances, and there is some small loss of polarization each time when the *unperturbed* spin tune $a\gamma$ crosses an integer. Partial snakes are not immune to these snake resonances either.

To see this more quantitatively, consider a partial type-1 snake with rotation around $\hat{z}$ by an angle ϕ. Spin precession can be described by Eq. (5.12) with

$$\vec{h} = -a\gamma\hat{y} + \phi\hat{z}\,\delta_p(\theta)\,. \tag{5.83}$$

where $\delta_p(\theta)$ is the periodic δ-function with period 2π. In spinor language, we have Eqs. (5.23–5.24) with

$$\vec{h} \cdot \vec{\sigma} \;=\; \begin{bmatrix} -a\gamma & \phi\delta_p(\theta) \\ \phi\delta_p(\theta) & a\gamma \end{bmatrix}.$$

From Eqs. (5.34–5.35), we then find the integer resonance strength

$$\epsilon \;=\; \frac{1}{2\pi} \int_0^{2\pi} d\theta\, e^{-iK\theta}\, \phi\delta_p(\theta) \;=\; \frac{\phi}{2\pi}\,.$$

Consider a full snake for example. The spin tune is $\frac{1}{2}$, meaning it is far from an integer resonance by a distance of $\frac{1}{2}$. However, the resonance strength $|\epsilon|$ is also very large, $\frac{\phi}{2\pi} = \frac{1}{2}$, comparable to the distance to the resonance. It is in fact incorrect to expect negligible influence on beam polarization during acceleration when the synchrotron is provided with a Siberian snake. On the contrary, the resonance strength is so large due to the snake that the polarization gets flipped each time an integer depolarization resonance is crossed. There is negligible loss of polarization because the resonance is overwhelmingly strong, not because the resonances are far away.

The snake does help, however, not because the resonances are far away, but because of the self-cancellation effects from one turn to the next due to the fact that the spin tune is equal to $\frac{1}{2}$. A full evaluation of the snake-induced depolarization resonances requires careful analysis.[30] One complication for example comes from the inevitable overlapping of resonances when their widths are so wide.

For an order of magnitude assessment of the magnitude of the effect, we may try to apply the Froissart–Stora equation to the integer resonances. For a snake with rotation angle ϕ, we find there is a loss of polarization each time when $a\gamma$ crosses an integer during acceleration,

$$\frac{S(\infty)}{S(-\infty)} \;=\; 2e^{-\phi^2/8\pi\Gamma} - 1\,. \tag{5.84}$$

With a full snake, we have $\phi = \pi$ and $|\epsilon| = \frac{1}{2}$. Strictly speaking, Eq. (5.84) is not really applicable because in this circumstance, the resonance width is so wide that the single resonance assumption that yields the Froissart–Stora formula breaks down. However, Eq. (5.84) is used here to give a qualitative indication. With a modest energy ramping speed, each crossing means the spin is flipped. The loss of polarization is very small because the spin flips are almost 100%. When the snake is partial, this may not be true any more, and one has to be careful not to cross the integer resonances too rapidly (not too slowly).

Homework 5.42 Froissart–Stora equation was derived earlier assuming the initial polarization is in the $\hat{y}$ direction. With a type-1 snake, the initial polarization is no longer along $\hat{y}$. Show that the Froissart–Stora equation still applies, as we assumed somewhat casually with Eq. (5.84).

---

[30]S.Y. Lee, AIP Conf. Proc. 187, 8th Int. Symp. High Energy Spin Physics, p. 1105 (1988); S.Y. Lee, Spin Dynamics and Snakes in Synchrotrons, World Scientific (1997).

Homework 5.43 This homework suggests a possibility of an oscillatory snake.

(a) Equation (5.83) applies when the snake strength does not change in time. What happens if the snake precession angle is oscillatory with $\phi \sin \nu_1 \theta \, \delta_p(\theta)$ where ν_1 is the snake modulation tune? Show that there are resonances at $a\gamma = K \pm \nu_1$. Identify a resonance strength.

(b) Consider a snake whose spin precession angle is time-dependent and is two-dimensional with $\phi(\hat{x} \cos \nu_1 \theta - \hat{z} \sin \nu_1 \theta)\delta_p(\theta)$. Show that there are resonances at $a\gamma = K + \nu_1$. What happens to the resonances $a\gamma = K - \nu_1$?

(c) What happens to the depolarization resonances if the snake has precession angle $\phi(\hat{x} \cos \nu_1 \theta + \hat{z} \sin \nu_1 \theta)\delta_p(\theta)$?

5.6.8   Snake design

Over the years, there has been invented a large number of snake designs by the "snake hunters". Most of them consist of interlacing horizontal and vertical bending dipoles whose net effect on spin is as prescribed for a Siberian snake of the type intended, while its net effect on orbit is made to vanish. Elegance and compactness become the name of the game.

Let us designate a bending magnet by $(\phi, \hat{\phi})$ where ϕ is the spin precession angle, and $\hat{\phi}$ is the spin precession axis which, in the case of a single dipole magnet, is the same as the magnetic field direction.[31] Then one possible type-1 snake design is (earlier elements on the right)

$$\left(\frac{\pi}{2}, \hat{x}\right) \left(\frac{\pi}{2}, -\hat{x}\right) \left(\frac{\pi}{2}, \hat{y}\right) \left(\frac{\pi}{2}, -\hat{x}\right) (\pi, -\hat{y}) \left(\frac{\pi}{2}, \hat{x}\right) \left(\frac{\pi}{2}, \hat{y}\right). \tag{5.85}$$

Figure 5.21 shows the design, where H and V means horizontal and vertical bending dipoles, and the orbital angle θ is chosen such that the spin precesses by $\frac{\pi}{2}$ relative to the beam trajectory. The x- and y-trajectories of the beam are shown. For protons, we have

$$\theta = \frac{B\ell}{(B\rho)} = \frac{\pi}{2a\gamma}, \qquad B\ell = 2.74211 \, \beta \text{ T-m}.$$

Note that, as emphasized before and exhibited in Fig. 5.2, the spin precession is relative to the rotating coordinate system. In absolute space, the spin precesses by $(a\gamma + 1)\theta$. Note also that (also mentioned before) the needed snake magnet strengths are independent of the beam energy, at least when $\beta \approx 1$.

The condition of no net orbital effect, however, is not always necessary for a Siberian snake. To save space for the accelerator, for example, it may be useful to incorporate part of the horizontal arc bending into the snake. We mention here one type-2 snake design by Derbenev[32] which contains a net horizontal

---

[31] A subtlety is to be noted here. Strictly speaking, $\hat{\phi}$ refers to the direction of the magnetic field in the rest frame of the moving particle. So we need to be careful with this statement if $\hat{\phi}$ contains a $\hat{z}$-component. See a discussion of this point on page 405.

[32] Ya.S. Derbenev, UM HE 97-07 (1997).

Figure 5.21: A design of type-1 snake. The design is sketched on top. The vertical and horizontal orbital excursions inside the snake are shown as the second and the third entries below the design sketch, respectively.

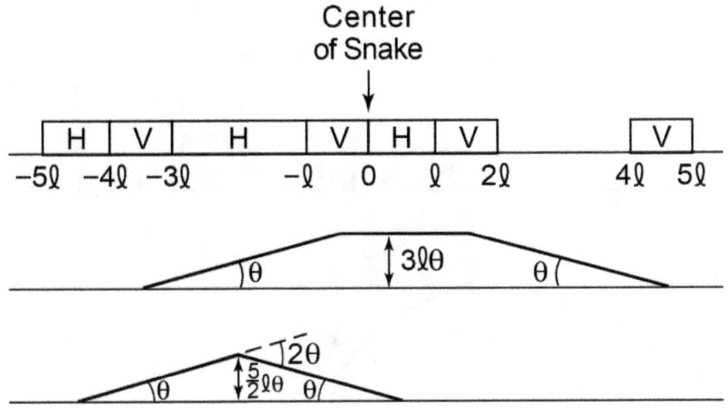

bending. The design consists of four magnets,

$$(\phi, \hat{n}_+)(\phi, \hat{n}_-)(\phi, \hat{n}_-)(\phi, \hat{n}_+),$$
$$\hat{n}_\pm = \pm\hat{x}\sin\alpha + \hat{y}\cos\alpha. \tag{5.86}$$

These magnets are tilted in the x-y plane by angles $\pm\alpha$ as sketched in Fig. 5.22. Their strengths are all the same, precessing the spin by an angle ϕ. The net spin rotation is

$$M_{\text{tot}} = e^{-\frac{i}{2}\vec{\sigma}\cdot\phi\hat{n}_+}e^{-\frac{i}{2}\vec{\sigma}\cdot\phi\hat{n}_-}e^{-\frac{i}{2}\vec{\sigma}\cdot\phi\hat{n}_-}e^{-\frac{i}{2}\vec{\sigma}\cdot\phi\hat{n}_+}.$$

Let Φ and $\hat{n}$ be the net precession angle and axis, then Eq. (5.21) gives

$$\cos\frac{\Phi}{2} = \cos^2\phi - \cos 2\alpha\sin^2\phi, \tag{5.87}$$

$$\hat{n}\sin\frac{\Phi}{2} = 2\cos\alpha\sin\phi\left[-\hat{x}\sin^2\frac{\phi}{2}\sin 2\alpha + \hat{y}\left(\cos^2\frac{\phi}{2} - \cos 2\alpha\sin^2\frac{\phi}{2}\right)\right].$$

To make a full Siberian snake of type 2, this design needs to provide $\Phi = \pi$ and $\hat{n} = -\hat{x}$. This means we must choose α and ϕ so that

$$\cot^2\phi = \cot^2\frac{\phi}{2} = \cos 2\alpha.$$

One solution is

$$\phi = \frac{2\pi}{3}, \qquad \alpha = \frac{1}{2}\cos^{-1}\frac{1}{3} = 35°15'. \tag{5.88}$$

Homework 5.44 Show that the design (5.85) constitutes a full type-1 Siberian snake.

Figure 5.22: A compact type-2 snake with net horizontal bending. The design is sketched on top. The horizontal and vertical orbital excursions inside the snake are shown as the second and the third entries below the design sketch respectively.

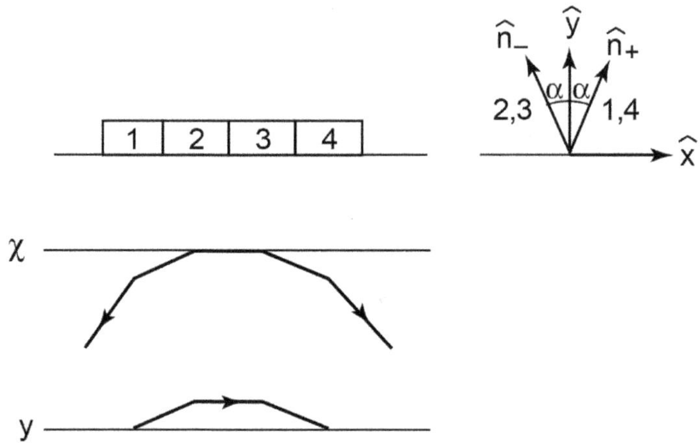

Homework 5.45 Use spinor algebra to show that the following configuration produces a type-1 snake,[33]

$$\left(\frac{\pi}{4},\hat{x}\right)\left(\frac{\pi}{4},\hat{y}\right)\left(\frac{\pi}{2},-\hat{x}\right)\left(\frac{\pi}{2},-\hat{y}\right)\left(\frac{\pi}{2},\hat{x}\right)\left(\frac{\pi}{2},\hat{y}\right)\left(\frac{\pi}{2},-\hat{x}\right)\left(\frac{\pi}{4},-\hat{y}\right)\left(\frac{\pi}{4},\hat{x}\right).$$

Homework 5.46 Use spinor algebra to show that the following configuration produces a type-2 snake,

$$\left(\frac{\pi}{2},-\hat{x}\right)\left(\frac{\pi}{2},\hat{y}\right)\left(\frac{\pi}{2},\hat{x}\right)\left(\pi,-\hat{y}\right)\left(\frac{\pi}{2},\hat{x}\right)\left(\frac{\pi}{2},\hat{y}\right)\left(\frac{\pi}{2},-\hat{x}\right).$$

Homework 5.47
   (a) Verify Eq. (5.87) for the type-2 snake of Fig. 5.22.
   (b) Confirm that $\hat{n}$ is a unit vector.
   (c) Verify that Eq. (5.88) gives a solution for a type-2 snake.
   (d) What is the net bending angle for the beam orbital motion by this snake?

Homework 5.48 In the design of snakes, the following theorem comes handy sometimes. If the magnet arrangement has a mirror symmetry, and if the spin precession axes of these magnets are all perpendicular to some direction $\hat{k}$, then the axis of the total precession is also perpendicular to $\hat{k}$.
   (a) Prove this theorem.
   (b) Use this theorem to explain why design (5.86) cannot be a type-1 snake no matter how α and ϕ are chosen.

[33]K. Steffen, DESY Laboratory Report DESY 83-062 (1983).

Chapter 6

Spin Dynamics of Electron

Spin dynamics for electrons is very different from that for protons. The main cause of the difference lies in synchrotron radiation. Synchrotron radiation is such an overwhelming mechanism for electrons while negligible for protons that these two areas of spin dynamics evolved along entirely different paths. One might note that there is little overlap of participating researchers in these two parts of the accelerator community, not to mention the very different languages and notations they developed.

This perhaps should not be too surprising. After all, even in orbital dynamics, protons and electrons behave differently, again due to synchrotron radiation. While a proton's orbital motion is determined from its initial conditions when injected into a synchrotron or a storage ring, an electron's orbital motion is determined by a balance between radiation damping and quantum excitation, both of which originate from synchrotron radiation. The same situation occurs for their spin dynamics.

When analyzing spin motion of protons in the preceding chapter, we first derived the Thomas–BMT equation. We then introduced a single-resonance approximation, and derived a Froissart–Stora formula that applied to a beam crossing a depolarization resonance during acceleration. To facilitate these resonance crossings for high energy synchrotrons, we then introduced a series of clever tricks of resonance jumping, resonance harmonic compensation, and then ultimately the invention called Siberian snakes. As we will see in this chapter, the only part of these developments relevant to electrons is the Thomas–BMT equation. The two analyses diverge right afterwards. In particular, we note that for protons, the main issue to be addressed involves synchrotrons where the proton beam's energy is being accelerated, while for electrons, we are more interested in storage rings where the electron beam's energy is fixed. We will also note that the key quantity to be calculated is the resonance strength ϵ for the proton case, while it is an entirely different quantity called spin chromaticity $\gamma \frac{\partial \hat{n}}{\partial \gamma}$ for the electron case.[1] While the complex quantity ϵ is used in

---

[1] Yes, they are connected underneath their appearances. For example, as will be discussed,

conjunction with the 2×2 spinor algebra, the real vector quantity $\gamma \frac{\partial \hat{n}}{\partial \gamma}$ is used in a language of 3×3 rotations of 3-D vectors. The diverging physical mechanisms also have evolved into diverging languages adopted. The underlying physics — the Thomas–BMT equation — however, remains the same.

6.1   Some quantum aspects of synchrotron radiation

Synchrotron radiation by a point charge is a familiar subject in classical electrodynamics. Results can be found in textbooks. Perhaps less familiar are some quantum mechanical corrections to the classical results. In this section, we briefly describe some of those quantum aspects of synchrotron radiation. We leave their detailed derivations to later sections.

As an electron travels in a circular accelerator, it is accelerated sideways and radiates electromagnetic waves known as the synchrotron radiation. The instantaneous power radiated by a relativistic electron of energy E in the classical limit is well-known,

$$\mathcal{P}_{\text{class}} = \frac{2}{3} \frac{r_0 \gamma^4 m c^3}{\rho^2}, \tag{6.1}$$

with r_0 the classical radius of the electron, $\gamma = \frac{E}{mc^2}$ the Lorentz factor, m the electron rest mass, c the speed of light and ρ the instantaneous bending radius. The frequency spectrum of the radiation is somewhat complicated. It covers more or less all frequencies up to a critical frequency defined as

$$\omega_c = \frac{3\gamma^3 c}{2\rho}, \tag{6.2}$$

which is essentially γ^3 times the revolution frequency of the electron. The impressive large factor of γ^3 is due to a Doppler effect compressing the radiation field during the radiation process.

Quantum effect   The above results assume that the electron follows a prescribed circular trajectory which is unperturbed by its radiation. This is a good approximation if the radiation can be regarded as being continuously emitted rather than being emitted as quantized photons as dictated by quantum mechanics. A more accurate picture is in fact to imagine an electron emitting sudden and discrete photons as it circulates along. The synchrotron radiation is a doubly-random process. Firstly, the times at which photons are emitted obey a certain average rate but are otherwise random. Secondly, the photon energies are typically around or below the value $\hbar\omega_c$ but their exact values are also otherwise unpredictable. These random and quantized nature of synchrotron

---

they connect when harmonic spin matching is addressed.

radiation in turn yields important quantum corrections to the well-known classical results. See also a discussion of the noise properties of synchrotron radiation on page 36.

As a quantum is emitted, the electron receives a recoil. The effective energy of the electron during the emission process is thus not E, but slightly lower than E by an amount comparable to the energy of the quantum. Assuming $\hbar\omega_c \ll E$, this slight reduction in the effective electron energy means the synchrotron radiation power is slightly reduced from expression (6.1). A quantum mechanical calculation shows in fact (see Sec. 6.3.4)

$$\mathcal{P} = \mathcal{P}_{\text{class}} \left(1 - \frac{55}{8\sqrt{3}} \frac{\hbar\omega_c}{E} \right). \tag{6.3}$$

The correction term to the classical expression (6.1) is of the order of $\frac{\hbar\omega_c}{E}$. The fact the correction involves the Planck constant $\hbar$ is a telling sign of quantum mechanical considerations. Equation (6.3) has included only the leading term linear in $\hbar$; higher order terms have been ignored.[2]

In practical electron storage rings, $\frac{\hbar\omega_c}{E}$ is very small. Take $E = 5$ GeV and $\rho = 25$ m for instance, we find $\frac{\hbar\omega_c}{E} = 2.2 \times 10^{-6}$. This means the quantum correction in Eq. (6.3) is not easy to detect. However, the discreteness of quantum emissions does have another easily detectable effect in practical accelerators. The equilibrium emittances of a bunched beam of electrons in a storage ring, for example, is determined by the balance between a damping effect (the "radiation damping," which is a purely classical phenomenon) and a diffusion effect (the "quantum excitation," which is a quantum mechanical phenomenon) of the electron trajectories.[3] See Fig. 6.1.

Should all electrons radiate continuously, the electron bunch will eventually shrink into a point bunch of zero dimensions due to radiation damping. The discreteness of photon emissions introduces noise into the electron trajectories and causes the beam dimensions to grow by diffusion, which counteracts and eventually balances the damping at equilibrium. The fact that the beam does have a finite size in a storage ring is therefore a quantum mechanical effect.

One of the beam size dimensions is the energy spread of the beam. Indeed, if we take the rms energy spread of the beam for example, it does contain a factor of $\hbar$,

$$\left(\frac{\sigma_E}{E} \right)^2 = \frac{55}{96\sqrt{3}} \frac{\hbar\omega_c}{E}.$$

The energy spread is thus of the order of the *square root* of $\frac{\hbar\omega_c}{E}$. The reason of the square root is due to the diffusion nature of the noise effect and should be apparent. Although $\frac{\hbar\omega_c}{E}$ is minuscule, its square root might not be. With $\frac{\hbar\omega_c}{E} = 2.2 \times 10^{-6}$ as in the example mentioned above, we find $\frac{\sigma_E}{E} = 0.9 \times 10^{-3}$, which

---

[2]An inspection of Eq. (6.1) leads to an expectation that the coefficient $\frac{55}{8\sqrt{3}}$ in Eq. (6.3) should lie between 0 and 4, and it should carry a minus sign. Equation (6.3) confirms this expectation.

[3]M. Sands, SLAC Report SLAC-121 (1970).

Figure 6.1: A cartoon showing the difference between the classical and the quantum mechanical pictures of synchrotron radiation.

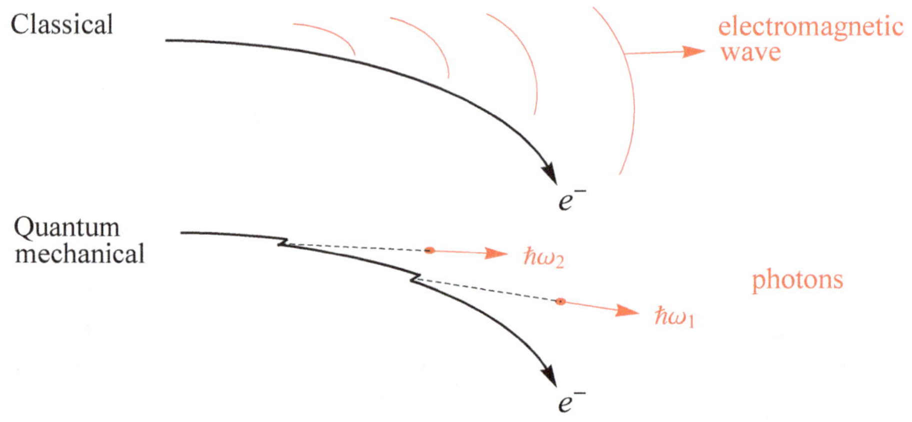

is now easily detectable. The beam's energy spread, a quantity of quantum mechanical origin therefore is small and yet macroscopic.

In addition to the discreteness of photon emissions, there is another quantum mechanical aspect of synchrotron radiation, namely that associated with the spin of the electron. Since spin is a quantity with its magnitude in units of $\hbar$, all spin effects of synchrotron radiation involve $\hbar$ and are necessarily of quantum mechanical origin. The problem becomes complicated when spin is taken into account because one now has to distinguish between two cases whether the electron spin stays in its initial state or flips over after emitting the synchrotron photon.

Let $\hat{n}$ be the spin orientation in the electron's rest frame before photon emission. When spin is taken into account, the main contribution to the synchrotron radiation power is still given by (6.3), but the $\hbar$ correction term now has an additional spin-dependent term (see Sec. 6.3.5 for derivation),

$$\mathcal{P} \; = \; \mathcal{P}_{\text{class}} \left[1 - \left(\frac{55}{8\sqrt{3}} + \hat{n} \cdot \hat{y} \right) \frac{\hbar\omega_c}{E} \right], \tag{6.4}$$

where $\hat{y}$ is the direction of magnetic field that bends the electron. If we average over all spin orientations $\hat{n}$, we recover Eq. (6.3) as it should for an unpolarized beam. Again, this spin correction term is very small in practice and is difficult to observe experimentally.

A comment is to be made concerning Eq. (6.4). Note that in the same spirit as the Thomas–BMT equation, $\hat{n}$ refers to the spin in the rest frame of the electron, even though all other quantities are taken in the laboratory frame. Note also that $\hat{n}$ is the spin *before* photon emission. After the emission, there is a small probability that it is flipped to $-\hat{n}$.

However, as discovered by Sokolov and Ternov,[4] there is another spin effect involving spin flip radiations that does have an easily observable effect on the electron beam. One can of course calculate the instantaneous power (energy radiated per unit time) radiated with spin flip and compare it with the overall instantaneous power (spin flip plus non-spin-flip), as done in Eq. (6.4), and conclude that this ratio is minuscule, but the more relevant quantity here is the instantaneous transition rate (number of photons emitted per unit time) that involves spin flip (see Sec. 6.3.6 for more discussions and derivation),

$$W = \frac{5\sqrt{3}}{16} \frac{\alpha_F \gamma^5 \hbar^2}{m^2 c \rho^3} \left[1 - \frac{2}{9}(\hat{n} \cdot \hat{z})^2 + \frac{8}{5\sqrt{3}} \hat{n} \cdot \hat{y} \right], \qquad (6.5)$$

where $\alpha_F = \frac{e^2}{4\pi\epsilon_0 \hbar c} = \frac{1}{137.036}$ is the fine structure constant and $\hat{z}$ is the unit vector in the direction of motion of the electron. We note this spin flip transition rate is proportional to $\hbar$, i.e. the entire term is quantum mechanical.[5]

This spin flip transition rate (6.5) is very slow. The overall rate of photon emissions is of the order of the radiation power divided by the photon energy, i.e. we expect that $W_{\text{overall}} \sim \frac{\mathcal{P}_{\text{class}}}{\hbar\omega_c}$. A more careful calculation in fact yields

$$W_{\text{overall}} = \frac{15\sqrt{3}}{8} \frac{\mathcal{P}_{\text{class}}}{\hbar\omega_c} = \frac{5\alpha_F}{2\sqrt{3}} \frac{\gamma c}{\rho} \approx \frac{\gamma}{95} \frac{c}{\rho}. \qquad (6.6)$$

This overall transition rate is of course dominated, in fact overwhelmed, by radiation events without spin flips. Compared with the overall rate (6.6), the rate that involves spin flips, Eq. (6.5), is smaller by a factor $\sim \frac{1}{6}(\frac{\hbar\omega_c}{E})^2 \sim 0.8 \times 10^{-12}$. Spin flip radiation events are extremely rare. And yet, we will explain below that these extremely rare spin flip radiations will contribute to a macroscopic effect on the net polarization of the beam. Indeed the discovery of spontaneous polarization of stored electron beams is to the credit of Sokolov and Ternov who courageously persevered on solving the Dirac equation exactly in spite of the overwhelming discouraging factor of $\frac{1}{6}(\frac{\hbar\omega_c}{E})^2$ to their disfavor.

In a storage ring, the guiding magnetic field is in the vertical direction $\hat{y}$. If we specify $\hat{n}$ to be either along the field (the up state) or against the field (the down state), we find that the transition rate from up state to down state, $W_{\uparrow\downarrow}$, is larger than that from down state to up state, $W_{\downarrow\uparrow}$. Noting that $\hat{n} \cdot \hat{z} = 0$ in this case, Eq. (6.5) gives

$$W_{\uparrow\downarrow} = \frac{5\sqrt{3}}{16} \frac{\alpha_F \gamma^5 \hbar^2}{m^2 c \rho^3} \left(1 + \frac{8}{5\sqrt{3}} \right)$$

$$W_{\downarrow\uparrow} = \frac{5\sqrt{3}}{16} \frac{\alpha_F \gamma^5 \hbar^2}{m^2 c \rho^3} \left(1 - \frac{8}{5\sqrt{3}} \right). \qquad (6.7)$$

Again, as mentioned, both $W_{\uparrow\downarrow}$ and $W_{\downarrow\uparrow}$ are very much smaller than W_{overall}.

---

[4]A.A. Sokolov and I.M. Ternov, Sov. Phys. Dokl. 8, 1203 (1964); A.A. Sokolov and I.M. Ternov, Synchrotron Radiation, Pergamon Press (1968); I.M. Ternov, Yu.M. Lokutov and L.I. Korovina, Sov. Phys. JETP 14, 921 (1962).

[5]The fact that Eq. (6.5) appears to be quadratic in $\hbar$ is because $\alpha_F \propto \frac{1}{\hbar}$.

Due to the fact that $W_{\uparrow\downarrow} \neq W_{\downarrow\uparrow}$, if we inject into a storage ring an unpolarized electron beam, an imbalance between the two spin flip transition rates would cause the beam to accumulate slowly a net polarization in the direction against the guiding field. One then observes that $W_{\uparrow\downarrow}$ and $W_{\downarrow\uparrow}$ are not only different but also so very different that the net equilibrium polarization reached is surprisingly high,

$$
\begin{aligned}
P_0 &= \frac{W_{\downarrow\uparrow} - W_{\uparrow\downarrow}}{W_{\downarrow\uparrow} + W_{\uparrow\downarrow}} \\
&= -\frac{8}{5\sqrt{3}} = -92.38\%.
\end{aligned}
\tag{6.8}
$$

The negative sign refers to the fact that the polarization built up by an electron beam is against the direction of the magnetic field. A positron beam would have the other sign.

We of course expect this equilibrium polarization to be reached very slowly because of the very weak transition rates. However, the time constant that this equilibrium polarization is approached by the initially unpolarized beam, in spite of being proportional to $\hbar^{-1}$ ($\hbar^{-2}$ if holding α_F fixed), is short enough to be practical. The polarization time constant is

$$
\begin{aligned}
\tau_0 &= (W_{\downarrow\uparrow} + W_{\uparrow\downarrow})^{-1} \\
&= \left(\frac{5\sqrt{3}}{8} \frac{\alpha_F \gamma^5 \hbar^2}{m^2 c \rho^3} \right)^{-1}.
\end{aligned}
\tag{6.9}
$$

Taking again $E = 5$ GeV and $\rho = 25$ m, the time constant is found to be 8 minutes. In a practical expression, we have

$$
\tau_0 = 99 \sec \frac{R[m]\, \rho[m]^2}{E[\mathrm{GeV}]^5},
$$

where $2\pi R$ is the ring circumference, and a factor $\frac{R}{\rho}$ has been included to take into account the filling factor of the dipole magnets.

One can now imagine the excitement when it was realized that the electron beam would spontaneously polarize itself to such a high level of 92% and all we have to do is to inject an unpolarized beam into a storage ring and wait 15 minutes or so. For once, we seem to be getting something free from mother nature. The value of an electron storage ring as a high energy physics research tool increased considerably due to this effect.

Radiative polarization build-up   It may be instructive to pause a moment to provide a proof of Eqs. (6.8) and (6.9) here. Consider a beam that consists of N electrons. Let $N_\uparrow$ electrons be in the up state and $N_\downarrow$ in the down state. Obviously we want $N_\uparrow + N_\downarrow = N$. Given the various transition rates, $N_\uparrow$ and $N_\downarrow$ will vary with time, and it should be easy to see that

$$
\begin{aligned}
\dot{N}_\uparrow &= W_{\downarrow\uparrow} N_\downarrow - W_{\uparrow\downarrow} N_\uparrow, \\
\dot{N}_\downarrow &= W_{\uparrow\downarrow} N_\uparrow - W_{\downarrow\uparrow} N_\downarrow.
\end{aligned}
$$

Figure 6.2: In an ideal planar electron storage ring, polarization builds up exponentially $P(t) = P_0(1 - e^{-t/\tau_0})$ towards $P_0 = -92\%$ after injecting an unpolarized beam.

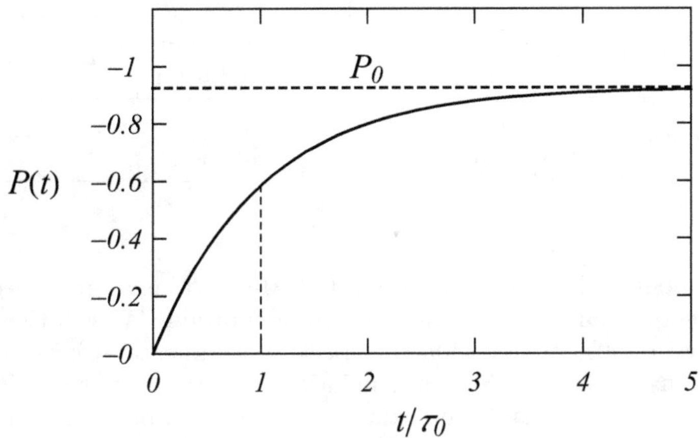

It follows that $\dot{N}_\downarrow + \dot{N}_\uparrow = 0$ as it should. The beam polarization is given by

$$P = \frac{N_\uparrow - N_\downarrow}{N_\uparrow + N_\downarrow} = \frac{N_\uparrow - N_\downarrow}{N}.$$

The polarization will change with time. It then follows with a small amount of algebra that

$$\dot{P} = -(W_{\downarrow\uparrow} + W_{\uparrow\downarrow})P + (W_{\downarrow\uparrow} - W_{\uparrow\downarrow}).$$

Equations (6.8) and (6.9) are then readily obtained by solving this equation for $P(t)$. In particular, $\tau_0^{-1} = W_{\downarrow\uparrow} + W_{\uparrow\downarrow}$ may be considered a total rate of spin flips (down-to-up plus up-to-down). Figure 6.2 illustrates the buildup of beam polarization when a beam is injected unpolarized [see also Eq. (6.50)].

Analogy between orbital and spin dynamics   If we draw an analogy between the electron's orbital motion and its spin motion and recall how equilibrium emittances are established in the orbital motion in a storage ring, saying the beam will polarize to 92% due to spin flip radiation is the same as saying the beam will shrink to a dimensionless size due to radiation damping. What we still need to take into account is the fact that the discrete photon emissions have also introduced noise into the system, and when taken into account, there is a diffusion effect of synchrotron radiation on both the emittances and the spin orientations of the electrons. The equilibrium value of beam polarization, just like the emittances, must be determined by a balance between the radiative polarizing effect of spin flip radiation (that we just discussed) and the depolarizing effect of quantum diffusion (that we will discuss).

Table 6.1: Analogy between the mechanism for equilibrium beam emittances and mechanism for equilibrium level of spin polarization.

	Damping $\longleftrightarrow$ Diffusion		Beam property
Orbital motion	Radiation damping	$\longleftrightarrow$ Quantum excitation	Emittances
Spin motion	Radiative polarization	$\longleftrightarrow$ Spin diffusion	Polarization

This analogy is illustrated in Table 6.1. In particular, it is necessary to calculate the quantum diffusion rate of spin orientation. We will then find that the ideal level of 92% polarization is subject to detrimental effects due to some depolarization resonances. Near a depolarization resonance, the spin diffusion rate becomes significant and the beam polarization can be reduced from 92%.

Several formulae mentioned in this section are given without proof. A rigorous proof can be found by solving Dirac equation in a magnetic field (see footnote 4). A semiclassical derivation using quantum electrodynamics without invoking the Dirac equation has also been developed (see Sec. 6.3). The price to pay in making a semiclassical approximation is that the results will be valid only to 1st order in $\hbar$. We shall follow this semiclassical approach.

As alluded to earlier, quantum mechanical corrections are characterized by the parameter $\frac{\hbar\omega_c}{E}$. A hierarchy of time scales can be observed by four quantities: the radiative polarization time τ_0, the longitudinal radiation damping time τ_{rad}, the accelerator revolution time T_0, and the overall synchrotron radiation emission rate W_{overall}. The hierarchy reads

$$\tau_0^{-1} \ll \tau_{\text{rad}}^{-1} \ll T_0^{-1} \ll W_{\text{overall}}.$$

Among the three quantities $\tau_0^{-1}, \tau_{\text{rad}}^{-1}$ and W_{overall}, we note that each is smaller by a factor $\sim \frac{\hbar\omega_c}{E}$. We can recap our findings by the following scalings,

$$
\begin{aligned}
\frac{\sigma_E}{E} &= \sqrt{\frac{55}{96\sqrt{3}}} \sqrt{\frac{\hbar\omega_c}{E}}, \\
\frac{\tau_0^{-1}}{\tau_{\text{rad}}^{-1}} &= \frac{5\sqrt{3}}{8} \frac{\hbar\omega_c}{E}, \\
\frac{\tau_0^{-1}}{W_{\text{overall}}} &= \frac{1}{3} \left(\frac{\hbar\omega_c}{E}\right)^2.
\end{aligned}
\tag{6.10}
$$

Homework 6.1 Consider an isomagnetic storage ring for which the bending radius ρ is uniform around the ring. From orbital dynamics, we know that the longitudinal radiation damping time is given by

$$\tau_{\mathrm{rad}}^{-1} = \frac{\mathcal{P}_{\mathrm{class}}}{E}.$$

The polarization rate is slower than the radiation damping rate by a large factor as given by Eq. (6.10). Derive Eq. (6.10) and show also that

$$\frac{\tau_0^{-1}}{\tau_{\mathrm{rad}}^{-1}} = \frac{36}{11}\left(\frac{\sigma_E}{E}\right)^2.$$

6.2 Spin precession — A recap

Spin motion in an electromagnetic field is described by the Thomas–BMT equation as discussed in Chapter 5,

$$\frac{d\vec{S}}{dt} = \vec{\Omega} \times \vec{S}, \tag{6.11}$$

$$\vec{\Omega} = -\frac{e}{m}\left[\left(a+\frac{1}{\gamma}\right)\vec{B} - \frac{a\gamma}{\gamma+1}\vec{\beta}(\vec{\beta}\cdot\vec{B}) - \left(a+\frac{1}{\gamma+1}\right)\frac{\vec{\beta}}{c}\times\vec{E}\right].$$

For accelerator applications, we make a change of time variable from t to $s = \beta ct$. For electrons, we assume the beam is relativistic, $\beta \approx 1$.

Having established the Thomas–BMT equation, the subject of spin dynamics bifurcates clearly into two paths — a proton path and an electron path. This unfortunate bifurcation is due to two camps of researchers, and the respective approaches and notations developed and evolved by each. In principle, they are speaking of the same spin dynamics, but in practice they have diverged too significantly for a possible reunification. The situation and the two approaches are sketched in Fig. 6.3. One obvious question, for example, is how are the two key quantities ϵ in the proton language and the $\gamma\frac{\partial \hat{n}}{\partial \gamma}$ in the electron language related.

In a storage ring, we apply several types of electric and magnetic fields to confine and to manipulate the motion of the electrons. These fields as seen by a circulating electron are periodic in s with the period equal to the circumference $2\pi R$ of the storage ring. Many of the applied fields, such as those provided by quadrupole and sextupole magnets, have effects on a particle only if its trajectory deviates from the designed orbit. An ideal electron traveling along the designed orbit sees only the guiding dipole magnetic field and the accelerating electric field. The accelerating electric field does not cause spin precession on the ideal electron because the field is parallel (or anti-parallel, rather, for negatively charged electrons) to the velocity $\vec{\beta}$ and the precession is, according to Eq. (6.11), proportional to $\vec{\beta}\times\vec{E}$. The guiding dipole field $\vec{B} = B_0(s)\hat{y}$, with $B_0(s+2\pi R) = B_0(s)$, on the other hand, does give rise to a precession,

$$\frac{d\vec{S}}{ds} = -\frac{eB_0(s)}{mc}\left(a+\frac{1}{\gamma}\right)\hat{y}\times\vec{S}. \tag{6.12}$$

Figure 6.3: The dynamics of proton spin and electron spin take on different paths after the Thomas–BMT equation.

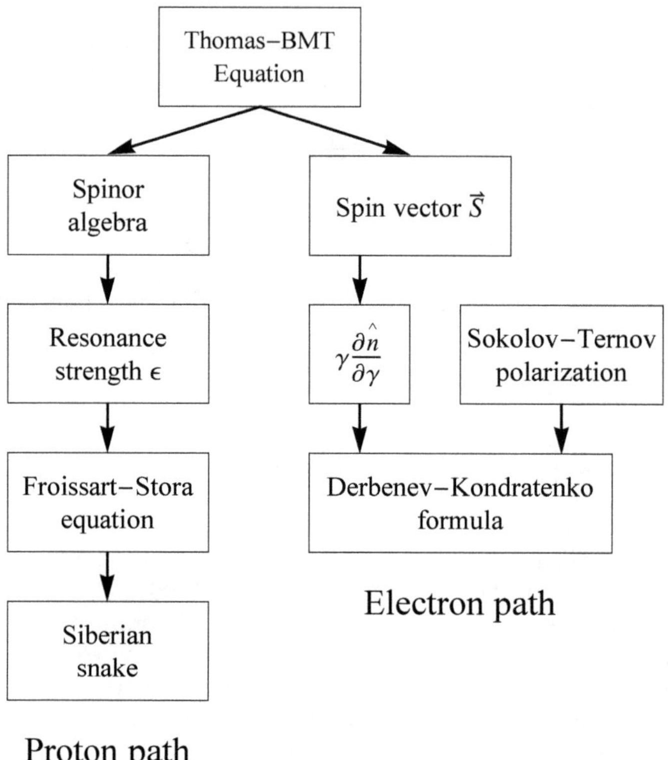

With the precession axis along $\hat{y}$, the y-component of spin S_y is preserved. If we adopt the coordinate system $(\hat{x}, \hat{y}, \hat{z})$ that rotates with the ideal electron with $\hat{z}$ along the electron's velocity and $\hat{x}$ along the horizontal direction, the other two spin components S_x and S_z rotate with the angular speed $a\gamma\frac{B_0}{B\rho}$ which is $a\gamma$ times the speed that the coordinate system rotates. As the electron completes one revolution, the coordinate system rotates by 2π and the spin has precessed around $\hat{y}$ by an angle $2\pi a\gamma$ relative to the rotating coordinate frame. In analogy to the definitions of tunes ν_x, ν_y, and ν_s for the horizontal, the vertical and the longitudinal motions, we define[6]

$$\text{spin tune} \;=\; a\gamma.$$

Consider an electron beam polarized initially in a certain direction. As the beam circulates around, only the polarization projection along $\hat{y}$ is preserved.

<hr>

[6]Deuterons are a curious case; its spin precession relative to the rotating coordinate system is backwards since $a < 0$.

Components perpendicular to $\hat{y}$ precess around $\hat{y}$ and, since different particles precess with different phases and very likely also with somewhat different speeds, rapidly smear out. As a result, if the beam is polarized at all, its net polarization at equilibrium can only be in the $\hat{y}$ direction,

$$\text{equilibrium beam polarization direction } \hat{n} \ = \ \hat{y}\,.$$

It is worthwhile to highlight the fact that the spin tune involves not the gyromagnetic ratio g but only the anomalous part of g, i.e., $g-2$. This fact can be traced back to result from the Thomas precession.

The fact that spin tune is equal to $a\gamma$ has an important practical application. Since a is a fundamental physical constant calculated and measured accurately to 9 digits, this property allows an accurate way to measure beam energy in a storage ring by trying to measure the spin tune.[7]

Perhaps it is helpful to view the situation in a different angle. Let us say the electron with spin $\vec{S}$ enters a magnetic field in the direction $\hat{b}$. Note that $\vec{S}$ is a vector in the electron's rest frame while $\hat{b}$ is in the laboratory frame. The first thing to do is to find the magnetic field in the rest frame by a Lorentz transformation. Let the rest-frame magnetic field be in the direction $\hat{b}_R$. The spin then precesses around $\hat{b}_R$ in its rest frame.

In general, $\hat{b}_R$ is not the same as $\hat{b}$. However, it turns out that they are equal in two cases. Case 1 is when the magnetic field is in the longitudinal $\hat{z}$-direction and electron is moving in the $\hat{z}$-direction. Then $\hat{b} = \hat{b}_R = \hat{z}$. Case 2 is when the magnetic field lies in the transverse x-y plane and electron is moving in the $\hat{z}$-direction. Then $\hat{b}$ and $\hat{b}_R$ are in the x-y plane and are equal. In general, $\hat{b}$ and $\hat{b}_R$ are not equal, but in these two cases, covering dipoles, quadrupoles, and solenoids,[8] they are equal, and together they give an impression that the spin in the rest frame always precesses around the direction of the magnetic field in the laboratory frame. The reader is advised to keep in mind that this is not true in general.

Homework 6.2

(a) Spin precesses ahead of the coordinate system by a factor of $a\gamma$ if we bend the trajectory by a magnetic field. What if we bend by an electric field? Consider an electron storage ring with purely electrical bending $\vec{E} = E_0\hat{x}$, what will be the spin tune?

(b) This electric storage ring still has synchrotron radiation. Along which direction will the Sokolov–Ternov radiative polarization be?

---

[7]See for example Ya.S. Derbenev, et al., Part. Accel. 10, 177 (1980); A.S. Artamonov, et al., Phys. Lett. 118B, 225 (1982); L. Arnaudon, et al., CERN SL/94-71 (BI) (1994); J.F. Zhang, et al., Proc. Part. Accel. Conf. Vancouver (2009).

[8]The reader might also recall and add some of the Siberian snake designs to the list.

6.3　Semiclassical description of spin effect on synchrotron radiation

Although spin effects are necessarily quantum mechanical, it is possible to derive most of the results of the previous Sec. 6.1 semiclassically provided we start with an effective Hamiltonian that includes a term that describes the interaction between electron spin and electromagnetic fields.[9] These derivations will be given in this section. The more rigorous quantum mechanical derivation by solving the Dirac equation can be found in the literature (see footnote 4).

6.3.1　The Hamiltonian

For a nonrelativistic electron in a magnetic field $\vec{B}$, the Hamiltonian is

$$H(\vec{r},\vec{p},t) \;=\; \frac{1}{2m}(\vec{p}-e\vec{A})^2 - \vec{\mu}\cdot\vec{B}\,, \tag{6.13}$$

where $\vec{A}$ is the vector potential associated with $\vec{B}$, and

$$\vec{\mu} \;=\; \frac{ge}{2m}\hbar\vec{S}$$

is the magnetic dipole moment for an electron with spin $\hbar\vec{S}$.

We then expand the term $(\vec{p}-e\vec{A})^2$ in Eq. (6.13). In the semiclassical calculation of electromagnetic radiation, one needs only the part of Hamiltonian $H(\vec{r},\vec{p},t)$ that describes the interaction between the electron and the field $\vec{B}$,

$$H_{\text{int}} \;=\; -ec\vec{\beta}\cdot\vec{A} - \vec{\mu}\cdot\vec{B}\,, \tag{6.14}$$

where we have dropped the term $\frac{\vec{p}^2}{2m}$ that describes a free electron and the term $\frac{e^2\vec{A}^2}{2m}$ that describes the negligible two-photon processes. Equation (6.14) is our starting point of the semiclassical treatment.

Equation (6.14) is the Hamiltonian in the nonrelativistic limit. To extend it to a relativistic electron, we need its relativistic generalization. A rigorous derivation should be obtained by making canonical transformations on the Dirac equation but we shall content ourselves with a lesser approach.

The first term in Eq. (6.14) does not require a change; it remains the same relativistically. To see that, we note that the Hamiltonian term $\frac{1}{2m}\left(\vec{p}-e\vec{A}\right)^2$ is to be replaced by the relativistic counterpart $[m^2c^4 + c^2(\vec{p}-e\vec{A})^2]^{1/2}$ which, up to 1st order in $e\vec{A}$, can be written as a free particle term $[m^2c^4 + c^2\vec{p}^2]^{1/2}$ plus an interaction term $-ec\vec{\beta}\cdot\vec{A}$, which is what we have in Eq. (6.14).

To generalize the second term of Eq. (6.14), we first rewrite it as

$$\hbar\vec{S}\cdot\vec{\Omega}\,, \qquad \text{with} \qquad \vec{\Omega} \;=\; -\frac{ge}{2m}\vec{B}\,. \tag{6.15}$$

---

[9] Ya.S. Derbenev and A.M. Kondratenko, Sov. Phys. JETP 37, 968 (1973); J.D. Jackson, Rev. Modern Phys. 48, 417 (1976); A.W. Chao, Summer School on High Energy Particle Accelerators, Fermilab (1981), AIP Conf. Proc., No. 87 (1983).

Generalization is then obtained by replacing $\vec{\Omega}$ by its relativistic extension (6.11). The relativistic interaction Hamiltonian is then found to be

$$H_{\text{int}} = -ec\vec{\beta}\cdot\vec{A} - \frac{e\hbar}{m}\vec{S}\cdot\left[\left(a+\frac{1}{\gamma}\right)\vec{B} - \frac{a\gamma}{\gamma+1}\vec{\beta}(\vec{\beta}\cdot\vec{B}) - \left(a+\frac{1}{\gamma+1}\right)\frac{\vec{\beta}}{c}\times\vec{E}\right].$$

$$(6.16)$$

It reduces to (6.14) in the nonrelativistic limit as it should.

Homework 6.3 Show that the Hamiltonian (6.15), together with $\dot{\vec{S}} = \frac{i}{\hbar}[H,\vec{S}]$ and

$$[S_i, S_j] = i\sum_k \epsilon_{ijk}S_k ,$$

gives the precession equation

$$\frac{d\vec{S}}{dt} = \vec{\Omega}\times\vec{S}.$$

Show that this is true whether $\vec{\Omega}$ is nonrelativistic (6.15) or has been generalized to (6.11).

6.3.2   Power and transition rate of synchrotron radiation

To describe synchrotron radiation, we let $\vec{A}, \vec{E}, \vec{B}$ in the interaction Hamiltonian to contain, in addition to the external applied fields, the field due to the radiation. The interaction Hamiltonian then contains two terms, a time-independent term due to external fields and a time-varying term due to the radiation. The external-field term is grouped with the free-particle term to from an unperturbed Hamiltonian H_0. We regard H_0 to be time-independent as we consider the photon emission event. The total Hamiltonian is then written as $H_0 + H_{\text{int}}$, where H_{int} is given by Eq. (6.16) with the understanding that $\vec{A}, \vec{E}, \vec{B}$ contain only the radiation field,

$$\vec{A} = \hat{\epsilon}\left(\frac{2\mu_0\hbar c}{k\mathcal{V}}\right)^{1/2}\frac{1}{2}\left(e^{i\vec{k}\cdot\vec{r}-i\omega t} + \text{c.c.}\right), \qquad (6.17)$$

where $\hat{\epsilon}, \omega, \vec{k}$ are the polarization, the frequency, and the wave vector of the emitted synchrotron photon, respectively. The normalization constant of $\vec{A}$ is chosen such that there is one such photon in a 3-D volume of space $\mathcal{V}$ under consideration — $\mathcal{V}$ is a dummy quantity that will be canceled out later. See Homework 6.4. Complex conjugate of $\vec{A}$ will be dropped because it contributes to photon absorption process that does not concern us here. The consideration of H_0 being constant in time is permissible because it does not change during the very rapid photon emission process.

From Maxwell equations, we have

$$\vec{E} = ikc\vec{A},$$

$$\vec{B} = \frac{1}{c}\hat{k}\times\vec{E}. \qquad (6.18)$$

To find the power and the transition rate of synchrotron radiation, we use the standard technique of quantum mechanics used to deal with time-dependent perturbations.[10] Let $|n(t)\rangle$ be the n-th eigenstate of the unperturbed Hamiltonian H_0 that evolves in time according to $\exp(-\frac{i}{\hbar}E_n t)$. Let the electron be initially in the state $|i(t)\rangle$. The time-dependent perturbation theory says that the probability amplitude that the electron is found in the state $|f(t)\rangle$ after completing the perturbation process, to first order of the perturbation strength, is given by

$$C_{fi} \;=\; \frac{1}{i\hbar} \int_{-\infty}^{\infty} dt \, \langle f(t)|H_{\text{int}}(t)|i(t)\rangle \;. \tag{6.19}$$

Inserting Eqs. (6.17) and (6.18) into Eq. (6.16), we have

$$H_{\text{int}} \;=\; \left(-ec\hat{\epsilon}\cdot\vec{\beta} - i\frac{e\hbar k}{m}\vec{S}\cdot\vec{V}\right)\left(\frac{\mu_0 \hbar c}{2k\mathcal{V}}\right)^{1/2} e^{i\vec{k}\cdot\vec{r}-i\omega t}, \tag{6.20}$$

where we have followed Jackson to define

$$\vec{V} \;=\; \left(a+\frac{1}{\gamma}\right)\hat{k}\times\hat{\epsilon} - \frac{a\gamma}{\gamma+1}\vec{\beta}(\vec{\beta}\cdot\hat{k}\times\hat{\epsilon}) - \left(a+\frac{1}{\gamma+1}\right)\vec{\beta}\times\hat{\epsilon}. \tag{6.21}$$

The first term in Eq. (6.20) describes a spinless point charge radiation and is independent of $\hbar$ aside from a normalization constant. The second term involves spin and is linear in $\hbar$.

In these expressions, we understand $\vec{\beta}, \vec{r}, \vec{S}$ and H_{int} are quantum mechanical operators. In our semiclassical calculations, however, they will be substituted by their classical values. Consequently, we avoid most of the troubles in taking expectation values between $|i\rangle$ and $|f\rangle$ and in keeping record of the ordering when the various operators appear — all commutators are ignored because they involve higher orders in $\hbar$. The only exception will be for the spin $\vec{S}$ when spin flips are involved, of which we will take care by using the 2×2 Pauli matrices.

The differential probability that a photon of polarization $\hat{\epsilon}$ is emitted with wave vector between $\vec{k}$ and $\vec{k}+d\vec{k}$ is

$$dp \;=\; |C_{fi}|^2 \frac{\mathcal{V}\,d^3\vec{k}}{(2\pi)^3}\,,$$

where the factor $\frac{\mathcal{V}d^3\vec{k}}{(2\pi)^3}$ is the number of free-photon states per polarization filling the volume of space $\mathcal{V}$. The power $d\mathcal{P}$ is given by the probability dp times the photon energy $\hbar\omega$, times the instantaneous frequency of revolution $\frac{c\beta}{2\pi\rho}$. Summed over the two photon polarizations, this gives the instantaneous power radiated per unit solid angle, per unit frequency interval,

$$\frac{d^2\mathcal{P}}{d\omega d\Omega} \;=\; \frac{\hbar\omega^3 \mathcal{V}}{(2\pi)^4 c^2 \rho}\sum_{\hat{\epsilon}} |C_{fi}|^2 \;.$$

---

[10]See for example, L.I. Schiff, Quantum Mechanics, McGraw-Hill (1968).

We have identified $d^3\vec{k} = k^2 d\Omega dk$. This subtle change of notable means now we have $k > 0$ or equivalently the frequency $\omega = kc > 0$. As mentioned, the dummy normalization factor $\mathcal{V}$ is self-canceled out.

Substituting explicitly from Eqs. (6.19) and (6.20), we find

$$\frac{d^2\mathcal{P}}{d\omega d\Omega} = \frac{\omega^2}{2(2\pi)^4\epsilon_0\rho} \sum_{\hat{\epsilon}} |I_1 + I_2|^2 \,, \qquad (6.22)$$

where we have defined a spin-independent integral

$$I_1 = \int_{-\infty}^{\infty} dt \left\langle f(t) \left| \left(-e\hat{\epsilon}\cdot\vec{\beta}\right) e^{i\vec{k}\cdot\vec{r}-i\omega t} \right| i(t) \right\rangle,$$

and a spin-dependent integral

$$I_2 = \int_{-\infty}^{\infty} dt \left\langle f(t) \left| \left(-\frac{ie\hbar k}{mc}\vec{S}\cdot\vec{V}\right) e^{i\vec{k}\cdot\vec{r}-i\omega t} \right| i(t) \right\rangle.$$

We have now obtained the differential power spectrum. We can also obtain the transition rate spectrum,

$$\frac{d^2W}{d\omega d\Omega} = \frac{1}{\hbar\omega} \frac{d^2\mathcal{P}}{d\omega d\Omega}.$$

Homework 6.4 Show the normalization constant in Eq. (6.17) is such that it represents one photon $\vec{k}$ in the 3-D space of volume $\mathcal{V}$.

Solution Let $\vec{A}$ be given by the real part of $\hat{\epsilon}Ne^{i\phi}$ with $\phi = \vec{k}\cdot\vec{r} - \omega t$ and normalization constant N. The field energy filling the volume $\mathcal{V}$ is $U = (\frac{\epsilon_0}{2}\vec{E}^2 + \frac{1}{2\mu_0}\vec{B}^2)\mathcal{V}$. With $\vec{E} = kcN\hat{\epsilon}\sin\phi, \vec{B} = \vec{k}\times\hat{\epsilon}N\sin\phi$, and letting $\langle U \rangle = \hbar\omega$ leads to the expression for N.

An equivalent derivation can be obtained using the Poynting vector by demanding $\langle \frac{\mathcal{V}}{\mu_0}\vec{E}\times\vec{B} \rangle = \hbar c^2\vec{k}$.

6.3.3   The classical limit

The classical result is obtained by ignoring all terms involving $\hbar$'s. The integral I_2 is therefore dropped from Eq. (6.22), and if we do not care about the electron motion after photon emission, the integral I_1 can be replaced by

$$I_1 = -e\hat{\epsilon}\cdot\int_{-\infty}^{\infty} dt\,\vec{\beta}(t)\,e^{i\vec{k}\cdot\vec{r}(t)-i\omega t} \,, \qquad (6.23)$$

with the operators $\vec{\beta}$ and $\vec{r}$ now replaced by their classical values $\vec{\beta}(t)$ and $\vec{r}(t)$. Calculation of I_1 using Eq. (6.23) can be found in textbooks but we reproduce it below using our notations. We first note the identity

$$\sum_{\hat{\epsilon}}(\hat{\epsilon}\cdot\vec{Z}_1)(\hat{\epsilon}\cdot\vec{Z}_2) = (\hat{k}\times\vec{Z}_1)\cdot(\hat{k}\times\vec{Z}_2) \,, \qquad (6.24)$$

Figure 6.4: Relative orientations of the coordinate system, the electron trajectory and the wave vector $\vec{k}$ of an emitted synchrotron photon. The bending magnetic field is in the $\hat{y}$ direction. The polar angle θ is defined relative to $\hat{z}$.

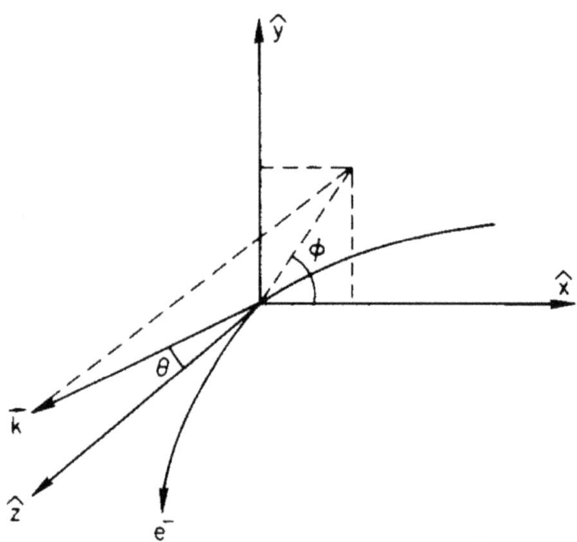

for any complex vectors $\vec{Z}_1$ and $\vec{Z}_2$. If we consider both $\vec{Z}_1$ and $\vec{Z}_2$ to be the integral that appears in (6.23), we realize that the quantity to be evaluated in the classical limit is

$$\hat{k} \times \int_{-\infty}^{\infty} dt\, \vec{\beta}(t)\, e^{i\vec{k}\cdot\vec{r}(t)-i\omega t} . \tag{6.25}$$

The coordinate system is shown in Fig. 6.4. The bending field is along $\hat{y}$; θ and ϕ define the direction of photon emission,

$$\hat{k} \;=\; \hat{z}\cos\theta + \hat{x}\sin\theta\cos\phi + \hat{y}\sin\theta\sin\phi . \tag{6.26}$$

In the classical limit, the electron motion is unperturbed by radiation and follows a prescribed circular path,

$$\begin{aligned}
\vec{\beta}(t) &= \beta\left(\hat{z}\cos\frac{\beta c t}{\rho} + \hat{x}\sin\frac{\beta c t}{\rho}\right), \\
\vec{r}(t) &= \rho\left[\hat{z}\sin\frac{\beta c t}{\rho} + \hat{x}\left(1 - \cos\frac{\beta c t}{\rho}\right)\right].
\end{aligned} \tag{6.27}$$

We recall that synchrotron radiation by a relativistic electron is strongly collimated in the direction of electron motion. The angle between the directions of motion of the electron and the photon is of the order of $\frac{1}{\gamma}$, i.e. we expect $\theta \lesssim \frac{1}{\gamma}$. Also, for a given $\hat{k}$, the time interval that takes an electron to emit a photon in the $\hat{k}$ direction can only last for a short time, $\left|\frac{\beta ct}{\rho}\right| \lesssim \frac{1}{\gamma}$. What we do is then straightforward, substitute Eqs. (6.26) and (6.27) into Eq. (6.25), keeping only leading terms in $\frac{1}{\gamma}$.

The phase factor to order $\mathcal{O}(\frac{1}{\gamma^3})$ is given by

$$i\vec{k}\cdot\vec{r}(t) - i\omega t = \frac{i\rho k}{2}\left[\theta\cos\phi\left(\frac{ct}{\rho}\right)^2 - \frac{ct}{\rho\gamma^2} - \frac{1}{3}\left(\frac{ct}{\rho}\right)^3\right] + \mathcal{O}\left(\frac{1}{\gamma^4}\right),$$

which in turn gives

$$\int_{-\infty}^{\infty} dt\, e^{i\vec{k}\cdot\vec{r}(t) - i\omega t} = \frac{2\rho}{\sqrt{3}\,c\gamma^2}(1+t^2)^{1/2}K_{1/3}(\eta),$$

$$\int_{-\infty}^{\infty} dt\, e^{i\vec{k}\cdot\vec{r}(t) - i\omega t}\left(\frac{ct}{\rho}\right) = -i\frac{2\rho}{\sqrt{3}\,c\gamma^2}(1+t^2)K_{2/3}(\eta).$$

The quantity in Eq. (6.25) is then found to be

$$\frac{2\rho}{\sqrt{3}\,c\gamma^2}(1+t^2)^{1/2}\left[\hat{x}tK_{1/3}(\eta) - i\hat{y}(1+t^2)^{1/2}\,K_{2/3}(\eta)\right]. \tag{6.28}$$

It follows that the classical differential radiation power is

$$\frac{d^2\mathcal{P}_{\text{class}}}{d\omega d\Omega} = \frac{e^2\rho\omega^2}{24\,\pi^4\epsilon_0 c^2\gamma^4}(1+t^2)\left[t^2 K_{1/3}^2(\eta) + (1+t^2)K_{2/3}^2(\eta)\right], \tag{6.29}$$

where we have defined

$$t = \theta\gamma\sin\phi,$$
$$\eta = \frac{\omega}{2\omega_c}(1+t^2)^{3/2},$$

with ω_c the synchrotron radiation critical frequency (6.2).

Geometrically, $\frac{t}{\gamma}$ is the angle between $\hat{k}$ and the orbital plane of the electron. The modified Bessel functions $K_{1/3}$ and $K_{2/3}$, together with some useful integrals involving them, are given in Table 6.2.

Integrating Eq. (6.29) over $\int_0^\infty d\omega$ gives the angular distribution of instantaneous power,

$$\frac{d\mathcal{P}_{\text{class}}}{d\Omega} = \frac{mc^3 r_0\gamma^5}{32\pi\rho^2}\frac{7+12t^2}{(1+t^2)^{7/2}}. \tag{6.30}$$

Making a change of variable

$$\int d\Omega = 2\pi\int_{-\infty}^{\infty}\frac{dt}{\gamma},$$

one can integrate (6.30) over solid angles. The result is of course just Eq. (6.1).[11]

---

[11] A few integrals over t have been used here and in the later text,

$$\int_{-\infty}^{\infty} dt\,\frac{t^{2m}}{(1+t^2)^n} = \pi\frac{(2m-1)!!(2n-2m-3)!!}{(2n-2)!!},$$

Table 6.2: Definition and some useful integrals of the modified Bessel functions $K_{1/3}$ and $K_{1/3}$.

$$\int_{-\infty}^{\infty} du\, e^{iz_0 u + \frac{i}{3}u^3} = 2\left(\frac{z_0}{3}\right)^{1/2} K_{1/3}\left(\frac{2}{3}z_0^{3/2}\right)$$

$$\int_{-\infty}^{\infty} du\, u\, e^{iz_0 u + \frac{i}{3}u^3} = \frac{2i}{\sqrt{3}} z_0 K_{2/3}\left(\frac{2}{3}z_0^{3/2}\right)$$

$$\int_0^{\infty} dx\, x^{-\lambda} K_\mu(x) K_\nu(x) = \frac{1}{2^{2+\lambda}\Gamma(1-\lambda)} \Gamma\left(\frac{1-\lambda+\mu+\nu}{2}\right) \Gamma\left(\frac{1-\lambda-\mu+\nu}{2}\right)$$

$$\times \Gamma\left(\frac{1-\lambda+\mu-\nu}{2}\right) \Gamma\left(\frac{1-\lambda-\mu-\nu}{2}\right), \qquad (\lambda < 1 - |\mu| - |\nu|)$$

$$\int_0^{\infty} dx\, x^2 K_{1/3}^2(x) = \frac{5\pi^2}{144}$$

$$\int_0^{\infty} dx\, x^2 K_{2/3}^2(x) = \frac{7\pi^2}{144}$$

$$\int_0^{\infty} dx\, x^3 K_{1/3}^2(x) = \frac{16\pi}{81\sqrt{3}}$$

$$\int_0^{\infty} dx\, x^3 K_{1/3}(x) K_{2/3}(x) = \frac{35\pi^2}{864}$$

$$\int_0^{\infty} dx\, x^3 K_{2/3}^2(x) = \frac{20\pi}{81\sqrt{3}}$$

Homework 6.5 Verify the identity (6.24). Make use of the fact that the two polarization vectors $\hat{\epsilon}_1$ and $\hat{\epsilon}_2$ are perpendicular to $\hat{k}$.

Homework 6.6

(a) Follow the outline in the text to derive Eq. (6.28) — the algebra is tedious but contains useful insight. In particular, note the need to calculate the radiation phase $i\vec{k}\cdot\vec{r}(t) - i\omega t$ to such a high accuracy of $\mathcal{O}(\frac{1}{\gamma^3})$. This homework illustrates the observation that all intricate dynamics occurs in the phase more than anything else.

(b) Derive Eq. (6.29) and use it and Table 6.2 to derive Eqs. (6.30) and (6.1).

6.3.4   Quantum correction for a spinless charge

By quantum correction here we mean correction to the classical results up to 1st order in $\hbar$. When we wrote the integrals I_1 and I_2 for the synchrotron radiation power, we were not too careful about the order in which the various

$$\int_{-\infty}^{\infty} dt\, \frac{7+12t^2}{(1+t^2)^{7/2}} = \frac{32}{3},$$

$$\int_{-\infty}^{\infty} dt\, \frac{5+9t^2}{(1+t^2)^5} = \frac{55\pi}{32},$$

$$\int_{-\infty}^{\infty} dt\, \frac{1}{(1+t^2)^{9/2}} = \frac{32}{35}.$$

operators appeared. Since noncommutability of operators are of the other of $\hbar$, this carelessness is acceptable for I_2, which is already first order in $\hbar$. It is, in fact, also acceptable for I_1 because I_1 is independent of spin and it is only the spin-dependent $\hbar$ correction that we are interested in.

Nevertheless, although not of prime concern for us, one can insist on doing the job right and obtain the quantum correction for a spinless charge, for which I_2 vanishes and we then want to calculate the spin-independent $\hbar$-correction term of I_1. This has been done by Schwinger.[12] It can be shown that the commutator relations can be cast (derivation omitted) in the form that says the first order $\hbar$ correction can be obtained by simply making a replacement

$$\omega \;\to\; \omega\left(1 + \frac{\hbar\omega}{E}\right)$$

in the classical result of the transition rate.[13] In other words,

$$\frac{d^2\mathcal{P}}{d\omega\, d\Omega} = \omega\left(\frac{1}{\omega}\frac{d^2\mathcal{P}_{\text{class}}}{d\omega\, d\Omega}\right)_{\omega\to\omega\left(1+\frac{\hbar\omega}{E}\right)}$$

$$= \frac{e^2 \rho \omega^2 \left(1 + \frac{\hbar\omega}{E}\right)}{24\,\pi^4 \epsilon_0 c^2 \gamma^4}\,(1+t^2)\left[t^2 K_{1/3}^2(\eta') + (1+t^2)K_{2/3}^2(\eta')\right], \quad (6.31)$$

where $\eta' = \frac{\omega}{2\omega_c}\left(1 + \frac{\hbar\omega}{E}\right)(1+t^2)^{3/2}$ and expanding to 1st order in $\hbar$ is understood.

One can then integrate the result over frequency to obtain the angular distribution

$$\frac{d\mathcal{P}}{d\Omega} = \frac{d\mathcal{P}_{\text{class}}}{d\Omega}\left[1 - \frac{128}{3\sqrt{3}\pi}\frac{\hbar\omega_c}{E}\frac{5+9t^2}{(1+t^2)^{3/2}(7+12t^2)}\right]. \quad (6.32)$$

Integrating again over solid angle then gives Eq. (6.3).

Homework 6.7
   (a) Use Eq. (6.31) to derive (6.32). Trick is to use integration by parts.
   (b) Use Eq. (6.32) to derive Eq. (6.3).

6.3.5   Radiation power without spin flip

Although the fact that g is not exactly equal to 2 plays an important role in how spin precesses in a storage ring, it is not so essential for the synchrotron radiation of the electron. In the rest of this and the next sections, we choose to ignore the slight difference between g and 2. Setting $g = 2$, or $a = 0$, we have

$$\vec{V} = \left(\frac{\hat{k}}{\gamma} - \frac{\vec{\beta}}{\gamma+1}\right)\times\hat{\epsilon}. \quad (6.33)$$

---

[12] J.S. Schwinger, Proc. Natl. Acad. Sci. USA 40, 132 (1954).
[13] Beware. This replacement is to be made on the transition rate, not on the radiation power.

We will briefly return to the case $g \neq 2$ in Eq. (6.51).

Just like I_1 can be approximated by a classical integral (6.23), a similar approximation can be made on I_2,

$$I_2 = -\frac{ie\hbar k}{mc} \int_{-\infty}^{\infty} dt \, \langle f|\vec{S}(t)|i\rangle \cdot \vec{V}(t) \, e^{i\vec{k}\cdot\vec{r}(t) - i\omega t} , \tag{6.34}$$

where $\vec{V}, \vec{r}$ are now classical quantities, $|i\rangle$ and $|f\rangle$ now only refer to the initial and final spin states of the electron. To find I_2, we need to evaluate $\langle f|\vec{S}(t)|i\rangle$.

Let the electron spin be instantaneously ($t = 0$) in the direction $\hat{n}$ and define angle θ_0 and ϕ_0 as shown in Fig. 6.5. Note that θ_0 is defined relative to $\hat{y}$, while θ of Fig. 6.4 is defined with respect to $\hat{z}$.

We distinguish between two cases according to whether there is a spin flip or not after photon emission. In case of no spin flip, $\langle f|\vec{S}(t)|i\rangle$ is easy to find. Knowing $\hat{n}$ at time $t = 0$, and knowing that, for $g = 2$, spin precesses with the same angular frequency as the electron circulates in the field $\vec{B}_0$, we find

$$\langle f|\vec{S}(t)|i\rangle = \frac{\hat{x}}{2} \sin\theta_0 \sin\left(\frac{\beta ct}{\rho} + \phi_0\right) + \frac{\hat{y}}{2}\cos\theta_0 + \frac{\hat{z}}{2}\sin\theta_0 \cos\left(\frac{\beta ct}{\rho} + \phi_0\right),$$
$$\tag{6.35}$$

where a factor $\frac{1}{2}$ is included on the right-hand-side because the electron spin is $\frac{\hbar}{2}$. Note that the case of no spin flip by a particle with spin is not to be confused with the case of a spinless particle.

We insert Eqs. (6.35) and (6.33) into I_2 to obtain

$$I_2 = -\frac{ie\hbar k}{2mc} \, \hat{\epsilon} \cdot \left(\cos\theta_0 \vec{u}_1 + \frac{1}{2}\sin\theta_0 e^{i\phi_0} \vec{u}_2 + \frac{1}{2}\sin\theta_0 e^{-i\phi_0} \vec{u}_3 \right), \tag{6.36}$$

where we have defined three additional symbols,

$$\vec{u}_1 = \hat{y} \times \int_{-\infty}^{\infty} dt \left[\frac{\hat{k}}{\gamma} - \frac{\vec{\beta}(t)}{\gamma + 1} \right] e^{i\vec{k}\cdot\vec{r} - i\omega t} ,$$

$$\vec{u}_{2,3} = (\hat{z} \mp i\hat{x}) \times \int_{-\infty}^{\infty} dt \left[\frac{\hat{k}}{\gamma} - \frac{\vec{\beta}(t)}{\gamma + 1} \right] e^{i\vec{k}\cdot\vec{r} - i\omega t \pm i\beta ct/\rho} . \tag{6.37}$$

In the expression for $\vec{u}_{2,3}$, the upper signs are for $\vec{u}_2$, the lower ones for $\vec{u}_3$. The reason we factor $\hat{\epsilon}$ outside of the parentheses in Eq. (6.36) is so that we can make use of the identity (6.24).

The $\hbar$ correction to synchrotron radiation power involves, from Eq. (6.22), the interference between the spin-independent amplitude and the spin-dependent amplitude. Explicitly, it involves the real part of $\sum_{\hat{\epsilon}}(I_1 I_2^*)$. Making use of Eq. (6.24), one finds

$$\sum_{\hat{\epsilon}}(I_1 I_2^*) = -i\frac{e^2\hbar k}{2mc}\left[\hat{k} \times \int_{-\infty}^{\infty} dt\vec{\beta}(t)\, e^{-i\vec{k}\cdot\vec{r}(t) + i\omega t} \right] \tag{6.38}$$

$$\cdot \left[\hat{k} \times \left(\cos\theta_0 \vec{u}_1^* + \frac{1}{2}\sin\theta_0 \, e^{-i\phi_0} \, \vec{u}_2^* + \frac{1}{2}\sin\theta_0 \, e^{i\phi_0} \, \vec{u}_3^* \right) \right] .$$

Figure 6.5: Relative orientations of the coordinate system, the electron trajectory and the instantaneous direction $\hat{n}$ of electron spin. The bending magnetic field is in the $\hat{y}$ direction. The polar angle θ_0 is defined relative to $\hat{y}$.

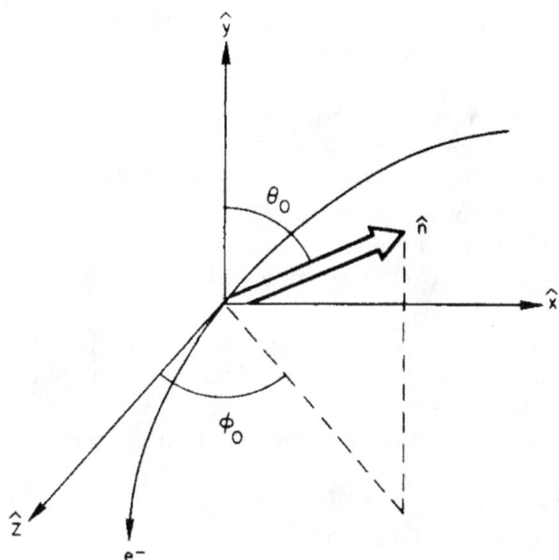

The quantity in the first square brackets has been evaluated before, given by Eq. (6.28). Similar steps also lead to, to leading order in $\frac{1}{\gamma}$,

$$\hat{k} \times \vec{u}_1^* = \frac{2}{\sqrt{3}} \frac{\rho \hat{y}}{\gamma^3 c} (1+t^2)^{1/2} K_{1/3}(\eta), \tag{6.39}$$

$$\hat{k} \times \vec{u}_{2,3}^* = -\frac{2}{\sqrt{3}} \frac{\rho}{\gamma^3 c}(1+t^2)^{1/2} \left[(t\hat{y} \mp i\hat{x})K_{1/3}(\eta) + i\hat{x}(1+t^2)^{1/2}K_{2/3}(\eta) \right].$$

Substituting into (6.38) gives the spin-dependent correction to synchrotron radiation power to 1st order in $\hbar$.

If we put together this result together with our previous spin-independent results, we obtain

$$\frac{d^2\mathcal{P}}{d\omega d\Omega} = \omega \left(\frac{1}{\omega} \frac{d^2\mathcal{P}_{\text{class}}}{d\omega d\Omega} \right)_{\omega \to \omega(1+\frac{\hbar\omega}{E})}$$

$$+ \frac{r_0 \rho \hbar \omega^3}{6\pi^3 c^2 \gamma^5} (1+t^2)^{3/2}(-\hat{n} \cdot \hat{y} + 2t\hat{n} \cdot \hat{z})K_{1/3}(\eta)K_{2/3}(\eta). \tag{6.40}$$

Integrating Eq. (6.40) over frequency gives

$$\frac{d\mathcal{P}}{d\Omega} = \frac{d\mathcal{P}_{\text{class}}}{d\Omega} \left[1 - \frac{128}{3\sqrt{3}\,\pi} \frac{\hbar\omega_c}{E} \frac{5+9t^2}{(1+t^2)^{3/2}(7+12t^2)} \right.$$

$$+ \frac{35}{3} \frac{\hbar \omega_c}{E} \frac{-\hat{n} \cdot \hat{y} + 2t\hat{n} \cdot \hat{z}}{(1 + t^2)(7 + 12t^2)} \Bigg] , \tag{6.41}$$

in which the spin-independent terms are those that appeared in Eq. (6.32). The second line in Eq. (6.41) gives the spin-dependent correction term; it vanishes if the spin orientation $\hat{n}$ is averaged over.

The t-dependence of Eq. (6.41) deserves a comment. For a longitudinally polarized electron, the term proportional to $\hat{n} \cdot \hat{z}$ gives rise to an up-down asymmetry of synchrotron radiation. With positive helicity $\hat{n} = \hat{z}$, there is more radiation in the upper plane, while with negative helicity $\hat{n} = -\hat{z}$, more radiation is found in the lower plane.

Integrating over the solid angles averages out the up-down asymmetry, the $\hat{n} \cdot \hat{z}$ term vanishes, but a term remains with connection to $\hat{n} \cdot \hat{y}$. We obtain Eq. (6.4).

We now see an asymmetry with respect to whether the spin is up ($\hat{n} = \hat{y}$) or down ($\hat{n} = -\hat{y}$); more energy is radiated if the electron spin points against the bending magnetic field. As we will see later in Sec. 6.6, the $\hat{n} \cdot \hat{y}$ term in Eq. (6.4) plays a role in determining the beam polarization in an electron storage ring.

Homework 6.8

    (a) Follow a similar step deriving Eq. (6.28) to derive Eq. (6.39).
    (b) Verify Eq. (6.40).
    (c) Integrate over ω to obtain Eq. (6.41).
    (d) Integrate again over t to verify Eq. (6.4).

Solution (a) First show that, to leading order in $\mathcal{O}(\frac{1}{\gamma})$,

$$\hat{k} \times (\hat{y} \times \vec{V}) = \frac{\hat{y}}{\gamma^2} + \mathcal{O}\left(\frac{1}{\gamma^3}\right) .$$

6.3.6　Transition rate with spin flip

The spin-independent integral I_1 does not contribute to spin flip radiation. To evaluate I_2 with spin flip, we first need to find the spin transition amplitude $\langle f | \vec{S}(t) | i \rangle$, but unlike Eq. (6.35) for the case without spin flip, this requires a bit more effort. We need to consider spin as operator, and we need Pauli matrices.

Let us choose the spin operator at time $t = 0$ to be $\vec{S}(0) = \frac{1}{2}\vec{\sigma}$ with $\vec{\sigma}$ the Pauli matrices defined in Eq. (5.18). The spin operator at other times can be obtained from $\vec{S}(0)$ by precession,

$$\begin{aligned}
S_x(t) &= \frac{1}{2}\left(\sigma_x \cos \frac{\beta ct}{\rho} + \sigma_z \sin \frac{\beta ct}{\rho} \right) , \\
S_y(t) &= \frac{1}{2}\sigma_y , \\
S_z(t) &= \frac{1}{2}\left(-\sigma_x \sin \frac{\beta ct}{\rho} + \sigma_z \cos \frac{\beta ct}{\rho} \right) .
\end{aligned} \tag{6.42}$$

Again let $\hat{n}$ be the direction of the electron spin before radiation. The initial and the final states $|i\rangle$ and $|f\rangle$ in the matrix representation are

$$|i\rangle \;=\; \begin{bmatrix} \cos\frac{\theta_0}{2}\, e^{-i\phi_0/2} \\ \sin\frac{\theta_0}{2}\, e^{i\phi_0/2} \end{bmatrix}, \qquad |f\rangle \;=\; \begin{bmatrix} -\sin\frac{\theta_0}{2}\, e^{-i\phi_0/2} \\ \cos\frac{\theta_0}{2}\, e^{i\phi_0/2} \end{bmatrix}, \qquad (6.43)$$

which are eigenstates of the operator $\hat{n}\cdot\vec{\sigma}$ with eigenvalues $+1$ and -1, respectively. Angles θ_0,ϕ_0 were the same as in Fig. 6.5. The matrix representation of $|f\rangle$ can be obtained from that of $|i\rangle$ by replacing $(\theta_0,\phi_0)\to(\pi-\theta_0,\pi+\phi_0)$. Matrix representation of $\langle f|$ is given by the transpose complex conjugate of $|f\rangle$. Note that we have assumed $\theta\sim\frac{1}{\gamma}\ll 1$ where θ was defined in Fig. 6.4, but here, θ_0 can be arbitrary; θ and θ_0 are not to be confused.

Having obtained Eqs. (6.42) and (6.43), it is straightforward by matrix multiplication to find

$$\langle f|\vec{S}(t)|i\rangle \;=\; -\frac{\hat{y}}{2}\sin\theta_0 + \cos^2\frac{\theta_0}{2}\left(\frac{\hat{z}}{2}+i\frac{\hat{x}}{2}\right)e^{-i\phi_0-i\beta ct/\rho}$$
$$-\sin^2\frac{\theta_0}{2}\left(\frac{\hat{z}}{2}-i\frac{\hat{x}}{2}\right)e^{i\phi_0+i\beta ct/\rho}.$$

We are now in a position to calculate I_2 — spin flip has been built in. Equation (6.34) gives

$$I_2 \;=\; -\frac{ie\hbar k}{2mc}\,\hat{\epsilon}\cdot\left(-\sin\theta_0\vec{u}_1 - \sin^2\frac{\theta_0}{2}e^{i\phi_0}\,\vec{u}_2 + \cos^2\frac{\theta_0}{2}e^{-i\phi_0}\,\vec{u}_3\right).$$

where $u_{1,2,3}$ were defined in Eq. (6.37).

The next step, by now familiar, is to sum over the polarizations $\hat{\epsilon}$ using the identity (6.24). After doing so, we get an expression that contains $\hat{k}\times\vec{u}_{1,2,3}$. Substituting from Eq. (6.39) then yields

$$\frac{d^2W}{d\omega d\Omega} \;=\; \frac{r_0\hbar\omega^3\rho}{24\pi^3 mc^4\gamma^6}(1+t^2)\Bigg\{ \sin^2\theta_0 K_{1/3}^2$$
$$+(1+t^2)\left(\frac{1+\cos^2\theta_0}{2}\right)\left(K_{1/3}^2+K_{2/3}^2\right)$$
$$+2\cos\theta_0(1+t^2)^{1/2}K_{1/3}K_{2/3}+t\cos\phi_0\sin 2\theta_0 K_{1/3}^2$$
$$-\frac{1}{2}\cos 2\phi_0\sin^2\theta_0\left[(1+t^2)K_{2/3}^2-(1-t^2)K_{1/3}^2\right]\Bigg\}. \quad (6.44)$$

We have given the transition rate rather than the power. The reason has been explained when we discussed Eq. (6.7).

Integrating over frequency gives

$$\frac{dW}{d\Omega} \;=\; \frac{2}{3\sqrt{3}\,\pi^2}\frac{r_0\hbar\gamma^6}{m\rho^3}\frac{1}{(1+t^2)^5}\Bigg\{ \sin^2\theta_0 + \frac{9}{8}(1+t^2)(1+\cos^2\theta_0) \qquad (6.45)$$
$$+\frac{105\sqrt{3}\pi}{256}\cos\theta_0(1+t^2)^{1/2}+t\cos\phi_0\sin 2\theta_0 - \frac{1}{8}\cos 2\phi_0\sin^2\theta_0(1+9t^2)\Bigg\}.$$

We have kept the five terms in the curly brackets the same order as the five terms in Eq. (6.44). The fourth term, being proportional to t, gives an up-down asymmetry to spin flip radiation. This asymmetry disappears if the spin direction is in the x-z plane or in the x-y plane. For example, one does not observe up-down asymmetry if $\hat{n}$ is along $\hat{x}$, or $\hat{y}$ or $\hat{z}$.

The total spin flip transition rate is obtained by integrating (6.45) over solid angles. Using the fact that $\cos\theta_0 = \hat{n} \cdot \hat{y}$ and $\sin\theta_0 \cos\theta_0 = \hat{n} \cdot \hat{z}$, we discover Eq. (6.5).

Homework 6.9 The text calculated the spin flip amplitude $\langle f|\vec{S}(t)|i\rangle$ following Eq. (6.43). A small modification can be applied to calculate $\langle f|\vec{S}(t)|i\rangle$ for the case without spin flip. Give this derivation. Your result should reconfirm Eq. (6.35).

Homework 6.10
  (a) Derive Eq. (6.44).
  (b) Integrate over ω to verify Eq. (6.45).
  (c) Integrate again over t to obtain Eq. (6.5).

Homework 6.11
  (a) The text focused on a calculation of the spin flip transition rate. Calculate the total radiation power that involves spin flips. Compare it with the radiation power without spin flips to see how much it is suppressed.
  (b) Does the sum of power with spin flip and the power without spin flip equal to the power of a spinless particle?

Solution (b) What answer did you give? You actually do not need the result in (a) to arrive at a negative answer. The spin flip power is quadratic in $\hbar$. To recover the spinless power, to 1st order in $\hbar$, you need to take the power without spin flip, and average it over the spin directions. In other words, the radiation power of a beam with spinless particles is the same as the radiation power of an unpolarized beam of particles with spin without spin flips.

6.4   Radiative polarization

6.4.1   Polarization buildup

We mentioned the Sokolov–Ternov mechanism for the beam to polarize itself in a storage ring when we discussed Eqs. (6.7–6.9), but so far we have been considering the radiation by a single electron. Polarization, on the other hand, is the net spin of a group of many electrons. Let $\vec{\zeta}$ be the polarization vector, i.e., its direction is along the direction of the net spin and its magnitude ζ ($\le 100\%$) is the degree of beam polarization. We now want to find an equation that determines the behavior of $\vec{\zeta}(t)$.

The equation of motion of $\vec{\zeta}$ contains, of course, the precession described by the Thomas–BMT terms, Eq. (6.11) or Eq.(6.12). It must also take into

account the polarizing effect of spin flip synchrotron radiation. In fact, it even
has to include the various depolarization effects so far not yet described. Here,
let us start by considering an idealistic situation in which we temporarily ignore
all quantum excitation diffusion, together with its depolarization effects, from
synchrotron radiation. To do this consistently, as revealed by Table 6.1, we
need to drop diffusion effects of both the orbital and the spin dynamics because
they have the same physical origin. In this idealized picture, the electrons form
a point bunch of zero emittances and no energy spread; all electrons follow
the designed trajectory and see only a guiding magnetic field in the vertical
direction $\hat{y}$. The only relevant terms are then the Thomas–BMT precession and
the Sokolov–Ternov spin flip transition rate, Eq. (6.5),

$$\frac{d\vec{\zeta}}{dt} = \frac{(a\gamma + 1)c}{\rho}\,\hat{y} \times \vec{\zeta} - \frac{1}{\tau_0}\left[\vec{\zeta} - \frac{2}{9}\hat{z}(\vec{\zeta}\cdot\hat{z}) + \frac{8}{5\sqrt{3}}\hat{y}\right], \qquad (6.46)$$

where the factor $a\gamma$, we recognize, is the spin tune; $\frac{c}{\rho}$ is the revolution frequency
of the electron and τ_0 was defined in Eq. (6.9). An additional factor of 2 appears
in the transition rate term because in one spin flip event, polarization changes
by two units of electron spin.

Admittedly Eq. (6.46) is somewhat awkward since in the first term that
describes precession, we have included, and indeed we must include, the fact
that $a \neq 0$. The second spin flip term, however, comes from Eq. (6.5); but we
have set $a = 0$ in the derivation of Eq. (6.5). The approximation $a = 0$ is good
for electrons and muons, but is clearly invalid for protons. Fortunately we most
likely will never need to worry about radiative polarization for protons. For
electrons, positrons and muons, taking into account of $a \neq 0$ in the second term
does not change our final result much, while the mathematics becomes much
more cumbersome. [See Eq. (6.51) later for the result when $a \neq 0$.]

Let us put the awkwardness of Eq. (6.46) in another way. Ironically, in the
first term in Eq. (6.46), we will claim $a\gamma \gg 1$ or at least $a\gamma \gtrsim 1$, while in the
second term, we claim $a \ll 1$. The justification partly lies in the fact that in
the first term, what we claim to be large is $a\gamma$, not a.

Let us rewrite Eq. (6.46) in terms of the three components ζ_x, ζ_y and ζ_z of
the polarization in a coordinate system rotating with the beam,

$$\begin{aligned}
\dot{\zeta}_x &= \frac{a\gamma c}{\rho}\zeta_z - \frac{1}{\tau_0}\zeta_x, \\
\dot{\zeta}_y &= -\frac{1}{\tau_0}\left(\zeta_y + \frac{8}{5\sqrt{3}}\right), \qquad\qquad (6.47)\\
\dot{\zeta}_z &= -\frac{a\gamma c}{\rho}\zeta_x - \frac{7}{9\tau_0}\zeta_z.
\end{aligned}$$

We see that ζ_x and ζ_z are coupled, while ζ_y is uncoupled from ζ_x and ζ_z. From
Eq. (6.47), we observe that at equilibrium when $\dot{\zeta}_x = \dot{\zeta}_y = \dot{\zeta}_z = 0$, we must
have $\zeta_x = \zeta_z = 0$ and $\zeta_y = -\frac{8}{5\sqrt{3}} = -92.38\%$.

To proceed to get a feeling about how this equilibrium polarization is reached
in time, let us simplify the problem by considering a uniform magnetic field; ρ

and τ_0 are then constants. We first note that ζ_y is readily solved,

$$\zeta_y(t) = \left[\zeta_y(0) + \frac{8}{5\sqrt{3}}\right] e^{-t/\tau_0} - \frac{8}{5\sqrt{3}} . \tag{6.48}$$

The vertical component of polarization thus approaches its equilibrium value $-\frac{8}{5\sqrt{3}}$ with time constant τ_0.

To find the time evolution for ζ_x and ζ_z, we first note that if we ignore spin precession, ζ_x will approach zero with a rate τ_0^{-1} while ζ_z will take a slightly lower rate, $\frac{7}{9}\tau_0^{-1}$, to reach its zero. Both rates are very slow compared with the rate $\frac{a\gamma c}{\rho}$ at which ζ_x and ζ_z rotate and mix into each other. It is therefore a good approximation if we replace $\frac{1}{\tau_0}$ in the $\dot{\zeta}_x$ equation and $\frac{7}{9\tau_0}$ in the $\dot{\zeta}_z$ equation by their average value $\frac{8}{9\tau_0}$. After doing so, we can solve ζ_x and ζ_z,[14]

$$\zeta_x(t) = \left[\zeta_x(0)\cos\frac{a\gamma ct}{\rho} + \zeta_z(0)\sin\frac{a\gamma ct}{\rho}\right] e^{-8t/9\tau_0} ,$$

$$\zeta_z(t) = \left[-\zeta_x(0)\sin\frac{a\gamma ct}{\rho} + \zeta_z(0)\cos\frac{a\gamma ct}{\rho}\right] e^{-8t/9\tau_0} . \tag{6.49}$$

Equations (6.48) and (6.49) describe the time evolution of polarization if we inject into a storage ring a beam with initial polarization $\vec{\zeta}(0)$.[15] In particular, if the injected beam is unpolarized, the spin flip synchrotron radiation will cause the beam to build up its polarization against the field,

$$\vec{\zeta}(t) = -\frac{8}{5\sqrt{3}}\,\hat{y}\left(1 - e^{-t/\tau_0}\right) . \tag{6.50}$$

This result was used to provide Fig. 6.2.

6.4.2   The case when $g \neq 2$

As mentioned, part of the analysis following Eq. (6.46) assumes a pure Dirac particle when the anomalous gyromagnetic ratio $a = \frac{g-2}{2}$ is set to zero. Our results are therefore approximately valid only for electrons and muons, and are invalid for protons and deuterons. In particular, when a is set to zero, we conclude that the equilibrium radiative polarization is 92.38% in the direction antiparallel to the magnetic field for electrons and parallel to the magnetic field for positrons.

One may pause a moment here and try to draw a more intuitive picture of this polarization buildup. For that, one imagines a magnetic moment $\vec{\mu}$ in a magnetic field $\vec{B}$. Quantum mechanics says that two quantum states are generated, one with $\vec{\mu}$ parallel to $\vec{B}$, another with $\vec{\mu}$ antiparallel to $\vec{B}$. Particles,

---

[14]V.N. Baier, Sov. Phys. Ups. **14**, 695 (1972).

[15]A cautious reader may rightfully question if Eqs. (6.48) and (6.49) together could yield an unphysical situation when $|\vec{\zeta}|^2 > 1$ at some time t. Indeed this is no longer guaranteed after the approximations made. But we are not going to pursue this slight inaccuracy.

one would then argue, prefer to stay in the lower energy state, namely the one with $\vec{\mu}$ parallel to $\vec{B}$. One then concludes that, due to the continuous excitation of the synchrotron radiation, electrons must polarize against $\vec{B}$ while positrons are polarized along $\vec{B}$. Some satisfaction can be drawn because this yields the correct answer.

The difficulty with such a picture has been discussed by Jackson.[16] The problem lies in the fact that the two states cannot be regarded as isolated states; orbital motion of the electron must be considered together with the spin as one coupled system. During the time interval it takes an electron to complete the process of emitting a photon, the electron has been bent by an angle $\sim \frac{1}{\gamma}$ in its orbital motion. In the mean time, the electron spin has precessed by an angle $a\gamma$ times as much, i.e., it has precessed by an angle $\sim a\gamma \times \frac{1}{\gamma} = a$. In order for the two quantum states to be regarded as being isolated, the spin must complete at least one full turn of precession during the photon emission process. This is true only if $|a| \gtrsim 2\pi$, or equivalently, $|g| \gtrsim 4\pi$. For electrons and positrons, this is far from being the case.

The above intuitive picture therefore breaks down, and it remains not too much more than a quick way to memorize the direction of polarization correctly for both electrons and positrons. In fact, even for this limited purpose, the fact that it does work is only accidental. According to this picture, electron polarization will be in the $-\hat{y}$ direction if $g > 0$ and $+\hat{y}$ direction if $g < 0$. The polarization direction switches sign at $g = 0$, and in either case the degree of polarization should be 100% when the final equilibrium is reached.

To reconcile the intuitive picture and our results for the case $a = 0$, we can refer to a general calculation performed for arbitrary a.[17] It is found there that the electron polarization switches direction between $-\hat{y}$ and $+\hat{y}$ not at $g = 0$ but at $g = 1.198$. Also the degree of polarization does not approach 100% until $|g| \gtrsim 4\pi$. More explicitly, for arbitrary a,

$$
\begin{aligned}
\frac{\tau_0(a)}{\tau_0} &= \left[\left(1 + \frac{41}{45}a - \frac{23}{18}a^2 - \frac{8}{15}a^3 + \frac{14}{15}a^4\right) e^{-\sqrt{12}|a|}\right. \\
&\quad - \frac{8}{5\sqrt{3}}\frac{a}{|a|}\left(1 + \frac{11}{12}a - \frac{17}{12}a^2 - \frac{13}{24}a^3 + a^4\right) e^{-\sqrt{12}|a|} \\
&\quad \left. + \frac{8}{5\sqrt{3}}\frac{a}{|a|}\left(1 + \frac{14}{3}a + 8a^2 + \frac{23}{3}a^3 + \frac{10}{3}a^4 + \frac{2}{3}a^5\right)\right]^{-1}, \\
P_0(a) &= -\frac{8}{5\sqrt{3}}\frac{\tau_0(a)}{\tau_0}\left(1 + \frac{14}{3}a + 8a^2 + \frac{23}{3}a^3 + \frac{10}{3}a^4 + \frac{2}{3}a^5\right). \quad (6.51)
\end{aligned}
$$

The reader should check that when a is set to zero, Eqs. (6.8) and (6.9) are recovered.

Plotted in Figs. 6.6(a) and 6.6(b) are the values of $\frac{\tau_0(a)}{\tau_0}$ and $P_0(a)$ versus a for a negatively charged particle. (For positively charged particle, the sign of $P_0(a)$

[16] J.D. Jackson, Rev. Modern Phys. 48, 417 (1976).

[17] Ya.S. Derbenev and A.M. Kondratenko, Sov. Phys. JETP 37, 968 (1973); J.D. Jackson, Rev. Modern Phys. 48, 417 (1976).

Figure 6.6: (a) The buildup time $\tau_0(a)$ for radiative polarization, normalized by τ_0 of Eq. (6.9), versus the magnetic moment anomaly a. (b) The equilibrium beam polarization $P_0(a)$ versus a. The red curves show the general results (6.51). The blue curves indicate what one would expect from an intuitive picture that is valid only for large $|g|$; see Eq. (6.52).

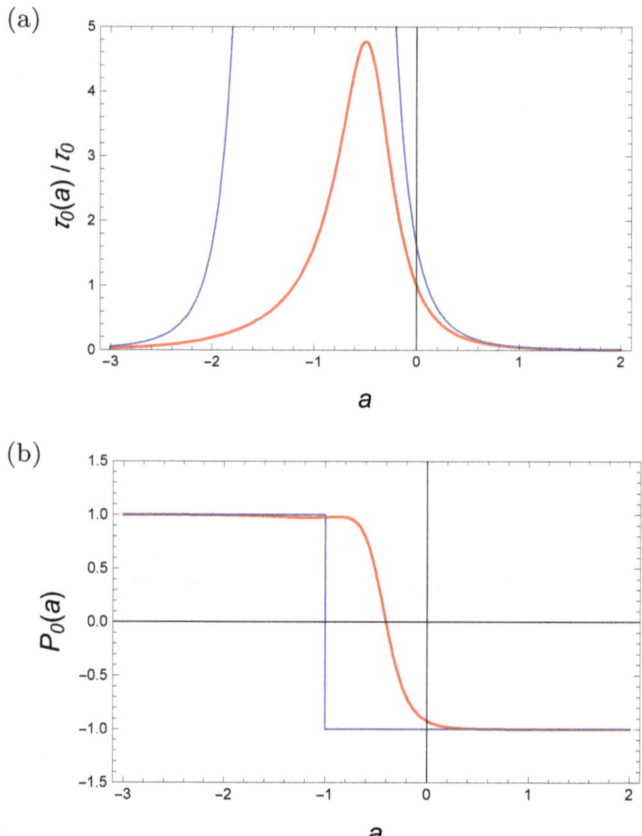

should be reversed.) For large $|a|$, the magnetic moment is large; the polarization time constant becomes short and the level of polarization approaches 100% as we would expect. Our earlier results, however, correspond only to $a = 0$. If we insist on using the right value of $a = 0.00116$, the equilibrium polarization would have been 92.44%, somewhat higher than the value 92.38% we have been talking about; and the polarization time constant would have been shorter by about half a percent.

Homework 6.12 Follow the arguments in the text to derive Eq. (6.46).

Solution Use $\dot{P} = (W_{\downarrow\uparrow} + W_{\uparrow\downarrow})P + (W_{\downarrow\uparrow} - W_{\uparrow\downarrow})$ and $\dot{P} = \hat{\zeta} \cdot \dot{\vec{\zeta}}$.

Homework 6.13 Since protons have $a = 1.793$, a proton beam will be almost 100% fully radiatively polarized. The problem is its polarization time. Find the polarization time for a 500-GeV proton storage ring of 1 km radius. Observe how the polarization time τ_0 more or less scales inversely with the 4th power of particle's mass.

Solution Use $\frac{\tau_0(1.793)}{\tau_0} = \frac{1}{116.5}$ from Eq. (6.51) and τ_0 from Eq. (6.9). The polarization time is 6×10^{10} years.

Homework 6.14 If the gyromagnetic ratio $|g| \gg 1$, show that

$$\frac{\tau_0(a)}{\tau_0} \approx \frac{16}{15\sqrt{3}} |a + 1|^5 ,$$

and Eq. (6.51) becomes

$$\tau_0^{-1}(a) = \frac{\alpha_F |g|^5 \hbar^2 \gamma^5}{48 m^2 c \rho^3} \quad \text{and} \quad P_0 = -\frac{g}{|g|} . \tag{6.52}$$

Show that these results agree with the intuitive picture discussed in connection to Eq. (6.51).

Note that the fine structure constant can be expressed as $\alpha_F = \frac{r_0 m c}{\hbar}$. That is why τ_0 in Eqs. (6.9) and (6.52) may appear being quadratic in $\hbar$.

6.4.3   Wiggler insertion

So far we have considered a planar storage ring with all bending magnets having the same sign, $\rho > 0$. If there are wiggler devices — a series of horizontal bending magnets with alternating positive and negative polarities — in the storage ring, we need to modify our results slightly. The ring bending magnets will still have the same sign $\rho > 0$ but in the wiggler insertion there are some bending magnets with $\rho < 0$. The "up" and the "down" spin states switch roles in a reversed bending magnet.

We need to keep track of the sign of ρ and the up and down spin states. To take into account the difference between electrons and positrons, let us also be careful about the sign of charge e. Here let us adopt the convention that e is positive for positrons and negative for electrons. Assuming the beam is injected unpolarized, then throughout the evolution of polarization, we have $\zeta_x = \zeta_z = 0$.

First, let us build some feeling by considering a simplified case in which the storage ring consists of a group of positive bending magnets with bending radius $\rho_1 > 0$ and a group of negative bending magnets with $\rho_2 < 0$. The polarization ζ_y will try to approach 92% value (consider positrons) in the positive bends with $\dot{\zeta}_y \propto (0.92 - \zeta_y)$. In the negative bends, because the magnetic field is now in the negative direction, ζ_y tries to approach -92% with $\dot{\zeta}_y \propto (-0.92 - \zeta_y)$. The equation governing ζ_y is therefore

$$\dot{\zeta}_y = \frac{1}{\tau_1}(0.92 - \zeta_y) + \frac{1}{\tau_2}(-0.92 - \zeta_y) ,$$

where τ_1 and τ_2 are the polarization times in the positive and the negative bends, respectively — they will have a positive sign regardless of the signs of the charge or the bending radius. It follows by solving the above equation that an initially unpolarized beam would have

$$
\begin{aligned}
\zeta_y(t) &= P_0\left(1 - e^{-t/\tau_0}\right), \\
\tau_0^{-1} &= \tau_1^{-1} + \tau_2^{-1}, \\
P_0 &= 92\%\left(\frac{\tau_1^{-1} - \tau_2^{-1}}{\tau_1^{-1} + \tau_2^{-1}}\right).
\end{aligned}
$$

We then consider the more general situation when bending radius ρ is a function of s with possibly positive and negative signs. Equation (6.47) then should read

$$
\dot\zeta_y = -\left(\frac{5\sqrt{3}}{8}\frac{\alpha_F\gamma^5\hbar^2}{m^2c|\rho^3|}\right)\zeta_y + \frac{e}{|e|}\frac{\alpha_F\gamma^5\hbar^2}{m^2c\rho^3}.
$$

Assuming ζ_y varies slowly, we can average this equation over s around the storage ring circumference. The final results give

$$
\begin{aligned}
\tau_0 &= \left(\frac{5\sqrt{3}}{8}\frac{\alpha_F\gamma^5\hbar^2}{m^2c}\frac{1}{2\pi R}\oint\frac{ds}{|\rho^3|}\right)^{-1}, \\
P_0 &= \frac{e}{|e|}\frac{8}{5\sqrt{3}}\frac{\oint ds/\rho^3(s)}{\oint ds/|\rho^3(s)|}.
\end{aligned}
\tag{6.53}
$$

Homework 6.15 Apply Eq. (6.53) to the following problem. Suppose your planar, ideal electron storage ring has too long a polarization time, and you wish to shorten it by inserting some strong wigglers. You are given three choices,

1. a wiggler consisting of three dipoles whose lengths are $\ell, 2\ell, \ell$ and bending radii are $-\rho_1, \rho_1, -\rho_1$,

2. a wiggler consisting of three dipoles whose lengths are ℓ, ℓ, ℓ and bending radii are $-\rho_1, \frac{\rho_1}{2}, -\rho_1$.

3. a wiggler consisting of three dipoles whose lengths are ℓ, ℓ, ℓ and bending radii are $\rho_1, -\frac{\rho_1}{2}, \rho_1$.

Which choice should you make?

Solution You should make choice 2, which can be referred to as an "asymmetric wiggler", and not to expect too much. You should note that a wiggler affects both P_0 and τ_0. Introducing a wiggler shortens τ_0 — the original intention — but also necessarily lowers P_0. In choice 1, $\zeta_y(t)$ with wiggler will never be higher than without wiggler at any time t. For choice 2, there could be a short

Figure 6.7: Behavior of polarization for the three cases of Homework 6.15, case 1 on the left, case 2 in the middle and case 3 on the right. The polarization $-\zeta_y(t)$ is plotted as a function of time $\frac{t}{\tau_0}$, where τ_0 refers to the polarization time without wiggler. The solid curves are when $\frac{\ell}{\rho_1^3}\frac{\rho_0^3}{2\pi R} = 0.1$, where ρ_0 is the uniform positive bending radius in the rest of the ring. The dashed curves are the case without wiggler for comparison.

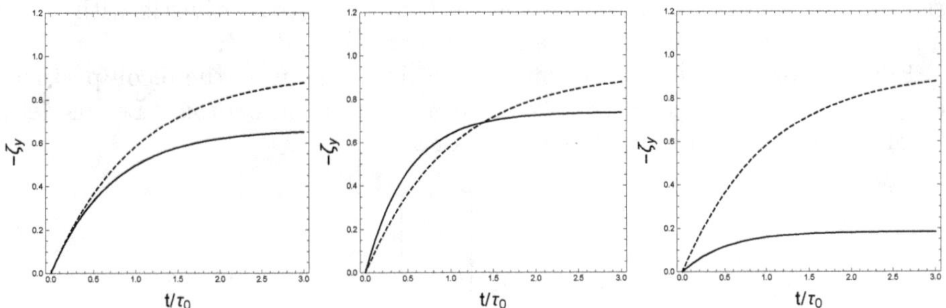

time period during which having wiggler could be beneficial. See Fig. 6.7 for an illustration of the situation. See also the discussion around Fig. 6.11 later.

As seen from Fig. 6.7, case 2 is the only case when there is a limited period of time when there is a gain of polarization with the wiggler. For case 1, the polarization has the same initial buildup slope as the case without wiggler, but never surpasses it. If wigglers are to be used to enhance polarization, the only way is to use case 2 initially and to switch it off at a certain time when the natural polarization becomes more favorable.

6.5   Polarization in a storage ring

6.5.1   Ideal storage ring

In Sec. 6.2, we mentioned that if an electron follows the designed trajectory exactly in a perfect planar storage ring, its spin will precess around the vertical direction $\hat{y}$; and if all electrons do so, the net beam polarization direction $\hat{n}$ will be along $\hat{y}$. We defined a spin tune as the number of spin precessions per revolution and found it is equal to $a\gamma$. We also concluded that under this same condition the radiative beam polarization will be 92%. To summarize, we have shown that

$$\hat{n} = \hat{y}, \tag{6.54}$$

$$\text{spin tune} = a\gamma, \tag{6.55}$$

$$P_0 = 92\%. \tag{6.56}$$

On the other hand, we know the assumption that all electrons follow the designed trajectory is never fulfilled because the beam distribution has a finite size. Even if we build a storage ring for which all electric and magnetic devices are constructed and installed perfectly, the designed trajectory is followed only by the center of beam distribution and not by all electrons. One urgent question to be answered is what happens to the polarization properties (6.54) to (6.56) if we take into account the finite size of the beam.

For this discussion, we assume that the perfect planar storage ring does not have skew quadrupole or sextupole fields. Those fields will be later included as error fields. We shall also assume the electron beam is relativistic with $\beta \approx 1$. In particular, we ignore the difference between $\frac{\Delta P}{P_0}$ and $\frac{\Delta E}{E_0}$.

Finite beam sizes in an electron storage ring come from the recoil perturbations that electrons receive as they radiate synchrotron photons. Let us define the orbital state of an electron by a vector

$$
X = \begin{bmatrix} x \\ x' \\ y \\ y' \\ z \\ \frac{\Delta E}{E_0} \end{bmatrix},
$$

where x, y and z are the displacements of the electron relative to the center of particle distribution; x' and y' are the corresponding conjugate momenta defined to be the slopes of the electron's trajectory; z is the longitudinal displacement of the electron from the reference synchronous electron; $\frac{\Delta E}{E_0}$ is the relative energy deviation. Immediately after radiating a photon, only the $\frac{\Delta E}{E_0}$-coordinate of the electron is perturbed by the event of radiation. As the electron continues to circulate around the storage ring, this perturbation in $\frac{\Delta E}{E_0}$ in general propagates into the other five orbital coordinates, giving the beam finite sizes in all six dimensions. In a perfect planar storage ring, however, the perturbation on $\frac{\Delta E}{E_0}$ propagates only into the x-, x'- and z-coordinates, leaving y- and y'-coordinates unperturbed. As a result, the beam distribution is an infinitely thin ribbon with finite width and length but zero height.

We will now try to reestablish properties (6.54) to (6.56) for a ribbon beam in the ideal storage rings as follows.

- A particle in such a ribbon beam experiences, in addition to the bending magnetic field and the RF accelerating electric field seen along the designed trajectory, the perturbing magnetic fields in the quadrupoles. The nice thing is, with $y = 0$, these quadrupole fields are all along $\hat{y}$. If we look at the Thomas–BMT equation (6.11), we find that the spin precession angular velocity $\vec{\Omega}$ is also along $\hat{y}$. [The second term in Eq. (6.11) vanishes, the other two terms are along $\hat{y}$.] This reestablishes (6.54) since any polarization components perpendicular to $\hat{y}$ will disappear rapidly due to the different precession phases of different particles.

- We also find from Eq. (6.11) that the contributions from quadrupoles and RF cavities oscillate between positive and negative values as x and x'

execute betatron and synchrotron oscillations — in general, x contains a betatron part and a synchrotron part. As a result, the average rate of spin precession is determined by the bending magnets alone. This reestablishes (6.55).

It is in fact a general result that the spin tune, on the average, is always determined from the electromagnetic field seen along the closed orbit. But we shall keep in mind that the actual spin precession angle per turn oscillates slightly, and deviates slightly from the average value $2\pi a\gamma$ by an amount that depends on the betatron and synchrotron motions of the electron. This effect of "frequency modulation," as we will see later, is in fact one of the mechanisms that depolarizes the beam.

- As to the level of radiative polarization, Eq. (6.56), it is also unaffected by the finite size of the ribbon beam. This is because the rate of spin flip transition, which we recall is the mechanism responsible for polarization buildup, is proportional to the magnetic field to the cubic power [see Eq. (6.9)]. The magnetic field in quadrupoles is too weak to have an appreciable effect on radiative polarization buildup.

We thus conclude that in a perfectly constructed planar storage ring, the beam has a ribbon distribution, and its polarization satisfies the nice properties (6.54) to (6.56).

A real storage ring will contain error fields. So we want to know what happens to beam polarization if the storage ring contains error fields. We will start by adding errors to our ideal storage ring.

Before we do so in the following sections, let us first consider two types of error fields: those due to sextupole magnets and those accidental dipole fields that cause a horizontal closed orbit distortion. An example of the latter type is when a quadrupole magnet is horizontally misaligned. These error fields are special because they do not cause particles to execute vertical motions and the beam keeps its ribbon distribution.

The perturbing magnetic fields seen by particles always point in the $\pm\hat{y}$ directions. Most of the previous discussions for a perfect ring still apply. In particular, (6.54) follows from the fact that $\vec{\Omega}$ is always along $\pm\hat{y}$ directions. Remembering that the beam distribution center always follows the closed orbit to rotate 2π radians per turn — no matter how distorted the closed orbit may be — and that spin precesses $a\gamma$ times faster than the coordinate does in a lock step manner, we find the spin tune is always equal to $a\gamma$, i.e., (6.55) is assured. As to (6.56), it again follows if the perturbing fields are much weaker than the main bending fields, which is satisfied for all practical cases.[18]

We thus conclude that as long as particles do not execute y-motions, the beam (a ribbon beam) will happily polarize itself according to (6.54) to (6.56).

---

[18]That is all except wigglers. If we insert a wiggler in the storage ring, the beam remains ribbon-shaped but the associated "error" fields are strong enough to have an appreciable effect on polarization level. In fact, it always lowers the equilibrium polarization level. See Eq. (6.53).

It is therefore not an overstatement if we blame (almost) all depolarization effects to some vertical motions of one cause or another. The reader should recall a similar situation when studying depolarization effects for protons in Chapter 5. Vertical motion is almost always the culprit.

Homework 6.16

(a) The text claims the level of polarization will reach 92% in a perfect planar ring with ribbon beams, arguing the effects due to the magnetic fields in the quadrupoles do not have much effect. Go through the argument more quantitatively and give an estimate of the effect being ignored, particularly on the ideal 92% polarization level and on the polarization time τ_0. Some hints can perhaps be obtained from the discussion of wiggler insertion, Sec. 6.4.3, and Homework 6.15.

(b) The text also claims the same when a horizontal closed orbit distortion is present. Give an expression of the expected change in the loss of polarization and effect on the polarization time in this case.

6.5.2   Integer resonance

Problems occur when we include error fields that cause y-excursions in particle motion. Our ribbon beam configuration breaks down. One might think these fields are weak and question why should they do much harm. After all, if an electron passes through a quadrupole of strength $G\ell = 5$ Tesla off-centered, say vertically, by 1 mm, the spin precesses by an angle of 0.3 mrad, which looks rather harmless. The answer to this question lies in two facts.

1. The electron passes through this quadrupole not just once but repeatedly as it circulates around. The innocent-looking 0.3 mrad may add up coherently every time the electron passes through the quadrupole. The conditions for those small spin rotations to add up coherently are referred to as depolarization resonance conditions.

2. Even more importantly, the strengths of some of those depolarization resonances are greatly enhanced due to the presence of a monstrous noise source — the synchrotron radiation. The enhancement factor involved, as will be explained, is typically as large as 10^6.

Mechanism 1 should be somewhat familiar, because it is the same mechanism that causes depolarization in a proton synchrotron (see Chapter 5). When the spin motion resonates with orbital motion, the spin motion is seriously perturbed. Analysis of this effect led to the calculation of resonance strength ϵ for proton synchrotrons. Mechanism 2, on the other hand, is new. It is associated with synchrotron radiation, and is absent in proton accelerators. This mechanism dominates — overwhelms, actually — in electron storage rings. Although synchrotron radiation does not directly generate a depolarization resonance, it can enhance an existing depolarization resonance by a large factor. The resonance has to be there first, but once there, synchrotron radiation magnifies its effects. The challenge is now how to formulate the problem and analyze both mechanisms for an electron storage ring with error fields.

Let us make another comment here. In storage rings, typical error fields, coming for example from magnet misalignments, are proportional to beam energy E because strengths of all magnets scale with E by design. When a particle passes through this error field, the factor of E in the error field cancels the factor of E in its rigidity, and as a result, the angular deflection θ in the particle's orbital trajectory is independent of E. The same thing does not happen for spin; the spin of the particle is perturbed by an angle $a\gamma\theta$ which is proportional to E. In other words, the higher the beam energy is, the more sensitive are the particle spins to magnet misalignments and therefore the more vulnerable is the beam polarization to the depolarization effects. Maintaining a good beam polarization in a high energy storage ring such as LEP (90 GeV) would be much more difficult than in a lower energy storage ring such as SPEAR (1.5 GeV).

One type of depolarization resonance occurs when the spin tune $a\gamma$ is close to an integer. To see that, let us start with Eq. (6.54), i.e., the nominal polarization direction $\hat{n}$ is along the vertical direction $\hat{y}$. The reason Eq. (6.54) is important is that the radiative polarization built up painstakingly by the spin flip synchrotron radiation is along $\hat{y}$. If $\hat{n} \neq \hat{y}$, the beam gets to keep only the polarization component along $\hat{n}$ and the net beam polarization will be reduced by a cosine factor $\hat{n} \cdot \hat{y}$. Clearly, one loses polarization if $\hat{n}$ deviates appreciably from $\hat{y}$.

Two short remarks might be in order at this point.

- Strictly speaking, such a loss of polarization is not a "depolarization" mechanism. It is rather a "lack of polarization."

- The reader may be reminded of the fact that $\hat{y}$ refers to the laboratory frame while $\hat{n}$ exists in the electron's rest frame. Calling $\hat{n} \cdot \hat{y}$ a "cosine factor" as if it is a geometric quantity admittedly will require a stretch of imagination.

Consider a particle at the center of beam distribution. It follows the closed orbit and sees an external electromagnetic field that is periodic in s with period $2\pi R$. Its spin therefore precesses with a periodic angular velocity $\vec{\Omega}$. Starting with s, we can integrate this angular velocity through one turn to obtain a net rotation on spin. If the spin is represented as a 3-D vector

$$\vec{S} = \begin{bmatrix} S_x \\ S_y \\ S_z \end{bmatrix},$$

the net rotation can be written as a 3×3 matrix $R(s)$. The beam polarization direction $\hat{n}(s)$ at position s is then given by the rotational axis of $R(s)$ (use right-hand rule),

$$R(s)\hat{n}(s) = \hat{n}(s).$$

In case particles do not execute y-motions (the ribbon beam case), $R(s)$ is simply a rotation about $\hat{y}$ by an angle $2\pi a\gamma$,

$$
R(s) = \begin{bmatrix} \cos 2\pi a\gamma & 0 & \sin 2\pi a\gamma \\ 0 & 1 & 0 \\ -\sin 2\pi a\gamma & 0 & \cos 2\pi a\gamma \end{bmatrix}. \tag{6.57}
$$

The rotational axis of this $R(s)$ is $\hat{n}(s) = \hat{y}$, which of course is just the property (6.54).

The calculation of $R(s)$ and $\hat{n}(s)$ for the general case will be described in detail later. Here let us consider an idealized case in which the spin precession from $s = 0$ to $s = 2\pi R$ is given by Eq. (6.57) but at $s = 0$ there is a perturbing magnetic field B_x along the horizontal direction. Such a field may come from a vertically misaligned quadrupole. Traversing this magnetic field B_x, the spin precesses around $\hat{x}$. The spin rotation across the perturbing field is described by

$$
\begin{bmatrix} 1 & 0 & 0 \\ 0 & \cos\theta & \sin\theta \\ 0 & -\sin\theta & \cos\theta \end{bmatrix}, \tag{6.58}
$$

with $\theta = (1 + a\gamma)\frac{B_x \ell}{B\rho}$. The total rotation matrix for one revolution is given by the product of (6.57) and (6.58).

It can be easily shown that the corresponding rotational axis $\hat{n}(s)$ everywhere outside the perturbing field region has the cosine factor

$$
\hat{n} \cdot \hat{y} = \frac{1}{\sqrt{1 + \tan^2 \frac{\theta}{2} \csc^2 \pi a\gamma}}. \tag{6.59}
$$

It follows that the beam polarization vanishes ($\hat{n} \cdot \hat{y} = 0$) on an "integer resonance," i.e., when the spin tune $a\gamma$ is equal to an integer k. Integer resonances show their depolarization effect by causing $\hat{n}$ to deviate from $\hat{y}$. See Fig. 6.8.

One can also calculate the width in $(a\gamma - k)$ within which beam polarization is significantly reduced. The width is found to be of the order of $\frac{\theta}{2\pi}$. Taking again $\theta = 0.3$ mrad as a typical value from our numerical example, the resonance width is about 0.5×10^{-4}, which is much narrower than the spacing between the integer resonances. This means integer depolarization resonances are easy to avoid. Furthermore, the integer resonances depolarize the beam through the cosine factor $\hat{n} \cdot \hat{y}$. Unlike the sideband resonances to be mentioned next, they are not enhanced by the noise due to synchrotron radiation. We thus expect that integer depolarization resonances are typically not a serious problem in electron storage rings.

Homework 6.17

(a) Verify Eq. (6.59).

(b) We used 3×3 matrices for spin precession in this section. One can equally use spinor algebra. Use spinor algebra to rederive Eq. (6.59).

Figure 6.8: The cosine factor $\hat{n} \cdot \hat{y}$ in Eq. (6.59) as a function of spin tune $a\gamma$. To make the integer resonances visible, the value of θ is set to an exaggerated value of 0.1.

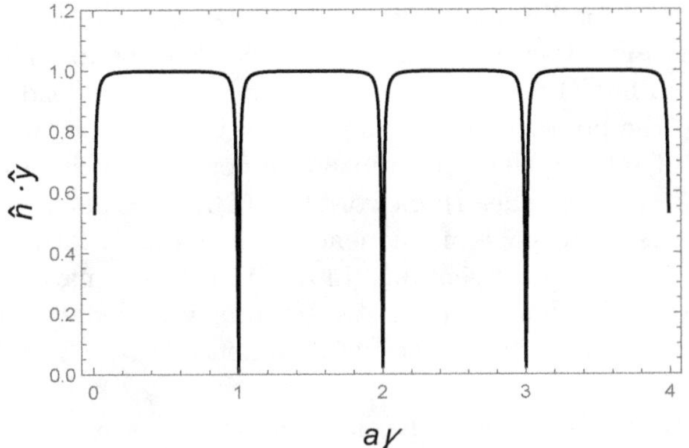

(c) Show that the width of the resonance, reasonably defined, in spin tune units is given by $\Delta \nu = \frac{|\theta|}{2\pi}$ when $|\theta| \ll 1$.

Homework 6.18

(a) Let $\vec{S}_n$ be the spin of an electron as it circulates the n-th revolution. Transform the spin according to

$$\vec{S}_{n+1} = M_1(a\gamma + \epsilon \cos n\nu) M_2(\theta) \vec{S}_n \,,$$

where $M_1(a\gamma)$ is the precession matrix (6.57); $M_2(\theta)$ is the perturbation matrix (6.58); $|\epsilon| \ll 1$ is a small parameter; ν is the frequency at which the spin precession frequency $a\gamma$ is modulated. Starting with $\vec{S}_0$ along the rotational axis of the transformation $M_1(a\gamma) M_2(\theta)$, show that $\vec{S}_n$ deviates from $\hat{y}$ significantly if $a\gamma \pm \nu$ is close to an integer. Estimate the resonance width. Do the analysis to 1st order in ϵ.

(b) Repeat the problem for the case when the spin is driven by an oscillating perturbation,

$$\vec{S}_{n+1} = M_1(a\gamma) M_2(\epsilon \cos n\nu) \vec{S}_n \,.$$

The two cases (a) and (b) describe two different depolarization mechanisms, (a) for a frequency modulation mechanism, (b) for a resonance driving mechanism. When synchrotron radiation is added, both mechanisms get excited each time a synchrotron radiation photon is emitted, as will be elaborated in the following section.

6.5.3   Sideband resonance

In the previous section, we have followed a particle at the beam center to obtain the polarization direction $\hat{n}$. Consider now instead a particle that executes a horizontal betatron oscillation x_β around the beam center. For this particle, the part of precession described by Eq. (6.57) needs to be modified; the rotation is still around $\hat{y}$, but the angle $2\pi a\gamma$ now contains an additional term that is proportional to the betatron amplitude $\hat{x}_\beta$ and is oscillatory with the betatron tune ν_x. One might say that the spin precession motion is "frequency modulated" by the x_β-motion. [See Homework 6.18(a).] A result of such a frequency modulation is the occurrence of frequency sidebands. In other words, to first order in $\hat{x}_\beta$, the system now contains, in addition to the natural frequency $a\gamma$, two more sideband frequencies $a\gamma \pm \nu_x$. If we now introduce the perturbation (6.58), the spin motion will be seriously influenced if $a\gamma \pm \nu_x$ is equal to an integer k.

A similar situation happens if the electron executes a synchrotron oscillation. The spin precession motion described by Eq. (6.57) is frequency modulated by the synchrotron motion at the synchrotron tune ν_s. Two sidebands at $a\gamma \pm \nu_s$ occur and the spin motion is seriously influenced by the perturbation (6.58) if $a\gamma \pm \nu_s = k$.

These resonances $a\gamma \pm \nu_x = k$ and $a\gamma \pm \nu_s = k$ are referred to as *sideband resonances* because they appear as sidebands of the *integer resonances* $a\gamma = k$.

Extending this discussion, it is expected that higher order sidebands $a\gamma \pm m_x\nu_x \pm m_s\nu_s = k$ are all excited in a Bessel function expansion. We shall not consider these nonlinear effects as they are outside of our present scope. But their existence should be kept in mind.

The spin motion is also seriously perturbed at the vertical betatron sidebands $a\gamma \pm \nu_y = k$. The mechanism, however, is different from the frequency modulation mechanism. In this idealized example, the source of the problem is now not Eq. (6.57) but Eq. (6.58). What happens now is that as the particle executes a y_β-oscillation, the magnetic field it experiences at the quadrupole contains two terms: the static B_x (coming from the vertical misalignment of the quadrupole magnet) that causes the spin to precess according to Eq. (6.58), and an additional B_x that oscillates with y_β. One might now say that the simple harmonic spin precession is "driven" by an oscillatory driving force every time the electron passes through the quadrupole with a vertical betatron displacement. [See Homework 6.18(b).] If the frequency ν_y of the driving force and the natural spin precession frequency $a\gamma$ satisfy the resonance condition $a\gamma \pm \nu_y = k$, we expect a strong response of spin to the driving. The frequency modulation mechanism is therefore replaced by a driving force mechanism, at least in this idealized example. Note that if there is a y_β-oscillation, in fact every quadrupole in the storage ring will contribute to a driving depolarization force, and all their contributions need to be included in an actual calculation.

Once we deviate from our idealized example, the situation rapidly becomes complicated. For example, if there is a skew quadrupole field somewhere, it produces a perturbation (6.58) when the electron has an x-displacement. The driving force mechanism then also applies to the horizontal betatron and the synchrotron sidebands. One can also imagine that the simple harmonic precession will be frequency modulated by y_β-motion if there are vertical bending dipoles in the storage ring. It is clear that studying these effects case by case will be cumbersome, if not impossible. What is needed is a general framework, which we will describe later in this chapter.

6.5.4 Spin diffusion due to synchrotron radiation

We have not yet explained the role of synchrotron radiation in enhancing the depolarization resonances. This we will do next.

Imagine an electron following the closed orbit with its spin $\vec{S}$ happily polarized along $\hat{n}$. Now suddenly it emits a photon of energy δE at time $t = 0$. After the emission, the electron starts to execute orbital oscillations around the closed orbit. The oscillations can be decomposed into three normal modes, which in an uncoupled accelerator are just the horizontal and vertical betatron modes and the synchrotron mode. In a coupled accelerator, these three nominal modes are simply to be replaced by the three eigenmodes of the accelerator optics. We know that these excited orbital oscillations are damped by radiation damping. The damping times τ_{rad} are somewhat different for the three modes but they are all comparable, typically about several milliseconds. A few τ_{rad}'s after the radiation, the electron's 3-D orbital motion damps to the closed orbit and quiets down again. Meanwhile, the spin $\vec{S}$ of the electron starts to precess away from $\hat{n}$ due to the perturbing electromagnetic fields seen away from the closed orbit. Similar to the orbital motion, this excited spin motion will also quiet down. The time constant, however, is not τ_{rad} but the polarization time constant τ_0 given by Eq. (6.9), which typically reads at least several minutes.

We illustrate in Fig. 6.9 the spin motion during this whole process. The perturbing electromagnetic field that acts on the spin from $t = 0$ to $t = $ a few τ_{rad}'s is oscillatory with frequencies ν_x, ν_y and ν_s.[19] This field perturbs the spin through both the frequency modulation and the resonance driving mechanisms mentioned before. In case the spin tune $a\gamma$ is such that one of the sideband conditions is fulfilled or nearly fulfilled, this photon emission event will significantly reduce the polarization of this electron. In Fig. 6.9, this means $\theta \to \infty$. The angle θ is to be considered the random walk step size in the spin diffusion due to synchrotron radiation.

Note that synchrotron radiation emissions do not excite integer depolarization resonances. On the other hand, in case the storage ring contains nonlinearities, the orbital disturbance after the photon emission contains combinations of multiples of the three frequencies $\nu_{x,y,s}$, and depolarization appears when

$$a\gamma = k \pm k_x \nu_x \pm k_y \nu_y \pm k_s \nu_s \,, \tag{6.60}$$

at all the nonlinear depolarization resonances with $k_{x,y,s}$ positive and negative integers. Synchrotron radiation excites and enhances all these linear and non-

---

[19]Or, in case of coupled optics, they might be designated $\nu_I, \nu_{II}, \nu_{III}$.

Figure 6.9: The motion of an electron spin $\vec{S}$ following the sudden emission of a synchrotron photon of energy δE. (a) Before the emission ($t < 0$), the electron is polarized with $\vec{S}$ along the direction $\hat{n}$ of the net beam polarization. (b) Photon emission at time $t = 0$ excites the orbital motions of the electron, which cause the electron to see perturbing electromagnetic fields. After the emission ($t > 0$) and before the orbital motions are radiation damped ($t \lesssim \tau_{\mathrm{rad}}$), $\vec{S}$ precesses according to the perturbing fields in some complicated manner (Thomas–BMT equation). The radiation damping time τ_{rad} is typically several milliseconds. (c) After the orbital motions are radiation damped ($t \gtrsim \tau_{\mathrm{rad}}$), $\vec{S}$ sees no perturbing fields and starts to execute a simple precession motion around $\hat{n}$. The angle θ is an important parameter that determines the strength of depolarization due to synchrotron radiation. If $\theta \gtrsim 10^{-6}$, one expects loss of polarization. (d) The precessing $\vec{S}$ slowly spirals in towards $\hat{n}$ due to the radiative polarizing effect of synchrotron radiation. Significant spiraling occurs after a time τ_0 given by the polarization time, typically at least several minutes. A few τ_0 later, $\vec{S}$ damps to $\hat{n}$. The excitation-damping process (a) to (d) is repeated every time a synchrotron photon is emitted.

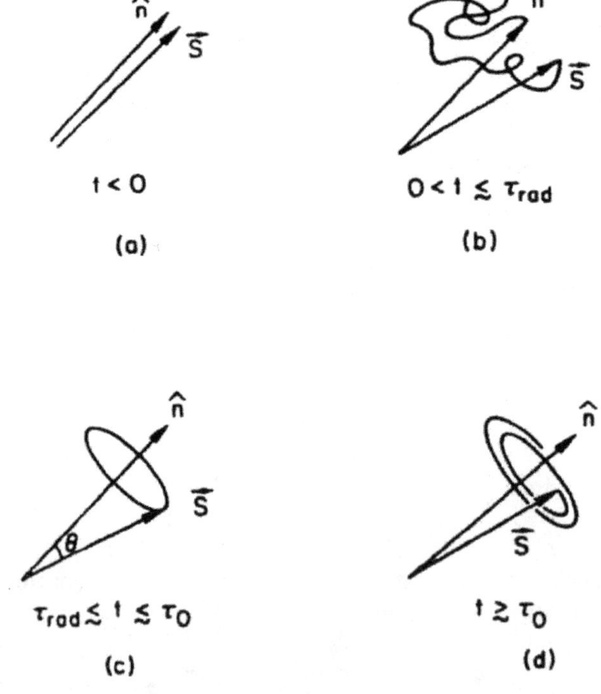

linear sideband resonances; it just does not excite the lone integer ones. These resonances (6.60) come from nonlinear field errors (error magnetic field being nonlinear in $x, y, \frac{\Delta E}{E}$) in the storage ring. Generally the nonlinear depolarization resonances tend to be weaker than the linear ones.

One can imagine doing a calculation of the widths of the sideband resonances just like we did for the integer resonances. One might argue that the resonance width should be defined so that within the width, the spin needs to deviate far away from $\hat{n}$, and the angle θ of Fig. 6.9 should be $\gtrsim 1$ rad. One then probably finds that the widths are very narrow and concludes that sideband resonances are not a serious problem for beam polarization. What happens, however, is that photons are constantly being emitted. Staying outside of a resonance width found this way does not guarantee a good polarization. For example, if each photon emission causes the spin to deviate from $\hat{n}$ by an angle θ of, say, 10^{-6} rad, then the spin will random-walk away from $\hat{n}$ in about 10^{12} emissions. For a 5 GeV storage ring of 25 m radius, this means a depolarization time of about 10^9 revolutions, since there are $\sim 10^3$ emissions per revolution, or about 10 minutes. To assure good polarization, the depolarization time must be much longer than the polarization time τ_0. If τ_0 is also about 10 minutes, it means one must stay away from the sideband resonances far enough so that θ is less than something like 10^{-6} rad rather than 1 rad. Synchrotron radiation thus greatly enhances, by a factor $\sim 10^6$, the sideband depolarization resonances.

Note that this enhancement of sideband depolarization resonances by synchrotron radiation applies only to electrons and positrons, and not protons. Note also that unlike what is more familiar in orbital dynamics, the integer resonances are *weaker* than the sideband resonances for spin dynamics in an electron storage ring because they do not get the 10^6 enhancement from synchrotron radiation.

If we plot the beam polarization as a function of beam energy in a given electron storage ring, the level of polarization should have more or less the following appearance. The nominal level of polarization should be 92%. But when the beam energy E is such that $a\gamma$ is close to an integer, which occurs at an even spacing of 440 MeV, the polarization dips substantially. These dips however are very narrow, almost invisible. Between every pair of adjacent integer resonances, there are 6 sideband resonances (considering linear sideband resonances only). Unless the storage ring is extremely flat and the beam is extremely ribbon-like, these sideband resonances tend to be much wider than the integer resonances. Furthermore, because ν_s usually is small, the two synchrotron sidebands $a\gamma \pm \nu_s = k$ tends to overlap, and overwhelm the integer resonance $a\gamma = k$ sitting between them, and all three resonances appear as a single resonance dip near the location $a\gamma = k$.[20]

Table 6.3 is a recap of our discussion concerning the integer and the sideband depolarization resonances. A few examples of such polarization plots will be shown in the next section.

Since τ_{rad} is so much shorter than τ_0, one can ignore the time period $0 <$

---

[20]Sometimes the three-resonance group is incorrectly identified as the integer resonance.

Table 6.3: Comparison between the integer and sideband depolarization resonances driven by synchrotron radiation in an electron storage ring.

	Driving mechanism	Enhanced by synch. rad.	Typical reso. width
Integer resonances $a\gamma = k$	$\hat{n} \cdot \hat{y} < 1$	no	$\sim 10^{-4}$
Sideband resonances $a\gamma = k \pm \nu_{x,y,s}$	$\left|\gamma \frac{\partial \hat{n}}{\partial \gamma}\right| \gtrsim 1$	yes	$\sim 10^{-1}$

$t \lesssim \tau_{\text{rad}}$ shown in Fig. 6.9(b) as far as spin polarization is concerned. For $t < 0$, we have $\vec{S} - \hat{n} = \vec{0}$. For $t > 0$, the deviation of $\vec{S}$ from $\hat{n}$ is proportional to the perturbation $\frac{\delta E}{E_0}$. If we extrapolate the spin precession motion of Fig. 6.9(c) backwards in time to the moment of emission $t = 0$, we can write

$$\vec{S} - \hat{n} = \frac{\delta E}{E_0} \cdot \gamma \frac{\partial \hat{n}}{\partial \gamma}, \qquad \text{at} \qquad t = 0^+, \qquad (6.61)$$

where we have defined a proportionality vector $\gamma \frac{\partial \hat{n}}{\partial \gamma}$.

The reader should note that this tracing back in time to $t = 0$ is the same as what we do in orbital dynamics. In treating orbital dynamics after a photon emission, Δx_β is traced back in time to give the equivalent orbital kicks $\Delta x_\beta = -\frac{\delta E}{E_0} D_x$ and $\Delta x'_\beta = -\frac{\delta E}{E_0} D'_x$. The same thing is being done here for the spin motion. The quantity $\gamma \frac{\partial \hat{n}}{\partial \gamma}$ plays the role of the dispersion functions D_x, D'_x for spin motion.

The vector $\gamma \frac{\partial \hat{n}}{\partial \gamma}$ is a key quantity in determining the radiative polarization of the beam. It is really a 2-D vector because it is always confined to be perpendicular to $\hat{n}$. For a ribbon beam, perturbations due to synchrotron radiation is decoupled from the spin motion and we have

$$\gamma \frac{\partial \hat{n}}{\partial \gamma} = \vec{0}, \qquad \text{(ribbon beam)}.$$

In general, it is a vector completely determined by the storage ring lattice, depending only on the location s where the photon is emitted, and is independent of the properties of synchrotron radiation spectrum or properties of the electron beam or its polarization.

The quantity $\gamma \frac{\partial \hat{n}}{\partial \gamma}$ is called the "spin chromaticity."[21] The notation used here follows that of Derbenev, Kondratenko and Skrinsky.[22] The angle θ shown

---

[21]Perhaps one should more accurately call it "spin dispersion function" in analogy to the orbital dispersion functions. Usually, dispersion function depends on the position s; chromaticity is a global quantity independent of s. The nomenclature has been adopted by our community, however.

[22]Ya.S. Derbenev, A.M. Kondratenko and A.N. Skrinsky, Part. Accel. **9**, 247 (1979).

in Fig. 6.9(c) is given by

$$\theta = \left| \frac{\delta E}{E} \gamma \frac{\partial \hat{n}}{\partial \gamma} \right|.$$

It specifies the random walk step-size of quantum diffusion on spin motion.

Perhaps it is worthwhile here to raise a cautionary flag. The notation $\gamma \frac{\partial \hat{n}}{\partial \gamma}$, although suggestive, is unfortunate. The spin diffusion is physically caused by photon emissions. The γ here has its physical origin accordingly; it refers to the energy of the randomly emitted photon instead of the energy of the emitting electron. It is the effect of the electron spin due to the photon emission that the quantity $\gamma \frac{\partial \hat{n}}{\partial \gamma}$ refers to, and $\gamma \frac{\partial \hat{n}}{\partial \gamma}$ needs to be derived in accordance to this physical process. It would be incorrect if for example, one calculates $\hat{n}$ of an electron as a function of the electron's energy, and then take a partial derivative of $\hat{n}$ with respect to the electron energy γ, and call that $\gamma \frac{\partial \hat{n}}{\partial \gamma}$. In some literatures, this in fact has been done. Beware!

As another illustration of this point, Homework 6.27 gives an example when a different noise source contributes to the depolarization diffusion. In that case, we still use the same notation $\gamma \frac{\partial \hat{n}}{\partial \gamma}$, but it then means a different quantity for a different physical process. The spin chromaticity $\gamma \frac{\partial \hat{n}}{\partial \gamma}$ should be defined according to the physical process of the noise source.

We mentioned the condition $\theta \lesssim 10^{-6}$ to assure a polarized beam. It may be useful to examine this criterion a bit closer. The overall photon emission rate W_{overall} was given in Eq. (6.6). To assure beam polarization, we must have the diffusion rate $\theta^2 W_{\text{overall}}$ smaller than the radiative polarization buildup rate τ_0^{-1}, i.e.,

$$\theta \lesssim \sqrt{\frac{\tau_0^{-1}}{W_{\text{overall}}}} \sim \frac{\hbar \omega_c}{E},$$

where use has been made of Eq. (6.10). Since the photon energy $\delta E \sim \hbar \omega_c$, this in turn then implies that the condition for polarization is

$$\left| \gamma \frac{\partial \hat{n}}{\partial \gamma} \right| \lesssim 1,$$

as claimed in Table 6.3.

Homework 6.19 We mentioned that nonlinear depolarization resonances occur at (6.60).

(a) Are there nonlinear resonances at $k_{\text{spin}} a\gamma = k \pm k_x \nu_x \pm k_y \nu_y \pm k_s \nu_s$ when k_{spin} being an integer larger than ± 1? Can you give a reason why? A similar situation occurs for the protons, as mentioned following Eq. (5.33).

(b) Consider the synchrotron sidebands driven by an energy oscillation of a modest amplitude $\hat{\delta}$ at the synchrotron tune ν_s. The text claims that this contributes to a frequency modulation depolarization mechanism and that higher order sidebands are weighted by Bessel functions. Complete this argument to find these Bessel function weights. Observe their dependencies on k_s and $\hat{\delta}$. Under what conditions may we claim the higher order sidebands are much weaker than the linear ones?

6.6   Derbenev–Kondratenko formula

We assume that the storage ring fields, including the closed-orbit distortion and all the error fields, are known. From this information, one can obtain the polarization direction $\hat{n}(s)$ and the spin chromaticity $\gamma\frac{\partial\hat{n}}{\partial\gamma}$ around the storage ring. See Fig. 6.10. A matrix formulation will be described for this purpose in the next section. Here let us assume $\hat{n}(s)$ and $\gamma\frac{\partial\hat{n}}{\partial\gamma}(s)$ are known and we will look for an expression of beam polarization in terms of these quantities. In particular, since we are no longer restricting ourselves to ribbon beams, $\gamma\frac{\partial\hat{n}}{\partial\gamma} \neq \vec{0}$ in general.

Figure 6.10: A schematic drawing of the direction of polarization $\hat{n}$ and the spin chromaticity $\gamma\frac{\partial\hat{n}}{\partial\gamma}$. The dashed curve indicates the designed trajectory. The solid curve is the distorted closed orbit. Note that the magnitude of $\hat{n}(s)$ is always 1 but the magnitude of $\gamma\frac{\partial\hat{n}}{\partial\gamma}(s)$ varies with s. The vectors $\hat{n}$ and $\gamma\frac{\partial\hat{n}}{\partial\gamma}$ are always perpendicular to each other.

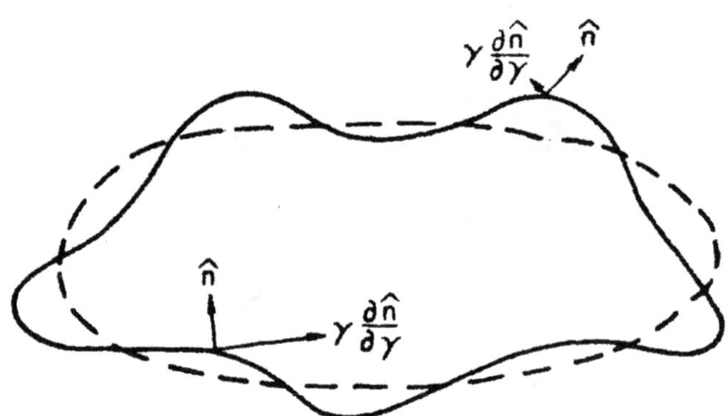

Consider an unpolarized electron beam stored at time $t = 0$. Due to synchrotron radiation, with all its polarizing as well as depolarizing effects, the beam slowly acquires a polarization $\zeta(t)\hat{n}$ along the direction $\hat{n}$. We expect $\zeta(t)$ to approach an equilibrium value P with a time constant τ. The ideal case has been worked out before to yield P_0 and τ_0. Here we want to find the general expressions for P and τ.

We start with Eq. (6.46). The first term in (6.46) describes the precession motion. For a polarization along $\hat{n}$, it can be dropped since $\hat{n}$ is the precession axis. The second term comes from spin flip radiation. It of course must be kept and we have

$$\dot{\zeta}(t) = -\frac{1}{\tau_0}\left[\zeta(t) - \frac{2}{9}(\hat{n}\cdot\hat{z})^2\zeta(t) + \frac{8}{5\sqrt{3}}\hat{n}\cdot\hat{y}\right], \qquad (6.62)$$

where $\hat{z}$ is along the beam motion, $\hat{y}$ is along the bending magnetic field.

As mentioned following Eq. (6.46), Eq. (6.62) has made the approximation that $a = 0$ considering the case of electrons.

Since we expect the polarization to be very slowly changing, it is a good approximation to average the right-hand-side of Eq. (6.62) over the circumference of the ring. Inserting τ_0 from Eq. (6.9), this gives

$$\dot{\zeta}(t) = -\frac{5\sqrt{3}}{8} \frac{\alpha_F \gamma^5 \hbar^2}{m^2 c} \frac{1}{2\pi R} \left\{ \zeta(t) \oint ds \frac{[1 - \frac{2}{9}(\hat{n} \cdot \hat{z})^2]}{|\rho|^3} + \frac{8}{5\sqrt{3}} \oint ds \frac{\hat{n} \cdot \hat{y}}{|\rho|^3} \right\}.$$
(6.63)

Equation (6.63) is incomplete; it contains only the polarizing effect. We will have to include more terms coming from the spin chromatic effects due to a nonzero $\gamma \frac{\partial \hat{n}}{\partial \gamma}$. Let us, however, ignore $\gamma \frac{\partial \hat{n}}{\partial \gamma}$ for a short moment. The equilibrium level of polarization would then be obtained by setting $\dot{\zeta} = 0$ in Eq. (6.63), which gives an equilibrium level of polarization

$$-\frac{8}{5\sqrt{3}} \frac{\oint ds\, (\hat{n} \cdot \hat{y}/|\rho|^3)}{\oint ds\, [1 - \frac{2}{9}(\hat{n} \cdot \hat{z})^2]/|\rho|^3}.$$

The factor $\hat{n} \cdot \hat{y}$ is the cosine reduction factor mentioned when we discussed integer depolarization resonances. It again reflects the fact that synchrotron radiation builds up the polarization along $\hat{y}$ while only its projection onto $\hat{n}$ survives. The (less important) factor $[1 - \frac{2}{9}(\hat{n} \cdot \hat{z})^2]$ in the denominator comes from the slight dependence of spin flip radiation on the z-component of electron spin. At this point, the only depolarization resonances playing a role are the integer resonances affecting the cosine factor $\hat{n} \cdot \hat{y}$.

The effects of spin chromaticity $\gamma \frac{\partial \hat{n}}{\partial \gamma}$ are associated with synchrotron radiation without spin flips. Consider an electron polarized along $\hat{n}$. As a photon of energy δE is emitted, its spin starts to precess around $\hat{n}$ with a small rotating deviation given by Eq. (6.61), which we designate as $\vec{d}$. It is easy to see from Eq. (6.61) that $\theta = |\vec{d}| \ll 1$. After emission, the electron has lost a polarization

$$\frac{\Delta P}{P} = -\frac{1}{2}|\vec{d}|^2 = -\frac{1}{2}\left| \frac{\delta E}{E_0} \gamma \frac{\partial \hat{n}}{\partial \gamma} \right|^2.$$

Let $\dot{N}$ be the number of photon emissions per unit time, we obtain the quantum diffusion rate on polarization,

$$\dot{\zeta}(t) = -\frac{1}{2} \frac{\zeta(t)}{2\pi R} \oint ds \left\langle \dot{N} \frac{\delta E^2}{E^2} \right\rangle \left| \gamma \frac{\partial \hat{n}}{\partial \gamma} \right|^2,$$
(6.64)

where we have averaged over s with $\left\langle \dot{N} \frac{\delta E^2}{E^2} \right\rangle$ and $\gamma \frac{\partial \hat{n}}{\partial \gamma}$ considered functions of s, and $\left\langle \dot{N} \frac{\delta E^2}{E^2} \right\rangle$ is given by integrating over the synchrotron radiation spectrum,[23]

$$\left\langle \dot{N} \frac{\delta E^2}{E^2} \right\rangle = \frac{55}{24\sqrt{3}} \frac{\alpha_F \hbar^2 \gamma^5}{m^2 c |\rho^3(s)|}.$$
(6.65)

---

[23]M. Sands, SLAC Report SLAC-121 (1970).

Equation (6.64) gives the contribution of the spin chromaticity to the evolution of the beam polarization $\zeta(t)$. The negative sign on the right-hand-side says it is depolarizing. The quadratic nature of its integrand reflects the mechanism is a diffusion effect.

But there is another more subtle effect on polarization due to the spin chromaticity. Consider now an electron that is not perfectly polarized before radiation. Let its spin be $\hat{n} + \vec{D}$ with $\vec{D}$ a small rotating vector perpendicular to $\hat{n}$. Now the electron emits a photon of energy δE. After emission, the spin acquires another rotating deviation $\vec{d}$. Let $\vec{D}_0$ and $\vec{d}_0$ be the values of $\vec{D}$ and $\vec{d}$ extrapolated to the moment of emission. If δE does not depend on $\vec{D}_0$, i.e., if the synchrotron radiation does not depend on the instantaneous spin, $\vec{d}_0$ is uncorrelated with $\vec{D}_0$ and we simply have observed a random walk in spin motion as has been considered so far. The story is different if δE does depend on $\vec{D}_0$. Then $\vec{d}_0$ correlates with $\vec{D}_0$ and the original amount of depolarization will decrease or increase according to how $\vec{d}_0$ and $\vec{D}_0$ are correlated. In the former case the correlation is polarizing, while in the latter case, depolarizing.

More quantitatively, the polarization of the electron before and after the radiation are equal to $1 - \frac{1}{2}|\vec{D}_0|^2$ and $1 - \frac{1}{2}|\vec{D}_0 + \vec{d}_0|^2$, respectively. Change of polarization due to the radiation is therefore

$$\frac{\Delta P}{P} = -\vec{D}_0 \cdot \vec{d}_0 - \frac{1}{2}|\vec{d}_0|^2 . \tag{6.66}$$

The second term in (6.66) is the random walk term already calculated. But now we have to also look at the first term. Summing over photon emission events, the contribution of the first term in (6.66) to the depolarization process is

$$\begin{aligned}
\dot{\zeta}(t) &= \langle -\dot{N}\vec{D}_0 \cdot \vec{d}_0 \rangle \\
&= -\left\langle \dot{N}\frac{\delta E}{E_0} \right\rangle \vec{D}_0 \cdot \gamma\frac{\partial \hat{n}}{\partial \gamma} .
\end{aligned} \tag{6.67}$$

where, as before, $\langle \ \rangle$ means averaging over the radiation spectrum. In addition, an overall averaging over s is understood.

Expression for $\langle \dot{N}\delta E \rangle$ has been obtained before in Eq. (6.4) — it is just the radiation power. The spin-independent terms in Eq. (6.4) do not concern us here. Keeping only the spin-dependent term gives

$$\langle \dot{N}\delta E \rangle = -\mathcal{P}_{\text{class}}\,\vec{S} \cdot \hat{y}\,\frac{\hbar\omega_c}{E_0} , \tag{6.68}$$

where the spin direction in Eq. (6.4) has been replaced by the instantaneous value $\hat{n} + \vec{D}_0$. Also since $\hat{n}$ is perpendicular to $\gamma\frac{\partial \hat{n}}{\partial \gamma}$, the vector $\vec{D}_0$ in Eq. (6.67) can be replaced by $\vec{S}$. Substituting Eq. (6.68) into Eq. (6.67) yields

$$\dot{\zeta}(t) = \frac{\hbar\omega_c}{E^2}\,\mathcal{P}_{\text{class}}\,(\vec{S} \cdot \hat{y})\left(\vec{S} \cdot \gamma\frac{\partial \hat{n}}{\partial \gamma}\right) . \tag{6.69}$$

Since $\vec{S}$ can in principle point in any arbitrary direction, the next step is to average over its solid angles, keeping its magnitude constant. When this is done, the factor $(\vec{S} \cdot \hat{y})\,(\vec{S} \cdot \gamma \frac{\partial \hat{n}}{\partial \gamma})$ becomes $\frac{1}{3}(\hat{y} \cdot \gamma \frac{\partial \hat{n}}{\partial \gamma})|\vec{S}|^2$. Now the question is what to use for $|\vec{S}|^2$. One may argue that since $\vec{S}$ is the unit spin direction, it obviously has $|\vec{S}|^2 = 1$. The correct answer, however, is $|\vec{S}|^2 = 3$, which comes from the fact that the magnitude of the electron spin must be determined from the quantum mechanical relation $\left|\frac{\hbar}{2}\vec{S}\right|^2 = \frac{1}{2}(\frac{1}{2} + 1)\hbar^2$.[24] Thus, Eq. (6.69) becomes, after averaging over s,

$$\dot{\zeta}(t) \;=\; \frac{1}{2\pi R} \oint ds\, \frac{\hbar \omega_c}{E^2}\, \mathcal{P}_{\text{class}} \left(\hat{y} \cdot \gamma \frac{\partial \hat{n}}{\partial \gamma} \right). \tag{6.70}$$

The subtle effect of correlation between spin-dependent radiation and the spin chromaticity has then made an additional contribution (6.70) to the time evolution of beam polarization. A reader who is familiar with orbital effect of synchrotron radiation should draw a close analogy of this subtle contribution to that of the radiation partition number, usually designated by the symbol $\mathcal{D}$, in the orbital dynamics.[25]

We have now obtained three separate contributions to $\dot{\zeta}(t)$; they are given by Eqs. (6.63), (6.64) and (6.70). Adding them up gives the final expression obtained by Derbenev and Kondratenko,[26]

$$\dot{\zeta}(t) \;=\; -\frac{5\sqrt{3}}{8} \frac{\alpha_F \gamma^5 \hbar^2}{m^2 c} \left[\alpha_+ \, \zeta(t) - \frac{8}{5\sqrt{3}} \alpha_- \right], \tag{6.71}$$

where

$$\alpha_+ \;=\; \frac{1}{2\pi R} \oint \frac{ds}{|\rho(s)|^3} \left[1 - \frac{2}{9}(\hat{n} \cdot \hat{v}) + \frac{11}{18} \left| \gamma \frac{\partial \hat{n}}{\partial \gamma} \right|^2 \right]_s,$$

$$\alpha_- \;=\; \frac{1}{2\pi R} \oint \frac{ds}{|\rho(s)|^3} \left[\frac{\dot{\hat{v}} \times \hat{v}}{|\dot{\hat{v}}|} \cdot \left(\hat{n} - \gamma \frac{\partial \hat{n}}{\partial \gamma} \right) \right]_s. \tag{6.72}$$

In Eq. (6.71), following Derbenev and Kondratenko, the symbol for the instantaneous direction of beam motion has been changed (hopefully for clarity) from $\hat{z}$ to $\hat{v}$; the symbol for the magnetic field direction has been changed from $\hat{y}$ to $-\frac{\dot{\hat{v}} \times \hat{v}}{|\dot{\hat{v}}|}$ with $\dot{\hat{v}}$ along the direction of acceleration. The latter change of symbol has the advantage that it also applies to positrons and to negative polarity of bending.

From Eq. (6.71), it follows that the polarization builds up according to

$$P(t) \;=\; P(1 - e^{t/\tau}).$$

---

[24]This is the only place where quantum mechanics is compelled in our semiclassical treatment. See also Homework 6.20.

[25]It is quite amazing that orbital dynamics and spin dynamics are so closely analogous and to this detailed level. Be reminded of Table 6.1 again.

[26]Ya.S. Derbenev and A.M. Kondratenko, Sov. Phys. JETP 37, 968 (1973). We have adopted a different semiclassical derivation, but mainly followed their notations.

The equilibrium beam polarization is equal to

$$P = \frac{8}{5\sqrt{3}} \frac{\alpha_-}{\alpha_+}, \tag{6.73}$$

and the e-folding time constant to reach the equilibrium is

$$\tau = \left(\frac{5\sqrt{3}}{8} \frac{\alpha_F \gamma^5 \hbar^2}{m^2 c} \alpha_+ \right)^{-1}. \tag{6.74}$$

In the case of a perfect planar storage ring, Eqs. (6.73) and (6.74) reduce to the results (6.8) and (6.9). For a perfect planar ring with isomagnetic bending ($\rho =$ constant), we have $\alpha_- = -\frac{1}{\rho^3}$ and $\alpha_+ = \frac{1}{\rho^3}$.

An inspection of Eqs. (6.71) and (6.72) shows that the spin chromaticity appears as a quadratic term in the denominator of the expression for P and only linearly in the numerator. Loss of polarization occurs if $|\gamma \frac{\partial \hat{n}}{\partial \gamma}| \gtrsim 1$. This means the angle θ of Fig. 6.9(c) will be $\gtrsim \frac{\delta E}{E_0}$, which is of the order of $\frac{\hbar \omega_c}{E_0}$.

Finally, skeptical readers who wonder if P, as given by Eq. (6.73), could end up being larger than unity (then something has obviously gone wrong) should work out Homework 6.21.

An approximation The exact expressions are given by Eqs. (6.72–6.74). We can also consider a couple of approximations as follows.

The spin chromaticity term $\hat{y} \cdot \gamma \frac{\partial \hat{n}}{\partial \gamma}$ in α_- is small for most practical cases. This follows from the fact that $\hat{n}$ is nearly equal to $\hat{y}$ and that $\gamma \frac{\partial \hat{n}}{\partial \gamma}$ is perpendicular to $\hat{n}$. If we further drop the noncritical term $\frac{2}{9}(\hat{n} \cdot \hat{v})^2$ in α_+, we obtain a simplified version of Eq. (6.72),

$$\alpha_+ \approx \frac{1}{2\pi R} \oint \frac{ds}{|\rho(s)|^3} \left[1 + \frac{11}{18} \left| \gamma \frac{\partial \hat{n}}{\partial \gamma} \right|^2 \right]_s,$$

$$\alpha_- \approx \frac{1}{2\pi R} \oint ds \frac{\hat{y} \cdot \hat{n}}{\rho(s)^3}. \tag{6.75}$$

Equation (6.75) shows explicitly that (i) the integer resonances affect mainly α_-; (ii) sideband resonances affect mainly α_+; and (iii) depolarization occurs mainly when $|\gamma \frac{\partial \hat{n}}{\partial \gamma}| \gtrsim 1$.

If we further assume $\rho(s) > 0$ and $\hat{n} \approx -\hat{y}$ for electrons, then Eq. (6.75) can be further simplified. In that case, let us define the Sokolov–Ternov polarization time,

$$\tau_0 = \left(\frac{5\sqrt{3}}{8} \frac{\alpha_F \gamma^5 \hbar^2}{m^2 c} \frac{1}{2\pi R} \oint \frac{ds}{\rho(s)^3} \right)^{-1},$$

and a "depolarization time" (recalling that depolarization is caused by the spin chromaticity $\gamma \frac{\partial \hat{n}}{\partial \gamma}$),

$$\tau_{\text{dep}} = \left(\frac{5\sqrt{3}}{8} \frac{\alpha_F \gamma^5 \hbar^2}{m^2 c} \frac{1}{2\pi R} \oint \frac{ds}{\rho(s)^3} \frac{11}{18} \left| \gamma \frac{\partial \hat{n}}{\partial \gamma} \right|^2 \right)^{-1},$$

then we have

$$
\dot{\zeta}(t) \approx -\left[\left(\frac{1}{\tau_0} + \frac{1}{\tau_{\text{dep}}}\right)\zeta(t) + \frac{8}{5\sqrt{3}}\frac{1}{\tau_0}\right],
$$

$$
P \approx -\frac{8}{5\sqrt{3}}\left(\frac{1/\tau_0}{1/\tau_0 + 1/\tau_{\text{dep}}}\right),
$$

$$
\tau \approx \left(\frac{1}{\tau_0} + \frac{1}{\tau_{\text{dep}}}\right)^{-1}. \tag{6.76}
$$

It follows that when there is a depolarization effect, the level of polarization P is reduced from the ideal level of $\frac{8}{5\sqrt{3}}$ by a factor $\frac{1/\tau_0}{1/\tau_0+1/\tau_{\text{dep}}}$. At the same time, the polarization buildup time τ is shorter than the Sokolov–Ternov time by the same factor.

At the first glance, it might sound peculiar that the depolarizing term makes the polarization time *shorter*. A moment's reflection, however, indicates that the initial rate of change of polarization at time $t = 0$ should be the same with and without depolarization, both yielding

$$
\dot{\zeta}(0) = -\frac{8}{5\sqrt{3}}\frac{1}{\tau_0},
$$

i.e., the initial slope of polarization buildup is independent of the depolarization — since the beam is initially unpolarized, there is nothing to depolarize. When depolarization is turned on, with a lower buildup level to reach in equilibrium, the e-folding time constants necessarily — and only superficially — gets shorter, as illustrated in Fig. 6.11. There is no benefit to gain with this shortened polarization time. See also Homework 6.15.

We have thus obtained the Derbenev–Kondratenko formula for the equilibrium level of beam polarization and the polarization time. What remains to be done is to calculate $\hat{n}(s)$ and $\gamma\frac{\partial\hat{n}}{\partial\gamma}(s)$ around the storage ring once the storage ring lattice, including its error fields, is specified. This calculation will be described in the next section using a SLIM formalism.

Homework 6.20 To obtain Eq. (6.70) from Eq. (6.69), we have used a semi-classical argument to replace the quantity $(\vec{S}\cdot\vec{A})(\vec{S}\cdot\vec{B})$ by $\vec{A}\cdot\vec{B}$ by averaging over the solid angles of spin $\vec{S}$, where $\vec{A}$ and $\vec{B}$ are arbitrary vectors. Prove the quantum mechanical spinor counterpart of this argument,

$$
\{(\vec{\sigma}\cdot\vec{A}),(\vec{\sigma}\cdot\vec{B})\} = \vec{A}\cdot\vec{B},
$$

where $\vec{\sigma}$ is a vector whose components are the Pauli matrices, $\{X,Y\} = \frac{1}{2}(XY + YX)$ is the symmetric anticommutator of two operators.

Homework 6.21

(a) If one can design a storage ring with arbitrary $\hat{n}$ and $\gamma\frac{\partial\hat{n}}{\partial\gamma}$, how should he/she choose these quantities so that the beam polarization given by Eq. (6.73)

Figure 6.11: Comparison of the behavior of polarization for a case with depolarization and a case without depolarization effect.

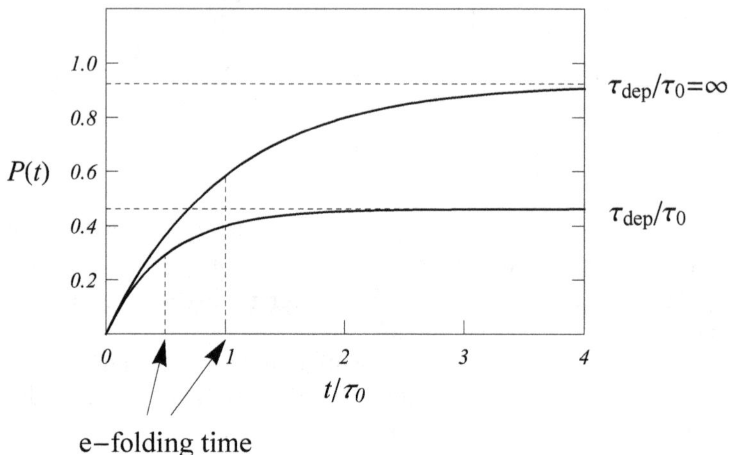

is maximized? Show that the answer is given by[27]

$$\hat{n} = \sqrt{\frac{7}{11}}\,\frac{\dot{\hat{v}} \times \hat{v}}{|\dot{\hat{v}}|} \pm \sqrt{\frac{4}{11}}\,\hat{v},$$

$$\gamma\frac{\partial \hat{n}}{\partial \gamma} = \frac{2\sqrt{7}}{11}\left(-\sqrt{\frac{4}{11}}\,\frac{\dot{\hat{v}} \times \hat{v}}{|\dot{\hat{v}}|} \pm \sqrt{\frac{7}{11}}\,\hat{v}\right).$$

(b) By substituting into Eq. (6.73), show that the maximum achievable polarization is

$$P_{\max} = \pm\frac{72\sqrt{21}}{113\sqrt{11} - 44} = \pm 99.75\%.$$

The value $\frac{8}{5\sqrt{3}}$ we reached with our perfect planar storage ring, although quite respectable, is not the best one can do. Also, it is intriguing that this optimization makes P to approach 1 ever so closely without violating $P < 1$.

Solution Let the optimal $\hat{n} = \hat{y}\sin\theta + \hat{v}\cos\theta$ and $\gamma\frac{\partial \hat{n}}{\partial \gamma} = A(-\hat{y}\cos\theta + \hat{v}\sin\theta)$. These parametrizations obey $\hat{n} \cdot \gamma\frac{\partial \hat{n}}{\partial \gamma} = 0$. Then,

$$P = \frac{8}{5\sqrt{3}}\frac{\sin\theta + A\cos\theta}{1 - \frac{2}{9}\cos\theta + \frac{11}{18}A^2}.$$

Look for solution that maximizes P. Looking only for solutions with real θ and A, the maximum occurs when $\cos\theta = \frac{2}{\sqrt{11}}$ and $A = \frac{2\sqrt{7}}{11}$. To visualize the situation, a contour plot of P in the (θ, A) space is shown in Fig. 6.12.

[27]Ya.S. Derbenev and A.M. Kondratenko, Sov. Phys. JETP 37, 968 (1973).

Figure 6.12: Contour plot of the achievable polarization P by varying the parameters θ and A as specified in Homework 6.21. The heights of the mountain and valley (the two eyes of the contour plot) in the contour plot are ± 0.9975.

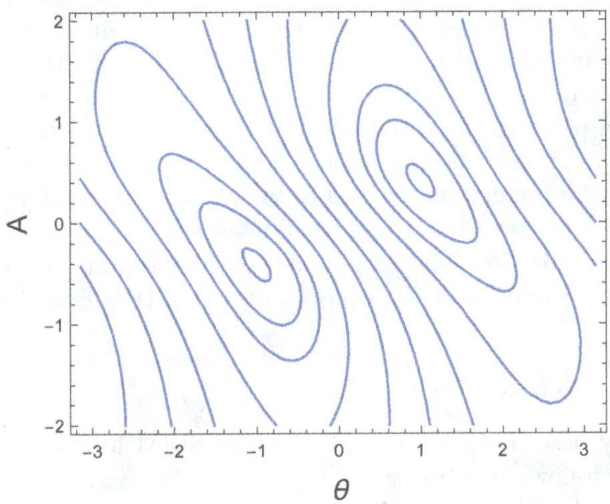

6.7 SLIM formalism

The existence of Sokolov–Ternov radiative spin polarization was confirmed experimentally in many existing electron storage rings.[28] The degree of polarization was often found to be lower than the expected 92% due to various depolarization effects. As elaborated in the previous sections 6.5 and 6.6, the very mechanism that gives rise to polarization, namely the synchrotron radiation, is also the main cause for depolarization. As an electron emits a photon during synchrotron radiation, it receives a recoil perturbation which excites its subsequent orbital oscillations. The electron then sees a perturbing electromagnetic field, which is modulated by these orbital oscillations, causing its spin to precess accordingly. Summing over the uncorrelated photon-emission events results in a diffusion of spin direction which becomes serious when the spin motion couples strongly to the oscillatory orbital motion. The physical process was illustrated in Fig. 6.9.

The achieved level of polarization is determined by a balance between the polarizing and the depolarizing effects of synchrotron radiation, as sketched in Table 6.1. The strength of the polarizing effect is given by the Sokolov–Ternov mechanism. The depolarization strength, on the other hand, depends on details of the storage ring operation conditions and is often difficult to analyze. For

---

[28]V.N. Baier, Sov. Phys. Uspekhi 14, 695 (1972); D. Potaux, Proc. 8th Int. Conf. High Energy Accel., CERN, p. 127 (1971); U. Camerini, et al., Phys. Rev. D12, 1855 (1975); SPEAR Polarization team, J. Johnson, R. Schwitters, et al., 3rd Int. Symp. High Energy Phys. with Polarized Beams and Targets, Argonne (1978).

a perfect storage ring with planar geometry, the spin-orbit coupling vanishes and the ideal 92% polarization is ensured. In the presence of imperfections or in storage rings designed with nonplanar configuration, depolarization strength must be known accurately in order to calculate the achievable polarization.

In Sec. 6.6, we formulated the problem and obtained expressions of the equilibrium polarization P and the polarization time τ through a Derbenev–Kondratenko formula. These expressions were given in terms of two quantities, the equilibrium polarization direction $\hat{n}(s)$ and the so-called spin chromaticity $\gamma\frac{\partial\hat{n}}{\partial\gamma}(s)$; both are functions of s, the position around the storage ring. We explained that the integer resonances are contained in $\hat{n}$ deviating from $\hat{y}$ and the sideband resonances (linear and nonlinear) are contained in $\gamma\frac{\partial\hat{n}}{\partial\gamma}(s)$ getting larger $\gtrsim 1$ in magnitude. We did not, however, elaborate on how to calculate $\hat{n}(s)$ and $\gamma\frac{\partial\hat{n}}{\partial\gamma}(s)$. In this section, we present a tool that fills in this step.

6.7.1　The formalism

Here we briefly describe the basic idea of the SLIM formalism.[29] Details are filled in by the following subsections.

In order to calculate the depolarization strength, one needs to know how the spin and orbital degrees of freedom of an electron couple among themselves. It is well known that in order to fully describe the orbital motion of an electron, one needs six phase-space coordinates $(x, x', y, y', z, \delta)$, where x, y and z are the horizontal, vertical and longitudinal displacements of a particle relative to the ideal designed trajectory; x', y' are the horizontal and vertical trajectory slopes; $\delta = \frac{\Delta P}{P_0}$ is the relative momentum error. In the linear approximation, evolution of the 6-D vector are described by 6×6 transport matrices. Since we are mostly interested in relativistic electron beams, $\beta \approx 1$, we also let $\delta = \frac{\Delta E}{E_0}$.

Why we must apply a full 3-D dynamics accurately to calculate the spin depolarization effects? The reason is that, first of all, the spin motion is very sensitive to the orbital motion especially at high energies when spin tune is large. Secondly, the electron motion after the emission of synchrotron photon generally involves all three orbital degrees of freedom, and when a depolarization perturbation is in action, there is not a priori reason that one degree of freedom plays a lesser role.

To emphasize the point that spin motion of a particle is intimately coupled to its orbital motion in all three degrees of freedom, perhaps one can consider a most typical sequence of events as follows. A photon emission excites the electron's synchrotron oscillation; the synchrotron oscillation couples to horizontal betatron oscillation by an energy dispersion; the horizontal betatron oscillation then couples to vertical betatron oscillation by x-y coupling; and finally the vertical betatron oscillation couples to spin motion by the Thomas–BMT equation and then depolarizes. All three orbital motions partake in this spin-orbital

---

[29]A.W. Chao, J. Appl. Phys. 50, 595 (1979); A.W. Chao, Nucl. Inst. Meth. 180, 29 (1981); H. Mais and G. Ripkin, DESY Laboratory Report 83-062 (1983); A.W. Chao, Lectures on Accelerator Physics, World Scientific (2020).

depolarization process. Note that, as mentioned before, the most direct culprit that depolarizes the beam is vertical motion, not horizontal or synchrotron motions.

In the absence of spin consideration, we know that the dynamics of a fully coupled 3-D (6-D phase space) linear system can be described by 6×6 matrices. Such a matrix formalism represents a generalization from the conventional 1-D Courant–Snyder analysis to 3-D. For this purpose, a SLIM algorithm was developed.

Following the SLIM algorithm, once the 6×6 one-turn map is obtained, one decomposes the map into three eigenmodes. Energy recoils due to quantum excitation are then decomposed into these eigenmodes. By statistically summing up these quantum excitations, one derives expressions for the three eigen-emittances of the orbital motion in an electron storage ring.

One important aspect of the SLIM algorithm lies in the fact that the three orbital motions are dealt with on equal footing. There is no assumption of a slow synchrotron oscillation. There is no assumption of strong or weak coupling among the three degrees of freedom. The three eigenmodes can be arbitrarily coupled, and the result is exact for any 3-D coupled linear storage ring system.

As all degrees of freedom are treated equally, there is a possibility to extend the dynamics to include other degrees of freedom, in particular, the spin. A proton or an electron actually has a 4th dimension in its dynamics. In addition to x-, y- and z-motions, the 4th dimension is a dynamics involving its spin. By adding this 4th dimension, and extending the eigenanalysis from 6-D to 8-D, the SLIM formalism also calculates the spin properties of the beam without much further ado in the SLIM algorithm using the same eigenanalysis package.

Spin motion can be included by adding two more spin coordinates (α, β) to form 8-D vectors,

$$
X = \begin{bmatrix} X_1 \\ X_2 \\ X_3 \\ X_4 \\ X_5 \\ X_6 \\ X_7 \\ X_8 \end{bmatrix} = \begin{bmatrix} x \\ x' \\ y \\ y' \\ z \\ \delta \\ \alpha \\ \beta \end{bmatrix}, \tag{6.77}
$$

and by extending the 6×6 transformations to 8×8.

Perhaps it should be mentioned here that the x and x' here refer to the total physical horizontal displacement and slope. They do not represent only the betatron components. Decomposing the orbital coordinates into "betatron" and "synchrotron" parts belongs to the job of eigenanalysis, which is contained internally in the SLIM calculation. In the fully coupled accelerator, "betatron" and "synchrotron" lose their meanings, the decomposition has to be done according to the three orbital eigenstates.

The 8×8 matrix, T, that transforms X for one revolution of the storage ring, has four eigenstates: three orbital x, y, z-states and one spin state. Each eigenstate is defined by a complex conjugate pair of eigenvectors of T. Any perturbation to the vector X, such as the recoil perturbation resulted from emitting a synchrotron photon, can be projected onto these four eigenstates. The projections onto the x, y, z-states give the contribution of this perturbation to the corresponding x, y, z-emittances, while the projection onto the spin state gives the contribution to spin diffusion. Since the same physical process of quantum emissions drives both the spin diffusion and the beam emittances of the electron beam, the SLIM formalism offers the possibility of obtaining the spin diffusion rate and the 21 beam distribution parameters $\langle X_i X_j \rangle$, $i, j = 1, \ldots, 6$, simultaneously in one single call of the eigenanalysis subroutine. How to develop those matrices and how to use them are subjects that we will cover below.

This formalism can be extended to noise sources other than synchrotron radiation recoils. When doing so, the same treatment and the same notation such as $\gamma \frac{\partial \hat{n}}{\partial \gamma}$ will be used. But each case will differ when the respective noise vector is decomposed into the four eigenstates. Generalization of the SLIM formalism to other noise sources can be straightforward.

The spin diffusion rate due to synchrotron radiation has been treated by several authors using different methods.[30] Since spin motion is intimately coupled to orbital motions, it is necessary to include all three orbital degrees of freedom in the analysis. Using 8×8 matrices to deal with this full complexity of spin-orbit coupling, treating the orbital and the spin motions of an electron on equal footing, then becomes a natural strategy to adopt.

One disadvantage of the matrix method is that nonlinear depolarization effects such as those caused by the nonlinear magnetic fields cannot be included in the SLIM formalism. Important advances that generalize to nonlinear spin dynamics have been developed, but not covered in this chapter, considered beyond the present scope.[31]

6.7.2   Determining $\hat{n}$

As mentioned, to calculate the spin polarization of a stored electron beam, the two key quantities to calculate are $\hat{n}(s)$ and $\gamma \frac{\partial \hat{n}}{\partial \gamma}(s)$. We will address $\hat{n}(s)$ in this section and $\gamma \frac{\partial \hat{n}}{\partial \gamma}(s)$ later in Sec. 6.7.5.

A comment is to be made here about the vector $\hat{n}$. Since we are considering a linear storage ring system, this $\hat{n}$ will be the equilibrium spin direction for all particles in the beam. If the orbital dynamics in the storage ring is nonlinear, different particles can have different $\hat{n}$, depending on the particle's orbital amplitudes A_x, A_y and A_δ. When that happens, the equilibrium spin direction $\hat{n}$ is no longer a single vector but a field of vectors centered around the central $\hat{n}$,

[30] V.N. Baier, Sov. Phys. Uspekhi 14, 695 (1972); Ya.S. Derbenev, A.M. Kondratenko and A.N. Skrinsky, Part. Accel. 9, 247 (1979); Ya.S. Derbenev and A.M. Kondratenko, Sov. Phys. JETP 37, 968 (1973); Ya.S. Derbenev and A.M. Kondratenko, Sov. Phys. JETP 35, 230 (1972).

[31] For further references, see D.P. Barber and G. Ripkin, Secs. 2.6.7 and 2.6.8, Handbook Accel. Phys. Eng., 2nd ed., World Scientific (2013), and the references therein.

which will acquire a more definitive notation $\hat{n}_0$.[32] The same complication also occurs to the other two base vectors $\hat{m}$ and $\hat{\ell}$ (to be mentioned momentarily). In our analysis below, we shall concentrate on linear systems, so we shall adopt the notations $\hat{n}, \hat{m}, \hat{\ell}$ without subscripts of 0's.

Given the electric and magnetic fields, spin precession of an electron is determined by the Thomas–BMT equation (6.11). Consider a storage ring in which all magnetic fields are transverse (e.g. in dipole, quadrupole and sextupole magnets, $\vec{B} = B_x\hat{x} + B_y\hat{y}$) and all electric fields are longitudinal (e.g. in accelerating RF cavities, $\vec{E} = E_z\hat{z}$).[33] We will simplify the expression for $\vec{\Omega}$ by linearizing it with respect to the coordinates of X. To do that, we first note that, to 1st order in the X coordinates, we have, assuming relativistic beam ($\beta \approx 1$),

$$\vec{\beta} \approx \hat{z} + x'\hat{x} + y'\hat{y}.$$

Substituting into Eq. (6.11) and linearize with respect to x', y' and $\delta = \frac{\Delta\gamma}{\gamma_0}$ gives

$$\frac{\vec{\Omega}}{c} \approx -\frac{1+\nu_0}{(B\rho)_0}\vec{B}$$
$$+ \frac{1}{(B\rho)_0}\left[\vec{B}\delta + (\nu_0+1)\frac{E_z}{c}(y'\hat{x} - x'\hat{y}) + \nu_0(x'B_x + y'B_y)\hat{z}\right], \quad (6.78)$$

where $\nu_0 = a\gamma_0$ is the nominal spin tune, $(B\rho)_0$ is the magnetic rigidity.

The first term on the right-hand-side of Eq. (6.78) is a leading term, containing terms zeroth order in the coordinates of X. The second term is a perturbation term, first order in X. The reason we need only to keep up to first order terms in X is because we are interested in depolarization resonances up to linear order.

Consider now an ideal electron that follows the closed-orbit turn after turn. By a closed-orbit here, we mean the 6-D closed-orbit defined relative to the ideal design trajectory, as described in the orbital part of the SLIM analysis,

$$X_0 = \begin{bmatrix} x_0 \\ x_0' \\ y_0 \\ y_0' \\ z_0 \\ \delta_0 \end{bmatrix},$$

in the presence of various perturbations. We assume X_0 has been obtained around the storage ring.

Equation (6.78) gives an expression of $\vec{\Omega}$ for an electron with coordinates X. In general, each component of X contains two parts. One part comes from

---

[32]D.P. Barber, J.A. Ellison, K. Heinemann, Phys. Rev. Special Topics – Accel. & Beams 7(12), 124002 (2004); G. Hoffstaetter, M. Vogt, D.P. Barber, Phys. Rev. Special Topics – Accel. & Beams 11(2), 114001 (1999).

[33]For cases not covered in this consideration (e.g. solenoids with longitudinal magnetic fields and crab cavities with transverse electric fields), the generalization, although not included here, is straightforward following the Thomas–BMT equation.

Table 6.4: Explicit expressions of $\frac{\vec{\Omega}(X_0)}{c}$ for various beamline elements. A nonlinear contribution for sextupole magnets is included.

Horizontal bending magnet	$-\frac{B_y}{(B\rho)_0}[(\nu_0 - \delta_0)\hat{y} - \nu_0 y_0'\hat{z}]$
Vertical bending magnet	$-\frac{B_x}{(B\rho)_0}[(\nu_0 - \delta_0)\hat{x} - \nu_0 x_0'\hat{z}]$
Quadrupole magnet	$-\frac{1+\nu_0}{(B\rho)_0}\frac{\partial B_y}{\partial x}[y_0\hat{x} + x_0\hat{y}]$
Skew quadrupole magnet	$-\frac{1+\nu_0}{(B\rho)_0}\frac{\partial B_y}{\partial y}[-x_0\hat{x} + y_0\hat{y}]$
RF cavity	$-\frac{1+\nu_0}{(B\rho)_0}\frac{E_z}{c}[x_0'\hat{y} - y_0'\hat{x}]$
Horizontal kicker	$-\frac{B_y}{(B\rho)_0}[(\nu_0 + 1 - \delta_0)\hat{y} - \nu_0 y_0'\hat{z}]$
Vertical kicker	$-\frac{B_x}{(B\rho)_0}[(\nu_0 + 1 - \delta_0)\hat{x} - \nu_0 x_0'\hat{z}]$
Sextupole magnet	$-\frac{1+\nu_0}{(B\rho)_0}\frac{\partial^2 B_y}{\partial x^2}[x_0 y_0\hat{x} + \frac{1}{2}(x_0^2 - y_0^2)\hat{y}]$

a closed-orbit contribution and the other part is a displacement relative to the closed-orbit. In other words, in all expressions up to now, X really should be written as $X_0 + X$, where X is the deviation from X_0. From here on, X will assume this new definition of a deviation from the 6-D closed-orbit X_0.

When we apply Eq. (6.78) to strong nonlinear magnets, we will include in their magnetic field $\vec{B}(X_0 + X)$ the nonlinear contributions due to X_0, but in the SLIM algorithm, linearization will always be performed with respect to X assuming X is small, although there is no linearization with respect to X_0 assuming X_0 is small.

Along the closed-orbit, the central electron then experiences well-defined electric and magnetic fields, $\vec{E}(X_0)$ and $\vec{B}(X_0)$, in each of the beamline elements. Given these fields, we obtain the spin precession angular velocity $\vec{\Omega}(X_0)$ seen by the central electron. Table 6.4 summarizes the explicit expressions of $\vec{\Omega}(X_0)$, obtained using Eq. (6.78), for various beamline elements. To find $\hat{n}$, the thing we want to do now is to analyze the spin motion of this central electron.

A horizontal bending magnet and a horizontal kicker really refer to the same physical devices of a dipole magnet with vertical magnetic field. They are different in Table 6.4, however, because, by its definition, the reference coordinate frame is rotated in a horizontal bending magnet but not in a horizontal kicker, leading to a difference in their angular velocities by $-\frac{B_y}{(B\rho)_0}\hat{y}$. Similarly, there is a difference between a vertical bending magnet and a vertical kicker.

We assume all beamline elements (other than drift spaces) are short enough so that X_0 and X do not change appreciably inside them. This is consistent with the thin-lens approximation we shall adopt. If the element's length is ℓ, then going through the element, the electron spin rotates by an angle $\vec{\Omega}(X_0)\frac{\ell}{c}$, i.e. by an angle of magnitude $\Omega\frac{\ell}{c}$ and around a rotation axis $\hat{\Omega}$. This rotation can be represented by a 3×3 matrix applied to the spin state

$$\vec{S} = \begin{bmatrix} S_x \\ S_y \\ S_z \end{bmatrix}.$$

Once $\vec{\Omega}(X_0)$ is known, the 3×3 spin rotation matrix going through the beamline element is given by

$$R = \begin{bmatrix} c_x^2(1-C)+C & c_x c_y(1-C)-c_z S & c_x c_z(1-C)+c_y S \\ c_x c_y(1-C)+c_z S & c_y^2(1-C)+C & c_y c_z(1-C)-c_x S \\ c_x c_z(1-C)-c_y S & c_y c_z(1-C)+c_x S & c_z^2(1-C)+C \end{bmatrix}, \quad (6.79)$$

where $c_{x,y,z}$ are the direction cosines $\hat{\Omega} \cdot \hat{x}, \hat{\Omega} \cdot \hat{y}$ and $\hat{\Omega} \cdot \hat{z}$ respectively, with $c_x^2 + c_y^2 + c_z^2 = 1$, and $C = \cos \frac{\Omega \ell}{c}, S = \sin \frac{\Omega \ell}{c}$. Derivation of Eq. (6.79) is given in Homework 6.22. For a drift space, spin is unaffected and R is a unit matrix.

Knowing the storage ring beamline, one multiplies all 3×3 matrices successively to obtain the total spin precession transformation $R_{\text{tot}}(s)$ for one revolution around position s. Being a rotation matrix, R_{tot} has three eigenvalues and one of them is necessarily 1. It is then easy to calculate the rotational axis of $R_{\text{tot}}(s)$ because it is given by the eigenvector of $R_{\text{tot}}(s)$ with eigenvalue 1. In other words, if we let this rotation axis be

$$\hat{n}(s) = \begin{bmatrix} n_x \\ n_y \\ n_z \end{bmatrix}, \quad \text{then} \quad R_{\text{tot}} \begin{bmatrix} n_x \\ n_y \\ n_z \end{bmatrix} = \begin{bmatrix} n_x \\ n_y \\ n_z \end{bmatrix}.$$

The unit vector $\hat{n}(s)$ has the property that it returns to itself after one revolution. In a sense, $\hat{n}(s)$ plays the role of a closed-orbit in spin motion.

Once $\hat{n}(s)$ is found, we look for two other unit vectors $\hat{m}(s)$ and $\hat{\ell}(s)$ so that the triplet $(\hat{n}, \hat{m}, \hat{\ell})$ forms a right-handed orthonormal base vectors.

Successive 3×3 transformations then bring this base to other positions with $2\pi R > s > 0$, where $2\pi R$ is the storage ring circumference. In one revolution, it gives

$$\begin{bmatrix} \hat{n} \\ \hat{m} \\ \hat{\ell} \end{bmatrix}_{s+2\pi R} = \begin{bmatrix} 1 & 0 & 0 \\ 0 & \cos 2\pi\nu & -\sin 2\pi\nu \\ 0 & \sin 2\pi\nu & \cos 2\pi\nu \end{bmatrix} \begin{bmatrix} \hat{n} \\ \hat{m} \\ \hat{\ell} \end{bmatrix}_s. \quad (6.80)$$

where $e^{\pm i 2\pi\nu}$ are the two other nontrivial eigenvalues of R_{tot}. The quantity ν gives the *spin tune*, i.e. the spin precession angle per turn divided by 2π.

A few comments concerning Eq. (6.80) are in order.

- For a planar ring without error fields, we found before that $\hat{n}$ coincides with the direction of the guiding magnetic field $\hat{y}$. In the presence of imperfections, this is no longer true. In particular, for example, as mentioned in the previous section, $\hat{n}$ may deviate from $\hat{y}$ noticeably if ν is close to an integer.

- Spin tune ν is defined around the ideal electron moving along the closed-orbit, while the closed-orbit can be distorted due to various error fields. (We have been considering an ideal electron in a nonideal storage ring.) So ν is the *perturbed* spin tune, and is not to be confused with the nominal spin tune $\nu_0 = a\gamma_0$ that applies only when all error fields are absent in a planar ring with a ribbon beam. In general, $\nu \neq \nu_0$.

- When trying to measure the beam energy in an electron storage ring by measuring its spin tune (see footnote 7), it is necessary to first show that ν does not deviate from ν_0 too much. Otherwise, the beam energy is actually not equal to the spin tune divided by a. In practice, however, the error is often small enough if integer resonances have been avoided.

- The base vectors $(\hat{n}, \hat{m}, \hat{\ell})$ defined this way around the ring has a discontinuity at the position s where the first definition of the vector base is chosen. The vector $\hat{n}$ will have no problem closing onto itself after one complete turn because it is the rotation axis of R_{tot}. But the other two components will have a discontinuity in their definitions. This discontinuity should not be a problem as long as it is kept in mind.

Homework 6.22 Verify Eq. (6.79).

Solution Consider a spin

$$\vec{S} \;=\; S_x\hat{x} + S_y\hat{y} + S_z\hat{z} \;=\; \tilde{S}\begin{bmatrix} \hat{x} \\ \hat{y} \\ \hat{z} \end{bmatrix},$$

where

$$S \;=\; \begin{bmatrix} S_x \\ S_y \\ S_z \end{bmatrix},$$

being rotated by an angle $\theta = \Omega\frac{\ell}{c}$ around axis $\hat{\Omega}$. A tilde means taking transpose of a matrix or a vector. We want to find a 3×3 matrix R that transforms the column matrix S to its new value after the rotation.

Consider an arbitrary unit vector $\hat{r}$ that is not parallel or anti-parallel to $\hat{\Omega}$. Define two unit vectors

$$\hat{a} \;=\; \frac{\hat{r} \times \hat{\Omega}}{|\hat{r} \times \hat{\Omega}|},$$

$$\hat{b} \;=\; \frac{\hat{a} \times \hat{\Omega}}{|\hat{a} \times \hat{\Omega}|} \;=\; \hat{a} \times \hat{\Omega}.$$

The three unit vectors $(\hat{a}, \hat{\Omega}, \hat{b})$ form a right-handed set of orthonormal base vectors. The choice of $\hat{r}$ is arbitrary for now, and is to be chosen by convenience later.

The transformations between $(\hat{x}, \hat{y}, \hat{z})$ and $(\hat{a}, \hat{\Omega}, \hat{b})$ are given by

$$
\begin{bmatrix} \hat{a} \\ \hat{\Omega} \\ \hat{b} \end{bmatrix} = \begin{bmatrix} \hat{a} \cdot \hat{x} & \hat{a} \cdot \hat{y} & \hat{a} \cdot \hat{z} \\ \hat{\Omega} \cdot \hat{x} & \hat{\Omega} \cdot \hat{y} & \hat{\Omega} \cdot \hat{z} \\ \hat{b} \cdot \hat{x} & \hat{b} \cdot \hat{y} & \hat{b} \cdot \hat{z} \end{bmatrix} \begin{bmatrix} \hat{x} \\ \hat{y} \\ \hat{z} \end{bmatrix} \equiv T \begin{bmatrix} \hat{x} \\ \hat{y} \\ \hat{z} \end{bmatrix},
$$

$$
\begin{bmatrix} \hat{x} \\ \hat{y} \\ \hat{z} \end{bmatrix} = \begin{bmatrix} \hat{x} \cdot \hat{a} & \hat{x} \cdot \hat{\Omega} & \hat{x} \cdot \hat{b} \\ \hat{y} \cdot \hat{a} & \hat{y} \cdot \hat{\Omega} & \hat{y} \cdot \hat{b} \\ \hat{z} \cdot \hat{a} & \hat{z} \cdot \hat{\Omega} & \hat{z} \cdot \hat{b} \end{bmatrix} \begin{bmatrix} \hat{a} \\ \hat{\Omega} \\ \hat{b} \end{bmatrix} \equiv T^{-1} \begin{bmatrix} \hat{a} \\ \hat{\Omega} \\ \hat{b} \end{bmatrix}.
$$

We note that $T^{-1} = \tilde{T}$.

Before we rotate $\vec{S}$, let us express it in the $(\hat{a}, \hat{\Omega}, \hat{b})$ coordinates,

$$
\vec{S} = \tilde{S} T^{-1} \begin{bmatrix} \hat{a} \\ \hat{\Omega} \\ \hat{b} \end{bmatrix}.
$$

Since $\hat{\Omega}$ is the rotation axis, when we rotate $\vec{S}$ in the $(\hat{a}, \hat{\Omega}, \hat{b})$ coordinates, we perform a rotation to its components by a matrix

$$
\Theta = \begin{bmatrix} \cos\theta & 0 & -\sin\theta \\ 0 & 1 & 0 \\ \sin\theta & 0 & \cos\theta \end{bmatrix}.
$$

The sign of the angle is such that it gives a right-handed rotation of angle θ around the $\hat{\Omega}$-axis. This means after the rotation, the components of the spin in the $(\hat{a}, \hat{\Omega}, \hat{b})$ coordinates is

$$
\tilde{S} T^{-1} \tilde{\Theta} \begin{bmatrix} \hat{a} \\ \hat{\Omega} \\ \hat{b} \end{bmatrix}.
$$

Then, expressed in the $(\hat{x}, \hat{y}, \hat{z})$ coordinates, the spin after rotation is

$$
\tilde{S} T^{-1} \tilde{\Theta} T \begin{bmatrix} \hat{x} \\ \hat{y} \\ \hat{z} \end{bmatrix}.
$$

The needed 3×3 matrix is therefore just

$$
R = \tilde{T} \Theta \tilde{T}^{-1} = \tilde{T} \Theta T.
$$

We still need to compute T. Take for example $\hat{r} = \hat{x}$. Defining $c_x = \hat{\Omega} \cdot \hat{x}, c_y = \hat{\Omega} \cdot \hat{y}, c_z = \hat{\Omega} \cdot \hat{z}$ with $c_x^2 + c_y^2 + c_z^2 = 1$, then we find

$$
\hat{a} = \frac{-c_z \hat{y} + c_y \hat{z}}{\sqrt{c_y^2 + c_z^2}}, \qquad \hat{b} = \frac{\hat{x} - c_x \hat{\Omega}}{\sqrt{c_y^2 + c_z^2}},
$$

which then gives

$$
T = \begin{bmatrix}
0 & -\dfrac{c_z}{\sqrt{c_y^2+c_z^2}} & \dfrac{c_y}{\sqrt{c_y^2+c_z^2}} \\[2ex]
c_x & c_y & c_z \\[2ex]
\sqrt{c_y^2 + c_z^2} & -\dfrac{c_x c_y}{\sqrt{c_y^2+c_z^2}} & -\dfrac{c_x c_z}{\sqrt{c_y^2+c_z^2}}
\end{bmatrix},
$$

which in turn gives, after some algebra,

$$
R = \begin{bmatrix}
c_x^2 + (c_y^2 + c_z^2)\cos\theta & c_x c_y(1-\cos\theta) - c_z\sin\theta & c_x c_z(1-\cos\theta) + c_y\sin\theta \\[1ex]
c_x c_y(1-\cos\theta) + c_z\sin\theta & c_y^2 + (c_x^2 + c_z^2)\cos\theta & c_y c_z(1-\cos\theta) - c_x\sin\theta \\[1ex]
c_x c_z(1-\cos\theta) - c_y\sin\theta & c_y c_z(1-\cos\theta) + c_x\sin\theta & c_z^2 + (c_x^2 + c_y^2)\cos\theta
\end{bmatrix}.
$$

Equation (6.79) then follows.

6.7.3   Spin motion by matrix

Once the base vectors $(\hat{n}, \hat{m}, \hat{\ell})$ are defined, the spin direction of a nearly polarized electron is expressed as

$$
\vec{S} \approx \hat{n} + \alpha\hat{m} + \beta\hat{\ell}, \qquad |\alpha, \beta| \ll 1. \tag{6.81}
$$

The quantities α and β thus describe the spin to a linear approximation and

$$
\frac{1}{2}(\alpha^2 + \beta^2)
$$

specifies the degree of depolarization of this electron. See Fig. 6.13. The assumption $|\alpha, \beta| \ll 1$ is acceptable since we will be interested in typical values of $|\alpha, \beta| \sim 10^{-6}$, as illustrated in Fig. 6.9. In contrast, $\hat{n}$ can deviate from $\hat{y}$ by an arbitrary angle; it is calculated exactly around the 6-D closed orbit.

So we describe the orbital motion with the coordinate base vectors $(\hat{x}, \hat{y}, \hat{z})$, and spin motion with $(\hat{n}, \hat{m}, \hat{\ell})$. In an ideal planar case, we have $\hat{n} = \hat{y}$. The base vectors $\hat{x}, \hat{z}$ rotate by 2π around the storage ring and close upon themselves. The base vectors $(\hat{m}, \hat{\ell})$ rotate by $2\pi a\gamma$ relative to $(\hat{x}, \hat{y}, \hat{z})$ and in general do not close upon themselves and an extra rotation is introduced at the end of one revolution to make them close.

In a nonideal storage ring, we still consider an ideal synchronous electron — an ideal electron in a nonideal storage ring — following the distorted closed orbit X_0 along which we have calculated the spin closed orbit $\hat{n}(s)$. The ideal electron has all its eight components of X vanish, including α and β.

We now consider an electron in general. Its spin motion is described by the Thomas–BMT equation,

$$
\frac{d\vec{S}}{dt} = c\frac{d\vec{S}}{ds} = \vec{\Omega}(X_0 + X) \times \vec{S}. \tag{6.82}
$$

Figure 6.13: Definitions of various parameters describing the spin motion of a single electron. Quantities (α, β) will form the spin phase space.

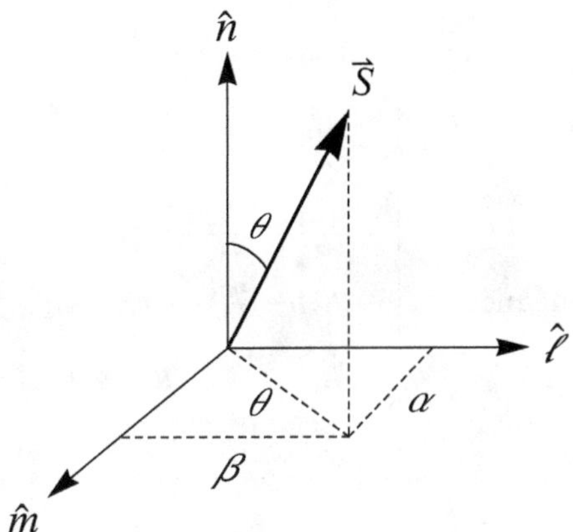

The precession angular velocity $\vec{\Omega}$ depends on the position of the electron, $X_0 + X$. It is given in Eq. (6.78), except that X in Eq. (6.78) needs to be represented by $X_0 + X$. In a linear approximation, $\vec{\Omega}(X_0 + X)$ can be decomposed into

$$\vec{\Omega}(X_0 + X) \; = \; \vec{\Omega}(X_0) + \vec{\omega}(X)\,,$$

where $\vec{\Omega}(X_0)$ has been given in Table 6.4. The perturbation term $\vec{\omega}$ is first order in the components in X and $|\vec{\omega}| \ll |\vec{\Omega}|$. Table 6.5 gives explicit expressions of $\vec{\omega}(X)$ for various beamline elements. This time, the bending magnets and kickers behave identically.

Noting that $\hat{n}, \hat{m}$ and $\hat{\ell}$ satisfy the precession equation along the closed-orbit,

$$\frac{d(\hat{n}, \hat{m}, \hat{\ell})}{ds} \; = \; \Omega(\vec{X}_0) \times (\hat{n}, \hat{m}, \hat{\ell})\,,$$

one obtains by substituting Eq. (6.81) into Eq. (6.82) that

$$\frac{d\alpha}{ds} \; = \; \vec{\omega}(X) \cdot \hat{\ell}\,,$$
$$\frac{d\beta}{ds} \; = \; -\vec{\omega}(X) \cdot \hat{m}\,. \tag{6.83}$$

In the derivation of Eq. (6.83), we have kept only terms linear in X. Unlike being done so far, however, this time the spin components α and β also participate in the linearization. Recalling that $\vec{\omega}(X)$ contains only terms linear in the *orbital* components of X, Eq. (6.83) describes how spin coordinates on the left-hand-side linearly couple to the orbital coordinates on the right-hand-side.

Table 6.5: Explicit expressions of $\vec{\omega}(X)$ for various beamline elements. Sextupoles are included by linearizing around the closed-orbit.

Horizontal bending magnet Horizontal kicker	$\frac{B_y}{(B\rho)_0}[\delta\hat{y} + \nu_0 y'\hat{z}]$
Vertical bending magnet Vertical kicker	$-\frac{B_x}{(B\rho)_0}[\delta\hat{x} + \nu_0 x'\hat{z}]$
Quadrupole magnet	$-\frac{1+\nu_0}{(B\rho)_0}\frac{\partial B_y}{\partial x}[y\hat{x} + x\hat{y}]$
Skew quadrupole magnet	$-\frac{1+\nu_0}{(B\rho)_0}\frac{\partial B_y}{\partial y}[-x\hat{x} + y\hat{y}]$
RF cavity	$-\frac{1+\nu_0}{(B\rho)_0}\frac{E_z}{c}[x'\hat{y} - y'\hat{x}]$
Sextupole magnet	$-\frac{1+\nu_0}{(B\rho)_0}\frac{\partial^2 B_y}{\partial x^2}[(xy_0 + yx_0)\hat{x} + (xx_0 - yy_0)\hat{y}]$

If we now form an 8-D state vector as in Eq. (6.77), the corresponding 8×8 transformation matrices that transform the state vector from one position to another in the storage ring would look like

$$M = \begin{bmatrix} \begin{array}{ccc|c} & 6 \times 6 & & \\ & \text{transport} & & 0 \\ & \text{matrix} & & \\ \hline & D & & 1 \end{array} \end{bmatrix}, \tag{6.84}$$

where the upper-left corner means the usual 6×6 transport matrix describing the transformation among the orbital coordinates; the upper-right corner is a 6×2 block filled by zeroes; the 2×6 block D is obtained from Table 6.5 and Eq. (6.83); the lower-right corner is a 2×2 unit matrix. The D elements are responsible for all the spin-orbit coupling effects. Explicit expressions of the 8×8 matrices, including the D elements, for some beamline elements are given in Sec. 6.7.4.

The fact that the upper-right corner is filled with zeroes reflects that the orbital motion of the electron is unperturbed by the particle's spin. This is true if we ignore spin-orbit quantum mechanical coupling contributions such as the Stern–Gerlach effect. These effects, being proportional to $\hbar$, are negligible in almost all cases unless when purposely being exploited in very special arrangements.

Ignoring the Stern–Gerlach effects, in writing down Eq. (6.84), we have assumed that the spin motion of an electron depends on its orbital motion, but its orbital motion disregards its spin motion.

One must not forget that, due to the discontinuous transition in the definition of the base vectors as the electron travels across $s = 2\pi R$ [see Eq. (6.80)], an extra transformation for the spin components is required at an infinitesimal distance before $s = 2\pi R$, by an angle $-2\pi\nu$,

$$\begin{bmatrix} \alpha \\ \beta \end{bmatrix}_{2\pi R} = \begin{bmatrix} \cos 2\pi\nu & \sin 2\pi\nu \\ -\sin 2\pi\nu & \cos 2\pi\nu \end{bmatrix} \begin{bmatrix} \alpha \\ \beta \end{bmatrix}_{2\pi R^-}.$$

We have assumed the position where we defined the initial set of base vectors $(\hat{n}, \hat{m}, \hat{\ell})$ is designated $s = 0$.

Starting from any position s we multiply all 8×8 transformation matrices successively around the beamline to obtain a transformation matrix for one revolution. It will be designated as $T(s)$. Let the eigenvalues and eigenvectors of $T(s)$ be λ_k and $E_k(s)$,

$$\begin{aligned} T(s)E_k(s) &= \lambda_k E_k(s), \\ \lambda_k^* &= \lambda_{-k}, \\ E_k^*(s) &= E_{-k}, \qquad k = \pm I, \pm II, \pm III, \pm IV. \end{aligned}$$

Eigenvectors at other positions, $E_k(s')$, are obtained from $E_k(s)$ by the transformation from s to s'.

Among the 8 eigenstates, due to its special form of the map (6.84), two of them will contain only spin components and no orbital components. These will be the spin eigenstates and are designated as the IV-th eigenstate. Its eigenvalues are given by $\lambda_{\pm IV} = e^{\pm i 2\pi\nu}$ with ν the perturbed spin tune of Eq. (6.80). The other six orbital eigenstates, on the other hand, carry with them some 7th and 8th spin components, and it is these components, if they turn out to be nonzero, that will cause depolarization to the electron beam due to synchrotron radiation.

It is not necessary to normalize the spin eigenstates since they are not used in later calculations. The orbital eigenstates are normalized by

$$\tilde{e}_k^* S e_k = \begin{cases} i, & \text{if } k = I, II, III, \\ -i, & \text{if } k = -I, -II, -III, \end{cases}$$

where e_k is the 6-D vector whose components are the six orbital components of E_k and

$$S = \begin{bmatrix} 0 & 1 & 0 & 0 & 0 & 0 \\ -1 & 0 & 0 & 0 & 0 & 0 \\ 0 & 0 & 0 & 1 & 0 & 0 \\ 0 & 0 & -1 & 0 & 0 & 0 \\ 0 & 0 & 0 & 0 & 0 & 1 \\ 0 & 0 & 0 & 0 & -1 & 0 \end{bmatrix}.$$

We now have prepared the eigenvalues and eigenvectors of the 8×8 matrices around the storage ring. They are to be used momentarily to calculate the spin chromaticity $\gamma \frac{\partial \hat{n}}{\partial \gamma}$.

Stern–Gerlach effect Let us make a slight detour here on how small the Stern–Gerlach forces — ignored in our treatment — are. Consider a vertically polarized electron with magnetic moment $\vec{\mu} = \mu\hat{y}$. For an electron, we have $\mu = \frac{ge\hbar}{4m}$.

Consider a quadrupole magnet with field gradient G and length ℓ. The force seen by the electron dipole moment in this magnet is given by

$$\vec{F} = -\nabla V,$$

where V is the potential

$$V = -\vec{\mu} \cdot \vec{B} = -\mu B_y = -\mu G x,$$

thus

$$\vec{F} = \mu G \hat{x}.$$

This means a quadrupole, acting as a focusing element for an electron's charge, now acts as a bending element for the electron's magnetic dipole moment.[34] In this case, the bending is in the horizontal plane and the bending angle is

$$\Delta x' = \frac{\mu G \ell}{E_0}.$$

Taking a typical quadrupole magnet with $\frac{G\ell}{B\rho} = 0.1\,\mathrm{m}^{-1}$, the kicking angle $\Delta x'$ is found to be about 0.2×10^{-13} rad.

Similarly, a sextupole magnet which produces a nonlinear field for an electron charge acts on $\vec{\mu}$ as a linear focusing element. The focal length is given by

$$\frac{1}{f} = \frac{2\mu\lambda\ell}{E_0},$$

with $\lambda = \frac{\partial^2 B_y}{\partial x^2}$. Taking a typical sextupole strength $\frac{\lambda\ell}{B\rho} = 1\,\mathrm{m}^{-2}$, we find a focal length of about 2.5×10^{12} m.

Both the bending by quadrupoles and focusing by sextupoles are indeed exceedingly weak. This is the basis of why the orbital beam dynamists do not have to worry about spin, in spite of the nonsymplecticity in making this approximation. On the other hand, spin dynamics do have to worry about, and in fact to take special care of all details of, the orbital dynamics because the D matrix is nonzero and sensitive.

Homework 6.23 Prove Eq. (6.83).

Homework 6.24 Derive at least a few of the entries of $\vec{\omega}(X)$ in Table 6.5.

6.7.4   Explicit expressions of SLIM 8 × 8 matrices

The generalized transport matrices for the state vector $(x, x', y, y', z, \delta, \alpha, \beta)$ are listed below for various beam-line elements. Thin-lens approximation has been taken.

---

[34]A curious reader is encouraged to contemplate a magnetic dipole moment as a tiny loop of electrical current and derive an expression of this force using this model.

Drift space $\ell = $ drift space length;

$$\begin{bmatrix}
1 & \ell & 0 & 0 & 0 & 0 & 0 & 0 \\
0 & 1 & 0 & 0 & 0 & 0 & 0 & 0 \\
0 & 0 & 1 & \ell & 0 & 0 & 0 & 0 \\
0 & 0 & 0 & 1 & 0 & 0 & 0 & 0 \\
0 & 0 & 0 & 0 & 1 & 0 & 0 & 0 \\
0 & 0 & 0 & 0 & 0 & 1 & 0 & 0 \\
0 & 0 & 0 & 0 & 0 & 0 & 1 & 0 \\
0 & 0 & 0 & 0 & 0 & 0 & 0 & 1
\end{bmatrix}.$$

Horizontal bending magnet or kicker $q = \dfrac{B_y \ell}{(B\rho)_0}$;

$$\begin{bmatrix}
1 & 0 & 0 & 0 & 0 & 0 & 0 & 0 \\
0 & 1 & 0 & 0 & 0 & q & 0 & 0 \\
0 & 0 & 1 & 0 & 0 & 0 & 0 & 0 \\
0 & 0 & 0 & 1 & 0 & 0 & 0 & 0 \\
-q & 0 & 0 & 0 & 1 & 0 & 0 & 0 \\
0 & 0 & 0 & 0 & 0 & 1 & 0 & 0 \\
0 & 0 & 0 & \nu_0 q \ell_z & 0 & q \ell_y & 1 & 0 \\
0 & 0 & 0 & -\nu_0 q m_z & 0 & -q m_y & 0 & 1
\end{bmatrix}.$$

Vertical bending magnet or kicker $q = \dfrac{B_x \ell}{(B\rho)_0}$;

$$\begin{bmatrix}
1 & 0 & 0 & 0 & 0 & 0 & 0 & 0 \\
0 & 1 & 0 & 0 & 0 & 0 & 0 & 0 \\
0 & 0 & 1 & 0 & 0 & 0 & 0 & 0 \\
0 & 0 & 0 & 1 & 0 & -q & 0 & 0 \\
0 & 0 & q & 0 & 1 & 0 & 0 & 0 \\
0 & 0 & 0 & 0 & 0 & 1 & 0 & 0 \\
0 & \nu_0 q \ell_z & 0 & 0 & 0 & q \ell_x & 1 & 0 \\
0 & -\nu_0 q m_z & 0 & 0 & 0 & -q m_x & 0 & 1
\end{bmatrix}.$$

Quadrupole $q = \dfrac{\ell}{(B\rho)_0} \dfrac{\partial B_y}{\partial x}$;

$$\begin{bmatrix}
1 & 0 & 0 & 0 & 0 & 0 & 0 & 0 \\
-q & 1 & 0 & 0 & 0 & q x_0 & 0 & 0 \\
0 & 0 & 1 & 0 & 0 & 0 & 0 & 0 \\
0 & 0 & q & 1 & 0 & -q y_0 & 0 & 0 \\
-q x_0 & 0 & q y_0 & 0 & 1 & 0 & 0 & 0 \\
0 & 0 & 0 & 0 & 0 & 1 & 0 & 0 \\
-(1+\nu_0) q \ell_y & 0 & -(1+\nu_0) q \ell_x & 0 & 0 & 0 & 1 & 0 \\
(1+\nu_0) q m_y & 0 & (1+\nu_0) q m_x & 0 & 0 & 0 & 0 & 1
\end{bmatrix}.$$

Skew quadrupole $q = \frac{\ell}{(B\rho)_0} \frac{\partial B_y}{\partial y}$;

$$
\begin{bmatrix}
1 & 0 & 0 & 0 & 0 & 0 & 0 & 0 \\
0 & 1 & -q & 0 & 0 & qy_0 & 0 & 0 \\
0 & 0 & 1 & 0 & 0 & 0 & 0 & 0 \\
-q & 0 & 0 & 1 & 0 & qx_0 & 0 & 0 \\
-qy_0 & 0 & -qx_0 & 0 & 1 & 0 & 0 & 0 \\
0 & 0 & 0 & 0 & 0 & 1 & 0 & 0 \\
(1+\nu_0)q\ell_x & 0 & -(1+\nu_0)q\ell_y & 0 & 0 & 0 & 1 & 0 \\
-(1+\nu_0)qm_x & 0 & (1+\nu_0)qm_y & 0 & 0 & 0 & 0 & 1
\end{bmatrix}.
$$

Rf cavity $q = \frac{e\hat{V}}{RE_0}\cos\phi_s$; $r = (1+\nu_0)\frac{e\hat{V}}{E_0}\sin\phi_s$; $\phi_s =$ synchronous phase; $\hat{V} =$ peak voltage;

$$
\begin{bmatrix}
1 & 0 & 0 & 0 & 0 & 0 & 0 & 0 \\
0 & 1 & 0 & 0 & 0 & 0 & 0 & 0 \\
0 & 0 & 1 & 0 & 0 & 0 & 0 & 0 \\
0 & 0 & 0 & 1 & 0 & 0 & 0 & 0 \\
0 & 0 & 0 & 0 & 1 & 0 & 0 & 0 \\
0 & 0 & 0 & 0 & q & 1 & 0 & 0 \\
0 & -r\ell_y & 0 & r\ell_x & 0 & 0 & 1 & 0 \\
0 & rm_y & 0 & -rm_x & 0 & 0 & 0 & 1
\end{bmatrix}.
$$

Sextupole $q = \frac{\ell}{(B\rho)_0}\frac{\partial^2 B_y}{\partial x^2}$; $r = (1+\nu_0)q$;

$$
\begin{bmatrix}
1 & 0 & 0 & 0 & 0 & 0 & 0 & 0 \\
-qx_0 & 1 & qy_0 & 0 & 0 & \frac{q}{2}(x_0^2 - y_0^2) & 0 & 0 \\
0 & 0 & 1 & 0 & 0 & 0 & 0 & 0 \\
qy_0 & 0 & qx_0 & 1 & 0 & -qx_0y_0 & 0 & 0 \\
-\frac{1}{2}q(x_0^2 - y_0^2) & 0 & qx_0y_0 & 0 & 1 & 0 & 0 & 0 \\
0 & 0 & 0 & 0 & 0 & 1 & 0 & 0 \\
-r(y_0\ell_x + x_0\ell_y) & 0 & -r(x_0\ell_x - y_0\ell_y) & 0 & 0 & 0 & 1 & 0 \\
r(y_0m_x + x_0m_y) & 0 & r(x_0m_x - y_0m_y) & 0 & 0 & 0 & 0 & 1
\end{bmatrix}.
$$

Homework 6.25 It is obligatory that as a homework at least some of the 8×8 matrices listed in this section be verified.

6.7.5   Determining $\gamma\frac{\partial\hat{n}}{\partial\gamma}$

The spin polarization in an electron storage ring approaches the equilibrium value P with a polarization time constant τ according to the Derbenev–Kondratenko formula, Eqs. (6.72–6.74). The polarization of an initially unpolarized beam builds up according to $P(1 - e^{-t/\tau})$.

In Eq. (6.72), an important quantity is the spin chromaticity $\gamma\frac{\partial\hat{n}}{\partial\gamma}$. In the SLIM language, it characterizes the projection of the recoil perturbation when emitting a synchrotron photon onto the spin eigenstate. A proper calculation of $\gamma \frac{\partial \hat{n}}{\partial \gamma}$ would take into full account the spin-orbit coupling after of the sudden emission of a synchrotron photon. In particular, the strongest spin-orbit coupling occurs near the depolarization resonances when the spin tune ν is close to $k \pm \nu_{x,y,s}$ for some integer k, where $\nu_{x,y,s}$ are the tunes for the horizontal betatron, the vertical betatron, and the synchrotron orbital motions.[35] Definition of $\gamma \frac{\partial \hat{n}}{\partial \gamma}$ needs to take into consideration the mechanism of these depolarization resonances.

Consider a fully polarized ideal electron. Let a photon be emitted at position s with energy δE. Immediately after the emission, the electron is left in the state

$$
\Delta X = \begin{bmatrix} 0 \\ 0 \\ 0 \\ 0 \\ 0 \\ -\frac{\delta E}{E_0} \\ 0 \\ 0 \end{bmatrix}.
$$

Decomposing into eigenstates at the emission position s, one has

$$
\Delta X = \sum_k A_k E_k(s)
$$

$$
= \sum_{k=\pm I, \pm II, \pm III} A_k E_k(s) + \begin{bmatrix} 0 \\ 0 \\ 0 \\ 0 \\ 0 \\ 0 \\ \bar{\alpha} \\ \bar{\beta} \end{bmatrix}_s , \tag{6.85}
$$

where we have used the property that the spin eigenstates $E_{\pm IV}$ contain no orbital coordinates.

Equation (6.85) contains eight unknowns $A_{\pm I, \pm II, \pm III}$, $\bar{\alpha}$, $\bar{\beta}$, and eight equations to determine them. The value for A_k for $k = \pm I, \pm II, \pm III$ can be obtained by equating the orbital components on both sides of Eq. (6.85). The solution can also be readily obtained using the orthogonality of the orbital eigenvectors, yielding (see footnotes 29)

$$
A_k = -i\frac{\delta E}{E_0} E_{5k}^*(s), \qquad k = I, II, III ,
$$

with $A_{-k} = A_k^*$ and E_{jk} the j-th component of vector E_k.

---

[35] More accurately, and it is important to point out, $\nu_{x,y,s}$ should be replaced by the three eigentunes $\nu_{I,II,III}$ for the three orbital eigenmodes. Also note that ν is the spin tune along the distorted closed-orbit; it is not necessarily equal to $a\gamma$.

Equating the last two spin components of Eq. (6.85) then yields

$$
\begin{bmatrix} \bar{\alpha} \\ \bar{\beta} \end{bmatrix} = -2 \frac{\delta E}{E_0} \sideset{}{'}\sum_k \begin{bmatrix} \mathrm{Im}(E_{5k}^* E_{7k}) \\ \mathrm{Im}(E_{5k}^* E_{8k}) \end{bmatrix}_s ,
\tag{6.86}
$$

where $\sum_k{}'$ means summation with k running over I, II and III only.

After the photon emission event, the orbital eigenstates are rapidly damped by the radiation damping, while the spin components precess with initial values given by Eq. (6.86). Ignoring the transient behavior of the spin, the spin right after the photon emission reads $\vec{S} = \hat{n} + \bar{\alpha}\hat{m} + \bar{\beta}\hat{\ell}$. Associating with the definition of the spin chromaticity $\gamma \frac{\partial \hat{n}}{\partial \gamma}$, Eq. (6.61), we then find

$$
\gamma \frac{\partial \hat{n}}{\partial \gamma}(s) = -2 \sideset{}{'}\sum_k \left[\mathrm{Im}(E_{5k}^* E_{7k})\hat{m} + \mathrm{Im}(E_{5k}^* E_{8k})\hat{\ell} \right]_s .
\tag{6.87}
$$

Knowing the eigenvectors $E_k(s)$ of the 8×8 transformation matrices around the storage ring thus allows a calculation of P and τ according to Eqs. (6.73) and (6.74).

Other noise sources   In the above analysis, the noise source is assumed to come from the random energies carried away by the synchrotron radiation photons. The notation $\gamma \frac{\partial \hat{n}}{\partial \gamma}$ for spin chromaticity is suggestive of its physical origin. It specifies the shock excitation to the spin vector when a synchrotron radiation photon is emitted and the electron energy receives a sudden recoil. However, noises do not have to come from synchrotron radiation alone. When electron dynamics is undergoing other noise sources, the same physics and the same SLIM analysis applies. In such cases, we will continue to adopt this same notation $\gamma \frac{\partial \hat{n}}{\partial \gamma}$ even if the noise is not applied to the electron energy.

Treating other noise sources following a similar line should be straightforward. The orbital angular noise coming from sideways synchrotron radiation (Homework 6.27), noise in the RF cavity's accelerating field, noise in an orbital feedback system, or noise due to beam-beam, intrabeam, or beam-residual-gas collisions, for example, can be treated similarly.

Homework 6.26 In the text, we calculated all eight eigenvectors, but only the six orbital eigenvectors are used in Eq. (6.87). The spin eigenvectors are not used and are thrown away. An orbital eigenvector satisfies the condition,

$$
\left[\begin{array}{c|c} M_{6\times6} & 0 \\ \hline D & C \end{array} \right] \left[\begin{array}{c} e_k \\ \hline v_k \end{array} \right] = \lambda_k \left[\begin{array}{c} e_k \\ \hline v_k \end{array} \right] .
$$

Show that the two spin components of the orbital eigenvector have closed form connection to its six orbital components as

$$v_k = (\lambda_k I - C)^{-1} De_k.$$

Homework 6.27 After emitting a sideways synchrotron radiation photon, the angular component of the recoil (consider vertical quantum excitation only) is given by

$$
\begin{bmatrix} 0 \\ 0 \\ 0 \\ -\frac{\delta E}{E_0}\theta_y \\ 0 \\ 0 \\ 0 \\ 0 \end{bmatrix}
=
\sum_{k=\pm I,\pm II,\pm III} A_k E_k(s) +
\begin{bmatrix} 0 \\ 0 \\ 0 \\ 0 \\ 0 \\ 0 \\ \bar{\alpha} \\ \bar{\beta} \end{bmatrix}.
$$

Find the corresponding expression for $\gamma\frac{\partial\hat{n}}{\partial\gamma}$.

The definition of spin chromaticity depends on the physical mechanism of how the noise couples and perturbs the spin motion. In this case, the perturbation is not executed in terms of energy, so the notation $\gamma\frac{\partial\hat{n}}{\partial\gamma}$ becomes inappropriate. The role it plays is a Green's function response to a sudden imposition of a noise kick, and therein lies the physical meaning of the quantity $\gamma\frac{\partial\hat{n}}{\partial\gamma}$.

6.7.6 Application

Computer code A computer code was prepared for the polarization and beam distribution calculations based on the SLIM formalism.[36] Applications to orbital dimensions are discussed elsewhere. In the rest of this section, we discuss its application to the calculation of polarization in electron storage rings.

SLIM is not a simulation code. No Monte-Carlo random number generation of a beam distribution is implemented. All calculations are analytical. Input to this code is the complete beamline data file of a storage ring. When a flag ISPIN is set to 1, the code will proceed to execute the polarization calculation; otherwise it executes only the orbital calculations. Thin-lens approximation was used initially. Extensions in several ways of the algorithm, including the extension to nonlinear effects, have also been developed (see footnote 31).

SPEAR The code is applied to estimate the achievable polarization for the electron storage ring SPEAR. The beamline elements for the ideal SPEAR lattice include horizontal bending magnets, quadrupole magnets, sextupole magnets, RF cavities, and drift spaces. Without field imperfections, the ideal lattice produces an equilibrium polarization of 92%. To simulate field imperfections, we introduce a random distribution of vertical orbit kickers. Horizontal orbit

[36] A.W. Chao, Proc. Int. Symp. High-Energy Physics with Polarized Beams and Polarized Targets, Lausanne (1980).

Figure 6.14: Estimated level of transverse polarization P versus beam energy E_0 for the electron storage ring SPEAR with a typical lattice configuration and with a vertical closed-orbit $\Delta y_{\rm rms} = 1.2$ mm. Depolarization resonances are indicated by the arrows above the figure.

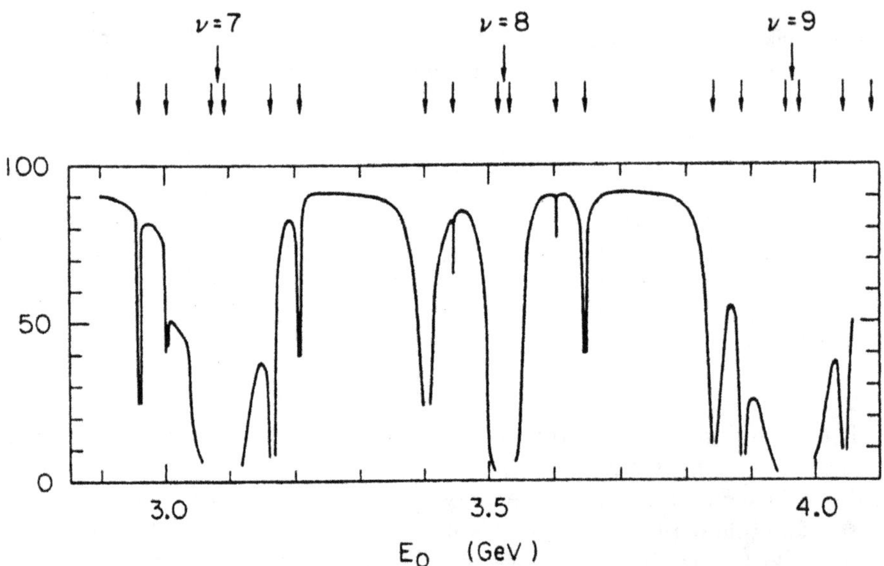

kickers should not produce too much depolarizing effects and are not included in the simulation. The resulting closed-orbit is then corrected by a set of orbit correctors. These orbit correctors contribute as additional vertical kickers in the lattice data set. The final vertical closed-orbit distortion makes sextupoles behave like skew quadrupoles and quadrupoles behave like additional vertical kickers.

In the presence of these field imperfections, the degree of polarization P is calculated and plotted in Fig. 6.14 as a function of the beam energy E_0. The SPEAR lattice used in Fig. 6.14 is specified by the lattice parameters $\nu_x = 5.28$, $\nu_y = 5.18$, $\nu_s = 0.022$, $\beta_x^* = 1.2$ m, $\beta_y^* = 0.10$ m and $D^* = 0$, where β_x^*, β_y^* and D^* are the usual unperturbed horizontal beta-function, vertical beta-function and the energy dispersion function at the interaction points. The strengths of the vertical kickers (both the error kickers and the corrector kickers) are normalized such that the rms closed-orbit distortion after orbit correction is $\Delta y_{\rm rms} = 1.2$ mm, which is typical for the SPEAR operation.

Expected locations of the depolarization resonances are indicated by arrows at the top of Fig. 6.14. Each integer resonance, $\nu =$ integer, is surrounded by six sideband resonances, $\nu \pm \nu_{x,y,s} =$ integer. Integer resonances are evenly spaced in beam energy at a spacing of 0.44065 GeV. Each integer resonance together with its two neighboring synchrotron sideband resonances overlap and

Figure 6.15: Comparison of calculation and measurements for SPEAR. Agreement near the two linear resonances $a\gamma - \nu_y = 3$ and $a\gamma - \nu_x = 3$ is reasonable. The third resonance $a\gamma - \nu_x + \nu_s = 3$ is nonlinear and is missed by the linearized SLIM. The measured width of the $\nu - \nu_y = 3$ resonance is wider than calculated, attributed to a spread of the betatron tune ν_y due to a known power supply ripple.

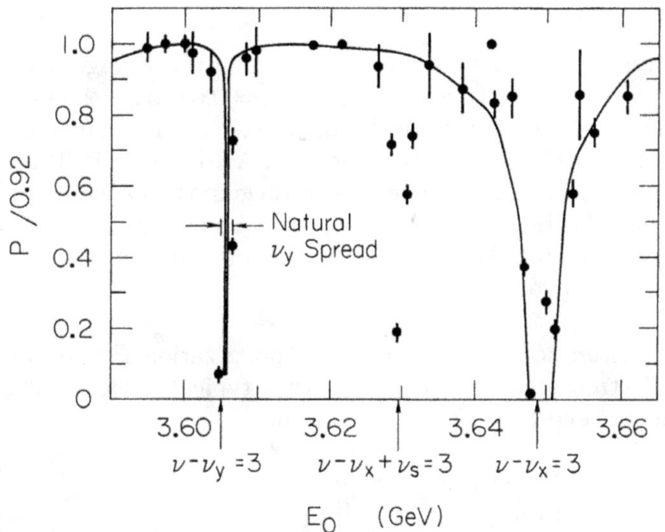

appear as single depolarization dip in Fig. 6.14. Otherwise, the width of the region covered by an integer resonance alone is typically much less than 10^{-3} in spin tune units — as mentioned before, depolarization is dominated by the sideband resonances. Here, ν should be taken as the perturbed spin tune given in Eq. (6.80). However, the shift of ν from its nominal value $\nu_0 = a\gamma$ is small, so in Fig. 6.14, the location of resonances indicated are given by $\nu_0 = k$ and $\nu_0 = k \pm \nu_{x,y,s}$.

For different distributions of vertical kickers whose strengths are normalized so that the orbit distortion after correction has $\Delta y_{\rm rms} = 1.2$ mm, the qualitative behavior of P versus E_0 does not change much from that shown in Fig. 6.14.

We have expanded the energy scale of Fig. 6.14 and plotted the result again in Fig. 6.15 in the region $E_0 \sim 3.6$ GeV. Superimposed are polarization measurements performed by the SPEAR polarization team.[37] The measurement involved back-scattering a polarized laser against a single positron beam to measure its transverse polarization in the expected $\hat{y}$-direction. The agreement between calculation and measurements is reasonable except that the calculation

[37]U. Camerini, et al., Phys. Rev. D12, 1855 (1975); SPEAR Polarization team, J. Johnson, R. Schwitters, et al., 3rd Int. Symp. High Energy Phys. with Polarized Beams and Targets, Argonne (1978).

has missed the nonlinear depolarization resonance located at $a\gamma - \nu_x + \nu_s = 3$. The SLIM program used here does not treat nonlinear resonances.

In actual operation, one must avoid the integer and the synchrotron sidebands by staying sufficiently away from integer multiples of 0.44065 GeV. As to the betatron sidebands, hopefully they can be avoided by properly choosing the betatron tunes. In addition, there will be some nonlinear depolarization resonances to avoid.

PEP   A similar example is shown in Fig. 6.16 for the storage ring PEP. Its lattice parameters are $\nu_x = 21.15$, $\nu_y = 18.75$, $\nu_s = 0.05$, $\beta_x^* = 2.8$ m, $\beta_y^* = 0.11$ m, $D^* = -0.49$ m. The rms vertical closed-orbit after correction is set to be $\Delta y_{\rm rms} = 0.6$ mm, half the value used for SPEAR because PEP has a more sophisticated orbit correction scheme. Nevertheless, the expected polarization is lower than that of SPEAR due to its higher beam energy. For a different distribution of errors yielding the same rms vertical orbit of 0.6 mm, the polarization pattern looks similar to Fig. 6.16.

Figure 6.16: Estimated level of transverse polarization P versus beam energy E_0 for the electron storage ring PEP with a typical lattice configuration and with a vertical closed-orbit $\Delta y_{\rm rms} = 0.6$ mm.

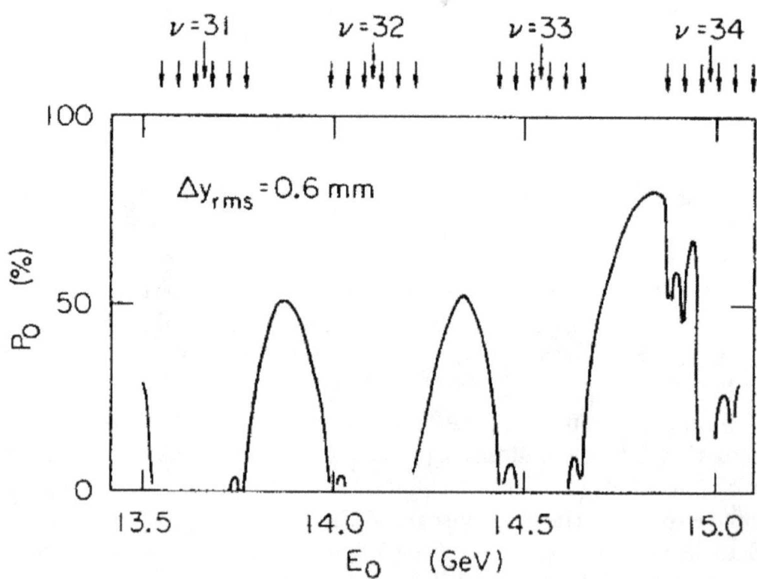

Rough scaling of reachable transverse polarization   An inspection of the polarization levels of SPEAR and PEP indicates their sensitive dependence on beam energy. Higher energy rings have stronger depolarization effects. This is expected because the disturbance to the spin direction for a given orbital angular disturbance is proportional to $a\gamma$. The higher the beam energy, the larger the precession angles. A similar behavior was emphasized for the proton polarization in Sec. 5.4.5.

Take the synchrotron sidebands for example. The enhancement effect of $a\gamma$ has been taken into account in the calculation of $\gamma\frac{\partial\hat{n}}{\partial\gamma}$, making its magnitude more or less proportional to the beam energy. The other energy dependencies such as the size of the synchrotron photons, the radiation damping rates, or the Sokolov–Ternov polarizing effects, are all embedded in the Derbenev–Kondratenko formula and included in the calculation.

In addition, we expect $\gamma\frac{\partial\hat{n}}{\partial\gamma}$ to scale linearly with $y_{\rm rms}$. If we now take the approximate formula (6.76), then very roughly, we expect the reachable level of transverse polarization to be approximately given by a pattern of resonances that is scaled by an overall scaling factor like

$$\hat{P}_0 \approx \frac{92\%}{1 + \zeta E_0^2 y_{\rm rms}^2},$$

with some fitting parameter ζ.

If we anchor the result so that $\hat{P}_0 = 60\%$ for PEP using Fig. 6.16, then

$$\hat{P}_0 \approx \frac{92\%}{1 + 0.5\left(\frac{E_0}{15\,{\rm GeV}}\right)^2\left(\frac{y_{\rm rms}}{0.6\,{\rm mm}}\right)^2}. \tag{6.88}$$

In case PEP is operated with $y_{\rm rms} = 1.2$ mm, the polarization drops to 30%.

For the collider LEP-I with $E_0 = 55$ GeV, and assume an rms vertical orbit distortion of 0.6 mm, it is expected that the vertical Sokolov–Ternov polarization to be around 12%.

This rough scaling (6.88) is meant to be applied when all strong resonances have been avoided and the storage ring is operated with closed-orbit well corrected. The ring is supposed to be nominally planar, and no strong solenoids or spin rotators are installed. To improve the polarization beyond this level, more dedicated operations will be required. In particular, one might apply harmonic correction schemes to be discussed in Sec. 6.8.

Another comment is to be made concerning the depolarization resonances at high energies. The growing synchrotron sidebands with beam energy can be cast in another way. The typical value of energy spread in an electron storage ring is $\sigma_\delta = 10^{-3}$. The energy spread is $\sigma_\delta E_0$. When this spread becomes large, a synchrotron sideband resonance can potentially cover a spin tune width of $\sigma_\delta a\gamma$. When the beam energy is high enough, this width becomes comparable to 1; all resonances begin to overlap and there is not room left for a good polarization. The linear portion of this effect has been included in the SLIM analysis and shows up automatically by the calculation. However, the higher order Bessel sidebands have been dropped. If we take a criterion like $\sigma_\delta a\gamma = \frac{1}{2}$ for the brick wall of losing all polarization, then the limit on highest beam energy is about 200 GeV. There is practically no polarization left also according to Eq. (6.88).

Nonlinear resonance One drawback of the linearized algorithm is that it does not treat the nonlinear resonances. Also mentioned, such nonlinear resonances have been observed at SPEAR as shown in Fig. 6.15. One effective way of studying the nonlinear effects is by numerical simulation. A simulation result for the storage ring PETRA is shown in Fig. 6.17(a) using a code SITROS.[38] It clearly shows a structure of nonlinear depolarization resonances. Figure 6.17(b) gives the linear SLIM calculation for the same lattice design. There is a general qualitative agreement between these results indicating the linear resonances play a dominant role.

Figure 6.17: Expected transverse beam polarization for the storage ring PETRA, (a) using a numerical simulation including nonlinear effects, and (b) using SLIM that includes only linear effects.

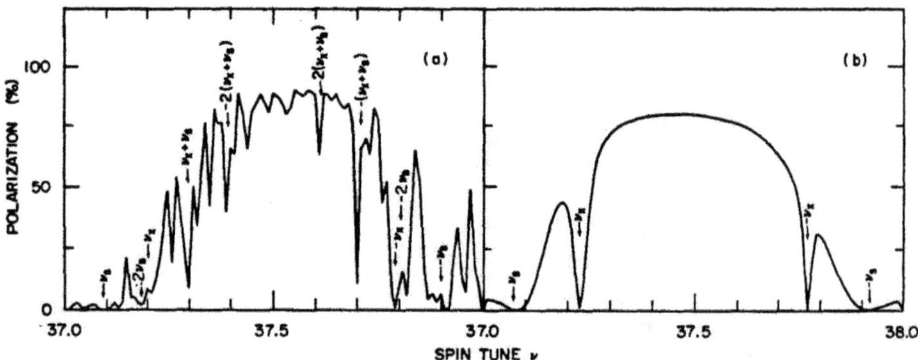

One advantage of computer simulation is its ability to include additional effects such as collective and beam-beam effects. Since, to a large extent, spin precession and orbital deflection are simply connected by a factor of $a\gamma$, it can be expected that when a collective effect affects the orbital motion of the electrons so that the beam emittance grows, it will also affect the polarization significantly. Similarly, if beam-beam interaction causes the beam size to grow, the beam polarization is expected to deteriorate as well as the luminosity.[39]

Spin rotator In the applications mentioned so far, depolarization resonances are all excited by field errors. Sometimes, however, there are situations when special-purpose beamline elements are installed that strongly affect the beam polarization even in the absence of errors. Particularly of importance are those that disrupt the planar geometry and/or the ribbon beam configuration of the storage ring. In those circumstances, the damage to polarization often is so strong that it dominates over the effect of the error fields. One example is the

---

[38] J. Kewish, DESY Lab. Report 83-032 (1983).

[39] A.M. Kondratenko, Sov. Phys. JETP 39, 592 (1974); J. Buon, Orsay Lab. Report LAL/RT/83-04 (1983).

Figure 6.18: A possible longitudinal spin rotator scheme. Shaded regions indicate vertical bending magnets. Each of the six bending magnets bends the spin direction by $\pm\frac{\pi}{2}$ around the $\hat{x}$ axis. The arrows indicate spin directions, blue for positron beam, red for electron beam. The positron beam moves to the right, the electron beam to the left. Collision point is in the middle.

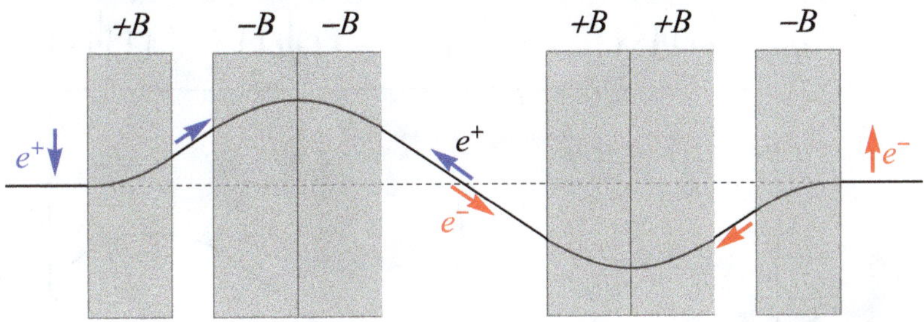

detector solenoids of a high-energy collider located around the collision point to analyze the collision secondary particles. Another example is the spin rotator insertions around the collision point to rotate the nominally vertical polarization into longitudinal for the most interested high-energy collision events. Both of these devices can strongly disrupt the beam polarization due to synchrotron radiation.

Generally any device that causes vertical motion of the beam depolarizes. A device like a spin rotator can be considered brutal in this sense. One possible spin rotator setup, perhaps the simplest one to conceive, consists of a set of vertical bending magnets (horizontal magnetic fields) as illustrated in Fig. 6.18. The nominal vertically polarized beams are spin rotated so that their polarizations become longitudinal at the collision point, and then restored to the nominal vertical direction before reentering the ring. The polarizations at the collision point of the two beams have opposite helicities.

It may seem this would work as far as spin rotation is concerned. But Fig. 6.19 shows the calculated level of polarization when a spin rotator of Fig. 6.18 is installed in an otherwise perfect PEP ring. The polarization is compromised primarily due to the $a\gamma \pm \nu_y = 6k$, where 6 is the superperiod of the PEP lattice design. When errors are included, we need to combine Figs. 6.16 and 6.19.

In this spin rotator scheme, the vertical bending magnets are strong, inducing strong synchrotron radiation at these magnets. Furthermore, they also generate large internal vertical dispersion, which means the synchrotron radiation excites directly vertical betatron oscillations, which in turn couples strongly to spin motion.

One might consider more sophisticated spin rotator schemes with other combinations of horizontal and vertical bending magnets and solenoids (see Homework 6.28 for one example). Siberian snakes or multiple-snake schemes, which have been successful devices for proton beams (no synchrotron radiation)

Figure 6.19: Expected level of polarization when a set of vertical bending magnets are used to produce a longitudinal polarization at one of the collision points in PEP.

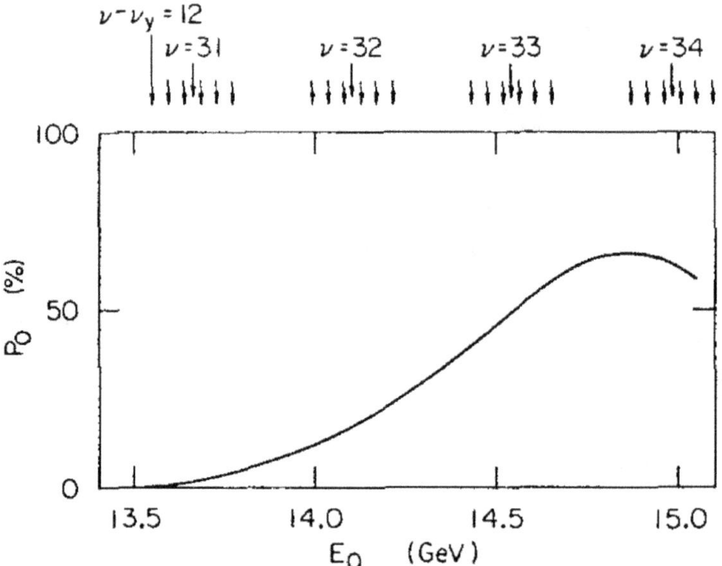

have also been considered. The polarization in all cases stay small. To overcome this difficulty, to design spin rotators that do not strongly depolarize, one idea using spin matching is to be discussed in the next section 6.8.

Homework 6.28 Show that the following arrangement can serve as a longitudinal spin rotator,[40]

$$\text{arc} \quad \leftarrow \quad (45°, -\hat{x}) \ (45°, -\hat{y}) \ (90°, \hat{x}) \ (90°, \hat{y}) \ (45°, -\hat{x}) \ (45°, -\hat{y}) \quad \rightarrow \quad \text{IP}$$

where a bending magnet is represented by $(\phi, \hat{\phi})$ with ϕ the spin rotation angle and $\hat{\phi}$ the rotation axis, and IP means the interaction point. To reverse the spin rotation, on the other side of the IP, the arrangements of the horizontal bending magnets are symmetric and the vertical bending magnets are antisymmetric.

Solution The spin is rotated by $\frac{4\pi}{3}$ around the axis $\frac{1}{\sqrt{3}}(\hat{x} + \hat{y} - \hat{z})$ from the arc to the IP. The total rotation across the collision region from arc to arc is by angle π around $\hat{y}$. As mentioned in the text, however, this rotator works for protons but does not work for electrons.

---

[40] J. Buon and K. Steffen, Nucl. Instr. Meth. Phys. Res. A245, 248 (1986).

6.8 Spin transparency

In the previous discussions, we studied the beam polarization when errors are added, or when additional insertions such as spin rotators or detector solenoids are installed. In either cases, we started with a nominal planar storage ring design and simply accepted the resulting polarization as the perturbations are added to the nominal design. In many cases, particularly for higher beam energies, we find serious limitations to the achieved level of polarization. The case of the spin rotator, needed to provide longitudinal beam polarization at the collision point of a collider storage ring, is of particular concern because all the rotator schemes we tried so far were found to depolarize the beam severely. We need to find the root cause of what is driving the depolarization — returning to first principles — and find a lattice design that is free of these driving terms. In this section, we will analyze this issue and introduce the concept of spin transparency.[41] A set of conditions are then introduced to make the rotator spin transparent. A spin rotator example will be designed obeying these spin transparency conditions as an example that potentially provides a high level of longitudinal polarization.

Consider an electron-positron storage ring collider. Our job is to insert a beamline section on each side of the collision point, whose net effect is to rotate the spin from vertical to longitudinal and restore it to vertical, as illustrated in Fig. 6.20. We shall call these two insertions the snake insertions. One special case of such a snake insertion was shown in Fig. 6.18 but we need to design the insertions that do not depolarize. In this section, we do not consider error fields. All depolarization effects come from the snake insertions.

Let us assume the storage ring, including all the snake insertion sections, consists of only horizontal bending magnets, vertical bending magnets and quadrupoles. In particular, there are no solenoids, skew quadrupoles or error fields. We assume no vertical bending magnets in the arc region and no quadrupoles in the snake insertions. We also assume that the interaction region (IR) between the snake insertions is a straight section with only quadrupoles.

As mentioned before, depolarization effects are contained in the two spin vectors $\hat{n}$ and $\gamma\frac{\partial\hat{n}}{\partial\gamma}$. Integer resonances are driven when $\hat{n}$ deviates from $\hat{y}$. Sideband resonances are driven by $\gamma\frac{\partial\hat{n}}{\partial\gamma}$. The integer resonances are narrow and can be avoided easily. They are not the main mechanism for loss of polarization. The spin transparency conditions we develop below address only the sideband resonances, excited by synchrotron radiation coupled to spin motion through the spin chromaticity $\gamma\frac{\partial\hat{n}}{\partial\gamma}$.

As mentioned, we will need to return to first principles. Synchrotron radiation occurs in the horizontal and vertical bending magnets. To minimize the depolarization effects, one must decouple the spin motion from the orbital

---

[41]A.W. Chao and K. Yokoya, KEK Laboratory Report 81-7 (1981); H. Mais and G. Ripkin, DESY report M-84-04 (1984); D.P. Barber, et al., Phys. Lett. B343, 436 (1995); D.P. Barber and G. Ripkin, Sec. 2.6.8, Handbook Accel. Phys. & Eng., 2nd ed., World Scientific (2013).

Figure 6.20: Spin rotator snake insertions considered to illustrate the spin transparency conditions. The red arrows indicated the directions of polarization.

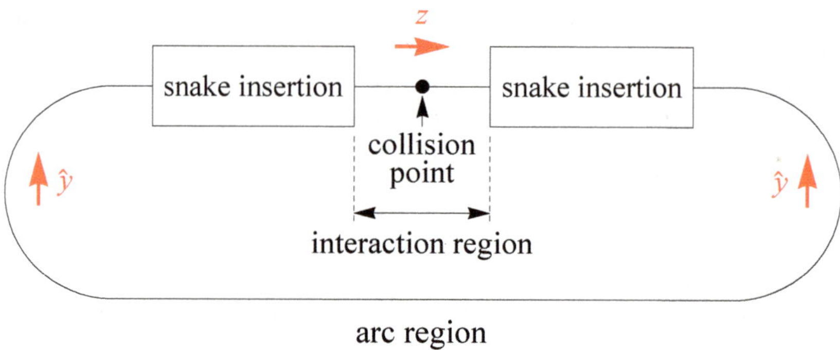

motions (horizontal betatron x_β, vertical betatron y_β and longitudinal synchrotron z) that are excited by photon emissions.

Emission in arc   First, let us consider a photon emission occurring in one of the normal bending magnets in the storage ring arc. After the photon emission, the electron starts to execute x_β- and z-oscillations (but no y_β-oscillation). We need to study the potential depolarization due to these oscillations. As this oscillating electron travels through the arc, the magnetic field it sees is always in the $\hat{y}$-direction, causing its spin to precess around $\hat{y}$. Since the polarization in the arc is also in the $\hat{y}$-direction, this does not cause any depolarization. As the electron traverses the snakes, the snake bending magnetic fields do not depolarize either because, without quadrupoles, they do not see any perturbation fields originating from the synchrotron-radiation induced orbital motions.

However, as the electron travels through the interaction straight section, it sees a magnetic field in the quadrupoles,

$$\vec{B}(s) \;=\; G(s)\big[x(s)\hat{y} + y(s)\hat{x}\big]\,,$$

where

$$\begin{aligned}
x(s) &= a\sqrt{\beta_x(s)}\sin\big[\psi_x(s) - \psi_{x0}\big] + bD_x(s)\,,\\
y(s) &= bD_y(s)\,,
\end{aligned} \tag{6.89}$$

with a and b some amplitudes proportional to the emitted photon energy, $\beta_{x,y}(s)$ the beta-functions, $D_{x,y}$ the dispersion functions, $\psi_{x,y}(s)$ the betatron phases. Note that in the IR, $D_y(s) \neq 0$ in general because we allow the snake insertions to contain vertical bends, and they are able to generate vertical dispersion in the IR.

Since the polarization in the interaction region is along $\hat{z}$, both the horizontal and vertical magnetic fields can cause depolarization unless we decouple the spin motion from the orbital motions after integrating through the passage. The decoupling condition is

$$\int_{\text{IR}} B_x \, ds \; = \; \int_{\text{IR}} B_y \, ds \; = \; 0 \,.$$

More specifically, the requirements read

$$\int_{\text{IR}} ds \, G(s) D_x(s) \; = \; 0 \,, \tag{6.90}$$

$$\int_{\text{IR}} ds \, G(s) D_y(s) \; = \; 0 \,, \tag{6.91}$$

$$\int_{\text{IR}} ds \, G(s) \sqrt{\beta_x(s)} \sin \psi_x(s) \; = \; 0 \,, \tag{6.92}$$

$$\int_{\text{IR}} ds \, G(s) \sqrt{\beta_x(s)} \cos \psi_x(s) \; = \; 0 \,. \tag{6.93}$$

Under these conditions, the IR looks transparent for spin motion of an electron executing motion (6.89).

Emission in snake We now consider a photon emission occurring in one of the snake bending magnets. The complication here is that the vertical dispersion in the snake magnet is in general nonzero; a photon emission therefore also excites y_β-oscillation of the electron. Consequently it is necessary to require two more conditions on the IR quadrupole magnets in order to decouple the spin motion,

$$\int_{\text{IR}} ds \, G(s) \sqrt{\beta_y(s)} \sin \psi_y(s) \; = \; 0 \,, \tag{6.94}$$

$$\int_{\text{IR}} ds \, G(s) \sqrt{\beta_y(s)} \cos \psi_y(s) \; = \; 0 \,. \tag{6.95}$$

In addition, the y_β-oscillation also introduces a depolarizing magnetic field in the arc region (recalling that the y-component of the magnetic field does not depolarize in the arcs),

$$B_x(s) \; = \; G(s) y(s) \,,$$

where

$$y(s) \; = \; a \sqrt{\beta_y(s)} \sin \left[\psi_y(s) - \psi_{y0} \right] \,, \tag{6.96}$$

and we have used the fact that $D_y(s) = 0$ in the arcs.

Let $\hat{\ell}(s)$ and $\hat{m}(s)$ be two unit vectors defined so that $(\hat{n}, \hat{m}, \hat{\ell})$ form a right-handed orthogonal vector system at position s and all three of them precess around the ring according to the bending magnetic fields. The condition that the arc looks transparent to the spin motion of an electron executing the motion (6.96) is

$$\int_{\text{arc}} ds \, \hat{\ell}(s) \cdot \hat{x}(s) \, B_x(s) \; = \; \int_{\text{arc}} ds \, \hat{m}(s) \cdot \hat{x}(s) \, B_x(s) \; = \; 0 \,,$$

which requires four conditions on the arc quadrupole magnets,

$$\int_{\text{arc}} ds\, \hat{\ell}(s) \cdot \hat{x}(s)\, G(s) \sqrt{\beta_y(s)}\, \sin\psi_y(s) \;=\; 0\,, \tag{6.97}$$

$$\int_{\text{arc}} ds\, \hat{\ell}(s) \cdot \hat{x}(s)\, G(s) \sqrt{\beta_y(s)}\, \cos\psi_y(s) \;=\; 0\,, \tag{6.98}$$

$$\int_{\text{arc}} ds\, \hat{m}(s) \cdot \hat{x}(s)\, G(s) \sqrt{\beta_y(s)}\, \sin\psi_y(s) \;=\; 0\,, \tag{6.99}$$

$$\int_{\text{arc}} ds\, \hat{m}(s) \cdot \hat{x}(s)\, G(s) \sqrt{\beta_y(s)}\, \cos\psi_y(s) \;=\; 0\,. \tag{6.100}$$

Transparency condition   We have thus obtained six IR conditions, Eqs. (6.90) to (6.95), and four arc conditions, Eqs. (6.97) to (6.100), that eliminates all the linear depolarization sideband resonances due to synchrotron radiation coupling to spin motion through $\gamma\frac{\partial\hat{n}}{\partial\gamma}$ for a spin rotator design of Fig. 6.20.

A few comments are in order.

- Conditions (6.90), (6.91), (6.92), (6.94) are trivially satisfied if the snake bending magnets and D_x, D_y, ψ_x, ψ_y are antisymmetric, and quadrupole magnets and β_x, β_y are symmetric about the collision point.

- With these ten conditions, all sideband resonances are no longer driven. In particular, conditions (6.90) and (6.91) eliminate the synchrotron sidebands; (6.92) and (6.93) eliminate the x_β sidebands; the remaining six conditions eliminate the y_β sidebands. Under these conditions, the spin chromaticity $\gamma\frac{\partial\hat{n}}{\partial\gamma}$ vanishes everywhere around the ring.

- In case it is already known that some of the sidebands are not important, then the number of transparency conditions to fulfill can be reduced accordingly, thus relieving partly the burden on the lattice design.

- In the arc conditions (6.97–6.100), $\hat{m}$ and $\hat{\ell}$ depend on the beam energy. This means the required quadrupole configuration in the arc changes with the beam energy. If for some reason, the arc quadrupoles cannot be varied with beam energy, conditions (6.97–6.100) should at least be satisfied at the energy for the nearest y_β sideband resonance.

- After the spin transparency conditions are fulfilled, there remains a residual depolarization due to $\hat{n}$ not being aligned to $\hat{y}$. This is not expected to cause a major problem as long as we avoid to be too close to the integer depolarization resonances.

- The transparency conditions are necessary and sufficient to eliminate the linear sideband depolarization resonances. Also, note that although they address spin transparency for a single passage, once fulfilled, spin transparency is assured for multiple revolutions of a stored beam.

Figure 6.21: Expected level of longitudinal polarization when spin rotators are included for four collision points around the TRISTAN ring. Ten spin transparency conditions are implemented in the lattice design at two specific energies: 20.235 GeV (solid curve) and 20.05 GeV (dashed curve). If the ten conditions are met at all energies, the polarization approaches 92% at all energies except for a very narrow region near the integer depolarization resonance at 20.27 GeV.

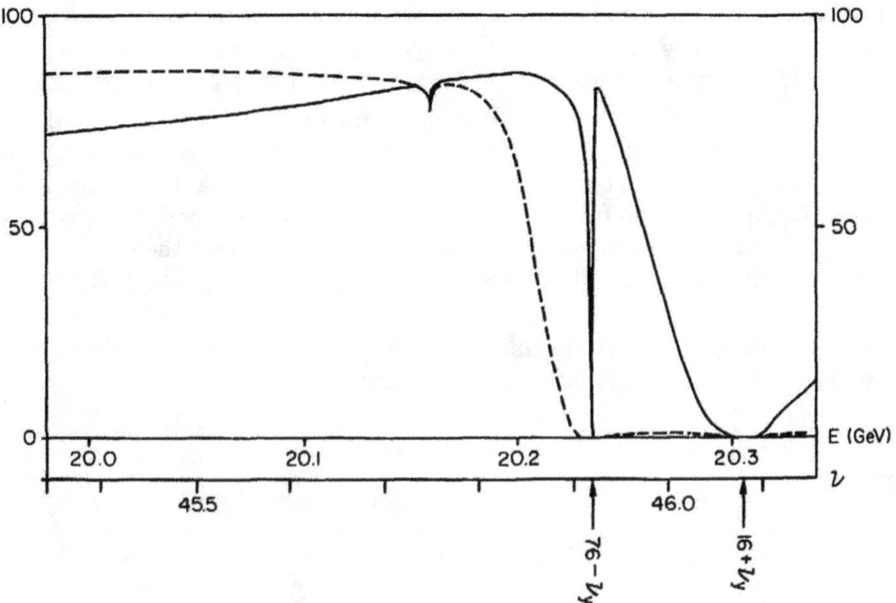

A design example was made for the TRISTAN collider. A spin rotator similar to that of Fig. 6.18 was first added to the collider lattice. Without implementing the spin transparency conditions, the achieved polarization was low. In this design, spin rotators were added to four collision points around the collider ring. SLIM code was used to calculate the polarization level. After matching the collider lattice fulfilling the ten conditions, the polarization is as shown in Fig. 6.21.

The dashed curve is when the transparency conditions are satisfied at a fixed energy of 20.05 GeV with $a\gamma = 45.50$. Solid curve is when the conditions are imposed at 20.235 GeV with $a\gamma$ at a y_β sideband resonance. Differences of these two curves at 20.05 GeV and near the neighboring resonances indicate the contribution of the deviations from the conditions (6.97) to (6.100). In particular, the width of the y_β sideband resonance is quite narrow in the case of the solid curve. Imposing the conditions (6.97) to (6.100) at each energy, if one wishes, can yield high level of polarization throughout the energy range.

General condition So far we obtained the transparency conditions based on a specific design configuration as sketched in Fig. 6.20. By extending the analysis, one can obtain a general expression for the spin chromaticity $\gamma\frac{\partial\hat{n}}{\partial\gamma}(s)$ as a function of s at least for the case when the storage ring contains only quadrupoles, and horizontal and vertical bending dipoles — in particular, there is no coupling elements. To make a dipole at position s spin transparent, we need to make $\gamma\frac{\partial\hat{n}}{\partial\gamma}(s) = \vec{0}$. In case all dipoles are spin transparent, then the whole ring is spin transparent, as we did for the special case of Fig. 6.20.

Consider a dipole at position s. A synchrotron radiation photon is emitted there with a spin chromaticity $\gamma\frac{\partial\hat{n}}{\partial\gamma}(s)$. Table 6.6 shows a decomposition of $\gamma\frac{\partial\hat{n}}{\partial\gamma}(s)$ into appropriate harmonic driving terms of each of the sideband resonances. Transparency of this dipole is achieved if all the integrals in Table 6.6 vanish. Since the driving terms are expressed as complex quantities, there are a total of ten conditions to fulfill. However, Table 6.6 offers a possibility of selectively eliminating the strongest sidebands. As mentioned, for the entire ring to be spin transparent, we need all dipoles to satisfy the transparency conditions.

Table 6.6: Depolarization sideband resonance and their corresponding driving terms for a synchrotron radiation photon emitted at location s.

Resonance	Driving term in $\gamma\frac{\partial\hat{n}}{\partial\gamma}(s)$	Harmonic correction knobs
$a\gamma+\nu_x = k$	$D_x(s)\times\int_s^{s+C} ds(\hat{m}+i\hat{\ell})\cdot\hat{y}\,G\sqrt{\beta_x}e^{i\psi_x}$	y_{COD} at $D_x\neq 0$
$a\gamma-\nu_x = k$	$D_x(s)\times\int_s^{s+C} ds(\hat{m}+i\hat{\ell})\cdot\hat{y}\,G\sqrt{\beta_x}e^{-i\psi_x}$	y_{COD} at $D_x\neq 0$
$a\gamma+\nu_y = k$	$D_y(s)\times\int_s^{s+C} ds(\hat{m}+i\hat{\ell})\cdot\hat{x}\,G\sqrt{\beta_y}e^{i\psi_y}$	x_{COD} at $D_y\neq 0$
$a\gamma-\nu_y = k$	$D_y(s)\times\int_s^{s+C} ds(\hat{m}+i\hat{\ell})\cdot\hat{x}\,G\sqrt{\beta_y}e^{-i\psi_y}$	x_{COD} at $D_y\neq 0$
$a\gamma\pm\nu_s = k$	$\int_s^{s+C} ds(\hat{m}+i\hat{\ell})\cdot G(D_x\hat{y}+D_y\hat{x})$	x_{COD} and y_{COD}

Each sideband in Table 6.6 contains an integer k. Note that all sidebands with different k's — not only the one nearest to the spin tune — are eliminated altogether by the corresponding transparency condition.

For an ideal planar ring, all driving terms vanish — in Table 6.6, the first two rows vanish because $\hat{m}$ and $\hat{\ell}$ are perpendicular to $\hat{y}$; the second two rows vanish because $D_y = 0$; the last row vanishes for the same reasons.

Table 6.6 says there are 10 conditions to fulfill for each dipole in the storage ring. Strictly speaking, a perfect spin transparency requires fulfilling $10\,N_B$ conditions, where N_B is the total number of dipoles. Imposing $10\,N_B$ matching conditions on the storage ring lattice is clearly impractical. On the other hand, for some special cases, such as the case of Fig. 6.20, these $10N_B$ conditions degenerate into a total of ten conditions. Furthermore, if approximate transparency is acceptable, there is an alternative to consider harmonic matching to be discussed below.

Harmonic matching   Consider for example the driving term for the $a\gamma + \nu_x = k$ resonance in Table 6.6. Strictly speaking, the driving terms depends on s, the location where synchrotron photon is emitted. However, one can observe that this driving term becomes independent of s if $a\gamma + \nu_x$ is exactly equal to a particular k, i.e. if we are exactly on a particular sideband resonance. If we consider a strict spin transparency only when the beam energy is exactly on a particular sideband, then there are only two conditions, instead of $2N_B$ conditions, to fulfill because once the transparency is fulfilled at one position, it will be fulfilled at all other locations around the ring.

When $a\gamma$ moves away from the exact value of the particular resonance, spin chromaticity no longer vanishes, but hopefully still stays small. This is obviously only an approximate spin transparency in general and may not be satisfactory for a strong spin rotator, but it might be acceptable for the weaker error fields. Similarly the same observation can be made on all the other sideband resonances.

This welcomed observation offers a way to make harmonic correction of depolarization effects for error field effects as follows.[42] We first understand that for each sideband resonance, there is a driving term represented as a complex quantity. This is the case even when the otherwise ideal planar ring is perturbed by error fields and even when the errors contains coupling elements. Exactly on the resonance, this quantity is independent of the observation point around the ring aside from a phase factor of unit magnitude. If we now introduce two sets of corrector knobs that make an additional contribution to the driving term, then we can vary the two knobs empirically to cancel the original driving term to zero. The sideband resonance is then eliminated, at least exactly at the resonance. By systematically eliminating all the stronger sideband resonances, an approximately spin transparent ring can be generated empirically.

In Table 6.6, we give a third column indicating the possible knobs of harmonic correction for each of the sideband resonances. Take a particular $a\gamma + \nu_x$ resonance for example. Each of the two knobs to perform harmonic correction against error fields can be a vertical closed orbit y_{COD} in a region where the horizontal dispersion is nonzero. When a vertical closed orbit is added, it generates components of $\hat{m}$ and $\hat{\ell}$ in the $\hat{y}$ direction. With $D_x \neq 0$, this knob contributes a driving term to the $a\gamma + \nu_x$ resonance. With two such knobs, we can compensate for the original driving term due to field errors. The other correction knobs in Table 6.6 can be obtained similarly.

Figure 6.22 shows an envisioned effort to improve the beam polarization for the PETRA storage ring. The level of expected transverse polarization, calculated using SLIM, is shown as the solid curve when error fields and orbital

---

[42]R. Rossmanith and R. Schmidt, Nucl. Instr. Meth. Phys. Res. A236, 231 (1985); A. Assmann, et al., Proc. 4th Euro. Part. Accel. Conf., London (1984); D.P. Barber and G. Ripkin, Sec. 2.6.8, Handbook Accel. Phys. & Eng., 2nd ed., World Scientific (2013); A.W. Chao, AIP Proc. No. 95, Brookhaven, p. 458 (1982).

Figure 6.22: Expected level of transverse polarization for PETRA. The solid curve is the case when the planar ring is perturbed by error fields. The dashed curve is after some effort on harmonic correction is made. [Courtesy Rudiger Schmidt (2021).]

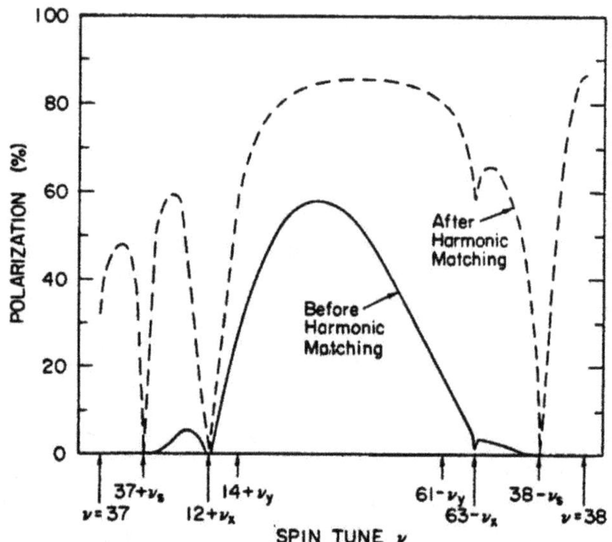

distortions have been included. When harmonic corrections are imposed, the polarization improves to become the dashed curve.[43] The required changes on the closed orbit is hardly noticeable.

---

[43]R. Schmidt, Workshop Polarized Electron Accel. & Storage, DESY Report M-82/09 (1982); D. Barber and H. Mais, DESY Report M-82/09 (1982).

Chapter 7

Echo

7.1 Echoes are everywhere

Echoes are a fascinating, curious, and common phenomenon. It occurs in many areas of physics,

- Acoustic echo, probably used to measure sound speed [ancient];
- Spin echo, in nuclear magnetic resonance [1950];[1]
- Echo in plasma [1969];[2]
- Echo in accelerators [1992].[3]

Echoes are prevalent in physics. In spite of their prevalence, however, each time when it emerged, it seemed to generate surprise and excitement. In retrospect, one concludes that echoes cause excitements because they are unexpected, and they are unexpected because they are hidden from our observation even when their presence is in fact rather pronounced in some hidden space. When they eventually emerge to be observed, they become surprises.

The echo effect is well known in plasma physics (see Fig. 7.1), and was first introduced to accelerator physics by Stupakov (1992).

In accelerator physics, echoes occur in multiple circumstances. Particle motion in an accelerator has four dimensions, and there are echo effects in each of these four dimensions, each with its own special characteristics,

- Transverse: horizontal x and vertical y;
- Longitudinal z;
- Spin: S.[4]

---

[1] E.L. Hahn, Phys. Rev. 80, 580 (1950).

[2] A.Y. Wong, D.R. Baker, Phys. Rev. 188, 326 (1969).

[3] G. Stupakov, SSC Laboratory Report 579 (1992).

[4] A.W. Chao and E.D. Courant, Phys. Rev. Special Topics – Accel. & Beams, 10, 014001 (2007).

Figure 7.1: Schematic of a plasma echo experiment.

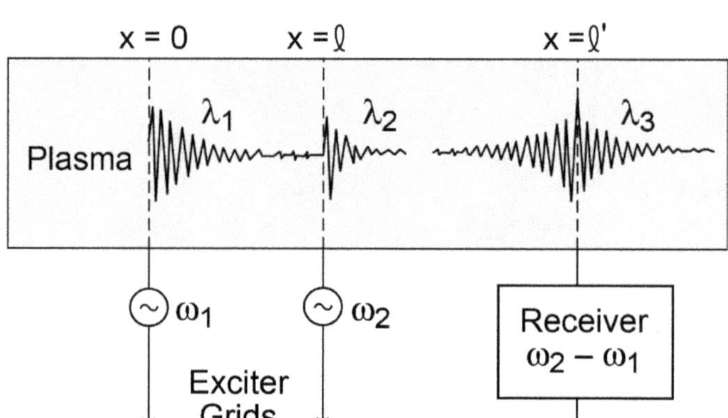

Echoes are everywhere in an accelerator. This is a rich subject in accelerator physics and technology, and is still evolving. After learning the subject, one might wonder why the storage ring vacuum chamber is not filled with many mysterious signals generated by the stored beam, each representing echoes of some beam manipulation of a long past.

A thought experiment   Consider a beam that has been stored for a long time in a storage ring and is circulating turn after turn in a quiet steady state. At $t = 0$ we kick this beam by a dipole kicker. The kicker is then quickly switched off. The beam's centroid then subsequently executes a free betatron oscillation, as detected by a good beam position monitor. Due to a spread in the betatron frequencies of the particles in the beam, the centroid signal detected by the position monitor decoheres in a relatively short time, say in 1 ms. The beam becomes quiet again after 1 ms according to our observation.

Long after the kick, say 1 s later, the centroid signal reads of course almost dead zero. At this point, we give the beam a second, this time quadrupole, kick. Such a kick does not affect the beam centroid, and thus the beam centroid signal stays zero. The beam stays motionless according to the beam position monitor.

The curious thing is that if we wait another 1 s — exactly 1 s — after the quadrupole kick, the beam centroid detector would receive a sudden and pronounced blip. The blip lasts approximately 1 ms long, and after that the beam centroid signal quiets down again. This sudden blip is called an echo, and is a result of the correlation and interplay between the two kicks and a long memory of the intricate beam dynamics in the phase space. A schematic of this gedanken experiment is sketched in Fig. 7.5 later, and will be discussed in detail then.

In passing, here we are seeing another manifestation of the relevance of the otherwise abstract concept of phase space. An appreciation of the echo effect must originate with an introduction of the phase space. In the event of an echo, even when there is no signal of anything going on in the real 3-D space, intricate information in particle motion is nevertheless registered, memorized, and evolved in the phase space. The curiosity of echoes is then further enhanced by the exceedingly long time this concealed memory lasts. Had we been creatures living in the 6-D phase space instead of the 3-D real space, echoes would not have been such a curiosity after all.

Another ingredient important to the echo effects is the Liouville theorem. The conservation of phase space area element is obeyed down to the finest details according to the theorem, thus allowing the echoes to occur in the first place.

Other echo effects in accelerator The above thought experiment addresses an echo in the beam's transverse motion. Echoes can also be observed in the longitudinal dimension. In case the stored beam is bunched, an RF phase shift and an RF amplitude jump play the roles of the dipole and quadrupole kicks, respectively. Experimentally, on the other hand, longitudinal echo has been first observed in the antiproton accumulator ring at Fermilab[5] and in the CERN SPS[6] for unbunched beams. Those experiments demonstrate that echoes can be effectively used for detecting extremely weak diffusion effects occurring inside the beam.

In the discussion of beam dynamics in accelerators, as pointed out in Chapters 5 and 6, in addition to the 3-dimensional transverse and longitudinal motions, particle motion also has a 4th dimension involving its spin. Indeed, there is also a spin echo effect. Furthermore, echo effect has also been explored[7] recently as a clever way to generate microbunching in beam distribution for the purpose of applying to free electron lasers.

As just mentioned, diffusion effects can significantly affect the echo signal. In the presence of diffusion, the precise details developed in an echo mechanism can be compromised when particles in the beam blurs its memory due to diffusion. As a result, the echo signal will be suppressed. We will include some discussions of the effect of diffusion in this chapter.

Quantum mechanics, together with its uncertainty principle, also dilutes the otherwise sharp phase space memory, and therefore affect the echo, but with a much longer time scale.[8]

Echoes are also influenced by collective effects. Unlike diffusion, a collective instability effect alone in principle does not cause phase space to smear; but it does mix up phase space elements by changing the beam dynamics. As a result, the echo signal will be affected in a complicated manner. Measurements of echoes should yield useful information about the wakefield and impedance

---

[5]L.K. Spentzouris, J.-F. Ostigy, P.L. Colestock, Phys. Rev. Lett. 76, 620 (1996).

[6]O. Brüning, et al., CERN SL-MD Note 217 (1996); O. Brüning, et al., Euro. Part. Accel. Conf. 1996.

[7]G. Stupakov, Phys. Rev. Lett. 102, 074801 (2009).

[8]A.W. Chao and B. Nash, 18th Advanced Beam Dynamics Workshop on Quantum Aspects of Beam Physics, p. 90 (2002).

of an accelerator. We do not pursue these collective effects in our discussions
however.

In general, due to its prevalent nature in the accelerator environment and
due to its extreme sensitivity to phase space disturbances, echoes are potentially
a sensitive diagnostic tool for accelerator applications. The challenge is to find
ways to use them.

7.2　Transverse echo

7.2.1　Transverse decoherence

Consider the setup for transverse echo mentioned earlier as a thought experiment. Let us first concentrate on the period from $t = 0$ when the beam receives
a sudden dipole kick but before it receives the second quadrupole kick. The
physics involved during this time period is *transverse decoherence*. Echo will be
discussed in the following section.

We describe the transverse dynamics as sketched in Fig. 7.2 using the action-
angle variables J and ϕ, with

$$x = \sqrt{2J\beta}\cos\phi, \qquad p = \beta x' + \alpha x = -\sqrt{2J\beta}\sin\phi,$$
$$J = \frac{x^2 + p^2}{2\beta}, \qquad \tan\phi = -\frac{p}{x},$$

where β, α are the usual Courant–Snyder functions.

Figure 7.2: Transverse phase space parameters to describe the decoherence effect.

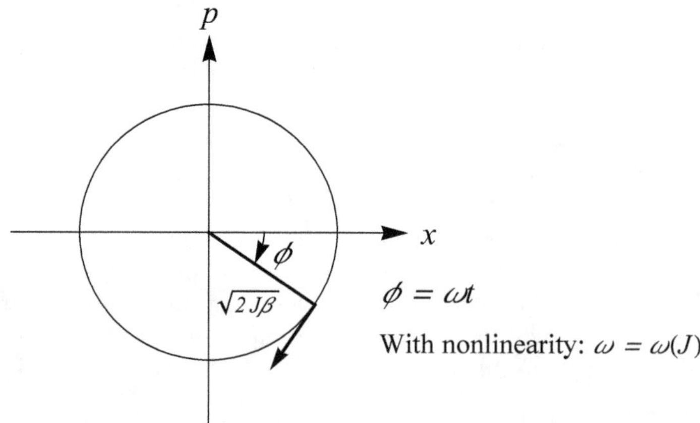

Before the transformation, x is the coordinate and p is the momentum.
After the transformation, ϕ is the coordinate and J is the momentum. This

transformation from (x, p) to (ϕ, J) has a Jacobian that is equal to β, not equal to 1. Strictly speaking, the transformation is not exactly canonical. But this will not affect our following analysis.

The beam receives a dipole kick at time $t = 0$. Let the kick be $\Delta x' = \theta$, which means $\Delta p = \beta\theta$. The kick is represented by the map

$$x_{\text{after}} = x_{\text{before}}, \qquad p_{\text{after}} = p_{\text{before}} + \beta\theta, \tag{7.1}$$

where "before" and "after" refer to $t = 0^-$ and $t = 0^+$, respectively.

The kick (7.1) is applied to all particles in the beam with the same kicking angle θ. As a result, the beam distribution immediately after the kick, $\psi_1(x, p)$, is related to the distribution immediately before the kick, $\psi_0(x, p)$, by

$$\psi_1(x, p) = \psi_0(x, p - \beta\theta).$$

Note it is $p - \beta\theta$, not $p + \beta\theta$, in the argument of ψ_0. Similar comment applies to Eq. (7.2) later.

Before the kick at $t = 0$, the beam's transverse distribution is in a steady-state. This means $\psi_0(x, p) = \psi_0(J)$ depends only on J and is independent of ϕ or t. We normalize the beam distribution by

$$\int_0^\infty dJ \int_0^{2\pi} d\phi \, \psi_0(J) = 2\pi \int_0^\infty dJ \, \psi_0(J) = 1.$$

It follows that

$$\int_{-\infty}^\infty dx \int_{-\infty}^\infty dp \, \psi_0 = \beta.$$

Immediately after the kick, the beam distribution is given by

$$\psi_1(x, p) = \psi_0\left(\frac{x^2 + (p - \beta\theta)^2}{2\beta}\right)$$

$$\implies \quad \psi_1(J, \phi) = \psi_0\left(J + \theta\sqrt{2J\beta}\sin\phi + \frac{1}{2}\beta\theta^2\right).$$

At time $t > 0$ after the kick, the motion of individual particles is described by

$$J_{\text{after}} = J_{\text{before}}, \qquad \phi_{\text{after}} = \phi_{\text{before}} + \omega(J_{\text{before}})t,$$

where "before" and "after" now refer to $t = 0^+$ and $t > 0$, respectively; $\omega(J)$ is the betatron oscillation frequency of a particle whose action is J. The dependence of ω on J may be weak, but it is important; it gives rise to a spread in the betatron frequencies and thus the decoherence effect. Later, it will also play an important role for the echo.

The mechanism of decoherence after an initial kick is illustrated below. The decoherence time is determined by the inverse of the beam particles' frequency spread. If for example, the beam has a betatron tune spread of 10^{-3}, then the decoherence time is of the order of 10^3 turns. In this mechanism, as illustrated

Figure 7.3: Decoherence mechanism after an initial kick. A beam with 11 particles are kicked at time $t = 0$. The particles have an oscillation frequency spread but all particles receive the same initial kick. (a) shows the trajectories of the 11 individual particles, (b) shows the centroid signal of the 11-particle beam. Individual particles oscillate with constant amplitudes, but the centroid signal decoheres with a decoherence time $\sim$(frequency spread)$^{-1}$.

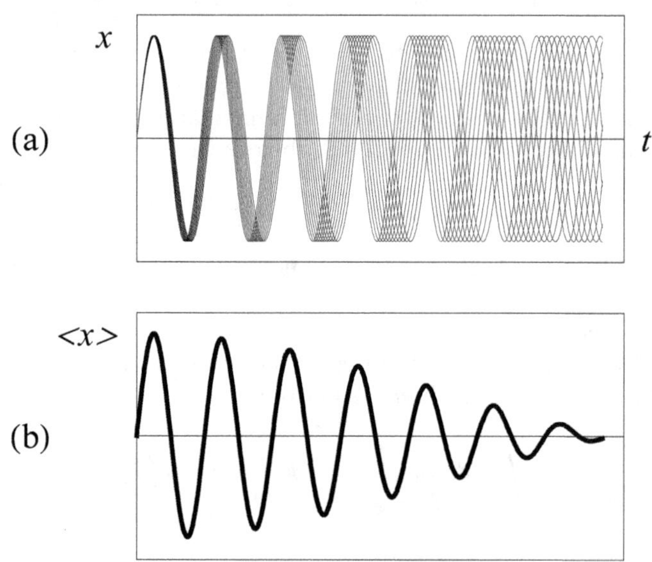

in Fig. 7.3, it is important to keep in mind that although the beam signal is observed to decohere, the oscillations of individual particles continue with constant amplitudes.

The distribution after the kick is given by

$$\psi_2(\phi, J, t) = \psi_1(\phi - \omega(J)t, J)$$
$$= \psi_0\left[J + \theta\sqrt{2J\beta}\sin(\phi - \omega(J)t) + \frac{1}{2}\beta\theta^2\right]. \qquad (7.2)$$

It is $\phi - \omega(J)t$, not $\phi + \omega(J)t$, that goes into Eq. (7.2).

The beam centroid signal after the kick is

$$\langle x\rangle(t > 0) = \int_0^\infty dJ \int_0^{2\pi} d\phi\, x\, \psi_2(\phi, J, t)$$
$$= \int_0^\infty dJ \int_0^{2\pi} d\phi\, \psi_2(\phi, J, t)\sqrt{2J\beta}\cos\phi. \qquad (7.3)$$

We are interested in the case when the beam distribution before the kick is

Gaussian,

$$\psi_0(J) \;=\; \frac{1}{2\pi J_0} e^{-J/J_0}\,, \tag{7.4}$$

and when the amplitude-dependent betatron frequency is given by

$$\omega(J) \;=\; \omega_0 + \omega' J\,, \tag{7.5}$$

where ω_0, the nominal betatron frequency, and ω', the betatron detuning parameter, are constants, independent of J. The quantity J_0 is the unperturbed rms beam emittance. We are mostly interested in the case when the beam's frequency spread $\omega' J_0$ is much smaller than ω_0.

Substituting Eqs. (7.2), (7.4) and (7.5) into Eq. (7.3) and changing variable from J to $a = \sqrt{2J\beta}$, we obtain

$$\langle x \rangle(t) \;=\; \frac{1}{\beta J_0} \int_0^\infty a^2 da \, \exp\left(-\frac{a^2}{2\beta J_0} - \frac{\beta\theta^2}{2J_0}\right) \sin\left(\omega_0 t + \frac{\omega' t}{2\beta} a^2\right) I_1\left(\frac{\theta a}{J_0}\right),$$

where we have integrated over ϕ using

$$\frac{1}{2\pi} \int_0^{2\pi} d\phi \cos\phi \, e^{x\cos\phi} \;=\; I_1(x)\,,$$

with $I_1(x)$ the Bessel function.

The integration over a can also be performed using

$$\int_0^\infty a^2 da \, e^{-Aa^2} I_1(Ba) \;=\; \frac{B}{4A^2} e^{B^2/4A}\,.$$

We then obtain

$$\begin{aligned}
\langle x \rangle(t) \;&=\; \beta\theta \, \mathrm{Im}\left[\frac{e^{i\omega_0 t}}{(1 - i\Theta)^2} \exp\left(\frac{\beta\theta^2}{2J_0} \frac{i\Theta}{1 - i\Theta}\right)\right] \\
&=\; \frac{\beta\theta}{1 + \Theta^2} \exp\left[-\frac{\beta\theta^2\Theta^2}{2J_0(1 + \Theta^2)}\right] \sin\left[\omega_0 t + \frac{\beta\theta^2\Theta}{2J_0(1 + \Theta^2)} + 2\tan^{-1}\Theta\right], \\
\Theta \;&\equiv\; \omega' J_0 t\,.
\end{aligned} \tag{7.6}$$

To analyze Eq. (7.6), we first observe that the nominal betatron frequency ω_0 is supposed to be much larger than the betatron frequency spread $\omega' J_0$ of the beam. This means the sine factor in Eq. (7.6) is rapidly oscillating because it contains $\omega_0 t$ in its argument. All terms involving Θ are also time-dependent but are varying slowly. With this observation, we can consider Eq. (7.6) as giving, in addition to a rapidly oscillating sine factor, a slowly varying oscillation amplitude

$$\langle x \rangle^{\mathrm{ampl}}(t) \;=\; \frac{\beta\theta}{1 + \Theta^2} \exp\left[-\frac{\beta\theta^2\Theta^2}{2J_0(1 + \Theta^2)}\right]. \tag{7.7}$$

This amplitude starts with the value $\beta\theta$ at time $t = 0$, but slowly decreases with time, and it is substantially reduced in a time determined by $\Theta^2 \approx 1$ and

$\frac{\beta\theta^2\Theta^2}{J_0} \approx 1$. This leads to consideration of a decoherence time,

$$\tau_{\text{decoh}} \approx \min\left[\frac{1}{\omega' J_0}, \frac{1}{\omega'\theta\sqrt{\beta J_0}}\right]. \tag{7.8}$$

Later in our analysis, we will concentrate on a first order calculation, and in so doing, we will assume $\beta\theta$ to be small. In that regard, only one of the results in Eq. (7.8) remains, and we will have only

$$\tau_{\text{decoh}} = \frac{1}{\omega' J_0},$$

such as claimed in the caption of Fig. 7.3.

Physically, decoherence occurs because different particles have slightly different oscillation frequencies. Although all individual particles continue to execute their own betatron oscillations without attenuation, their collective centroid signal smears out in time. The rate of smearing depends on the frequency spread. If $\omega = $ constant without a spread, the centroid oscillation after the kick will continue indefinitely, and there will be no decoherence.

Here we analyzed the decoherence assuming that the frequency spread originated from a slight dependence of the betatron frequency on the betatron amplitude of the oscillating particle. Frequency spread can also originate from other sources, such as a dependence of the betatron frequency on the particle's energy offset. Analysis of the decoherence as well as the echo effects will differ somewhat, but the end result, particularly the existence of a decoherence and echo will remain.

Figure 7.4 shows how the beam centroid signal decoheres. It shows $\frac{\langle x \rangle}{\beta\theta}$ versus Θ for $\omega_0 = 50\,\omega' J_0$, and various values of kick amplitude of

$$u \equiv \frac{\beta\theta^2}{2J_0}.$$

When $u \ll 1$, i.e. when the kick amplitude is much smaller than the unperturbed beam size, Eqs. (7.6) and (7.7) can be approximated as

$$\begin{aligned}
\langle x \rangle(t) &\approx \frac{\beta\theta}{1+\Theta^2}\sin(\omega_0 t + 2\tan^{-1}\Theta) \\
&= \frac{\beta\theta}{(1+\Theta^2)^2}[(1-\Theta^2)\sin\omega_0 t + 2\Theta\cos\omega_0 t], \\
\langle x \rangle^{\text{ampl}}(t) &\approx \frac{\beta\theta}{1+\Theta^2}.
\end{aligned} \tag{7.9}$$

In each panel of Fig. 7.4, there are two curves; the red curves are the first order approximation (7.9). When the kick amplitude is not small compared with the unperturbed beam size, Eq. (7.9) over-estimates the beam centroid motion. When $u \ll 1$, the decoherence time is $\tau_{\text{decoh}} \approx \frac{1}{\omega' J_0}$. When $u \gtrsim 1$, $\tau_{\text{decoh}} \approx \frac{1}{\omega' J_0}\frac{1}{\sqrt{2u}}$, according to Eq. (7.8).

Figure 7.4: Decoherence of the beam centroid signal after a kick, for various values of the kick amplitude $u = 0.5, 1, 2$. Black curves are the exact calculation. Red curves are approximation when $u \ll 1$; the red curves are the same in all three panels.

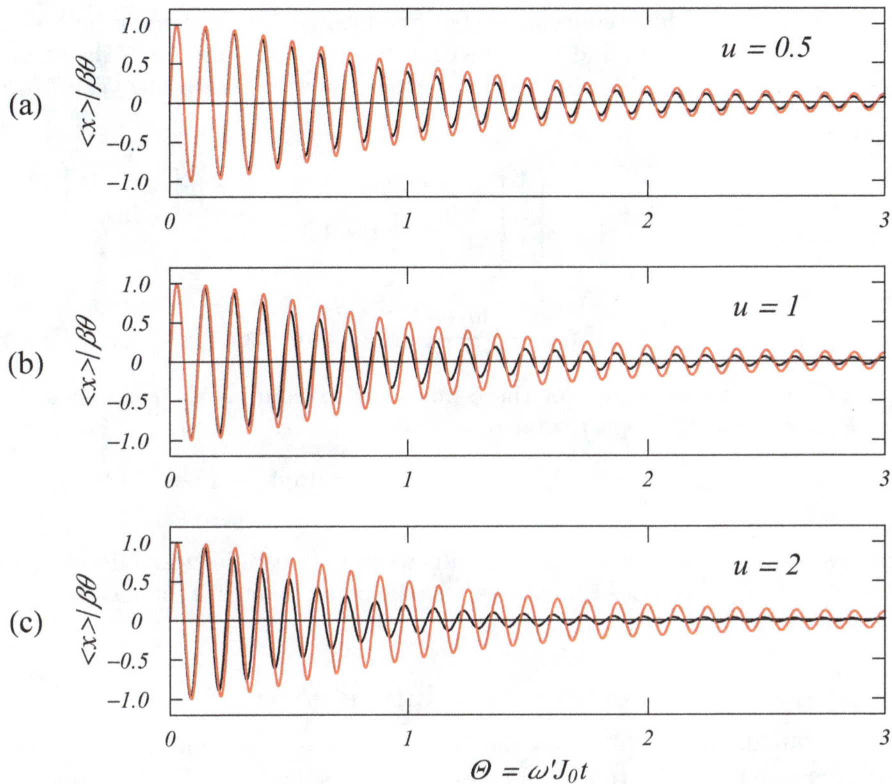

Homework 7.1 In our discussion, we assumed that the frequency spread in the beam particles comes from an amplitude-dependence of the oscillation frequency. As a result, ω' plays a role in our analysis. Does our analysis still apply if the frequency spread comes from some other parametric dependencies? What parameter would replace ω'? A special case when the frequency depends on the particle's energy deviation δ deserves special attention.

Homework 7.2 Equation (7.6) gives the centroid position as a function of time of a Gaussian beam after the kick. Follow similar steps to find the pairing expression for the centroid momentum.

Solution An intermediate step gives

$$\langle p \rangle(t) \;=\; \beta\theta \, \mathrm{Re}\left[\frac{e^{i\omega_0 t}}{(1 - i\Theta)^2} \exp\left(\frac{\beta\theta^2}{2J_0}\frac{i\Theta}{1 - i\Theta}\right)\right].$$

Homework 7.3 We have concentrated on the beam centroid signal after a kick. Extend the calculation to find the time evolution and decoherence of the second moments $\langle x^2 \rangle, \langle xp \rangle, \langle p^2 \rangle$. Apply to the special case with conditions (7.4–7.5).[9]

Solution

$$\begin{bmatrix} \langle x^2 \rangle \\ \langle xp \rangle \\ \langle p^2 \rangle \end{bmatrix} \;=\; \frac{\beta}{2}(2J_0 + \beta\theta^2)\begin{bmatrix} 1 \\ 0 \\ 1 \end{bmatrix} + \frac{\beta^2\theta^2 \exp\left[-\frac{2\beta\theta^2\Theta^2}{J_0(1+4\Theta^2)}\right]}{2(1+4\Theta^2)^{3/2}}\begin{bmatrix} -\cos\psi \\ \sin\psi \\ \cos\psi \end{bmatrix},$$

where

$$\psi \;=\; 2\omega_0 t - \frac{\beta\theta^2\Theta^2}{J_0(1+4\Theta^2)} + 3\tan^{-1} 2\Theta.$$

It may be instructive to plot these functions to examine their transient behavior after the kick. In particular note that

$$\langle x^2 \rangle + \langle p^2 \rangle \;=\; 2\beta J_0 + \beta^2\theta^2 \;=\; \text{constant in time}.$$

Homework 7.4 The kick we considered is when the beam is kicked in one turn, i.e. a δ-function. How does the decoherence behavior change if the kick is a step function?

7.2.2　Transverse echo — no diffusion

After the dipole kick, we now let the beam decohere for a time τ that is much longer than τ_{decoh}. The time τ is chosen intentionally long. For example, the parameter $\Theta = \omega' J_0 \tau$ could be chosen $\sim 10^3$. If so, a beam initially kicked to 1 mm oscillation would have a residual oscillation of ~ 1 nm at time τ.

By the time $t = \tau$, we give the now-very-quiet beam a second, one-turn, quadrupole kick,

$$x_{\text{after}} \;=\; x_{\text{before}}, \qquad p_{\text{after}} \;=\; p_{\text{before}} - qx_{\text{before}}, \tag{7.10}$$

where $q = \frac{\beta}{f}$ with f the focal length of the kicking quadrupole ($f > 0$ means focusing quadrupole).

For simplicity, we consider the dipole and the quadrupole kicks to be weak.[10] For a dipole kick to be weak, we require $u = \frac{\beta\theta^2}{2J_0} \ll 1$. The condition for a quadrupole kick to be considered weak will be given later in Eq. (7.21).

---

[9]See for example, M.A. Furman, Sec. 2.3.10, Handbook Accel. Phys. & Eng. 2nd ed., World Scientific (2013).

[10]Later in Sec. 7.4.3, we will consider another case when these kicks are made intentionally strong to explore their higher order effects. The reason weak kicks are more relevant here is due to the fact that we are only looking at the beam centroid signal for the echo.

For a weak dipole, let's first rederive Eq. (7.9) as follows,

$$\begin{aligned}
\psi_1 &= \psi_0(x, p - \beta\theta) \approx \psi_0(J) - \beta\theta\psi_0'(J)\frac{\partial J}{\partial p} \\
&= \psi_0(J) + \theta\psi_0'(J)\sqrt{2J\beta}\sin\phi, \\
\Longrightarrow \\
\psi_2 &\approx \psi_0(J) + \theta\psi_0'(J)\sqrt{2J\beta}\sin(\phi - \omega(J)t).
\end{aligned} \tag{7.11}$$

Note the appearance of the derivative term $\psi_0'(J)$ in the expression. As we will see, the echo signal is not driven by the whole beam distribution ψ_0, but by its derivative ψ_0'. Physically this eventually is due to the Liouville theorem of the phase space. The reader is suggested to think of the physical effect underlying this mathematical observation (Homework 7.6).

The first term $\psi_0(J)$ in Eq. (7.11) does not contribute to the beam centroid signal. The second term contributes

$$\begin{aligned}
\langle x\rangle(t) &\approx \theta\int_0^\infty dJ\int_0^{2\pi} d\phi\sqrt{2J\beta}\cos\phi\,\psi_0'(J)\sqrt{2J\beta}\sin(\phi - \omega(J)t) \\
&= -2\pi\beta\theta\int_0^\infty J\,dJ\,\psi_0'(J)\sin(\omega(J)t).
\end{aligned} \tag{7.12}$$

Equation (7.12) holds for arbitrary $\psi_0(J)$ and $\omega(J)$. If they are given by Eqs. (7.4) and (7.5), we recover Eq. (7.9).

At time $t = \tau$, the beam is given a weak quadrupole kick with "small q" [see Eqs. (7.20) and (7.21) later]. Immediately before the kick, we have

$$\psi_3 = \psi_0(J) + \theta\psi_0'(J)\sqrt{2J\beta}\sin(\phi - \omega(J)\tau). \tag{7.13}$$

Immediately after the kick, we have

$$\begin{aligned}
\psi_4(\phi, J) &\approx \psi_3(\phi, J) + qx\frac{\partial\psi_3}{\partial p} \\
&= \psi_3(\phi, J) + q\sqrt{2J\beta}\cos\phi\left[\frac{\partial J}{\partial p}\frac{\partial\psi_3}{\partial J} + \frac{\partial\phi}{\partial p}\frac{\partial\psi_3}{\partial\phi}\right] \\
&= \psi_3(\phi, J) - q\sqrt{2J\beta}\cos\phi\left[\sqrt{\frac{2J}{\beta}}\sin\phi\frac{\partial\psi_3}{\partial J} + \frac{\cos\phi}{\sqrt{2J\beta}}\frac{\partial\psi_3}{\partial\phi}\right].
\end{aligned} \tag{7.14}$$

From Eq. (7.13), we have

$$\begin{aligned}
\frac{\partial\psi_3}{\partial J} &= \psi_0'(J) + \theta\frac{\partial}{\partial J}[\psi_0'(J)\sqrt{2J\beta}]\sin(\phi - \omega(J)\tau) \\
&\quad -\theta\omega'(J)\tau\psi_0'(J)\sqrt{2J\beta}\cos(\phi - \omega(J)\tau), \\
\frac{\partial\psi_3}{\partial\phi} &= \theta\psi_0'(J)\sqrt{2J\beta}\cos(\phi - \omega(J)\tau).
\end{aligned} \tag{7.15}$$
$$\tag{7.16}$$

The three terms in Eq. (7.15) have the relative magnitudes of the order of

$$1 \ : \ \sqrt{u} \ : \ (\omega' J_0 \tau)\sqrt{u} \, . \tag{7.17}$$

We have already assumed $u \ll 1$. We will consider the case when the third term dominates the beam centroid signal. The first term, when substituted into Eq. (7.14), does not contribute to the centroid signal. Therefore, the third term dominates if

$$|\omega' J_0 \tau| \gg 1 \, . \tag{7.18}$$

The quantity $\omega' J_0 \tau$ is the betatron phase smear due to the frequency spread in a time period of τ. Equation (7.18) means the decoherence effect must be large, but that is just what we are interested in.

The term due to Eq. (7.16), relative to the three terms in Eq. (7.17), has a magnitude of $\sqrt{u}$, so is also negligible. We have thus obtained, under condition (7.18),

$$\psi_4 \ \approx \ \theta q \omega'(J)\tau J \psi_0'(J) \sqrt{2J\beta} \sin 2\phi \cos(\phi - \omega(J)\tau) \, . \tag{7.19}$$

If we return to Eq. (7.14), the condition of "small q" really requires $(\psi_4 - \psi_0) \ll (\psi_3 - \psi_0)$, or

$$|(\omega' J_0 \tau)q| \ \ll \ 1 \, . \tag{7.20}$$

Combining all the conditions for the validity of our analysis, we have

$$\left|\frac{1}{q}\right| \ \gg \ |\omega' J_0 \tau| \ \gg \ 1, \qquad \frac{1}{|\theta|}\sqrt{\frac{J_0}{\beta}} \ \gg \ 1 \, . \tag{7.21}$$

For $t > \tau$, after the quadrupole kick, the beam distribution is

$$\psi_5(\phi, J, t) \ = \ \psi_4(\phi - \omega(J)(t - \tau), J) \, . \tag{7.22}$$

This beam distribution will contain the echo signal. It may be useful to emphasize again here that the echo signal is rather pronounced and apparent in the phase space, as given by Eq. (7.22); however, it is well hidden and would be clueless to a 3-D observer.

Our observation is made through the detection of the beam centroid. The centroid of the beam distribution (7.22) is

$$\begin{aligned}
\langle x \rangle^{\text{echo}}(t) \ &= \ \int_0^\infty dJ \int_0^{2\pi} d\phi \sqrt{2J\beta} \cos\phi \, \psi_5(\phi, J, t) \\
&\approx \ 2\beta\theta q\tau \int_0^\infty J^2 dJ \omega'(J)\psi_0'(J) \\
&\quad \times \int_0^{2\pi} d\phi \cos\phi \sin[2\phi - 2\omega(J)t + 2\omega(J)\tau] \cos[\phi - \omega(J)t] \, .
\end{aligned} \tag{7.23}$$

Note that we are calling the centroid signal of ψ_5 the echo signal and designate it as $\langle x \rangle^{\text{echo}}$.

The integration over ϕ can be performed to yield

$$-\frac{\pi}{2}\sin[\omega(J)(t-2\tau)]\,. \tag{7.24}$$

Hidden behind this integration over ϕ, thus Eq. (7.24), is the secret to the echo phenomenon. Magically, the two kicks, the frequency spread, and the particle evolution in phase space, all conspire to produce a *recoherence* in the neighborhood of time $t = 2\tau$! — and yet, according to (7.24), exactly at $t = 2\tau$, the echo signal is zero because sine function is zero there.

Attention is to be paid to the fact that the recoherence, and the fact that it occurs at $t = 2\tau$, do not depend on the exact form of $\psi_0(J)$ or $\omega(J)$. One might consider the echo mechanism to be a result due to the fact that the second kick, being a δ-function kick, sends two beam responses, one continues to be winding forward in time and another proceeding with a reversed time. The component moving backward in time exactly unwinds the forward moving component following the first kick, thus resulting in an echo when the unwinding is completed. This winding and unwinding cancel each other perfectly regardless of the form of $\psi_0(J)$ and $\omega(J)$. For a further examination of this mechanism, see Homeworks 7.7 and 7.12.

After performing the ϕ-integral, we obtain

$$\langle x\rangle^{\mathrm{echo}}(t) \;=\; -\pi\beta\theta q\tau\int_0^\infty J^2 dJ\omega'(J)\psi_0'(J)\sin[\omega(J)(t-2\tau)]\,. \tag{7.25}$$

Equation (7.25) is our main result. As mentioned, regardless of the exact forms of $\psi_0(J)$ or $\omega(J)$, the echo signal clusters around the time $t = 2\tau$, although it vanishes when the time is exactly at $t = 2\tau$.

If $\omega(J)$ and $\psi_0(J)$ are given by Eqs. (7.4–7.5), the remaining integration over J can also be performed to yield

$$\langle x\rangle^{\mathrm{echo}}(t) \;=\; \beta\theta q\omega' J_0\tau\,\mathrm{Im}\left[\frac{e^{i\Phi}}{(1-i\xi)^3}\right]$$

$$\;=\; \beta\theta q\,\frac{\omega' J_0\tau}{(1+\xi^2)^3}\,[\xi(3-\xi^2)\cos\Phi+(1-3\xi^2)\sin\Phi]\,, \tag{7.26}$$

$$\xi \;=\; \omega' J_0(t-2\tau)\,,$$

$$\Phi \;=\; \omega_0(t-2\tau)\,. \tag{7.27}$$

Noting that Φ is a fast time variable and ξ is a slow time variable, the amplitude of the echo signal is found to be

$$\langle x\rangle^{\mathrm{echo\ ampl}}(t) \;=\; \beta\theta q\,\frac{\omega' J_0\tau}{(1+\xi^2)^{3/2}}\,. \tag{7.28}$$

Equation (7.28) contains a factor $\beta\theta q$, which is a measure of the product of the two kicks. It also contains a factor $\omega' J_0\tau$, which is the accumulated phase slippage during the waiting period. We note that $\beta\theta q$ is small, but $\omega' J_0\tau$ is large.

Figure 7.5: Schematic of a transverse echo experiment in a storage ring.

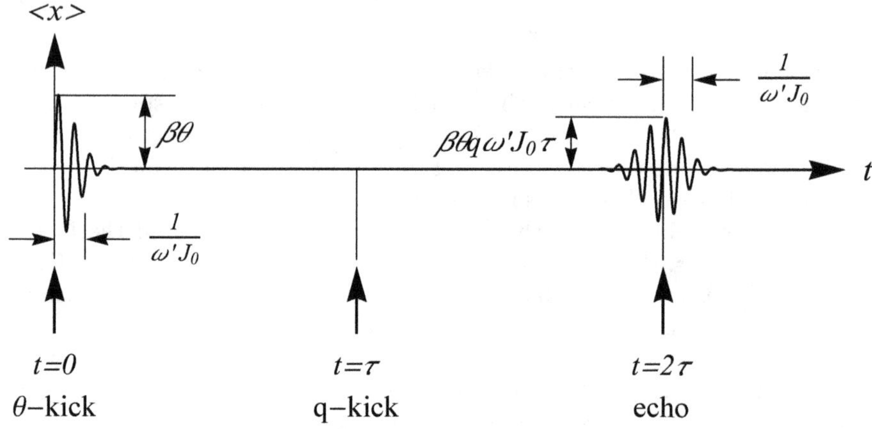

The echo signal clusters around $t = 2\tau$, as seen by the factor $(1 + \xi^2)^{3/2}$ in its denominator. The peak echo amplitude occurs at $\xi = 0$, i.e. $t = 2\tau$, and the maximum value is given by

$$\langle x \rangle^{\text{echo ampl max}} = \beta\theta q\omega' J_0 \tau . \tag{7.29}$$

Away from the peak, the echo amplitude decreases on both sides of the peak, proportionally to $\sim |t - 2\tau|^{-3}$. The full width at half maximum time duration around $t = 2\tau$ of the echo is

$$\Delta T^{\text{FWHM}} = \frac{2}{\omega' J_0} \sqrt{2^{3/2} - 1} \approx \frac{1.53}{\omega' J_0} . \tag{7.30}$$

This echo duration is comparable to the decoherence time τ_{decoh} after the dipole kick. One sees that both τ_{decoh} and ΔT^{FWHM} are much less than τ, while much longer than the beam betatron period $\frac{1}{\omega_0}$.

Equation (7.29) can be cast in a slightly different way,

$$\frac{\langle x \rangle^{\text{echo ampl max}}}{\sqrt{\beta J_0}} = \left(\theta\sqrt{\frac{\beta}{J_0}} \right) (q\omega' J_0 \tau) , \tag{7.31}$$

where the two factors on the right-hand-side are simply the dimensionless small parameters of the first and the second kicks, respectively [see Eq. (7.21)].

Figure 7.5 shows a schematic of the echo signal with the magnitudes of various physical parameters marked. It shows schematically what was expected of the thought experiment mentioned on page 480.

Figure 7.6 shows the echo signal near time $t = 2\tau$ according to Eq. (7.26), assuming $\omega_0 = 50\,\omega' J_0$. Note again that although the echo amplitude reaches maximum at $t = 2\tau$, the exact value of $\langle x \rangle^{\text{echo}}$ in fact vanishes there. Within

Figure 7.6: A close-up observation of the echo signal.

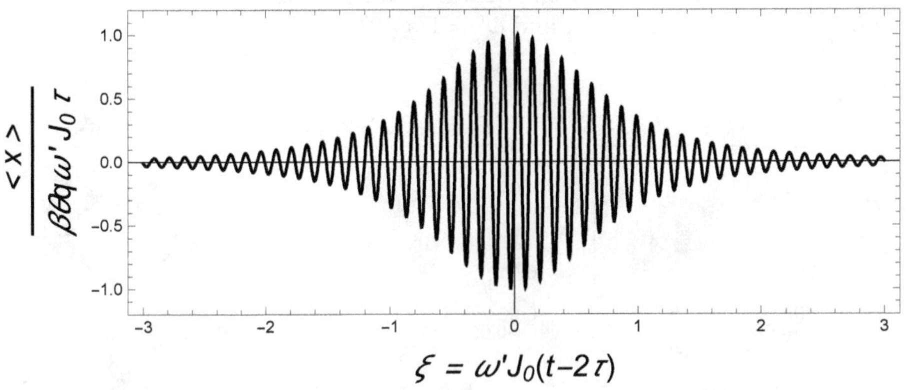

$$\xi = \omega' J_0(t - 2\tau)$$

the time period ΔT^{FWHM}, there are $\frac{\omega_0 \Delta T^{\mathrm{FWHM}}}{2\pi} \approx 12$ betatron oscillations as Fig. 7.5 confirms.

If we compare the maximum echo amplitude (7.29) to the initial kick amplitude of $\beta\theta$, we see that the echo amplitude is weaker by a factor of $q\omega' J_0\tau$, which, according to Eq. (7.20), is much less than 1 in order for our analysis to apply. One may want to push the parameters to see how to get as big an echo signal as possible. This occurs when $q\omega' J_0\tau$ is made to approach 1, when our analysis begins to break down.

In our analysis, we have made approximations summarized by Eq. (7.21). It is instructive to do a numerical simulation and observe what happens in detail in the phase space. Figure 7.7 shows some of the results.[11] The parameters used are exaggerated, with a very large initial kick and with a very large frequency spread, to illustrate the main features. Conditions (7.21) are therefore somewhat violated in this simulation. A Gaussian beam receives a dipole kick at $t = 0$ and a quadrupole kick at $t = \tau$. In the time between the two kicks, the beam decoheres. After the second kick and by the time $t = \frac{3}{2}\tau$, several "kinks" develop in the beam's phase space distribution. These kinks come from an interplay of the two kicks. Each of them occurs at a specific amplitude, which are most conspiring in the sense that, with the amplitude-dependent betatron frequency, they all come to coherence *simultaneously* at $t = 2\tau$ as they migrate in the phase space relative to one another, thus yielding an echo. The conspiring aspect is especially seen by the fact that the amplitudes at which the kinks occur must match exactly what is needed by the amplitude-dependence of the betatron frequency so that they can come into recoherence simultaneously at $t = 2\tau$. And this would have to be true for arbitrary $\omega(J)$.

The recoherence at time $t = 2\tau$ occurs with all kinks line up in the direction of momentum p. The beam centroid at that instant reads zero. The beam centroid signal reads maximum when $t = 2\tau \pm \frac{\pi}{2\omega_0}$, as also illustrated in Fig. 7.7.

---

[11]This should really be shown as a movie.

Figure 7.7: Simulation of a transverse echo experiment. [Courtesy Xiujie Deng (2017).] Parameters used are: $J_0 = 1 \times 10^{-6}$ m, $q = 0.25$, $\beta = 5$ m, $\theta = 2 \times 10^{-3}$, $\tau = 0.8$ ms, $\omega_0 = 1 \times 10^7$ s^{-1}, and $\omega' = 1 \times 10^9$ s^{-1}m^{-1}.

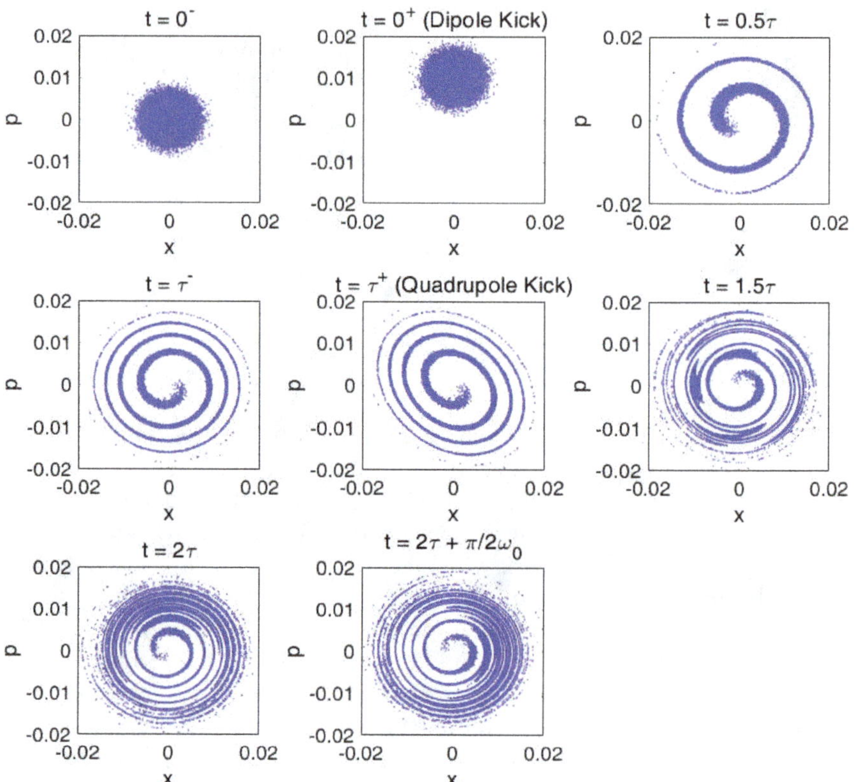

To illustrate the echo signal, on the other hand, we prefer to illustrate using parameters in the linear regime, i.e. with condition (7.21) fulfilled. One example by simulation is shown in Fig. 7.8. One can check that the simulation result agrees with the prediction using our linear theory.

Simulation also allows exploring parameters range beyond Eq. (7.21). For example, Fig. 7.9 shows what happens when q is varied. Cases 1 to 3 are for small q, in line with Eq. (7.21), and we obtain results consistent with Eq. (7.28). As q is increased further, cases 4 and 5, the echo signal develops a double hump structure. The maximum echo amplitude one can get from varying q is found to be about 40% of the initial kick amplitude (near case 3).

Following this section, we have included several homework problems to illustrate possible variations of transverse echo effects. They serve a purpose

Figure 7.8: Centroid signal of the simulation indicating an echo. [Courtesy Xiujie Deng (2017).] Parameters used are: $J_0 = 1 \times 10^{-6}$ m, $q = 0.08$, $\beta = 5$ m, $\theta = 2 \times 10^{-5}$, $\tau = 10$ ms, $\omega_0 = 1 \times 10^7$ s^{-1}, and $\omega' = 1 \times 10^9$ s^{-1}m^{-1}.

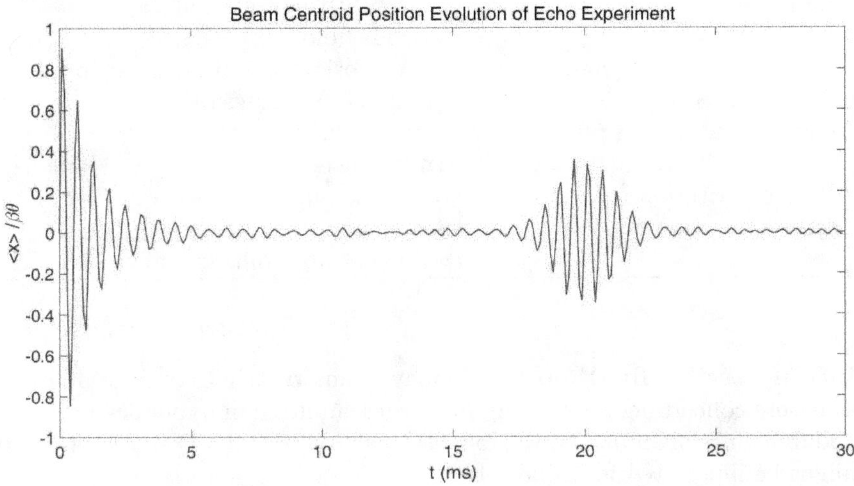

Figure 7.9: Echo amplitude signal obtained in a simulation. Curves 1 to 5 correspond to $q = 0.02, 0.03, 0.08, 0.2, 0.3$, respectively.

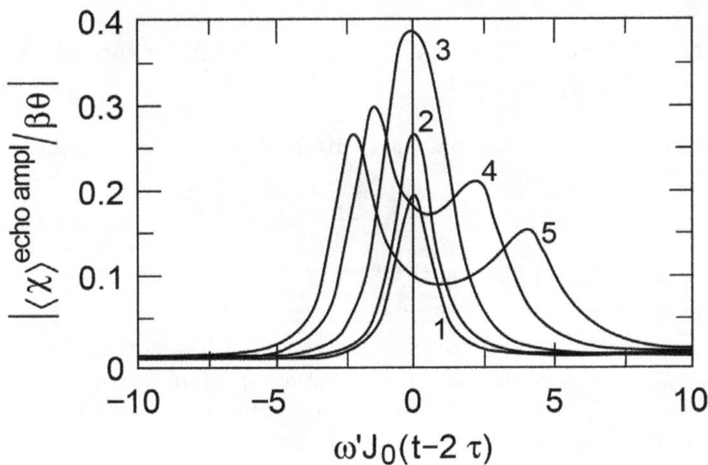

to echo the statement made earlier that there are echoes everywhere in an accelerator.

Table 7.1: Analogy between the Hamiltonian dynamics in the phase space and the incompressible fluid dynamics in the 3-D real space.

Hamiltonian dynamics	Fluid dynamics
Hamilton's equation	Navier–Stokes equation
Point particles	Viscous fluid
Particle motion in phase space	Fluid motion in real physical space
Liuoville theorem	Fluid is incompressible
Smooth contours in phase space	Laminar flow
Chaos	Turbulence
Chirikov criterion for chaos	Reynolds number for turbulence
Echo	See video
	www.youtube.com/watch?v=p08_KlTKP50

Incompressible fluid model   It may be instructive to make a detour discussion here concerning an analogy between Hamiltonian dynamics in phase space and dynamics of an incompressible viscous fluid in real 3-D space. A comparison might be illustrated in Table 7.1.

This insightful comparison was suggested by Lee Teng when he derived the Chirikov criterion of chaotic motion in phase space by drawing analogy with the turbulent motion in real space of a fluid.[12] The video[13] mentioned in Table 7.1 draws this analogy again between the echo effect in phase space with a fluid dynamics in real space. We will briefly return to this model when we discuss the Chirikov criterion for the beam-beam instability in Sec. 8.4.6.

Homework 7.5 Verify that Eq. (7.12) yields Eq. (7.9) for the special case when Eqs. (7.4) and (7.5) are applied.

Homework 7.6 Repeat the analysis (7.10–7.12) for a waterbag beam distribution

$$\psi_0(J) = \frac{1}{2\pi J_0} H(J_0 - J),$$

where $H(x)$ is the step function.

(a) Find $\langle x \rangle(t)$.

(b) The waterbag model contains particles with a spread of J. Together with an amplitude-dependent $\omega(J)$, the waterbag beam particles have a spread of frequencies. Yet, you will find there is no decoherence from (a) for weak kicks. Explain why there is no decoherence even though there is a frequency spread.

By this homework, bear in mind that an intuition that a beam with frequency spread necessarily imply decoherence after a kick is flawed — admittedly an issue exacerbated by Fig. 7.3. The key to this issue lies in the fact that it is $\psi'(J)$, not

---

[12]L.C. Teng, IEEE Trans. Nucl. Sci. NS-20, 843 (1973).

[13]University of New Mexico, Department of Physics and Astronomy, Youtube film on laminar flow (2007).

$\psi(J)$, that plays the key role, which in turn, as mentioned in the text, originates from the Liouville theorem.

Solution A simulation is shown in Fig. 7.10 to illustrate the comparison. Parameters used are the same as in Fig. 7.8, except that J_0 is now interpreted to be the maximum amplitude of the waterbag. The blue curve is for the Gaussian beam and is a reproduction of Fig. 7.8. The red curve is for the waterbag distribution. It shows no decoherence, and the echo is obscured.

Figure 7.10: There is no decoherence or echo effects for a waterbag beam distribution. The blue curve is the simulated response of a Gaussian beam to the double-kick experiment. The red curve is that for a waterbag beam. [Courtesy Xiujie Deng (2017).]

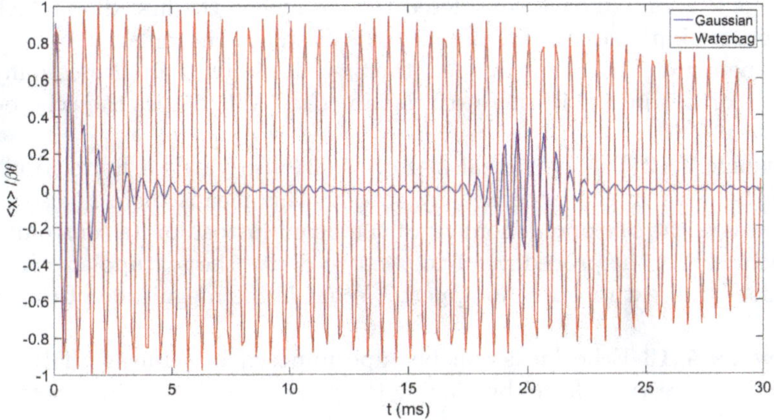

Homework 7.7 The integrand of the ϕ-integral in Eq. (7.23) contains four sinusoidal factors

$$\cos\phi \ \times \ \sin[\phi - \omega(J)t + \omega(J)\tau] \ \times \ \cos[\phi - \omega(J)t + \omega(J)\tau] \ \times \ \cos[\phi - \omega(J)t].$$

Trace back to find out where each factor originates from. This would give a feel for the intricate phase space dynamics that eventually leads to the echo.

Homework 7.8 We showed there is an echo at $t = 2\tau$. Are there multiple echoes, for example at $t = 4\tau$, etc.?

Homework 7.9 We have analyzed the problem with a quadrupole kick following a dipole kick. What happens if we reverse the order of these two kicks? Will there still be an echo?

Homework 7.10

(a) Consider a beam in a long linac. Give the beam at $s = 0$ a nonlinear angular kick of $\Delta x' = \theta_1 \sin k_1 x$. At $s = D$, kick the beam again by $\Delta x' = \theta_2 \sin k_2 x$. Assume the linac section under consideration is all drift spaces except the two kicks. Show that there is an echo if one observes the beam at location $s = \frac{k_1}{k_2 - k_1} D$. The echo signal this time is not in its center of charge $\langle x \rangle$ but in its transverse distribution.

(b) Can the setup be extended when the drift spaces contain also dipoles and quadrupoles? An experimental test can be considered if this generalized case can be established.

Solution (a) This homework prepares for the later discussion of longitudinal echo. One may consider this a transverse counterpart of the echo in longitudinal dimension of an unbunched beam.

Homework 7.11 We have considered the case when the first dipole kick gives the beam an initial angle θ. What if the first kick is replaced by an initial displacement x_0? This can be made if the first kick is given at the beam injection into the storage ring. Will there still be an echo? When does the echo occur?

Homework 7.12 The ϕ-integral in Eq. (7.23) also yields a finite result if we replace the $\cos \phi$ in the integrand by $\cos 3\phi$. This means there is also an echo if we measure the sextupole moment of the beam instead of its centroid (dipole moment). Follow the procedure of the text to derive the sextupolar echo signal. When does the sextupolar echo appear? What is the magnitude of the signal?

Homework 7.13 Echo kicks can be superimposed in a linear analysis. One can obtain interesting effects by playing the game of multiple kicks in producing echoes.

(a) Consider two sets of kicks

$$\text{(dipole kick } \theta_1 \text{ at } t = -2\tau_1, \text{ quadrupole kick } q_1 \text{ at } t = -\tau_1),$$
$$\text{(dipole kick } \theta_2 \text{ at } t = -2\tau_2, \text{ quadrupole kick } q_2 \text{ at } t = -\tau_2).$$

Show that the two echo signals simply add coherently near $t = 0$ and the net amplitude is proportional to $\theta_1 q_1 \tau_1 + \theta_2 q_2 \tau_2$.

(b) Apply dipole kicks θ_0 every other turn to the beam from $t = -2T$ to $t = 0$, and apply quadrupole kicks q_0 every turn from $t = -T$ to $t = 0$. There should be a very large echo signal at $t = 0$ because all these echoes add coherently. Find this coherent echo response.

(c) Same as (b) but when the dipole kicks are applied every turn (instead of every other turn) from $t = -2T$ to $t = 0$. The dipole and quadrupole kicks in this case are therefore effectively simple step-functions. Is there an echo signal in this case?

(d) Apply dipole ($-2T < t < 0$) and quadrupole ($-T < t < 0$) kicks every turn, but modulate the kick strength by $\theta = \theta_0 \sin \omega_1 t$ and $q = q_0 \sin 2\omega_1 t$, where

ω_1 is some modulation frequency. Find the echo response. Show that the echo amplitude is proportional to $\frac{\theta_0 q_0 T^2}{T_0}$, where T_0 is the revolution time.

Case (d) indicates that one can in principle apply very weak kicks — almost at a subliminal level — for multiple turns and obtain a large echo.

Homework 7.14 The echo set-up can be generalized. If the two kicks are m_1-th and m_2-th multipolar kicks ($m_2 > m_1$), we should observe an ($m_2 - m_1$)-th multipolar signal at time $t = \frac{m_1}{m_2 - m_1}\tau$.

Homework 7.15 What determines the number of phase space kinks in Fig. 7.7?

Homework 7.16 We call this effect an echo because the waiting period τ is so long so that, when the signal appears, it appears at an unexpectedly after all events seem to have disappeared. What happens if we shorten τ? When τ is shortened enough, the echo signal becomes mixed with the decoherence signal and no longer appears as an echo. However, that does not mean there is no echo signal.

To push this picture to an extreme, consider another thought experiment. Consider a linac section consisting of a dipole kicker, followed by an octupole magnet, followed by a quadrupole, and then another octupole magnet. Let the two octupoles be identical. By the time a beam goes through this linac section, the beam is heavily undergoing a transient and decoherence process. But is there hidden an echo signal in the beam's dipole motion?

7.2.3   Transverse echo — with diffusion

We now learned that the echo mechanism involves a long-term memory of the intricate structures of beam distribution in the phase space. One therefore expects that the echo will be easily affected by anything that even weakly perturbs the phase space distribution, such as a weak diffusion. This observation is then turned around and suggests a sensitive way to measure weak diffusion in storage rings. As mentioned earlier, question might also be raised whether the quantum mechanical uncertainty might play a role.

With diffusion, the beam distribution evolves according to the diffusion equation[14]

$$\frac{\partial \psi}{\partial t} + \omega(J)\frac{\partial \psi}{\partial \phi} \;=\; \frac{\partial}{\partial J}\left(D(J)J\frac{\partial \psi}{\partial J}\right), \tag{7.32}$$

where $D(J)$ is the diffusion coefficient. For now, let us keep the generality that $D(J)$ depends on the betatron amplitude J.

We will solve Eq. (7.32) in the limit of weak diffusion. Specifically, we assume that the diffusion has a small effect on a time scale during which the beam decoheres. The diffusion time is roughly equal to $\tau_{\text{diff}} \approx \frac{J}{D}$. Requiring $\tau_{\text{diff}} \gg \tau_{\text{decoh}}$, we get

$$D \;\ll\; \omega' J^2. \tag{7.33}$$

---

[14]G. Stupakov and A.W. Chao, Part. Accel. Conf. 1997.

In the limit of very strong diffusion, when the inequality opposite to (7.33) holds, the diffusion completely suppresses the echo effect. For this reason, echo effect is not relevant to electron storage rings due to the presence of the very strong diffusion caused by synchrotron radiation.

At time $t = 0$, the beam receives a small dipole kick. Immediately after the kick, the beam distribution is given by ψ_1 of Eq. (7.11). In the period $0 < t < \tau$, however, ψ_2 is not given by Eq. (7.11) but has to be found by solving Eq. (7.32) taking into account of the diffusion. Change variable from ϕ to

$$v = \phi - \omega(J)t.$$

Equation (7.32) gives

$$\frac{\partial \psi_2}{\partial t} = \left(\frac{\partial}{\partial J} - \omega'(J)t \frac{\partial}{\partial v} \right) \left[D(J)J \left(\frac{\partial \psi_2}{\partial J} - \omega'(J)t \frac{\partial \psi_2}{\partial v} \right) \right].$$

When $\omega' Jt \gg 1$, it then becomes

$$\frac{\partial \psi_2}{\partial t} \approx [\omega'(J)t]^2 D(J)J \frac{\partial^2 \psi_2}{\partial v^2}. \tag{7.34}$$

With initial condition $\psi_2(J, v, t = 0) = \psi_1(J, v)$, the solution of (7.34) for $0 < t < \tau$ is

$$\psi_2(J, v, t) = \theta \sqrt{2J\beta}\, \psi_0'(J)\, e^{-\frac{1}{3}D(J)J(\omega'(J))^2 t^3} \sin v, \tag{7.35}$$

where we have dropped the term $\psi_0(J)$ because it does not contribute to beam centroid signal.

At $t = \tau$, the beam receives a quadrupole kick, immediately before the kick, we have

$$\psi_3(J, \phi) = \theta \sqrt{2J\beta}\psi_0'(J)\, e^{-\frac{1}{3}D(J)J(\omega'(J))^2 \tau^3} \sin[\phi - \omega(J)\tau]. \tag{7.36}$$

Immediately after the kick, the perturbation ψ_4 is

$$\psi_4 \approx \theta q \omega'(J)\tau J \psi_0'(J)\sqrt{2J\beta}\; e^{-\frac{1}{3}D(J)J(\omega'(J))^2 \tau^3} \sin 2\phi \cos[\phi - \omega(J)\tau]. \tag{7.37}$$

Equations (7.35–7.37) are the same as (7.11), (7.13) and (7.19) except for the extra exponential factors due to diffusion.

In the period $t > \tau$, we make a change of variable from (ϕ, J, t) to (v_1, J, t), where

$$v_1 = \phi - \omega(J)(t - \tau),$$

in Eq. (7.32) to obtain

$$\frac{\partial \psi_5}{\partial t} = \left(\frac{\partial}{\partial J} - \omega'(J)(t - \tau)\frac{\partial}{\partial v_1} \right) \left[D(J)J \left(\frac{\partial \psi_5}{\partial J} - \omega'(J)(t - \tau)\frac{\partial \psi_5}{\partial v_1} \right) \right].$$

When $\omega' J(t - \tau) \gg 1$, we have

$$\frac{\partial \psi_5}{\partial t} \approx [\omega'(J)(t - \tau)]^2 D(J)J \frac{\partial^2 \psi_5}{\partial v_1^2}. \tag{7.38}$$

The solution of Eq. (7.38) with the initial condition $\psi_5(J, v_1 = \phi, t = \tau) = \psi_4(J, \phi)$, keeping only the term proportional to $\sin[v_1 + \omega(J)\tau]$, is

$$\psi_5 = \frac{1}{2}\theta q \omega'(J)\tau J\psi_0'(J)\sqrt{2J\beta}\sin[\phi - \omega(J)t + 2\omega(J)\tau]$$
$$\times \exp\left[-\frac{1}{3}D(J)J(\omega'(J))^2((t - \tau)^3 + \tau^3)\right]. \tag{7.39}$$

Substituting (7.39) into (7.23) and performing the ϕ-integration give for the echo signal,

$$\langle x\rangle^{\text{echo}}(t) = -\pi\beta\theta q\tau \int_0^\infty J^2 dJ\, \omega'(J)\psi_0'(J)\sin[\omega(J)(t - 2\tau)]$$
$$\times e^{-\frac{1}{3}D(J)J(\omega'(J))^2(\tau^3 + (t-\tau)^3)}. \tag{7.40}$$

The echo appears near $t \approx 2\tau$. Since we assume that diffusion is weak, the exponent in Eq. (7.40) is a slow function of time, and we can put $t = 2\tau$ in it, yielding

$$\langle x\rangle^{\text{echo}}(t) \approx \pi\beta\theta q\tau \int_0^\infty J^2 dJ\, \omega'(J)\psi_0'(J)\sin[\omega(J)(t - 2\tau)]e^{-\frac{2}{3}D(J)J(\omega'(J))^2\tau^3},$$
$$\tag{7.41}$$

which is a generalization of Eq. (7.25), and is our final echo result with diffusion.

If we now assume that $\omega(J)$ and $\psi_0(J)$ are given by Eqs. (7.4–7.5), and that the diffusion coefficient is a constant, $D(J) = D_0$, the integration in (7.41) can be performed explicitly. The result is

$$\langle x\rangle^{\text{echo}}(t) = \beta\theta q\omega' J_0\tau\, \text{Im}\left[\frac{e^{i\Phi}}{(\alpha - i\xi)^3}\right]$$
$$= \frac{\beta\theta q\omega' J_0\tau}{(\alpha^2 + \xi^2)^3}[\xi(3\alpha^2 - \xi^2)\cos\Phi + \alpha(\alpha^2 - 3\xi^2)\sin\Phi]. \tag{7.42}$$

where ξ and Φ were defined in Eq. (7.27), and

$$\alpha = 1 + \frac{2}{3}D_0(\omega')^2 J_0\tau^3. \tag{7.43}$$

In particular, ξ is a measure of the time, in units of the decoherence time, around $t = 2\tau$ when the echo signal appears. When $D_0 = 0$, this reduces to Eq. (7.26).

Echo amplitude   It also follows that

$$\langle x\rangle^{\text{echo ampl}}(t) = \beta\theta q\frac{\omega' J_0\tau}{(\alpha^2 + \xi^2)^{3/2}}. \tag{7.44}$$

Compared with Eq. (7.28), Eq. (7.44) is just to replace the quantity $1 + \xi^2$ by $\alpha^2 + \xi^2$ in the denominator. The difference between α and 1 can be written as

$$\alpha - 1 = \frac{2}{3}\left(\frac{D_0\tau}{J_0}\right)(\omega' J_0\tau)^2,$$

Figure 7.11: Reduction of echo amplitude when diffusion is included. Diffusion suppresses the echo amplitude but does not alter the time of the echo.

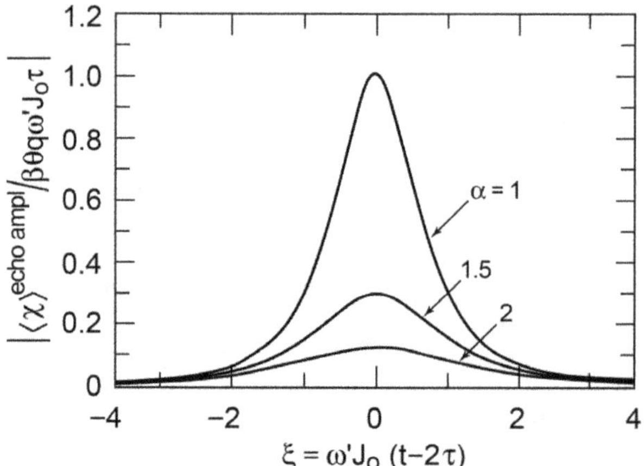

where the factor $\frac{D_0 \tau}{J_0}$ is just $\frac{\tau}{\tau_{\text{diff}}}$, i.e. the ratio of the time between the two kicks to the diffusion time. The factor $\omega' J_0 \tau$ is the betatron phase smear due to the frequency spread, and is $\gg 1$. We see that the difference between α and 1 is typically large, thus allowing a good opportunity to measure the diffusion strength. In fact, the measurement can in principle be so accurate that it is conceivable that one can explore the diffusion coefficient as a function of J using this method.

Figure 7.11 shows the echo amplitude as a function of ξ for various values of α. When $\alpha = 1$, the curve is just the envelope of Fig. 7.6. When $\alpha \neq 1$, the location of the echo is not changed, but the magnitude is reduced by a factor of α^3, indicating again the sensitive dependence on diffusion.

Optimizing echo amplitude by choosing the delay time τ   Given the diffusion coefficient D_0, one may want to maximize the echo signal by choosing τ. In the case without diffusion, the echo signal increases with τ indefinitely (within the linear approximation of weak dipole and quadrupole kicks). However, when diffusion is included, the echo signal would smear out if τ becomes too long. So there should be an optimum value of τ when the echo signal is optimized.

With

$$\langle x \rangle^{\text{echo ampl max}} = \frac{\beta \theta q \omega' J_0 \tau}{\alpha^3},$$

or, equivalently,

$$\frac{\langle x \rangle^{\text{echo ampl max}}}{\beta \theta q} = \frac{\Theta}{\left[1 + \frac{2}{3} \left(\frac{D_0}{\omega' J_0^2} \right) \Theta^3 \right]^3}, \tag{7.45}$$

where Θ, defined in Eq. (7.6), assumes the value $\omega' J_0 \tau$ here. The quantity $\frac{D_0}{\omega' J_0^2}$ in Eq. (7.45) is $\ll 1$ according to condition (7.33), while $\Theta \gg 1$.

Figure 7.12 shows $\langle x \rangle^{\text{echo ampl max}}$ as a function of τ. Without diffusion, $D_0 = 0$, the perturbation theory predicts a linear growth of the echo signal with the delay time τ. When $D_0 \neq 0$, the maximum value of $\langle x \rangle^{\text{echo ampl max}}$ is achieved at

$$\tau_{\max} = \left[\frac{16}{3} D_0 \, (\omega')^2 J_0 \right]^{-1/3}. \tag{7.46}$$

When D_0 is increased, it is expected that the echo observation should be made sooner. Equation (7.46) indicates that the observation time optimized for the largest echo signal scales as $D_0^{-1/3}$.

Figure 7.12: Dependence of the maximum echo amplitude when the time between the two kicks is varied. Three cases are for different strengths of the diffusion effect.

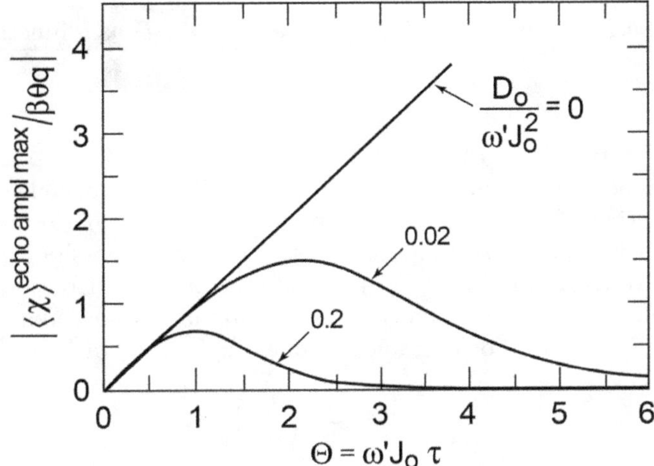

In terms of decoherence and diffusion times, the maximum is achieved at $\tau_{\max} \sim (\tau_{\text{diff}} \tau_{\text{decoh}}^2)^{1/3}$. When $\tau = \tau_{\max}$, we have $\alpha = \frac{9}{8}$ and the maximum echo amplitude is given by

$$\text{maximum} \left(\frac{\langle x \rangle^{\text{echo ampl max}}}{\beta\theta q} \right) = \left(\frac{8}{9} \right)^3 \left(\frac{16}{3} \frac{D_0}{\omega' J_0^2} \right)^{-1/3}. \tag{7.47}$$

The full width at half maximum time duration of the echo signal is [compare Eq. (7.30)]

$$\Delta T^{\text{FWHM}} = \frac{2\alpha}{\omega' J_0} \sqrt{2^{2/3} - 1} \approx 1.53 \frac{\alpha}{\omega' J_0}. \tag{7.48}$$

For $\tau = \tau^{\max}$, we have $\Delta T^{\text{FWHM}} = \frac{1.72}{\omega' J_0}$.

Measuring ω' and D_0 experimentally   Experimentally, both quantities ω' and D_0 can be found from the echo measurements. Measurement of ΔT^{FWHM} or the decoherence time τ_{decoh} can yield information on $\omega' J_0$. If the beam emittance J_0 is known, this allows one to determine ω'. After that, D_0 can be found by measuring $\tau_{\max}$ and using Eq. (7.46), or by measuring the maximum echo amplitude and using Eq. (7.47). See Table 7.2.

Table 7.2: Extracting information on the detuning parameter ω' and the diffusion coefficient D_0 from echo experiments.

Measurement	$\rightarrow$	information
τ_{decoh}	$\rightarrow$	$\omega' J_0$
height of echo	$\rightarrow$	$\frac{\omega' J_0 \tau}{\alpha^3}$
width of echo	$\rightarrow$	$\frac{\omega' J_0}{\alpha}$
$\tau_{\max}$	$\rightarrow$	$D_0 \omega'^2 J_0$
height of echo at $\tau = \tau_{\max}$	$\rightarrow$	$\frac{D_0}{\omega' J_0^2}$
shape of echo amplitude with respect to τ	$\rightarrow$	D as a function of J

Amplitude-dependent diffusion   In some cases (e.g. with beam-beam interaction, or near a nonlinear resonance), the diffusion coefficient may be a sensitive function of J, and it may not be a good approximation to treat it as a constant. It is conceivable that transverse echo gives a possibility not only to measure an average diffusion coefficient within the beam distribution, but also to obtain information on its dependence on betatron amplitude. With $D = D(J)$, we resort to Eq. (7.41). For example, we may model the diffusion coefficient as

$$D(J) \; = \; D_n \left(\frac{J}{J_0} \right)^n . \tag{7.49}$$

Equation (7.41) can be used to give

$$\langle x \rangle^{\mathrm{echo\ ampl\ max}} \; = \; -\pi \beta \theta q \tau \int_0^\infty J^2 dJ \; \omega'(J) \psi_0'(J) e^{-\frac{2}{3} D(J) J (\omega'(J))^2 \tau^3} , \tag{7.50}$$

where we have inserted $t = 2\tau$ to obtain the maximum echo amplitude.

If we assume Eqs. (7.4–7.5) for $\psi_0(J)$ and $\omega(J)$, then

$$\frac{\langle x \rangle^{\mathrm{echo\ ampl\ max}}}{\beta \theta q} \; = \; \frac{\Theta}{2} \int_0^\infty x^2 dx \; \exp\left[-x - \frac{2}{3} \left(\frac{D_n}{\omega' J_0^2} \right) \Theta^3 x^{n+1} \right] , \tag{7.51}$$

where $\Theta = \omega' J_0 \tau$. Figure 7.13 shows $\langle x \rangle^{\mathrm{echo\ ampl\ max}}$ as a function of Θ for various values of $\frac{D_n}{\omega' J_0^2}$; cases (a, b, c, d) are for $n = (-1, 1, 2, 3)$, respectively.[15] When $n = 0$, Fig. 7.12 is reproduced.

---

[15] Because of the polar coordinates used, the case $n = -1$, with its apparent divergence at $J = 0$, is not unphysical. It can result from kinematics. See Homework 7.17.

Figure 7.13: Same as Fig. 7.12, but the diffusion is now modeled as in Eq. (7.49). Cases (a), (b), (c), (d) correspond to $n = -1, 1, 2, 3$, respectively.

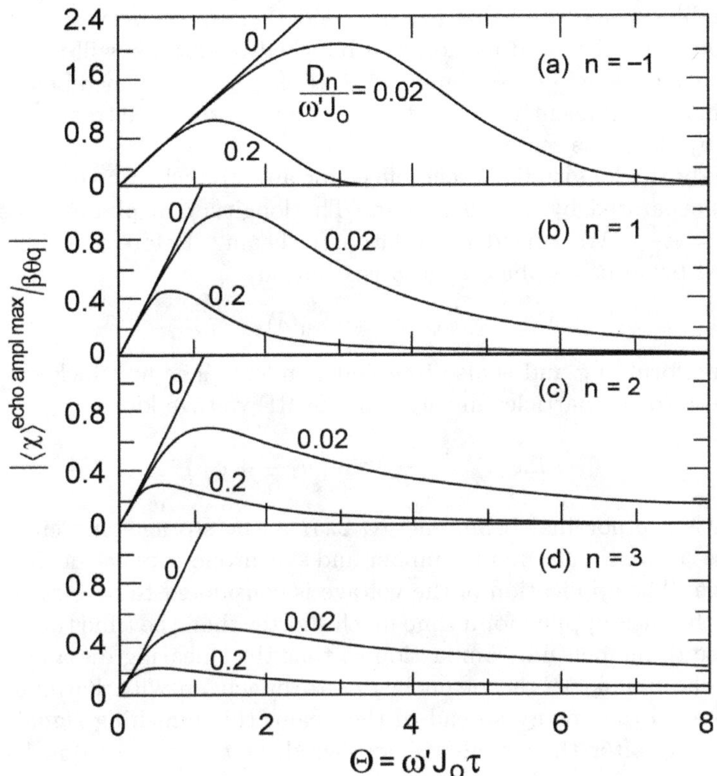

Homework 7.17

(a) Show that when $n = 0$, Eq. (7.51) reduces to (7.45).

(b) A special case occurs when $n = -1$, i.e. when the diffusion is stronger for smaller J than for larger J. Show that

$$\langle x \rangle^{\text{echo ampl max}} = \beta \theta q \omega' J_0 \tau \, e^{-\frac{2}{3} D_{-1} (\omega')^2 J_0 \tau^3}.$$

Solution (b) The fact that the $n = -1$ case is particularly simple can be traced to Eq. (7.32). Equation (7.32) becomes particularly simple when $D(J) \propto J^{-1}$. In fact, the longitudinal echo with diffusion has this same simplifying behavior. See Eq. (7.70) later.

Homework 7.18 Find what happens when diffusion is added to a waterbag model $\psi_0(J) = \frac{1}{2\pi J_0} H(J_0 - J)$ with arbitrary $\omega(J)$ and $D(J)$.

7.3　Longitudinal echo

The above analysis applies to the transverse motion of a bunched beam. With minor modifications, the same physics, and therefore a very similar analysis, applies to the longitudinal motion of a bunched beam. We will not repeat this similar analysis below — see Homework 7.21. For an unbunched beam, however, the dynamics is sufficiently different that requires a separate treatment, and will be the subject of this section.

To study the longitudinal echo effect for an unbunched beam, we also need two kicks separated by a long time τ. The longitudinal phase space is (z, δ), where $\delta = \frac{\Delta P}{P_0}$. We consider a relativistic beam. Before the first kick, the unbunched beam has a phase space distribution

$$\psi(z, \delta) = \psi_0(\delta),$$

which is uniform in z and is also time-independent. The first kick at time $t = 0$ is to impose to the particles in the beam an RF voltage kick,

$$\Delta\delta(z) = \frac{eV_1}{E_0} \sin\left(h_1 \frac{z}{R} + \phi_1\right),$$

where E_0 is the nominal beam energy, $2\pi R$ is the storage ring circumference, and h_1 are ϕ_1 are the harmonic number and synchronous phase of the oscillating RF voltage. The application of the voltage is considered to be instantaneous in the sense that it is applied for a time much shorter than the time the distribution smears due to momentum compaction so that the smearing distance $\Delta z \ll \frac{R}{h_1}$.

After the first kick, the beam begins to bunch up with harmonic number h_1, but due to the energy spread of the beam, this bunching signal decoheres in time. Long after the signal has decohered, at $t = \tau$, a second kick is then applied to the beam,

$$\Delta\delta(z) = \frac{eV_2}{E_0} \sin\left(h_2 \frac{z}{R} + \phi_2\right).$$

We choose a sign convention that $h_{1,2}$ are positive.

Another difference between the transverse (bunched) and the longitudinal (unbunched) echoes lies in the difference in the echo signals. In the transverse case, the echo signal lies in the centroid position $\langle x \rangle$, to be detected by a beam position monitor. On the other hand, the longitudinal echo signal lies in the beam current pattern as observed by a beam current monitor.

What we will show is that, at a much later, well-defined time, one will observe a sudden echo of longitudinal beam-current signal with harmonic number $h_2 - h_1$ if $h_1 < h_2$ and no echo if $h_1 \geq h_2$.

7.3.1　Longitudinal decoherence

We first consider the effect of decoherence after the first kick. The single-particle equations of motion are

$$\dot{z} = -\eta c\delta, \qquad \dot{\delta} = 0,$$

where η is the phase slippage factor. Immediately after the kick, the beam distribution is

$$
\begin{aligned}
\psi_1(z,\delta) &= \psi_0\left[\delta - \frac{eV_1}{E_0}\sin\left(h_1\frac{z}{R}+\phi_1\right)\right]\\
&\approx \psi_0(\delta) - \psi_0'(\delta)\frac{eV_1}{E_0}\sin\left(h_1\frac{z}{R}+\phi_1\right),
\end{aligned}
\tag{7.52}
$$

where we have assumed the kick is weak. Note the appearance of the derivative $\psi_0'(\delta)$.

After the kick, the particle coordinates are given by

$$
z_2 = z_1 - \eta ct\delta_1, \qquad \delta_2 = \delta_1.
$$

The beam distribution is therefore

$$
\begin{aligned}
\psi_2(z,\delta,t) &= \psi_1(z+\eta ct\delta,\delta)\\
&\approx \psi_0(\delta) - \psi_0'(\delta)\frac{eV_1}{E_0}\sin\left(h_1\frac{z+\eta ct\delta}{R}+\phi_1\right).
\end{aligned}
\tag{7.53}
$$

The beam current monitor measures

$$
\begin{aligned}
I_2(z,t) &= \int_{-\infty}^{\infty} d\delta\,\psi_2(z,\delta,t)\\
&\approx I_0 - \frac{eV_1}{E_0}\int_{-\infty}^{\infty} d\delta\,\psi_0'(\delta)\sin\left(h_1\frac{z+\eta ct\delta}{R}+\phi_1\right),
\end{aligned}
\tag{7.54}
$$

where $I_0 = \int_{-\infty}^{\infty} d\delta\,\psi_0(\delta)$ is the unperturbed beam current. We are of course interested only in the second term in Eq. (7.54). We shall drop the term I_0.

If we have a Gaussian δ-distribution,

$$
\psi_0(\delta) = \frac{I_0}{\sqrt{2\pi}\sigma_\delta}\,e^{-\delta^2/2\sigma_\delta^2},
\tag{7.55}
$$

Eq. (7.54) can be integrated to yield[16]

$$
I_2(z,t) \approx I_0\frac{eV_1 h_1\eta ct}{E_0 R}\cos\left(h_1\frac{z}{R}+\phi_1\right)\exp\left[-\frac{1}{2}\left(\frac{h_1\eta ct\sigma_\delta}{R}\right)^2\right].
\tag{7.56}
$$

At $t=0^+$ immediately after the kick, the beam current signal I_2 is zero. This is expected because the beam takes some time to become bunched after the energy modulation. In our linearized analysis, the signal will continue to grow indefinitely if not due to the energy spread of the beam which causes a

---

[16]Using

$$
\int_{-\infty}^{\infty} \delta d\delta\,e^{-\delta^2/2\sigma_\delta^2}\sin a\delta = \sqrt{2\pi}\,a\sigma_\delta^3 e^{-a^2\sigma_\delta^2/2}.
$$

decoherence. The decoherence time is the time it takes a particle with an energy error of $\delta = \sigma_\delta$ to move longitudinally by a distance $\frac{R}{h_1}$ relative to the reference on-momentum particle, i.e.

$$\tau_{\text{decoh}} = \frac{R}{h_1 \eta c \sigma_\delta}. \tag{7.57}$$

The fact that τ_{decoh} plays the role of a decoherence time is also clear from Eq. (7.56).

One obtains the amplitude of the signal by dropping the fast-oscillating cosine factor in Eq. (7.56),

$$I_2^{\text{ampl}}(t) \approx I_0 \frac{eV_1}{E_0 \sigma_\delta} \frac{t}{\tau_{\text{decoh}}} \exp\left(-\frac{t^2}{2\tau_{\text{decoh}}^2}\right).$$

The maximum value of I_2^{ampl} occurs when $t = \tau_{\text{decoh}}$ with

$$I_2^{\text{ampl max}} = I_0 \frac{eV_1}{E_0 \sigma_\delta} \exp\left(-\frac{1}{2}\right). \tag{7.58}$$

7.3.2　Longitudinal echo — no diffusion

The evolution of the coordinates of a single particle during the process is summarized by

$$
\begin{aligned}
&\text{At } t = 0^+, &&\begin{cases} z_1 = z_0, \\ \delta_1 = \delta_0 + \frac{eV_1}{E_0}\sin\left(h_1 \frac{z_0}{R} + \phi_1\right), \end{cases} \\
&\text{At } t = \tau^-, &&\begin{cases} z_3 = z_1 - \eta c \tau \delta_1, \\ \delta_3 = \delta_1, \end{cases} \\
&\text{At } t = \tau^+, &&\begin{cases} z_4 = z_3, \\ \delta_4 = \delta_3 + \frac{eV_2}{E_0}\sin\left(h_2 \frac{z_3}{R} + \phi_2\right), \end{cases} \\
&\text{At } t > \tau, &&\begin{cases} z_5 = z_4 - \eta c(t-\tau)\delta_4, \\ \delta_5 = \delta_4. \end{cases}
\end{aligned} \tag{7.59}
$$

What we need to do is to solve (z_0, δ_0) in terms of (z_5, δ_5), and substitute $\delta_0(z_5, \delta_5)$ into the initial distribution $\psi_0(\delta_0)$ to obtain the beam distribution at time $t > \tau$. Having obtained the beam distribution, the calculation of beam signal follows.

But this is a tedious calculation, and in the following, we shall be content with the perturbation approach again as we did in the transverse case. Keeping only the relevant terms for the echo effect, we find

$$\psi_3(z, \delta) \approx -\psi_0'(\delta)\frac{eV_1}{E_0}\sin\left(h_1 \frac{z + \eta c \tau \delta}{R} + \phi_1\right), \tag{7.60}$$

$$\psi_4(z, \delta) = \psi_3\left[z, \delta - \frac{eV_2}{E_0}\sin\left(h_2 \frac{z}{R} + \phi_2\right)\right]$$

$$\approx -\frac{eV_2}{E_0}\sin\left(h_2\frac{z}{R}+\phi_2\right)\frac{\partial\psi_3}{\partial\delta}$$

$$\approx \frac{eV_1}{E_0}\frac{eV_2}{E_0}\frac{h_1\eta c\tau}{R}\psi_0'(\delta)\sin\left(h_2\frac{z}{R}+\phi_2\right)\cos\left(h_1\frac{z+\eta c\tau\delta}{R}+\phi_1\right),$$

$$(7.61)$$

$$\psi_5(z,\delta,t) = \psi_4(z+\eta c(t-\tau)\delta,\delta)$$

$$\approx \frac{eV_1}{E_0}\frac{eV_2}{E_0}\frac{h_1\eta c\tau}{R}\psi_0'(\delta)\sin\left(h_2\frac{z+\eta c(t-\tau)\delta}{R}+\phi_2\right)$$

$$\times\cos\left(h_1\frac{z+\eta ct\delta}{R}+\phi_1\right).$$

Knowing ψ_5, the echo beam current signal is given by

$$I^{\text{echo}}(z,t) = \int_{-\infty}^{\infty} d\delta\,\psi_5(z,\delta,t). \qquad (7.62)$$

Substituting ψ_5 and assuming an initial Gaussian energy distribution (7.55), and performing the integration over δ, we obtain

$$I^{\text{echo}}(z,t) = -\frac{1}{2}I_0\frac{eV_1}{E_0}\frac{eV_2}{E_0}\frac{h_1\eta^2 c^2\tau}{R^2}$$

$$\times\left\{[(h_1+h_2)t-h_2\tau]\cos\left((h_1+h_2)\frac{z}{R}+\phi_1+\phi_2\right)\right.$$

$$\times\exp\left[-\frac{\eta^2 c^2\sigma_\delta^2}{2R^2}((h_1+h_2)t-h_2\tau)^2\right]$$

$$-[(h_1-h_2)t+h_2\tau]\cos\left((h_1-h_2)\frac{z}{R}+\phi_1-\phi_2\right)$$

$$\left.\times\exp\left[-\frac{\eta^2 c^2\sigma_\delta^2}{2R^2}((h_1-h_2)t+h_2\tau)^2\right]\right\}. \qquad (7.63)$$

An inspection of Eq. (7.63) shows that the echo signal occurs potentially at times $t^{\text{echo}} = \frac{h_2\tau}{h_1+h_2}$ and $t^{\text{echo}} = -\frac{h_2\tau}{h_1-h_2}$. However, we are interested only when $t^{\text{echo}} > \tau$. An echo signal occurs only if $h_2 > h_1$, and that the echo contribution to Eq. (7.63) can be rewritten as

$$I^{\text{echo}}(z,t) = -\frac{1}{2}I_0\frac{eV_1}{E_0}\frac{eV_2}{E_0}\frac{h_1\eta^2 c^2\tau}{R^2}[(h_2-h_1)t-h_2\tau]$$

$$\times\cos\left((h_2-h_1)\frac{z}{R}-\phi_1+\phi_2\right)$$

$$\times\exp\left[-\frac{\eta^2 c^2\sigma_\delta^2}{2R^2}((h_2-h_1)t-h_2\tau)^2\right]. \qquad (7.64)$$

Dropping the fast-oscillating cosine factor gives

$$I^{\text{echo ampl}}(t) = -\frac{1}{2}I_0\frac{eV_1}{E_0}\frac{eV_2}{E_0}\frac{h_1\eta c\tau}{R\sigma_\delta}\xi e^{-\xi^2/2}, \qquad (7.65)$$

$$\xi \;=\; \frac{\eta c \sigma_\delta (h_2 - h_1)}{R}\left(t - t^{\mathrm{echo}}\right),$$

$$t^{\mathrm{echo}} \;=\; \frac{h_2 \tau}{h_2 - h_1}\,.$$

Equation (7.65) for the longitudinal echo is the counterpart of Eq. (7.28) for the transverse echo.

One sees from Eqs. (7.64–7.65) that (a) the RF kick phases $\phi_{1,2}$ determine the phase of the current signal, but otherwise do not have a significant role; (b) the echo signal has a harmonic number $h_2 - h_1$; and (c) the echo amplitude vanishes at $t = t^{\mathrm{echo}}$. Observation (c) is in contrast to the transverse echo amplitude, which reaches maximum at $t = 2\tau$ (even though the *instantaneous signal* vanishes, the *amplitude* reaches a maximum).

The echo amplitude reaches maxima when $\xi = \pm 1$, or

$$t - t^{\mathrm{echo}} \;=\; \pm\frac{R}{\eta c \sigma_\delta (h_2 - h_1)}\,, \tag{7.66}$$

and the maximum value reached is

$$I^{\mathrm{echo\ ampl\ max}} \;=\; \pm\frac{1}{2} I_0 \frac{eV_1}{E_0} \frac{eV_2}{E_0} \frac{h_1 \eta c \tau}{R \sigma_\delta} \exp\left(-\frac{1}{2}\right). \tag{7.67}$$

Note that the beam current signal after the first kick and during the decoherence has harmonic number h_1, while the echo signal has harmonic number $h_2 - h_1$. So they are really different types of signals. In spite of this, however, one can compare the maximum echo amplitude (7.67) with the maximum decoherence signal (7.58) and obtain

$$\frac{I^{\mathrm{echo\ ampl\ max}}}{I_2^{\mathrm{ampl\ max}}} \;=\; \frac{1}{2}\frac{eV_2}{E_0}\frac{h_1 \eta c \tau}{R}\,. \tag{7.68}$$

Equation (7.68) is to be compared with Eq. (7.29).

The analogies between transverse (bunched beam) and longitudinal (unbunched beam) echoes are as follows,

$$\frac{eV_1}{E_0} \;\leftrightarrow\; \beta\theta$$

$$\frac{eV_2}{E_0} \;\leftrightarrow\; q$$

$$h_1 \;\leftrightarrow\; 1 \text{ (dipole kick)}$$

$$h_2 \;\leftrightarrow\; 2 \text{ (quadrupole kick)}$$

$$\frac{\eta c}{R} \;\leftrightarrow\; \omega' J_0$$

$$t - t^{\mathrm{echo}} \;\leftrightarrow\; t - 2\tau \tag{7.69}$$

Definition of ξ in Eq. (7.65) is analogous to the ξ defined in Eq. (7.27).

Echo experiment at AA ring Beautiful data, reproduced as Fig. 7.14, of longitudinal echo for an unbunched beam was obtained in the Antiproton Accumulator Ring at Fermilab (see footnote 5). Homework 7.19 intends to confront the analysis so far against these experimental observations. Additional AA data with diffusion effects are shown in Fig. 7.15 later.

Figure 7.14: A longitudinal echo experiment at the AA ring. The two decoherence signals after the two kicks, and the echo signal are clearly seen. There is also a high order echo at $t = 2\tau = 0.15$ s. [Courtesy Linda Spentzouris (2020).]

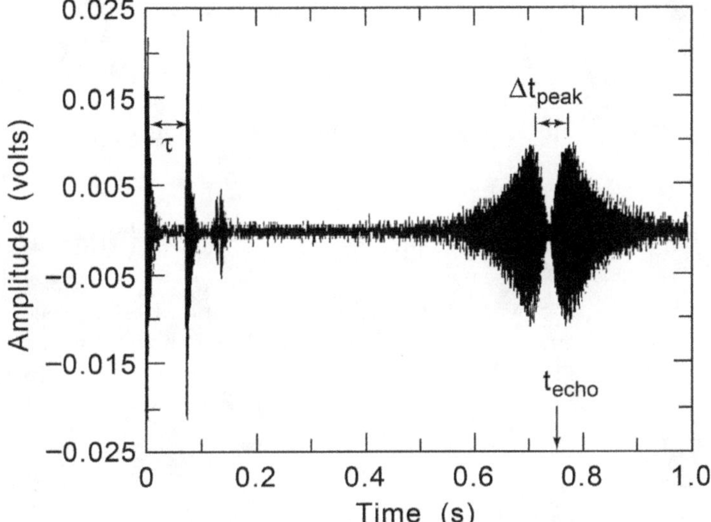

Homework 7.19 Compare the data in Fig. 7.14 with what is expected from Eqs. (7.65), (7.66) and (7.68). The parameters: $h_1 = 9$, $h_2 = 10$, $R = 70$ m, $\tau = 0.075$ s, $\eta = 0.023$, $E_0 = 8696$ MeV, $\Delta E = 3.2$ MeV.

Solution The following quantities can be checked: $t_{\text{peak}} = 10$, $\tau = 0.75$ s, $\Delta t_{\text{peak}} = \frac{2R}{\eta c \sigma_\delta} = 0.06$ s, $\tau_{\text{decoh}} = \frac{R}{h_1 \eta c \sigma_\delta} = 3$ ms, $\frac{I^{\text{echo ampl max}}}{I_2^{\text{ampl max}}} = 0.4$. Diffusion effect is not serious in this experiment. To detect diffusion effects, the waiting time τ needs to be longer.

7.3.3 Longitudinal echo — with diffusion

We next consider what happens to the longitudinal echo when there is diffusion. The analysis follows similar steps as in the transverse case. After the first kick, the beam distribution is still given by ψ_1 of Eq. (7.52). The distribution ψ_2, however, is not given by Eq. (7.53), but is to be solved with the diffusion equation

$$\frac{\partial \psi_2}{\partial t} - \eta c \delta \frac{\partial \psi_2}{\partial z} = \frac{\partial}{\partial \delta}\left[D(\delta) \frac{\partial \psi_2}{\partial \delta} \right]. \tag{7.70}$$

In Eq. (7.70) we have assumed that the diffusion originates from noise in δ. A noise in z is not going to have a significant effect on the echo.

Changing variable from z to

$$v = z + \eta ct\delta ,$$

and Let $\psi_2 = \psi_2(v, \delta, t)$, Eq. (7.70) becomes

$$\frac{\partial \psi_2}{\partial t} = \left(\frac{\partial}{\partial \delta} + \eta ct \frac{\partial}{\partial v} \right) \left[D(\delta) \left(\frac{\partial \psi_2}{\partial \delta} + \eta ct \frac{\partial \psi_2}{\partial v} \right) \right] .$$

When $\eta ct\delta \gg 1$, it simplifies,

$$\frac{\partial \psi_2}{\partial t} \approx (\eta ct)^2 D(\delta) \frac{\partial^2 \psi_2}{\partial v^2} . \tag{7.71}$$

The solution to Eq. (7.71), with the initial condition $\psi_2(v = z, \delta, t = 0) = \psi_1(z, \delta)$, is

$$\psi_2(v, \delta, t) \approx -\psi_0'(\delta) \frac{eV_1}{E_0} \exp\left[-\frac{1}{3} \left(\frac{\eta ch_1}{R} \right)^2 D(\delta) t^3 \right] \sin\left(h_1 \frac{v}{R} + \phi_1 \right) . \tag{7.72}$$

Having obtained ψ_2, the beam distributions before and after the second RF kick are found to be the same as Eqs. (7.60) and (7.61) when there was no diffusion, except that now they both have acquired an extra exponential diffusion factor

$$\exp\left[-\frac{1}{3} \left(\frac{\eta ch_1}{R} \right)^2 D(\delta) \tau^3 \right] .$$

After the second kick, the diffusion equation for ψ_5 is

$$\frac{\partial \psi_5}{\partial t} = \left(\frac{\partial}{\partial \delta} + \eta c(t - \tau) \frac{\partial}{\partial v_1} \right) \left[D(\delta) \left(\frac{\partial \psi_5}{\partial \delta} + \eta c(t - \tau) \frac{\partial \psi_5}{\partial v_1} \right) \right] ,$$

where we have changed variables to $\psi_5(v_1, \delta, t)$ with

$$v_1 = z + \eta c(t - \tau)\delta .$$

If $\eta c(t - \tau)\delta \gg 1$, we have

$$\frac{\partial \psi_5}{\partial t} \approx \eta^2 c^2 (t - \tau)^2 D(\delta) \frac{\partial^2 \psi_5}{\partial v_1^2} . \tag{7.73}$$

With the initial condition $\psi_5(v_1 = z, \delta, t = \tau) = \psi_4(z, \delta)$, the solution to Eq. (7.73) is

$$\psi_5(z, \delta, t) \approx \frac{1}{2} \frac{eV_1}{E_0} \frac{eV_2}{E_0} \frac{h_1 \eta c\tau}{R} \psi_0'(\delta)$$

$$\times \exp\left[-\frac{1}{3} \frac{\eta^2 c^2}{R^2} D(\delta) \left(h_1^2 \tau^3 + (h_2 - h_1)^2 (t - \tau)^3 \right) \right]$$

$$\times \sin\left[(h_2 - h_1) \frac{z}{R} + \phi_2 - \phi_1 + \frac{\eta c}{R} \delta((h_2 - h_1)t - h_2\tau) \right] , \tag{7.74}$$

where we have assumed $h_2 > h_1 > 0$ and have kept only the first-order echo contribution. The exponential factor in Eq. (7.74) can be replaced by its value near $t = \tau^{\text{echo}}$ with τ^{echo} given by Eq. (7.65) because it is slow function of t, i.e.

$$\exp\left[-\frac{1}{3}\frac{\eta^2 c^2}{R^2}D(\delta)\left(h_1^2\tau^3 + (h_2 - h_1)^2(t-\tau)^3\right)\right]$$

$$\approx \quad \exp\left[-\frac{1}{3}\frac{\eta^2 c^2}{R^2}D(\delta)\frac{h_1^2 h_2}{h_2 - h_1}\tau^3\right]. \tag{7.75}$$

Given ψ_5, the echo signal is given by Eq. (7.62). If we assume that the initial beam distribution is given by Eq. (7.55) and that the diffusion coefficient is a constant $D(\delta) = D_0$, we find that I^{echo} is given by the same expression (7.64) without diffusion, multiplied by the exponential factor (7.75).

It then follows that

$$I^{\text{echo ampl}} = \frac{1}{2}I_0\frac{eV_1}{E_0}\frac{eV_2}{E_0}\frac{h_1\eta c\tau}{R\sigma_\delta}\,\xi e^{-\xi^2/2}\exp\left[-\frac{1}{3}\frac{\eta^2 c^2}{R^2}D_0\frac{h_1^2 h_2}{h_2 - h_1}\tau^3\right], \tag{7.76}$$

where ξ was defined with Eq. (7.65).

The maxima of echo amplitude occur at $\xi = \pm 1$, with [compare Eq. (7.67)]

$$I^{\text{echo ampl max}} = \pm\frac{1}{2}I_0\frac{eV_1}{E_0}\frac{eV_2}{E_0}\frac{h_1\eta c\tau}{R\sigma_\delta}e^{-\frac{1}{2}}\exp\left[-\frac{1}{3}\frac{\eta^2 c^2}{R^2}D_0\frac{h_1^2 h_2}{h_2 - h_1}\tau^3\right], \tag{7.77}$$

and

$$\tau_{\max} = \left[\frac{(h_2 - h_1)R^2}{\eta^2 c^2 D_0 h_1^2 h_2}\right]^{1/3}.$$

Homework 7.20 Figure 7.15 reproduces the Fermilab AA ring data when the maximum echo amplitude is measured as τ is varied.

(a) Fit the data to Eq. (7.77) to find the value of D_0.

(b) Show that Eq. (7.70) implies the rms energy spread diffuses according to $\frac{d}{dt}\sigma_\delta^2 = 2D_0$.

(c) Use the result found in (a) to calculate how much time does it take the diffusion to contribute to an energy spread of 10^{-3}. This diffusion is weak, and yet it can be measured rather cleanly by the echo experiment.

Solution $D_0 = 1.3 \times 10^{-10}\,\text{s}^{-1}$. Proceed to show that this very weak diffusion causes σ_δ to grow by 10^{-3} in 1 hour. Note that the abscissa of Fig. 7.15 is time to echo, not τ.

Homework 7.21 Consider the longitudinal motion of a bunched beam. First kick the beam by an RF phase shift. Then, after a long waiting time τ, kick it again by an RF amplitude jump. Analyze the echo response of this beam. This problem is more similar to the transverse case than to the longitudinal unbunched beam case treated in the text.

Figure 7.15: Measured maximum echo amplitude as a function of t^{echo} in the AA ring. [Courtesy Linda Spentzouris (2000).] Diffusion is playing an important role. The solid curve is a fitting using Eq. (7.77), not just a smooth curve that goes through the data points.

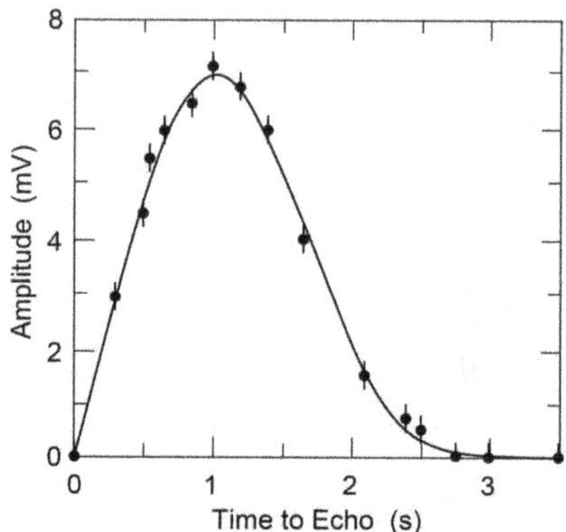

Homework 7.22 Find an expression for the echo amplitude if $D(\delta)$ is not a constant, but is $D(\delta) = D_n(\frac{\delta}{\sigma_\delta})^n$. Is there any indication from Fig. 7.15 that $D(\delta)$ is not a constant?

Homework 7.23 It was stated in the text after Eq. (7.70) that we have assumed the diffusion originates from noise in δ, while a noise in z does not affect the echo. Verify this statement.

Solution You first need to find the diffusion equation that replaces Eq. (7.70).

7.4   Echo-enabled harmonic generation

7.4.1   Harmonic generation

Consider an accelerator used as a radiation source. A charged-particle beam (say electrons) in the accelerator is made to pass through a radiator (say a dipole, an undulator, or an aluminum foil) to generate coherent electromagnetic radiation. As a beam passes through the radiator, its coherent radiation power depends sensitively on how the electron beam's longitudinal distribution is bunched. It is well known for example that if an electron beam is bunched to a length ℓ, its coherent radiation will concentrate up to around the frequency $\sim \frac{\ell}{c}$ or wavelength $\sim \ell$. A DC beam ($\ell = \infty$), for example, will never coherently radiate.[17]

---

[17]Put in another way: its radiation will have an infinite wavelength, i.e. it can only radiate the static Coulomb field.

Figure 7.16: Sketch of the layout of an echo-enabled harmonic generation scheme. [Courtesy Gennady Stupakov (2021).]

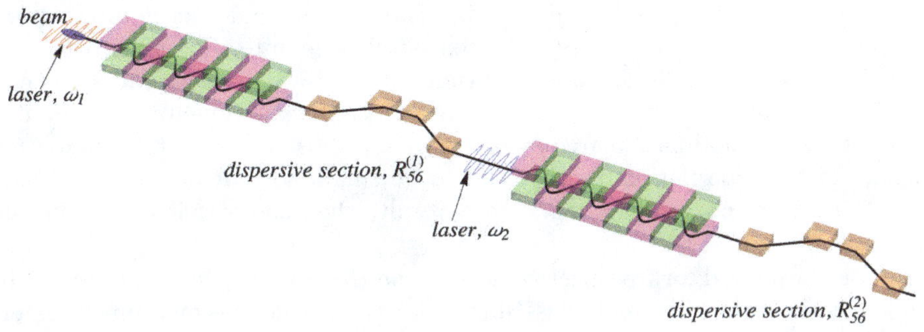

An important drive in the development of radiation sources has been the effort to push up on the radiation frequency. The higher the radiation frequency, the shorter the wavelength, and the higher its resolution in exploring the object under study using the radiation source. One way to push up the radiation frequency is by shortening the electron bunch length. However, this approach reaches a limit when the bunch length cannot be shortened beyond some point.

At this point, a more accessible approach is by harmonic generation. If an electron bunch of length ℓ can be modulated into a train of many smaller evenly spaced microbunches each of length $\ell_{\mathrm{microbunch}}$, then the coherent radiation will concentrate up to around a higher frequency with wavelength $\sim \ell_{\mathrm{microbunch}}$. This is referred to as harmonic generation. In case the original bunch is divided into M microbunches, then we have $\ell_{\mathrm{microbunch}} \lesssim \frac{\ell}{M}$; the harmonic generation is then by a factor of M, i.e. it is an M-th harmonic generation.

A free electron laser is an accelerator device that manipulates the longitudinal distribution of a charged-particle beam. In FEL applications, for example, there have been efforts to generate FEL radiation of shorter wavelength by various harmonic generation techniques. In this section, we shall mention one such technique using an ingenious application of the echo mechanism (see footnote 7). The setup of this echo-enabled harmonic generation (EEHG) is sketched in Fig. 7.16.

As seen in Fig. 7.16, there are two stages in the EEHG layout, each stage consisting of one laser-undulator section for energy modulation followed by one dispersion section for bunch compression. The basis of the EEHG mechanism is quite similar to that of the longitudinal echo of an unbunched beam; the first modulation is equivalent to the first kick, while the second modulation serves as the second kick. After a well-defined time periods, represented by the two dispersion sections, an echo emerges downstream. In the present case, the echo signal consists of a beam distribution that contains the harmonic we aim for.

The radiator (not shown in Fig. 7.16), is then located where the echo appears, downstream of the second dispersion section.

In a longitudinal echo in a circular accelerator, the beam energy is first modulated at frequency ω_1 and, after a long delay τ, a second energy modulation is applied at another frequency ω_2. Then, after a definite time related to τ, a sudden echo signal in the beam's longitudinal distribution occurs at frequency $m\omega_1 - k\omega_2$, where m and k are integers. In actual experiments in a circular accelerator, the modulation frequencies $\omega_{1,2}$ are most likely in the microwave range, and the modulation depths tend to be small, so that only the low harmonics, $m, k = \pm 1$, are excited. As a result, the echo signal is also in the microwave range.

To be considered for free electron lasers, the EEHG setup shown in Fig. 7.16 applies to the case of a single-pass linac. All time scales become much faster. Compared with the echo in circular accelerators, the two energy modulations and the echo all appear within the time frame of a single pass. The energy modulations are extended to laser frequencies, and the modulation depths considered will be deep. The reason the modulations have to be deep is because we intend to generate high harmonics, requiring high values of mode indices m, k; and as we will soon see, that in turn requires deeper modulations and exploits nonlinear effects the deep modulations provide. On the other hand, the basic physical process, the echo mechanism, is one and the same.

In Fig. 7.16, each energy modulation section is executed by beating the laser with an undulator tuned to resonate at the laser frequency. After passing the undulator, an electron beam copropagating with the resonating laser will develop an energy modulation with a modulation wavelength same as the laser wavelength.

Each dispersion section in Fig. 7.16 consists of a 4-dipole chicane-like beamline. After passing through a chicane, particles rearrange their longitudinal positions according to their energies. A particle with higher energy will move ahead while a particle with lower energy will fall behind compared with an on-energy particle ($R_{56}^{(1)}$ and $R_{56}^{(2)}$ are positive).

The two dispersion sections represent the two waiting periods in the case of longitudinal echo in circular accelerators. The long waiting time τ in the circular accelerator now becomes $R_{56}^{(1)}$ of the first dispersion section. To lengthen τ, we just increase the value of $R_{56}^{(1)}$. The length of the linac does not have to increase. Similarly, the second long waiting time between the second modulation and the echo is adjusted by the value of $R_{56}^{(2)}$. The radiator is then positioned right after the $R_{56}^{(2)}$ dispersion section.

In the application to free electron lasers, a single modulation-compression stage already serves as a harmonic generation scenario. It is called high gain harmonic generation (HGHG) scheme, invented by Lihua Yu.[18] In this section, we first briefly discuss the HGHG mechanism, and devote more of the contents to EEHG. Drawing analogy to our discussion in the earlier sections, the HGHG is

---

[18]L.H. Yu, Phys. Rev. A44, 5178 (1991).

analogous to an initial stage of the decoherence effect and the EEHG is analogous to the echo effect.

7.4.2 HGHG mechanism

Consider a layout consisting only of the first modulation-compression stage of Fig. 7.16. This is an arrangement for HGHG mechanism. For simplicity of analysis, let us consider the case when the energy modulation is done by a sawtooth waveform instead of the more conventional sinusoidal form.

Consider a beam with initial distribution that has zero energy spread and uniform longitudinal distribution. Let the energy modulation be sawtooth form with peak values $\pm\hat{\delta}_1$ and sawtooth wave length λ_1. Let the compressor have the transport R_{56} element given by R_1. The sawtooth energy modulation is first imprinted to the uniform beam after the laser-modulator. This energy imprint then transforms the beam into a train of microbunches, each with zero length and spaced by a distance of λ_1 after the beam passes through the compressor, provided the parameters are chosen to satisfy the HGHG condition,

$$\hat{\delta}_1 R_1 = \frac{\lambda_1}{2}. \tag{7.78}$$

The HGHG mechanism is illustrated in Fig. 7.17.

With this condition, harmonic generation is fulfilled. The initially uniform beam is now microbunched, and now because of its short microbunch length ($\ell_{\text{microbunch}} = 0$), radiates coherently strongly at all harmonic wavelengths $\frac{\lambda_1}{M}$ up to harmonics as high as $M \sim \frac{\lambda_1}{\ell_{\text{microbunch}}}$.

Figure 7.17: An illustration of the HGHG mechanism. The beam's phase space distribution is illustrated; (a) before entering the modulator, (b) after the modulator and before the compressor, and (c) after the compressor. In this idealized example, the beam is transformed from DC uniform to infinitely sharply microbunched.

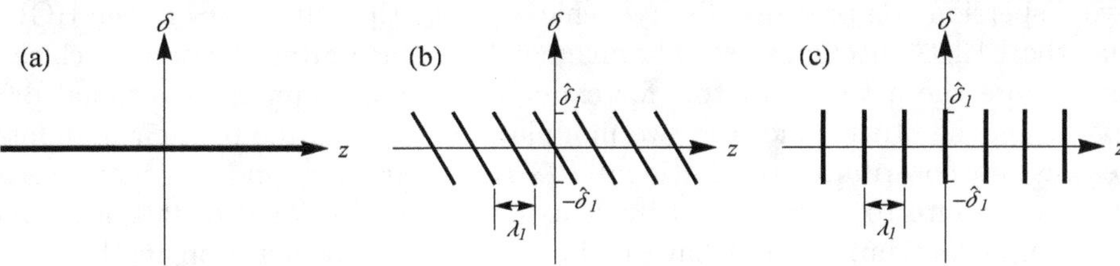

This calculation of beam distribution is for the case of a sawtooth modulation. The beam after HGHG consists of a train of infinitely sharp microbunches as shown. Later on page 525, we will calculate the beam distribution when the modulation has a sinusoidal waveform. The highest harmonics reachable by HGHG mechanism will be shown to be more limited.

7.4.3   EEHG mechanism

With two modulation-compression stages, the EEHG scheme offers a way to microbunch the initial beam with a spacing much finer than the HGHG scheme.

As mentioned, a single-staged modulation-compression scheme is similar to the decoherence effect, while the double-staged scheme is more analogous to the echo effect. In terms of echo language, the modulator of the first section acts as a first kicker. The chicane of the first section acts of the waiting period of τ. The modulator of the second section acts as the second kick. Then the final dispersion section acts as the second waiting period for the echo to form.

In the decoherence effect, as we learned in Sec. 7.3.1, the beam after the energy modulation takes a decoherence time τ_{decoh}, Eq. (7.57), to pick up a maximum bunching signal. During this time, the particle has accumulated a momentum compaction

$$R_{56} \;=\; \tau_{\text{decoh}}\eta c \;=\; \frac{R}{h_1 \sigma_\delta}.$$

For a particle with energy deviation of σ_δ, for example, its longitudinal position would shift by a distance $R_{56}\sigma_\delta = \frac{R}{h_1} = \lambda$ of the modulation wavelength. It is clear that the HGHG condition (7.78) corresponds to the condition when the beam bunching reaches a maximum in the single-stage HGHG scheme. If a radiator is installed at this position of maximum bunching, an HGHG radiation source can be constituted.

Perhaps a comment should be made here concerning the intuitive interpretation of what constitutes a "drift space". Being familiar with the transverse dynamics, a drift space is intuitively equated to a free space where there are no magnets. This is not so when dealing with longitudinal dynamics. In longitudinal dynamics, a free space without magnets ($R_{56} = 0$) does not constitute a drift space. A drift space has to consists of a space with R_{56}. In this context, the two dispersion compressors, the two chicanes, are the dift spaces in the HGHG and the EEHG mechanisms. The magnet free space after the second chicane and before the actual radiator, for example, do not count as additional drift space. On the other hand, the two modulation sections, due to their undulator designs, do contribute to additional R_{56}'s of their own, and for that reason, they contribute to additional drift spaces. These additional drift spaces may have important impact and have to be taken into consideration in the optics and beam dynamics designs.

The idea of EEHG is to generate an echo. To generate the echo, the procedure asks first for a long time delay after the first stage, and following the long delay, a second stage is applied. In the EEHG language, it translates to a condition when R_1 is much larger than that required for the HGHG, i.e. for the first modulation-compression stage of the EEHG, we have instead of (7.78),

$$\hat{\delta}_1 R_1 \gg \lambda.$$

In other words, the section is strongly over-compressed. The beam becomes hopelessly defocused from the point of view of HGHG, but is the regime where EEHG aims to apply.

We will analyze the two-stage EEHG mechanism in the following two sections, first for the case when the energy modulations are done by idealized sawtooth wave forms, and then repeat when the modulations are the more conventional sinusoidal wave forms.

In spite of the close analogy between EEHG in a linac and longitudinal echo for unbunched beam in a circular accelerator, however, for later reference, we also mention here a few details where they differ. These differences come from the fact that the modulation depths in EEHG are not small and, by intention, nonlinear effects enter. The longitudinal echo in a circular accelerator is linear in the modulation depth due to the typically weak modulation voltages applied, while for high harmonic generation, EEHG is intrinsically a nonlinear device. A few consequences of their difference appear as follows.

- Longitudinal echo requires $h_2 > h_1$. EEHG does not have this requirement.

- The location of the longitudinal echo does not depend on the amplitudes V_1 and V_2 of the two kicks. The location of the EEHG signal depends on the modulation depths $\hat{\delta}_1$ and $\hat{\delta}_2$.

- Longitudinal echo requires a time reversal after the second kick to reverse the winding of the phase space due to the first kick. This is not the mechanism of EEHG.

7.4.4 Sawtooth modulation

Let the energy modulations be sawtooth form with peak values $\pm\hat{\delta}_1$ and $\pm\hat{\delta}_2$, and sawtooth wave lengths λ_1 and λ_2, for the two stages respectively. Let the two compressors have the R_{56} elements given by R_1 and R_2. The EEHG nominally proposes to use sinusoidal energy modulations, but for illustration of its mechanism, we use an idealized sawtooth modulation first.[19] The sinusoidal case is given later.

As mentioned, the HGHG scheme requires only one modulation-compression stage. If the electron beam has no energy spread, a single-stage HGHG can provide very high harmonics by a sawtooth modulator under condition (7.78). In principle, after HGHG, the beam becomes infinitely sharp in longitudinal length, which in turn gives infinitely high harmonic generation. However, even with sawtooth modulation, the finite beam energy spread limits the highest

---

[19]Sawtooth modulation is also a way to increase the reach of even higher harmonics; D. Ratner and A.W. Chao, Proc. Free Electron Laser Conf., Shanghai, 2011, MOPB21; G. Stupakov and M. Zolotorev, ibid (2011), MOPB19.

Figure 7.18: Illustration of the sawtooth EEHG mechanism with the phase space distribution of the beam followed through the two stages of energy modulation-bunch compression until the echo signal appears. This illustration assumes sawtooth wave form for both energy modulations.

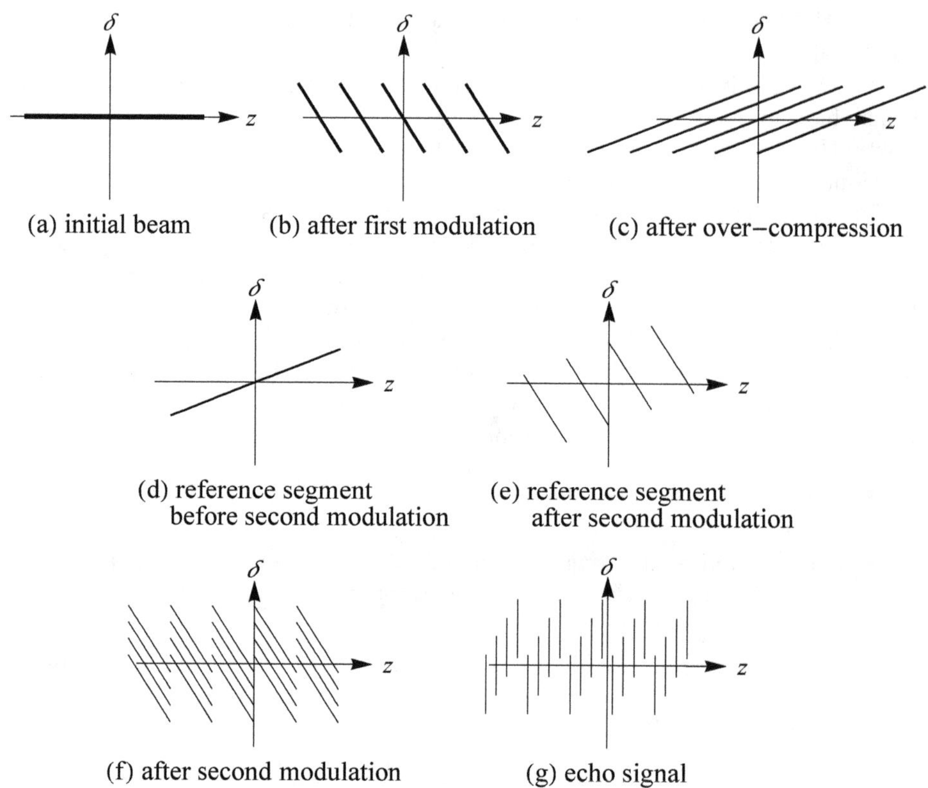

(a) initial beam   (b) after first modulation   (c) after over−compression

(d) reference segment
before second modulation   (e) reference segment
after second modulation

(f) after second modulation   (g) echo signal

harmonic reachable by the HGHG scheme. EEHG was invented to overcome this energy-spread limitation by having two modulation-compression stages. The trick is to apply the echo effect.

Figure 7.18 illustrates the sawtooth EEHG process. Consider a beam initially with a uniform longitudinal distribution and no energy spread. Its phase space distribution is shown as Fig. 7.18(a). Figure 7.18(b) shows the beam phase space after the first sawtooth modulation.

The beam then goes through a compression of the first stage. For echo bunching, unlike the case of HGHG, the first stage will over-compress by a large factor. After compression, there is a large number of zigzag filaments in the beam's distribution in phase space (z, δ), as shown in Fig. 7.18(c). At a given value of z, the number of filaments is

$$\frac{2\hat{\delta}_1 R_1}{\lambda_1} \gg 1\,.$$

For EEHG, we intend to have

$$\hat{\delta}_1 R_1 \gg \lambda_1, \lambda_2\,.$$

See Homework 7.24 for the parameters used in Fig. 7.18. In particular, $\lambda_2 = \lambda_1$ and $\hat{\delta}_1 R_1 = -\frac{5}{2}\lambda_1$ in this illustration.

Taking the single segment that goes through $(0,0)$ point for reference, reproduced as Fig. 7.18(d). Its distribution after the second modulator is shown in Fig. 7.18(e). We have assumed in this example that the second sawtooth modulation has shifted its phase by $\frac{\lambda_2}{2}$ relative to the first modulation. Figure 7.18(f) then shows the phase space distribution of the whole beam at that point when all segments are included.

The beam is then sent to go through the second compression so that all the segments "stand up" vertically, yielding a highly microbunched beam with 100% bunching factor if we choose the value of R_2 just right. After the second compression, the beam appears as Fig. 7.18(g). The beam is now microbunched, ready to radiate at a high harmonic determined by the microbunch spacing. In this case, the microbunch spacing is $\frac{\lambda_1}{4} = \frac{\lambda_2}{4}$, in agreement with Eq. (7.79) of Homework 7.25.

Harmonics much higher than either of the two seed lasers can thus be generated, and the final bunching factor is nominally 100% provided the beam initially has no energy spread and provided the modulations are truly sawtooth. If wave form is not sawtooth but is sinusoidal, the conclusions should not change substantially except that the final bunching factor then would have to be compromised. See next section.

It should be noted that what EEHG offers is to increase the harmonic number, i.e. the microbunches become more finely and evenly spaced compared with the HGHG mechanism for which the microbunches are more lumped. In case the beam has no energy spread and the modulation is sawtooth, both HGHG and EEHG give a train of δ-function microbunches, and both give infinitely high harmonics. For a given harmonic, the HGHG and the EEHG give the same radiation power. However, either a finite beam energy spread or a sinusoidal modulation waveform will broaden the δ-function microbunch lengths and therefore limit the highest harmonics that can be reached. In this situation, EEHG offers a much higher reachable harmonics than the HGHG because its microbunches have thinner width due to the long stretching of the first modulation-compression stage. The advantage of EEHG over HGHG lies in the first stretching and thinning of phase space in the first modulation-compression stage. The more stretching the first stage is, the higher the harmonic generation can potentially reach.

The stretching also means the beam is more vulnerable to energy fluctuations due to any noise — due to intrabeam scattering, residual gas or ion scattering or incoherent synchrotron radiation, for example — whose wavelength is comparable or shorter than λ_1 or λ_2.[20] Let the magnitude of this energy fluctuation at the end of the first stage be δ_{noise}, the echo microbunching mechanism requires that δ_{noise} be less than the fine energy spacing between the fine filaments, which is equal to $2\hat{\delta}_1 / \frac{2\hat{\delta}_1 R_1}{\lambda_1} = \frac{\lambda_1}{R_1}$. The noise criterion is therefore

$$\delta_{\text{noise}} \ \ll \ \frac{\lambda_1}{R_1} \, .$$

Homework 7.24 Follow the beam dynamics for the case illustrated in Fig. 7.18. Convince yourself that the beam is echo microbunched after the EEHG process as illustrated.

Solution This illustration in Fig. 7.18 has the following parameters: $\hat{\delta}_1 R_1 = \frac{5}{2}\lambda_1, \hat{\delta}_1 R_2 = \frac{1}{2}\lambda_2, \lambda_2 = \lambda_1, \frac{R_2}{R_1} = \frac{1}{5}, \frac{\hat{\delta}_2}{\hat{\delta}_1} = \frac{5}{4}$. The modulations have opposite signs as that shown for the HGHG case in Fig. 7.17 but this choice is nonessential to EEHG. The second sawtooth wave form is shifted by half wavelength at the origin $z = 0$ relative to the first sawtooth. The echo microbunch spacing is $\frac{\lambda_1}{4}$.

Homework 7.25 The example shown in Fig. 7.18 has $\lambda_2 = \lambda_1$. This does not have to be the case. Show that another case can be achieved when the two sawtooth modulations have $\lambda_2 \neq \lambda_1$ and the two wave forms do not contain a phase shift. Analyze the conditions for an echo signal to be produced under this scenario and show that under these conditions, the microbunch spacing is given by

$$\lambda_{\text{microbunch}} \ = \ \left| \frac{\lambda_2^2}{\lambda_1 + 2\hat{\delta}_1 R_1} \right| \, . \tag{7.79}$$

Solution Consider four particles.

<u>Particle 1</u> After the first modulation, consider the particle at $z = \frac{\lambda_1}{2}, \delta = \hat{\delta}_1$. After first dispersion section, it has $z = \frac{\lambda_1}{2} + R_1\hat{\delta}_1, \delta = \hat{\delta}_1$. We want

$$\frac{\lambda_1}{2} + R_1\hat{\delta}_1 \ = \ \frac{M\lambda_2}{2} \, ,$$

with M a large integer (M can be negative). Particles along this line segment has

$$\delta(z) \ = \ \frac{\hat{\delta}_1}{\frac{\lambda_1}{2} + R_1\hat{\delta}_1} \, z \, .$$

After the second modulation (assuming it has the same phase as the first modulation), the particle at $z = \frac{\lambda_2}{2}$ has a new energy deviation

$$\delta\left(\frac{\lambda_2}{2}\right) + \hat{\delta}_2 \ = \ \frac{\hat{\delta}_1 \lambda_2}{\lambda_1 + 2R_1\hat{\delta}_1} + \hat{\delta}_2 \, .$$

---

[20]G. Stupakov, Proc. FEL Workshop, New York, 2013, WEPSO68.

We want this particle to end up with $z = 0$ after the second dispersion section. This requires

$$\frac{\lambda_2}{2} + \left(\frac{\hat{\delta}_1 \lambda_2}{\lambda_1 + 2R_1 \hat{\delta}_1} + \hat{\delta}_2 \right) R_2 = 0,$$

$$\implies \quad \hat{\delta}_2 = -\frac{\lambda_2}{2R_2} \left(1 + \frac{2R_2 \hat{\delta}_1}{\lambda_1 + 2R_1 \hat{\delta}_1} \right).$$

<u>Particle 2</u> Consider particles along the same line segment $\delta(z)$ as particle 1. After the second modulation, consider the particle at $z = -\frac{\lambda_2}{2}$. It has a new energy deviation

$$\delta\left(-\frac{\lambda_2}{2} \right) - \hat{\delta}_2 = -\frac{\hat{\delta}_1 \lambda_2}{\lambda_1 + 2R_1 \hat{\delta}_1} - \hat{\delta}_2.$$

After the second dispersion section, it has

$$z = -\frac{\lambda_2}{2} - \left(\frac{\hat{\delta}_1 \lambda_2}{\lambda_1 + 2R_1 \hat{\delta}_1} + \hat{\delta}_2 \right) R_2 = 0,$$

as it should.

<u>Particle 3</u> We now consider a particle initially at $z = \frac{3\lambda_1}{2}$. After the first modulation, it has $\delta = \hat{\delta}_1$. After the first dispersion, it has $z = \frac{3\lambda_1}{2} + R_1 \hat{\delta}_1, \delta = \hat{\delta}_1$. Particles along this line segment has

$$\delta(z) = \frac{\hat{\delta}_1}{\frac{\lambda_1}{2} + R_1 \hat{\delta}_1} (z - \lambda_1).$$

After the second modulation, the particle at $z = \frac{\lambda_2}{2}$ has a new energy deviation

$$\delta\left(\frac{\lambda_2}{2} \right) + \hat{\delta}_2 = \frac{\hat{\delta}_1 (\lambda_2 - 2\lambda_1)}{\lambda_1 + 2R_1 \hat{\delta}_1} + \hat{\delta}_2.$$

After the second dispersion, this particle has

$$z = \frac{\lambda_2}{2} + \left(\frac{\hat{\delta}_1 (\lambda_2 - 2\lambda_1)}{\lambda_1 + 2R_1 \hat{\delta}_1} + \hat{\delta}_2 \right) R_2 = -\frac{2\hat{\delta}_1 \lambda_1 R_2}{\lambda_1 + 2R_1 \hat{\delta}_1}.$$

We want this z to be equal to $\pm\frac{\lambda_2}{M}$.

<u>Particle 4</u> We now consider particles along the same segment $\delta(z)$ as particle 3. After the second modulation, the particle at $z = -\frac{\lambda_2}{2}$ has an energy deviation

$$\delta\left(-\frac{\lambda_2}{2} \right) - \hat{\delta}_2 = \frac{\hat{\delta}_1 (-\lambda_2 - 2\lambda_1)}{\lambda_1 + 2R_1 \hat{\delta}_1} - \hat{\delta}_2.$$

After the second dispersion, this particle has

$$z \;=\; -\frac{\lambda_2}{2} + \left(\frac{\hat{\delta}_1(-\lambda_2 - 2\lambda_1)}{\lambda_1 + 2R_1\hat{\delta}_1} - \hat{\delta}_2 \right) R_2 \,,$$

which automatically is equal to $\pm\frac{\lambda_2}{M}$.

Summarizing, these conditions give

$$
\begin{aligned}
R_1\hat{\delta}_1 \;&=\; \frac{1}{2}(M\lambda_2 - \lambda_1)\,, \\[2mm]
|R_2\hat{\delta}_1| \;&=\; \frac{\lambda_2^2}{2\lambda_1}\,, \\[2mm]
R_2\hat{\delta}_2 \;&=\; -\frac{\lambda_2}{2} - \frac{R_2\hat{\delta}_1}{M}\,, \\[2mm]
\lambda_{\text{microbunch}} \;&=\; \frac{\lambda_2}{|M|} \;=\; \frac{\lambda_2^2}{|\lambda_1 + 2R_1\hat{\delta}_1|}\,.
\end{aligned}
$$

The example case in Fig. 7.18 satisfies these conditions with $M = -4$.

If $|M| \gg 1$, we have

$$R_1\hat{\delta}_1 \;\approx\; \frac{M}{2}\lambda_2, \qquad R_2\hat{\delta}_2 \;\approx\; -\frac{\lambda_2}{2}, \qquad R_2\hat{\delta}_1 \;=\; \pm\frac{\lambda_2^2}{2\lambda_1}\,.$$

7.4.5   Sinusoidal modulation

In practice, sawtooth wave forms are not easily available and one has to approximate them by superposing multiple wave frequencies. In case of only one frequency component, we have sinusoidal wave forms. The analysis for sinusoidal energy modulations can be found in Stupakov's original literature and is reproduced below (see footnote 7).

Consider an electron bunch interacting with a laser beam of frequency ω_1 in an undulator. The bunch length is much longer than the laser wavelength, so one can locally consider a longitudinally uniform electron beam. As mentioned earlier, the energy spread of the electron beam plays a role in HGHG and EEHG. In particular, if the energy spread is zero and if sawtooth wave forms are available, the HGHG mechanism already provides a way to reach very high harmonics, and there is not much incentive to consider the EEHG. In the analysis below, we will therefore include two additional considerations, a sinusoidal modulation and an electron beam with finite energy spread.

We assume the initial energy distribution of the beam is Gaussian around E_0 with the variance σ_E, i.e.

$$f(p) \;=\; \frac{N_0}{\sqrt{2\pi}}\, e^{-p^2/2}\,,$$

where $p = \frac{E - E_0}{\sigma_E}$ and N_0 is the number of particles per unit length of the beam. Normalization is $\int dp\, f(p) = N_0$.

Let us consider the first modulator-chicane stage first. As the beam passes through the first modulator, its particle energy is modulated with the amplitude ΔE_1, so that the final dimensionless energy deviation p' is related to the initial p by

$$p' = p + A_1 \sin \zeta ,$$

where

$$A_1 = \frac{\Delta E_1}{\sigma_E}, \qquad \zeta = \frac{\omega_1 z}{c} ,$$

and z is the longitudinal coordinate in the beam.

The distribution function after the interaction with the laser becomes

$$f(\zeta, p) = \frac{N_0}{\sqrt{2\pi}} \exp\left[-\frac{1}{2}(p - A_1 \sin \zeta)^2 \right],$$

where we have used (ζ, p) as the phase space.

Sending then the beam through a dispersive system with the dispersive strength $R_{56}^{(1)}$, converts the longitudinal z into z', $z' = z + R_{56}^{(1)} \frac{p\sigma_E}{E_0}$, and makes the distribution function to read

$$f(\zeta, p) = \frac{N_0}{\sqrt{2\pi}} \exp\left[-\frac{1}{2}(p - A_1 \sin(\zeta - B_1 p))^2 \right], \tag{7.80}$$

where

$$B_1 = R_{56}^{(1)} \frac{\omega_1}{c} \frac{\sigma_E}{E_0} .$$

Integration of f over p gives the distribution of the beam density N as a function of the coordinate ζ,

$$N(\zeta) = N_0 \int_{-\infty}^{\infty} dp\, f(\zeta, p) .$$

Noting that this density is a periodic function of ζ, one can expand it into Fourier series,

$$\frac{N(\zeta)}{N_0} = 1 + 2 \sum_{k=1}^{\infty} b_k \cos k\zeta ,$$

where the coefficient b_k is the Fourier amplitude of the harmonic k. Calculation with the Gaussian beam gives an analytical expression for b_k,

$$b_k = e^{-\frac{1}{2} B_1^2 k^2} J_k(A_1 B_1 k) , \tag{7.81}$$

where J_k is the Bessel function of order k. See Homework 7.26.

HGHG scheme In the HGHG scheme, Eq. (7.81) gives the bunching factor for the k-th harmonic radiation. For an HGHG application, we optimize the value of $|b_k|$ for the desired harmonic k. It is evident from Eq. (7.81) that one would choose B_1 to be small so that the factor $e^{-\frac{1}{2} B_1^2 k^2}$ is not small. The factor $|J_k(A_1 B_1 k)|$ then requires A_1 to be large. An inspection of the expression of

A_1 indicates that the energy modulation therefore has to be much larger than the initial energy spread of the electron beam. This is one of the practical limitations of the HGHG scheme.

One consequence of HGHG scheme is the fact that several harmonics are generated, starting with $k = 1$ to some highest reachable harmonic as given by Eq. (7.81). Depending on the intended harmonic k, quantities A_1 and B_1 are chosen to optimize the bunching factor $|b_k|$. Figure 7.19 shows $|b_k|$ as contour plots in the (A_1, B_1) space for three cases $k = 3, 5, 8$. The orders of magnitudes of the optimal values of A_1 and B_1 for an intended harmonic k can be read off these contour plots.

EEHG scheme   For the EEHG scheme, we do not optimize the value of $|b_k|$ in Eq. (7.81), and the beam is then sent through a second modulator-chicane section, and the optimization is done after both stages are completed. In general, the frequency ω_2 of the second laser beam can differ from the frequency of the first laser ω_1. As we will see, one advantage of the EEHG is that the energy modulations do not have to be too large, and it offers to reach a much higher harmonics. The disadvantage is that it is more complicated and it is more susceptible to noise effects because the microbunches are much thinner in phase space.

As mentioned, optimizing HGHG requires a small B_1 and a large A_1, while as we will see, optimizing EEHG requires a large B_1 and a small A_1. Again, keep in mind that B_1 is related to the momentum compaction R_{56} and A_1 is related to the energy modulation depth.

The final distribution function at the exit from the second dispersion section can be easily found by applying consecutively two more transformations to (7.80). The first of these two transformations corresponding to the modulation of the beam energy with dimensionless amplitude A_2 is $p' = p + A_2 \sin(\frac{\omega_2 z}{c} + \phi)$, where ϕ is a phase of the second laser beam. The second transformation corresponds to the passage through the second dispersive element, $z' = z + R_{56}^{(2)} \frac{p \sigma_E}{E_0}$. The resulting final distribution function is

$$f(\zeta, p) \;=\; \frac{N_0}{\sqrt{2\pi}} \exp\left[-\frac{1}{2}\left(p - A_2 \sin(K\zeta - KB_2 p + \phi)\right.\right.$$

$$\left.\left. - A_1 \sin(\zeta - (B_1 + B_2)p + A_2 B_1 \sin(K\zeta - KB_2 p + \phi))\right)^2\right], \quad (7.82)$$

where

$$\zeta \;=\; \frac{\omega_1 z}{c}, \qquad B_2 \;=\; R_{56}^{(2)} \frac{\omega_1}{c} \frac{\sigma_E}{E_0}, \qquad K \;=\; \frac{\omega_2}{\omega_1}.$$

To simplify the analysis, we first consider the limit of small energy modulations,

$$|A_1, A_2| \ll 1.$$

At the same time we assume that the product $A_2 B_1$ may not be small. We then expand the distribution function (7.82) keeping only linear terms in A_1 and A_2. It is easy to see that the result of such an expansion gives three terms: the zero

Figure 7.19: Contour plots of the HGHG bunching factor $|b_k|$ in the (A_1, B_1) space; (a) $k = 3$, (b) $k = 5$, (c) $k = 8$. The blue curves show the contours for the regions where $|b_k| > 0.3, 0.2, 0.15$, respectively for (a), (b) and (c).

(a)

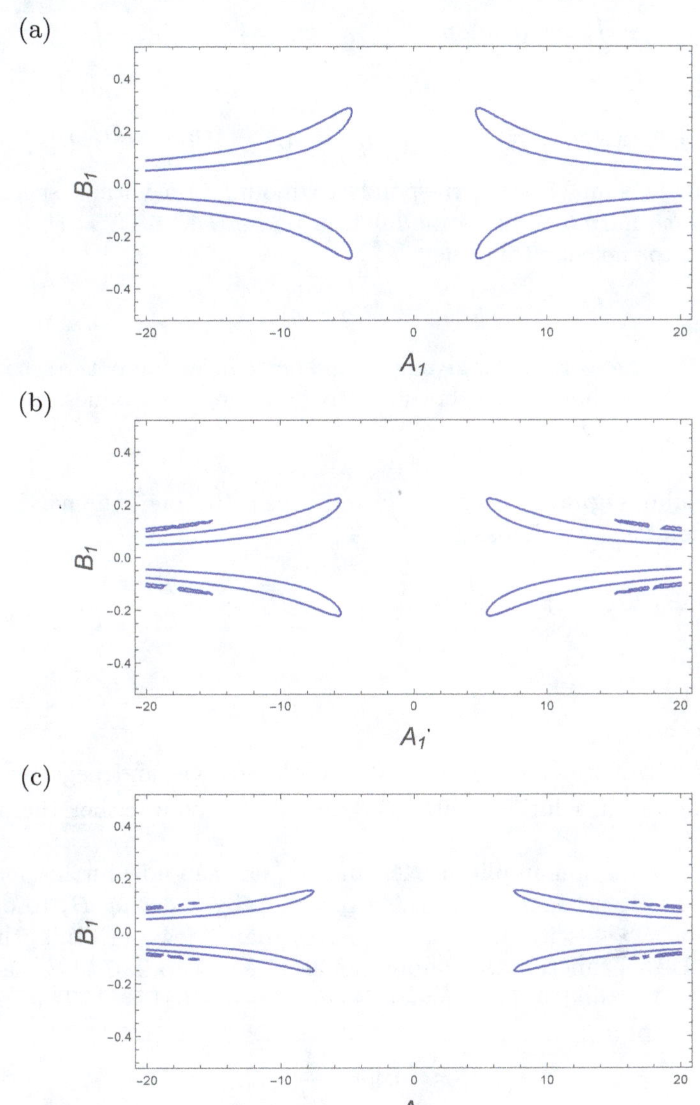

(b)

(c)

order term $\frac{N_0}{\sqrt{2\pi}} e^{-p^2/2}$, the term $\frac{N_0}{\sqrt{2\pi}} e^{-p^2/2} p A_2 \sin(K\zeta - K B_2 p + \phi)$ linear in A_2, and finally, the term, which we denote f_3,

$$f_3(\zeta, p) = \frac{N_0}{\sqrt{2\pi}} e^{-\frac{p^2}{2}} p A_1 \sin\left[\zeta - (B_1 + B_2)p + A_2 B_1 \sin(K\zeta - K B_2 p + \phi)\right].$$

$$(7.83)$$

This last term is responsible for the echo effect, so we will focus on this term only. Integrating Eq. (7.83) over p gives

$$\frac{N(\zeta)}{N_0} = \frac{1}{N_0}\int_{-\infty}^{\infty} f_3(\zeta,p)dp = \sum_{k=-\infty}^{\infty} c_k \cos\left(\zeta + kK\zeta + k\phi\right), \qquad (7.84)$$

with

$$c_k = -A_1[B_1 + B_2(kK+1)]J_k\left(B_1 A_2\right) e^{-\frac{1}{2}(B_1 + B_2(kK+1))^2}.$$

Each term in sum (7.84) corresponds to modulation with a specific harmonic. The k-th harmonic has a modulation phase $(kK+1)\zeta = (k\omega_2 + \omega_1)\frac{z}{c}$, corresponding to the echo frequency

$$\omega_{\text{echo}} = k\omega_2 + \omega_1. \qquad (7.85)$$

and we see that large values of k give up-frequency conversion of the second laser frequency ω_2. Note that k takes both positive and negative values; modulation wavelength is given by $\frac{2\pi c}{|\omega_{\text{echo}}|}$.

Optimum echo signal　For given k and given B_1, one can maximize the absolute value of $|c_k|$ by choosing

$$B_1 + B_2(kK+1) = \pm 1 \qquad \Longrightarrow \qquad B_2 = -\frac{B_1 \mp 1}{kK+1}, \qquad (7.86)$$

which gives

$$|c_k| = \frac{A_1}{\sqrt{e}}\left|J_k(A_2 B_1)\right|, \qquad (7.87)$$

where $e = 2.72$. Note that since k takes both positive and negative values, Eq. (7.86) allows for solutions with $R_{56}^{(1)}$ and $R_{56}^{(2)}$ having either the same or opposite signs.

For given modulation amplitudes A_1 and A_2, one can now further maximize the value of the Bessel function in this expression by choosing B_1 to optimize $J_k(A_2 B_1)$. For $k > 4$, with an accuracy better than a few per cents, the maximum of the Bessel function J_k is approximately equal to $0.67\,k^{-1/3}$, and it is attained when the value of its argument is equal to $k + 0.81\,k^{1/3}$. This gives the maximum value of $|c_k|$,

$$|c_k| \approx 0.406\,\frac{|A_1|}{|k|^{1/3}}, \qquad (7.88)$$

for

$$|B_1| \approx \frac{|k|}{A_2}\left(1 + 0.81\,|k|^{-2/3}\right). \qquad (7.89)$$

To summarize, for chosen modulation frequencies $\omega_{1,2}$ and a chosen harmonic $k \gg 1$, and given the two modulation depths $|A_1| = \left|\frac{\Delta E_1}{\sigma_E}\right| \ll 1$ and $|A_2| = \left|\frac{\Delta E_2}{\sigma_E}\right| \ll 1$, the optimal bunching factor is obtained by choosing the compression parameter $|B_1| = \left|R_{56}^{(1)}\frac{\omega_1}{c}\frac{\sigma_E}{E_0}\right|$ to be given by Eq. (7.89) and $|B_2| = \left|R_{56}^{(2)}\frac{\omega_1}{c}\frac{\sigma_E}{E_0}\right|$

given by Eq. (7.86). The echo frequency is then given by Eq. (7.85). The optimal bunching factor $|c_k|$ is given by Eq. (7.88). In all these expressions, σ_E is the initial unperturbed energy spread of the electron beam.

Since we assume $|A_2| \ll 1$, it follows from Eq. (7.89) that $|B_1|$ should be much larger than unity. The first compression therefore specifies — it in fact requires — a condition of over-compression. Also note that although our definitions of $A_{1,2}$ and $B_{1,2}$ involve the energy spread in the beam σ_E, the optimal settings for $R_{56}^{(1)}$ and $R_{56}^{(2)}$, as follows from Eqs. (7.89) and (7.86), do not depend on σ_E. They do depend however on the amplitude of the energy modulation $A_2 \sigma_E$ in the second modulator.

Equation (7.88) demonstrates a remarkable feature of the EEHG, namely a very slow decay $\propto |k|^{-1/3}$ with the harmonic number k. This is in sharp contrast with the Gaussian suppression of the HGHG bunching factor with k, as demonstrated by Eq. (7.81). Compared with the two-stage EEHG, the single-stage HGHG is much simpler, while the EEHG can be used to aim for much higher harmonics.

From a practical point of view, a particular case of interest is when $\omega_1 = \omega_2$. This case can be realized with a single laser system by splitting the laser light and sending it to both undulators. Remarkably, this case allows an analytical solution for arbitrary values of parameters A_1 and A_2.

One can show that in the general case of arbitrary values of parameters A_1 and A_2, the frequency for the echo signal is $\omega_{\text{echo}} = k\omega_2 + m\omega_1$, where k and m are arbitrary (positive or negative) integer numbers. The reason we concluded Eq. (7.85) with $m = \pm 1$ in our analysis so far is because we have linearized with respect to A_1.

Figure 7.20 illustrates one example of the evolution of the phase space of the beam as it travels through the sinusoidal EEHG system. A comparison can be made with the EEHG mechanism for a sawtooth modulations as shown schematically in Fig. 7.18. A large value of $R_{56}^{(1)}$ in the first modulator over-compresses the beam, leading to "shredding" of the beam phase space as shown in the top right picture. At this stage after the first modulation-compression, the beam has an almost uniform density distribution in its z-projection, while each beam shred has an energy spread much smaller than that of the original beam. With this much reduced energy spread, the second modulator has a much easier job to do using a relatively modest value of $R_{56}^{(2)}$.

Homework 7.26

(a) Derive Eq. (7.81) for the HGHG bunching factor starting from Eq. (7.80).

(b) Take the limit when the beam has no energy spread, $\sigma_E = 0$, to see the behavior of the bunching factors. Note that $A_1 B_1 \neq 0$ in this limit.

Solution (a) Change variables of the double integration, and then apply the identities

$$\frac{1}{\sqrt{2\pi}} \int_{-\infty}^{\infty} dp\, e^{-ikB_1 p} \exp\left[-\frac{1}{2}(p - A_1 \sin \zeta)^2 \right] = e^{-ikB_1 A_1 \sin \zeta - \frac{1}{2}k^2 B_1^2},$$

Figure 7.20: Simulation of phase space distribution of a beam going through an EEHG; (top left) after the first energy modulation, (top right) after the first over-compression, (lower left) after the second energy modulation, (lower right) after the second compression and emergence of echo. [Courtesy Gennady Stupakov (2020).]

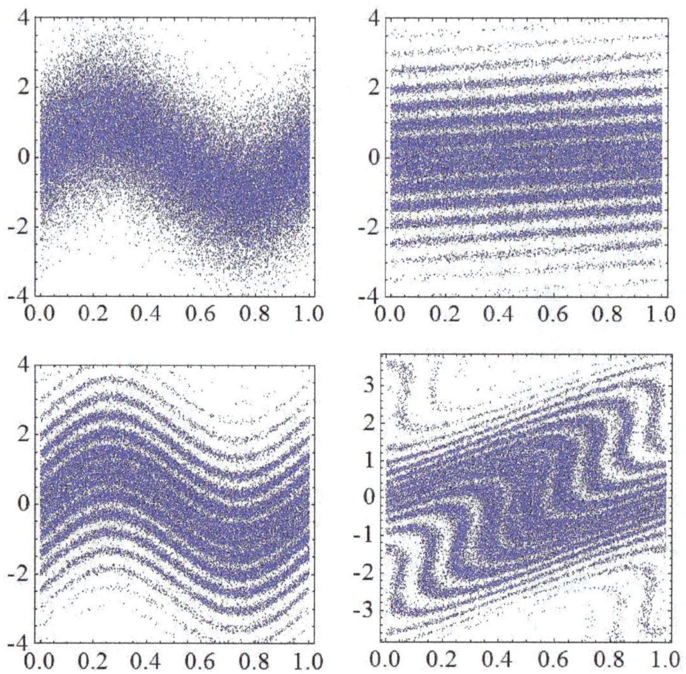

$$\frac{1}{2\pi} \int_0^{2\pi} d\zeta \, e^{-ik\zeta - ikB_1 A_1 \sin \zeta} \;=\; J_k(kB_1 A_1).$$

The result (7.81) is valid for arbitrary A_1, not only when it is small.

Homework 7.27 Derive Eq. (7.84) for the EEHG bunching factor starting from Eq. (7.83).

Solution Use the identities

$$\sin(\phi_1 + x \sin \phi_2) \;=\; \sum_{k=-\infty}^{\infty} J_k(x) \sin(\phi_1 + k\phi_2),$$

$$\int_{-\infty}^{\infty} dp \, p e^{-p^2/2} \sin \beta p \;=\; \sqrt{2\pi} \, \beta e^{-\beta^2/2}.$$

Homework 7.28 Our analysis of transverse echo was carried out for weak kicks. Our result, particularly Eq. (7.31), applies when the kick strengths are kept to the leading linear order. In this section we learned that higher order terms have their significant contributions under some circumstances. Try to contemplate the effects if the transverse echo analysis is carried out to higher orders. Are there more echoes? When do they appear and what are their relative amplitudes? Qualitative answers suffice.

7.5 Spin echo

As mentioned in Sec. 7.1, echoes are a widespread occurrence in accelerators. For each dimension of beam dynamics, there is a set of corresponding echo effects. In addition to the three orbital dimensions of beam dynamics, there is a fourth dimension involving spin. Naturally, there is a set of spin echo effects in accelerators.

Also as mentioned, echoes come about due to a long memory of intricate phase space interplays — Liouville theorem as their underlying foundation, so they do not apply to electron beams when synchrotron radiation overwhelms the dynamics and washes out the memory. For a discussion of spin echo, therefore, we consider the case of a polarized proton beam in a synchrotron, a subject we covered in Chapter 5.

7.5.1 Spin interference and spin echo — A recap

As elaborated in Chapter 5, when a polarized proton beam is accelerated through a depolarization resonance, its polarization is reduced by a well-defined calculable reduction factor. When the beam subsequently crosses a second resonance, the final beam polarization is considered to be reduced by the product of the two reduction factors corresponding to the two crossings, each calculated independently of the other. This is a good approximation when the spread of spin precession frequency $\Delta\nu_{\text{spin}}$ of the beam (particularly due to its energy spread) is sufficiently large that the spin precession phases of individual particles smear out completely during the time τ between the two crossings. This approximate picture, however, ignores two spin dynamical effects: an interference effect and a spin echo effect.

The interference effect occurs when $\Delta\nu_{\text{spin}}$ is too small, or when τ is too short, to complete the smearing process. In this case, the two resonance crossings interfere with each other, and the final polarization exhibits constructive or destructive patterns depending on the exact value of τ. Typically, the beam's energy spread is large and this interference effect does not appear. To study this effect, therefore, it is necessary to reduce the beam energy spread and to consider two resonance crossings very close to each other. The spin interference effect has been discussed in some detail on page 366.

The other mechanism, also due to the interplay between two resonance crossings, is the much more subtle effect of spin echo. It turns out that even when the precession phases appear to be completely smeared between the two crossings, there will still be a sudden and short-lived echo signal of beam polarization signal at a time τ after the second crossing; we will show later that the magnitude of the spin echo can be as large as 57% if the beam was initially 100% polarized. The surprise is that this echo signal exists even when the beam has a sizable energy spread and even when τ is large.

So in echo effect there could be a sensitive way to experimentally test the intricate spin dynamics, including a detection of possible weak spin diffusion in a synchrotron. By examining the orbital and the spin echoes, detailed and intricate orbital dynamics or spin dynamics can be explored.

In a planar synchrotron, the spin of a particle precesses rapidly around the vertical $\hat{y}$-axis with the spin tune $\nu_{\text{spin}} = a\gamma$, where $a = \frac{g-2}{2}$ with g the gyromagnetic ratio of the particle under consideration, and γ is the Lorentz energy factor. As the particle is accelerated or decelerated in the synchrotron, its γ changes and spin tune changes accordingly. As the spin tune varies, the spin motion of a particle will be strongly affected if the particle experiences perturbing electromagnetic fields as it executes orbital motion in the synchrotron, and if its spin tune comes close to, or crosses a depolarization resonance

$$a\gamma = \kappa, \tag{7.90}$$

where κ specifies the resonance location (κ = integer for imperfection resonances, κ = integer $\pm$ vertical betatron tune for intrinsic resonances). In this situation, the perturbation on spin motion can be characterized by a single complex quantity, the resonance strength ϵ, which can be expressed in terms of a Fourier harmonic of the perturbing fields around the accelerator.

One such analysis was obtained by Froissart and Stora when the spin tune crosses the resonance linearly in time starting and ending far from the resonance. Their result yields the neat Froissart–Stora formula, Eq. (5.50),

$$\frac{S_y(\infty)}{S_y(-\infty)} = 2e^{-\pi|\epsilon|^2/2\Gamma} - 1,$$

that relates the final polarization $S_y(\infty)$ after crossing to the initial polarization $S_y(-\infty)$ before crossing, where Γ is the resonance crossing speed.

As a generalization of the Froissart–Stora analysis, a matrix formalism has been developed that allows the calculation of polarization near a resonance when the crossing pattern of the spin tune consists of a combination of constant in time, linear in time, and sudden discrete jumps (see Sec. 5.5). The condition of being far from the resonance before and after crossing is also removed. This matrix formalism can be used, for example, to study multiple crossings of a resonance. Using this formalism, we are able to explore the constructive and destructive interference as well as the spin echo effects of these crossings.

Experimentally, the interference effects are expected to be most readily observable when the polarized beam has a small spread $\Delta\nu_{\text{spin}}$ in particles' spin tunes. To meet this requirement, the beam must have a small energy spread.

Ways to produce a beam with small energy spread should help greatly the exploration of the spin interference effects.

For a polarized beam of larger energy spreads, the multiple resonance crossings are far separated from each other and the interference effects are not readily observable. However, interference still emerges in the more subtle echo effect. In this echo experiment, two resonance crossings normally considered to be far separated can still interfere with each other to produce a spin echo signal at an unexpected long time after the second crossing. In Sec. 5.5, we showed the echo effect with Figs. 5.9 and 5.10 by numerical simulation. In the following section, we intend to give a more detailed derivation of the spin echo signal with two successive resonance crossings.

7.5.2 Beam with energy spread

We apply the matrix formalism, Sec. 5.5, to our analysis of the spin echo effect. Consider a case when the on-momentum particle of the beam is made to double jump-cross a resonance according to the prescription (5.64). In other words, the on-momentum particle's energy as a function of time θ is such that

$$a\gamma_0(\theta) = \kappa + \begin{cases} -A, & \text{if } \theta < \theta_1, \\ A, & \text{if } \theta_1 < \theta < \theta_2, \\ -A, & \text{if } \theta_2 < \theta. \end{cases}$$

For an off-momentum particle in the beam with energy deviation $\delta = \frac{\Delta\gamma}{\gamma_0}$, on the other hand, its spin tune will be given by

$$a\gamma(\theta) = \kappa + \begin{cases} -A + \kappa\delta, & \text{if } \theta < \theta_1, \\ A + \kappa\delta, & \text{if } \theta_1 < \theta < \theta_2, \\ -A + \kappa\delta, & \text{if } \theta_2 < \theta. \end{cases}$$

We will continue the analysis following Eq. (5.67), and assume $|\delta| \ll 1, |\kappa\delta| \ll 1$ and $|\kappa\delta| \ll A$. With $A > 0$, the first resonance crossing is a jump from below. Given these conditions, the polarization of the off-momentum particle is found after some algebra to be (see footnote 4)

$$P_y(\theta < \theta_1) \approx \frac{A}{\Omega},$$

$$P_y(\theta_2 > \theta > \theta_1) \approx \frac{A}{\Omega^3}\left\{ A^2 - |\epsilon_0|^2 + 2|\epsilon_0|^2 \cos\left[(\Omega + \frac{A\kappa\delta}{\Omega})(\theta - \theta_1)\right] \right\},$$

$$\begin{aligned} P_y(\theta > \theta_2) \approx \frac{A}{\Omega^5}\Bigg\{ &(A^2 - |\epsilon_0|^2)^2 + 2|\epsilon_0|^4 \cos\left[\Omega(\theta - \theta_1) + \frac{A\kappa\delta}{\Omega}(2\theta_2 - \theta - \theta_1)\right] \\ &- 2A^2|\epsilon_0|^2 \cos\left[\Omega(\theta + \theta_1 - 2\theta_2) - \frac{A\kappa\delta}{\Omega}(\theta - \theta_1)\right] \\ &+ 2|\epsilon_0|^2(A^2 - |\epsilon_0|^2) \cos\left[(\Omega - \frac{A\kappa\delta}{\Omega})(\theta - \theta_2)\right] \\ &+ 4A^2|\epsilon_0|^2 \cos\left[(\Omega + \frac{A\kappa\delta}{\Omega})(\theta_2 - \theta_1)\right] \Bigg\}. \end{aligned} \tag{7.91}$$

where $\Omega = \sqrt{A^2 + |\epsilon_0|^2}$, and we have used the fact that $\Omega_0 = \Omega_2 \approx \Omega - \frac{A\kappa}{\Omega}\delta$ and $\Omega_1 \approx \Omega + \frac{A\kappa}{\Omega}\delta$.

Equation (7.91) applies to the case of a single particle. For a beam of particles with a finite energy spread, an averaging on the result (7.91) over the beam's energy distribution will have to be performed. Assuming the energy distribution is Gaussian with rms σ_δ, the result is

$$P_y(\theta < \theta_1) \approx \frac{A}{\Omega},$$

$$P_y(\theta_2 > \theta > \theta_1) \approx \frac{A}{\Omega^3}\left\{A^2 - |\epsilon_0|^2 + 2|\epsilon_0|^2 e^{-\frac{A^2\kappa^2\sigma_\delta^2}{2\Omega^2}(\theta-\theta_1)^2}\cos\Omega(\theta-\theta_1)\right\},$$

$$\begin{aligned}
P_y(\theta > \theta_2) \approx \frac{A}{\Omega^5}\Big\{&(A^2 - |\epsilon_0|^2)^2 + 2|\epsilon_0|^4 e^{-\frac{A^2\kappa^2\sigma_\delta^2}{2\Omega^2}(2\theta_2-\theta-\theta_1)^2}\cos\Omega(\theta-\theta_1)\\[2mm]
&- 2A^2|\epsilon_0|^2 e^{-\frac{A^2\kappa^2\sigma_\delta^2}{2\Omega^2}(\theta-\theta_1)^2}\cos\Omega(\theta+\theta_1-2\theta_2)\\[2mm]
&+ 2|\epsilon_0|^2(A^2 - |\epsilon_0|^2)e^{-\frac{A^2\kappa^2\sigma_\delta^2}{2\Omega^2}(\theta-\theta_2)^2}\cos\Omega(\theta-\theta_2)\\[2mm]
&+ 4A^2|\epsilon_0|^2 e^{-\frac{A^2\kappa^2\sigma_\delta^2}{2\Omega^2}(\theta_2-\theta_1)^2}\cos\Omega(\theta_2-\theta_1)\Big\}.
\end{aligned} \tag{7.92}$$

The echo signal is contained in the expression of $P_y(\theta > \theta_2)$ of Eq. (7.92). We discuss some features of the result (7.92) below.

- In $P_y(\theta_2 > \theta > \theta_1)$, there is a sinusoidal oscillating term with oscillation frequency Ω. This term is the shock response of the beam polarization to the first resonance jump, its phase depending on $\theta - \theta_1$.

- In $P_y(\theta > \theta_2)$, there are four oscillating terms, all with oscillation frequency Ω. Each term has its own physical origin. The third oscillating term gives the shock response to the second resonance crossing, its phase depending on $\theta - \theta_2$. The fourth term describes the interference between the two crossings, its phase depending on $\theta_2 - \theta_1$. (This fourth term is independent of time θ, so strictly speaking, it is not an "oscillating" term.) The remaining two oscillating terms (the first and second terms) appear more mysterious; they give rise to the spin echo effect, while the first term dominates over the second term.

- Each of the four oscillating terms in $P_y(\theta > \theta_2)$ of (7.92) contains a Gaussian factor corresponding to the effect of phase smearing due to the finite beam energy spread. The rate the phase information is lost is such that each of the oscillating terms is damped in N_{smear} turns, where

$$N_{\text{smear}} \approx \frac{\sqrt{2}\,\Omega}{2\pi|A|\kappa\sigma_\delta}.$$

Because of these exponential smearing factors, all the oscillating terms will be significant only within a time span of the order of $\Delta\theta \sim 2\pi N_{\text{smear}}$

centered around specific values of time θ. The shock terms will center around $\theta = \theta_{1,2}$, while the echo term will center around $\theta = 2\theta_2 - \theta_1$. Physical meanings of each of these four terms are clear and distinct.

- The oscillatory terms are responsible for the transient, the interference, and the echo effects. After they damp out by their respective Gaussian smearing factors, the level of polarization is given by

$$
P_y = \begin{cases}
\frac{A}{\Omega}, & \text{if } \theta < \theta_1, \\[2mm]
\frac{A}{\Omega}\left(\frac{A^2 - |\epsilon_0|^2}{\Omega^2}\right), & \text{if } \theta_2 > \theta > \theta_1, \\[2mm]
\frac{A}{\Omega}\left(\frac{A^2 - |\epsilon_0|^2}{\Omega^2}\right)^2, & \text{if } \theta > \theta_2.
\end{cases}
$$

The initial level of polarization, $\frac{A}{\Omega}$, comes from the initial eigenstate. It is clear that the first jump contributes a polarization reduction factor $\left(\frac{A^2 - |\epsilon_0|^2}{\Omega^2}\right)$, while the second jump gives rise to the same loss factor. When the beam has a sufficiently large energy spread and the two resonance crossings are sufficiently separated in time, therefore, it does seem justified to consider the two jumps as two separate independent events, and ignore any interference effects. On the other hand, even under this condition, an echo remains unavoidable.

- To observe a significant interference effect, i.e., for the fourth oscillating term in $P_y(\theta > \theta_2)$ to be significant, it is necessary that

$$
\theta_2 - \theta_1 \lesssim \frac{\sqrt{2}\,\Omega}{|A|\kappa\sigma_\delta}. \tag{7.93}
$$

The quantity on the right-hand-side, therefore, specifies how long the memory of crossing a resonance lasts. As discussed earlier, when $\sigma_\delta = 0$, such as for a single particle, the interference will be remembered indefinitely.

- Our analysis assumes the resonances are crossed by sudden jumps in the spin tune. In practice, spin tune is varied at a finite speed. A "sudden" jump means the crossing is made in a time short enough that the polarization has not made a significant change. This requires

$$
N_{\text{jump}} \ll \frac{1}{\Omega},
$$

where N_{jump} is the number of turns it takes to complete the jump.

- As mentioned earlier, the polarization oscillation frequency Ω is much slower than the precession frequency κ. In fact, in order for the spinor equation of motion (5.54) to hold, we require another condition, namely, one needs the jump not to be too fast,

$$
N_{\text{jump}} \gg \frac{1}{\kappa}.
$$

This condition, however, is easily fulfilled in practice.

- The interference effect is pronounced only when Eq. (7.93) holds, i.e. only when the two jumps are sufficiently close in time. More specifically, interference occurs significantly when

$$\frac{1}{\kappa} \ll N_{\text{jump}} \ll \left\{ \begin{array}{c} \frac{1}{\Omega} \\ \frac{1}{2\pi}(\theta_2 - \theta_1) \end{array} \right\} \ll \frac{\sqrt{2}\,\Omega}{2\pi|A|\kappa\sigma_\delta}. \qquad (7.94)$$

- In comparison, the spin echo effect stands out and becomes most visible only when Eq. (7.93) does not hold. In that case, the two shock responses and the echo signal are all clearly separated in time. To examine the echo effect, we are interested mainly in the parameters regime

$$\frac{1}{\kappa} \ll N_{\text{jump}} \ll \left\{ \begin{array}{c} \frac{1}{\Omega} \\ \frac{\sqrt{2}\,\Omega}{2\pi|A|\kappa\sigma_\delta} \end{array} \right\} \ll \frac{1}{2\pi}(\theta_2 - \theta_1). \qquad (7.95)$$

In this regime, the interference term does not contribute, and can be dropped.

- In addition to (7.94) and (7.95), we should keep in mind that our analytic approximation (7.91) requires

$$\kappa\sigma_\delta \ll |A|, \qquad (7.96)$$

although the absolute validity of this condition may not be too critical as far as observing an echo is concerned.

- The calculation leading to Eq. (7.92) for the echo signal requires sufficiently small σ_δ. For small δ, we have

$$\begin{aligned} \Omega &= \sqrt{(A + \kappa\delta)^2 + |\epsilon_0|^2} \\ &\approx \sqrt{A^2 + |\epsilon_0|^2} + \frac{A\kappa\delta}{\sqrt{A^2 + |\epsilon_0|^2}} + \frac{|\epsilon_0|^2\kappa^2\delta^2}{2(A^2 + |\epsilon_0|^2)^{3/2}} + \cdots. \end{aligned}$$

Equation (7.92) has included the term linear in δ only. For its approximation to be valid, the accumulated spread of spin precession angle caused by the term second-order in δ over the long time period $\theta_2 - \theta_1$ must be much less than 2π. This requires

$$\frac{|\epsilon_0|^2\kappa^2\sigma_\delta^2}{2\Omega^3}(\theta_2 - \theta_1) \ll 2\pi. \qquad (7.97)$$

This condition puts a limit on the echo waiting time $\theta_2 - \theta_1$. The echo signal diminishes if $\theta_2 - \theta_1$ is too long due to the quadratic δ-dependence of the spin precession frequency.

7.5.3 Echo signal

The echo signal comes about because the shock response to the spin produced by the first resonance crossing contains a precessing term in its spinor representation, and when the second crossing at a time τ later produces a shock response that contains another precessing term of equal speed but opposite direction, the two terms cancel each other at a time τ after the second crossing. Although particles with different energy errors precess with different speeds, the time when cancelation occurs is exactly the same for all particles independent of their energy errors, at least to 1st order in the energy errors. An echo is then produced as a result.

We showed earlier a few examples of echo signals in Figs. 5.9 and 5.10 when we summed up the single-particle result over a beam with energy spread numerically. The same result are of course also be obtained analytically using Eq. (7.92) provided the condition (7.97) is satisfied, as both cases of Figs. 5.9 and 5.10 do.

Figure 5.9 shows the case when the two crossings interfere destructively. For completeness, Fig. 7.21 shows the case using Eq. (7.92) with the same parameters as Fig. 5.9 except for $\Omega(\theta_2 - \theta_1) = 15\pi$ instead of 16π for a constructive interference. It should be observed that although the interference patterns are distinctly differently, the echo signal looks basically the same.

Figure 7.21: Conditions are the same as Fig. 5.9 ($\theta_1 = 0$, $|\epsilon_0| = 0.08$, $\kappa = 4$, $A = 0.1$), except that this is for a case of constructive interference with $\Omega(\theta_2 - \theta_1) = 15\pi$. Three cases are shown, $\sigma_\delta = 0$ (red), $\sigma_\delta = 0.1\%$ (green), $\sigma_\delta = 0.3\%$ (blue). As σ_δ increases, the interference effect is suppressed while an echo signal becomes more apparent.

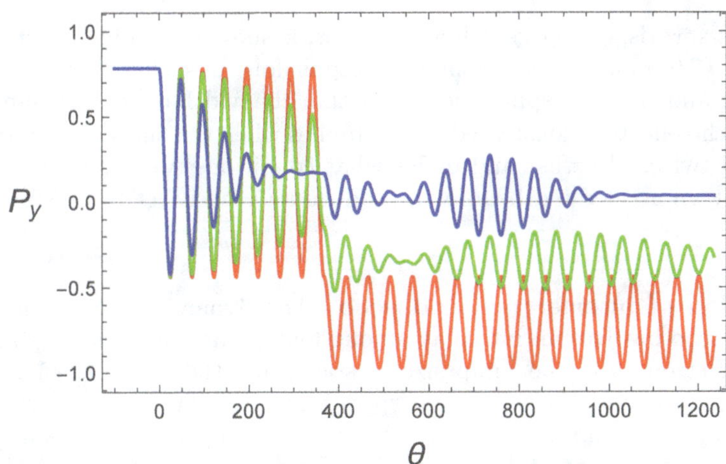

The case of a more pronounced echo, Fig. 5.10, is also reproduced using Eq. (7.92). Again as we mentioned when discussing Fig. 5.10, the same echo signal is produced whether the two resonance crossings are destructive or constructive.

In Eq. (7.95), the term $\frac{1}{\Omega}$ in the curly bracket represents the oscillatory motion of the polarization, while the term $\frac{\sqrt{2}\,\Omega}{2\pi|A|\kappa\sigma_\delta}$ corresponds to the damped behavior of the polarization due to phase smearing. Depending on the relative values of $\frac{1}{\Omega}$ and $\frac{\sqrt{2}\,\Omega}{2\pi|A|\kappa\sigma_\delta}$, the separated responses will appear damped-oscillatory or critically-damped. Figure 7.21 (green) appears damped-oscillatory, while Fig. 7.21 (blue) appears critically damped. When σ_δ is increased further from Fig. 7.21 (blue), the polarization will look basically the same except that the responses become increasingly sharply centered around θ_1, θ_2, and $2\theta_2 - \theta_1$ as depicted in Fig. 5.10.

Having identified the term in Eq. (7.92) responsible for the echo signal — the first oscillatory term in the expression for $P_y(\theta > \theta_2)$, it follows that the magnitude of the echo signal, relative to its background value is given by

$$P_{y,\text{echo}} = \frac{2A|\epsilon_0|^4}{\Omega^5}.$$

It follows that this echo signal is maximum when

$$|A|_{\text{max. echo}} = \frac{|\epsilon_0|}{2}. \tag{7.98}$$

When condition (7.98) is fulfilled, the maximum echo signal is

$$P_{y,\text{echo max.}} = \left(\frac{4}{5}\right)^{5/2} = 57\%, \tag{7.99}$$

a perhaps surprisingly large value. However, it should be pointed out also that condition (7.98) means the resonance jump is done in such a way that both the launching and the final spin tunes are within the width of the resonance.

The echo effect demonstrated above applies when the one and only resonance is crossed twice. Whether and under what conditions two separate resonances crossed by an accelerated beam will produce an echo is yet to be studied more closely.

Simulation of interference and echo   The dynamics of spin interference and spin echo involves an interplay among four parameters: the depolarization resonance strength ϵ, the jump size in spin tune, the time duration between the two jumps, and the spin tune spread of the beam. By adjusting these parameters, the beam polarization after the second jump exhibits a wealth of effects of constructive and destructive interferences and spin echo. A simulation code is written to study these effects and compare with the analysis using spinor algebra and the matrix formalism and for a beam with energy spread.[21]

---

[21]M. Bai, A. Chao, and T. Roser, AIP Conf. Proc. 1149, 785 (2009).

To observe a clear interference effect, we choose the condition (7.93),

$$\frac{1}{2\pi}(\theta_2 - \theta_1) \lesssim \frac{\sqrt{2}\,\Omega}{2\pi|A|\kappa\sigma_\delta}.$$

On the other hand, the condition for a prominent echo occurs when, from Eqs. (7.95) and (7.97),

$$\frac{\sqrt{2}\,\Omega}{2\pi|A|\kappa\sigma_\delta} \ll \frac{1}{2\pi}(\theta_2 - \theta_1) \ll \frac{2\Omega^3}{|\epsilon_0|^2\kappa^2\sigma_\delta^2}.$$

The tracking of the spin motion was carried out by multiplying the 2×2 one-turn spin transfer matrix for each orbital revolution. In the absence of the spin resonance, the one turn spin transfer matrix is described as

$$\psi_{n+1} = e^{-\frac{i}{2}a\gamma 2\pi}\psi_n,$$

where ψ_{n+1} and ψ_n are the spinors at the $(n+1)$-th and n-th orbital revolution, respectively. The spin resonance was simulated using a localized RF spin rotator which kicks the spin vector by an angle of $\tilde{\phi}_{\text{res}}$ around an axis which rotates in the horizontal plane at a frequency of ν_{res} times of orbital revolution frequency with ν_{res} the tune of the spin resonance, and the spin resonance strength is $\epsilon = \frac{\tilde{\phi}_{\text{res}}}{2\pi}$.

In the simulation, the beam energy was kept constant and the RF spin rotator tune was jumped cross the spin tune $\nu_{\text{spin}} = a\gamma$, for each spin resonance crossing. For all the simulations, the RF spin rotator was first adiabatically ramped from zero to its intended strength with its tune at a distance A below spin tune, i.e. $\nu_{\text{res}} = \nu_{\text{spin}} - A$ with $A > 0$. Its tune was then jump crossed to $\nu_s + A$ in one turn at n_1-th turn and then jumped back at the n_2-th turn.

The spin echo simulation was done with 1201 particles with Gaussian distributed momentum spread that corresponds to a spin tune spread 0.000022. The resonance strength amplitude as well as the resonance jump size were chosen to maximize the spin echo signal, i.e. $A = 0.001$, or from 4.599 to 4.601, and the resonance strength $\epsilon = 0.002$. A total of 40,000 orbital turns was between the two resonance jumps. Figure 7.22 shows the simulation result where an echo signal peaked up at 40,000 turns after the second resonance jump and the size of the echo signal is 0.57 as expected by Eq. (7.99).

The lower panel in Fig. 7.22 shows the case when the time between the two jumps went up to 300,000 turns, the spin echo was attenuated due to the higher order dependence of the spin precession frequency Ω on the beam's energy spread as dictated by Eq. (7.97).

Homework 7.29 The text discusses the double crossing result, Eq. (7.92) in some detail. It is a rich result containing the interference and echo effects. Spend some effort to derive Eq. (7.92) and try to follow the discussions. Assume $|\delta| \ll 1$ and keep only terms up to linear order in δ.

Homework 7.30 Verify the condition for maximum spin echo polarization is given by Eq. (7.98). Show that the maximum spin echo is $P_{y,\text{max}} = \left(\frac{4}{5}\right)^{5/2}$.

Figure 7.22: The two plots show the vertical component of average spin of 1201 particles as a function of orbital turns. Both simulations were done in similar conditions, namely, the spin resonance strength was first ramped from 0 to 0.002 in the first 20,000 turns with its tune set below spin tune by 0.001. This tune was then jumped to 0.001 above spin tune at 25,000th turn. The spin resonance tune then made a second jump back to 0.001 below spin tune. The upper plot corresponds to the interval between the two resonance jumps is 40,000 turns, and an echo signal was then peaked up at 40,000 turns after the second jump as expected. The echo signal has been optimized to the level of 57% as predicted. The lower plot shows when the two resonance jumps were spaced with 300,000 turns, the amplitude of the echo signal was significantly reduced due to higher order smearing by the energy spread.

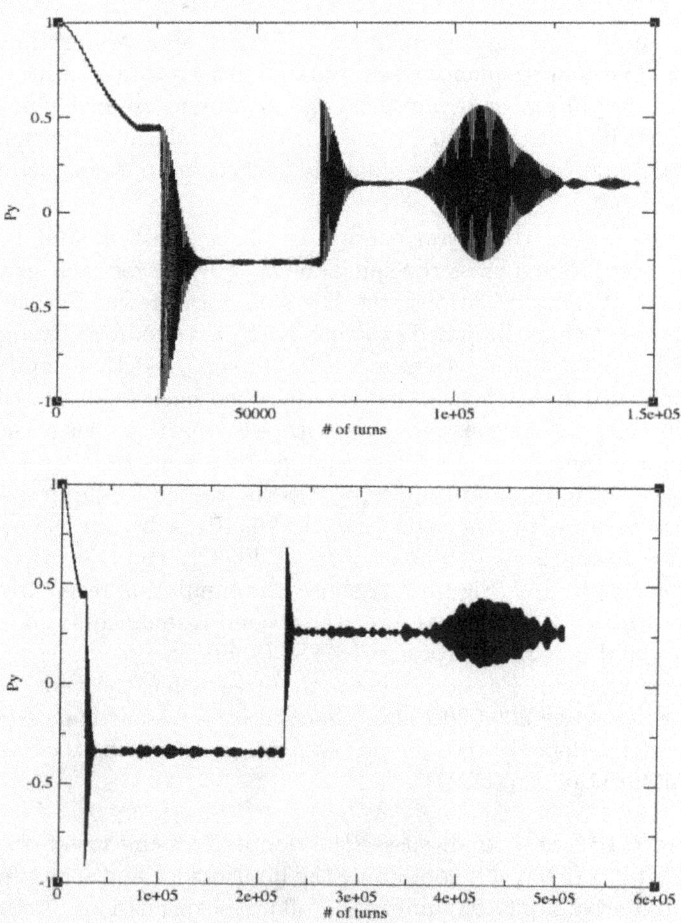

Chapter 8

Beam-Beam Interaction

The subject of beam-beam interaction has been a focus of studies since the advent of the colliding beam storage rings. Its study has led to an impressive continuing increase of the collider luminosities in a wide range of storage ring colliders.[1] One of the main limitations to achieving ever higher luminosities in these storage ring colliders is the beam-beam limit caused by beam-beam perturbation. When two beams collide in a storage ring collider, the particles in one beam sees mutually the electromagnetic fields carried by the on-coming beam. The fields perturb the motion of the particles, and when the beam intensity is high enough, the beam motion becomes unstable. As the high luminosity requires as strong a beam-beam perturbation as can be tolerated, the design of colliders requires pushing the beam-beam strength to its very limit. This limit is referred to as the beam-beam limit, and the challenge is to understand what mechanisms determine its behavior.

Beam-beam interaction has evolved into a rather deep subject. This chapter intends to introduce a part of this subject covering the stability issues of beam motion as the beams undergo beam-beam interaction. An emphasis is to introduce a range of physical models to describe the beam-beam instability mechanisms. More specifically, the important issue of beam manipulation and storage ring operation to optimize the luminosity is considered outside the present scope. Other important topics considered beyond the scope include at least the following,

- beam-beam effects in linear colliders;
- simulation efforts;
- crab waist operation;
- beamstrahlung;
- parasitic collisions.

---

[1]See for example, K. Hirata, M. Zobov, P. Chen, D. Schulte, J.M. Jowett, Secs. 2.5.1 to 2.5.4, and M.A. Furman, M.S. Zisman, Sec. 4.1, Handbook Accel. Phys. & Eng., 2nd ed., World Scientific (2013). See also Nonlinear Dynamics and the Beam-beam Interaction, AIP Proc. No. 57, BNL (1979).

Figure 8.1: The weak-strong (left) and strong-strong (right) pictures of the beam-beam interaction. The weak-strong picture addresses the stability of a single particle in the weak beam in a static nonlinear environment of the strong beam which stays intact. The strong-strong picture studies the mutual interaction of the two beams. IP means the interaction point.

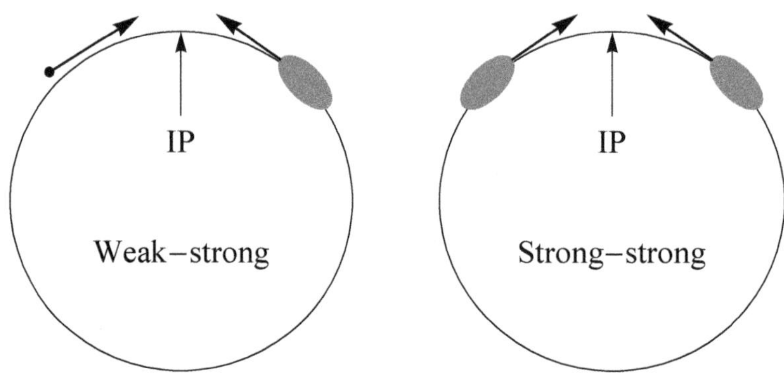

In the following, we divide the subject into two categories, based on two physical pictures that describe the beam-beam interaction: a weak-strong, and a strong-strong picture as depicted in Fig. 8.1. Both pictures play a role in determining the beam-beam behavior. A more traditional effort on this subject has been mostly addressing a weak-strong picture and its corresponding single-particle nonlinear dynamics. Coherent beam-beam effects associated with the strong-strong picture, on the other hand, constitute an important aspect that must be taken into account. In this chapter, we will address both pictures.

8.1　The luminosity

Before we enter the subject of beam-beam interaction, we need to introduce a quantity called the luminosity. The luminosity is the main performance quality parameter of a collider.

8.1.1　Head-on luminosity

Consider a bunch of particles in beam 1 colliding with a bunch of particles in beam 2. Let the two beam bunches have charge distributions given by

$$\rho_1(x,y) \;=\; \frac{N_1 e}{L_z}\, n_1(x,y)\,, \qquad \rho_2(x,y) \;=\; -\frac{N_2 e}{L_z}\, n_2(x,y)\,,$$

where $N_{1,2}$ are the number of particles in the two bunches, L_z is the longitudinal length of both bunches. We assume the particles in the two beams have opposite charges. The transverse distribution densities are normalized as

$$\int_{-\infty}^{\infty} dx \int_{-\infty}^{\infty} dy\, n_{1,2}(x,y) \;=\; 1\,.$$

Consider a particle in beam 1 traversing beam 2 in a collision. Let the particle have a transverse displacements x and y, and assuming the collision is strictly head on so that x and y do not change during the collision, the probability that the particle finds a partner in beam 2 to collide with is

$$\mathcal{P}(x,y) \;=\; N_2 n_2(x,y)\Sigma\,,$$

where Σ is the cross-section of the collision event, a universal constant that depends on the type of collision of interest.

The number of collision events per crossing $\mathcal{N}$ is then obtained by integrating $\mathcal{P}(x,y)$ over the distribution of beam 1,

$$\begin{aligned}
\mathcal{N} \;&=\; N_1 \int_{-\infty}^{\infty} dx \int_{-\infty}^{\infty} dy\, \mathcal{P}(x,y) n_1(x,y) \\
&=\; N_1 N_2 \Sigma \int_{-\infty}^{\infty} dx \int_{-\infty}^{\infty} dy\, n_1(x,y) n_2(x,y)\,.
\end{aligned}$$

The number of events per second is obtained by multiplying $\mathcal{N}$ by f, the rate of collisions per second, i.e.

$$\dot{\mathcal{N}} \;=\; f\mathcal{N}\,.$$

Obviously $\dot{\mathcal{N}}$ is proportional to Σ. The luminosity is defined to be the proportionality coefficient,

$$\mathcal{L} \;=\; \frac{\dot{\mathcal{N}}}{\Sigma}\,.$$

It follows that for head-on collisions,

$$\mathcal{L} \;=\; N_1 N_2 f \int_{-\infty}^{\infty} dx \int_{-\infty}^{\infty} dy\, n_1(x,y) n_2(x,y)\,. \tag{8.1}$$

If both beams have a bi-Gaussian distribution

$$n_{1,2}(x,y) \;=\; \frac{1}{2\pi\sigma_x\sigma_y}\, e^{-\frac{x^2}{2\sigma_x^2}-\frac{y^2}{2\sigma_y^2}}\,,$$

then

$$\mathcal{L} \;=\; \frac{N_1 N_2 f}{4\pi\sigma_x\sigma_y}\,.$$

If for example, $N_1 = N_2 = 10^{12}$, $f = 10^5$ s^{-1}, $\sigma_x = 300$ μm, $\sigma_y = 30$ μm, we have $\mathcal{L} = 10^{32}$ cm^{-2}s^{-1}. If the collision event of interest has a cross-section $\Sigma = 10^{-32}$ cm^2, then the expected rate of events of the collider is 1 event per second.

Equation (8.1) holds under the following conditions,

- The transverse beam distributions $n_{1,2}(x,y)$ stay the same along the lengths of the beam bunches.

- The beams and all their particles move strictly in the $\pm\hat{z}$-directions.

- The traversal of collision is strictly head-on; there is no crossing angle and no x' or y' slopes of the individual particles.

Homework 8.1

(a) Calculate the luminosity when the two bi-Gaussian beams have different beam sizes σ_{x1}, σ_{y1} for beam 1 and σ_{x2}, σ_{y2} for beam 2.

(b) Calculate the luminosity when the two beams have the same beam sizes but one is misaligned by a transverse displacements Δx and Δy at the collision point.

(c) Calculate the luminosity when the two beams have the same beam sizes but one is misaligned by a tilt angle θ_1 and the other beam with angle θ_2 in the x-y plane.

Solution

(a)

$$\mathcal{L} = \frac{N_1 N_2 f}{2\pi\sqrt{(\sigma_{x1}^2 + \sigma_{x2}^2)(\sigma_{y1}^2 + \sigma_{y2}^2)}} \, .$$

(b)

$$\mathcal{L} = \frac{N_1 N_2 f}{4\pi\sigma_x\sigma_y} \exp\left(-\frac{\Delta x^2}{4\sigma_x^2} - \frac{\Delta y^2}{4\sigma_y^2} \right) \, .$$

(c) Having obtained (a), you may try to guess at the answer; it is given by

$$\mathcal{L} = \frac{N_1 N_2 f}{4\pi\Sigma_x\Sigma_y} \, ,$$

$$\Sigma_x^2 = \sigma_x^2 \cos^2\left(\frac{\theta_1 - \theta_2}{2}\right) + \sigma_y^2 \sin^2\left(\frac{\theta_1 - \theta_2}{2}\right) \, ,$$

$$\Sigma_y^2 = \sigma_x^2 \sin^2\left(\frac{\theta_1 - \theta_2}{2}\right) + \sigma_y^2 \cos^2\left(\frac{\theta_1 - \theta_2}{2}\right) \, .$$

8.1.2   Hour-glass luminosity

The luminosity expression (8.1) needs to be modified when we want to take into consideration that some particles enter the beam-beam collision with finite slopes x' and y', which is unavoidable if the beams have finite transverse emittances.

When trajectory slopes are taken into account, the beam sizes no longer can be constant throughout the collision. The quantities σ_x and σ_y depend on the location of observation. Away from the collision point, the beam distribution remains bi-Gaussian but its sizes change according to

$$\sigma^2(s) = \sigma^{*2}\left[1 + \frac{(s - s^*)^2}{\beta^{*2}}\right] \, ,$$

where s is the longitudinal distance of the observation point from the collision point, s^* is the location of the waist, and β^* is the Courant–Snyder β-function at the waist. We define the coordinate s so that the collision point has $s = 0$. In case the waist position coincide with the collision point, then $s^* = 0$.

Since there are two dimensions in x and y, and two beams 1 and 2, there are four degrees of freedom. For each degree of freedom, we have three variables σ^*, β^*, s^*. There are therefore 12 possible variables to describe the optics in the collision region and the corresponding luminosity. We shall assume $s^* = 0$ and assume the same β^* for the two beams, so we have four variables σ_x^*, σ_y^*, β_x^*, β_y^* in our treatment below.

The beam assumes an hour-glass shape with its waist occurring at the collision point. To reach higher luminosity, there tends to be an effort to squeeze the β-functions β_x^*, β_y^* as small as possible at the collision point. In doing so, and as they become low enough to be comparable to the bunch length L_z, the hour-glass effect becomes significant as we will illustrate below.

Let the 3-D beam distributions be

$$\rho_1(x,y,z) = N_1 e\, m_1(z) n_1(x,y), \qquad \rho_2(x,y,z) = -N_2 e\, m_2(z) n_2(x,y),$$

where $m_{1,2}(z)$ are the longitudinal distribution density normalized by

$$\int_{-\infty}^{\infty} dz\, m_{1,2}(z) \;=\; 1.$$

Consider a longitudinal slice of beam 1 at $z = z_1$ and a slice of beam 2 at $z = z_2$, where $z_{1,2}$ refer to the respective centers of each beam bunch. These two slices collide at a longitudinal position $s = \frac{z_1 - z_2}{2}$. So at the point where these two slices collide, we have

$$\sigma_x^2 \;=\; \sigma_x^{*2}\left[1 + \frac{(z_1 - z_2)^2}{4\beta_x^{*2}}\right], \qquad \sigma_y^2 \;=\; \sigma_y^{*2}\left[1 + \frac{(z_1 - z_2)^2}{4\beta_y^{*2}}\right],$$

and therefore for those two slices,

$$n_1(x,y) \;=\; n_2(x,y) \;=\; \frac{\exp\left\{-\dfrac{x^2}{2\sigma_x^{*2}\left[1 + \frac{(z_1 - z_2)^2}{4\beta_x^{*2}}\right]} - \dfrac{y^2}{2\sigma_y^{*2}\left[1 + \frac{(z_1 - z_2)^2}{4\beta_y^{*2}}\right]}\right\}}{2\pi\sigma_x^*\sigma_y^*\left[1 + \frac{(z_1 - z_2)^2}{4\beta_x^{*2}}\right]^{1/2}\left[1 + \frac{(z_1 - z_2)^2}{4\beta_y^{*2}}\right]^{1/2}}. \tag{8.2}$$

These two bunch slices contribute to a number of collision events,

$$d\mathcal{N} \;=\; [N_1 m_1(z_1) dz_1]\,[N_2 m_2(z_2) dz_2]\Sigma \int_{-\infty}^{\infty} dx \int_{-\infty}^{\infty} dy\, n_1(x,y) n_2(x,y).$$

Integrating over z_1 and z_2 of all the slices, we obtain the luminosity

$$\mathcal{L} \;=\; N_1 N_2 f \int_{-\infty}^{\infty} dz_1 \int_{-\infty}^{\infty} dz_2\, m_1(z_1) m_2(z_2) \int_{-\infty}^{\infty} dx \int_{-\infty}^{\infty} dy\, n_1(x,y) n_2(x,y).$$

We now assume a Gaussian longitudinal distribution

$$m_1(z) = m_2(z) = \frac{1}{\sqrt{2\pi}\sigma_z} e^{-\frac{z^2}{2\sigma_z^2}}.$$

Together with Eq. (8.2), we then have

$$\mathcal{L} = \frac{N_1 N_2 f}{8\pi^3 \sigma_z^2 \sigma_x^{*2} \sigma_y^{*2}} \int_{-\infty}^{\infty} dz_1 \int_{-\infty}^{\infty} dz_2 \int_{-\infty}^{\infty} dx \int_{-\infty}^{\infty} dy$$

$$\times \frac{\exp\left\{ -\frac{z_1^2 + z_2^2}{2\sigma_z^2} - \frac{x^2}{\sigma_x^{*2}\left[1 + \frac{(z_1-z_2)^2}{4\beta_x^{*2}}\right]} - \frac{y^2}{\sigma_y^{*2}\left[1 + \frac{(z_1-z_2)^2}{4\beta_y^{*2}}\right]} \right\}}{\left[1 + \frac{(z_1-z_2)^2}{4\beta_x^{*2}}\right]\left[1 + \frac{(z_1-z_2)^2}{4\beta_y^{*2}}\right]}$$

$$= \frac{N_1 N_2 f}{8\pi^2 \sigma_z^2 \sigma_x^* \sigma_y^*} \int_{-\infty}^{\infty} dz_1 \int_{-\infty}^{\infty} dz_2 \frac{\exp\left(-\frac{z_1^2 + z_2^2}{2\sigma_z^2}\right)}{\left[1 + \frac{(z_1-z_2)^2}{4\beta_x^{*2}}\right]^{1/2} \left[1 + \frac{(z_1-z_2)^2}{4\beta_y^{*2}}\right]^{1/2}}$$

$$= \frac{N_1 N_2 f}{8\pi^2 \sigma_x^* \sigma_y^*} \int_{-\infty}^{\infty} du_1 \int_{-\infty}^{\infty} du_2 \frac{\exp\left(-\frac{u_1^2 + u_2^2}{2}\right)}{\left[1 + \frac{\sigma_z^2}{4\beta_x^{*2}}(u_1-u_2)^2\right]^{1/2} \left[1 + \frac{\sigma_z^2}{4\beta_y^{*2}}(u_1-u_2)^2\right]^{1/2}}.$$

By a change of variables $v_1 = u_1 - u_2$, $v_2 = u_1 + u_2$, the double integral becomes a single integral,

$$\mathcal{L} = \frac{N_1 N_2 f}{4\pi \sigma_x^* \sigma_y^*} \frac{1}{2\sqrt{\pi}} \int_{-\infty}^{\infty} dv_1 \frac{e^{-\frac{v_1^2}{4}}}{\left(1 + \frac{\sigma_z^2}{4\beta_x^{*2}} v_1^2\right)^{1/2} \left(1 + \frac{\sigma_z^2}{4\beta_y^{*2}} v_1^2\right)^{1/2}}.$$

Let us define the nominal case without hour-glass effect as

$$\mathcal{L}_0 = \frac{N_1 N_2 f}{4\pi \sigma_x^* \sigma_y^*}.$$

Compared with the nominal case, the hour-glass effect has reduced the luminosity by a reduction factor[2]

$$R_{hg}(h_x, h_y) = \frac{1}{2\sqrt{\pi}} \int_{-\infty}^{\infty} dv_1 \frac{e^{-v_1^2/4}}{\left(1 + \frac{h_x^2}{4} v_1^2\right)^{1/2} \left(1 + \frac{h_y^2}{4} v_1^2\right)^{1/2}}. \tag{8.3}$$

This reduction factor depends on the values of

$$h_x = \frac{\sigma_z}{\beta_x^*}, \qquad h_y = \frac{\sigma_z}{\beta_y^*}.$$

One sees $R_{hg}(0,0) = 1$ as expected. Figure 8.2 gives the contour plot of $R_{hg}(h_x, h_y)$.

---

[2]S.Y. Lee, Accelerator Physics, 3rd ed., World Scientific (2011).

Figure 8.2: Contour plot of the hour-glass luminosity reduction factor $R_{hg}(h_x, h_y)$ in the (h_x, h_y) plane, where $h_x = \frac{\sigma_z}{\beta_x^*}$, $h_y = \frac{\sigma_z}{\beta_y^*}$. The contour lines are for $R_{hg}(h_x, h_y) = 0.9, 0.8, 0.7, 0.6, 0.5, 0.4$, respectively. The origin has $R_{hg}(0, 0) = 1$.

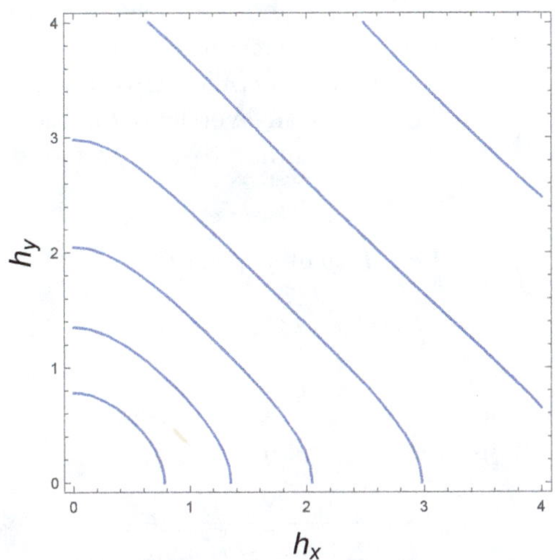

Round beam One special case is when the lattice design around the collision point is round in the sense that $\beta_x^* = \beta_y^* = \beta^*$ (the beam sizes do not have to be round because the horizontal and vertical emittances are not specified). In that case, letting $h = \frac{\sigma_z}{\beta^*}$,

$$R_{hg}(h, h) \;=\; \frac{\sqrt{\pi}}{h}\, e^{1/h^2}\, \mathrm{erfc}\left(\frac{1}{h}\right), \tag{8.4}$$

where $\mathrm{erfc}(x)$ is the error function.

Flat beam Another special case occurs when the lattice optics is flat in such a way that $\beta_x^* \gg \sigma_z$ while only the vertical dimension observes the hour-glass effect. In this case, $h_x \to 0$, we have

$$R_{hg}(0, h_y) \;=\; \frac{1}{\sqrt{\pi}\, h_y}\, e^{1/2h_y^2}\, K_0\left(\frac{1}{2h_y^2}\right), \tag{8.5}$$

where $K_0(x)$ is the Bessel function.

Homework 8.2
    (a) Follow the text to derive Eq. (8.3).
    (b) Use Eq. (8.3) to derive the round optics result (8.4).
    (c) Use Eq. (8.3) to derive the flat optics result (8.5).

8.1.3 Crossing angle luminosity

In the previous section, we dealt with the modification needed to calculate the luminosity when particles in the beams are not strictly moving in the $\pm\hat{z}$-directions due to the hour-glass effect, i.e. due to the fact that the beams have finite transverse emittances. There is another possibility when particles do not move in the strict $\pm\hat{z}$-directions even without the hour-glass effect, and that is when two beams are purposely made to collide with a crossing angle.

Let the crossing angle be $\pm\alpha$ in the vertical y-plane. A reasoning of integrating the colliding beam slices similar to the previous section leads to an expression of the luminosity,

$$\mathcal{L} \;=\; N_2 N_2 f \int dz_1 \int dz_2 \int dx \int dy\, m_1(z_1\cos\alpha - y\sin\alpha)\, m_2(z_2\cos\alpha - y\sin\alpha)$$
$$\times\, n_1(x, z_1\sin\alpha + y\cos\alpha)\, n_2(x, z_2\sin\alpha + y\cos\alpha)\,. \tag{8.6}$$

Hour-glass effect is ignored here.

Assuming the two beams have the same tri-Gaussian distribution, the four-fold integration can be performed, yielding

$$\mathcal{L} \;=\; \frac{N_1 N_2 f}{4\pi\sigma_x\sqrt{\sigma_y^2\cos^2\alpha + \sigma_z^2\sin^2\alpha}}\,. \tag{8.7}$$

One minor subtlety for the cases when particles do not collide head on in the $\pm\hat{z}$-directions concerns the cross-section Σ. The cross-section is defined for a given center of mass collision energy. Strictly speaking, when particles collide at an angle, its center of mass energy changes. Here we have ignored this subtlety and assumed Σ to be constant over a range of center of mass energies of interest.

We will return to the subject of luminosity, crossing angle, and hour-glass effect after we introduce the subject of beam-beam limit.

Homework 8.3
 (a) The reasoning behind Eq. (8.6) deserves to be carried through.
 (b) Derive Eq. (8.7) using Eq. (8.6).

8.2 Beam-beam perturbation

8.2.1 Beam-beam potential

The beam-beam interaction is one of the strongest perturbations in a collider and has a pronounced influence on beam stability. Consider two beams colliding head-on. Let the two beams have the same transverse bi-Gaussian distribution at the collision point,

$$\rho(x, y) \;=\; \frac{Ne}{2\pi\sigma_x\sigma_y L_z}\,\exp\left(-\frac{x^2}{2\sigma_x^2} - \frac{y^2}{2\sigma_y^2}\right), \tag{8.8}$$

where L_z is the longitudinal beam bunch length.

Let the particle under consideration have a transverse position (x, y) as it collides with the on-coming beam. We need to calculate the beam-beam kick delivered to this particle due to the beam-beam interaction. To do so, we need to consider the electromagnetic potential generated by the on-coming beam.

In the relativistic limit, the electromagnetic fields carried by the on-coming beam are transverse; all longitudinal fields vanish because of Lorentz contraction. The fields therefore can be calculated in a 2-D configuration as if the on-coming beam has a transverse distribution (8.8) but with an infinite longitudinal length. To calculate the 2-D electric field carried by the on-coming beam, we apply the Gauss law and the Poisson equation,

$$\vec{E}(x, y) = -\nabla \Phi(x, y),$$
$$\nabla^2 \Phi(x, y) = -\frac{\rho(x, y)}{\epsilon_0}.$$

Our first job is to solve for $\Phi(x, y)$ together with the boundary conditions that $\Phi(x, y) \to 0$ at $|x, y| \to \infty$. The result we obtain applies when the beam collision point is in free space without wall boundaries. We first define the Fourier transforms[3]

$$\rho(x, y) = \frac{1}{4\pi^2} \int_{-\infty}^{\infty} dk_x \int_{-\infty}^{\infty} dk_y \, \tilde{\rho}(k_x, k_y) \, e^{ik_x x + ik_y y},$$
$$\Phi(x, y) = \frac{1}{4\pi^2} \int_{-\infty}^{\infty} dk_x \int_{-\infty}^{\infty} dk_y \, \tilde{\Phi}(k_x, k_y) \, e^{ik_x x + ik_y y}.$$

From the bi-Gaussian beam distribution, we have

$$\tilde{\rho}(k_x, k_y) = \frac{Ne}{L_z} \exp\left(-\frac{\sigma_x^2 k_x^2}{2} - \frac{\sigma_y^2 k_y^2}{2} \right).$$

The Poisson equation then yields

$$-(k_x^2 + k_y^2)\, \tilde{\Phi}(k_x, k_y) = -\frac{\tilde{\rho}(k_x, k_y)}{\epsilon_0}$$
$$\implies \quad \tilde{\Phi}(k_x, k_y) = \frac{\tilde{\rho}(k_x, k_y)}{\epsilon_0(k_x^2 + k_y^2)} = \frac{Ne}{\epsilon_0(k_x^2 + k_y^2)L_z} \exp\left(-\frac{\sigma_x^2 k_x^2}{2} - \frac{\sigma_y^2 k_y^2}{2} \right).$$

We need to make an inverse Fourier transformation to obtain $\Phi(x, y)$. It requires a double integral. However, a trick can be used to make it a single integral by using the identity

$$\frac{1}{k_x^2 + k_y^2} = \frac{1}{2} \int_0^{\infty} dt \, e^{-\frac{t}{2}(k_x^2 + k_y^2)}.$$

[3]S. Kheifets, DESY report PETRA-119 (1976); S.Y. Lee, Accelerator Physics, 3rd ed., World Scientific (2011).

Figure 8.3: Contour plots of the beam-beam potential $\Phi(x, y)$ for three cases of bi-Gaussian beam distributions. The upper row shows $\rho(x, y)$; the lower row shows $\Phi(x, y)$. The three columns from left to right are for the cases $\frac{\sigma_y}{\sigma_x} = 1, 0.3$ and 0.001, respectively.

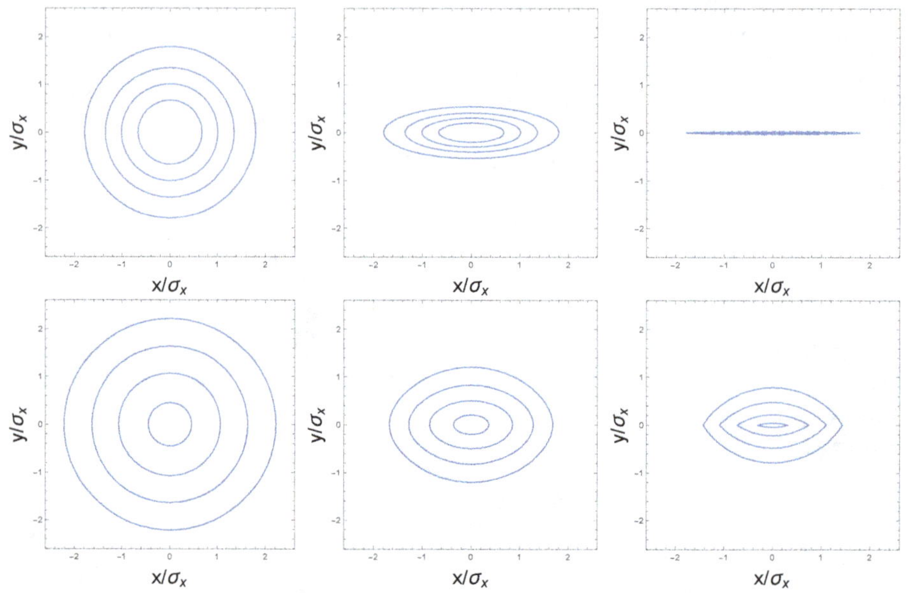

Using this trick, the static 2-D potential is found to be

$$\Phi(x, y) \;=\; \frac{Ne}{4\pi\epsilon_0 L_z} \int_0^\infty dt \; \frac{\exp\left[-\frac{x^2}{2(\sigma_x^2 + t)} - \frac{y^2}{2(\sigma_y^2 + t)}\right] - 1}{\sqrt{(\sigma_x^2 + t)(\sigma_y^2 + t)}}. \tag{8.9}$$

In Eq. (8.9), an additional term of -1 is added to the numerator of the integrand to remove a logarithmic divergence. The potential is thus shifted by a constant in such a way that $\Phi(0, 0) = 0$. Also we have $\Phi \to 0$ when $|x, y| \to \infty$ as we intend to accomplish.

Figure 8.3 shows a few contour plots of the potential $\Phi(x, y)$ as a function of $\frac{x}{\sigma_x}$ and $\frac{y}{\sigma_x}$ (not $\frac{y}{\sigma_y}$) for three transverse beam distributions $\rho(x, y)$. The case $\frac{\sigma_y}{\sigma_x} = 1$ represents a round beam, and the case $\frac{\sigma_y}{\sigma_x} = 0.001$ represents an extra flat beam.

Homework 8.4 Go through the algebra as outlined in the text to derive Eq. (8.9).

Homework 8.5

(a) Start with Eq. (8.9) to find expressions for the electric and the magnetic fields carried by the bi-Gaussian beam. Assume the beam motion is relativistic.

(b) A beam with 10^{12} electrons is focused to $\sigma_x = 300\,\mu$m, $\sigma_y = 30\,\mu$m at the interaction point. The beam bunch length is $L_z = 1$ cm. Find the electric and magnetic fields seen by a test charge colliding with and traversing this beam bunch head-on at a position $x = 0$, $y = 1\ \sigma_y$.

8.2.2 Beam-beam kick

As a particle with transverse displacements (x, y) traverses the beam-beam potential (8.9) of the on-coming beam, it receives a transverse kick in $\Delta x'$ and $\Delta y'$. The electric field seen by the particle is $\vec{E} = -\nabla\Phi(x, y)$. Let us assume the charge of the two beams are opposite. Then the electric force experienced by the particles is

$$\vec{F}_e \;=\; -e\vec{E} \;=\; e\nabla\Phi(x, y)\,.$$

There is also a magnetic force. The on-coming beam also generates a magnetic field by Ampere's law. The corresponding magnetic force is given by

$$\vec{F}_m \;=\; \beta^2 \vec{F}_e\,, \tag{8.10}$$

where $\beta = \frac{v}{c}$ with v the beam's moving speed.

To see Eq. (8.10), consider the on-coming beam as a collection of infinitely thin wires of moving charges. The electric and magnetic fields it generates can be obtained by superposition of these wires. For each wire, its fields are such that Eq. (8.10) is fulfilled, so by superposition the total forces fulfill the same condition. We have taken into account the fact that the on-coming beam moves in opposite direction (with velocity $-\beta c$) to the particle under consideration (with velocity βc).

The total force experienced by the particle as it traverses the on-coming beam is therefore

$$\vec{F} \;=\; \vec{F}_e + \vec{F}_m \;=\; (1 + \beta^2)\, e\nabla\Phi(x, y)\,.$$

The transverse impulse received by the particle is

$$\Delta\vec{p}_\perp \;=\; \vec{F}\Delta t\,,$$

where Δt is the time duration of the beam-beam force acting on the particle. Assuming the two colliding bunches are of the same length L_z, the time duration is

$$\Delta t \;=\; \frac{L_z}{2\beta c}\,,$$

where a factor of 2 is due to the fact that both beam are moving.

The transverse beam-beam kicks are then found to be

$$\Delta x'\hat{x} + \Delta y'\hat{y} \;=\; \frac{\Delta\vec{p}_\perp}{m\beta\gamma c} \;=\; \frac{(1 + \beta^2)eL_z}{2mc^2\beta^2\gamma}\,\nabla\Phi(x, y)\,.$$

For relativistic beams, $\beta \approx 1$. We summarize the result,

$$\Delta x' = -\frac{\partial U(x,y)}{\partial x},$$

$$\Delta y' = -\frac{\partial U(x,y)}{\partial y},$$

$$U(x,y) = -\frac{Nr_0}{\gamma} \int_0^\infty dt \, \frac{\exp\left[-\frac{x^2}{2(\sigma_x^2+t)} - \frac{y^2}{2(\sigma_y^2+t)}\right] - 1}{\sqrt{(\sigma_x^2+t)(\sigma_y^2+t)}}, \qquad (8.11)$$

where r_0 is the classical radius of the particle, N is the total number of particles in the on-coming beam bunch.

Note that we have assumed the two beams have opposite charges, opposite directions of motion, but the same β and γ. In case of colliding an asymmetric beam configuration, the factors of e, β, γ and N that appear in Eq. (8.11) need to be sorted out carefully to see which beam does each factor refer to.

Equation (8.11) considers the beam-beam collision as thin-lens kicks. The beam-beam perturbation is being reduced to that of a nonlinear thin-lens. In other words, in this picture, therefore,

$$\text{beam-beam problem } = \text{ a single-particle nonlinear mapping problem}. \quad (8.12)$$

This is an important formulation of the beam-beam problem. However, it should be kept in mind that it also is based on one rather restricted view. Obviously, Eq. (8.11) applies only if the transverse beam distribution is an upright bi-Gaussian. But more seriously, there are also a few additional limitations as follows.

- The factor L_z of the bunch length drops out in the calculation of beam-beam kicks when we integrate the beam-beam force over the length of the interaction. This simplification assumes the transverse beam distribution does not change during the collision and is held fixed as Eq. (8.8). Several complications, such as the hour-glass effect, synchrobetatron effects and the crossing angle effects, ensue if this approximation is removed, as we will discuss.

- In Eq. (8.11), we have made the approximation that x and y do not change appreciably during the collision. In case the interaction is so strong as to cause beam disruption within a single beam-beam passage, our approximation breaks down. Such a beam-beam disruption effect tends not to occur in storage ring colliders but can occur when pushed in linear colliders.

- It is clear from Eq. (8.11) that the beam-beam force can be written as the gradient of an electromagnetic potential. The source of this potential is due to the collective Coulomb force of a continuum of charge distribution with total charge Ne rather than a collection of N discrete charges. Effects that do not come from a collective Coulomb interaction such as beamstrahlung, or effects that are consequences of the interaction between individual particles such as beam-beam diffusion, are thus excluded from this treatment.

- Finally, Eq. (8.11), as mentioned, treats the beam-beam interaction as a static nonlinear thin-lens. It is appropriate only if we consider a weak-strong picture and consider the stability only of the weak beam. The strong beam stays unaffected by the beam-beam interaction. As a result, Eq. (8.11) also excludes coherent beam-beam effects where both beams are disturbed and their disturbances mutually bootstrap to grow by the beam-beam interaction.

We shall elaborate on some of these issues later in the chapter.

Taylor expansion Let us return to Eq. (8.11). For particles in the core of the beam distribution, $|x| \lesssim \sigma_x$ and $|y| \lesssim \sigma_y$, we can Taylor expand the expression of $U(x, y)$ and obtain[4]

$$U(x, y) \approx \frac{N r_0}{\gamma} \left[\frac{x^2}{\sigma_x(\sigma_x + \sigma_y)} + \frac{y^2}{\sigma_y(\sigma_x + \sigma_y)} \right.$$
$$\left. - \frac{x^4}{12} \frac{2\sigma_x + \sigma_y}{\sigma_x^3(\sigma_x + \sigma_y)^2} - \frac{x^2 y^2}{2\sigma_x \sigma_y(\sigma_x + \sigma_y)^2} - \frac{y^4}{12} \frac{\sigma_x + 2\sigma_y}{\sigma_y^3(\sigma_x + \sigma_y)} + \cdots \right]. \tag{8.13}$$

Only even powers of x and y appear in the Taylor expansion.

It follows from Eq. (8.13) that if we take only the quadratic x^2 and y^2 terms of the potential $U(x, y)$, valid when $|x| \ll \sigma_x, |y| \ll \sigma_y$, the beam-beam lens gives a linear orbital kick whose x and y components are

$$\Delta x' = -\frac{2 N r_0}{\gamma \sigma_x(\sigma_x + \sigma_y)} x \,,$$
$$\Delta y' = -\frac{2 N r_0}{\gamma \sigma_y(\sigma_x + \sigma_y)} y \,. \tag{8.14}$$

Equation (8.14) constitutes the linear thin-lens model of the beam-beam interaction. Both $\Delta x'$ and $\Delta y'$ kicks are focusing as a result of our assumption that the two beams have opposite charges. In the linear thin-lens model, the x- and y-motions are decoupled as a result of the assumption that the beam distribution is upright in the x-y plane. In this model, x-y beam-beam coupling comes only from nonlinear beam-beam terms.

Bassetti–Erskine formula Equation (8.11) can be cast in an elegant closed-form, Bassetti–Erskine formula, in terms of the error function.[5] This closed-form

---

[4]B.W. Montague, Nucl. Instr. Meth. 187, 335 (1981).

[5]M. Bassetti and G.A. Erskine, CERN report CERN-ISR-TH/80-06 (1980); R. Talman, AIP Proc. 153, p. 827 (1987); K. Hirata, Sec. 2.5.1 and J.M. Jowett, Sec. 2.5.4, Handbook Accel. Phys. & Eng., 2nd ed., World Scientific (2013).

formula is particularly convenient in numerical calculations. The beam-beam kick received by a particle with transverse displacements (x, y) in an electron-positron collision is given by

$$\Delta y' + i\Delta x' = -\frac{Nr_0}{\gamma} \begin{cases} \Pi(x, y, \sigma_x, \sigma_y), & (\sigma_x > \sigma_y, y > 0), \\ -\Pi^*(x, -y, \sigma_x, \sigma_y), & (\sigma_x > \sigma_y, y < 0), \\ i\,\Pi^*(y, x, \sigma_y, \sigma_x), & (\sigma_x < \sigma_y, x > 0), \\ -i\,\Pi(y, -x, \sigma_y, \sigma_x), & (\sigma_x < \sigma_y, x < 0), \end{cases} \tag{8.15}$$

where

$$\Pi(x, y, \sigma_x, \sigma_y) = \sqrt{\frac{2\pi}{\sigma_x^2 - \sigma_y^2}} \left[w\left(\frac{x + iy}{\sqrt{2(\sigma_x^2 - \sigma_y^2)}}\right) - e^{-\frac{x^2}{2\sigma_x^2} - \frac{y^2}{2\sigma_y^2}} w\left(\frac{\frac{\sigma_y x}{\sigma_x} + i\frac{\sigma_x y}{\sigma_y}}{\sqrt{2(\sigma_x^2 - \sigma_y^2)}}\right) \right],$$

and

$$w(z) = e^{-z^2}\left[1 - \mathrm{erf}(-iz)\right].$$

For a round beam case, Eq. (8.15) reduces to

$$\Delta y' + i\Delta x' = -\frac{2Nr_0}{\gamma}\left(\frac{ix + y}{x^2 + y^2}\right)\left(1 - e^{-\frac{x^2 + y^2}{2\sigma^2}}\right),$$

where $\sigma_x = \sigma_y = \sigma$. This result agrees with Eq. (8.17) of Homework 8.7.

Homework 8.6 Fill in the derivation of Eq. (8.13).

Solution A few integral identities might be useful,

$$\int_0^\infty \frac{dt}{(\sigma_x^2 + t)^{3/2}(\sigma_y^2 + t)^{1/2}} = \frac{2}{\sigma_x(\sigma_x + \sigma_y)},$$

$$\int_0^\infty \frac{dt}{(\sigma_x^2 + t)^{5/2}(\sigma_y^2 + t)^{1/2}} = \frac{2(2\sigma_x + \sigma_y)}{3\sigma_x^3(\sigma_x + \sigma_y)^2},$$

$$\int_0^\infty \frac{dt}{(\sigma_x^2 + t)^{3/2}(\sigma_y^2 + t)^{3/2}} = \frac{2}{\sigma_x\sigma_y(\sigma_x + \sigma_y)^2}. \tag{8.16}$$

It may be instructive to make a contour plot of this Taylor expanded $U(x, y)$ and compare with Fig. 8.3. If you do so, keep in mind that this expression is applicable only when $|x| \lesssim \sigma_x$ and $|y| \lesssim \sigma_y$. In contrast, note that the contour plots in Fig. 8.3 have courageously extended to $y = \pm 2.5\sigma_x$ which in some cases is way out of the vertical beam core, and that is permitted because we did not assume $|y| \lesssim \sigma_y$ there.

Homework 8.7 Equation (8.11) can be used to yield a simpler result for a round beam with $\sigma_x = \sigma_y = \sigma$. The same result can also be obtained directly from the Gauss law and the Ampere law and use the property of cylindrical symmetry. Show that, for a round beam,

$$U(x, y) = -\frac{Nr_0}{\gamma}\int_0^{r^2/2\sigma^2} \frac{du}{u}\left(e^{-u} - 1\right),$$

$$\Delta x' \;=\; \frac{2Nr_0}{\gamma}\,\frac{x}{r^2}\left(e^{-r^2/2\sigma^2}-1\right),$$

$$\Delta y' \;=\; \frac{2Nr_0}{\gamma}\,\frac{y}{r^2}\left(e^{-r^2/2\sigma^2}-1\right),\tag{8.17}$$

where $r^2 = x^2 + y^2$, and in the linear thin-lens limit,

$$\Delta x' \;=\; -\frac{Nr_0}{\gamma\sigma^2}\,x,\qquad \Delta y' \;=\; -\frac{Nr_0}{\gamma\sigma^2}\,y,\qquad \text{(linear thin-lens)}.$$

Homework 8.8 Apply Eq. (8.11) to the case of a flat bi-Gaussian beam with $\sigma_x \gg \sigma_y$, and consider a particle with $|x| \ll \sigma_x$. Show that

$$U(x,y) \;=\; -\frac{2Nr_0}{\gamma\sigma_x}\,y\int_0^{y/\sigma_y}\frac{du}{u}\left(e^{-u^2/2}-1\right),$$

$$\Delta y' \;=\; -\frac{2Nr_0}{\gamma\sigma_x}\int_0^{y/\sigma_y}du\,e^{-u^2/2},$$

$$\Delta y' \;=\; -\frac{2Nr_0}{\gamma\sigma_x\sigma_y}\,y,\qquad \text{(linear thin-lens)},$$

Homework 8.9 Given the Bassetti–Erskine formula for the case $\sigma_x > \sigma_y, y > 0$ of Eq. (8.15), go over a geometrical reorientation to obtain the results for
  (a) the case $\sigma_x > \sigma_y, y < 0$,
  (b) the case $\sigma_x < \sigma_y, x > 0$,
  (c) the case $\sigma_x < \sigma_y, x < 0$,
  (d) and the case for $\sigma_x = \sigma_y$ by taking the limit $\sigma_x \to \sigma_y$.

8.3 Linear thin-lens model

The thin-lens model of beam-beam interaction, valid for collision of two short bunches with upright transverse bi-Gaussian distribution is the most basic beam-beam model we will discuss. In particular, the motion of small amplitude particles is described by the linear uncoupled Eq. (8.14). This case is most conveniently analyzed by using a matrix technique. In this section, we will push the linear thin-lens model, in spite of its apparent over-simplicity, to explore several beam-beam mechanisms,

- weak-strong stability limit;

- dynamic-β^* effect;

- flip-flop effect;

- synchrobetatron coupling due to crossing angle;

- synchrobetatron coupling due to dispersion at the collision point.

The fact that we have linearized the beam-beam kick in x and y limits our treatment to particles with small displacements x, y, i.e. we are limited to the dynamics of the beam core. On the other hand, a linearized dynamics is consistent with the fact that we insist on a bi-Gaussian distribution. With linear perturbations, a bi-Gaussian distribution remains bi-Gaussian even if distorted. Since the linear thin-lens kicks are x-y decoupled, the bi-Gaussian distribution will stay upright. The linear thin-lens model is internally self-consistent in this regard.

8.3.1   Beam-beam tune shift parameter

Since the x- and the y-motions are decoupled, we can concentrate on one of the dimensions, say the y-dimension, below. The motion of the particle is characterized by the state vector

$$\begin{bmatrix} y \\ y' \end{bmatrix}.$$

The transformation of this state vector across the collision thin-lens is, from Eq. (8.14),

$$\begin{bmatrix} 1 & 0 \\ -\frac{1}{f} & 1 \end{bmatrix}, \qquad \frac{1}{f} = \frac{2Nr_0}{\gamma \sigma_y(\sigma_x + \sigma_y)},$$

which is the same as that of a thin-lens quadrupole of focal length f.

Let there be one collision point per revolution around the storage ring collider, and let ν be the unperturbed betatron tune and β_0^* be the unperturbed β-function at the interaction point, the transformation from the middle of the interaction point around one revolution of the ring can be written as

$$\begin{bmatrix} \cos 2\pi(\nu + \Delta\nu) & \beta^* \sin 2\pi(\nu + \Delta\nu) \\ -\frac{1}{\beta^*} \sin 2\pi(\nu + \Delta\nu) & \cos 2\pi(\nu + \Delta\nu) \end{bmatrix}$$
$$= \begin{bmatrix} 1 & 0 \\ -\frac{1}{2f} & 1 \end{bmatrix} \begin{bmatrix} \cos 2\pi\nu & \beta_0^* \sin 2\pi\nu \\ -\frac{1}{\beta_0^*} \sin 2\pi\nu & \cos 2\pi\nu \end{bmatrix} \begin{bmatrix} 1 & 0 \\ -\frac{1}{2f} & 1 \end{bmatrix}, \qquad (8.18)$$

where $\Delta\nu$ is the tune shift due to the beam-beam thin-lens, β^* is the perturbed β-function at the middle of the collision point. We have assumed the unperturbed α-function $\alpha_0^* = 0$. In the thin-lens model, we will assume that the transverse beam size of the weak beam scales with the square root of the perturbed β-function when the beam-beam effect is taken into account.[6]

Solving Eq. (8.18) yields

$$\cos 2\pi(\nu + \Delta\nu) = \cos 2\pi\nu - \frac{\beta_0^*}{2f} \sin 2\pi\nu,$$

---

[6]An alert reader may rightfully question this statement as it assumes the beam emittance is unaffected by the beam-beam distorted linear lattice. This scaling $\sigma_y \propto \sqrt{\beta_y^*}$ could be applicable if the colliding beams are protons or antiprotons but is not strictly valid for electrons or positrons. However, the effect is small and not a prominent qualitative feature; we shall set aside this issue here as we focus on the beam-beam dynamics.

$$\frac{\beta^*}{\beta_0^*} = \frac{\sin 2\pi\nu}{\sin 2\pi(\nu + \Delta\nu)}. \tag{8.19}$$

It has been hoped, and seriously studied theoretically and experimentally, that one or two dimensionless scaling parameters may characterize the beam-beam interaction completely or almost completely. In the linear-lens model, it is clear from Eq. (8.19) that two such parameters are the unperturbed betatron tune ν and a beam-beam strength parameter defined as[7]

$$\xi_0 = \frac{\beta_0^*}{4\pi f}, \tag{8.20}$$

or

$$\xi_x = \frac{\beta_{x0}^* N r_0}{2\pi\gamma\sigma_x(\sigma_x + \sigma_y)}, \qquad \xi_y = \frac{\beta_{y0}^* N r_0}{2\pi\gamma\sigma_y(\sigma_x + \sigma_y)}. \tag{8.21}$$

If the two beams have the same sign of charge, these expressions will contain an extra minus sign. In particular, then $\xi_{x,y} < 0$.

With this definition of ξ_0, it follows from Eq. (8.19) that

$$\frac{\beta^*}{\beta_0^*} = \frac{1}{\sqrt{1 - (2\pi\xi_0)^2 + 4\pi\xi_0 \cot 2\pi\nu}}. \tag{8.22}$$

The factor of 4π in Eq. (8.20) is included so that when $2\pi\xi_0$ is small, the beam-beam tune shift $\Delta\nu$ is equal to ξ_0,

$$\begin{aligned} \Delta\nu &\approx \xi_0, \\ \frac{\beta^*}{\beta_0^*} &\approx 1 - 2\pi\xi_0 \cot 2\pi\nu. \end{aligned} \tag{8.23}$$

For this reason, the strength parameter ξ_0 is often referred to as the beam-beam tune shift parameter, although its meaning goes actually beyond the tune shift. Note that for Eq. (8.23) to be valid, we need $|2\pi\xi_0| \ll 1$, not $|\xi_0| \ll 1$.

A subtle point is to be made here. From the derivation of Eq. (8.19), it needs to be noted that it is the unperturbed β_0^* that enters the definition of ξ_0. In contrast, the $\sigma_{x,y}$ that enter the focal lengths $f_{x,y}$ comes from the actual perturbed beam size. So Eq. (8.19) actually mixes perturbed and unperturbed quantities into a single equation. Later when we analyze the perturbed beam-beam parameters, we shall insert the perturbed $\sigma_{x,y}$ into $\xi_{x,y}$, but we must not replace their $\beta_{x0,y0}^*$ by the perturbed β-functions. Readers beware.

The factor β_0^* can be thought of as a magnification factor which, when applied to a perturbation, gives the effectiveness of the perturbation on the particle's betatron motion. To reduce the effect of the beam-beam perturbation, it is advantageous to reduce the value of β_0^*. This leads to the idea of low-β^*

---

[7]F. Amman and D. Ritson, Proc. Int. Conf. High Energy Accel., Brookhaven, 471 (1961); D. Ritson and J. Rees, SLAC report SLAC-TN-65-39 (1965); M. Bassetti, V-th Int. Conf. High Energy Accel., Frascati, 708 (1965).

insertion that has been successfully used to increase the luminosity of many storage rings.[8] We shall return to this topic later on page 634.

Homework 8.10

(a) Follow the text to derive Eq. (8.19).
(b) Use Eq. (8.19) to show Eq. (8.22).
(c) Derive Eq. (8.23) when $|2\pi\xi_0| \ll 1$.

8.3.2 Stability condition

Stability of small-amplitude betatron motion is determined by the requirement that $|\cos 2\pi(\nu + \Delta\nu)| \le 1$. The stability region is best conveniently represented by a region in a (ν, ξ_0) plot, where ν and ξ_0, as mentioned, are the two dimensionless scaling parameters of this model. The instability boundary is given by

$$\nu + \Delta\nu \;=\; \frac{m}{2}, \qquad m \;=\; \text{integer},$$

or

$$\xi_{\text{limit}} \;=\; \begin{cases} -\frac{\tan \pi\nu}{2\pi}, & \text{if } m = \text{even}, \\ \frac{\cot \pi\nu}{2\pi}, & \text{if } m = \text{odd}. \end{cases} \tag{8.24}$$

The case $m =$ even applies when ν is below an integer; the case $m =$ odd is when ν is above an integer. For the case above integer when $\frac{1}{2} > [\nu] > 0$ ($[\nu]$ is the fractional part of ν), for example, the stability boundary is $\xi_{\text{limit}} = \frac{\cot \pi\nu}{2\pi}$ as indicated by the black curve in Fig. 8.4. The shaded region indicates unstable motion. The plot is shown only for the range $0 < \nu < \frac{1}{2}$ since it is periodic in ν with period $\frac{1}{2}$.

Physically, the condition $\nu + \Delta\nu = \frac{m}{2}$ simply states that the beam-beam shifted tune has reached a half-integer. Since $\Delta\nu > 0$ (for the case the two beams have opposite signs of charge), the half-integer reached by the shifted tune is the nearest half-integer above ν, and it is a region below that half-integer where the instability region lies. The beam is stable when ν is above a half-integer, even when ν is close to the half-integer value as Fig. 8.4 illustrates.

The width of this unstable region for a given ξ_0 is

$$\delta\nu \;=\; \frac{1}{\pi}\tan^{-1} 2\pi\xi_0\,.$$

When $|2\pi\xi_0| \ll 1$,

$$\delta\nu \;\approx\; 2\xi_0\,. \tag{8.25}$$

The total resonance width — the stopband — covers from $\nu = \frac{1}{2} - \delta\nu$ up to $\frac{1}{2}$.

In the stable region, the upper panel of Fig. 8.4 shows a few contours of constant $\Delta\nu$. For the case when $\Delta\nu = 0.05$, for example, it shows that $\Delta\nu \approx \xi_0$ over a wide range of ν except when ν approaches 0 or 0.5. This is when Eq. (8.23) holds. When $\Delta\nu = 0.2$, however, this approximate relationship breaks down.

---

[8]K.W. Robinson and G.-A. Voss, Cambridge Electron Accelerator Laboratory report CEAL-TM-149 (1965); P.L. Morton and J. Rees, IEEE Trans. Nucl. Sci. NS-14, 630 (1967).

Figure 8.4: Stability region of betatron motion in the thin-lens beam-beam model in the (ν, ξ_0) plane. Shaded region is unstable. The pattern repeats in ν with period of $\frac{1}{2}$. Upper panel shows the contours of constant $\Delta\nu$ in the stable region. The lower panel shows the contours of $\frac{\beta^*}{\beta_0^*}$.

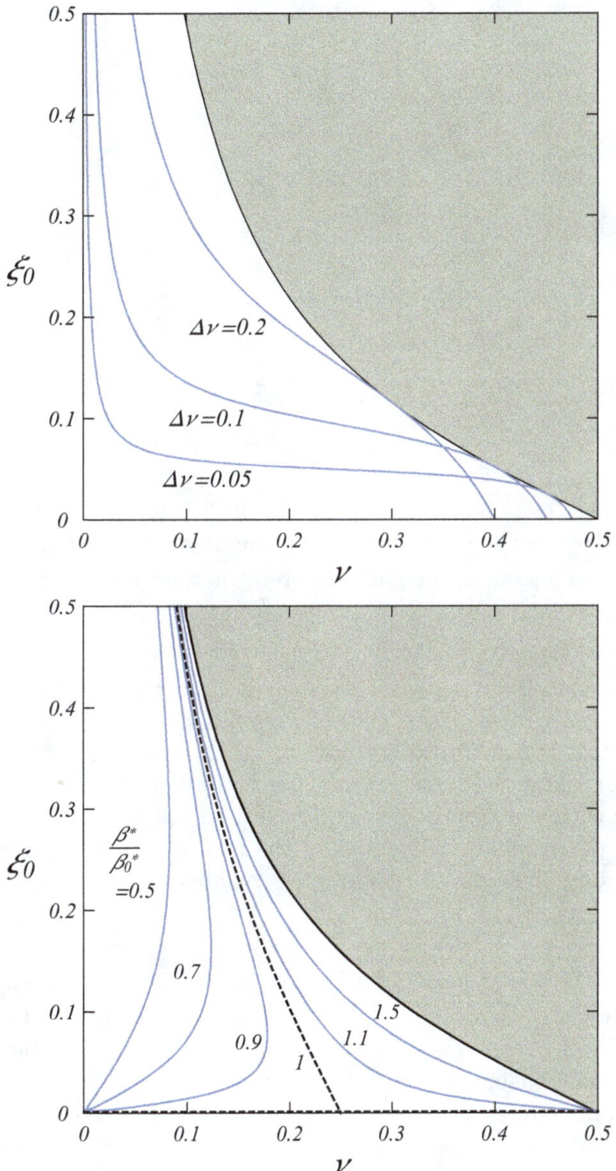

Figure 8.5: The beam-beam pinch changes β^*. Figure shows $r = \frac{\beta^*}{\beta_0^*}$ as a function of ξ_0 for colliding round beams. The blue, red, green, brown curves are for the cases $\nu = 0.05, 0.1, 0.2, 0.3$, respectively.

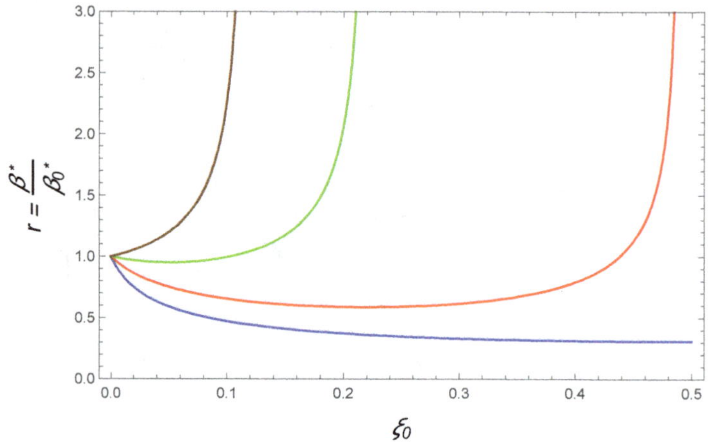

The lower panel of Fig. 8.4 shows a few contours of $\frac{\beta^*}{\beta_0^*}$. It may be instructive to observe this plot to see how the second member of Eq. (8.23) is fulfilled.

Figure 8.4 can also be shown in a different manner as in Fig. 8.5, where $\frac{\beta^*}{\beta_0^*}$ is plotted as a function of ξ_0 using Eq. (8.22) for four values of $\nu = 0.05, 0.1, 0.2, 0.3$. When ξ_0 approaches the instability value from below, the $\frac{\beta^*}{\beta_0^*}$ diverges — the red, green and brown curves each diverges at its own respective stability limit value of ξ_0. As we will see later, these divergences do not occur any more when dynamic-β^* effect is taken into account.

It should be reminded that our analysis is applicable only for particles near the beam core with small betatron amplitudes. In the unstable region of Fig. 8.4, it only means particles are driven out of the beam core. It does not necessarily mean they extend their motion to infinity. Nonlinear effects will be considered later in the chapter.

β^* **pinch**   In the lower panel of Fig. 8.4, a few contours of constant $\frac{\beta^*}{\beta_0^*}$ are plotted. It shows a dividing line — indicated by the black dashed curve — between the cases $\beta^* > \beta_0^*$ and the cases $\beta^* < \beta_0^*$. The beam-beam kick being assumed is focusing, so the β-function is pinched by the beam-beam kick. However, on one side of the parameter space, the β-function at the collision point actually becomes larger. The dividing line when $\beta^* = \beta_0^*$ is

$$\xi_{\beta^* = \beta_0^*} = \frac{\cot 2\pi\nu}{\pi}. \qquad (8.26)$$

To optimize the luminosity, we want the beam size to be small at the collision point. It is therefore expected that we have a strong preference to choose the beam parameters so that $\beta^* < \beta_0^*$ to benefit from the pinch effect. If we take the approximate expression (8.23), then we will choose to operate with

$$\xi_0 \cot 2\pi\nu \, > \, 0.$$

The preferred operation parameters from this point of view is then,

$$\text{beams have the same sign of charge } (\xi_0 < 0) \quad \rightarrow \quad \frac{1}{4} < \nu < \frac{1}{2}\,,$$

$$\text{beams have opposite signs of charge } (\xi_0 > 0) \quad \rightarrow \quad 0 < \nu < \frac{1}{4}\,. \tag{8.27}$$

The ν here applies for both ν_x and ν_y. Electron-positron or proton-antiproton colliders might prefer $0 < \nu < \frac{1}{4}$ per collision point, while proton-proton colliders might prefer $\frac{1}{4} < \nu < \frac{1}{2}$.

Homework 8.11

(a) Derive Eq. (8.24).

(b) How is Eq. (8.24), the beam-beam limit according to the linear thin-lens model, to be modified if the two beams have the same sign of charge?

8.3.3 The dynamic-β^* effect

Self-consistent dynamic-β^* In the thin-lens model so far, we have been considering a particle in weak beam under the influence of an on-coming strong beam. The strong beam is unaffected by the beam-beam interaction; its beam size stays at $\sigma_{x,y}$ independent of the beam-beam perturbation. The perturbation to the particle in the weak beam under consideration is described by Eq. (8.19) and Figs. 8.4 and 8.5. An approximate expression is given by (8.23).

What happens if the two beams have equal intensity? In this case, the on-coming beam is affected in the same way as the beam under consideration. Its beam sizes $\sigma_{x,y}$ will change just like those of the weak beam. The beam-beam lens affects the β-functions which in turn affect the beam sizes, and the beam sizes then affect the beam-beam strengths. It is clear that a self-consistent model needs to be established. We turn to this question in this section. The fact that the β-functions at the collision point are changed by the mutual thin-lens beam-beam kicks is sometimes called the dynamic-β^* effect. Dynamic-β^* is a strong-strong linear thin-lens beam-beam effect.

For a more quantitative evaluation of this issue, for simplicity, let us consider the case when the unperturbed beam is round, with $\sigma_{0x} = \sigma_{0y} = \sigma$ and $\beta_{x0}^* = \beta_{y0}^* = \beta_0^*$ at the collision point. Let us assume $\xi_x = \xi_y$ and denote it by ξ. Let ξ_0 designate the beam-beam parameter before taking into account of the dynamic-β^* effect.

We first note that $\xi \propto \frac{\beta_0^*}{\sigma^2}$. Here β_0^* assumes the unperturbed value; but σ is the actual perturbed transverse beam size, therefore it scales with the perturbed

β-function and is proportional to $\sqrt{\beta^*}$ [see footnote 6, also see the subtle point made after Eq. (8.23)]. It follows that

$$\xi = \frac{\beta_0^*}{\beta^*} \xi_0 .\tag{8.28}$$

We then are in a situation where the perturbed β^* depends on ξ according to Eq. (8.19), while ξ depends on β^* according to Eq. (8.28). Assuming equal beam intensities, the two beams experience the same perturbation, a self-consistent condition then requires

$$\frac{\beta^*}{\beta_0^*} = \frac{\sin 2\pi\nu}{\sin 2\pi(\nu + \Delta\nu)} = \frac{\sin 2\pi\nu}{\sqrt{1 - \left(\cos 2\pi\nu - 2\pi\xi_0 \frac{\beta_0^*}{\beta^*} \sin 2\pi\nu\right)^2}}$$

$$= \frac{1}{\sqrt{1 + 4\pi\xi_0 \cot 2\pi\nu \left(\frac{\beta_0^*}{\beta^*}\right) - (2\pi\xi_0)^2 \left(\frac{\beta_0^*}{\beta^*}\right)^2}} ,$$

$$\implies \quad r^2 + (4\pi\xi_0 \cot 2\pi\nu)r - (2\pi\xi_0)^2 - 1 = 0, \quad \text{where} \quad r = \frac{\beta^*}{\beta_0^*} ,$$

$$\implies \quad r = -2\pi\xi_0 \cot 2\pi\nu + \sqrt{1 + 4\pi^2\xi_0^2 \csc^2 2\pi\nu} .\tag{8.29}$$

Equation (8.29) gives the dynamic-β^* of a round beam due to the mutual beam-beam perturbation in an equal-beam strong-strong case.

Figure 8.6 shows a contour plot of constant $r = \frac{\beta^*}{\beta_0^*}$ in the (ν, ξ_0) space. It replaces the lower panel of Fig. 8.4, which is valid only for the weak-strong case. The red curves in Fig. 8.6 are the contour when $r = 1$, i.e. the β-function is unperturbed by the linear thin-lens beam-beam interaction. They are the same as the dashed curves shown in Fig. 8.4. Unlike Fig. 8.4 however, here we extend the plot also to the region $\xi_0 < 0$.

The red curves in Fig. 8.6 divide the (ν, ξ_0) space into four quadrants. The northwest and southeast regions are when the beam-beam interaction pinches the beam size at the collision point and helps the luminosity. The southwest and northeast regions are unfavorable to the luminosity.

It should be noted that, unlike Fig. 8.4, there is no unstable regions in Fig. 8.6. This striking difference results because when ξ_0 approaches the unstable limit ξ_{limit}, the beam size has grown in such a way that ξ stays always below the limit. As a result, when dynamic-β^* effect is taken into consideration, there is no beam-beam instability in the linear thin-lens model.

Figure 8.6 can also be shown in a different manner as in Fig. 8.7. Both Figs. 8.6 and 8.7 show that, for small $2\pi\xi_0$, a pinch effect $\beta^* < \beta_0^*$ occurs when $\nu < \frac{1}{4}$ while antipinching occurs if $\frac{1}{4} < \nu < \frac{1}{2}$. The pattern repeats with period ν of $\frac{1}{2}$.

Figure 8.7 is to be compared with Fig. 8.5 for a weak-strong case without dynamic-β^*. It can be seen in Fig. 8.7, particularly the blue curve, that to benefit from the β^* pinch, it is advisable to choose ν close to and above an integer.

Figure 8.6: Dynamic-β^* effect showing, for the case of colliding two round beams, the contour plot of constant $r = \frac{\beta^*}{\beta_0^*}$ in the (ν, ξ_0) space. Quantity ξ_0 is the unperturbed beam-beam strength parameter before the dynamic-β^* effect is taken into account, ν is the unperturbed tune between collision points.

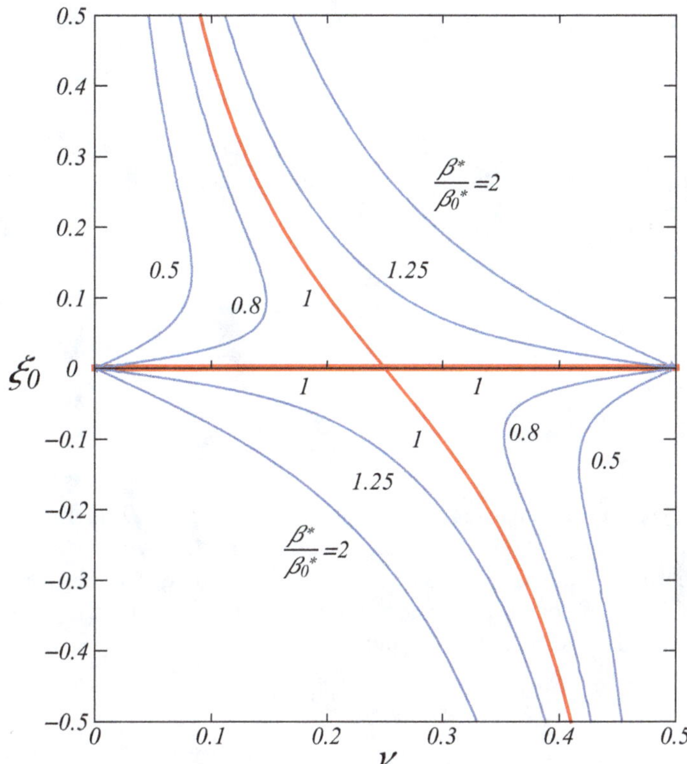

Luminosity   The dynamic-β^* also affects the luminosity. The expression for the luminosity for round beams is

$$\mathcal{L} = \frac{N^2 f}{4\pi\sigma^2} \,,$$

where σ is the rms size of the round beam at the collision point, f is the collision frequency. As the beam size changes by dynamic-β^*, so does the beam-beam parameter,

$$\xi = \frac{\beta_0^*}{4\pi} \frac{N r_0}{\gamma\sigma^2} \,.$$

Be reminded again that here β_0^* is the unperturbed β-function while σ is the perturbed beam size.

Figure 8.7: Dynamic-β^* effect showing $r = \frac{\beta^*}{\beta_0^*}$ as a function of ξ_0 for round beams. The blue, red, green, brown are for the cases $\nu_0 = 0.05, 0.1, 0.2, 0.3$, respectively.

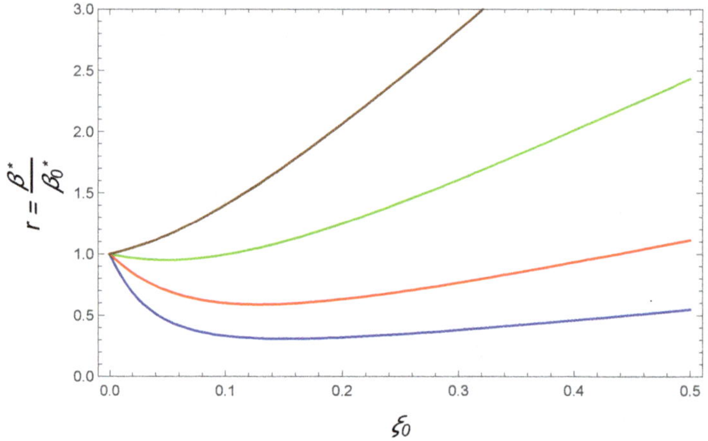

Consider the case without dynamic-β^* effect first. The beam size stays constant at σ_0. The luminosity increases quadratically with N until the beam becomes unstable when ξ reaches the unstable point ξ_{limit}, Eq. (8.24). To be specific, let us consider the case $\xi > 0$ for colliding beam of opposite charges and unperturbed tune $0 < \nu < \frac{1}{2}$, then

$$\xi_{\text{limit}} = \frac{\cot \pi \nu}{2\pi}.$$

If we define N_{limit} and $\mathcal{L}_{\text{limit}}$ to be the beam intensity and the luminosity at the point when ξ reaches ξ_{limit}, again without taking into account of dynamic-β^*,

$$N_{\text{limit}} = \frac{2\gamma \sigma_0^2}{\beta_0^* r_0} \cot \pi \nu, \qquad \mathcal{L}_{\text{limit}} = \frac{f \sigma_0^2 \gamma^2}{\pi \beta_0^{*2} r_0^2} \cot^2 \pi \nu,$$

and define

$$x = \frac{N}{N_{\text{limit}}},$$

then we have

$$\xi_0 = \begin{cases} \xi_{\text{limit}} x, & \text{if } x < 1, \\ \text{unstable}, & \text{if } x > 1, \end{cases}$$

$$\mathcal{L}_0 = \begin{cases} \mathcal{L}_{\text{limit}} x^2, & \text{if } x < 1, \\ \text{unstable}, & \text{if } x > 1. \end{cases}$$

The situation is shown in Fig. 8.8 as the black curve. The end point of the curve indicates where the instability sets in and beam becomes unstable.

When dynamic-β^* is taken into account, as mentioned earlier, there is no instability any more. The beam size grows when instability approaches in such a way that the perturbed ξ never exceeds ξ_{limit}. However, the luminosity suffers by the blown-up beam size,

$$\xi = \frac{\xi_0}{r}, \qquad \mathcal{L} = \frac{\mathcal{L}_0}{r}, \qquad \text{where} \qquad r = \frac{\beta^*}{\beta_0^*} = \frac{\sigma^2}{\sigma_0^2}. \qquad (8.30)$$

With dynamic-β^*, the value of r depends on ξ_0 and ν. The value of r in Eq. (8.30) is obtained by Eq. (8.29) with $\xi_0 = \xi_{\text{limit}}x$. We then obtain

$$\mathcal{L} = \frac{\mathcal{L}_{\text{limit}}x^2}{\sqrt{1 + 4\pi^2\xi_{\text{limit}}^2 x^2 \csc^2 2\pi\nu - 2\pi\xi_{\text{limit}}x \cot 2\pi\nu}}.$$

As mentioned, this expression holds even when $x > 1$, i.e. when $N > N_{\text{limit}}$.

Figure 8.8 shows the expected dependence of luminosity on the beam intensity. The horizontal axis is $x = \frac{N}{N_{\text{limit}}}$. The vertical axis is $\frac{\mathcal{L}}{\mathcal{L}_{\text{limit}}}$. We have normalized the axes by N_{limit} and $\mathcal{L}_{\text{limit}}$, so note that these normalizations depend on ν. The luminosity is expected to benefit from the β^* pinching when ν is close to and above an integer (the blue curve).

When $0 < \nu < \frac{1}{4}$, the luminosity is higher than $\mathcal{L}_0$ initially due to the pinching effect. However, when N continues to increase, $\mathcal{L}$ drops below the quadratic behavior as pinching cannot overcome the beam-beam blow-up. When $\frac{1}{4} < \nu < \frac{1}{2}$, the beam-beam lens is always antipinching if the two beams have the same sign of charge.

The beam size growth with beam intensity and the beneficial pinching effect when ν is close to an integer are general model-independent features to be expected of the beam-beam effects.

As we will discuss in the rest of this chapter, there are other beam-beam mechanisms that prohibits the beam-beam strength parameter to reach the limit such as predicted by Eq. (8.24) of the linear thin-lens model. However, as the first simplest beam-beam model, this linear thin-lens model already yields useful conclusions.

A quantitative example   It may be instructive to give a quantitative example. Consider an e^+e^- storage ring collider. Assume $\gamma = 10^4$ (5 GeV), a round beam with unperturbed rms beam size $\sigma_0 = 0.2$ mm, $\beta_0^* = 0.4$ m, and a collision rate of $f = 5 \times 10^5$ s^{-1}. Table below gives the relevant quantities as functions of ν.[9]

---

[9]This quantitative example illustrates the dynamic-β^* effect under a linearized thin-lens model. The values of ξ_{limit} in the table are exaggerated because linearized thin-lens is a rather optimistic model. Actually achieved limits come nowhere close to some of the values listed in the table.

Figure 8.8: Expected luminosity as a function of N, the number of particles per bunch. The black curve refers to the case without dynamic-β^* effect. It stops at $x = 1$ with the end point marking the instability limit. The blue, red, and brown curves correspond to $\nu = 0.05, 0.1$ and 0.3, respectively, when dynamic-β^* effect is considered.

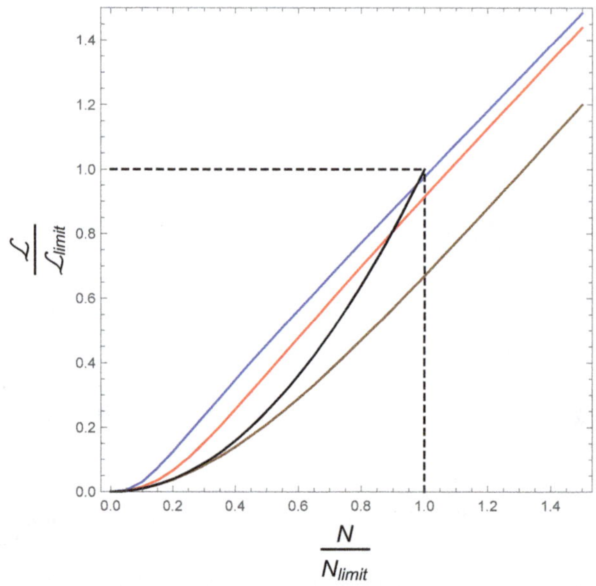

ν	0.05	0.1	0.3
$\xi_{\text{limit}} = \frac{\cot \pi \nu}{2\pi}$	1.005	0.49	0.115
$N_{\text{limit}} = \xi_{\text{limit}} \frac{4\pi}{\beta_0^*} \frac{\gamma \sigma_0^2}{r_0}$	9.02×10^{12}	4.4×10^{12}	1.04×10^{12}
$\mathcal{L}_{\text{limit}} = \frac{N_{\text{limit}}^2 f}{4\pi \sigma_0^2} \ [\text{cm}^{-2}\text{s}^{-1}]$	8.1×10^{33}	1.9×10^{33}	1.07×10^{32}

Be reminded that in this table, $\xi_{\text{limit}}, N_{\text{limit}}, \mathcal{L}_{\text{limit}}$ refer to values before taking into consideration of the dynamic-β^* effect. Figure 8.9 shows the behavior of the perturbed $\frac{\beta^*}{\beta_0^*}$, the perturbed beam-beam parameter ξ and the luminosity $\mathcal{L}$ as functions of the beam intensity N for the three cases of $\nu = 0.05, 0.1, 0.3$ when the dynamic-β^* is considered. The β^* pinch of the case $\nu = 0.05$ is clearly visible.

Note that following the tradition, the unit of luminosity $\mathcal{L}$ is $\text{cm}^{-2}\text{s}^{-1}$, not $\text{m}^{-2}\text{s}^{-1}$. The black dashed curves are references when dynamic-β^* effect is not considered. They are plotted over the entire range of beam intensity but they actually stop at the value N_{limit} (see the above table for the values of N_{limit} for the three cases of ν) when instability sets in. The colored curves however do not stop.

Figure 8.9: A quantitative example of dynamic-β^* effect. The black dashed curves are the reference case when dynamic-β^* is ignored. The blue, red, and brown curves correspond to $\nu = 0.05, 0.1$ and 0.3, respectively. All three plots are in log-log scales.

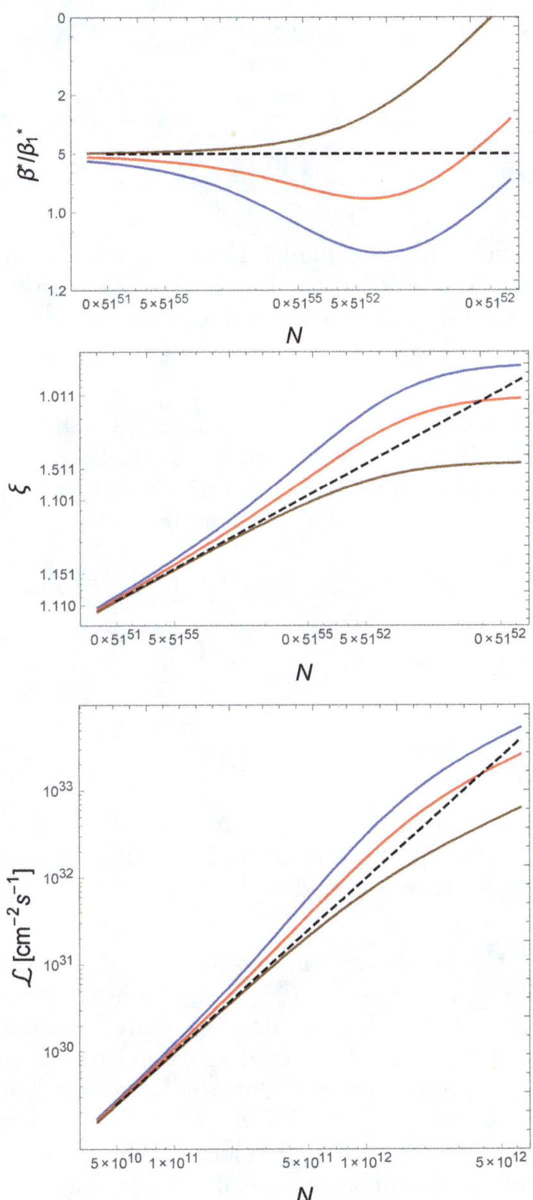

Also seen by the behavior of ξ is the fact that it approaches a constant value given by the instability threshold for the specified ν. As N increases, ξ approaches the limiting value but does not cross it. The asymptotic behavior when N reaches the saturation level is (Homeworks 8.13 and 8.14)

$$\frac{\beta^*}{\beta_0^*} \;\to\; \frac{\beta_0^* N r_0}{2\gamma\sigma_0^2}\tan\pi\nu \,,$$

$$\xi \;\to\; \frac{\cot\pi\nu}{2\pi}\,,$$

$$\mathcal{L} \;\to\; \frac{Nf\gamma}{2\pi\beta_0^* r_0}\cot\pi\nu\,. \tag{8.31}$$

According to the linear thin-lens model, therefore, when the beam intensity N increases approaching saturation, the beam-beam parameter saturates, the β-function grows linearly with N, the beam size grows as $\sqrt{N}$, and the luminosity grows linearly (not quadratically) with N.

Tune shift   We have been paying attention to the beam-beam perturbation to β^* so far mainly because the luminosity depends on it.   There is also a perturbation to the tune from ν to $\nu + \Delta\nu$ which we now turn attention to.

With dynamic-β^* effect, $\Delta\nu$ is determined by

$$\cos 2\pi(\nu + \Delta\nu) \;=\; \cos 2\pi\nu - 2\pi\xi_0\frac{\beta_0^*}{\beta^*}\sin 2\pi\nu\,.$$

We then use Eq. (8.29) to obtain

$$\cos 2\pi(\nu + \Delta\nu) \;=\; \cos 2\pi\nu - \frac{2\pi\xi_0\sin 2\pi\nu}{\sqrt{1 + 4\pi^2\xi_0^2\csc^2 2\pi\nu} - 2\pi\xi_0\cot 2\pi\nu}\,. \tag{8.32}$$

A contour plot of constant $\Delta\nu$ is shown in Fig. 8.10. The figure illustrates how the tune shifts when $|2\pi\xi_0|$ is small and how the instability is approached. See Homework 8.16 for more discussions on Fig. 8.10.

Flat beam   So far we simplified the problem assuming a round beam at the collision point with $\sigma_{x0} = \sigma_{y0}$ and $\beta_{x0}^* = \beta_{y0}^*$. Since the two dimensions are decoupled, it is straightforward to extend the analysis to the case of elliptical beams. In fact, it is also straightforward to extend to the case when the two beams have different intensities or even different beam energies. But to illustrate this flexibility, let us consider only the case of flat beams. We assume in this case that $\sigma_x \gg \sigma_y$ and $\xi_x \ll \xi_y$ and consider only the y motion of the particles. We again assume the two beams have equal intensity and equal energy.

The unperturbed ξ_{y0} is given by

$$\xi_{y0} \;=\; \frac{\beta_{y0}^* N r_0}{2\pi\gamma\sigma_{y0}(\sigma_{x0} + \sigma_{y0})} \;\approx\; \frac{\beta_{y0}^* N r_0}{2\pi\gamma\sigma_{y0}\sigma_{x0}}\,.$$

Figure 8.10: Contour plot of constant tune shift $\Delta\nu$ for a round beam with dynamic-β^* taken into account. The divergences appear when $\nu + \Delta\nu \to \frac{1}{2}$.

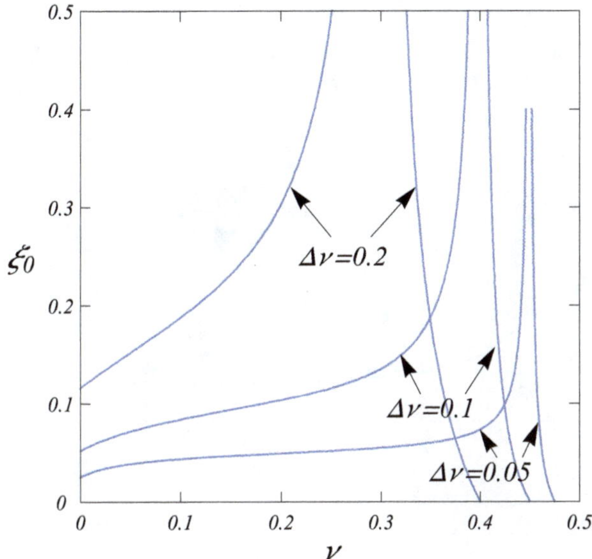

The beam-beam parameter perturbed by the dynamic-β^* is to replace σ_{y0} in the denominator by σ_y, and as a result, we have

$$\xi_y = \xi_{y0}\sqrt{\frac{\beta_0^*}{\beta_y^*}},$$

because $\sigma_y \propto \sqrt{\beta_y^*}$ (see footnote 6).

Following similar steps as Eq. (8.29), the self-consistent solution for a flat beam is found to be given by

$$r^2 - (2\pi\xi_{y0})^2 r + 4\pi\xi_{y0}r\sqrt{r}\cot 2\pi\nu_y = 1, \quad \text{where} \quad r = \frac{\beta_y^*}{\beta_{y0}^*}. \tag{8.33}$$

The dynamic-β^* effect of a flat beam on β_y is shown in Fig. 8.11. It is to be compared with Fig. 8.5, the case when beam-beam thin-lens is applied but for a weak-strong case without dynamic-β^*, and Fig. 8.7 when dynamic-β^* is included but for round Gaussian beams. The dynamic-β^* effect reacts more strongly for flat beams than for round beams for given beam-beam strength parameter.

Homework 8.12

(a) Refer to Fig. 8.4, what is the condition for $\beta^* = \beta_0^*$? Show that there are two solutions. One of them is the trivial case when $\xi_0 = 0$. The other is

Figure 8.11: Dynamic-β^* effect showing $r = \dfrac{\beta_y^*}{\beta_{y0}^*}$ as a function of ξ_{y0} for flat beams. The blue, red, green, brown are for the cases $\nu_{y0} = 0.05, 0.1, 0.2, 0.3$, respectively.

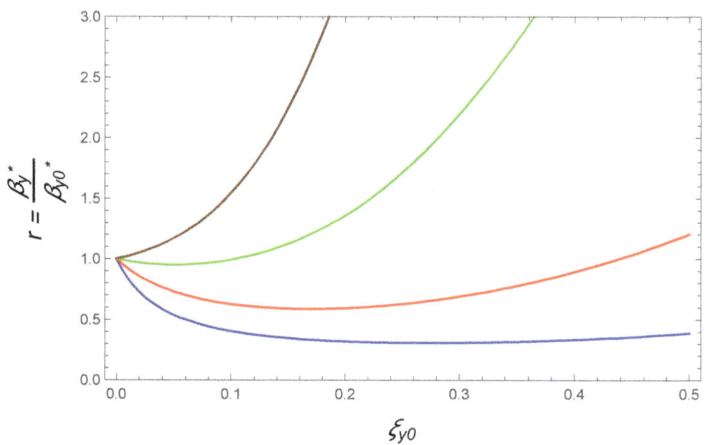

given by Eq. (8.26). Both solutions are indicated by dashed curves in the lower panel of Fig. 8.4.

(b) What is the tune shift when the beam-beam perturbation does not change the β-function, i.e. when $\beta^* = \beta_0^*$ and $\xi_0 \neq 0$?

Solution (b) You might want to guess at the answer first. The answer is $\Delta\nu = \frac{1}{2} - 2\nu$ when $\frac{1}{2} > \nu > 0$.

Homework 8.13

(a) Verify Eq. (8.29) for round beams.

(b) Verify Eq. (8.33) for flat beams.

(c) Show that for $|2\pi\xi_0| \ll 1$, the relevant solutions of both Eq. (8.29) and (8.33) are give by $r \approx 1 - 2\pi\xi_0 \cot 2\pi\nu$. Note that this is the same as Eq. (8.23), i.e. to 1st order in ξ_0, the beam-beam perturbed β^* is the same with or without dynamic-β^* effect. Physically this is expected because the dynamic-β^* effect, requiring both beams to have a finite ξ_0, necessarily has its leading order $\sim \mathcal{O}(\xi_0^2)$.

(d) In the other extreme when $|2\pi\xi_0| \gg 1$, show that

$$\frac{\beta^*}{\beta_0^*} \approx \begin{cases} 2\pi\xi_0 \tan \pi\nu, & \text{for round beams,} \\ (2\pi\xi_0 \tan \pi\nu)^2, & \text{for flat beams.} \end{cases}$$

(e) Use the result in (d) to verify Eq. (8.31).

Homework 8.14 The asymptotic behavior (8.31) applies for a round beam. In case of a flat beam, we obtain a different scaling with N. Find the asymptotic scaling of $\xi_{x,y}$, $\sigma_{x,y}$ and $\mathcal{L}$ with N. As one would expect, the vertical beam size blows up first.

Solution As $N \to \infty$, the linear thin-lens dynamic-β^* model gives $\xi_x \propto N$ (provided it stays well below the instability condition), $\sigma_x \to$ constant, $\xi_y \to$ constant, $\sigma_y \propto N$, and $\mathcal{L} \propto N$.

Homework 8.15 When $\beta^* = \beta_0^*$, or $r = 1$, one expects the difference between the cases with and without dynamic-β^* to disappear because the perturbed beam size is equal to the unperturbed beam size. We have now calculated r for four cases, round beams with and without dynamic-β^*, and flat beams with and without dynamic-β^*. Show that all four cases have the same condition for $r = 1$, namely

$$\xi_0 = \frac{\cot 2\pi\nu}{\pi}.$$

Homework 8.16

(a) Use Eq. (8.32) to confirm that $\Delta\nu \approx \xi_0$ when $|2\pi\xi_0| \ll 1$.
(b) Show that $\Delta\nu$ when $|2\pi\xi_0| \gg 1$ is such that $\nu + \Delta\nu \to \frac{1}{2}$.
(c) Confirm both of your answers in (a) and (b) with Fig. 8.10.

8.3.4 Flip-flop effect

Recap We started with a weak-strong picture and a linear thin-lens model and established the physical model represented by Eq. (8.19) and Figs. 8.4 and 8.5. In this model, the beam-beam perturbation is applied only to the weak beam particles. It must be kept in mind that, in Eq. (8.19), ν and β_0^* refer to the unperturbed tune and the unperturbed β-function of the weak beam, while the focal length f is determined by the beam sizes of the strong beam. In this weak-strong model, f is fixed while ν and β_0^* are perturbed by it.

We then introduced a complication to consider the case when the two beams are mutually perturbed in a strong-strong configuration, and we introduced the dynamic-β^* effect, represented by Eq. (8.29) and Figs. 8.6 and 8.7 (round beam case). In this picture, the two beams mutually perturb each other while maintaining to behave the same way because we assume the two beams have equal intensity. The analysis then follows by a self-consistency condition.

The linear thin-lens model will be extended another step in this section. This time we repeat the dynamic-β^* analysis, but keep track of which parameters come from which beam by a more careful bookkeeping. The goal of the analysis is to distinguish a possibility when the two beams end up in different states even if we assume they have the same intensity and in the same initial configuration. In particular, we will look for the possibility of a steady state when one beam is blown up while the other beam stays focused. If that happens, it is called a flip-flop effect. There is no a priori reason which beam becomes blown up as they started identical; the blown-up beam effectively becomes the weak beam, the other beam becomes the strong beam and stays minimally perturbed.

Self-consistency condition As mentioned, the analysis proceeds by a more careful bookkeeping. To be specific, let us consider a case of an electron-positron storage ring collider. We then refer to these two beams the $-$ and the $+$ beams. We will still consider linearized beam-beam forces (linear thin-lens model), but now allow the two beams to behave differently. For simplicity, we consider the case of two round beams.

Using a similar derivation to Eq. (8.29) but keeping track of the fact that the perturbation of one beam is due to the beam-beam force from the other beam, we obtain[10]

$$r_+ = \frac{1}{\sqrt{1 + 4\pi\xi_0 \cot 2\pi\nu \left(\frac{1}{r_-}\right) - (2\pi\xi_0)^2 \left(\frac{1}{r_-}\right)^2}},$$

$$r_- = \frac{1}{\sqrt{1 + 4\pi\xi_0 \cot 2\pi\nu \left(\frac{1}{r_+}\right) - (2\pi\xi_0)^2 \left(\frac{1}{r_+}\right)^2}}, \qquad (8.34)$$

where $r_\pm = \left(\frac{\beta^*}{\beta_0^*}\right)_\pm$. As always, ξ_0 and β_0^* denote the unperturbed values. Equation (8.34) describes how the two beams couple to each other through the beam-beam interaction.

One obvious solution to Eq. (8.34) is when $r_+ = r_-$, i.e. Eq. (8.29). But now there is another set of solutions with $r_+ \neq r_-$, which is found to be (Homework 8.17)

$$r_\pm = \frac{4\pi^2\xi_0^2 - 1}{16\pi^2\xi_0^2 \cot^2 2\pi\nu - (4\pi^2\xi_0^2 - 1)^2} \times \left\{ 2\pi\xi_0(4\pi^2\xi_0^2 + 1)\cot 2\pi\nu \right.$$

$$\left. \pm \frac{1}{\sqrt{2}}\sqrt{(4\pi^2\xi_0^2 + 1)\csc^2 2\pi\nu \left[1 - 20\pi^2\xi_0^2 + 32\pi^4\xi_0^4 - (4\pi^2\xi_0^2 + 1)\cos 4\pi\nu\right]} \right\}.$$

$$(8.35)$$

In case the beams stay in this solution, one beam is statically blown up (r_+), the other one pinches (r_-). This constitutes a bi-stable state of the beams because switching the blown-up beam and the pinched beam is also a solution. This situation can be a model for the flip-flop phenomenon observed in storage ring colliders. Note that this is not the only possible explanation of the flip-flop effect. We will mention another possible explanation later in Sec. 8.5.1.

Figure 8.12 shows the steady-state dynamic-β^* flip-flop solutions as functions of the unperturbed beam-beam parameter ξ_0 for three values of unperturbed tune ν. As mentioned, the equal-beam solution (8.29) always exists for all values of ξ_0 and is indicated by the red curves. The flip-flop solution (8.35) exists only over a finite range of ξ_0 and are indicated by the black curves.

For the flip-flop solution (8.35) to exist, the condition on ξ_0 is

$$\xi_{0,\text{threshold}} < \xi_0 < \frac{\cot \pi\nu}{2\pi}. \qquad (8.36)$$

---

[10] A.W. Chao, SSCL Laboratory report SSCL-346 (1991), Lecture Notes Phys. 460, p. 363 (1992).

Figure 8.12: The flip-flop solution of dynamic-β^* effect. The three panels show the cases $\nu = 0.05, 0.1, 0.15$, respectively from left to right. The flip-flop solution, indicated by the black curves, exists only in a range of ξ_0. The end point of the range occurs when one of the beam becomes infinitely large while the other beam reaches $r = 1$. The red curves represent the third equal-beam solution without flip-flop. The flip-flop states exist only when $\nu < \frac{1}{4}$.

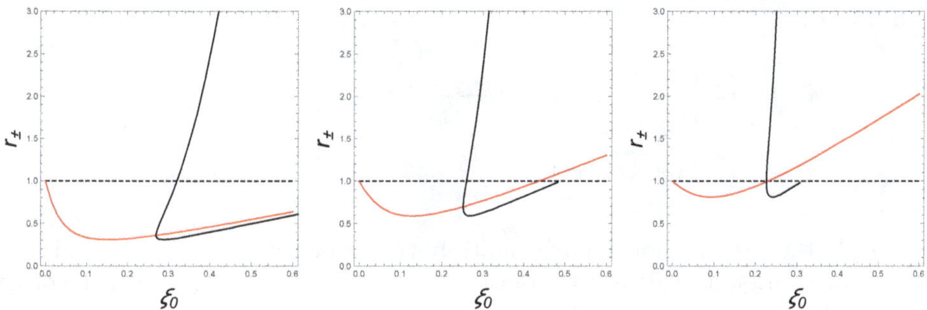

where the flip-flop threshold occurs at

$$\xi_{0,\text{threshold}} = \frac{1}{4\pi}\sqrt{5 + \cos 4\pi\nu + \cos 2\pi\nu\sqrt{34 + 2\cos 4\pi\nu}}\,.$$

In order for this range to exist, ν has to be less than $\frac{1}{4}$,

$$\nu < \frac{1}{4},$$

i.e. the beams must pinch each other to begin with. At the low end of the range (8.36), the two beams have equal sizes. At the high end of the range, one of the beams is blown up to an infinite beam size; the other pinched beam in this case sees no beam-beam force from the blown-up beam, and its beam size is unperturbed, $r = 1$. These features can be seen in Fig. 8.12.

Note that not all steady-states are stable. In particular, in the region (8.36), we will show later that the equal-beam steady-state solution, indicated by the red curves, is in fact unstable; any infinitesimal perturbation away from it will force the beams to converge to a flip-flop state.

Homework 8.17

(a) Derive Eq. (8.35) starting with Eq. (8.34).

(b) In your derivation, verify the validity condition (8.36).

(c) Show that at the low end of the validity range, the two beams have equal sizes. At the high end, show that one beam has $r \to \infty$ and the other beam has $r = 1$.

(d) Show that there is no flip-flop solution if $\nu > \frac{1}{4}$ regardless of the value of ξ_0.

(e) Show that the widest range of flip-flop occurs when ν is slightly above a half-integer, and in that case, the flip-flop occurs only when ξ_0 exceeds a threshold value,

$$\xi_0 \;>\; \frac{\sqrt{3}}{2\pi}\,.$$

Solution This homework requires a bit of straightforward algebra. Rewrite Eq. (8.34),

$$r_+^2 \left[1 + 4\pi\xi_0 \cot 2\pi\nu\,\frac{1}{r_-} - (2\pi\xi_0)^2\frac{1}{r_-^2} \right] \;=\; 1\,,$$

$$r_-^2 \left[1 + 4\pi\xi_0 \cot 2\pi\nu\,\frac{1}{r_+} - (2\pi\xi_0)^2\frac{1}{r_+^2} \right] \;=\; 1\,.$$

Multiply the first member by r_-^2, multiply the second member by r_+^2, and subtract the results, it leads to two solutions. One of them says $r_+ = r_-$; the other solution says

$$\frac{1}{r_+} + \frac{1}{r_-} \;=\; \frac{4\pi\xi_0 \cot 2\pi\nu}{4\pi^2\xi_0^2 - 1}\,.$$

(Do not worry about the apparent divergence at $2\pi\xi_0 = 1$.) It then follows that $r_\pm$ are the two solutions to a quadratic equation, which can be solved, yielding Eq. (8.35). In the solution process, it also will show the condition Eq. (8.36).

8.3.5   Linear synchrobetatron coupling with crossing angle

Although the linear thin-lens beam-beam interaction does not couple the x- and y-motions for an upright beam cross-section, as indicated by Eq. (8.14), it can still couple the x- or the y-motions to the longitudinal synchrotron degree of freedom if the beams collide with a crossing angle.[11]

Consider a weak-strong configuration with two beams crossing vertically at an angle 2α. We will look for the betatron and synchrotron tune shifts as well as changes in particle distribution of the weak beam under the influence of the collisions with the strong beam. To do this, we consider a particle in the weak beam with no horizontal displacement, and with vertical displacement y and longitudinal displacement z relative to the beam center. In this model, x-motion is still decoupled from y and z, but as we will see, y and z are coupled.

To analyze the problem, let us first consider the case when synchrotron oscillation is very slow, with synchrotron tune $\nu_s \ll 1$, so that z of the particle does not change much per revolution. The coordinate z can be considered quasistatic. With a crossing angle, the weak-beam particle receives a vertical kick that depends on z as it traverses the strong beam. If z stays quasistatic,

---

[11]B. Richter and D. Ritson, Int. Conf. High Energy Accel., Dubna, p. 461 (1963); L.E. Augustin, Orsay Laboratory note 35-69 (1969); A.W. Chao, AIP Conf. Proc. 57, Brookhaven (1979).

the beam-beam kicks generate a vertical closed orbit distortion $\Delta y_{\rm COD}$ around the collider ring. In the linear thin-lens model, $\Delta y_{\rm COD}$ is proportional to z. As a result, the weak beam acquires a y-z tilt as Fig. 8.13 illustrates.

Figure 8.13: The envisioned weak-strong configuration when the beams collide with a crossing angle. The two beam trajectories are tilted due to the crossing angle geometry. The weak beam's orientation will be tilted in addition relative to its tilted trajectory.

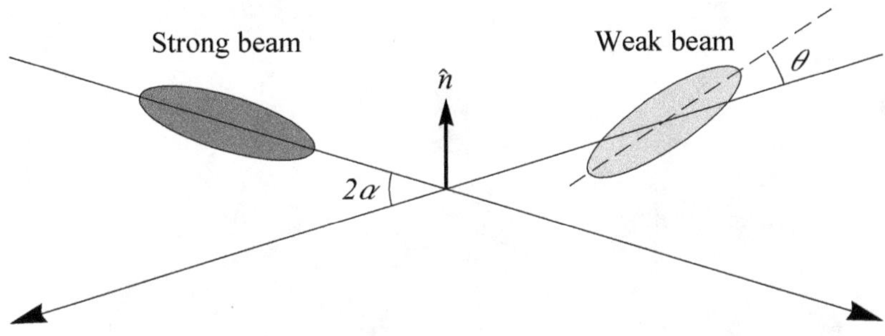

On the other hand, with a crossing angle, the beam-beam interaction also generates a kick in the z-direction. This component of the kick changes the particle's energy, which in turn affects its subsequent z motion. It is clear that a complete treatment of the problem should take into account of the synchrotron motion. The coupling between the vertical and the synchrotron motions generates synchrobetatron resonances.

Beam-beam kick The particle under consideration has coordinates (y, z) relative to the center of the weak beam as illustrated in Fig. 8.14.

Let the on-coming strong beam have the distribution in the y-z plane,

$$\rho(y, z) \;=\; \frac{Ne}{2\pi\sigma_y\sigma_z L_x}\, e^{-\frac{y^2}{2\sigma_y^2} - \frac{z^2}{2\sigma_z^2}},$$

where $L_x = \sqrt{2\pi}\,\sigma_x$ is the horizontal beam width and y and z here are coordinates relative to the center of the strong beam. Let us assume $\sigma_x \gg \sigma_y$ for simplicity, so here we are considering a flat beam configuration.

The impulse received by the particle of displacements y and z relative to the center of the weak beam, in the relativistic limit, is given by

$$
\begin{aligned}
\frac{\Delta\vec{p}}{P_0} &= -G\hat{n} \int_0^{y\cos\alpha - z\sin\alpha} dt\,\exp\left(-\frac{t^2}{2\Sigma^2}\right), \\
\Sigma^2 &= \sigma_z^2\sin^2\alpha + \sigma_y^2\cos^2\alpha, \\
G &= \frac{2Nr_0}{\gamma\sigma_x\Sigma}.
\end{aligned}
\tag{8.37}
$$

Figure 8.14: Coordinates (y, z) of a particle in the weak beam in a collision with a strong beam with a crossing angle 2α. The red arrows indicate the electric force, the magnetic force, and the direction $\hat{n}$ of the net force.

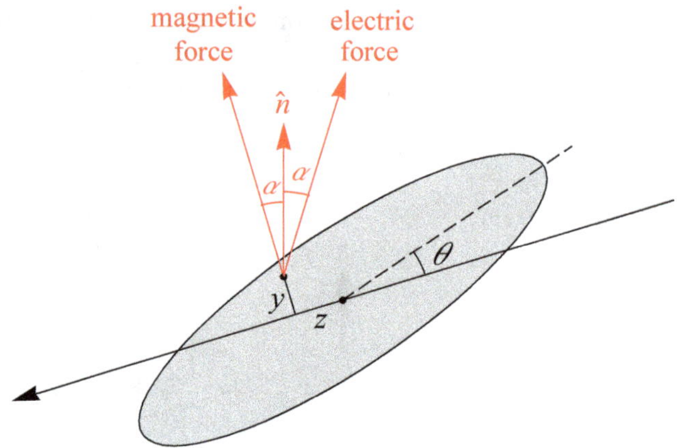

The reason this impulse is in the $\hat{n}$-direction is because the electric and the magnetic beam-beam forces, unlike the case of head-on collision, are not in the same direction, and their total force is in the $\hat{n}$-direction. See Fig. 8.14.

If $|y| \ll \sigma_y$ and $|z| \ll \sigma_z$, then $|y\cos\alpha - z\sin\alpha| \ll \Sigma$, we can linearize the beam-beam kick,

$$\frac{\Delta\vec{p}}{P_0} = -G\hat{n}\,(y\cos\alpha - z\sin\alpha)\,,$$

or, decomposing $\hat{n}$ into y- and z-components,

$$\hat{n} = \hat{y}\cos\alpha - \hat{z}\sin\alpha\,,$$
$$\Delta y' = -G(y\cos\alpha - z\sin\alpha)\cos\alpha\,,$$
$$\Delta\delta = G(y\cos\alpha - z\sin\alpha)\sin\alpha\,,$$

where $\delta = \frac{\Delta E}{E_0}$ is the relative energy deviation of the particle.

Defining the 4-D state vector of the weak-beam particle as

$$\begin{bmatrix} y \\ y' \\ z \\ \delta \end{bmatrix}\,,$$

the beam-beam kick can be represented by a matrix map,

$$T_{bb} = \begin{bmatrix} 1 & 0 & 0 & 0 \\ -G\cos^2\alpha & 1 & G\sin\alpha\cos\alpha & 0 \\ 0 & 0 & 1 & 0 \\ G\sin\alpha\cos\alpha & 0 & -G\sin^2\alpha & 1 \end{bmatrix}\,. \qquad (8.38)$$

This matrix map is symplectic. When $\alpha = 0$, it reduces to the 1-D thin-lens model. When $\alpha = \frac{\pi}{2}$, it switches the roles of σ_y and σ_z as one would expect.

Resonance stopband The map T_{bb} couples the y- and z-motions of particles in the weak beam. Let the rest of the collider ring be described by the uncoupled map

$$
T_0 = \begin{bmatrix}
\cos 2\pi\nu_y & \beta_0^* \sin 2\pi\nu_y & 0 & 0 \\
-\frac{1}{\beta_0^*} \sin 2\pi\nu_y & \cos 2\pi\nu_y & 0 & 0 \\
0 & 0 & \cos 2\pi\nu_s & -\eta C \\
0 & 0 & \frac{\sin^2 2\pi\nu_s}{\eta C} & \cos 2\pi\nu_s
\end{bmatrix},
$$

where η is the phase slippage factor and C is the circumference of the collider ring. The total one-turn map around the entrance point to the beam-beam collision point is

$$
T_{\text{tot}} = T_0 T_{bb} . \tag{8.39}
$$

The unperturbed tunes are ν_y and ν_s. The perturbed eigentune ν satisfies the eigenvalue equation,

$$
\det(T_{\text{tot}} - \lambda I) = 0, \quad \text{and} \quad \lambda + \frac{1}{\lambda} = 2\cos 2\pi\nu ,
$$

which yields, after some algebra,

$$
\left(\cos 2\pi\nu - \cos 2\pi\nu_s - \frac{\eta CG}{2} \sin^2 \alpha \right) \left(\cos 2\pi\nu - \cos 2\pi\nu_y + \frac{\beta_0^* G}{2} \cos^2 \alpha \sin 2\pi\nu_y \right)
$$

$$
= -\frac{\eta C \beta_0^* G^2}{16} \sin^2 2\alpha \sin 2\pi\nu_y . \tag{8.40}
$$

When $\alpha = 0$, the system is decoupled, and the two eigentunes are $\nu = \nu_s$ and

$$
\cos 2\pi\nu = \cos 2\pi\nu_y - \frac{\beta_0^* G}{2} \sin 2\pi\nu_y ,
$$

as they should be. If $\alpha \neq 0$, solution of Eq. (8.40) gives the two shifted eigentunes.

Away from synchrobetatron resonances, the two eigentunes are given by

$$
\cos 2\pi\bar{\nu}_s = \cos 2\pi\nu_s + \frac{\eta CG}{2} \sin^2 \alpha ,
$$

$$
\cos 2\pi\bar{\nu}_y = \cos 2\pi\nu_y - \frac{\beta_0^* G}{2} \cos^2 \alpha \sin 2\pi\nu_y . \tag{8.41}
$$

For the beam to be stable, we need both these expressions to have absolute values ≤ 1. This condition generates half-integer stopbands around $\nu_s = \frac{m}{2}$ and around $\nu_y = \frac{m}{2}$. The stopbands depend on whether m is even or odd, and also depend on the sign of G and η ($G > 0$ if two beams have opposite charges, $G < 0$ if they have the same charge; $\eta > 0$ above transition, $\eta < 0$ below transition). For the ν_s resonance, let $\nu_s = \frac{m}{2} + \Delta$, the stopband occurs when

	$\eta G > 0$	$\eta G < 0$
$m = $ even	$\|\Delta\| < \frac{\|\sin\alpha\|}{2\pi}\sqrt{\eta C G}$	stable
$m = $ odd	stable	$\|\Delta\| < \frac{\|\sin\alpha\|}{2\pi}\sqrt{-\eta C G}$

Since ν_s is usually small, most likely only the $m = 0$ case applies.

For the ν_y resonances, let $\nu_y = \frac{m}{2} + \Delta$, the stopband occurs when

	$G > 0$	$G < 0$
$m = $ any integer	$0 > \Delta > -\frac{\beta_0^* G}{2\pi}\cos^2\alpha$	$-\frac{\beta_0^* G}{2\pi}\cos^2\alpha > \Delta > 0$

This betatron resonance stopband is just Eq. (8.25) obtained earlier if we set $\alpha = 0$.

Our attention actually turns more to the synchrobetatron coupling resonances when $\bar\nu_y \pm \bar\nu_s \approx m$.[12] Equation (8.40) can be solved for $\cos 2\pi\nu$. With the definitions of $\bar\nu_{y,s}$ given by Eq. (8.41), the solution reads

$$\cos 2\pi\nu = \frac{\cos 2\pi\bar\nu_y + \cos 2\pi\bar\nu_s}{2}$$
$$\pm\sqrt{\left(\frac{\cos 2\pi\bar\nu_y - \cos 2\pi\bar\nu_s}{2}\right)^2 - \frac{\eta C \beta_0^* G^2}{16}\sin^2 2\alpha \sin 2\pi\nu_y}\ .$$

Let $\bar\nu_y \pm \bar\nu_s = m + \Delta$. Stability requires the quantity inside the square root to be positive. The stopband occurs when

	$\eta\sin 2\pi\nu_y > 0$	$\eta\sin 2\pi\nu_y < 0$
$m = $ any integer	$\|\Delta\| < \frac{\|G\sin 2\alpha\|}{4\pi}\sqrt{\frac{\eta C \beta_0^*}{\sin 2\pi\nu_y}}$	stable

In case $\nu_s \ll 1$, $\sin 2\pi\nu_y > 0$ for difference resonances $\nu_y - \nu_s = m$, and $\sin 2\pi\nu_y < 0$ for sum resonances $\nu_y + \nu_s = m$. It follows that there are synchrobetatron sum resonance stopbands if $\eta < 0$ (below transition), and there are difference resonance stopbands if $\eta > 0$ (above transition).

In contrast, to 1st order in the coupling perturbation, the betatron coupling resonances always have stopbands only for the sum resonances and not for the difference resonances. The reason there is a flip of roles above transition here is due to the negative longitudinal mass effect above transition.

As a numerical example, let us take $\beta_0^* G = 0.6$ (two beams have opposite signs of charge), $\alpha = 1$ mrad, $\nu_y = 0.1$, $\eta C = 2$ m (above transition), and $\beta_0^* = 0.2$ m. In this case, $\eta G > 0$, the betatron half-integer resonances hav a total width of $\Delta = 0.1$ below $\nu_y = \frac{m}{2}$. The synchrotron resonances occur around $\nu_s = m$ with a width $\pm 4 \times 10^{-4}$. The synchrobetatron stopband occurs around $\nu_y - \nu_s = m$ and has a width $\Delta = \nu_y - \nu_s - m = \pm 4 \times 10^{-4}$.

---

[12]A. Piwinski, DESY Laboratory report DESY77/18 (1977).

Beam tilt   We mentioned by Fig. 8.13 that one consequence of beams colliding with a crossing angle is a tilt of the beam distribution in the y-z plane. The tilt angle is contained in the one-turn map T_{tot} obtained in Eq. (8.39). To find it, we first look for the eigenvectors of T_{tot}.

For simplicity of algebra, we assume $G\beta_0^* \ll 1$. For our purpose we need only to find one of the four eigenvectors — the one that is slightly beam-beam perturbed from the original longitudinal mode. It is found to be (Homework 8.20)

$$V = \begin{bmatrix} -\dfrac{G\beta_0^*}{2} \dfrac{\sin\alpha\cos\alpha\sin 2\pi\nu_y}{\cos 2\pi\nu_y - \cos 2\pi\nu_s} \\[2ex] -\dfrac{G}{2}\sin\alpha\cos\alpha \dfrac{\cos 2\pi\nu_y - e^{-i2\pi\nu_s}}{\cos 2\pi\nu_y - \cos 2\pi\nu_s} \\[2ex] 1 \\[2ex] -i\dfrac{\sin 2\pi\nu_s}{\eta C} \end{bmatrix}. \tag{8.42}$$

We need to find the distribution tilt of the eigenvector V. Let its corresponding eigenvalue be $e^{i2\pi\nu_1}$. We know that T_{tot} is symplectic, so V^* is also an eigenvector with eigenvalue $e^{-i2\pi\nu_1}$. Let us consider a particle initially in the state

$$X(0) = V + V^*.$$

The state of this particle k revolutions later will be

$$X(k) = Ve^{i2\pi k\nu_1} + V^*e^{-i2\pi k\nu_1}.$$

Let us designate the vector V as

$$V = \begin{bmatrix} v_{1r} + iv_{1i} \\ v_{2r} + iv_{2i} \\ v_{3r} + iv_{3i} \\ v_{4r} + iv_{4i} \end{bmatrix}.$$

The y- and z-components of the particle motion then reads

$$\begin{aligned} y(k) &= 2v_{1r}\cos 2\pi k\nu_1 - 2v_{1i}\sin 2\pi k\nu_1, \\ z(k) &= 2v_{3r}\cos 2\pi k\nu_1 - 2v_{3i}\sin 2\pi k\nu_1. \end{aligned}$$

As k progresses from turn to turn, this particle traces out an ellipse in the y-z plane. The ellipse has a tilt angle θ_{yz} given by

$$\tan 2\theta_{yz} = \frac{2(v_{1r}v_{3r} + v_{1i}v_{3i})}{v_{3r}^2 + v_{3i}^2 - v_{1r}^2 - v_{1i}^2}.$$

Substituting the various quantities using Eq. (8.42), we obtain

$$\tan\theta_{yz} = v_{1r} = \frac{G\beta_0^*}{2}\frac{\sin\alpha\cos\alpha\sin 2\pi\nu_y}{\cos 2\pi\nu_s - \cos 2\pi\nu_y}. \tag{8.43}$$

Since $G\beta_0^* \ll 1$, we have $\theta_{yz} \approx \tan\theta_{yz}$. The presence of synchrobetatron resonance — either sum or difference resonance — is obvious from the denominator $\cos 2\pi\nu_y - \cos 2\pi\nu_s$.

To see the effect of this beam tilt on luminosity, we need to compare θ_{yz} to the value of $\frac{\sigma_y}{\sigma_z}$. To avoid losing luminosity, we need

$$\theta_{yz} \ll \frac{\sigma_y}{\sigma_z},$$

or

$$|\nu_y \pm \nu_s - m| \gg \frac{G\beta_0^*}{4\pi} \frac{\sigma_z}{\sigma_y} \sin\alpha \cos\alpha.$$

If we take $\frac{\sigma_z}{\sigma_y} = 600$, $G\beta_0^* = 0.6$, $\alpha = 1$ mrad, we find we need to stay away from the synchrobetatron coupling resonances by $|\nu_y \pm \nu_s - m| \gg 0.03$.

We just showed that the beam distribution has a tilt in the y-z plane. Obviously the analysis can be extended to calculate the tilts in the y'-z and the δ-z planes. The head and the tail can have different values of y' and δ. In particular, since $v_{3i} = 0$ and $v_{4r} = 0$, there is no correlation between δ and z and therefore no head-tail asymmetry in the δ-z distribution. But the correlation between y' and z will show up as a head-tail asymmetry in the y'-z distribution.

Quasistatic treatment　　We mentioned at the beginning of this section a quasistatic picture in the limit when $\nu_s \to 0$. If we take that limit, the y and y' of the particle become closed orbit distortions as functions of z because z varies only slowly with time in the quasistatic limit. Our result (8.43) then says

$$y_{\text{COD}} = z \frac{G\beta_0^*}{2} \frac{\sin\alpha \cos\alpha \sin 2\pi\nu_y}{1 - \cos 2\pi\nu_y} = z \frac{G\beta_0^*}{2} \sin\alpha \cos\alpha \cot \pi\nu_y.$$

According to the closed orbit distortion analysis, the closed orbit at the position of a kicking angle θ is given by $y_{\text{COD}} = \frac{\theta\beta_y}{2} \cot \pi\nu_y$. In the present case, the kicking angle is $\theta = Gz\cos\alpha \sin\alpha$. Other than the synchrobetatron resonance, the quasistatic picture agrees with our result.

The quasistatic picture also predicts that the closed orbit slope immediately before the kick is $-\frac{\theta}{2}$ when the Courant–Snyder α-function is zero as we have assumed. In this case our analysis says

$$y'_{\text{COD}} = zv_{2r} = -z\frac{G}{2} \sin\alpha \cos\alpha.$$

Indeed, this also agrees with the quasistatic picture. Immediately after the beam-beam collision, $y'_{\text{COD}} = z\frac{G}{2} \sin\alpha \cos\alpha$.

Homework 8.18 Use the geometry of Fig. 8.14, convince yourself of Eq. (8.37).

Homework 8.19 Show that the map (8.38) is symplectic and (therefore) has unit determinant.

Homework 8.20 Verify the perturbed longitudinal eigenvector is given by Eq. (8.42) to 1st order in $|G\beta_0^*| \ll 1$.

Solution Do not try to solve for the eigenvectors by brute force. To lowest order in $G\beta_0^*$, all you need to realize is that we need to keep the first order terms in $G\beta_0^*$ in all cross-plane (off-diagonal) terms and only zeroth order in in-plane (diagonal) terms. This means T_{tot} is simplified to read $T_{\text{tot}} \approx T_{\text{tot1}}$, where

$$
T_{\text{tot1}} = \begin{bmatrix} \mathcal{O}(1) & \mathcal{O}(G\beta_0^*) \\ \hline \mathcal{O}(G\beta_0^*) & \mathcal{O}(1) \end{bmatrix}
$$

$$
= \begin{bmatrix}
\cos 2\pi\nu_y & \beta_0^* \sin 2\pi\nu_y & G\beta_0^* \cos\alpha \sin\alpha \sin 2\pi\nu_y & 0 \\
-\dfrac{\sin 2\pi\nu_y}{\beta_0^*} & \cos 2\pi\nu_y & G\cos\alpha \sin\alpha \cos 2\pi\nu_y & 0 \\
-\eta C G \cos\alpha \sin\alpha & 0 & \cos 2\pi\nu_s & -\eta C \\
G\cos\alpha \sin\alpha \cos 2\pi\nu_s & 0 & \dfrac{\sin^2 2\pi\nu_s}{\eta C} & \cos 2\pi\nu_s
\end{bmatrix}.
$$

After this approximation, T_{tot1} is still symplectic. This treatment amounts to paying attention to the synchrobetatron resonances $\nu_y \pm \nu_s = m$ but ignores the in-plane resonances $\nu_y = \frac{m}{2}$ and $\nu_s = \frac{m}{2}$.

Also, since we are looking for the perturbed synchrotron mode eigenvector, we approximate V to be

$$
V = \begin{bmatrix} \mathcal{O}(G\beta_0^*) \\ \hline \mathcal{O}(1) \end{bmatrix}.
$$

With these approximations valid to 1st order in $G\beta_0^*$, Eq. (8.42) follows. Furthermore, it can be shown that the eigenvalue associated with V is $e^{i2\pi\nu_s}$.

Homework 8.21 A counterpart of y-z coupling due to a crossing angle has an equivalent case for x-y coupling. In the linear thin-lens model, the x- and y-motions are decoupled when the on-coming beam has an upright bi-Gaussian transverse cross-section. In case the bi-Gaussian cross-section of the on-coming beam is tilted by an angle $\pm\alpha$ in the x-y plane at the collision point, we need to treat the problem with a 2-D dynamics. Modify our analysis to find the stability condition for this problem. You have already expected that this beam-beam interaction excites x-y coupling resonances of the types $\nu_x \pm \nu_y = m$.

8.3.6 Synchrobetatron coupling due to dispersions at the RF and the collision point

The text so far examined one mechanism of synchrobetatron resonances driven by linear thin-lens beam-beam force. Another possible candidate of beam-beam induced synchrobetatron mechanism is when the collision point is dispersive. In this case, however, in the weak-strong linearized thin-lens model, the linearized beam-beam force is equivalent to a quadrupole. Having a dispersion at the collision point is therefore similar to the very common case when a quadrupole is located at a dispersive location. It disturbs the quadrupole focusing, thus affecting the integer and half-integer resonances. It does not affect the synchro-betatron resonances.

Synchrobetatron resonances, however, are still driven if the RF cavity is located at a dispersive location.[13] So we ask a curiosity question what happens when both the collision point and the RF are dispersive in this section.

Again assuming a weak-strong linear thin-lens model. This time let us assume the collision point has a horizontal dispersion $D^* \neq 0$. The single-particle state vector is (x, x', z, δ). The beam-beam kick to a weak-beam particle is

$$T_{bb} \;=\; \begin{bmatrix} 1 & 0 & 0 & 0 \\ -\frac{1}{f_x} & 1 & 0 & 0 \\ 0 & 0 & 1 & 0 \\ 0 & 0 & 0 & 1 \end{bmatrix}, \qquad \frac{\beta_x^*}{4\pi f_x} \;=\; \xi \, .$$

To assure stability in the longitudinal dimension and to allow the study of synchrobetatron coupling resonances, we need to introduce an RF focusing cavity. The map for the cavity is

$$T_{\mathrm{cav}} \;=\; \begin{bmatrix} 1 & 0 & 0 & 0 \\ 0 & 1 & 0 & 0 \\ 0 & 0 & 1 & 0 \\ 0 & 0 & k_{rf} & 1 \end{bmatrix}, \qquad k_{rf} \;=\; \frac{e\hat{V}_{rf}\omega_{rf}}{E_0 c} \, .$$

The transfer map from one collision point to the next is

$$T_0 \;=\; T_2 T_{\mathrm{cav}} T_1 \, ,$$

where T_1 is the map from the interaction point to the cavity, T_2 is the map from the cavity to the next interaction point, both are given by the Courant–Snyder maps that involve the β-functions at the interaction point and the cavity, and involve the dispersion function D^*. For simplicity, however, we assume the cavity is located at the collision point, so that $T_1 = I$ and

$$T_2 \;=\; \begin{bmatrix} \cos\mu_x & \beta^* \sin\mu_x & 0 & D^*(1 - \cos\mu_x) \\ -\frac{\sin\mu_x}{\beta^*} & \cos\mu_x & 0 & \frac{D^*}{\beta^*}\sin\mu_x \\ -\frac{D^*}{\beta^*} & D^*(\cos\mu_x - 1) & 1 & -A \\ 0 & 0 & 0 & 1 \end{bmatrix} \, ,$$

where $A = \eta C - \frac{D^{*2}}{\beta^*}\sin\mu_x$, η is the phase slippage factor, C is the ring circumference. Synchrotron phase μ_s is determined by $\cos\mu_s = 1 - \frac{k_{rf}A}{2}$ with $\mu_s < \pi$.

The total map for one superperiod (half a revolution assuming two collision points) is

$$T_{\mathrm{tot}} \;=\; T_2 T_{\mathrm{cav}} T_{bb} \, .$$

---

[13]A. Piwinski, 11th Int. Conf. High Energy Accel., 638 (1980); A. Piwinski, Sec. 2.3.4, Handbook Accel. Phys. & Eng., 2nd ed., World Scientific (2013).

For small full-turn synchrotron tune $\nu_s = \frac{\mu_s}{\pi}$, we have $\nu_s \approx \frac{1}{\pi}\sqrt{k_{rf}A}$. For phase stability, we need $k_{rf}A > 0$.

The eigenfrequency μ is determined by the equation

$$\det\left(T_{\text{tot}} + T_{\text{tot}}^{-1} - 2\cos\mu\, I\right) \;=\; 0$$

$$\Longrightarrow$$

$$\cos^2\mu - (\cos\mu_x + \cos\mu_s - 2\pi\xi\sin\mu_x)\cos\mu + \left[\cos\mu_x\cos\mu_s - 2\pi\xi\cos\mu_s\sin\mu_x\right.$$

$$\left. + \frac{8D^{*2}}{A\beta^*}\sin^2\frac{\mu_s}{2}\sin^3\frac{\mu_x}{2}\left(\cos\frac{\mu_x}{2} - 2\pi\xi\sin\frac{\mu_x}{2}\right)\right] \;=\; 0. \tag{8.44}$$

The synchrobetatron resonance strength scales with two parameters, ξ and

$$h \;=\; \frac{D^{*2}}{A\beta^*}. \tag{8.45}$$

Typically, we would have $|h| \ll \frac{1}{\mu_x}$.

In Eq. (8.44), one observes that the beam-beam effect (characterized by ξ) and the dispersive cavity effect (characterized by h) are mostly decoupled. The only term potentially caused by a dispersive beam-beam interaction is the contribution proportional to the product $h\xi$. Homework 8.22 illustrates these resonances by an explicit example. It confirms that synchrobetatron resonances are mainly driven by the dispersive cavity. The dispersive collision point plays a role but is minor. We will encounter a similar situation when we discuss the synchrobetatron instability of coherent beam-beam instability in Sec. 8.7.

Homework 8.22

(a) Verify Eq. (8.44).

(b) Investigate the synchrobetatron resonance behavior by plotting the instability stopband using Eq. (8.44).

Solution (b) It should be noted that with only one RF per superperiod as we assumed, μ_s must not exceed π. Figure 8.15 shows the instability stopband in the $(\frac{\mu_x}{2\pi}, \frac{\mu_s}{2\pi})$ space. The four cases are (a) $h = 0.01$, $\xi = 0$; (b) $h = -0.01$, $\xi = 0$; (c) $h = 0.01$, $\xi = 0.05$; (d) $h = -0.01$, $\xi = 0.05$. Cases (a), (b) are without beam-beam effect; the stopbands are driven purely by the dispersive cavity. Case (a) is above transition, driving only the difference resonance. Case (b) is below transition, driving only the sum resonance. Cases (c), (d) includes the dispersive beam-beam effect. It is seen that the dispersive beam-beam effect disturbs only the integer and the half-integer resonances, leaving the synchrobetatron resonance basically intact.

Figure 8.15: Synchrobetatron resonance due to a dispersive cavity and dispersive collision point. For simplicity, the RF cavity is located at the collision point. Parameters $\mu_{x,s}$ are the betatron and synchrotron phase advance between collision points. Various resonance stopbands are plotted in black. The red curves highlight the synchrobetatron resonances. The four cases are (a) $h = 0.01$, $\xi = 0$; (b) $h = -0.01$, $\xi = 0$; (c) $h = 0.01$, $\xi = 0.05$; (d) $h = -0.01$, $\xi = 0.05$.

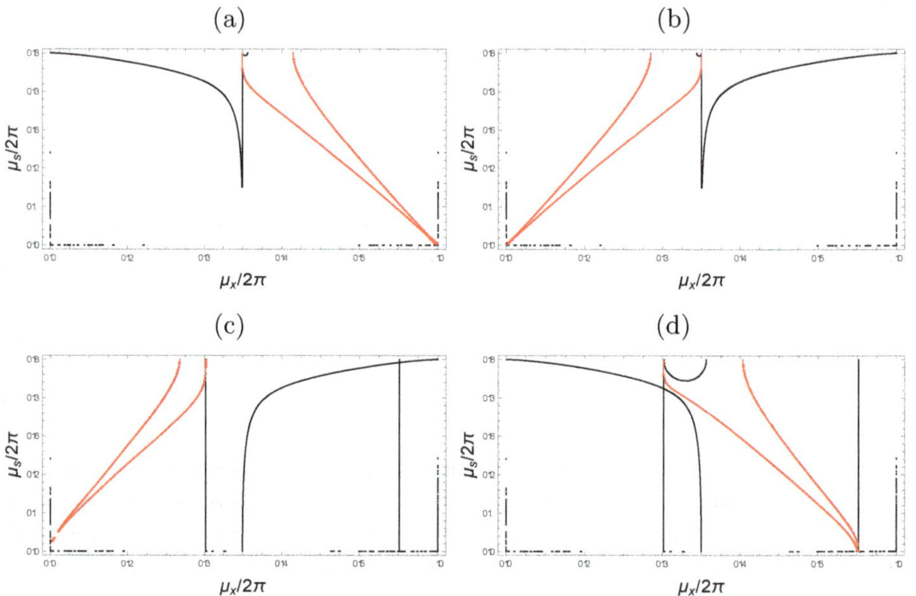

8.4   Weak-strong nonlinear beam-beam effect

8.4.1   Weak-strong beam-beam resonance

To discuss the weak-strong beam-beam effects on particle dynamics, we need to bring focus on beam-beam induced resonances. In the linear thin-lens model of the previous section, we already encountered resonances of the types $2\nu_x = m$, $2\nu_y = m$, and $\nu_y \pm \nu_s = m$. When the full beam-beam potential is taken into account, many more resonances can be excited.

In the weak-strong picture, the strong beam is unperturbed by the weak beam. It is regarded as a rigid nonlinear lens at the beam-beam interaction point (IP). Beam dynamics in this picture is that of an incoherent single-particle motion in the presence of a static nonlinear lens — the lens is still considered thin. The beam-beam problem thus reduces to a nonlinear mapping problem, Eq. (8.12). A beam-beam limit sets in when the nonlinearity strength increases and particle motion becomes unstable at some point.

In the weak-strong picture, the beam-beam limit comes from excitation of nonlinear resonances,

$$n_x \nu_x + n_y \nu_y + n_s \nu_s = m, \tag{8.46}$$

for all integers n_x, n_y, n_s, m. As the resonance indices $|n_{x,y,s}|$ increase, higher order resonances come into action.

In a storage ring, there are many possible sources of nonlinearities. The beam-beam interaction is only one of them, but its notoriety in a storage ring collider has a few particular reasons.

1. The effort to increase the luminosity necessarily means pushing for the highest possible colliding beam currents. The beam-beam limit is always to be reached by definition. We must understand the beam-beam limit in order to push up the luminosity.

2. Unlike the nonlinearities of a magnet, the beam-beam potential has a natural profile conforming to the beam size, rather than the size of the vacuum chamber pipe. That means the beam-beam force is extremely nonlinear and the higher order terms become significant as particle amplitudes become comparable or larger than the rms beam size. This means resonances of very high orders are potentially excited. Consider a resonance $\nu = \frac{m}{n}$, this means excitation up to very high values of n.

3. The beam collisions take place over a short distance of the colliding bunch length. This δ-function like nonlinear kicks have the potential of exciting very high harmonics of each of the resonances. Consider a resonance $\nu = \frac{m}{n}$, this means excitation up to very high values of m.

The first reason is pretty obvious. Let us address the other two reasons a bit more below.

To address the second reason, let us consider the betatron resonances $\nu_y = \frac{m}{n_y}$ for example and ask the question how high the order of resonances can go, i.e. how high can the index n_y go? Since the beam-beam force starts to become nonlinear when the particle's betatron amplitude exceeds σ_y and we demand stability of particle motion up to $k\sigma_y$, the maximum order of resonance $\hat{n}$ should at least roughly be such that the corresponding Taylor expansion term in Eq. (8.13) becomes negligible,

$$\frac{1}{\hat{n}!} \left(\frac{k^2}{2}\right)^{\hat{n}} \lesssim 1.$$

For $\hat{n} \gg 1$, we use the Stirling's formula,[14]

$$\hat{n}! \approx \sqrt{2\pi\hat{n}} \left(\frac{\hat{n}}{e}\right)^{\hat{n}}, \tag{8.47}$$

---

[14]The Stirling's formula is rather amazing. It says that, other than the comparatively small correction factor of $\sqrt{2\pi\hat{n}}$, we have

$$1 \cdot 2 \cdot 3 \cdot 4 \cdots \hat{n} \approx \frac{\hat{n}}{e} \cdot \frac{\hat{n}}{e} \cdot \frac{\hat{n}}{e} \cdot \frac{\hat{n}}{e} \cdots \frac{\hat{n}}{e}.$$

where $e = 2.72$, to obtain

$$\hat{n} \approx e\frac{k^2}{2}. \tag{8.48}$$

The expression (8.48) gives approximately the highest resonance orders $n_{x,y}$ that can be excited by the beam-beam perturbation. For example, if $k = 6$, i.e. if we care about particles up to 6σ amplitudes, then $\hat{n} \approx 49$. All resonances up to the 49th order will potentially need to be avoided when choosing the working point in the (ν_x, ν_y) tune space.

The third cause on the list is addressed by a highest value of m of resonances that can be excited by the beam-beam interaction. The resonance (8.46) is driven by the m-th Fourier component of the beam-beam perturbation. Let the collider circumference be $2\pi R$. Let us assume both of the colliding beam bunches have a length L_z. The length of the collision region is L_z. The beam-beam perturbation therefore has basically a white Fourier spectrum all the way up to

$$\hat{m} \approx \frac{2\pi R}{L_z}, \tag{8.49}$$

Since $L_z \ll 2\pi R$, very high order resonances can be excited.

The total number of betatron resonances potentially excited by the beam-beam perturbation is therefore approximately

$$\hat{n}^2 \hat{m} = \left(e\frac{k^2}{2}\right)^2 \frac{2\pi R}{L_z}.$$

Each of these resonances can also have synchrotron Bessel function sidebands. See Homework 8.23(b).

Beam-beam limit   The criterion, Eqs. (8.48) and (8.49), serves only as a rough indicator of the resonance orders to be concerned with. It for example does not require the knowledge of the beam-beam strength ξ. It also does not depend on the radiation damping for the case of electron-positron colliders. So we have not yet touched upon the beam-beam limit to be specified by a value for ξ_{limit}. In the rest of this chapter, we will mention a few estimates of ξ_{limit}, each estimate given when a specific beam-beam mechanism is considered. In particular, we will mention the beam-beam limits set

- by fitting the weak-strong beam-beam tune footprint into a resonance-free region of tune space on page 594;
- by using the weak-strong Chirikov criterion in Sec. 8.4.5;
- by drawing an analogy to incompressible viscous fluid and Navier-Stoke's equation in Sec. 8.4.6;
- as a result of a diffusion model in Sec. 8.4.8;
- with a potential role of synchrotron radiation damping;
- as a result from strong-strong coherent beam-beam blow-up model in Sec. 8.8.3.

A weak-strong simulation The linear thin-lens model focuses on particles with small betatron amplitudes in the weak beam. For particles with larger amplitudes, the linearization breaks down and the beam-beam force needs to be taken in its full glory. Such a nonlinear mapping problem is difficult to handle analytically. Before we venture into the following sections, and to have an initial look at the beam-beam resonances, we introduce some results of a numerical computer simulation here.

To simplify our consideration, let us assume round beams. The beam-beam potential and the beam-beam kick are given by Eq. (8.17). Let us consider the case $x = 0, x' = 0$, the particle then moves strictly in the y-plane.

Consider a weak-beam particle launched with an initial condition (y_0, y_0') at the position immediately before the beam-beam kick. Going through the beam-beam nonlinear lens, it receives a kick

$$\Delta y = 0, \qquad \Delta y' = -\frac{2Nr_0}{\gamma y}\left(1 - e^{-y^2/2\sigma^2}\right).$$

After the beam-beam kick, the particle circulates around the ring with a map

$$\begin{bmatrix} \cos 2\pi\nu & \beta^* \sin 2\pi\nu \\ -\frac{1}{\beta^*} \sin 2\pi\nu & \cos 2\pi\nu \end{bmatrix}.$$

In the simulation, the beam-beam kick and the ring map are repetitively applied to the particle motion. The turn-by-turn particle coordinates are then recorded and plotted in the normalized phase space ($u = \frac{y}{\sigma}, v = \frac{\beta^* y'}{\sigma}$).

Figure 8.16 shows our result. In Fig. 8.16, four cases are presented. We assume $4\pi\xi = 0.3$ and a tune close to a 4th order resonance with $\nu = 0.239$. In each case, five particles are tracked $(u_0, v_0) = (0.5, 0), (1, 0), (2, 0), (1.5, 1), (3, 0)$. Case (a) is when the beam-beam force is ignored. The particles trace out perfect circles. Case (b) is the linear thin-lens approximation. Since we stay in the region where linear thin-lens model is stable, the particles trace out ellipses. The ellipses are tilted in the phase space because we observe at the entrance to, instead of the middle of, the beam-beam kick.

We then included nonlinear effects. Case (c) is when the beam-beam kick takes into account of the next order octupole term from the Taylor expansion of the beam-beam kick. The phase space is seen to be rather distorted. There are 4 islands formed at some amplitude; they reflect the fact that we are close to a 4th order resonance. In addition, there is a notable chaos in the trajectory of the outer most particle. In fact, although all other particles have been tracked for 1000 turns, this last 3σ-particle is lost in 350 turns. By including the octupole nonlinearity of the beam-beam kick, the beam has thus become unstable.

Had we stopped here, we would say that we have found the explanation for the beam-beam instability. But Fig. 8.16(d) shows the case when the complete beam-beam kick is included. What is striking is that the stable islands we saw in case (c) are still there, but the stochastic behavior of the chaotic motion seems to have disappeared and all particles survive for 1000 turns again.

This means the nonlinear beam-beam force alone does not destabilize the beam, at least for the parameters being considered. This is in sharp contrast to

Figure 8.16: Simulation of the motion of a weak-beam particle under the beam-beam influence. The trajectory of a particle is plotted in the normalized phase space (u, v) when $4\pi\xi = 0.3$, $\nu = 0.239$. Case (a) ignores the beam-beam force; (b) linearized; (c) includes the octupole term; (d) complete beam-beam force. The same five initial conditions are used in all cases.

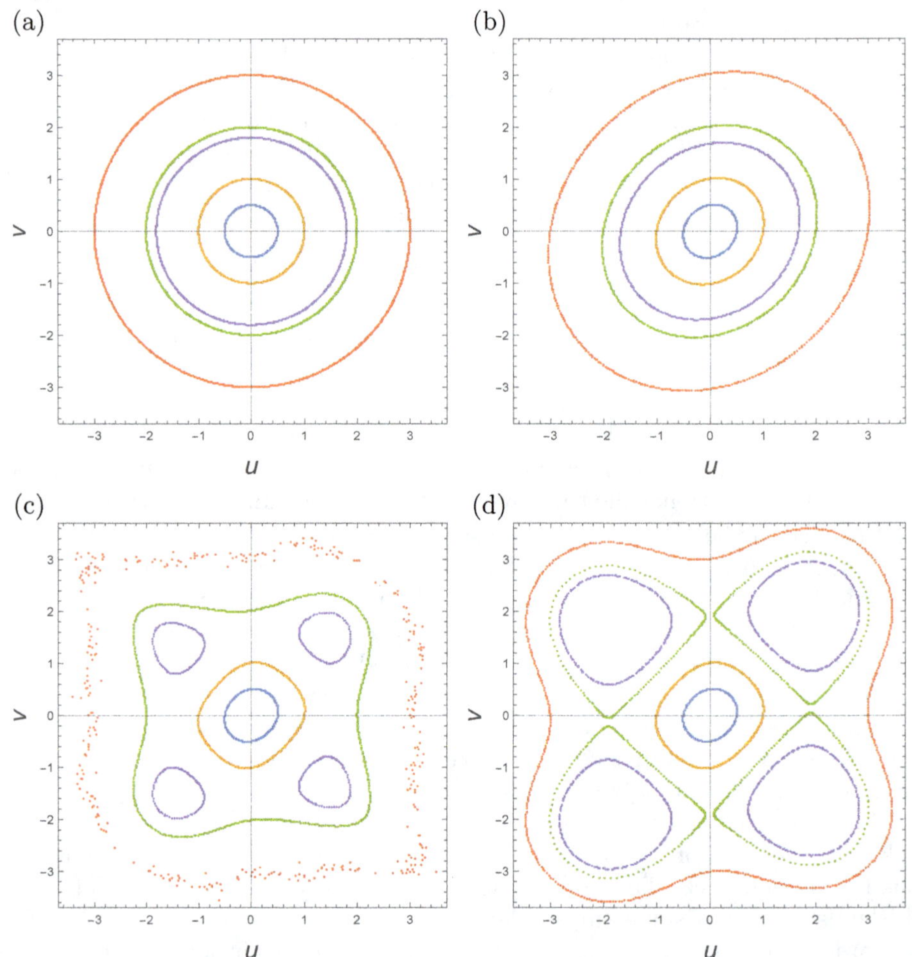

the resonance instabilities driven by magnetic field imperfections, as Fig. 8.16(c) could be an example of. The reason for such a behavior is because the beam-beam force, although very nonlinear in the beam core, diminishes quickly once the weak-beam particle acquires an amplitude larger than the core size of the strong beam. In the limit of very large amplitudes, the weak-beam particle acts as if unperturbed. The beam-beam force therefore produces islands in the phase space but not an instability. We briefly conclude that the beam-beam instability will necessarily invoke mechanisms beyond the one that is just suggested.

Homework 8.23

(a) Verify the somewhat qualitative result of Eqs. (8.48) and (8.49).

(b) Extend the rough analysis leading to Eqs. (8.48) and (8.49) to include the synchrotron sidebands by considering a Bessel function frequency-modulation mechanism.

Solution (a) Use the Stirling formula (8.47), and that

$$\lim_{M\to\infty} M^{1/M} = 1\,.$$

8.4.2 Detuning and tune spread

In an earlier section, we simplified the beam-beam kick Eq. (8.11) by linearizing it for particles in the beam core with small amplitudes, $|x| \ll \sigma_x, |y| \ll \sigma_y$. When nonlinear terms are included, as mentioned in the previous section, one expects significant changes, such as island formation in the phase space, resonances, chaos and instabilities.

In this section, we will examine an effect that the betatron tunes start to shift as the particle coordinates $|x|$ and $|y|$ increase from the beam center. To do so, we consider an idealistic case when the tunes are away from any significant resonances. We will incorporate the resonance effects in the following sections by adding terms on top of what we obtain in this section.

Hamiltonian   The kick equation (8.11) can be represented by a Hamiltonian

$$H(x, p_x, y, p_y, s) = \frac{1}{2}\left[p_x^2 + K_x(s)x^2\right] + \frac{1}{2}\left[p_y^2 + K_y(s)y^2\right] + U(x, y)\delta(s)\,, \qquad (8.50)$$

where $K_{x,y}(s)$ are the focusing functions of the storage ring, s is the distance along the design trajectory serving as the time variable of the dynamical system, $\delta(s)$ is the periodic δ-function, and $U(x, y)$ was given in Eq. (8.11). The Hamilton equations of motion give

$$\frac{d^2 x}{ds^2} + K_x(s)x = -\frac{\partial U(x, y)}{\partial x}\delta(s)\,,$$

$$\frac{d^2 y}{ds^2} + K_y(s)y = -\frac{\partial U(x, y)}{\partial y}\delta(s)\,.$$

A Courant–Snyder canonical transformation changes the canonical coordinates from (x, p_x, y, p_y) to $(\phi_x, J_x, \phi_y, J_y)$ and the time variable from s to $\theta = \frac{s}{R}$ where $2\pi R$ is the ring circumference, and

$$x = \sqrt{2\beta_x^* J_x}\cos\phi_x\,, \qquad y = \sqrt{2\beta_y^* J_y}\cos\phi_y\,,$$

with $\beta^*_{x,y}$ the β-functions at the collision point $s = 0$. The new Hamiltonian is

$$K(\phi_x, J_x, \phi_y, J_y, \theta) = \nu_x J_x + \nu_y J_y + U(\sqrt{2\beta^*_x J_x}\,\cos\phi_x, \sqrt{2\beta^*_y J_y}\,\cos\phi_y)\delta(\theta),$$

(8.51)

where $\nu_{x,y}$ are the unperturbed betatron tunes.

Smooth approximation So far the transformation is exact. The next step is a critical one called the smooth approximation. What it aims to do is to remove all fast oscillating terms and keep only the slowly varying terms in the Hamiltonian.[15] Away from all resonances, this is equivalent to replacing the term $U(\sqrt{2\beta^*_x J_x}\,\cos\phi_x, \sqrt{2\beta^*_y J_y}\,\cos\phi_y)\delta(\theta)$ by its averaging over ϕ_x, ϕ_y and θ,

$$\int_0^{2\pi} \frac{d\phi_x}{2\pi} \int_0^{2\pi} \frac{d\phi_y}{2\pi} \int_0^{2\pi} \frac{d\theta}{2\pi}\, U(\sqrt{2\beta^*_x J_x}\,\cos\phi_x, \sqrt{2\beta^*_y J_y}\,\cos\phi_y)\delta(\theta)\,. \quad (8.52)$$

In the smooth approximation away from resonances, only this average term contributes significantly to the dynamics. When there is a resonance nearby, there will be additional slowly varying terms to be included in the smoothed Hamiltonian, as will be considered in the next section.

Substituting $U(x, y)$ from Eq. (8.11) gives the smoothed new Hamiltonian

$$K(\phi_x, J_x, \phi_y, J_y, \theta) = \nu_x J_x + \nu_y J_y + K_1(\phi_x, J_x, \phi_y, J_y, \theta)\,,$$

where

$$K_1(\phi_x, J_x, \phi_y, J_y, \theta) = -\frac{N r_0}{2\pi\gamma} \int_0^\infty dt \int_0^{2\pi} \frac{d\phi_x}{2\pi} \int_0^{2\pi} \frac{d\phi_y}{2\pi}$$
$$\times \left[\frac{\exp\left(-\frac{J_x \beta^*_x \cos^2\phi_x}{\sigma_x^2 + t} - \frac{J_y \beta^*_y \cos^2\phi_y}{\sigma_y^2 + t}\right) - 1}{\sqrt{(\sigma_x^2 + t)(\sigma_y^2 + t)}} \right].$$

The integrations over ϕ_x and ϕ_y can be performed using the identity

$$\int_0^{2\pi} \frac{d\phi}{2\pi}\, e^{-2\alpha\cos^2\phi} = e^{-\alpha} I_0(\alpha)\,,$$

yielding the result

$$K_1(\phi_x, J_x, \phi_y, J_y, \theta) = -\frac{N r_0}{2\pi\gamma} \int_0^\infty dt \left[\frac{e^{-\frac{\alpha_x \sigma_x^2}{\sigma_x^2 + t} - \frac{\alpha_y \sigma_y^2}{\sigma_y^2 + t}} I_0\left(\frac{\alpha_x \sigma_x^2}{\sigma_x^2 + t}\right) I_0\left(\frac{\alpha_y \sigma_y^2}{\sigma_y^2 + t}\right) - 1}{\sqrt{(\sigma_x^2 + t)(\sigma_y^2 + t)}} \right],$$

---

[15]A. Schoch, CERN report 57-23 (1958); P.A. Sturrock, Ann. Phys. 3, 113 (1958); G. Guignard, CERN report 76-06 (1976); A.W. Chao, Lectures on Accelerator Physics, World Scientific (2020).

where

$$\alpha_x \;=\; \frac{J_x \beta_x^*}{2\sigma_x^2}, \qquad \alpha_y \;=\; \frac{J_y \beta_y^*}{2\sigma_y^2}\,.$$

This smoothed Hamiltonian is independent of θ, meaning it is a constant of the motion. In addition, it is independent of ϕ_x and ϕ_y, meaning the Courant–Snyder actions J_x and J_y are also constants of the motion. The remaining two Hamilton equations read

$$\frac{d\phi_x}{d\theta} \;=\; \frac{\partial K}{\partial J_x} \;=\; \nu_x + \frac{\partial K_1}{\partial J_x} \;=\; \nu_x + \Delta\nu_x(\alpha_x,\alpha_y)\,,$$

$$\frac{d\phi_y}{d\theta} \;=\; \frac{\partial K}{\partial J_y} \;=\; \nu_y + \frac{\partial K_1}{\partial J_y} \;=\; \nu_y + \Delta\nu_y(\alpha_x,\alpha_y)\,,$$

where[16]

$$\Delta\nu_x(\alpha_x,\alpha_y) = \frac{N r_0 \beta_x^*}{4\pi\gamma} \int_0^\infty dt \, \frac{e^{-\frac{\alpha_x\sigma_x^2}{\sigma_x^2+t}-\frac{\alpha_y\sigma_y^2}{\sigma_y^2+t}}\left[I_0(\frac{\alpha_x\sigma_x^2}{\sigma_x^2+t})-I_1(\frac{\alpha_x\sigma_x^2}{\sigma_x^2+t})\right]I_0(\frac{\alpha_y\sigma_y^2}{\sigma_y^2+t})}{(\sigma_x^2+t)^{3/2}(\sigma_y^2+t)^{1/2}}\,,$$

$$\Delta\nu_y(\alpha_x,\alpha_y) = \frac{N r_0 \beta_y^*}{4\pi\gamma} \int_0^\infty dt \, \frac{e^{-\frac{\alpha_x\sigma_x^2}{\sigma_x^2+t}-\frac{\alpha_y\sigma_y^2}{\sigma_y^2+t}}I_0(\frac{\alpha_x\sigma_x^2}{\sigma_x^2+t})\left[I_0(\frac{\alpha_y\sigma_y^2}{\sigma_y^2+t})-I_1(\frac{\alpha_y\sigma_y^2}{\sigma_y^2+t})\right]}{(\sigma_x^2+t)^{1/2}(\sigma_y^2+t)^{3/2}}\,. \tag{8.53}$$

We have used the identity

$$\frac{d}{dx}\left[e^{-x}I_0(x)\right] \;=\; -e^{-x}\left[I_0(x)-I_1(x)\right]\,.$$

Equation (8.53) can also be expressed to read

$$\frac{\Delta\nu_x}{\xi_x} = \frac{1+a}{2}\int_0^\infty dt \, \frac{e^{-\frac{\alpha_x}{1+t}-\frac{\alpha_y a^2}{a^2+t}}\left[I_0(\frac{\alpha_x}{1+t})-I_1(\frac{\alpha_x}{1+t})\right]I_0(\frac{\alpha_y a^2}{a^2+t})}{(1+t)^{3/2}(a^2+t)^{1/2}}\,,$$

$$\frac{\Delta\nu_y}{\xi_y} = \frac{1+a}{2}\int_0^\infty dt \, \frac{e^{-\frac{\alpha_x}{1+a^2 t}-\frac{\alpha_y}{1+t}}I_0(\frac{\alpha_x}{1+a^2 t})\left[I_0(\frac{\alpha_y}{1+t})-I_1(\frac{\alpha_y}{1+t})\right]}{(1+a^2 t)^{1/2}(1+t)^{3/2}}\,,$$

where $\xi_{x,y}$ are the beam-beam strength parameters introduced earlier,

$$\xi_x \;=\; \frac{N r_0 \beta_x^*}{2\pi\gamma\sigma_x(\sigma_x+\sigma_y)}, \qquad \xi_y \;=\; \frac{N r_0 \beta_y^*}{2\pi\gamma\sigma_y(\sigma_x+\sigma_y)}\,,$$

and we have defined an aspect ratio of beam sizes,

$$a \;=\; \frac{\sigma_y}{\sigma_x}\,.$$

---

[16]A. Lejcic and L. Le Duff, Proc. VII Conf. High Energy Accel., Stanford, p. 394 (1974); E. Keil, CERN report CERN/ISR-TH/72-7 and CERN/ISR-TH/72-25 (1972); A.W. Chao, AIP Conf. Proc. 57, Brookhaven, p. 42 (1979).

The two expressions in (8.53) represent the beam-beam perturbed betatron tune shifts as functions of α_x and α_y, which we already know are constants of the motion. We have thus designated them as $\Delta\nu_x(\alpha_x, \alpha_y)$ and $\Delta\nu_y(\alpha_x, \alpha_y)$. Once the Courant–Snyder amplitudes J_x and J_y, and therefore α_x and α_y, of the initial conditions of the particle are given, its tune shifts are determined by Eq. (8.53). This effect of tune shifts with amplitudes is referred to as the detuning effect.

It should be kept in mind that we have made a smooth approximation in the particle dynamics. Validity of this approximation applies only to 1st order in the beam-beam strengths. Our results are valid only to the order $\sim \mathcal{O}(\xi_{x,y})$ and only away from all resonances.

We now work out a few special cases of beam-beam detuning below.

Linear lens When $\alpha_x \ll 1$ and $\alpha_y \ll 1$, keeping only up to first order terms in $\alpha_{x,y}$ (quadrupole terms in x and y), we obtain

$$K_1 \approx \frac{Nr_0}{\pi\gamma(\sigma_x + \sigma_y)}(\sigma_x\alpha_x + \sigma_y\alpha_y),$$

where use has been made of Eq. (8.16). It follows that the small-amplitude particles have the tune shifts,

$$\Delta\nu_x \approx \xi_x, \qquad \Delta\nu_y \approx \xi_y.$$

Beam core with $\alpha_x \lesssim 1, \alpha_y \lesssim 1$ We may extend the linear-lens model to include up to 2nd order in $\alpha_{x,y}$ (octupole terms in x and y). The result is

$$K_1 = \frac{Nr_0}{\pi\gamma}\left[\frac{\sigma_x\alpha_x + \sigma_y\alpha_y}{\sigma_x + \sigma_y} - \frac{\sigma_x\sigma_y}{(\sigma_x+\sigma_y)^2}\alpha_x\alpha_y - \frac{\sigma_x(2\sigma_x+\sigma_y)}{4(\sigma_x+\sigma_y)^2}\alpha_x^2 - \frac{\sigma_y(\sigma_x+2\sigma_y)}{4(\sigma_x+\sigma_y)^2}\alpha_y^2\right],$$

and

$$\Delta\nu_x = \xi_x - \frac{\xi_x}{2(\sigma_x + \sigma_y)}[(2\sigma_x + \sigma_y)\alpha_x + 2\sigma_y\alpha_y],$$

$$\Delta\nu_y = \xi_y - \frac{\xi_y}{2(\sigma_x + \sigma_y)}[2\sigma_x\alpha_x + (\sigma_x + 2\sigma_y)\alpha_y]. \tag{8.54}$$

The octupole terms lower the tune shifts from the small-amplitude values $\xi_{x,y}$ as the amplitudes increases.

A particle with $\alpha_y \ll 1$, arbitrary α_x Consider a particle with small vertical amplitude $|y| \ll \sigma_y$ and arbitrary horizontal amplitude, and consider its vertical tune shift as a function of α_x. The result is

$$\Delta\nu_y = \xi_y e^{-\alpha_x} I_0(\alpha_x). \tag{8.55}$$

The tune shift is ξ_y when α_x is small because the particle stays close to the beam center where it sees the full beam density of the strong beam. When α_x increases, the tune shift decreases as it begins to see the weaker beam tails.

Round beam Here we call this a round beam case when $\sigma_x = \sigma_y$ at the collision point. There is no requirement that $\beta_x^* = \beta_y^*$ or $\nu_x = \nu_y$, so it is not a case that degenerates into 1-D cylindrically symmetry. It is still a full blown 2-D case particularly since we must stay away from the $\nu_x = \nu_y$ resonance. The result is given by

$$
\begin{aligned}
K_1 &= \frac{Nr_0}{2\pi\gamma} \int_0^1 \frac{dv}{v} \left[1 - e^{-(\alpha_x+\alpha_y)v} I_0(\alpha_x v) I_0(\alpha_y v)\right], \\
\Delta\nu_x &= \xi_x \int_0^1 dv\, e^{-(\alpha_x+\alpha_y)v}[I_0(\alpha_x v) - I_1(\alpha_x v)]I_0(\alpha_y v), \\
\Delta\nu_y &= \xi_y \int_0^1 dv\, e^{-(\alpha_x+\alpha_y)v} I_0(\alpha_x v)[I_0(\alpha_y v) - I_1(\alpha_y v)].
\end{aligned}
\tag{8.56}
$$

If $\alpha_x = 0$, the vertical tune shift has a simpler form,

$$
\Delta\nu_y = \frac{\xi_y}{\alpha_y}\left[1 - e^{-\alpha_y} I_0(\alpha_y)\right].
\tag{8.57}
$$

Again the tune shift becomes smaller than ξ_y when $\alpha_x \neq 0$ because the particle then samples also the weaker beam-beam kicks in the horizontal tails.

Flat beam Consider a flat beam, meaning $\sigma_x \gg \sigma_y$ and we consider a particle with $|x| \ll \sigma_x$, i.e. $\alpha_x \to 0$. We consider only the vertical tune shift, which can be written as

$$
\Delta\nu_y(\alpha_y) = \frac{Nr_0\beta_y^*}{4\pi\gamma\sigma_x} \int_0^\infty dt\, \frac{e^{-\frac{\alpha_y\sigma_y^2}{\sigma_y^2+t}}\left[I_0(\frac{\alpha_y\sigma_y^2}{\sigma_y^2+t}) - I_1(\frac{\alpha_y\sigma_y^2}{\sigma_y^2+t})\right]}{(\sigma_y^2+t)^{3/2}}.
$$

Change variable from t to $u = \frac{\alpha_y\sigma_y^2}{\sigma_y^2+t}$, the integration can be performed,

$$
\Delta\nu_y(\alpha_y) = \xi_y e^{-\alpha_y}\left[I_0(\alpha_y) + I_1(\alpha_y)\right],
\tag{8.58}
$$

where use has been made of the identity

$$
\int \frac{du}{\sqrt{u}}\, e^{-u}\left[I_0(u) - I_1(u)\right] = 2\sqrt{u}\, e^{-u}[I_0(u) + I_1(u)].
$$

As expected, $\Delta\nu_y = \xi_y$ when $\alpha_y \to 0$.

For completeness, the perturbation Hamiltonian also has a closed form for a flat beam, by integrating $\Delta\nu_y$ over J_y,

$$
\begin{aligned}
K_1 &= \frac{2\sigma_y^2}{\beta_y^*} \int_0^{\alpha_y} d\alpha_y\, \Delta\nu_y(\alpha_y) \\
&= \frac{Nr_0\sigma_y}{\pi\gamma\sigma_x} \left\{e^{-\alpha_y}\left[(1 + 2\alpha_y)I_0(\alpha_y) + 2\alpha_y I_1(\alpha_y)\right] - 1\right\}.
\end{aligned}
\tag{8.59}
$$

Tune footprint   We now learned that under the weak-strong beam-beam perturbation, the weak-beam particle's betatron tunes are shifted. Particles with small betatron amplitudes suffer the maximum tune shifts $\Delta\nu_{x,y} = \xi_{x,y}$. As the amplitudes increase, its tune shifts decrease as it meets the strong beam more in its weaker tail. When $\alpha_x, \alpha_y \to \infty$, the tune shifts vanish. As a result, when particles of all amplitudes are considered, the beam acquires tune spreads, and

$$\text{tune spreads} \;=\; \text{small-amplitude tune shifts} \;=\; \xi_{x,y}\,. \tag{8.60}$$

So we expect the strong beam will keep its unperturbed working point (ν_x, ν_y) in the tune space, but the weak beam will spread out its tunes into a distribution which we refer to as the tune footprint. The footprint is expected to stay inside a rectangle from the unperturbed (ν_x, ν_y) to $(\nu_x + \xi_x, \nu_y + \xi_y)$. The point $(\nu_x + \xi_x, \nu_y + \xi_y)$ is occupied by the particle at the beam center, the point (ν_x, ν_y) by particles far in the beam tail.

But not the entire rectangle is occupied. The footprint has a particular shape depending on the beam aspect ratio a. Figure 8.17 gives the shape of the footprint plotted in normalized coordinates $\frac{\Delta\nu_x}{\xi_x}$ and $\frac{\Delta\nu_y}{\xi_y}$ for three cases $a = 1, 0.1, 0.01$, respectively.

In each of Figs. 8.17(a), (b) and (c), two black solid curves indicate the boundaries of the working point footprint. One of them is determined by the condition $\alpha_x = 0$ while scanning α_y from 0 to ∞; the other is scanning with $\alpha_y = 0$. Interior of the footprint is determined by various combinations of α_x and α_y values. Black dots indicate the unperturbed working point where the strong beam firmly sits. As mentioned, for a bi-Gaussian strong-beam distribution as we have assumed, the weak-beam footprint stays inside the rectangle from (ν_x, ν_y) to $(\nu_x + \xi_x, \nu_y + \xi_y)$. We have assumed $\xi_{x,y} > 0$, meaning the two beams have opposite signs of charges.

The footprint is of course not evenly populated. The dashed curves in Fig. 8.17 indicate the contours on the footprint when $\alpha_x + \alpha_y = n$ with $n = 1, 2, 3, 4$. They are marked to give a feel of the beam population on the footprint. In particular, the shifted point $(\nu_x + \xi_x, \nu_y + \xi_y)$ is much more populated than the original working point (ν_x, ν_y).

Beam-beam limit   To maintain a good stable beam, it is necessary to fit the tune footprint in the tune space without hitting dangerous low-order resonances, as illustrated in Fig. 8.17(d). To do so, we first want to identify a good resonance free region in the tune space. We obviously want to locate the unperturbed working point at a southwest corner of this space (assuming $\xi_{x,y} > 0$) as is done in Fig. 8.17(d). According to this beam-beam model, the beam-beam limit is reached when the footprint can no longer be fitted into a resonance-free tune space.

In electron-positron colliders, it is most likely that $\sigma_x > \sigma_y$, i.e. $a < 1$. In this case, the footprint acquires a shape that bulges in the southeast direction as seen in Figs. 8.17(b) and (c). If there is a difference coupling resonance of importance to avoid — they often need to be avoided for luminosity reasons even

Figure 8.17: Tune footprint. Panels (a), (b), (c) are for the cases $a = \frac{\sigma_y}{\sigma_x} = 1, 0.1, 0.01$, respectively. Black dots indicate the unperturbed working point. Panel (d) shows an attempt to fit the footprint into a resonance free region in the tune space, and it shows a case when the beam-beam limit is about to be reached.

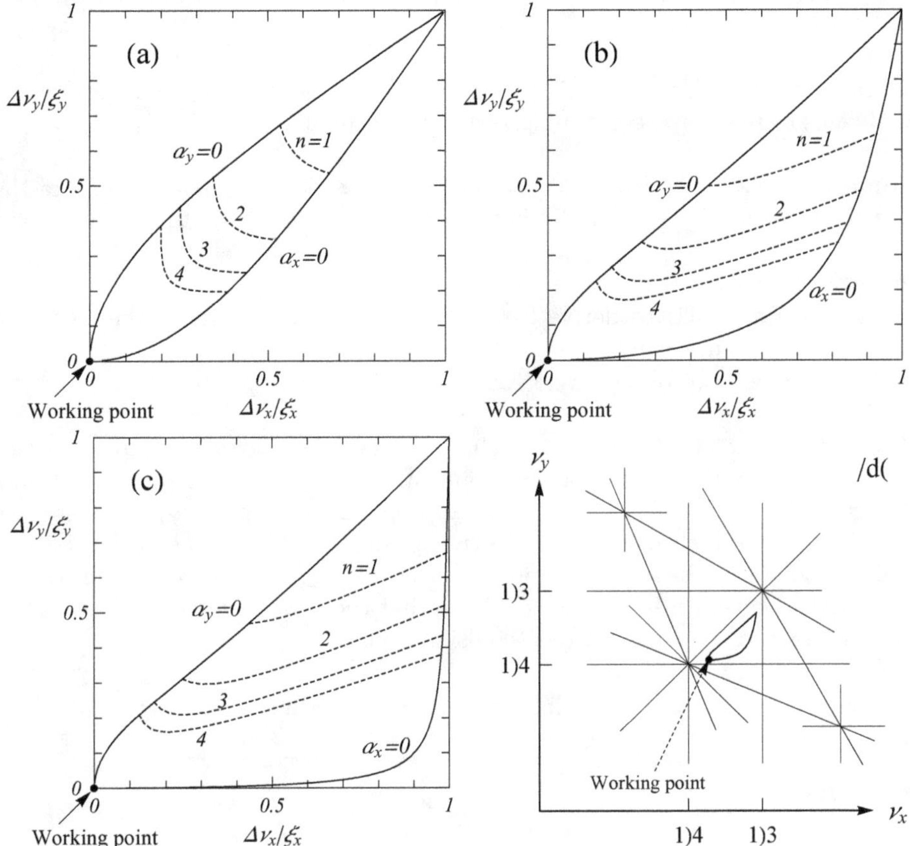

though, being difference resonances, they directly do not cause instability — it is advisable to locate the footprint on the southeast side to the resonance line, i.e. the working point should be chosen below any important coupling resonance lines. This situation of an optimal choice of working point is illustrated in Figs. 8.17(d). If the $\nu_x = \nu_y$ resonance is a concern, for example, this requires the fractional part of ν_y be slightly less than the fractional part of ν_x.

The footprint calculated above is when the strong beam has an upright bi-Gaussian distribution. It is conceivable that if the strong beam has a uniform disk distribution, the weak beam's tune footprint can be reduced to a point

(Homework 8.27). If both beams are made to have uniform distributions by some beam shaping technique, therefore, according to this conceived picture, one can minimize the nonlinear beam-beam resonance effects and thus mitigate the beam-beam limit threshold. This viewpoint, however, must also take into consideration that it is conceived only from the single-particle dynamics of a weak-strong picture. It does not apply when coherent beam-beam effects (Sec. 8.5) are included.

Homework 8.24 Use Eq. (8.53) to derive Eq. (8.54) for the tune shifts when the beam-beam octupole detuning terms are included.

Homework 8.25 Use Eq. (8.53) to derive Eq. (8.55) for the tune shifts of the case when the vertical amplitude is small but horizontal amplitude is arbitrary.

Homework 8.26

(a) Use Eq. (8.53) to derive Eq. (8.56) for the tune shifts of the case of a round strong beam with $\sigma_x = \sigma_y$.

(b) Then use Eq. (8.56) to derive Eq. (8.57).

Homework 8.27 Perhaps we put what we learned so far to a simple practice by this homework. Which working point, among those shown by the red dots in Fig. 8.18, would you consider most appropriate for a proton-proton storage ring collider? Which one for an electron-positron collider? Let there be two collision points in the collider, and $\nu_{x,\mathrm{ring}}$ and $\nu_{y,\mathrm{ring}}$ refer to the tunes of the whole ring. Consult also Eq. (8.27). Note that the ν we have been using in this section refers to the tune between collision points.

Figure 8.18: Proposed possible working point for a storage ring collider.

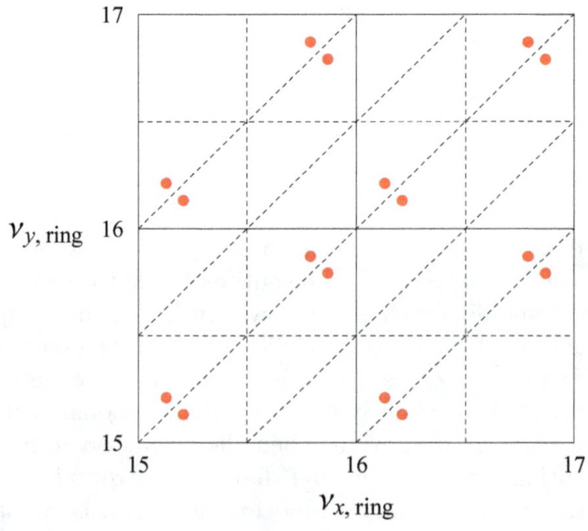

Homework 8.28 This homework deserves some consideration. It seems to point a way to mitigate the beam-beam problem. However, as emphasized in the text, this suggestion applies only if one considers weak-strong effects.

(a) The text calculated in detail for the case of an upright bi-Gaussian strong beam. What happens to the weak beam's tune footprint if the strong beam has a uniform round disk distribution? You might try to guess the answer.

(b) Try to repeat the question for a uniform elliptical disk distribution.

8.4.3 Single resonance model

In a linear thin-lens model, the highly nonlinear beam-beam force is linearized. The particle motion can be analyzed exactly using a matrix technique. We found that the x- and the y-dimensions are decoupled and, for each dimension, the single particle motion is completely described by two parameters, the betatron tune per superperiod ν and the beam-beam strength parameter ξ. We also found that when ν is sufficiently close to $\frac{m}{2}$, the particle motion becomes unstable with a certain stopband width.

When nonlinearity of the beam-beam force is taken into account, particle motion is significantly affected, particularly near nonlinear resonances. Take 1-D case for example, it occurs when the tune is close to a resonance $\nu = \frac{m}{n}$. The linear thin-lens model has addressed only the case when $n = 2$. Unfortunately, it is not possible to perform an exact analysis for $n \neq 2$ when nonlinear terms are included. To proceed, we make a drastic assumption.

$$\text{There is one and only one resonance } \nu = \text{rational number } \frac{m}{n} \text{ that}$$
$$\text{dominates the single-particle motion.} \tag{8.61}$$

This assumption is drastic because, first of all, there must be infinitely more possible rational numbers that are closer to ν than $\frac{m}{n}$ as long as ν is not exactly equal to $\frac{m}{n}$. One then argues that perhaps only smaller values of n, say $n < 10$, are worth considering because resonances of higher orders than ~ 10 do not affect particle motion too much. But this argument is subject to be challenged. First of all, there are reasons to expect a role played by resonances up to very high orders — see Eqs. (8.48) and (8.49). Secondly, when n increases, even though the resonance strength weakens, the number of resonances increases factorially as well. So there will be an infinite number of infinitely weak resonances around any choice of ν.

There is another reason why (8.61) is a drastic assumption. The analysis that allows our treatment of the problem under the assumption (8.61), the smooth approximation, will exclude some important physical phenomena from being studied, of particular concern is the existence of chaos in the phase space as we already saw in the simulation work of Fig. 8.16(c).

In spite of these reservations, in the following, we shall insist on the assumption (8.61). The method of analysis is the smooth approximation, whose validity applies to first order of the nonlinear perturbation. In our case, the results will be valid to 1st order in the beam-beam parameter ξ.

In a sense, we have been self-consistent. The effects due to the interplay of multiple resonances are expected to be generated by the product of the strengths of the various resonances involved. Since each resonance strength is proportional to ξ, these effects, including chaos, are necessarily higher order in ξ. As we declared, these higher order effects are excluded from our analysis here, including the smooth approximation to be discussed next.

On the other hand, as we will soon see, according to this model, particle motion is always stable against the beam-beam perturbation even though it can be strongly perturbed in the core region of the phase space.

To explain the beam-beam instability, the assumption (8.61) needs to be relieved or supplemented by considering multiple resonances or by numerical simulation studies of the chaotic layer as we did in Fig. 8.16(c). Another possibility is to modify the present static model so that ν or ξ become time-dependent. We shall briefly mention this latter possibility in Sec. 8.4.7.

Canonical transformation   In the previous section, the smooth approximation was accomplished by Eqs. (8.51) and (8.52) and results were obtained for the beam-beam detuning $\Delta\nu_x(\alpha_x,\alpha_y), \Delta\nu_y(\alpha_x,\alpha_y)$ away from all resonances. In this section, the smooth approximation needs to be performed when there is in addition the presence of a resonance nearby (see footnote 15).

Before proceeding, we need to define the resonance. Let the unperturbed tunes $\nu_{x,y}$ be such that

$$\nu_x \approx \nu_{x0}, \quad \nu_y \approx \nu_{y0}, \qquad \text{where} \quad n_x\nu_{x0} + n_y\nu_{y0} = m.$$

The reference tunes $\nu_{x0,y0}$ define the resonance. Clearly their exact definitions are not unique, but as we will see, this ambiguity does not matter. All that is important is that they define the resonance by $n_x\nu_{x0} + n_y\nu_{y0} = m$ and that the actual unperturbed tunes are close to these reference tunes. Physically, we are defining a rotating reference frame and as a result will be observing the dynamics in a stroboscopic system designed to extract the slow parts of the dynamics.

Basically we start with the Hamiltonian (8.50) and make three successive canonical transformations on it, a Courant–Snyder transformation, an action-angle transformation, and a slow-variable transformation. These three transformations can be combined into a single transformation with a generating function

$$G(x,y,\psi_x,\psi_y) = -\frac{x^2}{2\beta_x(s)}\left[\tan\phi_x - \frac{\beta'_x(s)}{2}\right] - \frac{y^2}{2\beta_y(s)}\left[\tan\phi_y - \frac{\beta'_y(s)}{2}\right],$$

with

$$\phi_z = \psi_z + \nu_{z0}\frac{s}{R} + \int_0^s ds'\left(\frac{1}{\beta_z(s')} - \frac{\nu_z}{R}\right), \qquad z = x, y, \qquad (8.62)$$

where $\beta_{x,y}(s)$ are the β-functions, $2\pi R$ is the storage ring circumference, $\phi_{x,y}$ are the betatron phases. The old canonical variables are (x, p_x, y, p_y), the new ones are $(\psi_x, J_x, \psi_y, J_y)$. These two sets of coordinates are related by

$$
\begin{aligned}
z &= \sqrt{2J_z\beta_z}\cos\phi_z\,, \\
p_z &= -\sqrt{\frac{2J_z}{\beta_z}}\left(\sin\phi_z - \frac{\beta_z'}{2}\cos\phi_z\right), \qquad z = x,y\,.
\end{aligned}
$$

What motivated these complicated transformations? In the presence of beam-beam perturbation, and to 1st order in the perturbation strength, we expect the actions $J_{x,y}$ to change slowly in time. The betatron phases $\phi_{x,y}$, however, vary quickly. To filter out the slow components, we need to observe the dynamics first by extracting the contributions of $\nu_{x0,y0}$ from the phases $\phi_{x,y}$ like what is done in Eq. (8.62), so that the remaining residual phases $\psi_{x,y}$ become slowly varying. The aim is to define the new canonical variables $(\psi_x, J_x, \psi_y, J_y)$ so that they all are slowly varying.

The reader will note the "extra" term $\int_0^s ds'\left(\frac{1}{\beta_z(s')} - \frac{\nu_z}{R}\right)$ in Eq. (8.62). It comes from a need when carrying out the detailed derivation so that we can use s, a uniformly progressing parameter, as our time variable. It is needed because the betatron phase does not advance uniformly like s does. The reader should note that this "extra" term is not going to be too critical because it resets to zero every turn of revolution and does not accumulate from turn to turn. For the beam-beam problem in particular, it does not contribute at all because the perturbation occurs only at $s=0$ and this term vanishes at all revolutions.

After this canonical transformation, the new Hamiltonian reads

$$
K(\psi_x, J_x, \psi_y, J_y, \theta) = (\nu_x - \nu_{x0})J_x + (\nu_y - \nu_{y0})J_y + K_1(\psi_x, J_x, \psi_y, J_y, \theta),
\tag{8.63}
$$

where

$$
K_1(\psi_x, J_x, \psi_y, J_y, \theta) = \delta(\theta)\,U\left[\sqrt{2J_x\beta_x^*}\cos(\psi_x + \nu_{x0}\theta), \sqrt{2J_y\beta_y^*}\cos(\psi_y + \nu_{y0}\theta)\right].
$$

In this new Hamiltonian, we have also changed time variable from s to $\theta = \frac{s}{R}$ (which can be considered a more trivial fourth canonical transformation in our derivation). Only the β-function at the collision point enters the expression due to the presence of $\delta(\theta)$.

Smooth approximation So far no approximations have been made. We now make the assumption (8.61) that there is one and only one resonance $n_x\nu_x + n_y\nu_y \approx m$ in the neighborhood. We then decompose the perturbation Hamiltonian K_1 in Fourier components,

$$
K_1 = \sum_{\alpha,\beta=-\infty}^{\infty} f_{\alpha\beta}(J_x, J_y)\, e^{i\alpha(\psi_x + \nu_{x0}\theta) + i\beta(\psi_y + \nu_{y0}\theta)} \cdot \frac{1}{2\pi}\sum_{\gamma=-\infty}^{\infty} e^{i\gamma\theta}\,.
\tag{8.64}
$$

The fact that this Fourier decomposition is possible is because K_1 is periodic in $\psi_x + \nu_{x0}\theta$, $\psi_y + \nu_{y0}\theta$ and θ, all with period 2π.

The Fourier coefficients are given by

$$f_{\alpha\beta} \;=\; \int_0^{2\pi} \frac{d\theta_x}{2\pi} \int_0^{2\pi} \frac{d\theta_y}{2\pi}\, e^{-i\alpha\theta_x - i\beta\theta_y}\, U\!\left[\sqrt{2J_x\beta_x^*}\cos\theta_x,\, \sqrt{2J_y\beta_y^*}\cos\theta_y\right].$$

So far our result is still exact. Our next critical step is to extract only the slowly varying terms. We already expect ψ_x, ψ_y, J_x, J_y to be slowing varying. The issue is therefore to examine the factor

$$e^{i(\alpha\nu_{x0} + \beta\nu_{y0} + \gamma)\theta}$$

in the expression (8.64) to see when it is slowly varying.

Away from resonances, the only way for this factor to be slowly varying is when $\alpha = \beta = \gamma = 0$ because θ varies with time too quickly. In this case, we obtain

$$K_1 \;=\; \frac{f_{00}}{2\pi}.$$

Here we have just recovered Eq. (8.52).

Near a resonance, we have, in addition to the f_{00} term, some other resonance-driving slow terms appearing when

$$\alpha \;=\; kn_x, \qquad \beta \;=\; kn_y, \qquad \gamma \;=\; -km,$$

where k is any integer. Keeping all slow terms, we then obtain

$$K_1 \;=\; \frac{1}{2\pi} \sum_{k=-\infty}^{\infty} f_{(kn_x)(kn_y)}(J_x, J_y)\, e^{ik(n_x\psi_x + n_y\psi_y)}.$$

All terms are slowly varying in this expression. The term $k = 0$ is the term contributing when away from all resonances. The other terms are resonance driving. Note that our treatment has included all higher order resonances,

$$k(n_x\nu_x + n_y\nu_y) \;=\; km$$

with all k's.

We now apply the smooth approximation to the beam-beam potential. The result is

$$K_1 \;=\; -\frac{Nr_0}{\gamma} \int_0^{2\pi} \frac{d\theta_x}{2\pi} \int_0^{2\pi} \frac{d\theta_y}{2\pi} \sum_{s=-\infty}^{\infty} \delta(n_x\psi_x + n_y\psi_y - n_x\theta_x - n_y\theta_y + 2\pi s)$$

$$\times \int_0^{\infty} dt\, \frac{\exp\left(-\dfrac{J_x\beta_x^*\cos^2\theta_x}{\sigma_x^2+t} - \dfrac{J_y\beta_y^*\cos^2\theta_y}{\sigma_y^2+t}\right) - 1}{\sqrt{(\sigma_x^2 + t)(\sigma_y^2 + t)}}, \tag{8.65}$$

where we have applied the identity

$$\frac{1}{2\pi} \sum_{k=-\infty}^{\infty} e^{iku} \;=\; \sum_{s=-\infty}^{\infty} \delta(u + 2\pi s).$$

The δ-function in Eq. (8.65) can in principle be used to remove one of the triple integrations.

We have now obtained the Hamiltonian (8.63) with a smoothed perturbation K_1 given by Eq. (8.65).

Invariant   The smoothed Hamiltonian, unlike the original one before smoothing, is now s-independent. The δ-function beam-beam kicks in the original Hamiltonian are now smoothed to become a time-independent potential well when we transform to the slow dynamical variables $(\psi_x, J_x, \psi_y, J_y)$. It follows that the smoothed Hamiltonian,

$$K \; = \; \text{invariant of the motion.}$$

This is of course an important result.

But the smooth approximation also gives a second invariant of the motion. An inspection of (8.65) shows that K depends on ψ_x and ψ_y not separately but always in their combination $n_x\psi_x + n_y\psi_y$. A degree of freedom is therefore lacking in the Hamiltonian. When there appears a lack of degree of freedom in the Hamiltonian, it always implies the existence of a certain invariant. That is why a symmetry in physical laws always implies a conservation law — time invariance implies energy conservation; translational invariance implies momentum conservation; rotational symmetry implies angular momentum conservation, etc. Here we have a similar situation.

Lacking the θ-degree of freedom, we already see K is an invariant. Now lacking anther degree of freedom in $\psi_{x,y}$ leads to another invariant. More specifically, the Hamilton equations read

$$\frac{d\psi_x}{d\theta} \; = \; \nu_x - \nu_{x0} + \frac{\partial K_1}{\partial J_x},$$
$$\frac{d\psi_y}{d\theta} \; = \; \nu_y - \nu_{y0} + \frac{\partial K_1}{\partial J_y},$$
$$\frac{dJ_x}{d\theta} \; = \; -\frac{\partial K_1}{\partial \psi_x},$$
$$\frac{dJ_y}{d\theta} \; = \; -\frac{\partial K_1}{\partial \psi_y}.$$

It is easy to see that the fact that K depends on ψ_x and ψ_y only by the combined variable $n_x\psi_x + n_y\psi_y$ implies

$$n_y\frac{dJ_x}{d\theta} - n_x\frac{dJ_y}{d\theta} \; = \; 0,$$

or equivalently,

$$n_y J_x - n_x J_y \; = \; \text{another invariant of the motion.} \tag{8.66}$$

Under the smooth approximation, we have therefore found two invariants of the motion in this 2-D dynamical system. This is quite an accomplishment by the smooth approximation.

On the other hand, as mentioned before, the smooth approximation has one important weakness, namely it fails to find the chaotic motion in a nonlinear dynamical system. On the contrary, having found two invariants in a 2-D system means the system is fully integrated, and there is no more room for chaotic motion. The smooth approximation actually precludes any existence of chaos.

Sum and difference resonances   Equation (8.66) leads to another important result. When n_x and n_y are of opposite signs, the resonance is called a difference resonance for obvious reason. When they are of the same sign, it is called a sum resonance. Since J_x and J_y are necessarily positive quantities, the theorem (8.66) says both of them will have to be bounded in the presence of a difference resonance and the motion is necessarily stable, at least to 1st order in the resonance strength.

Near a sum resonance, J_x and J_y can grow indefinitely without violating the theorem. So the particle motion can potentially be unstable. In particular, if J_x grows, it predicts that J_y has to grow as well to maintain the validity of the theorem. Near a sum resonance, either both J_x and J_y are bounded, or both grow.

Scaling parameters   The Hamiltonian (8.65) can be written as $\frac{Nr_0}{\gamma}$ times a dimensionless factor that depends on the dimensionless quantities

$$n_x, \quad n_y, \quad n_x\psi_x + n_y\psi_y, \quad \frac{J_x\beta_x^*}{2\sigma_x^2}, \quad \frac{J_y\beta_y^*}{2\sigma_y^2}, \quad \frac{\sigma_y}{\sigma_x}.$$

The quantities $n_x\psi_x + n_y\psi_y, \frac{J_x\beta_x^*}{2\sigma_x^2} = \alpha_x, \frac{J_y\beta_y^*}{2\sigma_y^2} = \alpha_y$ are the dynamical variables describing the particle's coordinates. The equations of motion, derived from the Hamiltonian K in terms of $n_x\psi_x + n_y\psi_y, \alpha_x, \alpha_y$, contains these dimensionless parameters,

$$\frac{Nr_0\beta_x^*}{\gamma\sigma_x^2}, \quad \frac{Nr_0\beta_y^*}{\gamma\sigma_y^2}, \quad \frac{\sigma_y}{\sigma_x}, \quad n_x, \quad n_y, \quad n_x\nu_x + n_y\nu_y - m.$$

The first three are replaceable by

$$\xi_x = \frac{Nr_0\beta_x^*}{2\pi\gamma\sigma_x(\sigma_x + \sigma_y)}, \quad \xi_y = \frac{Nr_0\beta_y^*}{2\pi\gamma\sigma_y(\sigma_x + \sigma_y)}, \quad a = \frac{\sigma_y}{\sigma_x}.$$

The remaining three are determined by ν_x and ν_y. We thus conclude that the single-resonance model is completely characterized by the scaling parameters

$$\xi_x, \xi_y, a, \nu_x, \nu_y.$$

Homework 8.29 It should be noted that we have applied the smooth approximation to the beam-beam problem with a particular form of $U(x, y)$. This does not have to be the case. Follow similar steps to analyze the case of a thin-lens sextupole at location $s = 0$ that gives kicks

$$\Delta x' = -S(x^2 - y^2), \qquad \Delta y' = 2Sxy.$$

Homework 8.30 Similar to Homework 8.29, this time analyze the case of a thin-lens quadrupole at location $s = 0$ that gives kicks

$$\Delta x' = -Gx, \qquad \Delta y' = Gy.$$

This perturbation is linear. The problem can be solved exactly using the matrix technique. However, this homework asks to do the analysis using the smooth approximation, valid to 1st order in G.

8.4.4 Single-resonance model in 1-D

In this section, we treat a simpler problem ignoring the x-y coupling and considering only the vertical y motion of the weak-beam particle. We illustrate this by the case of a flat beam. The Hamiltonian of this case is obtained by setting $\sigma_x \gg \sigma_y$ and $n_x = 0$. The resonant terms included in the single-resonance model are

$$\nu_y = \frac{\pm m}{\pm n}, \frac{\pm 2m}{\pm 2n}, \frac{\pm 3m}{\pm 3n}, \cdots . \tag{8.67}$$

Although the contributions from $\frac{\pm 2m}{\pm 2n}, \frac{\pm 3m}{\pm 3n}, \cdots$ actually belong to higher order resonances (of order $2n, 3n, \cdots$), they are driven by the same resonance $\nu_y \approx \frac{m}{n}$ and are included in our analysis.

The Hamiltonian is obtained from Eq. (8.65),

$$K(\psi_y, J_y, \theta) = \left(\nu_y - \frac{m}{n}\right)J_y - \frac{\xi_y \sigma_y^2}{\beta_y^*}\frac{1}{n}\sum_{s=0}^{n-1}\int_0^\infty dt \, \frac{\exp\left[-\frac{2\alpha_y}{1+t}\cos^2\left(\psi_y - \frac{2\pi s}{n}\right)\right] - 1}{\sqrt{1+t}},$$

$$\tag{8.68}$$

where $\alpha_y = \frac{J_y \beta_y^*}{2\sigma_y^2}$. As we did before, we designate the second term on the right-hand-side as the perturbation Hamiltonian K_1. The perturbation Hamiltonian is independent of m. This is due to the assumption of a δ-function beam-beam kick, which has a white spectrum.

Instantaneous detuning The particle's instantaneous tune is given by

$$\frac{d\psi_y}{d\theta} = \frac{\partial K}{\partial J_y} = \nu_y - \frac{m}{n} + \frac{\partial K_1}{\partial J_y},$$

where the term

$$\frac{\partial K_1}{\partial J_y} = \frac{\xi_y}{n}\sum_{s=0}^{n-1}\cos^2\left(\psi_y - \frac{2\pi s}{n}\right)\int_0^\infty \frac{dt}{(1+t)^{3/2}}\exp\left[-\frac{2\alpha_y}{1+t}\cos^2\left(\psi_y - \frac{2\pi s}{n}\right)\right]$$

$$= \frac{\xi_y}{n}\sum_{s=0}^{n-1}\cos^2\left(\psi_y - \frac{2\pi s}{n}\right)\int_0^1 du \exp\left[-2\alpha_y u^2 \cos^2\left(\psi_y - \frac{2\pi s}{n}\right)\right], \tag{8.69}$$

has the meaning of an instantaneous tune shift.

Note that this tune shift depends on J_y and ψ_y. Near a resonance, both J_y and ψ_y are slowly time varying. For a given J_y, the instantaneous tune shift spans a range because ψ_y ranges from 0 to 2π. The center of this range is the detuning $\Delta\nu_y(\alpha_y)$, Eq. (8.58).

When $J_y \to 0$, by keeping only the linear term in α_y in Eq. (8.68), the instantaneous tune shift is found to be

$$
\begin{aligned}
\frac{\partial K_1}{\partial J_y} &= \frac{2\xi_y}{n} \sum_{s=0}^{n-1} \cos^2\left(\psi_y - \frac{2\pi s}{n}\right) \\
&= \begin{cases} \xi_y(1 + \cos 2\psi_y), & \text{if } n = 1,2, \\ \xi_y, & \text{if } n \neq 1,2. \end{cases}
\end{aligned}
\tag{8.70}
$$

It follows that the small amplitude detuning is given by ξ_y for all orders of resonances. However, all high order resonances have no resonance spanning at small amplitude except for the integer and half-integer resonances for which case, the tune spans a range $\pm\xi_y$ due to a dependence on $\cos 2\psi_y$. We will return to this subject on page 610 when discussing the resonance width. See also Homework 8.31.

Fourier decomposition　　We now Fourier decompose the expression of the perturbation Hamiltonian K_1 in Eq. (8.68),

$$
K_1(\psi_y, J_y) = \sum_{k=0}^{\infty} G_k(J_y) \cos k\psi_y \, .
$$

It follows that

$$
\begin{aligned}
G_k(J_y) &= \frac{1}{\pi(1 + \delta_{k0})} \int_0^{2\pi} d\psi_y \cos k\psi_y \, K_1(\psi_y, J_y) \\
&= -\frac{\xi_y \sigma_y^2}{\pi \beta_y^*(1 + \delta_{k0})} \int_0^{\infty} \frac{dt}{\sqrt{1+t}} \int_0^{2\pi} d\theta \left[\exp\left(-\frac{2\alpha_y}{1+t}\cos^2\theta\right) - 1\right] \\
&\quad \times \frac{1}{n} \sum_{s=0}^{n-1} \cos\left(k\theta + k\frac{2\pi s}{n}\right),
\end{aligned}
$$

where $\delta_{k0} = 1$ if $k = 0$ and 0 otherwise, and we have made a change of variable from ψ_y to $\theta = \psi_y - \frac{2\pi s}{n}$ in the integration.

The quantity in the last line can be calculated,

$$
\frac{1}{n} \sum_{s=0}^{n-1} \cos\left(k\theta + k\frac{2\pi s}{n}\right) = \begin{cases} \cos pn\theta, & \text{if } k = pn \text{ is a multiple of } n, \\ 0, & \text{otherwise.} \end{cases}
$$

This means G_k vanishes unless $k = pn$ is an integer multiple of n, and

$$
G_{pn} = -\frac{\xi_y \sigma_y^2}{\pi \beta_y^*(1 + \delta_{p0})} \int_0^{\infty} \frac{dt}{\sqrt{1+t}} \int_0^{2\pi} d\theta \left[\exp\left(-\frac{2\alpha_y}{1+t}\cos^2\theta\right) - 1\right] \cos pn\theta \, .
$$

The integral over θ can also be performed analytically using the identity, for any A and any integer ℓ,

$$\frac{1}{\pi}\int_0^{2\pi} d\theta \, \cos\ell\theta \, e^{2A\cos^2\theta} = \begin{cases} 0, & \text{if } \ell = \text{odd}, \\ 2e^A \, I_{\frac{\ell}{2}}(A), & \text{if } \ell = \text{even}. \end{cases}$$

We then obtain

$$G_{pn} = \begin{cases} -\frac{2\xi_y\sigma_y^2}{\beta_y^*(1+\delta_{p0})}\int_0^\infty \frac{dt}{\sqrt{1+t}}\left[e^{-\frac{\alpha_y}{1+t}}I_{\frac{pn}{2}}\left(-\frac{\alpha_y}{1+t}\right) - \delta_{p0}\right], & \text{if } pn = \text{even}, \\ 0, & \text{otherwise}. \end{cases}$$

Finally, the integration over t can be performed by making a change of variable from t to $x = \frac{1}{1+t}$ and by using the identity

$$\int dx \, \frac{e^{-x}}{x\sqrt{x}}I_n(x) = \frac{2}{4n^2-1}\frac{e^{-x}}{\sqrt{x}}\left[(1+2x)I_n(x) + xI_{n+1}(x) + xI_{n-1}(x)\right].$$

The result is, when $pn = \text{even}$,

$$G_{pn} = -\frac{2\xi_y\sigma_y^2}{\beta_y^*}\left[g_{pn}(\alpha_y) + \delta_{p0}\right],$$

where the term δ_{p0} is a result of derivation and has no physical consequences,[17] and we have defined

$$g_{pn}(\alpha_y) = \frac{2(-1)^{\frac{pn}{2}}e^{-\alpha_y}}{(1+\delta_{p0})(p^2n^2-1)}\left[(1+2\alpha_y)I_{\frac{pn}{2}}(\alpha_y) + \alpha_y I_{\frac{pn}{2}+1}(\alpha_y) + \alpha_y I_{\frac{pn}{2}-1}(\alpha_y)\right].$$

$$(8.71)$$

The perturbation Hamiltonian is then found to be

$$K_1 = -\frac{2\xi_y\sigma_y^2}{\beta_y^*}\left[1 + \sum_{p=0}^{\infty}{}' g_{pn}(\alpha_y) \cos pn\psi_y\right], \qquad (8.72)$$

where $\sum_p'$ means the summation is over p when pn is even. In case n is even, then the summation is over all p's from 0 to infinity. In case n is odd, then only even p's are summed over.

Having obtained the Hamiltonian, the Hamilton equations then give the single-particle dynamics,

$$\frac{d\psi_y}{d\theta} = \nu_y - \frac{m}{n} + \frac{\partial K_1}{\partial J_y},$$

$$\frac{d\alpha_y}{d\theta} = \xi_y \sum_{p=0}^{\infty}{}' pn \, g_{pn}(\alpha_y) \sin pn\psi_y, \qquad (8.73)$$

---

[17]This term is needed to define the potential without a possible divergence. But being a constant term in a Hamiltonian, it has no physical consequences.

where the instantaneous tune shift is

$$\frac{\partial K_1}{\partial J_y} = -\xi_y \sum_{p=0}^{\infty}{}' g'_{pn}(\alpha_y) \cos pn\psi_y \,, \tag{8.74}$$

with $g'_{pn}(\alpha_y)$ the derivative of $g_{pn}(\alpha_y)$,

$$g'_{pn}(\alpha_y) = \frac{2(-1)^{\frac{pn}{2}}}{(1+\delta_{p0})\,(p^2 n^2 - 1)} \tag{8.75}$$

$$\times\, e^{-\alpha_y}\left[\left(1+\frac{p^2 n^2}{2\alpha_y}\right) I_{\frac{pn}{2}}(\alpha_y) + \frac{1}{2}I_{\frac{pn}{2}+1}(\alpha_y) + \frac{1}{2}I_{\frac{pn}{2}-1}(\alpha_y)\right].$$

Equations (8.71–8.75) are our final results for the flat-beam 1-D case under the smooth approximation near the resonance $\nu_y \approx \frac{m}{n}$. The particle motion is completely specified by the two scaling parameters ν_y (and therefore n and $\nu - \frac{m}{n}$) and ξ_y.

Two special cases deserve to be checked.

1. Away from resonances, we shall keep only the $p = 0$ term, and we recover the flat beam results (8.58) and (8.59).

2. For small amplitude particles $\alpha_y \ll 1$ but arbitrary n, we recover Eq. (8.70).

There are a few peculiarities resulting from the beam-beam summations $\sum_p{}'$ in Eqs. (8.72) and (8.74). If n is odd, the beam-beam perturbation does not drive the $\frac{m}{n}$ resonance. Instead, it drives the $\frac{2m}{2n}$ resonance. As a result, the integer resonance is driven identically as the 2nd order one, and a 3rd order resonance is driven identically as a 6th order one. The lowest order nonlinear resonance is therefore 4th order. Physically this traces to the fact that we have assumed a strong beam distribution that is up-down symmetric and the collision is head-on.

Resonance island, resonance width   We often depict the resonant particle motion by its trajectory in the polar $(\sqrt{\alpha_y}, \psi_y)$ space, as we did in several of our earlier plots, particularly those exhibiting island structures. It would be convenient to define a resonance width associated with those islands in the $(\sqrt{\alpha_y}, \psi_y)$ space — we shall somewhat loosely call this the phase space. For each resonance of order n, we designate this width by the island full width in the radial direction, designated as $(\delta\sqrt{\alpha_y})^{(n)}$ (in units of $\sqrt{\alpha_y}$).

We now want to calculate $(\delta\sqrt{\alpha_y})^{(n)}$. To do so, let us first prepare constant Hamiltonian contours in the phase space. The Hamiltonian is given by

$$\frac{\beta^*}{2\sigma_y^2}K(\psi_y, \alpha_y) = \left(\nu_y - \frac{m}{n}\right)\alpha_y - \xi_y\left[1 + \sum_{p=0}^{\infty}{}' g_{pn}(\alpha_y)\cos pn\psi_y\right],$$

where it is reminded that ν_y is the unperturbed betatron tune.

Take a 4th order resonance for example, $n = 4$. Figure 8.19 shows what happens to the phase space when the unperturbed tune is scanned through the resonance of strength $\xi_y = 0.2$ with $\nu_y - \frac{1}{4} = 0.05, 0.02, 0, -0.02, -0.05, -0.1$.

It is noted from Fig. 8.19 that when ν_y is -0.02 below the resonant value of $\frac{1}{4}$, a set of four islands are formed, indicating the moment of resonance crossing. The resonance has a strong influence of the particle motion at this point. Let us focus our attention to these islands.

First let us find out the location of these islands. The islands are generated at a particular amplitude $\sqrt{\alpha_0}$. A moment's reflection indicates that α_0 must be such that the shifted tune together with detuning, $\nu_y + \Delta_y(\alpha_0)$, is equal to the resonance value $\frac{m}{n}$, i.e.

$$\nu_y + \Delta\nu_y(\alpha_0) \approx \frac{m}{n},$$

or, using Eq. (8.58),

$$-\xi_y e^{-\alpha_0}[I_0(\alpha_0) + I_1(\alpha_0)] \approx \nu_y - \frac{m}{n}. \tag{8.76}$$

Equation (8.76) indicates that, for $\xi_y > 0$, resonance occurs only when ν_y is below $\frac{m}{n}$ while not to be lower than $\frac{m}{n} - \xi_y$. If ν_y is outside the range $(\frac{m}{n}, \frac{m}{n} - \xi_y)$, there is no solution for α_0 and there will be no islands formed. Physically this is because the beam-beam effect shifts the tune upward, so to hit a resonance, the original unperturbed tune needs to be below the resonance. This behavior is confirmed by Fig. 8.19.

In Fig. 8.19(d), the island location confirms Eq. (8.76). The location of the four islands agrees with Eq. (8.76) when $\nu_y - \frac{1}{4} = -0.02$.

Furthermore, an inspection of Eq. (8.76) indicates that α_0 decreases and moves towards the origin as ν_y moves away from $\frac{m}{n}$. This is confirmed by Fig. 8.19(e). We will return to this point in Fig. 8.29 and Homework 8.36.

When ν_y is away from the range between $\frac{1}{4}$ and $\frac{1}{4} - \xi_y$, the phase space evolves back to the nominal circular shape, again consistent with Fig. 8.19.

A similar resonance phase space contour plot can be prepared for other resonances. For example, the $\frac{1}{3}$ and the $\frac{1}{6}$ resonances have the same contour plot. An example is shown in Fig. 8.20, where a set of six islands is seen for the case $\xi_y = 0.1, \nu_y - \frac{1}{6} = -0.013$. The panel on the right shows more clearly the boundary of the islands, also known as the separatrix.

The resonance amplitude width $(\delta\sqrt{\alpha_y})^{(n)}$ is meant to be the total width of an island in the $(\sqrt{\alpha_y}\cos\psi_y, \sqrt{\alpha_y}\sin\psi_y)$ space in the radial direction. To calculate $(\delta\sqrt{\alpha_y})^{(n)}$, we need to return to the Hamiltonian and find its contours. For this purpose, let us take an approximate expression, taking only into account of the $p = 0$ and $p = 1$ terms (assuming $n = $ even),

$$\frac{\beta_y^*}{2\sigma_y^2}K \approx \left(\nu_y - \frac{m}{n}\right)\alpha_y - \xi_y\left[1 + g_0(\alpha_y) + g_{1n}(\alpha_y)\cos n\psi_y\right]. \tag{8.77}$$

The dropping of the $p \neq 0, 1$ terms is equivalent to keeping only the $\frac{\pm m}{\pm n}$ resonances and dropping all higher order contributions from $\frac{\pm 2m}{\pm 2n}, \frac{\pm 3m}{\pm 3n}, \cdots$.

Figure 8.19: Constant Hamiltonian contours as a flat-beam 1-D 4th order beam-beam resonance is crossed by sweeping the unperturbed betatron tune. The six cases are for $\nu_y - \frac{1}{4} = 0.05, 0.02, 0, -0.02, -0.05, -0.1$, respectively. The phase space is $(\sqrt{\alpha_y}\cos\psi_y, \sqrt{\alpha_y}\sin\psi_y)$. The beam-beam strength is $\xi_y = 0.2$.

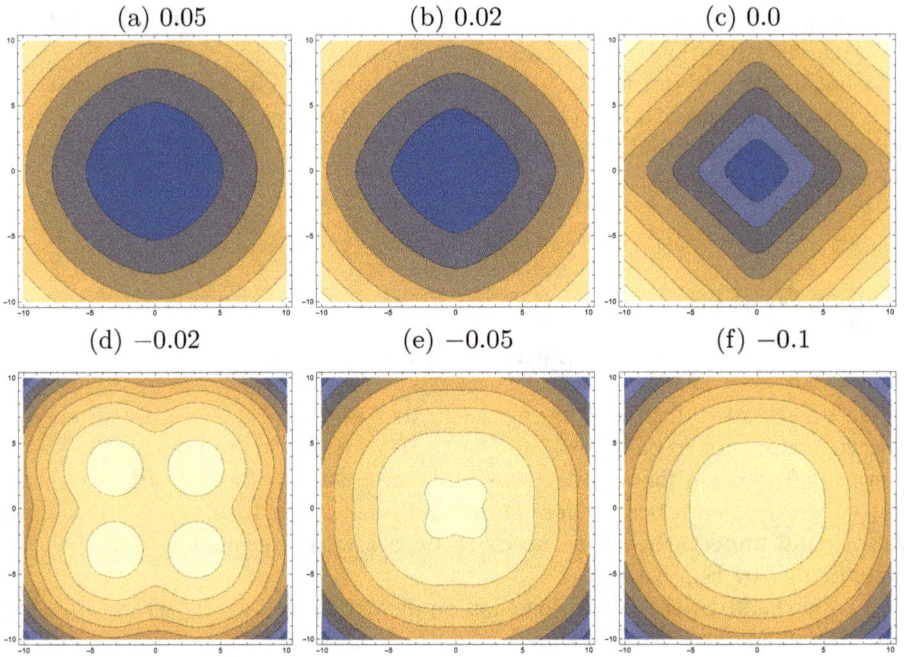

From Eqs. (8.71) and (8.75), we also have

$$
\begin{aligned}
g_0(\alpha_y) &= -e^{-\alpha_y}\left[(1+2\alpha_y)I_0 + 2\alpha_y I_1\right], \\
g_{1n}(\alpha_y) &= \frac{2(-1)^{\frac{n}{2}}}{n^2-1}e^{-\alpha_y}\left[(1+2\alpha_y)I_{\frac{n}{2}} + \alpha_y I_{\frac{n}{2}+1} + \alpha_y I_{\frac{n}{2}-1}\right], \\
g_0'(\alpha_y) &= -e^{-\alpha_y}(I_0 + I_1), \\
g_{1n}'(\alpha_y) &= \frac{2(-1)^{\frac{n}{2}}}{n^2-1}e^{-\alpha_y}\left[\left(1+\frac{n^2}{2\alpha_y}\right)I_{\frac{n}{2}} + \frac{1}{2}I_{\frac{n}{2}+1} + \frac{1}{2}I_{\frac{n}{2}-1}\right], \\
g_0''(\alpha_y) &= \frac{1}{2}e^{-\alpha_y}(I_0 - I_2),
\end{aligned}
$$

where all Bessel functions are evaluated at α_y.

We then look for the fixed points in the phase space. The location of the fixed point (ψ_0, α_0) are determined by the conditions

$$
\frac{\partial K}{\partial \alpha_y} = 0, \qquad \frac{\partial K}{\partial \psi_y} = 0.
$$

Figure 8.20: Constant Hamiltonian contours as a 3rd order or 6th order beam-beam resonance with $\nu_y - \frac{1}{6}$ or $\nu_y - \frac{1}{3} = -0.013$. The phase space is $(\sqrt{\alpha_y}\cos\psi_y, \sqrt{\alpha_y}\sin\psi_y)$. The beam-beam strength is $\xi_y = 0.1$. The panel on the right is the same case but singles out only the contour that outlines the boundary of the islands — the separatrix.

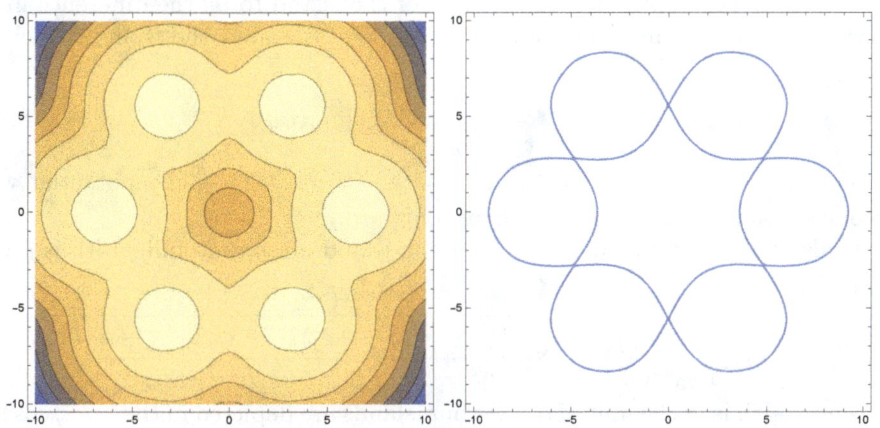

The condition $\frac{\partial K}{\partial \psi_y} = 0$ immediately gives

$$\sin n\psi_0 = 0 \quad \Longrightarrow \quad \psi_0 = \pm \frac{k\pi}{n}, \quad k = 0, 1, 2, \cdots, (n-1).$$

There are $2n$ fixed points with ψ_y given by the $2n$ values of ψ_0 above. Out of these $2n$ fixed points, n of them are stable and n are unstable. When $\frac{n}{2}$ is even, the fixed points at $\psi_0 = $ even multiples of $\frac{\pi}{n}$ are unstable, the ones at $\psi_0 = $ odd multiples of $\frac{\pi}{n}$ are stable. When $\frac{n}{2}$ is odd, the opposite holds.

We also need to find α_0, the radial location of the fixed points. The condition $\frac{\partial K}{\partial \alpha_y} = 0$ gives the result. But here a reasonable approximation can be made by dropping the term g_{1p}. We then obtain

$$\nu_y - \frac{m}{n} - \xi_y g_0'(\alpha_0) = 0, \tag{8.78}$$

which we recognize simply reproduces Eq. (8.76) obtained earlier. We will return to this equation in Homework 8.36.

Having obtained the locations of the islands, we now proceed to calculate their amplitude width and their potential well depth. To do so, let us concentrate over the region around the islands, i.e. we consider α_y in the neighborhood of α_0. The Hamiltonian can be approximated by Taylor expansion,

$$\frac{\beta_y^*}{2\sigma_y^2} K \approx \left(\nu_y - \frac{m}{n}\right)\alpha_y - \xi_y \left[1 + g_0(\alpha_0) + g_0'(\alpha_0)(\alpha_y - \alpha_0) \right.$$
$$\left. + \frac{1}{2}g_0''(\alpha_0)(\alpha_y - \alpha_0)^2 + g_{1n}(\alpha_0)\cos n\psi_y\right].$$

One can drop a constant term without affecting the Hamiltonian. Then applying Eq. (8.78), we find

$$\frac{\beta_y^*}{2\sigma_y^2} K \approx -\xi_y \left[\frac{1}{2} g_0''(\alpha_0)(\alpha_y - \alpha_0)^2 + g_{1n}(\alpha_0) \cos n\psi_y \right]. \qquad (8.79)$$

We define the depth of potential well of the island to be the difference in K between the stable and the unstable fixed points around the island, which in turn is equal to

$$\delta K_{\text{island}} = \frac{2\sigma_y^2}{\beta_y^*} \times 2\xi_y |g_{1n}(\alpha_0)|. \qquad (8.80)$$

This potential well depth scales as $\frac{1}{\sqrt{\alpha_0}}$ when $\alpha_0 \gg 1$. The potential well becomes shallower as the islands move outward.

It also follows from Eq. (8.79) that the island amplitude full width is given by

$$\delta\alpha_y^{(n)} = 4\sqrt{\left| \frac{g_{1n}(\alpha_0)}{g_0''(\alpha_0)} \right|} = 8\sqrt{\left| \frac{1}{n^2 - 1} \frac{(1 + 2\alpha_0)I_{\frac{n}{2}} + \alpha_0 I_{\frac{n}{2}+1} + \alpha_0 I_{\frac{n}{2}-1}}{I_0 - I_2} \right|}.$$

This width is in the unit of α_y. As our islands are depicted in the $(\sqrt{\alpha_y} \cos \psi_y, \sqrt{\alpha_y} \sin \psi_y)$ space, we may want to convert it to the $\sqrt{\alpha_y}$ units,

$$(\delta\sqrt{\alpha_y})^{(n)} = \frac{\delta\alpha_y^{(n)}}{2\sqrt{\alpha_0}}. \qquad (8.81)$$

Figure 8.21 shows $(\delta\sqrt{\alpha_y})^{(n)}$ as a function of $\sqrt{\alpha_0}$ for resonances $n = 2, 4, 6$. Once the resonance location α_0 is known from Eq. (8.78), Eq. (8.81) gives the full resonance amplitude width in $\sqrt{\alpha_y}$ units as a function of $\sqrt{\alpha_0}$. For the case $n = 2$, the resonance has a finite width $(\delta\sqrt{\alpha_y})^{(2)} = 2\sqrt{2}$ at $\sqrt{\alpha_0} = 0$ (see Fig. 8.21). We will return to Eq. (8.81) in Homework 8.36.

One more phase space quantity to be calculated is the island area in the $(\sqrt{\alpha_y} \cos \psi_y, \sqrt{\alpha_y} \sin \psi_y)$ space. Approximating each island to an elliptical shape, its area is approximately

$$\Phi_{\text{island}}^{(n)} \approx \pi \frac{(\delta\sqrt{\alpha_y})^{(n)}}{2} \frac{\pi}{n}. \qquad (8.82)$$

Resonance tune width   It would be desirable to convert the island amplitude width $\delta\alpha_y^{(n)}$ into tune units to see how wide these resonances occupy in the tune plane. Since the resonance overlapping refers to island overlapping in the phase space, it is natural that we define the resonance tune width by connecting it to the island width (see Homework 8.31),

$$\delta\nu_y^{(n)}(\alpha_y) = |\delta\nu_y'(\alpha_0)| \, \delta\alpha_y^{(n)} = 4\xi_y \sqrt{|g_{1n}(\alpha_0)g_0''(\alpha_0)|}$$

$$= 4\xi_y \sqrt{\frac{e^{-2\alpha_0}}{n^2-1} [(1+2\alpha_0)I_{\frac{n}{2}} + \alpha_0 I_{\frac{n}{2}+1} + \alpha_0 I_{\frac{n}{2}-1}](I_0 - I_1)}, \qquad (8.83)$$

where all Bessel functions are evaluated at α_0, the island location.

Figure 8.21: Amplitude full width $(\delta\sqrt{\alpha_y})^{(n)}$ of resonance islands in the $(\sqrt{\alpha_y}\cos\psi_y, \sqrt{\alpha_y}\sin\psi_y)$ space for resonance orders $n = 2$ (black), 4 (red) and 6 (blue).

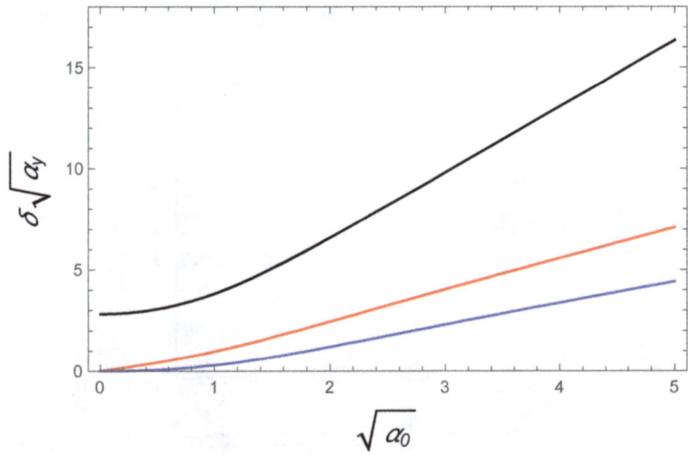

The situation is illustrated in Fig. 8.22. Each of the pink resonance band indicates the region $\frac{m}{n} + \Delta\nu_y \pm \frac{\delta\nu_y^{(n)}}{2}$ as a function of α_0, where $\Delta\nu_y$ is the detuning contribution given by (8.58) and $\frac{\delta\nu_y^{(n)}}{2}$ is the resonance tune width (8.83). The rational tune values of the resonances are indicated by blue straight lines.

The detuning (8.58) starts with ξ_y at $\alpha_y = 0$ and decays with increasing α_y only very slowly,

$$\Delta\nu_y(\alpha_y) \approx \frac{2\xi_y}{\sqrt{2\pi\alpha_y}}, \qquad \alpha_y \gg 1.$$

This means the tune bandwidths (pink bands) converge to their corresponding resonant values (blue straight lines) only very slowly.

The resonance tune widths vanish at small amplitudes for all resonances including the integer and half-integer ones. As mentioned earlier, the resonance $\frac{1}{3}$ behaves identically to the resonance $\frac{1}{6}$. The lowest order nonlinear resonance is in fact the $\frac{1}{4}$ resonance. In particular, it may be noted that the $\frac{1}{4}$ resonance is wider than the $\frac{1}{3}$ resonance. The $\frac{1}{2}$ resonance is considered linear.

One way to use Fig. 8.22 is as follows. To see if the beam-beam interaction would significantly affect particle motion, first decide on the choice of the unperturbed tune ν_y. The motion would be affected if the horizontal straight line of ν_y meets any of the pink resonance bands. The location where they meet indicates the amplitude at which the beam motion is affected. The bandwidth at that location then determines how wide the resonance is in tune space.

Figure 8.22: Beam-beam resonance tune widths as a function of betatron amplitude α_0 of the island location, for the $\frac{1}{2}, \frac{1}{3}, \frac{1}{4}$ and $\frac{1}{6}$ resonances. Beam-beam strength is $\xi_y = 0.1$.

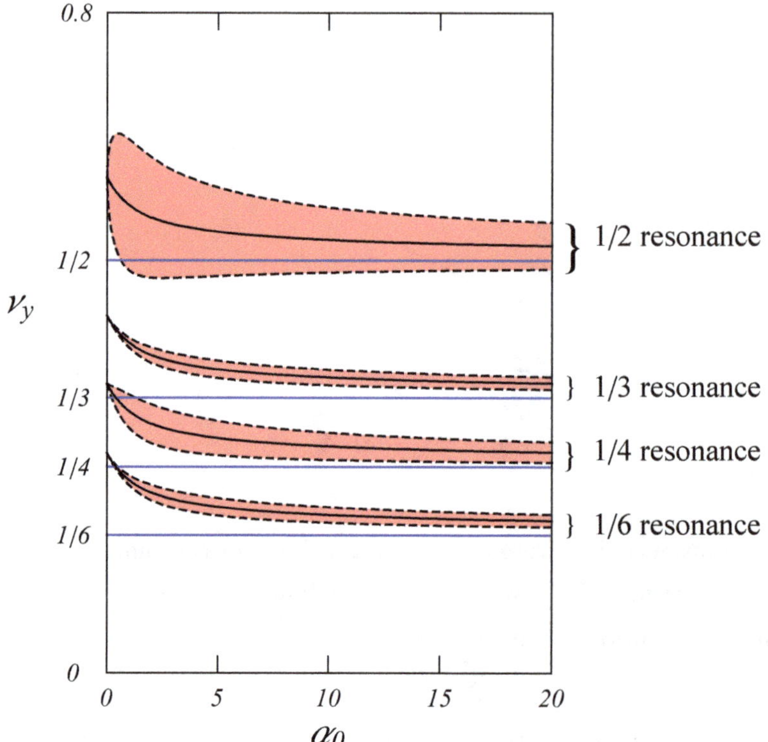

Observation of beam-beam resonances   A vivid manifestation of weak-strong beam-beam resonances can be seen in an early experiment at ACO, as shown in Fig. 8.23.[18] In such a weak-strong configuration, only the weak beam is affected by the beam-beam interaction. Figure 8.23 clearly shows the weak beam blows up every time the vertical tune approaches a fractional value. Note that the strong beam also blows up near the resonances $\nu_y = \nu_x$ and $\nu_y = \frac{2}{3}$. These blow-ups are not induced by the beam-beam interaction but by the residual field errors in the storage ring.

In early days of PETRA, colliding beams exhibited distributions shown in Fig. 8.24 in the neighborhood of a 6th order horizontal resonance.[19] The beam distributions were distorted strongly into a set of islands, a behavior that is unmistakably that caused by a nonlinear resonance.

---

[18]H. Zyngier, AIP Conf. Proc. 57, p. 137 (1979).
[19]A. Piwinski, AIP Conf. Proc. 57, p. 115 (1979); A. Piwinski, IEEE Trans. Nucl. Sci. NS-26, 4268 (1979).

Figure 8.23: ACO colliding-beam profiles as vertical tune is varied. Strong beam profile on the left column, weak beam profile on the right. Weak beam profile exhibits resonances when vertical tune is near a fractional value (see footnote 18).

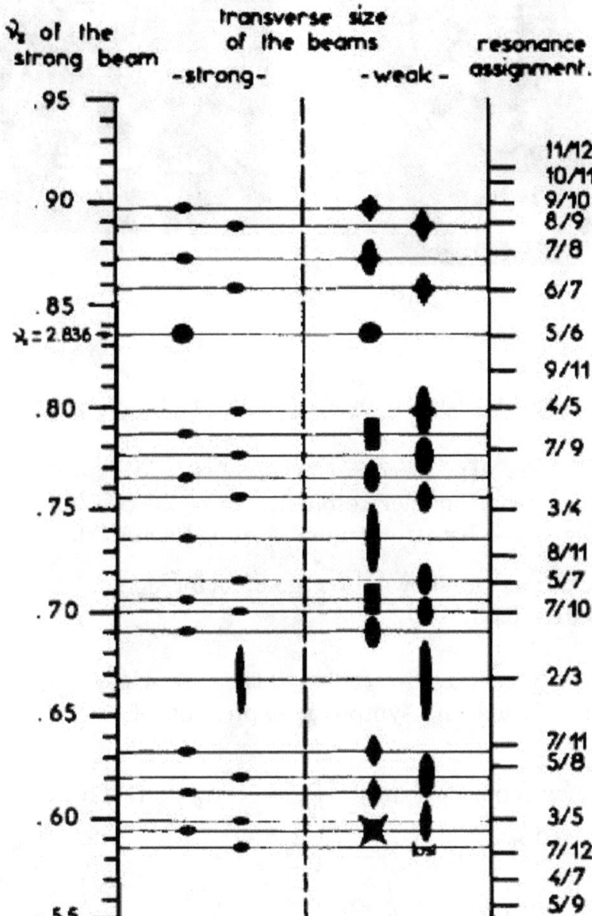

Homework 8.31

(a) Verify Eq. (8.69), the general expression of the instantaneous tune shift.

(b) Specialize Eq. (8.69) to the case when $J_y \to 0$ to obtain Eq. (8.70).

(c) Specialize Eq. (8.69) to the case when $n = 2$ but allow for arbitrary J_y.

It should be pointed out that this instantaneous tune (8.69) spans a region as ψ_y varies from 0 to 2π, and it might seem that it can serve as another definition of tune spread. However, this tune span is not connected to the phase space island, and is not to be confused with the resonance tune width $\delta\nu_y^{(n)}$ defined in the text.

Figure 8.24: PETRA colliding beams observed from a synchrotron light monitor.
The horizontal tune is close to a 6th order resonance (see footnote 19).

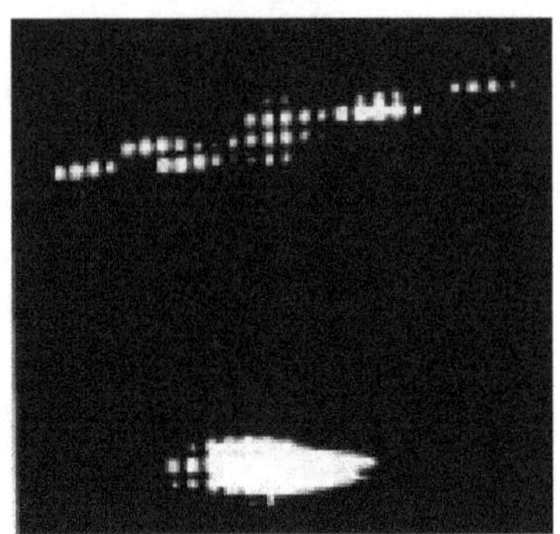 

Solution

(c) $\frac{\partial K_1}{\partial J_y} = 2\xi_y \cos^2 \psi_y \int_0^1 du \exp(-2\alpha_y u^2 \cos^2 \psi_y)$.

Homework 8.32 As a practice, make a constant-Hamiltonian contour plot
of Eq. (8.68) near a half-integer resonance, $n = 2$. See how the phase space
topology changes as the betatron tune is scanned across the resonance.

Solution An example is shown in Fig. 8.25 for $\xi_y = 0.04$ and $\nu_y - \frac{m}{2} =$
$0.03, 0.01, 0, -0.01, -0.03, -0.1$.

Homework 8.33 Specialize Eq. (8.69) to the case when $J_y \to \infty$ for arbitrary
resonance order n. Find an asymptotic expression of the tune shift far in the
beam tail.

Solution You may replace $\int_0^1 du$ by $\int_0^\infty du$ if $\alpha_y \gg 1$. Then,

$$\frac{\partial K_1}{\partial J_y} = \frac{\xi_y}{n} \sqrt{\frac{\pi}{8\alpha_y}} \sum_{s=0}^{n-1} \left| \cos\left(\psi_y - \frac{2\pi s}{n}\right)\right|.$$

If information on ψ_y is not too important, you may replace it by an average
expression

$$\frac{\partial K_1}{\partial J_y} \approx \frac{2\xi_y}{\pi} \sqrt{\frac{\pi}{8\alpha_y}}.$$

The instantaneous tune shift scales like $\frac{1}{\sqrt{\alpha_y}}$.

Homework 8.34 The text made a substantial effort to analyze the case for a
flat bi-Gaussian beam with $\sigma_x \gg \sigma_y$ and considered a particle executing only
y-motion. Follow similar steps to analyze the problem for the case of a round
beam $\sigma_x = \sigma_y$. Consider again the case of a particle executing only y-motion.

Figure 8.25: A flat-beam 1-D beam-beam phase space when the beta-tron tune is scanned through a half-integer resonance. Horizontal axis is $\sqrt{\alpha_y}\cos\psi_y$. Vertical axis is $\sqrt{\alpha_y}\sin\psi_y$. The six panels correspond to $\nu_y - \frac{m}{2} = 0.03, 0.01, 0, -0.01, -0.03, -0.1$, respectively.

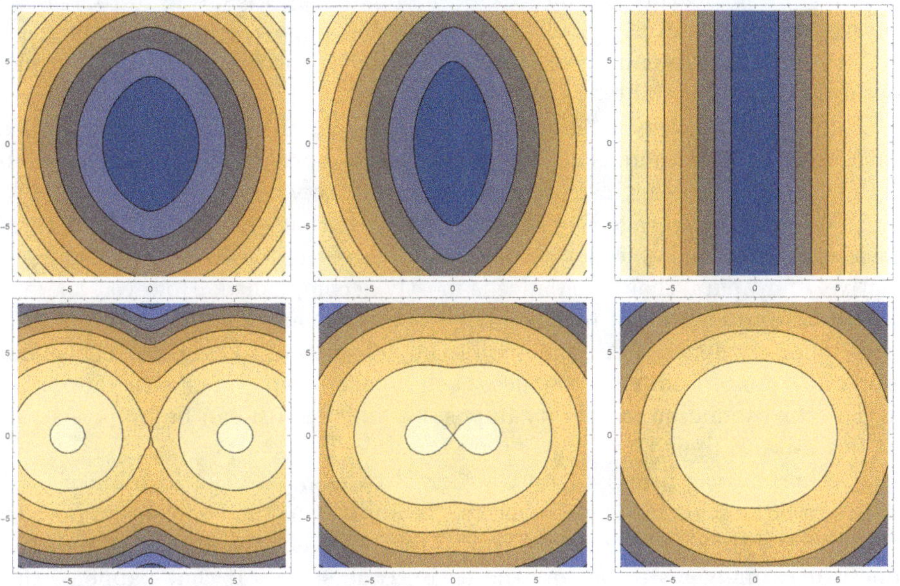

Homework 8.35 Analyze the problem for the case when a weak-beam particle executing y-motion colliding with a strong beam with a flat uniform disk distribution of dimensions $w_x \gg w_y$. There should be much less resonance structures if the particle's oscillation amplitude is less than w_y.

8.4.5 Chirikov criterion

One important observation for the weak-strong beam-beam perturbation lies in the fact that the effect of the perturbation Hamiltonian weakens when the beta-tron amplitude $\alpha_y \to \infty$. Physically this originates because, unlike the case of a magnet nonlinear multipole, the beam-beam force diminishes towards the tail of the strong-beam distribution. Because of this special beam-beam property, it follows that, according to the weak-strong single-resonance model, the beam-beam perturbation eventually does not cause instabilities. Mathematically, it follows from the fact that the perturbation Hamiltonian

$$K_1 \approx 4\xi_y \sqrt{\frac{\sigma_y^2 J_y}{\pi \beta_y^*}} \qquad J_y \to \infty\,,$$

so the unperturbed linear term $(\nu_y - \frac{m}{n})J_y$ eventually always dominates as long as ν_y is not exactly on resonance, and the far tail phase space trajectories become circles. At sufficiently large amplitudes, the particle motion eventually becomes stabilized even when nonlinear resonances mess up the beam distribution in the beam core. This feature has already been apparent in Figs. 8.19, 8.20, and 8.25 and was alluded to in Fig. 8.16. As much as the smooth approximation gives a bucketful of beam dynamics of the beam-beam interaction, it does not explain the beam-beam instability.

This picture, however, assumes the validity of the smooth approximation when we insist that there is at most one resonance present. In actuality, we also at the same time know this can never really be true.

To operate the storage ring collider, it is clear that one chooses the betatron tunes to avoid the low order resonances. Since the low-order resonances are relatively far apart, their avoidance is mostly accomplished. However, there are also an infinite number of higher order resonances, and they populate the tune space infinitely densely. At some point they are bound to overlap. The smooth approximation of isolated resonances then breaks down and chaotic motion sets in. The corresponding instability imposes a limit, which can be proposed to be the beam-beam limit.

A criterion was proposed by Chirikov[20] to associate the instability to the onset of a large scale resonance overlap. To apply it to the beam-beam problem, it is proposed that a fundamental limit is reached when the sum of all resonance tune widths manages to occupy the entire tune space. If that happens, there is no room left to choose the working point of the betatron tune, a large scale chaos occurs and the beam becomes unstable.

The total tune width of all beam-beam resonances is given by

$$\delta\nu_{y,\text{tot}}(\alpha_y) = \sum_{n=2,\text{even}}^{\infty} (n-1) \times \delta\nu_y^{(n)}(\alpha_y),$$

where the single resonance tune width $\delta\nu_y^{(n)}(\alpha_y)$ is given in Eq. (8.83). We have excluded the $n = $ odd terms to avoid double counting of the resonances. The additional factor $n - 1$ is to take into account that there are $n - 1$ resonances of order n in the tune space from 0 to 1.[21]

Figure 8.26 shows $\delta\nu_{y,\text{tot}}$ as a function of α_y. The few dashed curves indicate the first few terms in the summation over n. The red curve is the total sum up to $n = 50$.

One observation to be made with Fig. 8.26 is the fact that when α_y is large, meaning for particles in the beam tail, the total tune width, and therefore the mechanism of beam-beam instability, involves increasingly higher order of resonances. When $\alpha_y = 20$ (particle amplitude is $2\sqrt{20}\,\sigma$), for example, as shown

---

[20]G. Zaslavsky and B. Chirikov, Dokl. Akad. Nauk. SSR 169, 306 (1964); B.V. Chirikov, Phys. Reports 52, 263 (1979).

[21]Strictly, one needs to exclude contributions from resonance $\frac{m}{n}$ where m and n have common divisible factors. But we shall not be worried about that.

Figure 8.26: Total resonance tune width $\frac{\delta\nu_{y,\text{tot}}(\alpha_y)}{\xi_y}$ as a function of α_y. The dashed curves are the first few terms in the summation representing the contributions of the lower order resonances. The red curve is a summation over n up to 50 to assure convergence over the range of α_y in the plot. Many higher order resonances constitute the bulk of the total sum.

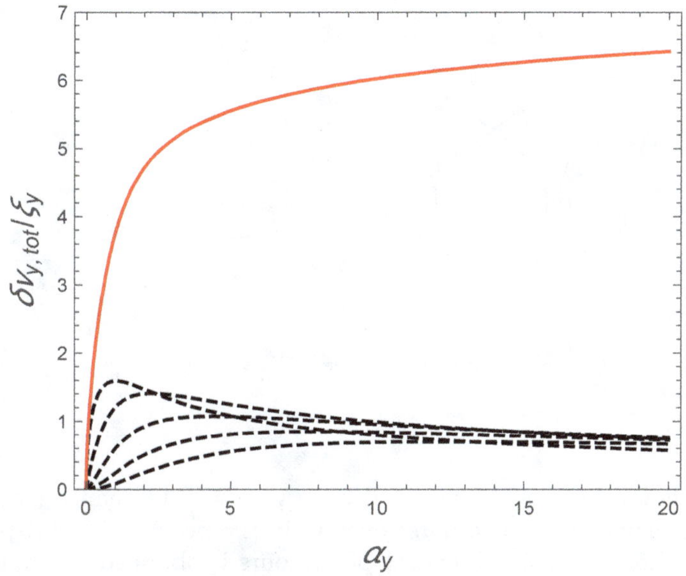

in Fig. 8.26, it is necessary to consider resonances up to $n = 50$ for convergence. The convergence requires even more terms as α_y increases. [Consult also Eq. (8.48).] Asymptotically,

$$\delta\nu_{y,\text{tot}} = 7.6\,\xi_y, \qquad \text{when} \qquad \alpha_y \to \infty\,.$$

Chirikov criterion then states that large scale chaos happens at amplitude α_y when

$$\delta\nu_{y,\text{tot}}(\alpha_y) \; > \; 1\,.$$

If we take the asymptotic value with $\alpha_y \to \infty$, then the Chirikov stability requires

$$\xi_y \; < \; \frac{1}{7.6} \; = \; 0.13\,. \tag{8.84}$$

According to the thin-lens weak-strong beam-beam Chirikov model, the beam-beam limit is reached at $\xi_y = 0.13$ for a flat-beam configuration.

As mentioned, Fig. 8.26 indicates that overlapping of higher order resonances contribute significantly to the Chirikov instability. Rather exaggerated parameters are needed for low order resonances to overlap. As an illustration,

Figure 8.27: Overlapping of a $\frac{2}{5}$ resonance and a $\frac{3}{8}$ resonance in an exaggerated case of $\xi_y = 0.5$ shown in the polar $(\sqrt{\alpha_y}, \psi_y)$ phase space.

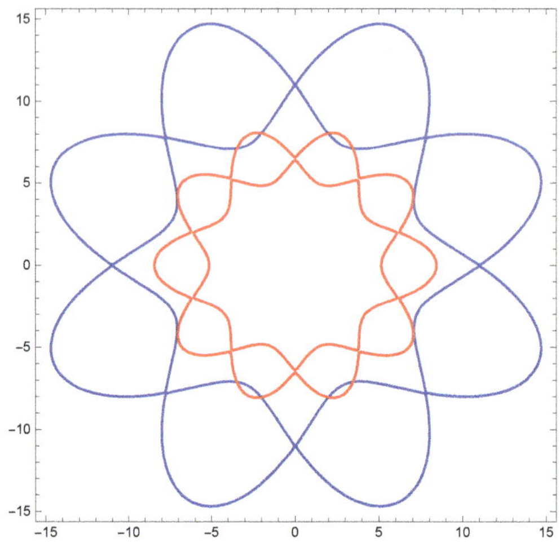

Fig. 8.27 shows a rather exaggerated case when a $\frac{2}{5}$ resonance (red contours) and a $\frac{3}{8}$ resonance (blue contours) overlap in the polar $(\sqrt{\alpha_y}, \psi_y)$ phase space when $\xi_y = 0.5, \nu_y = 0.34$. Each set of contours is obtained assuming one and only one resonance is playing the role by the smooth approximation disregarding the presence of the other resonance. When both resonances are present and the contours overlap, the smooth approximations break down; both contours break, resulting in chaotic particle trajectory in the overlapped region of the phase space.

Particle loss When the beam becomes unstable per the Chirikov criterion, it is most likely when the working point has been optimized away from all low-order resonances but there are multiple high-order resonances forcing the particle trajectory to enter a large-scale chaotic motion, causing the particle to be lost from the beam.

There remains a catch, however. The Chirikov criterion addresses the onset of large scale chaos in the phase space. It applies well to the cases of magnetic multipole nonlinearities. Applying to the beam-beam perturbation, however, it does not really address the beam-beam limit as an eventual particle loss mechanism. This is again because the beam-beam perturbation vanishes as the particle amplitude increases and at $\alpha_y \to \infty$, it is the unperturbed term that dominates the Hamiltonian, and the particle motion, even if chaotic, must stay bounded. The chaotic layers become extra thin and limited to form circles. The presence of a large scale chaos may explain a beam blow-up but still falls short of explaining a beam-beam instability.

Figure 8.28: Particle motion remains bounded even when large scale chaos occurs according to the Chirikov criterion for the beam-beam perturbation. The normalized coordinates are $u = \frac{y}{\sigma_y}, v = \frac{\beta_y^* y'}{\sigma_y}$.

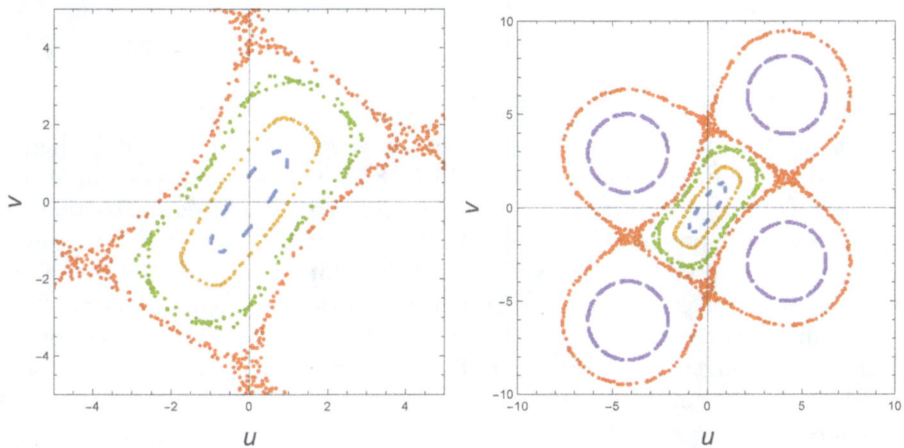

Figure 8.28 illustrates this point by a numerical simulation of a flat-beam case with $\xi_y = 0.2, \nu_y = 0.193$ and several particles with various initial conditions. The left panel views more of the beam core and observes chaotic behavior. However, when the view is expanded to larger scales, the same plot shows bounded motion even for the chaotic trajectories. There is not a particle loss mechanism in the weak-strong nonlinear lens model.

A further extension along this line of reasoning then introduces multidimensional dynamics involving a mechanism when chaotic layers get connected into a web in the multidimensional phase space. Particle loss then occurs by negotiating through channeling from layers to layers in an apparent diffusion behavior.[22] This effect is sometimes referred to as Arnold diffusion.

8.4.6 Incompressible fluid model

An interesting alternative view of the Chirikov criterion was suggested by Teng.[23] The idea starts with an analogy between particle motion in the phase space of a Hamiltonian system and the motion of a viscous incompressible fluid in the real physical space — after all, according to the Liouville theorem, phase space volume is incompressible. We mentioned this model earlier in Sec. 7.2.2 when we discussed the echo effect.

By writing down the Hamilton equations on one hand and the fluid equation

---

[22]V. Arnold, Soviet Math. 5, 581 (1964); D. Neuffer, A. Riddiford and A.G. Ruggiero, IEEE Trans. Nucl. Sci. 28, 3, 2494 (1981).
[23]L.C. Teng, IEEE Tran. Nucl. Sci. NS-20, 843 (1973).

on the other, it is possible to establish the analogy as was given in Table 7.1. The Chirikov criterion is then equivalent to the Reynolds condition in fluid dynamics that the viscosity must be large enough to prevent turbulence from occurring. This offers an alternative view as well as derivation of the Chirikov criterion.

8.4.7 Island trapping model

We mentioned a weakness of the beam-beam single-resonance model when we discussed the beam loss mechanism and Fig. 8.28, i.e., as much as it can distort the beam core distribution, a single beam-beam resonance lacks a particle loss mechanism, in apparent conflict with observations. We remedied it by considering multiple overlapping resonances with the Chirikov model, and then found necessary to supplement it by an Arnold diffusion mechanism of chaotic layers in multidimensional phase space. If this seems contrived, in this section, we return to the single-resonance model but modify it into another model, called the island trapping model,[24] with an aim to focus on an alternative particle loss mechanism.

The single-resonance model described so far assumes all parameters such as the tune ν and the beam-beam strength ξ stay constant in time. In the island trapping model, let us consider the tune being modulated more or less sinusoidally in time with a certain slow frequency and, while doing so, periodically maneuvers around a resonance $\frac{m}{n}$. Unlike the static single-resonances, this provides a mechanism which continuously brings particles from small to large amplitudes. If particles are moved to amplitudes large enough, a physical aperture limitation on the amplitude then enhances the particle loss mechanism and potentially explains the observed lifetime limitation in colliding beams.

In the static model, a particle moves along a constant Hamiltonian contour and some of the contours form islands in the phase space. The distance of the islands from the phase space origin depends on the distance of the tune from the resonance, $\nu - \frac{m}{n}$. In particular, we mentioned when we derived Eq. (8.76) that $\alpha_0 = \frac{\beta_y^* J_y}{2\sigma_y^2}$, the normalized amplitude of the island centers from the origin, increases and moves away from the origin as ν_y moves away from $\frac{m}{n}$, as illustrated in Fig. 8.29 from the black islands to red and to blue ones.

When the tune slowly modulates near a resonance, therefore, a set of islands then breathes in and out in phase space. Phase space area elements, together with particles trapped in them, are distorted and transported by this island motion. Not only the island position and area, but also the depth of the island potential well also modulate together with the tune modulation.

This tune modulation provides a mechanism for transporting particles from the beam core to the beam tail. Islands at lower amplitudes tend to have a deeper potential well, and more likely to trap particles as they move outward in phase space, while at larger amplitudes, they become shallow and more likely

---

[24]A.W. Chao and M. Month, Nucl. Instr. Meth. 121, 129 (1974); M. Month, IEEE Trans. Nucl. Sci. NS-22, 1376 (1975); A. Chao, AIP Conf. Proc. 693, p. 291 (2003).

Figure 8.29: Trapping model for a 6th order resonance depicted in the polar $(\sqrt{\alpha_y}, \psi_y)$ space. Three sets of islands are shown corresponding to $\nu_y - \frac{m}{6} = -0.007$ (black), -0.013 (dashed red) and -0.021 (dotted blue). The beam-beam strength parameter is $\xi_y = 0.1$. If the tune ν_y ripples, the islands breathe in and out in the phase space, providing a mechanism to transport particles from the beam core to the beam tail.

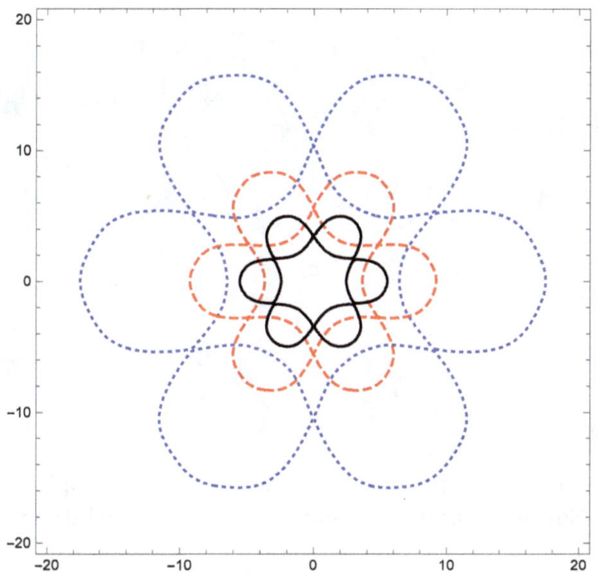

to become leaky and drop off particles. How many particles are trapped and transported depends on the resonance strength and the speed and adiabaticity of island motion. The process is sketched in Fig. 8.30.

One possible source of tune modulation comes from the synchrotron oscillation of the particle's energy coupled with a chromatic dependence of the betatron tune. In this case, the tune is modulated by the synchrotron frequency. Another possible source of tune modulation could be a result of power supply ripples.

Phase space cleaning   As an extension of the application of resonance island trapping, it is conceivable that an appropriate choice of a resonance driven by an intentional multipole magnet can be used to clean the beam distribution, remove the beam halos, or perform beam shaping by an optimized program of tune modulation and resonance strength. This possibility is then a special case of a phase space displacement manipulation technique.

Homework 8.36 There are three sets of islands in Fig. 8.29 corresponding to $\nu_y - \frac{1}{6} = -0.007, -0.013, -0.021$. Here it is a good opportunity to check the accuracy of Eqs. (8.76) and (8.81).

Figure 8.30: A sketch of an island trapping particle loss mechanism.

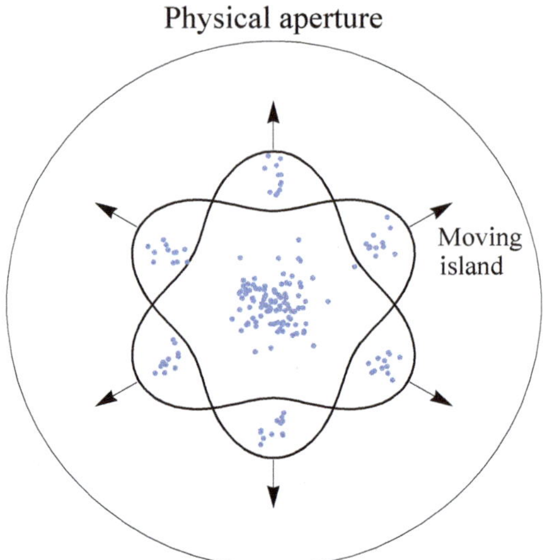

(a) Show that the locations of these three sets of islands agree with what is expected by Eq. (8.76) and the resonance amplitude widths reasonably agree with Eq. (8.81).

(b) As seen in Fig. 8.29, as the islands move outward the phase space, their area increases. At the same time, however, their potential-well depth shallows. Estimate their depths (not shown in Fig. 8.29) using Eq. (8.80).

8.4.8   Diffusion model

We now consider another variation of beam-beam model. This time the parameters are time-independent, but accompanying the beam-beam perturbation we now impose a diffusion effect on particle motion. Although as emphasized before, a beam-beam induced single resonance does not directly cause particle loss, it does enhance the diffusion process and cause particle losses indirectly.[25] We briefly discuss this subject in this section. The diffusion mechanism can be unrelated to the beam-beam interaction, or it can be beam-beam induced. In either case, the presence of beam-beam single-resonances enhances their diffusion effects.

Only a small fraction of the beam-beam kick can be noisy   Let us imagine a certain beam-beam induced diffusion. Let us assume that the beam-beam kick contains a random part in the sense that, for some reason, this part

---

[25]H.G. Hereward, CERN/ISR-DI/72-26 (1972); J. Le Duff, CERN/ISR-AS/74-53 (1974); D. Neuffer and A. Ruggiero, Proc. Beam-beam Int. Seminars, SLAC-PUB-2624, p. 332 (1980).

of the beam-beam kick assumes a stochastic nature from one kick to the next. Clearly the linearized thin-lens beam-beam kicks are totally correlated from one collision to the next and are not a candidate here. The same applies to kicks after a smooth approximation in the presence of one and only one resonance. In those cases, the beam-beam kick contains zero noise.

But there could still be some noise, for example in the Chirikov limit. As to noises unrelated to beam-beam, they can originate from quantum excitation or intrabeam scattering, and can contribute even in the case of a smooth approximation.

Let us now assume that the noise is beam-beam induced and that a fraction f of the beam-beam kick is stochastic. Let us consider the weak-strong case of a flat beam to be specific. Consider a typical particle with $y \sim \sigma_y$ getting a beam-beam kick. The kicking angle is

$$\Delta y' \sim -4\pi\xi_y \frac{y}{\beta_y^*} \sim -4\pi\xi_y \frac{\sigma_y}{\beta_y^*},$$

or

$$\frac{\Delta y'}{\sigma_{y'}} \sim -4\pi\xi_y,$$

where $\sigma_{y'}$ is the rms angular divergence of the beam at the interaction point.

It should first be noted that this is a very large kicking angle. If $\xi_y = 0.1$, for example, the typical beam-beam kicking angle is larger than the rms beam divergence. If these kicks are uncorrelated from turn to turn, there is not the least of hope to store such a beam.[26]

Let us consider a phenomenological model for an electron-positron collider.[27] Let us assume this beam-beam kick contains a small fraction f that is truly stochastic,

$$\frac{\Delta y'_{\text{noise}}}{\sigma_{y'}} \sim -4\pi\xi_y f.$$

The beam-beam random kicks contribute to a diffusion in the beam size just like the random contribution from the synchrotron radiation noise. Their contributions are counteracted by radiation damping and in balance this gives an equilibrium rms beam size,

$$\sigma_y^2 = \sigma_0^2 \left[1 + \frac{\tau_y N_{\text{IP}}}{T_0} (4\pi\xi_y f)^2 \right], \tag{8.85}$$

where τ_y is the radiation damping time, T_0 is the revolution period, N_{IP} is the number of interaction points in the collider (i.e. it is the number of beam-beam collisions per revolution), and σ_0 is the nominal beam size in the absence of the beam-beam collisions. We take ξ_y to be the value of the nominal beam-beam strength parameter because we are considering a weak-strong case.

---

[26]The same numerical exercise also says how surprisingly well the collider physicists have done their job when they have managed to handle such hefty a beam-beam perturbation!

[27]A.W. Chao, Accel. Summer School, Brookhaven, AIP Conf. Proc. 127, p. 201 (1983).

There are two situations when a beam-beam limit is reached. In one case, it is reached when there is a significant drop in luminosity. Another case occurs when the beam suffers from a loss of lifetime imposed by an aperture limit. For a very rough estimate, we might take the limit to be set by the conditions

$$\begin{cases} \frac{\tau_y N_{\mathrm{IP}}}{T_0}(4\pi\xi_y f)^2 = 1, & \text{luminosity limited,} \\ \sigma_0^2 \frac{\tau_y N_{\mathrm{IP}}}{T_0}(4\pi\xi_y f)^2 = \left(\frac{A}{10}\right)^2, & \text{aperture quantum lifetime limited,} \end{cases}$$

where A is the physical aperture for the y-motion.

In the luminosity limited case, for example, the beam-beam limit occurs when

$$f\xi_{\mathrm{limit}} = \frac{1}{4\pi}\sqrt{\frac{T_0}{\tau_y N_{\mathrm{IP}}}}.$$

If the limit occurs at $\xi_{\mathrm{limit}} = 0.1$, and if $\frac{\tau_y N_{\mathrm{IP}}}{T_0} = 2000$, we find the percentage of the beam-beam kick that can be considered stochastic to be $f \sim 2\%$.

So we conclude that if the beam-beam limit is caused by some stochastic diffusion process, roughly, a small but perhaps significant portion of the beam-beam kick should be stochastic to explain the observed beam-beam limit.

Beam-beam limit scaling, two distinctly opposite models However, it must be kept in mind that the predictions based on this phenomenological diffusion model can be very different from the predictions based on the nonlinear resonance model.

In the nonlinear resonance model, the instability mechanism is the myriad of resonances. In the Chirikov model, for example, the particle amplitude grows as a result of the build-up from turn to turn due to overlapping resonances. The instability threshold is sensitively dependent on the betatron tune, but is basically independent of the radiation damping time. If we try to fit the nonlinear resonance model into the present diffusion model, the quantity f will have to be rather complicated.

In a simplistic version of the diffusion model treating f as a universal constant, in contract, the instability comes from stochastic diffusion between collisions. There is no dependence on the tune. It also predicts a dependence on the radiation damping time τ_y as well as the number of interaction points N_{IP}.

So we see that the physical bases of these two models are two extreme opposites, and the parameters scaling of these two models has to be distinctly different. In the nonlinear resonance model, beam-beam kicks are assumed to be 100% correlated from one collision to next. After all, the basic physical mechanism in this model is the resonances, and keeping track of turn-by-turn correlation is how resonances work.

To the contrary is the diffusion model. The stochastic portion of the beam-beam kicks are assumed to be 0% correlated between collisions. No resonances play any role in this model, and yet the synchrotron radiation damping plays a pivotal role, in sharp contrast to the resonance-based models. We will pick up the discussion of the role played by synchrotron radiation damping later on page 642.[28]

More specifically, assuming $f = $ constant, and taking into account $\tau_y \propto \gamma^{-3}$ and $\sigma_0 \propto \gamma\sqrt{\beta_y^*}$. We conclude that, for a given electron storage ring with fixed T_0, the luminosity-limited beam-beam limit, in the diffusion model, has the scaling

$$\xi_{\text{limit}} \propto \frac{1}{\sqrt{\tau_y N_{\text{IP}}}} \propto \gamma^{3/2} N_{\text{IP}}^{-1/2}. \tag{8.86}$$

Since $\xi_y \propto \frac{N\beta_y^*}{\gamma\sigma_{y0}(\sigma_{x0}+\sigma_{y0})}$, we then have, for a fixed β_y^*,

$$N_{\text{limit}} \propto \gamma^{9/2} N_{\text{IP}}^{-1/2}. \tag{8.87}$$

Since the luminosity per interaction point $\mathcal{L} \propto \frac{N^2}{\sigma_x \sigma_y}$, it follows that

$$\mathcal{L}_{\text{limit}} \propto \gamma^7 N_{\text{IP}}^{-1}. \tag{8.88}$$

Attention is to be drawn particularly to the very steep γ^7 dependence of luminosity on beam energy.

For the case of aperture-limited beam-beam interaction, we find a somewhat different scaling,

$$\begin{aligned}
\xi_{\text{limit}} &\propto \gamma^{1/2} N_{\text{IP}}^{-1/2}, \\
N_{\text{limit}} &\propto \gamma^{7/2} N_{\text{IP}}^{-1/2}, \\
\mathcal{L}_{\text{limit}} &\propto \gamma^5 N_{\text{IP}}^{-1}.
\end{aligned} \tag{8.89}$$

In addition to the tell-tale independence on the betatron tune, two scaling properties deserve some attention. The first is the scaling with the number of interaction points. Its presence is straightforward and can be anticipated by the stochastic mechanism. The other is the sensitive scaling with beam energy γ. Note that we have assumed a weak-strong picture here. A slight complication (Homework 8.37) can be added to deal with a strong-strong picture, yielding slightly different scaling results.

Attempt to find f from resonances　So far our phenomenological model has not addressed the issue where and how the beam-beam noise comes from. If we take the simplistic diffusion model seriously, one might ask if there can be an analysis to estimate the diffusion coefficient f somehow. Here we suggest[29] one such an attempt based on the smooth approximation in the presence of overlapping resonances. In the following section 8.4.10, we will give another more concrete model for the beam-beam noise based on binary Coulomb scatterings.

---

[28]Admittedly, it is unsatisfactory that after decades of in-depth study of the beam-beam problem, we have not yet resolved its very basic physical mechanism, and the applicability of two extreme opposite mechanisms with distinctly different scaling laws are still being resolved.

[29]To be taken with a sizable grain of salt.

Based on single resonances, the beam-beam kick comes from Eq. (8.77),

$$\frac{dJ_y}{d\theta} \;=\; -\frac{\partial K_1}{\partial \psi_y},$$

where ($n =$ even)

$$K_1 \;=\; -\frac{2\sigma_y^2 \xi_y}{\beta_y^*}\left[1 + g_0(\alpha_y) + g_{1n}(\alpha_y)\cos n\psi\right].$$

The change in J_y per turn is

$$\Delta J_y \;=\; -2\pi \frac{2\sigma_y^2 \xi_y}{\beta_y^*} g_{1n}(\alpha_y) n \sin\psi_y \,.$$

Translating into kicking angle, it gives

$$\Delta y' \;=\; \frac{\Delta J_y}{2\sigma_y} \;=\; -2\pi\sigma_{y'}\xi_y g_{1n}(\alpha_y) n \sin\psi_y \,.$$

To accommodate the diffusion model, we need to associate this kick to an angular noise. If we take a view along the Chirikov diffusion, we might consider this kick $\Delta y'$ in its entirety to be the noise while taking $n \gg 1$ for the possible overlap of high order resonances. The resonance order n now represents the higher order of the resonance when Chirikov overlap becomes unavoidable.

In the previous consideration, the noise was based on the typical kicking angle of $-4\pi\xi_y\sigma_{y'}$. In the present consideration, we need to make the replacement in Eq. (8.85),

$$(4\pi\xi_y f)^2 \;\rightarrow\; (4\pi\xi_y)^2 \,\frac{n^2}{8}\, g_{1n}^2(\alpha_y)\,.$$

If we further take $\alpha_y \sim 1$ for the typical particle to be in a diffusion region, we have

$$g_{1n} \;\approx\; \frac{8(-1)^{\frac{n}{2}}}{n^2 - 1}\sqrt{\frac{1}{2\pi}}\,.$$

The replacement is then approximately

$$(4\pi\xi_y f)^2 \;\approx\; (4\pi\xi_y)^2 \frac{4\,n^2}{\pi(n^2-1)^2} \;\approx\; (4\pi\xi_y)^2 \frac{4}{\pi n^2}\,.$$

or

$$f \;\approx\; \frac{2}{\sqrt{\pi}\,n}\,.$$

In case we take seriously our earlier estimate of $f = 2\%$, we conclude that the stochastic resonances are ~ 56-th order. [Consult Eq. (8.47).]

Homework 8.37 In Eq. (8.85), we treated the diffusion model for the simpler case of weak-strong picture. To consider a strong-strong case, the parameter ξ_y is then determined not by σ_0 but by σ_y and a self-consistency condition needs to be imposed upon Eq. (8.85). Try to extend the analysis to a strong-strong case and derive the modified scaling laws of ξ_{limit}, N_{limit}, $\mathcal{L}_{\text{limit}}$ with respect to γ accordingly.

8.4.9 Quantum lifetime reduction

In electron storage rings, the diffusion caused by synchrotron radiation quantum noise and the radiation damping provided at the acceleration cavities counteract each other. In the absence of any physical aperture limit in particle's amplitudes, an equilibrium beam distribution is reached when the two effects are in balance. When an aperture is present, however, the beam suffers a steady loss of particles. The beam lifetime is then determined by the particle diffusion rate across the physical aperture.

Clearly the distortion of phase space by the presence of a single resonance will also distort the equilibrium distribution. The formation of phase space islands pushes the particle distribution outwards. As a result, the effective stable aperture is reduced from the physical aperture by an amount in such a way that the phase space area enclosed within the stable region must exclude the areas occupied by the islands,

$$\text{(stable phase space area)} \approx \text{(phase space area provided by physical aperture)}$$
$$- \sum \text{(phase space area of all islands)}. \qquad (8.90)$$

The beam lifetime is then shortened accordingly. An analysis by solving the Fokker–Planck equation and applying the smooth approximation to first order of the perturbation yields the reduced quantum lifetime.[30]

We can take Eq. (8.90) for a phenomenological estimate as follows. Consider a case when there is a vertical aperture limiting the beam's quantum lifetime. Let the aperture limit be given by

$$\alpha_y < \hat{\alpha} .$$

The corresponding physical limit in y is $|y| < 2\sigma_y \sqrt{\hat{\alpha}}$. The phase space area provided by this physical aperture and available for the beam is

$$\Phi_0 = \pi \hat{\alpha} .$$

The quantum lifetime in the absence of beam-beam (or other) nonlinear perturbations has been calculated before in Eq. (1.113) of Chapter 1. In the present notation with aperture defined by $\hat{\alpha}$, we rewrite it as

$$\tau_{\text{quantum}} = \tau_{\text{rad}} \frac{e^{2\hat{\alpha}}}{4\hat{\alpha}} ,$$

where τ_{quantum} means the quantum lifetime and τ_{rad} means the vertical radiation damping time. In terms of the available phase space area, we write

$$\tau_{\text{quantum}} = \tau_{\text{rad}} \frac{\pi e^{2\Phi_0/\pi}}{4\Phi_0} . \qquad (8.91)$$

Validity range of this expression requires $\Phi_0 \gg 1$.

---

[30]A.W. Chao, Phys. Rev. Special Topics — Accel. & Beams, 6, 094001 (2003).

On the other hand, the islands occupy a total area of $n\Phi_{\text{island}}^{(n)}$, where $\Phi_{\text{island}}^{(n)}$ is given by Eq. (8.82). This island area needs to be excluded from that available to the stable beam according to Eq. (8.90). To take into account of this loss of phase space, we rewrite the quantum lifetime as

$$\tau_{\text{quantum}} = \tau_{\text{rad}} \frac{\pi e^{2\Phi/\pi}}{4\Phi},$$

where now $\Phi = \Phi_0 - n\Phi_{\text{island}}^{(n)}$.

For small island areas, this quantum lifetime is shortened by a factor approximately

$$\exp\left(\frac{2n}{\pi}\,\Phi_{\text{island}}^{(n)}\right). \tag{8.92}$$

The exponent quantity in Eq. (8.92) is basically the total island phase space area measured in terms of the strong beam's rms beam emittance. A similar expression of reduction factor can be applied to other sources of nonlinearities in the storage ring.

For proton rings, candidates for the diffusion effects include intrabeam scattering, residual gas scattering, power supply noise, RF noise, etc. These diffusion effects will also be enhanced by the presence of single resonances, and in particular the beam-beam resonances.

Homework 8.38 It should be instructive to look into the expected loss of quantum lifetime due to the beam-beam resonances by plotting the quantum lifetime reduction factor (8.92) as a function of α_0, the expected island position in phase space, for a number of low order resonances up to $n = 6$. Consult Fig. 8.22.

8.4.10   Coulomb diffusion model

So far in all our considerations, the beam-beam perturbation is facilitated by an on-coming beam that generates an electromagnetic potential. The beam-beam kicks are described by Eq. (8.11) with an electromagnetic potential $U(x, y)$. This electromagnetic potential was derived under the assumption that the beam has a continuum distribution in space, in particular, a bi-Gaussian distribution.

The actual beam, however, consists of a collection of discrete point charges, not a continuum distribution, as illustrated in Fig. 8.31. Equation (8.11) faithfully gives the beam-beam kick averaged over the collective fields of the on-coming beam, but it fails to consider accidental short-range collision events between individual discrete particles. These short-range discrete collisions are expected to contribute to noise effects to the beam-beam interaction in addition to the smooth kicks of (8.11). In this section, we take a brief look of possible effects of these short-range collisions (see footnote 27).

The system will still be based on the smooth kicks (8.11), but on top of this continuum model, we shall add binary collisions when specifically two and only two particles interact by the Coulomb fields. This binary picture supplements

Figure 8.31: Beam-beam continuum model versus a discrete model. The continuum model has been assumed for the description of beam-beam interaction by nonlinear lens; the discrete model is assumed for beam-beam binary collisions.

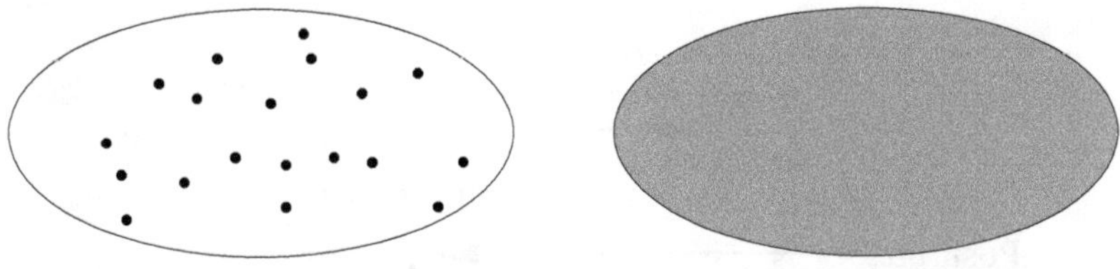

The discrete beam   The continuum model

the continuum model, and takes into consideration of the shortest-range fields. Medium-range interactions taking into account of discrete multiple particles in one collision event are still not included.

Impact parameter   Consider two particles, say one positron and one electron, engaging in a binary collision event as illustrated in Fig. 8.32. The positron carries with it a pancake field,

$$\vec{E} \;=\; \frac{e}{2\pi\epsilon_0 r}\,\delta(s-ct)\hat{r}, \qquad \vec{B} \;=\; \frac{e}{2\pi\epsilon_0 c r}\,\delta(s-ct)\hat{\theta}\,.$$

The electron moves in the $-\hat{s}$ direction with speed c; its position at time t is $s = -ct$. The positron delivers a force to the electron,

$$\begin{aligned}
\vec{f} \;&=\; (-e)(\vec{E}-c\hat{s}\times\vec{B}) \\
&=\; -\frac{e^2}{\pi\epsilon_0 r}\hat{r}\delta(s-ct)\Big|_{s=-ct} \;=\; -\frac{e^2}{\pi\epsilon_0 r}\hat{r}\delta(-2ct)\,.
\end{aligned}$$

The impulse received by the electron as it passes by the positron in this binary encounter is

$$\frac{\Delta\vec{P}}{P} \;=\; \frac{1}{mc\gamma}\int_{-\infty}^{\infty} dt\,\vec{f} \;=\; -\frac{2r_0}{\gamma r}\hat{r}\,,$$

with r_0 the classical radius of the electron or positron, $\vec{r}$ is the transverse displacement from the positron to the electron, and r is the impact parameter designated by b in Fig. 8.32.

Beam lifetime due to direct Coulomb scattering   If the two particles collide with a very small impact parameter, the collision will be violent. If b is small enough, the scattering angle of both particles will exceed the tolerance of the storage ring acceptance and both particles will be lost. The minimum impact parameter for the particle to stay in the respective beams is given by

$$\frac{2r_0}{\gamma b_{\min}} \;=\; \hat{\Theta}, \qquad \text{and} \qquad b_{\min} \;=\; \frac{2r_0}{\gamma\hat{\Theta}}\,,$$

Figure 8.32: A binary collision between a positron and an electron with impact parameter b. Both particles are assumed to be relativistic. The pancake electric (red) and magnetic (blue) fields carried by the positron are indicated.

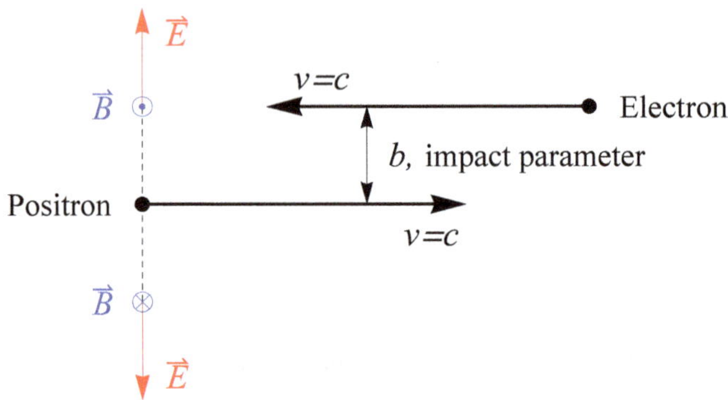

where $\hat{\Theta}$ is the angular aperture of the storage ring acceptance. If for example, the ring has a physical aperture $\hat{X}$ or $\hat{Y}$ at a position in the ring with β-functions β_x, β_y, then

$$\hat{\Theta}_x \;=\; \frac{\hat{X}}{\sqrt{\beta_x \beta_x^*}}, \qquad \text{or} \qquad \hat{\Theta}_y \;=\; \frac{\hat{Y}}{\sqrt{\beta_y \beta_y^*}},$$

where $\beta_{x,y}^*$ are the β-functions at the collision point.

Let us simplify the consideration to assume the angular aperture is $\hat{\Theta}$ regardless of in the x or the y direction. The cross section of violent Coulomb binary collision events causing direct particle loss is

$$\Sigma \;=\; \pi b_{\min}^2 \;=\; \pi \left(\frac{2r_0}{\gamma \hat{\Theta}} \right)^2.$$

The rate of particle loss in each beam is then given by

$$\dot{N} \;=\; -\mathcal{L}\Sigma,$$

where $\mathcal{L}$ is the luminosity of the collider.

Let us take a collider with $\mathcal{L} = 10^{32}$ cm^{-2}s^{-1}, $\gamma = 10^4$ (5 GeV beam energy), and $\Theta = 5$ mrad, then we find $b_{\min} = 1.1 \times 10^{-16}$ m and $\dot{N} = -4 \times 10^4$ s^{-1}. In case there are $N = 10^{12}$ electrons in the beam, then the Coulomb lifetime is $\tau = -\frac{N}{\dot{N}} \approx 7000$ hours.

Coulomb diffusion   Particles undergoing violent Coulomb scatterings are immediately removed from the two beams. After their removal, the remaining beams still experience a diffusion effect causing emittances to grow.

As mentioned, we model the beam-beam interaction now as a superposition of a continuum beam-beam kick supplemented by an additional contribution of binary collisions. The continuum contribution is given by Eq. (8.11). We now consider the additional binary contribution.

The on-coming positron beam has a transverse distribution

$$n(x, y) = \frac{1}{N} \sum_{i=1}^{N} \delta(x - x_i)\delta(y - y_i) \,.$$

The kick received by an electron traversing this positron beam with transverse displacements (x, y) is

$$\frac{\Delta \vec{P}}{P}(x, y) = -\frac{2r_0}{\gamma} \sum_{i=1}^{N} \frac{(x - x_i)\hat{x} + (y - y_i)\hat{y}}{(x - x_i)^2 + (y - y_i)^2} \,,$$

or

$$\Delta x' = -\frac{2r_0}{\gamma} \sum_{i=1}^{N} \frac{x - x_i}{(x-x_i)^2 + (y-y_i)^2}, \qquad \Delta y' = -\frac{2r_0}{\gamma} \sum_{i=1}^{N} \frac{y - y_i}{(x-x_i)^2 + (y-y_i)^2},$$

The diffusion part of the angular kicks are

$$\langle \Delta x'^2 \rangle = \left(\frac{2r_0}{\gamma}\right)^2 \sum_{i=1}^{N} \frac{(x - x_i)^2}{[(x-x_i)^2 + (y-y_i)^2]^2} \,,$$

$$\langle \Delta y'^2 \rangle = \left(\frac{2r_0}{\gamma}\right)^2 \sum_{i=1}^{N} \frac{(y - y_i)^2}{[(x-x_i)^2 + (y-y_i)^2]^2} \,.$$

Given the assumed continuum distribution $n(x_i, y_i)$, for example an upright bi-Gaussian distribution,

$$n(x_i, y_i) = \frac{1}{2\pi\sigma_x\sigma_y} e^{-\frac{x_i^2}{2\sigma_x^2} - \frac{y_i^2}{2\sigma_y^2}} \,,$$

we have

$$\langle \Delta x'^2 \rangle(x, y) = N \left(\frac{2r_0}{\gamma}\right)^2 \int_{-\infty}^{\infty} dx_i \int_{-\infty}^{\infty} dy_i \, n(x_i, y_i) \frac{(x - x_i)^2}{[(x-x_i)^2 + (y-y_i)^2]^2} \,,$$

$$\langle \Delta y'^2 \rangle(x, y) = N \left(\frac{2r_0}{\gamma}\right)^2 \int_{-\infty}^{\infty} dx_i \int_{-\infty}^{\infty} dy_i \, n(x_i, y_i) \frac{(y - y_i)^2}{[(x-x_i)^2 + (y-y_i)^2]^2} \,.$$

Here we have made a physical assumption that the positron beam returns to the collision point each time with the same bi-Gaussian average distribution but the exact location of individual positrons have completely rearranged stochastically. This must be a good approximation because we are considering binary collisions, so the particle collides with a different set of nearby particles at each passage.

These diffusive kicks are to be superimposed onto the continuum kicks of Eq. (8.11) and they are functions of (x, y), the transverse position of the electron under consideration. The regions

$$\frac{|x - x_i|}{(x - x_i)^2 + (y - y_i)^2} > \frac{1}{b_{x,\min}}, \quad \text{or} \quad \frac{|y - y_i|}{(x - x_i)^2 + (y - y_i)^2} > \frac{1}{b_{y,\min}}$$

are to be excluded from the integration because once that happens, the particles involved are no longer eligible to stay in the collider.

For an estimate of this Coulomb diffusion, let us consider the case of a round bi-Gaussian beam and let us consider an electron at the beam center with $x = y = 0$. We then have

$$
\begin{aligned}
\langle \Delta x'^2 + \Delta y'^2 \rangle(0,0) &= \frac{4Nr_0^2}{\gamma^2} \int_{-\infty}^{\infty} dx_i \int_{-\infty}^{\infty} dy_i \frac{1}{2\pi\sigma^2} \frac{e^{-\frac{x_i^2 + y_i^2}{2\sigma^2}}}{x_i^2 + y_i^2} \\
&= \frac{4Nr_0^2}{\gamma^2\sigma^2} \int_{b_{\min}}^{\infty} \frac{dr}{r} e^{-\frac{r^2}{2\sigma^2}} \approx \frac{4Nr_0^2}{\gamma^2\sigma^2} \ln\frac{\sigma}{b_{\min}}.
\end{aligned}
$$

We have approximated the violent collision events to be defined by $x_i^2 + y_i^2 < b_{\min}^2$ and they are excluded in the integration, and we have assumed $b_{\min} \ll \sigma$.

The Coulomb logarithm $\ln\frac{\sigma}{b_{\min}}$ is an expected factor due to the divergent Coulomb force when the beam size σ cuts off the force at long distance, the closest impact parameter $b_{\min}$ cuts off at short distance, and the Coulomb log takes the logarithm of their dynamic range.

If we take this beam-beam angular diffusion as statistical and assume it to add statistically in each beam-beam encounter, then it contributes to a growth rate of the transverse beam emittance according to

$$\frac{d\epsilon_x}{dt} = \frac{\beta_x^*}{T_0} \frac{2Nr_0^2}{\gamma^2\sigma^2} \ln\frac{\sigma}{b_{\min}}, \qquad \frac{d\epsilon_y}{dt} = \frac{\beta_y^*}{T_0} \frac{2Nr_0^2}{\gamma^2\sigma^2} \ln\frac{\sigma}{b_{\min}}, \qquad (8.93)$$

where T_0 is the time between collisions. In case of one collision per turn, then T_0 is the revolution period.

The beam-beam induced Coulomb emittance growth, Eq. (8.93), can then be applied to electron-positron colliders and proton-proton colliders as follows.

- For an electron-positron collider, this emittance growth (8.93) is to be balanced against radiation damping, leading to growths of the equilibrium emittances of

$$\Delta\epsilon_x = \frac{\beta_x^*\tau_x}{T_0} \frac{Nr_0^2}{\gamma^2\sigma^2} \ln\frac{\sigma}{b_{\min}}, \qquad \Delta\epsilon_y = \frac{\beta_y^*\tau_y}{T_0} \frac{Nr_0^2}{\gamma^2\sigma^2} \ln\frac{\sigma}{b_{\min}},$$

where $\tau_{x,y}$ are the radiation damping times.

Take for example, $N = 10^{12}, \sigma = 0.3$ mm, $\gamma = 10^4$ and $b_{\min} = 1.1 \times 10^{-16}$ m as in the previous numerical example, then the Coulomb logarithmic factor is $\ln\frac{\sigma}{b_{\min}} = 28$, and $\sqrt{\langle \Delta x'^2 \rangle + \langle \Delta y'^2 \rangle}(0,0) = 1 \times 10^{-8}$ rad.

If we further take $\frac{\tau_{x,y}}{T_0} = 2000, \beta^*_{x,y} = 1$ m, it follows that the emittance growth due to beam-beam binary Coulomb scattering is $\Delta\epsilon_{x,y} = 0.5 \times 10^{-13}$ m, which is a small contribution.

- For a proton-proton or proton-antiproton collider, this contribution of Coulomb diffusion (8.93) does not have radiation damping to balance it. It can contribute to a beam-beam induced emittance growth and the subsequent degradation of luminosity. Its effect needs to be evaluated.

Higher moments, beam tail The emittance growth (8.93), together with its Coulomb logarithm, deals with the beam core. There remains the possibility that the $\frac{1}{r}$-singular Coulomb force contributes more to a beam tail, to be represented by the higher moments. We extend the above calculation to give an estimate again for a round positron beam and consider the electron at $x = y = 0$. For the $2m$-th moment $(m \geq 2)$, we have

$$\langle(\Delta x'^2 + \Delta y'^2)^m\rangle(0,0) = \left(\frac{2r_0}{\gamma}\right)^{2m}\frac{N}{\sigma^2}\int_{b_{\min}}^{\infty}\frac{rdr}{r^{2m}}e^{-\frac{r^2}{2\sigma^2}}$$

$$\approx \left(\frac{2r_0}{\gamma}\right)^{2m}\frac{N}{\sigma^2}\frac{1}{2(m-1)b_{\min}^{2m-2}} = \frac{2Nr_0^2}{\gamma^2\sigma^2(m-1)}\hat{\Theta}^{2m-2},$$

where $\hat{\Theta}$ is the angular acceptance of the storage ring.

If we let $\Delta x'^2 = \Delta y'^2$, then

$$\langle\Delta x'^{2m}\rangle = \frac{1}{2^m}\langle(\Delta x'^2 + \Delta y'^2)^m\rangle = \frac{Nr_0^2}{2\gamma^2\sigma^2(m-1)}\left(\frac{\hat{\Theta}}{2}\right)^{2m-2}, \qquad (m \geq 2).$$

If we then compare these high order contributions with the rms beam divergence $\sigma' = \frac{\sigma}{\beta^*}$ at the collision point, we obtain

$$\left\langle\left(\frac{\Delta x'}{\sigma'}\right)^{2m}\right\rangle = \frac{Nr_0^2}{2\gamma^2\sigma^2\sigma'^2(m-1)}\left(\frac{\hat{\Theta}}{2\sigma'}\right)^{2m-2}, \qquad (m \geq 2).$$

Again taking the numerical example as above, $\langle(\frac{\Delta x'}{\sigma'})^4\rangle = 3.4 \times 10^{-10}$ and $\langle(\frac{\Delta x'}{\sigma'})^6\rangle = 1.2 \times 10^{-8}$. The violent collision events have been excluded from contributing.

Similar to the case of the second moments, these contributions to higher moments make small contributions to the beam tail distribution for an electron-positron collider but may be more significant for a proton-proton collider.

8.4.11 Beam-beam limit by ξ_{limit}

The beam-beam strength parameters ξ_x and ξ_y, as explained in Eq. (8.60), have the meaning of beam-beam induced tune spreads and the small-amplitude tune shifts. In Eq. (8.84), it is also used to set the stochastic limit. These and other results we obtained so far indicate that ξ has the meaning of being the dimensionless scaling parameter of the beam-beam problem. It is the parameter that specifies the linear as well as the nonlinear beam-beam perturbation strengths. This observation leads to a hypothesis that the beam-beam limit is imposed through some universal limitation on the beam-beam strength parameter ξ — as Eq. (8.84) already implicitly assumed. In this section, we pursue some considerations along this hypothesis.

It should be noted however that this basic hypothesis that ξ plays the role of setting the beam-beam limit is immediately in contradiction with the diffusion models of Secs. 8.4.8 and 8.4.10. As we will see, this beam-beam limit model yields scaling behaviors different from those predicted by the diffusion models. As mentioned earlier, the main difference is of a fundamental nature, that this hypothesis based on ξ assumes the beam-beam limit comes purely from a 100%-correlated dynamics, while the diffusion models assume the limit comes from a 100%-uncorrelated stochastic noise effects.

Low-β^* insertion   The hypothesis of ξ playing the determining role has led to one important practical consequence, namely the invention of low-β^* insertions.[31] According to this scaling property, it would be beneficial to make ξ as small as possible and an inspection of Eq. (8.21) shows that a small β^* does the job.

In a low-β^* insertion, a few strong quadrupoles are inserted in the interaction region to pinch β^* to a small value. Today, low-β^* insertions are implemented in all colliding beam storage rings. As a result, luminosities have generally increased by one or two orders of magnitude. Figure 8.33 illustrates the difference between a low-β^* insertion and a normal cell structure.

The low-β^* insertion quadrupoles cannot be too close to the collision point since the detector solenoid and its compensation solenoids on its both sides occupy the space. This puts a limit on the smallest β^* achievable at the collision point. Low-β^* insertion, together with the sophisticated lattice insertion that compensates for its nonlinear optics, is a highly evolved skill in collider lattice design. It is considered outside the scope of the present discussion.[32]

Scaling behavior   It is incorrect to say that the benefit of low-β^* comes from the fact that the beam size at the collision point is pinched to a smaller value. Although the luminosity does benefit when beam size is made smaller, the idea of low-β^* actually, somewhat counter-intuitively, tends to ask for a large beam size. The benefit of small β^* comes from minimizing the beam's dynamical response to the beam-beam perturbation; it does not come from a geometric effect of a smaller beam size.

Put it in another way, one of the many difficulties of a low-β^* insertion design for a collider lies in the conflicting requirements of a small β^* but at the

---

[31]K.W. Robinson and G.-A. Voss, Cambridge Electron Accel. report CEAL-TM-149 91965); P.L. Morton and J.R. Rees, IEEE Trans. Nucl. Sci. NS-14, 630 (1967).

[32]See for example, E. Keil, Sec. 2.2.3; M. Zobov, F. Zimmermann, Sec. 4.13; J.P. Koutchouk, V. Shiltsev, Sec. 4.14, Handbook Accel. Phys. & Eng. 2nd edition, World Scientific (2013).

Figure 8.33: Difference of the optics between (a) a normal cell structure and (b) a low-β^* insertion. In each case, two typical particle trajectories are drawn. In (a), the effects of the focusing and the defocusing magnets tend to cancel each other. The net effect is focusing but the focal length is long and the displacement of a particle changes sign only after the particle passes through several magnets. In (b), a strong quadrupole "overfocuses" the particle trajectories so that all displacements flip sign near the collision point. Such an overfocused configuration is usually to be avoided in a normal cell structure.

(a) Normal cell structure

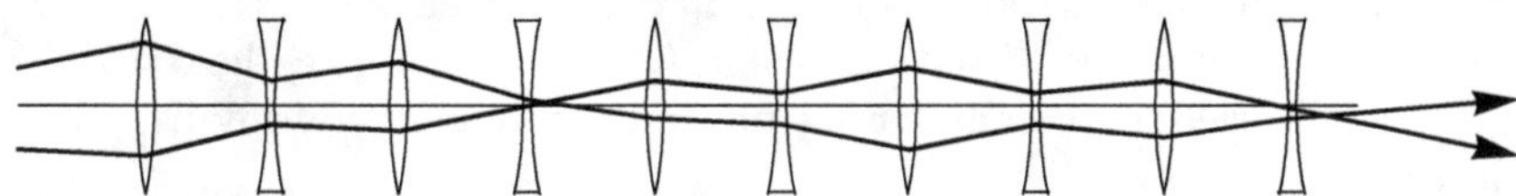

(b) Low$-\beta^*$ insertion

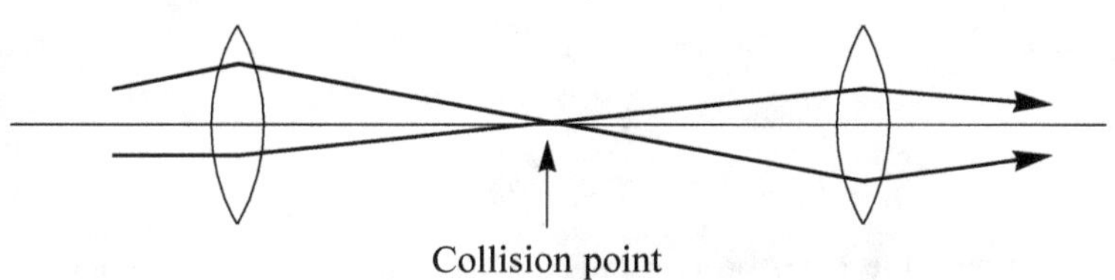

same time also a large beam size at the collision point. The beam size is to be controlled not by β^* but by the emittance ϵ. The reason is described as follows.

Let us consider a flat-beam configuration with $\sigma_x \gg \sigma_y$, and the beam-beam limit occurs in the y-motion of the particles when ξ_y reaches a certain universal value ξ_{limit} per the hypothesis we are making. The two key parameters of interest to us here are

$$\xi_y = \frac{N r_0 \beta_y^*}{2\pi \gamma \sigma_x \sigma_y}, \qquad \mathcal{L} = \frac{N^2 f}{4\pi \sigma_x \sigma_y}.$$

When the beam-beam limit is reached, the beam intensity is

$$N_{\text{limit}} = \frac{2\pi \xi_{\text{limit}}}{r_0} \frac{\gamma \sigma_x \sigma_y}{\beta_y^*}, \tag{8.94}$$

and the luminosity is

$$\mathcal{L}_{\text{limit}} = \frac{\pi \xi_{\text{limit}}^2 f}{r_0^2} \frac{\gamma^2 \sigma_x \sigma_y}{\beta_y^{*2}}, \tag{8.95}$$

where f is the collision rate.

It should be noted that in this model, since the limit ξ_{limit} is held fixed, the beam intensity is varied to match this limit without other limitations. Implicitly it has assumed that there is no other collective instability issues standing in the way.

To examine the scaling behavior under this beam-beam-limit model, we further note that with a given storage ring, the beam sizes $\sigma_{x,y}$ are proportional to γ. Let us express

$$\sigma_x = \sigma_{x0}\frac{\gamma}{\gamma_0}, \qquad \sigma_y = \sqrt{\beta_y^* \epsilon_{y0}}\,\frac{\gamma}{\gamma_0},$$

where $\sigma_{x0}, \epsilon_{y0}, \gamma_0$ are reference quantities not of critical importance to us here — they can assume the values at the top design energy of the collider for example, but we have explicitly introduced the $\sqrt{\beta_y^*}$ dependence of σ_y.

We now consider the situation when $f, \xi_{\text{limit}}, \sigma_{x0}, \gamma_0$ are held fixed while $\gamma, \epsilon_{y0}, \beta_y^*$ are varied in a manner as mutually independent knobs. We then obtain the scaling laws,

$$
\begin{aligned}
\xi_y &\sim \text{constant}, \\
\sigma_x &\sim \gamma, \\
\sigma_y &\sim \gamma \epsilon_{y0}^{1/2} \beta_y^{*\,1/2}, \\
N_{\text{limit}} &\sim \gamma^3 \epsilon_{y0}^{1/2} \beta_y^{*\,-1/2}, \\
\mathcal{L}_{\text{limit}} &\sim \gamma^4 \epsilon_{y0}^{1/2} \beta_y^{*\,-3/2}.
\end{aligned}
\tag{8.96}
$$

This scaling behavior is to be compared with the behavior predicted using other beam-beam-limit models, for example Eqs. (8.86–8.88) for the diffusion model with luminosity limitation and Eq. (8.89) for the diffusion model with beam lifetime limitation. In particular, if β_y^* and ϵ_{y0} are held fixed, the luminosity depends on the beam energy sensitively as

$$\mathcal{L}_{\text{limit}} \sim \gamma^4.$$

In comparison, the diffusion models have even deeper energy dependencies, mainly because diffusion models invoke the radiation damping and radiation damping time depends sensitively on beam energy with $\tau_y \sim \gamma^{-3}$.

To maximize luminosity for given beam energy, according to this scaling behavior (8.96), we want small β_y^* and large ϵ_{y0}, and a small β_y^* is particularly a sensitive parameter. On the other hand, both reducing β_y^* and increasing ϵ_{y0} have their limits. In the optimization process, a small β_y^* reduces the vertical beam size, while increasing ϵ_{y0} increases it.

Optimum ϵ_{y0} Once β_y^* is optimized, asking for a large ϵ_{y0} is equivalent to asking for as big a beam size as possible. The reason this is helpful, in spite of generating a larger beam size, is so that a higher beam intensity can be accommodated before reaching the beam-beam limit.

There are potentially several ways to increase ϵ_{y0}, for example, for the case of flat beams,

- weaker focusing in the collider cells, i.e. lower vertical tune;

- adding vertical wigglers in the collider storage ring;[33]
- mismatching dispersions in the storage ring lattice;[34]
- adding a vertical dispersion at the collision point;
- adding x-y betatron coupling.

Among these possibilities, the first three involve artificially enlarging the beam emittance. This means the beam size in the whole ring, not only at the collision point, is enlarged. Clearly, these ideas are restricted by the limitation when vertical aperture is reached. The fourth idea does not have this problem, as strictly speaking it affects the beam size without enlarging the emittance, but there is the issue that the beam-beam force excites the undesirable synchrobetatron resonances.[35]

The fifth idea involves an intentional x-y coupling that splits the horizontal emittance between x- and y-dimensions. Another limitation along this line then occurs when the beam-beam effect in the horizontal dimension begins to emerge.[36]

Optimum x-y coupling   In the above analysis, we have assumed a flat beam configuration. On the other hand, regardless of how β_y^* and ϵ_{y0} are controlled, as these two quantities are optimized for luminosity paying attention solely to the vertical dimension, and as the required beam intensity increases to fulfill the increasing luminosity, at some point, the horizontal dimension begins to take a role.

When both the horizontal and vertical beam-beam limits play the role, it is conceived that the optimal choice is to impose the condition that the horizontal and the vertical limits are reached simultaneously, i.e.

$$\xi_x \;=\; \xi_y, \qquad \text{or} \qquad \frac{\beta_x^*}{\sigma_x} \;=\; \frac{\beta_y^*}{\sigma_y}. \tag{8.97}$$

Condition (8.97) is consistent with our assumption of a flat beam if $\beta_x^* \gg \beta_y^*$.

Although this condition is not necessarily achieved by adjusting the x-y coupling, it is conventionally referred to as the optimum coupling condition. The scaling laws under this condition is discussed in Homework 8.39.

Optimal β_y^* due to hour-glass   We have concluded by our scaling laws that a higher luminosity can be reached by a smaller β_y^*. One question concerns what determines the lowest value of β_y^* when the scaling laws cease to be applicable. One answer, among other answers coming from lattice design issues, lies in the hour-glass effect when the colliding beams have finite bunch lengths, as illustrated in Fig. 8.34.

---

[33] J.M. Paterson, J. Rees, and H. Wiedemann, SLAC note PEP-125 (1975).

[34] R.H. Helm, M.J. Lee, and J.M. Paterson, Proc. IXth Int. Conf. High Energy Accel., Stanford (1974).

[35] A. Piwinski, DESY report 77/18 (1977); S. Myers, Nucl. Instr. Meth. 211, 263 (1983).

[36] J. Rees and L. Rivkin, Proc. Summer School on High Energy Accel., SLAC (1982).

Figure 8.34: A sketch of an hour-glass collision of beam #1 and beam #2. Each beam bunch has length $2L$. The sketch shows an exaggerated (nonoptimized) case when $\beta_y^* = \frac{L}{2\sqrt{5}}$.

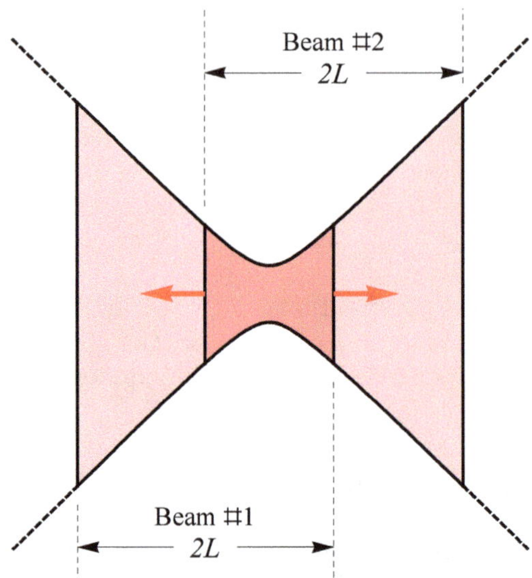

Let us consider a simplified case in which the bunches have uniform longitudinal distributions with a full length $2L$. We assume a flat beam with Fig. 8.34 showing only the vertical projection, i.e. we are making β_y^* as small as possible while leaving β_x^* unchanged. We will show that in this model, the optimum choice of β_y^* is approximately $\frac{1}{5}$ of the full bunch length $2L$. A lower value of β_y^* beyond this point begins to hurt the luminosity.

When we discussed the subject of luminosity, we introduced a luminosity reduction factor for a flat beam configuration, Eq. (8.5), due to the hour-glass effect. The result is

$$\mathcal{L} = R_{hg}\mathcal{L}_0, \qquad \mathcal{L}_0 = \frac{N^2 f}{4\pi\sigma_x^*\sigma_y^*},$$

$$R_{hg} = \frac{1}{\sqrt{\pi}\,h_y}\, e^{1/2h_y^2}\, K_0\left(\frac{1}{2h_y^2}\right),$$

where $h_y = \frac{\sigma_z}{\beta_y^*}$.

This luminosity reduction factor is based on evaluating the geometric overlapping of the two colliding bunches. But the hour-glass geometry affects not only the luminosity; it also affects the beam-beam strength ξ. So we now also need to evaluate ξ under the influence of the hour-glass geometry.

As we move away from the collision point by a distance s, the β-function increases hyperbolically according to

$$\beta_y(s) \;=\; \beta_y^* \left[1 + \left(\frac{s}{\beta_y^*} \right)^2 \right].$$

Since $\sigma_y \propto \sqrt{\beta_y}$, a modified beam-beam strength parameter, integrated over the collision range from $s = -L$ to $s = L$ for a uniform beam over a total bunch length of $2L$ reads

$$\xi_y \;=\; \xi_y^* \times \frac{1}{2L} \int_{-L}^{L} ds \, \sqrt{1 + \left(\frac{s}{\beta_y^*} \right)^2}.$$

It follows that the beam-beam strength parameter is enhanced due to the hour-glass effect,

$$\xi_y \;=\; \xi_y^* \times \eta_{hg}\,,$$

by an enhancement factor

$$\eta_{hg} \;=\; \frac{1}{2} \left[\sqrt{1 + u^2} + \frac{1}{u} \ln \left(u + \sqrt{1 + u^2} \right) \right], \qquad u \;=\; \frac{L}{\beta_y^*} \;=\; \frac{\sqrt{3}\,\sigma_z}{\beta_y^*}\,,$$

where we replace L by the rms value $\sqrt{3}\,\sigma_z$. The quantity ξ_y^* here is the beam-beam parameter without taking into account of the hour-glass effect, i.e.

$$\xi_y^* \;=\; \frac{N r_0 \beta_y^*}{2 \pi \gamma \sigma_x^* \sigma_y^*}\,,$$

with $\sigma_{x,y}^*$ the beam sizes at the collision point without hour-glass.

In the absence of hour-glass, the nominal beam-beam parameter and luminosity are given by Eqs. (8.94) and (8.95),

$$N_{\text{limit}}^* \;=\; \frac{2 \pi \xi_{\text{limit}}}{r_0} \frac{\gamma \sigma_x^* \sigma_y^*}{\beta_y^*}\,,$$

$$\mathcal{L}_{\text{limit}}^* \;=\; \frac{\pi \xi_{\text{limit}}^2 f}{r_0^2} \frac{\gamma^2 \sigma_x^* \sigma_y^*}{\beta_y^{*2}}\,.$$

With the hour-glass effect, we have a reduced beam intensity and a reduced luminosity,

$$N_{\text{limit}} \;=\; \frac{1}{\eta_{hg}} N_{\text{limit}}^*, \qquad \mathcal{L}_{\text{limit}} \;=\; \frac{R_{hg}}{\eta_{hg}^2} \mathcal{L}_{\text{limit}}^*$$

We now need to optimize the luminosity by setting $\sigma_y^* = \sqrt{\beta_y^* \epsilon_y}$ and varying β_y^* while holding $\gamma, \sigma_x^*, \epsilon_y, \xi_{\text{limit}}$ and σ_z fixed. This requires maximizing the function

$$F(h_y) \;=\; h_y^{3/2} \frac{R_{gh}}{\eta_{hg}^2}$$

by varying $h_y = \frac{\sigma_z}{\beta_y^*}$.

Figure 8.35: The hour-glass luminosity geometric reduction factor R_{hg}, the beam-beam enhancement factor η_{hg}, and the optimization factor $F(h_y)$ as functions of $h_y = \frac{\sigma_z}{\beta_y^*}$. A relatively broad optimal luminosity occurs at $h_y = 1.4$, where $R_{hg} = 0.8, \eta_{hg} = 1.65, F(h_y) = 0.48$.

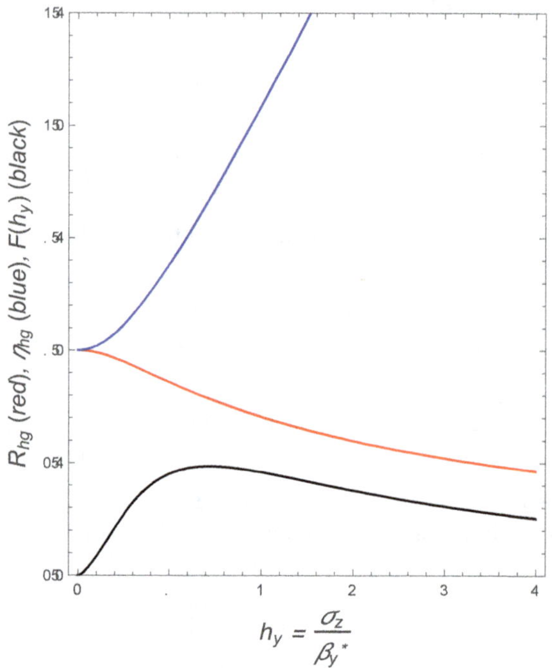

Figure 8.35 shows the luminosity reduction factor R_{hg}, the beam-beam enhancement factor η_{hg}, and the optimization factor $F(h_y)$ as functions of h_y. For optimal luminosity, one chooses

$$h_y = 1.4, \qquad \text{or} \qquad \beta_y^* = 0.7\,\sigma_z.$$

The corresponding beam intensity is

$$N = 7.5\,\frac{\xi_{\text{limit}}\gamma\sigma_x^*}{r_0}\sqrt{\frac{\epsilon_y}{\sigma_z}}, \tag{8.98}$$

and the optimal luminosity is

$$\mathcal{L} = 2.6\,\frac{\xi_{\text{limit}}^2\gamma^2\sigma_x^* f\sqrt{\epsilon_y}}{r_0^2\sigma_z^{3/2}}. \tag{8.99}$$

The optimum is relatively broad.

It should be pointed out that we have considered only the geometric implications of a small β^*. There are other considerations. One is that a small β^*

necessarily implies a large β at the neighboring strong quadrupoles in the low-β^* insertion. This means these quadrupoles need to have large apertures to clear the beam passage, which in turn means they need to be very strong in order to provide the required field gradient. In addition, a large β at these quadrupoles means extreme sensitivity to errors in their construction and installation.

Synchrobetatron coupling due to hour-glass Another effect associated with small β^* is the hour-glass induced synchrobetatron coupling and the possibility of exciting synchrobetatron resonances (see footnote 35).

To illustrate the mechanism, consider a weak-strong flat beam configuration. Consider the vertical motion of a weak-beam particle with $x = 0$. The weak beam has a finite bunch length σ_z, thus allowing an hour-glass to have an effect. To simplify the consideration, we assume the strong beam has zero length, so it functions simply as a thin-lens, and it always dutifully arrives the collision point $s = 0$ at the repetitive revolution time.

Due to a longitudinal displacement $z \neq 0$ of the weak-beam particle, however, the particle collides with the strong beam not at $s = 0$ but at $s = -\frac{z}{2}$. Due to hour-glass, the strong beam has a larger beam size $\sigma_y = \sigma_y^* \sqrt{1 + \frac{z^2}{4\beta_y^{*2}}}$ there. Take the linear thin-lens model for example, the kick delivered to the weak-beam particle is given by

$$\Delta y' \;=\; -4\pi \xi_y^* \, \frac{y}{\left(\beta_y^{*2} + \frac{z^2}{4}\right)^{1/2}}\,, \qquad \xi_y^* \;=\; \frac{N r_0 \beta_y^*}{2\pi\gamma\sigma_x^*\sigma_y^*}\,.$$

As the particle executes synchrotron oscillation in the weak beam, its transverse beam-beam kick is then modulated by its z-oscillation. If β_y^* is reduced, this modulation is enhanced and becomes a source of synchrobetatron coupling. In the case of a linear thin-lens model, for example, synchrobetatron sideband resonances are excited at

$$\nu_y + \xi_y^* \;\approx\; k \pm 2\nu_s\,, \tag{8.100}$$

for all integers k. These resonances have resonance width of

$$\delta\nu_y \;=\; \xi_y^* \left(\frac{z_0}{2\beta_y^*}\right)^2\,, \tag{8.101}$$

where z_0 is the amplitude of the synchrotron oscillation of the particle under consideration.

Since these beam-beam hour-glass driven sideband resonances (8.100) are spaced by $2\nu_s$ apart, it follows that these resonances will overlap, and a large scale chaos will occur in phase space, when the resonance width $\delta\nu_y$ approaches or exceeds $2\nu_s$. A synchrobetatron beam-beam limit is then expected to be given by

$$\xi_{\text{limit}}^* \;=\; 2\nu_s \left(\frac{2\beta_y^*}{\sigma_z}\right)^2\,. \tag{8.102}$$

Equation (8.102) can be considered a case of Chirikov condition for beam-beam hour-glass induced synchrobetatron resonances. It can become significant when the bunch length σ_z approaches β_y^*.

Role of synchrotron radiation damping　We mentioned earlier that there were basically two physical mechanisms considered to explain the beam-beam instability; one is based on various linear and nonlinear resonances assuming the particle motion is 100% correlated from one collision to the next; the other is based on identifying some diffusion effects assuming 0% correlation between collisions (see Sec. 8.4.8). One key difference between these two physical mechanisms lies in the role of synchrotron radiation damping time $\tau_{\rm rad}$ — resonance models are insensitive to $\tau_{\rm rad}$ while the diffusion models depend on $\tau_{\rm rad}$ rather directly.

Experimentally, a strong evidence that radiation damping does play a pivotal role is provided by the observation that all proton (or hadron) colliders have been limited by a value of $\xi_{\rm limit}$ an order of magnitude smaller than that of the electron colliders. Typical value of $\xi_{\rm limit}$ would be around 0.005 for protons and 0.05 for electrons. Even in the case of electron-proton colliders, it seems that the proton ring is found to withstand a much lower value of $\xi_{\rm limit}$ than its electron partner. One possible explanation of this observation is that the protons do not have radiation damping to help them fight the beam-beam perturbations. This almost readily available explanation, however, necessarily imply the radiation damping plays a pivotal role, and the diffusion models must be taken seriously. A possible model combining the resonance and the diffusion models was attempted on page 625 through the Chirikov condition, but in-depth study has been lacking.

Homework 8.39 The scaling (8.96) refers to the case of a flat beam where only the vertical motion is playing the role imposing the beam-beam limit. Add the horizontal motion also into consideration, but take into account of the condition of optimal coupling, Eq. (8.97), to obtain an updated set of scaling laws.

Solution

$$N_{\rm limit} \sim \gamma^3 \epsilon_{y0} \beta_x^* \beta_y^{*-1},$$
$$\mathcal{L}_{\rm limit} \sim \gamma^4 \epsilon_{y0} \beta_x^* \beta_y^{*-2}.$$

Homework 8.40 It might be instructive to touch base with a numerical example. Let us assume a colliding beam setup with $\xi_{\rm limit} = 0.05, \gamma = 10^4, f = 10^6\,{\rm s}^{-1}, \sigma_x^* = 1\,{\rm mm}, \epsilon_y = 5 \times 10^{-8}\,{\rm m}, \sigma_z = 5\,{\rm cm}$.

(a) Calculate the hour-glass limited beam intensity and luminosity.

(b) Compare the hour-glass limited luminosity you obtain with what the same $\xi_{\rm limit}$ yields without hour-glass effect.

Solution (a) $N = 1.3 \times 10^{12}, \mathcal{L} = 1.7 \times 10^{32}\,{\rm cm}^{-2}{\rm s}^{-1}$.

Homework 8.41 Follow the discussion outlined in the text on the hour-glass induced synchrobetatron coupling effect. Convince yourself that

(a) the mechanism drives resonances (8.100), and that

(b) the resonance width is given by Eq. (8.101).

8.5 Coherent beam-beam effect

We mentioned in Fig. 8.1 that there are two pictures to describe the beam-beam effect, a weak-strong picture and a strong-strong picture. So far we have mostly concentrated in a weak-strong picture when only the particles in the weak beam are perturbed while the strong beam stays intact. The weak-strong beam-beam interaction picture has provided useful insights into the mechanism of the beam-beam instability. It is, however, only a limited view. In fact, the identification of the beam-beam instability with instabilities in a periodic nonlinear mapping, as proposed by Eq. (8.12), cannot completely represent the physics of two strong colliding beams.

We did venture into a strong-strong regime when we discussed the dynamic-β^* (Sec. 8.3.3) and the flip-flop (Sec. 8.3.4) effects. However, in those discussions, the dynamics involves only the static or quasistatic single-particle incoherent effects. The beams as a whole do not move, even though their distributions may be distorted by the mutual beam-beam interaction in a static manner. This picture then misses the wide range of possibilities when two strong beams execute coupled coherent motion and the additional instability mechanisms connected to them. In this and the following sections, we address these coherent beam-beam effects.

Strong-strong picture In the strong-strong picture, both beams are perturbed. To study the coherent beam-beam effects, the procedure in principle follows two steps.

1. Let the steady-state distribution be Ψ_0. For example, Ψ_0 may be Gaussian in the absence of the beam-beam interaction. With strong-strong perturbation, Ψ_0 must be solved self-consistently, and the resulting perturbed steady-state Ψ_0 will in general no longer be Gaussian. An example of this steady state distortion is given by the dynamic-β^* and the flip-flop effects as we did in Secs. 8.3.3 and 8.3.4, except that what we did then took into account only the linear beam-beam force, and we need to extend the calculation to nonlinear beam-beam force.

2. Given the perturbed, generally non-Gaussian, steady-state Ψ_0, let the two beams have distributions

$$\begin{aligned}
\Psi_1 &= \Psi_0 + \Delta\Psi_1, \\
\Psi_2 &= \Psi_0 + \Delta\Psi_2,
\end{aligned} \tag{8.103}$$

where $\Delta\Psi_{1,2}$ are now time-oscillatory. Beam 1 sees the beam-beam force from beam 2, which is determined by the distribution Ψ_2 of beam 2, while beam 2 sees a beam-beam force similarly determined by Ψ_1. The beam-beam force thus provides a coupling between the two beam distributions in a dynamical manner. The question is then if the perturbations $\Delta\Psi_{1,2}$ initially infinitesimal would stay infinitesimal or if they, under some unfavorable conditions, bootstrap each other into an unstable growth.

Coherent beam-beam modes do exist   In the following sections, we intend to introduce some of the efforts that formulate the coherent beam-beam problem. Before we start the discussion, a comment is in order at the outset concerning the coherent beam-beam modes. An observation has sometimes been made that the only observed coherent beam-beam mode so far has been the dipole modes — see Figs. 8.39 and 8.40 later — and there does not seem to have an evidence of higher modes been observed. It is then argued that it is questionable that coherent beam-beam modes actually exist. We have a different view.

The observation of dipole modes is a strong evidence that these coherent beam-beam modes do exist. As to the modes higher order than dipole, they are expected to be strongly Landau damped due to the large beam-beam tune spread ξ. The point is that the ℓ-th coherent mode is Landau damped with a rate of $\ell\xi$. Higher modes with $\ell > 1$ are therefore strongly suppressed by Landau damping and not observed as their behavior, once Landau damped, becomes indistinguishable from incoherent motions. However, not easily observable does not mean they do not play a role. We shall comment on this point along the way and a summary comment is given on page 707.

8.5.1   Steady-state with nonlinear beam-beam force

In the earlier treatment of dynamic-β^* and flip-flop effects, we assumed a linearized beam-beam force. With this assumption, the perturbed steady-state remains Gaussian, and only the rms beam size gets changed. However, if the beam distribution is indeed Gaussian, the beam-beam force cannot be strictly linear. (Only a uniform elliptical distribution gives strictly linear beam-beam force, and even then only inside the elliptical region of the distribution.) We therefore have an internal inconsistency here.

One way to proceed is to take into consideration of the nonlinear beam-beam force for Gaussian beams, but insist that the beam-beam force acts only on the second moments of the beam distributions in such a way that the distributions are describable by equivalent Gaussians. This is a slight improvement but is still not self-consistent. We shall not pursue this approach except for some intermittent comments.

In general, beam-beam forces and beam distributions are to be found self-consistently. Assuming flat beams for simplicity, the two beam bunch distributions are coupled through the Vlasov equation,

$$\frac{\partial\Psi_1}{\partial s} + y'\frac{\partial\Psi_1}{\partial y} - F_2(y,s)\frac{\partial\Psi_1}{\partial y'} \;=\; 0,$$

$$\frac{\partial \Psi_2}{\partial s} + y' \frac{\partial \Psi_2}{\partial y} - F_1(y,s) \frac{\partial \Psi_2}{\partial y'} \; = \; 0, \tag{8.104}$$

where

$$
\begin{aligned}
F_{1,2}(y,s) \;&=\; K(s)y + \mathrm{BB\ force}\\[4pt]
\mathrm{BB\ Force} \;&=\; \frac{2\pi N r_0}{L_x \gamma} \delta_P(s) \int_{-\infty}^{\infty} d\bar{y}\, H(y-\bar{y})\rho_{1,2}(\bar{y})\\[4pt]
&=\; \frac{2\pi N r_0}{L_x \gamma} \delta_P(s) \int_{-\infty}^{\infty} d\bar{y}\, H(y-\bar{y}) \int_{-\infty}^{\infty} d\bar{y}'\, \Psi_{1,2}(\bar{y},\bar{y}',s),
\end{aligned}
\tag{8.105}
$$

with $H(x)$ the sign-function, $H(x) = -1$ if $x < 0$ and $+1$ when $x > 0$. We have assumed the two beams have opposite charges. Admittedly, Eq. (8.104) is slightly awkward because the coordinates s in the two equations point in opposite directions, but we trust this does not introduce confusion.

The external focusing force provided by the accelerator is given by the term $K(s)y$ in Eq. (8.105). The beam-beam force occurs only at the interaction point as described by $\delta_P(s)$, the periodic δ-function with period $s = C$, the collider circumference if there is one collision per revolution. Quantity L_x is the horizontal total width of the beam assuming a uniform distribution in x.

The equal-beam, steady-state self-consistent distribution satisfies

$$\frac{\partial \Psi_0}{\partial s} + y' \frac{\partial \Psi_0}{\partial y} - F_0(y,s) \frac{\partial \Psi_0}{\partial y'} \; = \; 0, \tag{8.106}$$

where

$$F_0(y,s) \;=\; K(s)y + \frac{2\pi N r_0}{L_x \gamma} \delta_P(s) \int_{-\infty}^{\infty} d\bar{y}\, H(y-\bar{y}) \int_{-\infty}^{\infty} d\bar{y}'\, \Psi_0(\bar{y},\bar{y}',s). \tag{8.107}$$

In the dynamic-β^* analysis, we linearized the beam-beam force in Eq. (8.107). With linearization, the beam-beam problem is analytically soluble, as discussed in Sec. 8.3.3.

Mutually pinched uniform beams Curiously, it turns out that there exists another interesting analytically soluble case in the nonlinear regime. Consider two relativistic, unbunched, round beams of opposite charges penetrating each other, not only at the interaction point but all along the distance from $s = -\infty$ to $s = \infty$. Let both beams have a distribution that is Gaussian in the transverse momenta x' and y' and algebraic in coordinates x and y as follows,

$$\Psi_0(x,x',y,y') \;=\; \frac{e\lambda_0 \exp\left(-\frac{x'^2 + y'^2}{2\sigma'^2}\right)}{2\pi^2 \sigma'^2 A^2 \left(1 + \frac{x^2 + y^2}{A^2}\right)^2}, \tag{8.108}$$

where λ_0 is the line density of the two beams, and

$$\sigma'^2 \;=\; \frac{\lambda_0 r_0}{\gamma}.$$

It can be shown[37] (Homework 8.42) using the Vlasov equation — counterpart of Eq. (8.104) for the round geometry — that if the two beams have the distribution (8.108) with the specific $\sigma' = \sqrt{\frac{\lambda_0 r_0}{\gamma}}$ but arbitrary beam size A, they will mutually pinch each other with their beam-beam forces in such a way that they both maintain the steady-state (8.108).

Take for example two beams with $\lambda_0 = 10^{12}\,\mathrm{m}^{-1}, \gamma = 10^4$, the required beam divergence for mutual pinching is $\sigma' = 0.5$ mrad.

In the dynamic-β^* model, we noted that a localized linearized beam-beam force drives half-integer resonances $\nu_0 = \frac{m}{2}$. The problem was analytically soluble. We now note that in case the beam-beam force is nonlinear but smooth (non-localized, independent of s), as in the case of two mutually pinching unbunched beams, a steady-state (8.108) can be found without indication of resonances. Analytical solutions can be found either if the beam-beam perturbation is localized but linearized, or if it is nonlinear but nonlocalized.

Localized and nonlinear beam-beam perturbation   We now return to the discussion following Eq. (8.107). In general, unfortunately, the beam-beam force is both localized and nonlinear, i.e.

$$\text{BB force} \;\propto\; \delta_P(s) \times (\text{nonlinear function of } y)\,.$$

In this case, like the case in single-particle nonlinear dynamics, the problem becomes very difficult to solve. In particular, resonances are driven when ν_0 is close to a rational number $\frac{m}{n}$. A self-consistent strong-strong distribution is yet to be found in this case. The closest soluble case is that of the weak-strong case when there is only one prominent resonance playing a role as we did using the smooth approximation.

Drawing analogy with the longitudinal potential well distortion due to coherent wakefields (Sec. 1.2.3) in a storage ring, one may say that in the present case the steady-state solution Ψ_0 is "transverse potential well distorted" by the beam-beam force at the interaction point. One important difference, however, is the fact that the longitudinal motion is slow enough that the coherent wakefields can be approximated as an averaged nonlocal effect. After applying this adiabatic approximation, the problem becomes independent of s and is analytically accessible, while in the beam-beam problem, the perturbation is truly localized. See Homework 8.43.

It seems the two most available tricks we have developed for analysis have been the use of matrices to deal with linear localized systems and the use of adiabatic approximation to deal with nonlocalized nonlinear systems. In passing, the use of smooth approximation near a nonlinear resonance is in a sense a way to identify the slow variables so that adiabatic approximations can apply.

Setting the difficulty aside, assuming Eq. (8.106) is solved, it gives the solution when both beams have the same distribution Ψ_0. This equal-beam solution

---

[37]S. Kheifets and A.W. Chao, SLAC report PEP-325 (1979).

describes a nonlinear dynamic-β^* effect. One could also consider other nonlinear flip-flop steady-state solutions when the two beams behave differently. All these solutions will be sensitive to the proximity of the tune ν_0 to rational values $\frac{m}{n}$ just like the dynamic-β^* and the flip-flop solutions in the linear lens model is sensitive to $\nu_0 = \frac{m}{2}$. A systematic search for general solutions of Ψ_0 using Eqs. (8.106–8.107) will be rather involved and intricate. In what follows, however, we will bypass this step, simply assuming a steady-state Ψ_0 has been given.

Once the steady-state Ψ_0 is known, whether it is the equal-beam or flip-flop solutions, the next question to ask is whether the beam motion is stable against small perturbations from it. The way to consider the problem was introduced in Eq. (8.103). In the following sections, we address this question by introducing a few simplified models illustrating the mechanisms of coherent beam-beam interaction.

Homework 8.42 Verify what is claimed in the text that two beams with distribution (8.108) will mutually pinch each other by the beam-beam forces in a steady state.

Solution First show that with the beam distribution (8.108), the beam-beam force gives

$$y'' = -\frac{4\lambda_0 r_0}{\gamma} \frac{y}{x^2 + y^2 + A^2}.$$

The rest is to verify that

$$y' \frac{\partial \Psi_0}{\partial y} + y'' \frac{\partial \Psi_0}{\partial y'} = 0,$$

for a steady-state solution, provided $\sigma'^2 = \frac{\lambda_0 r_0}{\gamma}$.

Homework 8.43 In the text, we argued that the steady state beam distribution Ψ_0 under beam-beam perturbation cannot be considered in the context of a "transverse potential well distortion" because its δ-function perturbation is too amiable to resonances. On the other hand, if we are assured to be "away from all resonances", one might be able to draw some simplified conclusions. To explore this unrealistic but idealistic circumstance, we look for a solution of a time-independent Ψ_0 in the limit when the beam-beam interaction is smoothed out over the storage ring circumference, so that Eqs. (8.106–8.107) can be simplified by letting

$$N \delta_P(s) \rightarrow \lambda_0,$$

where λ_0 is the line particle density as in Eq. (8.108). Show that with this idealistic smoothing, Eq. (8.108) is a solution to Ψ_0 of a beam-beam potential well distortion problem when cylindrical symmetry is assumed. With this smoothing, a potential well distortion picture then does apply. In a sense, we are making an adiabatic approximation in the transverse dynamics.

Homework 8.44

(a) Generalize Eq. (8.108) by adding a uniform linear external focusing in the derivation.

(b) Find the counterpart of Eq. (8.108) for the case of flat beams.

8.5.2   Rigid dipole model

As mentioned, the dynamic-β^* effect is the simplest strong-strong model, which involves static beam distortions. The simplest coherent beam-beam model that involves dynamics (moving beams) is the rigid dipole model.[38] In this model, each beam bunch is assumed to have a rigid distribution except that its center-of-charge is allowed to move. All bunches are then coupled through the beam-beam kicks to form a dynamic system in which all bunches oscillate in time.

Note that no such coherent motion is allowed in the weak-strong picture. This in fact is one serious limitation of a weak-strong picture, especially since the coherent motions, as we will soon see, tend potentially to set tighter stability limits than the incoherent motion.

To illustrate the model, let us assume the unperturbed beam distribution Ψ_0 is flat at the collision point. The horizontal beam size is taken to be $L_x = \sqrt{2\pi}\,\sigma_x$. We assume $\sigma_x \gg \sigma_y$ and consider only the y-motion of the rigid bunches.

Coherent beam-beam kick   In the rigid beam model, the kicks given to the bunches are computed by averaging the single-particle kicks over the beam distribution. The kick received by a single particle at displacement y is

$$\Delta y'(y) \;=\; -\frac{2\pi N r_0}{L_x \gamma} \int_{-\infty}^{\infty} d\bar{y}\, \rho_0(\bar{y}) H(y - \bar{y}) \,,$$

with $H(x)$ the sign-function and $\rho_0(y)$ being the rigid beam's y-distribution, $\int_{-\infty}^{\infty} dy\, \rho_0(y) = 1$.

Let the center-of-charge coordinates of the two beams be designated by Y. The kick received by the rigid beam is then

$$\begin{aligned}
\Delta Y' \;&=\; \int_{-\infty}^{\infty} dy\, \rho_0(y - \Delta Y) \Delta y'(y) \\[1mm]
&=\; -\frac{2\pi N r_0}{L_x \gamma} \int_{-\infty}^{\infty} dy\, \rho_0(y - \Delta Y) \int_{-\infty}^{\infty} d\bar{y}\, \rho_0(\bar{y}) H(y - \bar{y}) \,,
\end{aligned}$$

where ΔY is the displacement of the kicked beam relative to the kicking beam at the collision point. Both beams are assumed to have the same rigid distribution $\rho_0(y)$.

---

[38]A. Piwinski, Proc. VIII Int. Conf. High Energy Accel., CERN, 357 (1971); A. Chao and E. Keil, CERN/ISR-TH/79-31 or SLAC report PEP-300 (1979); E. Keil, Nucl. Instr. Meth. 188, 9 (1981); K. Hirata, Nucl. Instr. Meth. Phys. Res. A269, 7 (1988); K. Yokoya and H. Kioso, Part. Accel. 27, 181 (1990); J.D. Bjorken, Workshop on p-pbar Options for the Super Collider, Fermilab Conf-84/29-THY (1984).

For $|\Delta Y|$ smaller than the beam size so that the beam-beam force can be linearized with respect to ΔY. We then replace

$$\rho_0(y - \Delta Y) \approx \rho_0(y) - \rho_0'(y)\Delta Y \,.$$

Note we are linearizing with respect to ΔY, i.e. we assume the two beams' centers-of-charge are only slightly separated; the beam-beam force remains non-linear.

It is reasonable to assume an up-down symmetry of the distribution $\rho_0(y)$. In follows then that

$$\Delta Y' = \frac{2\pi N r_0}{L_x \gamma} \Delta Y \int_{-\infty}^{\infty} dy\, \rho_0'(y) \int_{-\infty}^{\infty} d\bar{y}\, \rho_0(\bar{y}) H(y - \bar{y}) \,.$$

It is easy to show that (Homework 8.45)

$$\int_{-\infty}^{\infty} dy\, \rho_0'(y) \int_{-\infty}^{\infty} d\bar{y}\, \rho_0(\bar{y}) H(y - \bar{y}) = -2 \int_{-\infty}^{\infty} dy\, \rho_0^2(y) \,. \tag{8.109}$$

As a result, we obtain

$$\Delta Y' = -\frac{4\pi N r_0}{L_x \gamma} \Delta Y \int_{-\infty}^{\infty} dy\, \rho_0^2(y) \,.$$

We may facilitate a comparison with the incoherent beam-beam effect by expressing our result in terms of the familiar focal length of the incoherent motion,

$$\frac{1}{f_y} = \frac{2N r_0}{\gamma \sigma_y (\sigma_x + \sigma_y)} \,.$$

For a flat beam, $\sigma_x \gg \sigma_y$, this gives in terms of f_y,

$$\Delta Y' = -G \frac{\Delta Y}{f_y} \,,$$

where

$$G = \sqrt{2\pi}\, \sigma_y \int_{-\infty}^{\infty} dy\, \rho_0^2(y) \tag{8.110}$$

is the coherent dipole kick form factor.

For a Gaussian beam,

$$G = \frac{1}{\sqrt{2}} \,. \tag{8.111}$$

For a uniform distribution with beam height $L_y = \sqrt{2\pi}\sigma_y$, we find

$$G = 1 \,. \tag{8.112}$$

Coupled dipole motion   We now continue with the case of flat Gaussian beams with form factor (8.111). Assuming small center-of-charge motions, the coherent dipole beam-beam kicks on the two beams are

$$
\Delta Y_1' = -\frac{1}{\sqrt{2}}\frac{Y_1 - Y_2}{f_y},
$$

$$
\Delta Y_2' = -\frac{1}{\sqrt{2}}\frac{Y_2 - Y_1}{f_y}. \tag{8.113}
$$

Equation (8.113) can be written in a matrix form. Defining the vector

$$
\begin{bmatrix} Y_1 \\ Y_1' \\ Y_2 \\ Y_2' \end{bmatrix},
$$

the beam-beam kick is represented by the matrix map,

$$
T_{bb} = \begin{bmatrix} 1 & 0 & 0 & 0 \\ -\frac{G}{f_y} & 1 & \frac{G}{f_y} & 0 \\ 0 & 0 & 1 & 0 \\ \frac{G}{f_y} & 0 & -\frac{G}{f_y} & 1 \end{bmatrix}, \qquad G = \frac{1}{\sqrt{2}}. \tag{8.114}
$$

Consider the case of two beams of opposite charges, with each beam consisting of one bunch, colliding in a single storage ring. The beams will collide at two diagonally opposite collision points. After the beam-beam collision, each beam is transformed from one interaction point to the next by the matrix map,

$$
T_0 = \begin{bmatrix} t_0 & 0 \\ 0 & t_0 \end{bmatrix}, \qquad t_0 = \begin{bmatrix} \cos\mu & \beta_0^* \sin\mu \\ -\frac{1}{\beta_0^*}\sin\mu & \cos\mu \end{bmatrix}, \tag{8.115}
$$

where $\mu = \pi\nu$ is the betatron phase advance between the two interaction points, and ν is the betatron tune of the storage ring for a whole revolution.

Let the product of these two matrices be

$$
T_{\text{tot}} = T_0 T_{bb}. \tag{8.116}
$$

The four eigenvalues of T_{tot} form two complex conjugate pairs; each corresponds to a coherent beam-beam dipole mode. The coherent dipole motion is stable if all eigenvalues of T_{tot} have absolute values of 1.

In the case of two equal bunches, the two modes are a "0-mode" and a "π-mode". In the 0-mode, the two bunches move up and down together. It is always stable because as the two bunches move up and down together with $Y_1 = Y_2$ at the interaction point, there is no beam-beam force acting on the bunch centers regardless of the beam-beam strength. In particular, the 0-mode frequency is always equal to the unperturbed frequency μ.

In the π-mode, the two bunches move out of phase with $Y_1 = -Y_2$ at the interaction point. The π-mode is stable if (Homework 8.47)

$$
\xi_y < \frac{1}{2\sqrt{2}\,\pi}\cot\frac{\mu}{2}. \tag{8.117}
$$

Note that this is $\sqrt{2}$ times more stringent than the weak-strong incoherent stability limit (8.24) when μ is set to be the betatron phase advance between the two interaction points. This is because in the π-mode, the effective coherent beam separation is $2G$ times that in the incoherent case (a factor of G from the form factor, a factor of 2 due to both beams moving out of phase). The π-mode frequency therefore follows the incoherent analysis with the replacement $\xi_y \to 2G\xi_y$. A general expression for arbitrary form factor G, a generalization of Eq. (8.117), is given by

	$0 < \mu < \pi$	$\pi < \mu < 2\pi$
Two beams opposite charges	$0 < \xi_y < \frac{1}{4G\pi} \cot \frac{\mu}{2}$	$0 < \xi_y < -\frac{1}{4G\pi} \tan \frac{\mu}{2}$
Two beams same charge	$0 > \xi_y > -\frac{1}{4G\pi} \tan \frac{\mu}{2}$	$0 > \xi_y > \frac{1}{4G\pi} \cot \frac{\mu}{2}$

$$(8.118)$$

where μ is the betatron phase between interaction points. With $G = \frac{1}{\sqrt{2}}$ for flat Gaussian beams, the strong-strong rigid dipole instability is a factor $\sqrt{2}$ more stringent than the incoherent weak-strong picture. For uniform beams, $G = 1$, the instability is a factor of 2 more stringent.

In the unstable region, except when ξ_y is very close to the stability boundary, the instability growth rate obtained from the imaginary part of the eigenvalues of the π-mode is large, which means it would be difficult to fight this dipole instability by a typical feedback system.

Another comment should be made to emphasize the fact that in spite of the similar appearance of the coherent instability limit Eq. (8.117) and the incoherent limit (8.24), their instability mechanisms are quite different. Figure 8.36 illustrates the difference as viewed in the phase space.

Multiple bunches per beam   Still one more comment addresses the superperiodicity structure of these instabilities. The weak-strong and the strong-strong instabilities have very different superperiodicity structures, in spite of the similarity between Eqs. (8.117) and (8.24). We now address this point.

We see that strong-strong coherent beam-beam effect imposes stronger limit than the limit imposed by the weak-strong picture due to the fact that in one of the modes — the π-mode — the effective beam separation is doubled because both beams move. This result is obtained for the case of one bunch per beam, and there are two collision points in the collider. We now show that this situation (that coherent stability requires tighter conditions than incoherent stability) is much further exacerbated if there are multiple bunches per beam, mainly because there are many additional coherent beam-beam modes and they all need to be stable.

Let us consider the case of two bunches per beam, bunches 1 and 3 in one beam and bunches 2 and 4 in the other beam. The bunches collide at four locations A, B, C, D around the collider as illustrated in Fig. 8.37. We

Figure 8.36: Distributions in phase space of two colliding beams for four cases: (a) incoherent weak-strong case, (b) dynamic-β^* effect, (c) coherent dipole 0-mode, and (d) coherent dipole π-mode. (b), (c) and (d) are strong-strong cases, the two beams designated as $+$ and $-$ beams. The solid and dashed curves indicate the unperturbed and the beam-beam perturbed distributions, respectively. The perturbed distribution moves with time in (c) and (d) but are static in (a) and (b). The beam dynamics behavior are different in all cases, even though their stability criteria look similar.

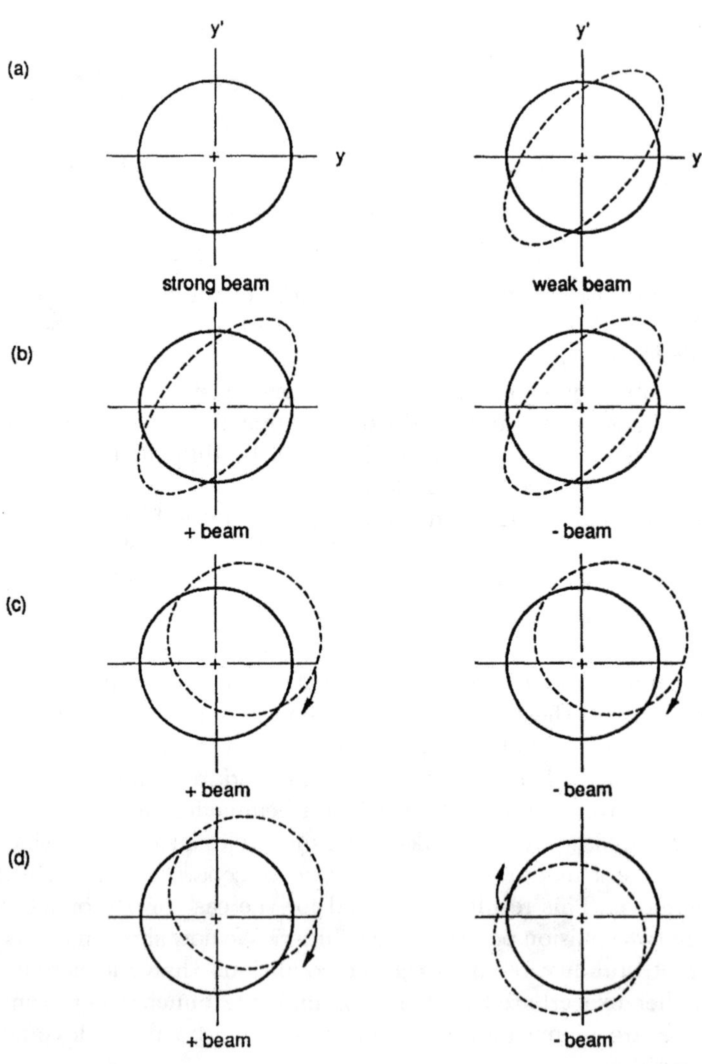

Figure 8.37: A configuration for the case of two bunches per beam and four interaction points.

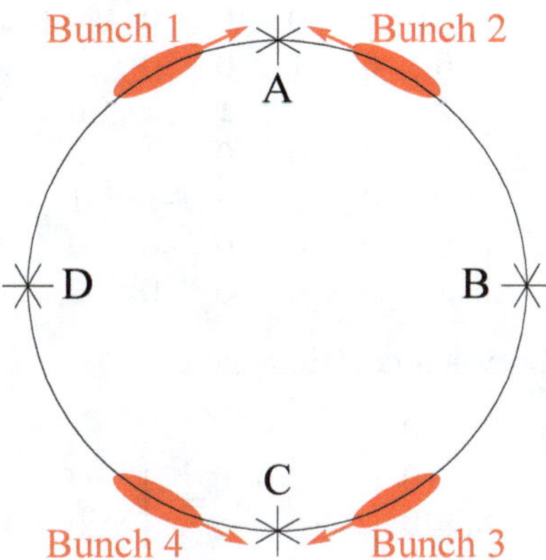

again consider the case of flat Gaussian beams and consider only the linearized coherent rigid dipole motion (small center-of-charge displacements) in the y-dimension.[39]

The sequence of events is as follows. First the bunches 1 and 2 collide at position A and bunches 3 and 4 collide at position C. All bunches then propagate by transport to the next collision when bunch 1 collide with bunch 4 at position B and bunch 2 and bunch 3 collide at position D. After the next transport propagation, the collision events repeat. The net effect is the dynamics repeats every half a revolution time.

Describing the dynamics with the vector $(Y_1, Y_1', Y_2, Y_2', Y_3, Y_3', Y_4, Y_4')$, the matrix map for the first collision is, letting $x = \frac{1}{\sqrt{2}\, f_y}$,

$$
T_{bbAC} = \begin{bmatrix}
1 & 0 & 0 & 0 & 0 & 0 & 0 & 0 \\
-x & 1 & x & 0 & 0 & 0 & 0 & 0 \\
0 & 0 & 1 & 0 & 0 & 0 & 0 & 0 \\
x & 0 & -x & 1 & 0 & 0 & 0 & 0 \\
0 & 0 & 0 & 0 & 1 & 0 & 0 & 0 \\
0 & 0 & 0 & 0 & -x & 1 & x & 0 \\
0 & 0 & 0 & 0 & 0 & 0 & 1 & 0 \\
0 & 0 & 0 & 0 & x & 0 & -x & 1
\end{bmatrix}.
$$

---

[39]Unless stated otherwise, we maintain $+$ and $-$ as indices for the two beams, and use $1, 2, 3, \cdots$ for the individual bunches.

The next collision is described by the map

$$
T_{bbBD} = \begin{bmatrix}
1 & 0 & 0 & 0 & 0 & 0 & 0 & 0 \\
-x & 1 & 0 & 0 & 0 & 0 & x & 0 \\
0 & 0 & 1 & 0 & 0 & 0 & 0 & 0 \\
0 & 0 & -x & 1 & x & 0 & 0 & 0 \\
0 & 0 & 0 & 0 & 1 & 0 & 0 & 0 \\
0 & 0 & x & 0 & -x & 1 & 0 & 0 \\
0 & 0 & 0 & 0 & 0 & 0 & 1 & 0 \\
x & 0 & 0 & 0 & 0 & 0 & -x & 1
\end{bmatrix}.
$$

The transport map T_0 between collisions is

$$
T_0 = \begin{bmatrix}
t_0 & 0 & 0 & 0 \\
0 & t_0 & 0 & 0 \\
0 & 0 & t_0 & 0 \\
0 & 0 & 0 & t_0
\end{bmatrix}, \qquad
t_0 = \begin{bmatrix}
\cos\mu & \beta_0^* \sin\mu \\
-\frac{1}{\beta_0^*}\sin\mu & \cos\mu
\end{bmatrix},
$$

where $\mu = \frac{\pi\nu}{2}$ is the betatron phase advance between the two interaction points, and ν is the betatron tune of the storage ring for a whole revolution.

The total map for half a revolution is

$$
T_{\text{tot}} = T_0 T_{bbBD} T_0 T_{bbAC}.
$$

As before, we then calculate the four eigenvalues of T_{tot}. The stability condition is that all eigenvalues have absolute values equal to 1. The instability boundaries are given by (Homework 8.48),

$$
\xi_y < \frac{1}{2\sqrt{2}\, f_y}\cot\mu, \qquad \text{and} \qquad \xi_y < \frac{1}{2\sqrt{2}\, f_y}\cot\frac{\mu}{2}. \tag{8.119}
$$

The analysis for two bunches per beam, $M = 2$, can be extended to higher values of M. Figure 8.38 shows the stability limit ξ_0 as a function of ν, the betatron tune in the whole collider ring. With each beam containing M bunches and two beams colliding at $2M$ collision points, the betatron phase between collision points is $\mu = \frac{\pi\nu}{M}$.

Figure 8.38(a) is for the case $M = 1$. The dashed curve gives the weak-strong stability limit (8.24). The solid curve represents (8.117). As mentioned, these two curves differ by a factor of $\sqrt{2}$.

For the case $M = 2$, as shown in Eq. (8.119) and Homework 8.48, there are two instability cuts. Figure 8.38(b) shows the situation. Also shown is a dashed curve that represents the corresponding weak-strong limit when there are four collision points; it has a single cut of instability band at $\nu = M$.

For the general case of M bunches per beam, the transformation matrices will be $4M \times 4M$ and there are $2M$ coherent dipole modes. Stability of the system requires all modes be stable, i.e. all eigenvalues have absolute values of 1. One of the coherent modes remains the unperturbed 0-mode. Results for the cases $M = 3$ and $M = 4$ are shown in Figs. 8.38(c) and (d), respectively. We refer to Fig. 8.38 as the sawtooth stability diagram for obvious reason.

Figure 8.38: Stability diagram with ξ_0, the limiting value of ξ_y, plotted versus ν, the total betatron tune per revolution. The rigid dipole form factor is taken to be $G = \frac{1}{\sqrt{2}}$. Cases are for $M = 1, 2, 3, 4$ bunches per beam. The betatron phase advance between interaction points is $\mu = \frac{\pi\nu}{M}$. The figure repeats with a period of $\nu = M$. The dashed curves give the stability region for the weak-strong case for comparison. The rigid dipole instability is much more stringent compared with the weak-strong single-particle picture.

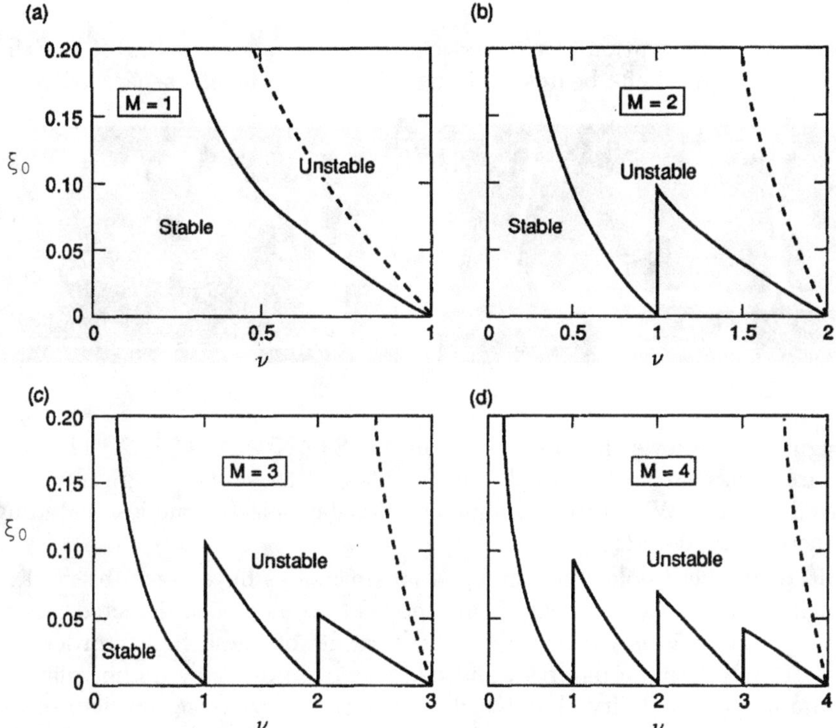

Clearly the strong-strong rigid dipole picture has a much more stringent stability requirement compared with that of the weak-strong picture, not only because of the additional factor of $2G$ but also because of the introduction of additional coherent modes. In particular, the incoherent limit obeys the superperiodicity and there is an instability cut only when $\nu = M$, while coherent dipole modes cut an instability at every $\nu = $ integer. For multiple bunches per beam, the superperiodicity suppresses the incoherent effects, but not the coherent effects. Coherent effects thus impose stricter limit on beam stability most notably for multibunch operations. See also a discussion on page 707.

Observation of coherent dipole beam-beam modes   Coherent beam-beam dipole modes have been observed in several collider storage rings such as SPEAR,[40] PETRA (see footnote 19), and TRISTAN.[41] Figure 8.39(left) shows the observation of the 0- and the π-modes in PETRA with $M = 1$ using a spectrum analyzer around the betatron frequency. The peak at the lower frequency corresponds to the 0-mode, the one at higher frequency is the π-mode. In Fig. 8.39(right), the π-mode frequency is slightly shifted because, although separated, the beams still couple slightly.

Figure 8.39: Observation of the vertical 0- and π-modes in PETRA with one bunch per beam. (Left) beams colliding and (Right) beams separated.

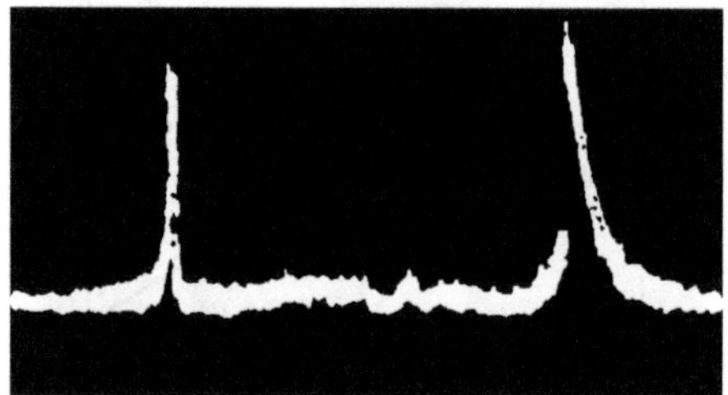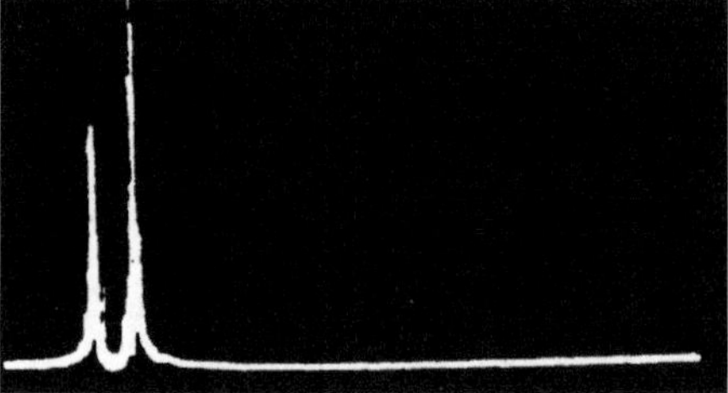

Figure 8.40 shows the observation in TRISTAN with $M = 2$. The parameters are such that all four modes are stable and these modes are observed by external driving. With proper frequency filtering, selected modes, including an intermediate mode, are observed.

Although the dipole coherent beam-beam modes have been observed readily, higher order coherent beam-beam modes have not been detected. As mentioned earlier, this is considered to be a result of these higher order modes being strongly Landau damped, and it does not imply they do not play a role in beam-beam instability. Landau damping is a very strong mechanism when the beam-beam interaction induces such a large tune spread specified by ξ, especially for higher order modes.

Homework 8.45

(a) Verify Eqs. (8.109) and (8.110).

(b) Derive Eq. (8.111) of beam-beam coherent kick for the case of flat Gaussian-beam coherent kick and Eq. (8.112) for the case of flat uniform-disk beam.

Solution (a) The interesting identity (8.109) follows from an integration by parts.

[40]SPEAR Group, Proc. 9th Int. Conf. High Energy Accel., Stanford, p. 37 (1974).

[41]T. Ieiri, T. Kawamoto, and K. Hirata, Nucl. Instr. Meth. A265, 364 (1988); K. Satoh, Proc. 6th Symp. Accel. Tech., Tokyo (1987), p. 18 (1987); T. Ieiri and K. Hirata, Proc. IEEE Trans. Part. Accel. Conf., Chicago, p. 709 (1989).

Figure 8.40: Observation of coherent dipole modes at TRISTAN with 2 bunches per beam. (a) Beam spectrum with four modes superimposed. The 0-mode and the π-mode stand out. By proper filtering, one could observe (b) the 0-mode, (c) an intermediate mode.

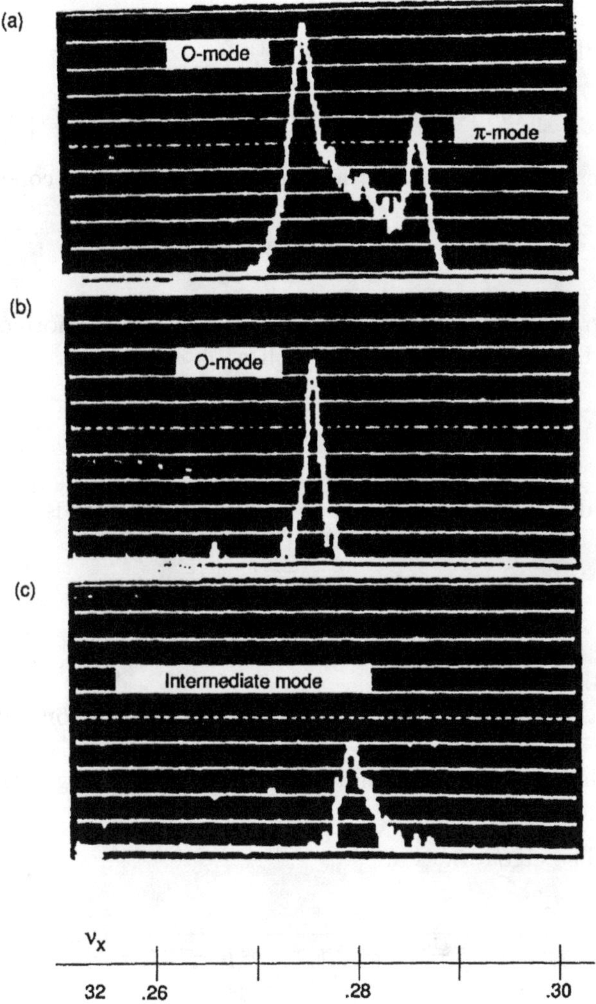

Homework 8.46 Extend the analysis of the coherent dipole kick form factor to the case of two round beams — the text gave the result of two flat beams. Apply your result first to a Gaussian distribution, then to the case of a uniform disk distribution.

Solution Let the round beam have transverse rigid distribution $\rho_0(r)$ with $\int_0^\infty 2\pi r\, dr\, \rho_0(r) = 1$. The beam-beam kick to a particle with transverse displacement (x, y), $r^2 = x^2 + y^2$, is

$$\Delta y'(x, y) = -\frac{4\pi N r_0}{\gamma} \frac{y}{r^2} \int_0^r r'\, dr'\, \rho_0(r')\,.$$

A vertically displaced beam has a distribution

$$\rho_0\!\left(\sqrt{x^2 + (y - \Delta Y)^2}\right) dx\, dy \;\approx\; \left[\rho_0(r) - \Delta Y\, \frac{y}{r}\rho_0'(r)\right] dx\, dy\,.$$

Integrating over the distribution, the displaced beam sees a coherent kick

$$\Delta Y' = -\frac{4\pi^2 N r_0}{\gamma}\, \Delta Y \int_0^\infty r\, dr\, \rho_0^2(r)\,,$$

where an integration by parts like Eq. (8.109) has been performed.

For a round Gaussian distribution with $\sigma_x = \sigma_y = \sigma$,

$$\Delta Y' = -\frac{1}{2}\frac{\Delta Y}{f_y}\,, \qquad \frac{1}{f_y} = \frac{N r_0}{\gamma \sigma^2}\,. \tag{8.120}$$

For a uniform disk beam of radius $\sqrt{x^2 + y^2} = R$, the kick is

$$\Delta Y' = -\frac{2 N r_0}{\gamma R^2}\, \Delta Y\,.$$

Homework 8.47

(a) Follow the text to calculate T_{tot} in Eq. (8.116) and compute its eigenvalues.

(b) Show that the stability condition for the coherent rigid dipole motion is given by Eq. (8.117).

Solution (a) The eigenvalues are

$$\begin{cases} e^{\pm i\mu}, & \text{0-mode,} \\[2mm] \cos\mu - \dfrac{\beta_0^*}{\sqrt{2}f_y}\sin\mu \pm i\sqrt{\dfrac{\sin\mu}{2f_y^2}\left[2\sqrt{2}\beta_0^* f_y\cos\mu + (2f_y^2 - \beta_0^{*2})\sin\mu\right]}\,, & \pi\text{-mode.} \end{cases}$$

Homework 8.48 Follow the text to analyze the case of rigid dipole beam-beam instability when there are two bunches per beam colliding at four locations in the collider.

(a) Obtain the four eigenvalues of the total map T_{tot}. Show that there is a 0-mode whose eigenvalues are the unperturbed $e^{\pm 2i\mu}$. Show that two of the remaining modes are degenerate.

(b) Verify the stability condition (8.119). The first of the condition refers to the stability of the two degenerate modes. The second condition refers to the last mode. Both conditions need to be fulfilled for the beam to be stable. The 0-mode is of course always stable.

Solution

(a) The two degenerate modes have the eigenvalues, with $x = \frac{1}{\sqrt{2}\,f_y}$,

$$\cos 2\mu - \beta_0^* x \sin 2\mu \pm i\sqrt{\sin 2\mu\left[2\beta_0^* x \cos 2\mu + (1 - \beta_0^{*2} x^2)\sin 2\mu\right]}\,.$$

The fourth mode has eigenvalues

$$\beta_0^{*2} x^2 + (1 - \beta_0^{*2} x^2)\cos 2\mu - 2\beta_0^* x \sin 2\mu$$
$$\pm 2i\sqrt{\sin\mu\,(\cos\mu - \beta_0^* x \sin\mu)^2\left[2\beta_0^* x \cos\mu + (1 - \beta_0^{*2} x^2)\sin\mu\right]}\,.$$

When $x = 0$, all four modes have eigenvalues $e^{\pm 2i\mu}$.

8.5.3 Asymmetric beams

The rigid dipole instabilities have been studied in the literature more extensively than the other coherent beam-beam effects. Several extensions of the basic model described so far have been examined. These include

- unequal beams ($N_+ \neq N_-$);[42]
- unequal tunes ($\nu_+ \neq \nu_-$);[42]
- asymmetric arcs;[43]
- error in phase advances between interaction points;[43]
- unequal circumferences ($C_+ \neq C_-$);[44]
- four compensating beams;[45]
- coherent beam-beam synchrobetatron resonance;[46]
- spontaneous beam separation;[47]
- long-range beam-beam interactions;[48]
- head-tail modes with crossing angle.[49]

---

[42]K. Hirata and E. Keil, CERN report CERN/LEP-TH/89-76 (1989); K. Hirata, AIP Proc. 214, Workshop on Beam Dynamics Issues of High Luminosity Asymmetric Colliders, LBL, 175 (1990).

[43]A. Chao and E. Keil, CERN report CERN/ISR-TH/79-31 or SLAC report PEP-300 (1979); E. Keil, CERN report LEP-226 (1980).

[44]K. Hirata and E. Keil, CERN reports CERN/LEP-TH/89-54 and CERN/LEP-TH/89-76 (1989); K. Hirata and E. Keil, Phys. Lett. B232, 413 (1989).

[45]E. Keil, CERN report CERN LEP-TH/89-37; Proc. 3rd Advanced ICFA Beam Dynamics workshop on Beam-Beam Effects in Circular Colliders, Novosibirsk, 85 (1989).

[46]Y. Kamiya and A.W. Chao, SLAC report SLAC/AP-10 (1983).

[47]K. Hirata and E. Keil, CERN report CERN/LEP-TH/89-76 (1989); K. Hirata, KEK Laboratory report KEK report TN-880010 (1989).

[48]M. Furman and A. Chao, IEEE Trans. Nucl. Sci., NS-32, 2297 (1985); E. Forest, SSC Central Design Group report SSC-51 (1986); E. Forest and M. Furman, SSC Central Design Group report SSC-32 (1985); M.A. Furman, SSC Central Design Group report SSC-62 (1986); K. Hirata, AIP Proc. 214, Workshop on Beam Dynamics Issues of High Luminosity Asymmetric Colliders, LBL, 441 (1990).

[49]K. Ohmi, et al., Phys. Rev. Lett. 119, 134801 (2017); N. Kuroo, et al., Phys. Rev. Accel. Beams 21, 031002 (2018).

One in fact makes the observation that for each incoherent beam-beam effect, there is most likely a coherent counterpart, and vice versa. Some of these extensions are reviewed below in this section.

Unequal beam Consider the case of one bunch per beam with two interaction points, but now the two bunches have different intensities, $\xi_+ \neq \xi_-$, and the two beams are in separate storage rings with different phase advances between interaction points, $\mu_+ \neq \mu_-$. In this case, the map (8.114) becomes

$$
T_{bb} = \begin{bmatrix} 1 & 0 & 0 & 0 \\ -\dfrac{G}{f_-} & 1 & \dfrac{G}{f_-} & 0 \\ 0 & 0 & 1 & 0 \\ \dfrac{G}{f_+} & 0 & -\dfrac{G}{f_+} & 1 \end{bmatrix}, \qquad \text{where} \qquad \begin{cases} \dfrac{1}{f_-} = \dfrac{2N_- r_0}{\gamma_+ \sigma_{x-}\sigma_{y-}}, \\[2mm] \dfrac{1}{f_+} = \dfrac{2N_+ r_0}{\gamma_- \sigma_{x+}\sigma_{y+}}, \end{cases}
$$

with G the rigid dipole form factor, and the map (8.115) becomes

$$
T_0 = \begin{bmatrix} t_+ & 0 \\ 0 & t_- \end{bmatrix}, \qquad t_\pm = \begin{bmatrix} \cos\mu_\pm & \beta_\pm^* \sin\mu_\pm \\ -\dfrac{1}{\beta_\pm^*}\sin\mu_\pm & \cos\mu_\pm \end{bmatrix}.
$$

By forming the total map $T_{\text{tot}} = T_0 T_{bb}$, the eigenmode frequencies $\mu_{1,2}$ are given by (Homework 8.49) (see footnote 42),

$$
2\cos\mu_{1,2} = \cos\mu_+ + \cos\mu_- - 2\pi G\xi_+ \sin\mu_+ - 2\pi G\xi_- \sin\mu_- \pm \sqrt{Q}, \qquad (8.121)
$$

and

$$
Q = (\cos\mu_+ - \cos\mu_- - 2\pi G\xi_+ \sin\mu_+ + 2\pi G\xi_- \sin\mu_-)^2 + 16\pi^2 G^2 \xi_+ \xi_- \sin\mu_+ \sin\mu_-,
$$

where

$$
\xi_+ = \frac{N_- r_0 \beta_+^*}{2\pi\gamma_+ \sigma_{x-}\sigma_{y-}}, \qquad \xi_- = \frac{N_+ r_0 \beta_-^*}{2\pi\gamma_- \sigma_{x+}\sigma_{y+}}.
$$

Note that this analysis allows other differences in the two beams, e.g. differences in N, γ, β_0^*, and $\sigma_{x,y}$ can all be incorporated into ξ. In general, there is no longer a 0-mode and a π-mode.

Unequal tune Equation (8.121) also includes the effect of unequal tunes. Figure 8.41 shows the stability region in the (ν_{0+}, ν_{0-}) plane (see footnote 42). Note that in addition to the expected integer and half-integer resonances, a new sum resonance $\nu_{0+} + \nu_{0-} = m$ is introduced, where $\nu_{0\pm} = \frac{\mu_\pm}{2\pi}$ are the tune advances between adjacent interaction points in the two storage rings (total tune is $\nu = 2\nu_0$). Strictly speaking, only this sum resonance is a true coupled-beam mode. Coherent coupled beam-beam modes are strongly excited at the sum resonances.

In passing, we note that splitting ν_{0+} and ν_{0-} did not improve the instability. The possible expectation that splitting the tune would introduce Landau damping, and therefore stability, to the system is incorrect. If anything, it is made worse due to the possibility of exciting the sum resonances.

Figure 8.41: Stability diagram in the (ν_{0+}, ν_{0-}) space. Shaded regions indicate instability. It particularly shows the excitation of sum resonances $\nu_{0+} + \nu_{0-} = m$. The horizontal ν_{0+} axis is from 0 to 1. The vertical ν_{0-} axis is from 0 to $\frac{1}{2}$. The pattern repeats if $\nu_{0\pm}$ are replaced by $\nu_{0\pm} + \frac{1}{2}$. [Courtesy Eberhard Keil (2020).]

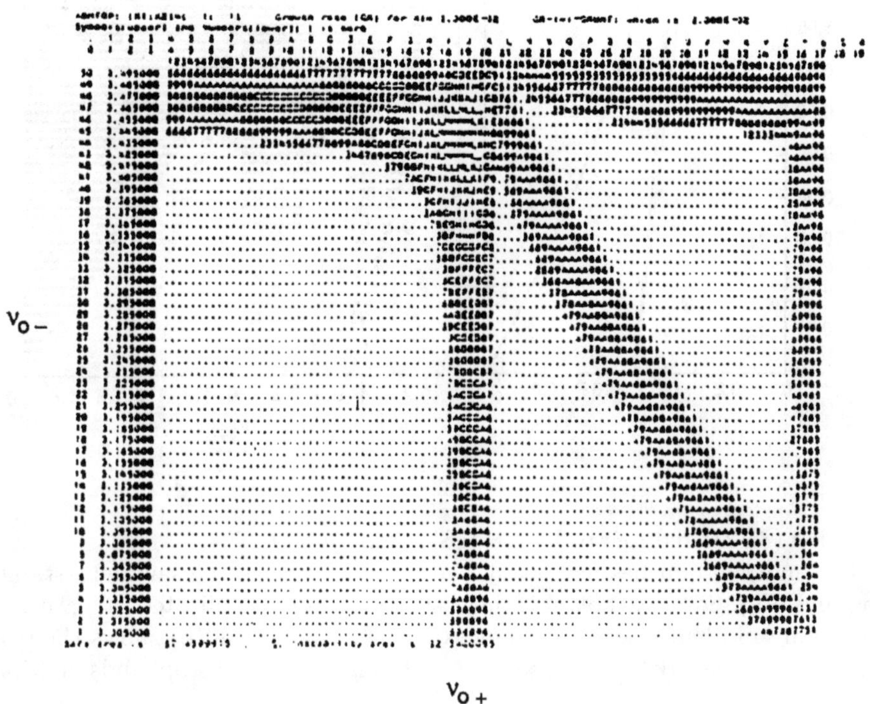

Phase advance error Consider again the case with one bunch per beam. In the previous analysis, the two interaction points are separated by two arcs with equal betatron phase advances. The analysis can be extended to unequal phase advances parametrized as $\mu_0 \pm \delta\mu$ while keeping both beams to have the same total tune $\nu = 2\nu_0 = \frac{\mu_0}{\pi}$ (see footnote 43). For simplicity, assume β_0^* at the two interaction points be the same. The result is shown in Fig. 8.42(a). The most pronounced new feature is the emergence of a stopband when the total tune ν is close to $\frac{1}{2}$. The stopband edges are found to be (Homework 8.50)

$$\xi_{\text{stopband}} = \frac{\cot \mu_0}{2\pi G} \frac{1 \pm \sin \delta\mu}{\cos^2 \delta\mu}. \tag{8.122}$$

On the other hand, the sawtooth part of the stability boundary reads

$$\xi_{\text{sawtooth}} = \frac{\sin \mu_0}{4\pi G(\cos \delta\mu - \cos \mu_0)}. \tag{8.123}$$

Figure 8.42: Effect of phase advance errors in the storage ring arcs on the sawtooth diagram. (a) $M = 1, \frac{\delta\mu}{2\pi} = 0.02$. (b) $M = 2, \frac{\delta\mu}{2\pi} = 0.03$. The form factor is taken to be $G = 1$. [Courtesy Kohji Hirata (2020).]

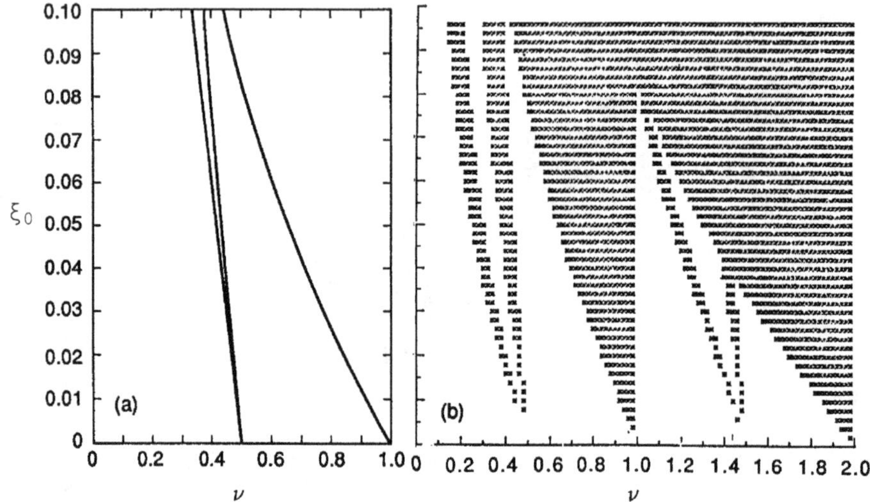

Equation (8.123) reduces to Eq. (8.118) when $\delta\mu = 0$.

For multiple bunches per beam, the results resemble the sawtooth stability diagrams Fig. 8.38, except that new stopbands of complicated structure are created around half-integer resonances $\nu = \frac{m}{2}$. Figure 8.42(b) shows the result obtained numerically for the case of $M = 2$. The number of stopbands multiplies in this case.

Unequal circumference   Consider the case when the two beams are stored in separate storage rings of circumferences $C_\pm$. The number of bunches $M_\pm$ are so as to have a fixed bunch spacing $\frac{C_+}{M_+} = \frac{C_-}{M_-}$. The two-ring system has a superperiod $= M_+ C_- = M_- C_+$. The coherent dipole beam-beam coupling excites, in addition to the respective single-ring half-integer resonances, also a set of coupled sum resonances in this system,

$$M_+ \nu_{0-} + M_- \nu_{0-} = m, \tag{8.124}$$

where $\nu_{0\pm}$ are the tune advances between adjacent interaction points in the two storage rings. Since there is only one interaction point for both rings, $\nu_{0\pm}$ are also the respective betatron tunes of the two rings. With this new set of resonances (8.124), there are now many more stopbands, which makes it undesirable to have unequal circumferences for the two beams.

The case we discussed earlier on unequal tunes can be considered a special case when $M_+ = M_- = 1$. Figure 8.43 shows a schematic layout of a collider

Figure 8.43: Schematics of the arrangement of unequal circumferences for the case $M_+ = 2, M_- = 3$.

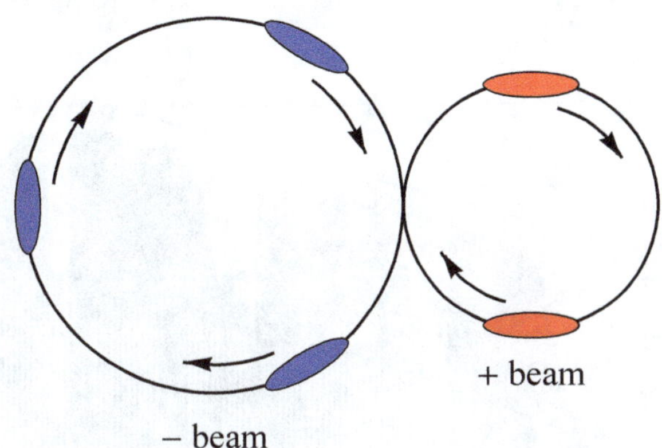

configuration for the case $M_+ = 2, M_- = 3$. The stability result is shown in Fig. 8.44 (see footnote 44). Note particularly the three sum resonances strongly driven by the beam-beam coupling. In comparison, the single-beam resonances are driven only when $\nu_{0\pm} = \frac{m}{2}$.

Homework 8.49

(a) Verify Eq. (8.121).

(b) Show that Eq. (8.121) reduces to the equal-beam result for both the 0- and the π-modes in Homework 8.47 when $\mu_+ = \mu_-$ and $\xi_+ = \xi_-$.

(c) When $\xi_+ = 0$, show that the $-$ beam is unperturbed, while the $+$ beam frequency shift $= G \times$ (the incoherent shift). This is the coherent dipole mode in the weak-strong limit.

(d) When $\mu_+ = \mu_-$ but $\xi_+ \neq \xi_-$, show that the two eigenmodes are still described by a 0-mode and a π-mode, and that the mode frequencies are given by

$$\cos \mu_{1,2} = \cos \mu_0 - \pi G \sin \mu_0 \begin{cases} 0, & \text{0-mode}, \\ 2(\xi_+ + \xi_-), & \pi\text{-mode}. \end{cases}$$

Solution (a) the eigenvalues λ satisfy the equation

$$\left(\lambda + \frac{1}{\lambda}\right)^2 - 2(\cos \mu_- + \cos \mu_+ - 2\pi G \xi_- \sin \mu_- - 2\pi G \xi_+ \sin \mu_+)\left(\lambda + \frac{1}{\lambda}\right)$$
$$+ (4\cos \mu_+ \cos \mu_- - 8\pi G \xi_- \cos \mu_+ \sin \mu_- - 8\pi G \xi_+ \cos \mu_- \sin \mu_+) = 0,$$

where $2\cos \mu = \lambda + \frac{1}{\lambda}$.

Homework 8.50 Consider the case when there is one bunch per beam with two interaction points, and the two arcs have a phase advances $\mu_0 \pm \delta\mu$. The

Figure 8.44: Maximum stable ξ plotted in the $(1 - \nu_{0+}, \nu_{0-})$ plane for the case $M_+ = 2, M_- = 3$. In addition to the single-ring half-integer resonances, it shows three coupling sum resonances $2\nu_{0-} + 3\nu_{0+} = 1, 2, 3$. [Courtesy Kohji Hirata and Eberhard Keil (2020).]

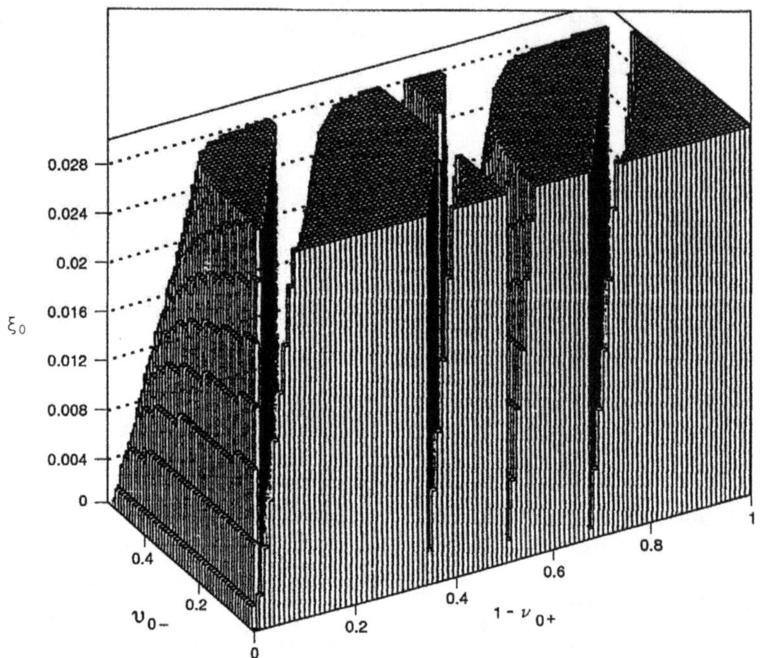

instability boundary consists of a sawtooth structure together with a stopband at half-integer total tune, as shown in Fig. 8.42(a).

(a) Perform the eigenanalysis of the system. Is there still a 0-mode and a π-mode?

(b) Use the result of (a) to verify that the stopband edges are given by the condition (8.122).

(c) Verify (8.123) for the sawtooth portion of the stability diagram.

Solution

(a) This time the total map is for the full revolution because the two arcs are now different. Find the 4th order equation for the eigenvalue λ. Let $Y = \lambda + \frac{1}{\lambda}$ to obtain a quadratic equation for Y.

(b) Stopbands occur when the condition that for stability, Y must be real.

(c) The sawtooth boundary comes from the condition that $|Y| = 2$.

8.5.4 Four-beam compensation

A point was made earlier that the strong-strong coherent beam-beam instabilities tend to assert more stringent stability conditions than that required for a weak-strong picture because

1. the effective beam-beam strength is larger due to the fact that both beams move; and

2. when there are multiple bunches per beam, there are many modes and each requires a stability condition of its own.

Item 1 makes the stopbands wider, while item 2 generates more stopbands. This point was best illustrated by the sawtooth diagram Fig. 8.38. We will drive this point a step further in this section by a discussion of a four-beam compensation scheme.

One conceived way to compensate for the beam-beam effect — therefore greatly increasing the luminosity — is the four-beam idea of DCI.[50] In this idea, two beam bunches $+Ne$ and $-Ne$ are made to collide with two other beam bunches $+Ne$ and $-Ne$ at the collision point, as sketched in Fig. 8.45. The colliding bunches have neutral net charges and thus presumably produce no beam-beam forces — if we consider only incoherent beam-beam interaction. Unfortunately, this interesting idea had not worked at DCI. If anything, the maximum reached beam-beam parameter was lower than that reached with a two-beam scheme.

Figure 8.45: Schematics of the interaction region of a four-beam compensation scheme.

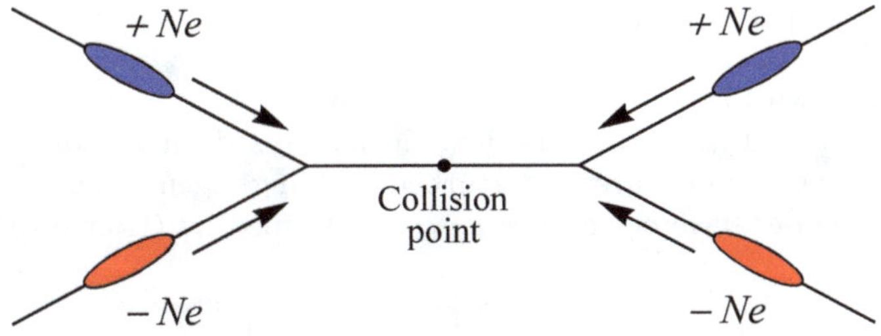

The problem is that what has been eliminated is the incoherent beam-beam force. Coherent beam-beam forces are not dealt with, because infinitesimal deviations from the steady-state can still grow as coherent modes. In fact, this four-beam scheme has worsened the coherent beam-beam effect because it introduces many more modes. The net result is that one has traded one problem with another, i.e. eliminating incoherent effects for worse coherent effects.

---

[50] J.E. Augustin, et al., Proc. 7th Int. Conf. High Energy Accel., Yerevan, Vol. 2, 113 (1970); G. Arzelier, et al., Proc. 8th Int. Conf. High Energy Accel., CERN, 150 (1971).

To analyze the four-beam dipole modes, let us designate the state vector to be $(Y_{1+}, Y_{1+}', Y_{1-}, Y_{1-}', Y_{2+}, Y_{2+}', Y_{2-}, Y_{2-}')$, and we need 8×8 matrices for the transfer maps. Take the $1+$ beam bunch for example. Its beam-beam kick is given by

$$
\begin{aligned}
\Delta Y_{1+}' &= -\frac{G}{f_y}(Y_{1+} - Y_{2-}) + \frac{G}{f_y}(Y_{1+} - Y_{2+}) \\
&= \frac{G}{f_y}(Y_{1+} - Y_{1-}).
\end{aligned}
$$

This kick is independent of Y_{1+} due to the cancellation of the neutralized oncoming beam. Combining the kicks to the four bunches, for the beam-beam interaction, we have

$$
T_{bb} = \begin{bmatrix}
1 & 0 & 0 & 0 & 0 & 0 & 0 & 0 \\
0 & 1 & 0 & 0 & -x & 0 & x & 0 \\
0 & 0 & 1 & 0 & 0 & 0 & 0 & 0 \\
0 & 0 & 0 & 1 & x & 0 & -x & 0 \\
0 & 0 & 0 & 0 & 1 & 0 & 0 & 0 \\
-x & 0 & x & 0 & 0 & 1 & 0 & 0 \\
0 & 0 & 0 & 0 & 0 & 0 & 1 & 0 \\
x & 0 & -x & 0 & 0 & 0 & 0 & 1
\end{bmatrix}, \qquad
x = \frac{G}{f_y} = \frac{2GNr_0}{\gamma \sigma_x \sigma_y}.
$$

The map for the collider ring, assuming one four-beam collision point in the ring, is

$$
T_0 = \begin{bmatrix}
t_0 & 0 & 0 & 0 \\
0 & t_0 & 0 & 0 \\
0 & 0 & t_0 & 0 \\
0 & 0 & 0 & t_0
\end{bmatrix}, \qquad
t_0 = \begin{bmatrix}
\cos \mu & \beta_0^* \sin \mu \\
-\frac{1}{\beta_0^*} \sin \mu & \cos \mu
\end{bmatrix},
$$

where $\mu = 2\pi\nu$ with ν the unperturbed tune.

The total map $T_{\text{tot}} = T_0 T_{bb}$ is then eigenanalyzed for its four eigenmodes. Two of these eigenmodes are unperturbed yielding eigenvalues $e^{\pm i\mu}$. The remaining two modes have eigentunes $\mu_{I,II}$ determined by (Homework 8.51)

$$
\begin{aligned}
\cos \mu_{I,II} &= \cos \mu \pm \beta_0^* x \sin \mu \\
&= \cos \mu \pm 4\pi\xi \sin \mu, \qquad \xi = \frac{GNr_0 \beta_0^*}{2\pi\gamma\sigma_x\sigma_y}.
\end{aligned}
\tag{8.125}
$$

Figure 8.46 shows the sawtooth stability diagram for the four-beam case (see footnote 45). The stability boundary consists of two segments, separated at $\nu = \frac{1}{4}$. The segment below $\nu = \frac{1}{4}$ is what one would obtain with colliding $-Ne$ and $-Ne$ (or colliding $+Ne$ and $+Ne$), while the segment above $\nu = \frac{1}{4}$ is that for colliding $+Ne$ and $-Ne$. The coherent dipole instability therefore is a result of the worse of the two combinations, as an inspection of Eq. (8.125) also indicates, by the fact that it holds for both signs of ξ. The maximum tolerable ξ is $\frac{1}{4\pi}$ when $\nu = \frac{1}{4}$.

Figure 8.46: Sawtooth stability diagram for a four-beam scheme when rigid-dipole coherent instability is taken into consideration. The shaded region is stable.

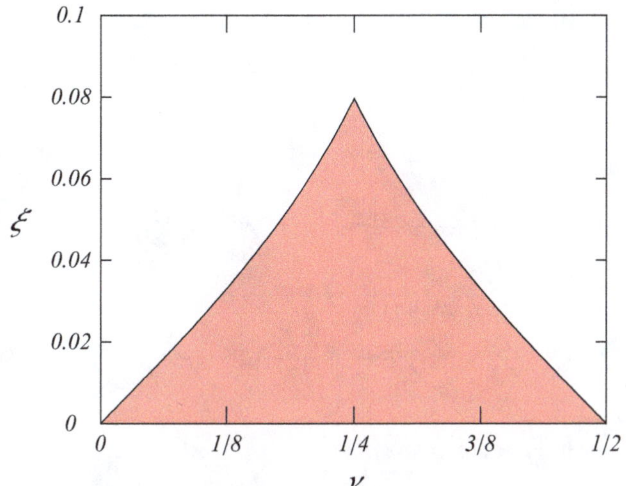

The situation would change if the four colliding beams are provided by linac colliders instead of storage rings, in which case the beams are discarded after collision. There will be no coherent beam-beam effects (at least not the multiturn resonant type of interest here), and the four-beam idea could indeed apply. In fact, one of the four beams can still come from a storage ring.

We have addressed only the rigid-dipole modes of the four-beam scheme. Analyses have been carried out also to higher order modes using Vlasov techniques.[51] We will return to the Vlasov technique for higher order modes later in Sec. 8.8 for the traditional two-beam collider case.

Homework 8.51

(a) Verify that the stability condition for the four-beam scheme is given by Eq. (8.125).

(b) According to Eq. (8.125) the stability region is divided into two segments with a dividing line at $\nu = \frac{1}{4}$. Show that the two segments are expressed by $\xi = \frac{\tan \pi \nu}{4\pi}$ and $\xi = \frac{\cot \pi \nu}{4\pi}$, respectively.

(c) It is instructive to examine the eigenvectors of the four modes exhibiting the interesting dancing of the four bunches. Convince yourself that the two unperturbed modes are those when the $+$ and the $-$ bunches move up and

[51]N.N. Chau and D. Potaux, Orsay Tech. report 5-74 (1974) and Orsay Tech. report 2-75 (1975); Ya.S. Derbenev, Proc. III All Union Part. Accel. Conf., Moscow, 382 (1973); N.S. Dikansky and D.V. Pestrikov, Part. Accel. 12, 27 (1982); Ya.S. Derbenev, N.S. Dikansky, D.V. Pestrikov, Prepront 7-75, INPh, Siberian Division, USSR Academy of Sciences, 1972.

Figure 8.47: Sketch of the four rigid-dipole modes of a four-beam scheme in the stable region.

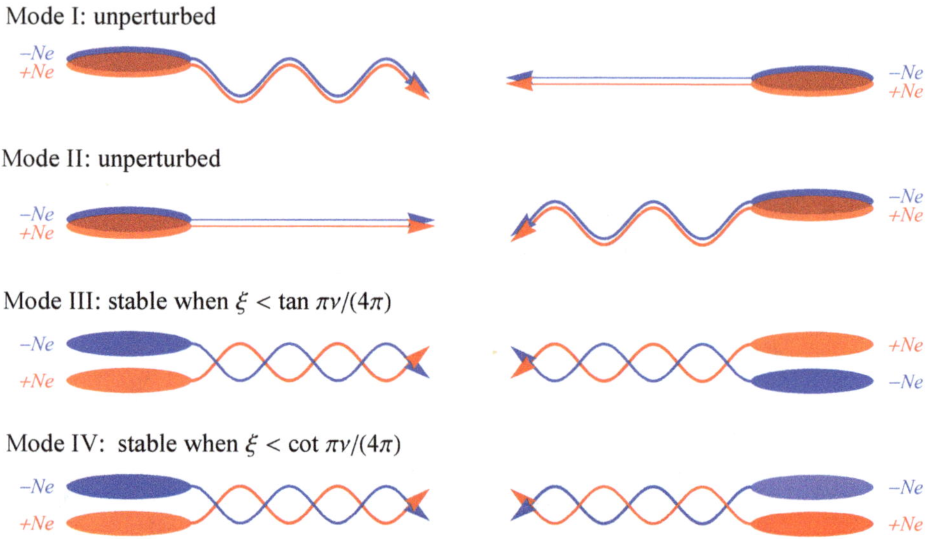

down together in either beam 1 or beam 2. The two perturbed modes are when the $+$ and $-$ bunches within each beam move out of phase while the two beams move in 0- or π-mode relative to each other.

Solution (c) The four mode patterns look like the sketch in Fig. 8.47.

8.5.5   Spontaneous beam separation

So far in analyzing the coherent dipole effects, we have linearized the beam-beam kicks with respect to the center-of-charge displacements.[52] This is valid only for center-of-charge motions small compared with the beam size σ. It suffices if one is interested only in the effect of beam-beam interaction on luminosity because even small beam motion must be avoided for luminosity purposes. The rigid-dipole motion in the presence of nonlinear center-of-charge kicks has been simulated numerically. Figure 8.48 shows one such result.[53]

With nonlinear center-of-charge kicks, the magnitude of center-of-charge motion saturates when beam separation becomes comparable to σ. In fact, in the linearly unstable region, the beams do not get lost. Instead, as beams separate far enough, they can find a steady-state separation to stay in if condition allows.

---

[52]A reminder: what has been linearized is with respect to the center-of-charge displacements. The beam-beam force is not linearized.

[53]E. Keil, Nucl. Instr. Meth. 188, 9 (1981).

Figure 8.48: The value of ξ_y as a function of the total vertical tune ν_y for the case of three bunches per beam ($M = 3$) and fixed $\nu_x = 0.2$. Rigid center-of-charge motions are simulated with nonlinear kicks. Solid and dotted curves represent the minimum and average values of ξ_y over 200 turns simulated. Sawtooth curve is that predicted by a linear theory. [Courtesy Eberhard Keil (2020).]

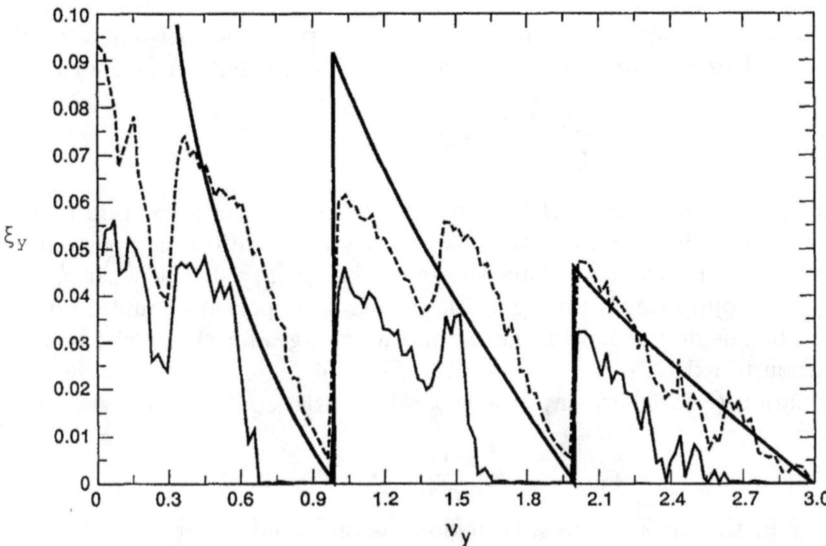

Take for example the case of one bunch per beam with round, equal-intensity beams. Let us consider a situation when the two beams reach steady-state locations with a separation distance of Δ. The linearized rigid-dipole kicks, valid when $|\Delta| \ll \sigma$, are specified by Eq. (8.120),

$$\Delta Y' = -\frac{G}{f}\Delta, \qquad \frac{1}{f} = \frac{N r_0}{\gamma \sigma^2}, \qquad G = \frac{1}{2} \text{ if round Gaussian}.$$

For the case $\xi > 0$ and $\frac{1}{2} < \nu < 1$, for example, we have the stable region specified in Eq. (8.118), required for the stability of the π-mode,

$$0 < 4\pi G \xi < -\tan \pi\nu, \qquad \xi = \frac{N r_0 \beta_0^*}{4\pi \gamma \sigma^2}.$$

When Δ is not small compared with σ, we need to resort to the corresponding nonlinear kick,

$$\Delta Y' = -\frac{2 G N r_0}{\gamma \Delta}\left(1 - e^{-\frac{\Delta^2}{2\sigma^2}}\right).$$

Here we have kept the form factor G even when Δ is not small, recognizing that when Δ becomes much larger than σ, the factor G is to be removed.

With this kick $\Delta Y'$ at the collision point, the beam executes a closed orbit distortion around the collider ring. The closed orbit at the location of the kick, according to the closed orbit analysis, is given by

$$\Delta Y^*_{\text{COD}} \;=\; \frac{\beta^*_0}{2}\Delta Y' \cot \pi \nu \,.$$

But this closed orbit ΔY^*_{COD} must be equal to $\frac{\Delta}{2}$ to be consistent with the assumption of steady-state beam separation of Δ. This self-consistency condition then yields

$$\frac{2\sigma^2}{\Delta^2}\left(1 - e^{-\frac{\Delta^2}{2\sigma^2}}\right) \;=\; -\frac{\tan \pi \nu}{4\pi G \xi}\,. \tag{8.126}$$

An inspection of Eq. (8.126) shows that there exists a solution only when $\xi \tan \pi \nu < 0$. This means a steady-state beam separation happens only when $0 < \nu < \frac{1}{2}$ if the two beams have the same charge ($\xi < 0$) and when $\frac{1}{2} < \nu < 1$ if they have opposite charges ($\xi > 0$). Incidentally, perhaps counter-intuitively, the two beams do not have to have opposite charges for the spontaneous beam separation to exist.

Equation (8.126) also says the separation exists only beyond the threshold when

$$-\frac{4\pi G \xi}{\tan \pi \nu} > 1\,,$$

i.e. only in the linearly unstable region, as one would expect. In the linearly stable region, below the threshold, the small-amplitude motion is stable and the beams do not spontaneously separate.

With beams separated, the luminosity is reduced by a factor,

$$\frac{\mathcal{L}}{\mathcal{L}_0} \;=\; \exp\left(-\frac{\Delta^2}{4\sigma^2}\right),$$

as worked out in Homework 8.1. Figure 8.49 shows the behavior of $\frac{\Delta}{\sigma}$ and luminosity reduction factor as functions of the stability parameter $x = -\frac{4\pi G \xi}{\tan \pi \nu}$.

One may study the behavior of small perturbations around a separated-beam configuration. The analysis will yield the corresponding mode frequencies and the stability conditions. The case of unequal beams has also been studied. It was found that the behavior of beam separation becomes complicated near a sum resonance $\nu_{0-} + \nu_{0+} = m$.[54]

Homework 8.52
    (a) Derive Eq. (8.126).
    (b) Show that if $|\Delta| \ll \sigma$, then

$$\frac{\Delta^2}{4\sigma^2} \;\approx\; 1 + \frac{\tan \pi \nu}{4\pi G \xi}\,.$$

---

[54]K. Hirata and E. Keil, CERN report CERN/LEP-TH/89-76 (1989); K. Hirata, KEK laboratory report TN-880010 (1989).

Figure 8.49: Steady-state beam separation due to nonlinear beam-beam force in the linearly unstable region. The beam separation $\frac{\Delta}{\sigma}$ and the luminosity reduction factor are plotted as functions of $x = -\frac{4\pi G\xi}{\tan \pi\nu}$. Occurrence of the beam spontaneous separation requires $x \geq 1$.

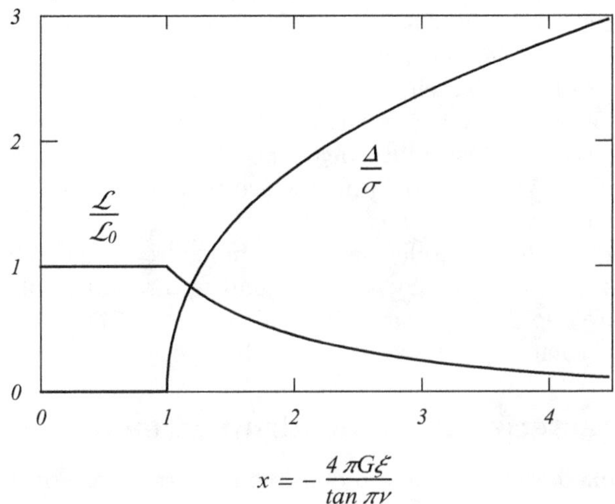

Confirm this behavior with Fig. 8.49.

Homework 8.53 The text treated the case of two equal bunches. Extend the result to the case when the two bunches are nonequal,

(a) when $\xi_+ \neq \xi_-$.

(b) when $\nu_+ \neq \nu_-$.

8.6 Quadrupole mode

In the dipole model we have been discussing so far, the beam distribution is assumed to be rigid; only center-of-charge motions are allowed. In general, coherent beam motion is to be described as a superposition of all modes, not only dipole modes. If we consider the transverse coherent beam-beam instabilities, all transverse modes need to be considered.

For small beam-beam strength parameters, coupling among different transverse modes is weak, particularly when

$$\xi \ll \text{fractional part of } \nu_y, \tag{8.127}$$

it is possible to study the coherent beam-beam effects by considering each transverse mode separately. Our calculation of the dipole coherent instability then still applies. In case of M bunches per beam, the dipole mode degenerates into

$2M$ modes in a 1-D case. In particular, the stability region will be a sawtooth diagram with instability bands starting when the total tune is near $\nu =$ integer, as shown in Fig. 8.38.

One can also perform a stability analysis on, say, the quadrupole mode and focus on its stability limit. With M bunches per beam, the quadrupole mode degenerates into $3M$ modes in a 1-D case. (See Homework 8.60.) In a quadrupole mode, it is envisioned the beam bunches execute transverse quadrupole oscillations. The center of each bunch does not move and all bunches collide head-on, but the shape of each bunch oscillates in time, leading to loss of luminosity. As we will see, there is instability near the resonances $\nu_x = \frac{m}{2}$, $\nu_y = \frac{m}{2}$, or $\nu_x + \nu_y = m$. Similarly, sextupole modes are unstable near the 3rd order uncoupled or coupled resonances, etc.

The degenerated dipole modes have their mode frequencies clustered around the betatron oscillation frequency ν_y. The quadrupole modes will cluster around $2\nu_y$.[55] They are separated cleanly if condition (8.127) holds, justifying our analysis to treat them separately and individually.

8.6.1   Characterizing the quadrupole mode

To study the coherent quadrupole modes, consider the case of two beams with one bunch each colliding head-on (i.e. no dipole motion). There are two collision points per revolution. Let both beams have bi-Gaussian distributions but their beam sizes oscillate around some equilibrium values in a quadrupole mode oscillation. Each beam is characterized by a Σ-matrix whose elements are the second moments,

$$
\Sigma =
\begin{bmatrix}
\langle x^2 \rangle & \langle xx' \rangle & 0 & 0 \\
\langle xx' \rangle & \langle x'^2 \rangle & 0 & 0 \\
0 & 0 & \langle y^2 \rangle & \langle yy' \rangle \\
0 & 0 & \langle yy' \rangle & \langle y'^2 \rangle
\end{bmatrix}.
$$

We have assumed the two beams are upright in x-y plane and they maintain so even undergoing the quadrupole oscillation.

There is a Σ-matrix for each beam. Designate the two Σ-matrices by $\Sigma_\pm$. Let there be the equilibrium Σ-matrix designated as Σ_0, and on top of Σ_0, each bunch has a small time-dependent perturbation, i.e. let

$$
\begin{aligned}
\Sigma_+ &= \Sigma_0 + \Delta\Sigma_+ , \\
\Sigma_- &= \Sigma_0 + \Delta\Sigma_- .
\end{aligned}
\tag{8.128}
$$

To study the problem of the stability limit under the quadrupole beam-beam effect, we consider $\Delta\Sigma_\pm$ infinitesimally small. Due to this assumption, our result will be valid to yield the instability threshold when $\Delta\Sigma_\pm$ barely begin to start

---

[55]More accurately, this statement is excluding the dynamic-β^* mode in the 1-D flat beam case (see Sec. 8.6.3) and the two invariants in the 2-D x-y coupled case (see Sec. 8.6.4), which are static quantities with zero frequency. All other dynamical modes are clustered around $2\nu_y$ in the 1-D case and $2\nu_x, 2\nu_y, \nu_x \pm \nu_y$ in the 2-D case. See Homework 8.60

their exponential growths. It is not applicable, for example, if we want to study any instability saturation effects beyond the instability threshold when $\Delta\Sigma_\pm$ become significant.

The transformations through the arcs are

$$(\Sigma_\pm)_{\text{out}} = T_0 \, (\Sigma_\pm)_{\text{in}} \, \tilde{T_0} \,, \tag{8.129}$$

with

$$T_0 = \begin{bmatrix} t_x & 0 \\ 0 & t_y \end{bmatrix}, \qquad t_{x,y} = \begin{bmatrix} \cos\mu_{x,y} & \beta^*_{x,y}\sin\mu_{x,y} \\ -\dfrac{1}{\beta^*_{x,y}}\sin\mu_{x,y} & \cos\mu_{x,y} \end{bmatrix},$$

where $\mu_{x,y} = \pi\nu_{x,y}$ are the betatron phase advances between the two interaction points, and $\nu_{x,y}$ are the betatron tunes of the storage ring for a whole revolution. For simplicity, however, later we will consider only the case when the beams are round, with $\beta^*_x = \beta^*_y = \beta^*_0$ and $\mu_x = \mu_y = \mu_0$.

As emphasized earlier, nonlinear dynamics plays a fundamental role in the coherent beam-beam dynamics. Unlike the single-particle nonlinear dynamics, however, even a linearized beam-beam force is capable of driving nonlinear resonances because the linearized focal lengths depend on the beam sizes. The transformation of the Σ-matrices through the interaction point due to a linearized beam-beam force is given by

$$(\Sigma_+)_{\text{out}} = T_{bb-} \, (\Sigma_+)_{\text{in}} \, \tilde{T}_{bb-} \,,$$
$$(\Sigma_-)_{\text{out}} = T_{bb+} \, (\Sigma_-)_{\text{in}} \, \tilde{T}_{bb+} \,,$$

where

$$T_{bb+} = \begin{bmatrix} 1 & 0 & 0 & 0 \\ -\dfrac{1}{f_{x-}} & 1 & 0 & 0 \\ 0 & 0 & 1 & 0 \\ 0 & 0 & -\dfrac{1}{f_{y-}} & 1 \end{bmatrix}, \qquad T_{bb-} = \begin{bmatrix} 1 & 0 & 0 & 0 \\ -\dfrac{1}{f_{x+}} & 1 & 0 & 0 \\ 0 & 0 & 1 & 0 \\ 0 & 0 & -\dfrac{1}{f_{y+}} & 1 \end{bmatrix}, \tag{8.130}$$

with the focal lengths $f_{x\pm}, f_{y\pm}$ given by

$$\frac{1}{f_{x\pm}} = \frac{2Nr_0}{\gamma\sqrt{\langle x^2\rangle_\pm}\left(\sqrt{\langle x^2\rangle_\pm} + \sqrt{\langle y^2\rangle_\pm}\right)} \,,$$

$$\frac{1}{f_{y\pm}} = \frac{2Nr_0}{\gamma\sqrt{\langle y^2\rangle_\pm}\left(\sqrt{\langle x^2\rangle_\pm} + \sqrt{\langle y^2\rangle_\pm}\right)} \,.$$

The fact that the focal lengths $f_{x\pm}$ and $f_{y\pm}$ depend on the elements of the matrices $\Sigma_\pm$ contributes to the source of nonlinearity of the beam-beam map. Furthermore, they couple the $+$ and the $-$ beams. Such couplings are absent in the map T_0 around the rest of the ring.

The problem we are dealing with is to consider a beam-beam force that is linearized with respect to x and y (that is why we can use matrices for the beam dynamics), and then consider the dynamics of the Σ-matrices. We decompose the Σ-matrices according to Eq. (8.128) and then linearize with respect to $\Delta\Sigma_\pm$. The two linearizations, one being the linearization of the beam-beam force with respect to x and y, the other being the linearization of the dynamics with respect to $\Delta\Sigma$, are not to be confused. The dynamics of $\Delta\Sigma_\pm$, as we will see, contains nonlinear resonances when $\nu = \frac{m}{2}$.

Since there are two beam bunches, each having its own Σ-matrix, and each Σ-matrix has six independent elements, we have a general problem of a 12×12 matrix to describe the dynamics. The vector X for the dynamical state of the two beams is

$$\tilde{X} \;=\; (\Delta\Sigma_{+11}, \Delta\Sigma_{+12}, \Delta\Sigma_{+22}, \Delta\Sigma_{+33}, \Delta\Sigma_{+34}, \Delta\Sigma_{+44},$$
$$\Delta\Sigma_{-11}, \Delta\Sigma_{-12}, \Delta\Sigma_{-22}, \Delta\Sigma_{-33}, \Delta\Sigma_{-34}, \Delta\Sigma_{-44}). \quad (8.131)$$

In writing down Eq. (8.131), we have assumed the two beams do not execute any motion in the coupled x-y phase space, i.e. the beams stay upright at all times in the x-y plane. If tilting is allowed, the dynamical state of the two beams will be 20-dimensional.

The job is to linearize the beam-beam maps (8.130) with respect to the elements in X, and find the 12×12 map M that transforms X for one period from one interaction point to the next,

$$X_{\text{after}} \;=\; MX_{\text{before}}. \quad (8.132)$$

The stability of the system is then specified by the condition that all twelve eigenvalues of M must be equal to 1.

In the following sections 8.6.2, 8.6.3 and 8.6.4, we apply this algorithm to a few circumstances.

8.6.2   Dynamic-β^* as a static quadrupole mode

We mentioned a certain steady-state Σ_0 but did not elaborate on it. Before we analyze quadrupole oscillations and their instabilities, we now fill in the gap solving for Σ_0.

This steady state Σ_0 is supposed to be the equilibrium Σ-matrix of both beams around which quadrupole modes oscillate. We consider the case of two equal flat bi-Gaussian beams with $\sigma_x \gg \sigma_y$, and we ignore x-motion. Assume the two beams have opposite charges and there are no dipole center-of-charge motions.

In the flat-beam case, and assuming no x-motion, the problem reduces to 1-D. In this case, both beams have

$$\Sigma = \begin{bmatrix} \langle y^2 \rangle & \langle yy' \rangle \\ \langle yy' \rangle & \langle y'^2 \rangle \end{bmatrix}.$$

Specify the beam-beam strength by the usual single-particle parameter

$$\xi_0 \;=\; \frac{\beta^*}{4\pi f_0}, \qquad \frac{1}{f_0} \;=\; \frac{2Nr_0}{\sigma_{x0}\sigma_{y0}},$$

where β^* is the unperturbed vertical β-function at the collision point. The beam-beam kick is

$$T_{bb0} = \begin{bmatrix} 1 & 0 \\ -\frac{1}{f_0} & 1 \end{bmatrix}.$$

The map in the collider ring from one collision point to the next is

$$T_0 = \begin{bmatrix} \cos\mu & \beta^* \sin\mu \\ -\frac{1}{\beta^*}\sin\mu & \cos\mu \end{bmatrix},$$

where μ is the unperturbed vertical betatron phase between collision points. The total map for one superperiod is then

$$T_{\text{tot}} = T_0 T_{bb0}.$$

The steady-state Σ_0 is determined by the self-consistency condition that it reproduces itself after one superperiod, i.e.,

$$\Sigma_0 = T_{\text{tot}}\Sigma_0\tilde{T}_{\text{tot}}.$$

Solving for the elements of Σ_0, we obtain

$$(\Sigma_0)_{11} = \frac{\beta^{*2}f_0\sin\mu}{\beta^*\cos\mu + f_0\sin\mu}(\Sigma_0)_{22},$$

$$(\Sigma_0)_{12} = \frac{\beta^{*2}\sin\mu}{2(\beta^*\cos\mu + f_0\sin\mu)}(\Sigma_0)_{22}. \tag{8.133}$$

Equation (8.133) then defines the steady-state beam distribution when the linearized beam-beam effect is present. We can translate it to a language of β-functions, and the corresponding perturbed β-function is given by (Homework 8.55)

$$\bar{\beta}^* = \frac{(\Sigma_0)_{11}}{(\text{Det}\,\Sigma_0)^{-1/2}} = \frac{\beta^*\sin\mu}{\sqrt{\sin\mu\left[\frac{\beta^*}{f_0}\cos\mu - \frac{\beta^{*2}}{4f_0^2}\sin\mu + \sin\mu\right]}}. \tag{8.134}$$

The reader should note that this result (8.134) is exactly the beam-beam perturbed β-function described by Eq. (8.22) for a weak-strong case if f_0 is specified by the strong beam. In the present strong-strong case, when we demand the two beams have the same distribution Σ_0 and f_0 be specified by the perturbed beam sizes, then the equal-beam self-consistency condition leads to the dynamic-β^* solution (8.29). In the present strong-strong equal-beam analysis, therefore, the Σ_0 corresponds to the dynamic-β^* state as described by Eq. (8.29).

If we further allow the two beams to have different equilibrium beam distributions, then it leads to the flip-flop solution as discussed in Sec. 8.3.4 and Eq. (8.35). Later in Sec. 8.6.5, we will discuss the stability of quadrupole motion around the dynamic-β^* and the flip-flop solutions. We will see that the equal-beam dynamic-β^* solution and the flip-flop solution are mutually exclusive; when one is stable, the other must be unstable, and the beams must choose one or the other (or neither) to be their stable equilibrium state.

Homework 8.54 This homework is a curiosity quiz. We know the matrix Σ is symmetric. Suppose it is given an expression

$$\Sigma = A,$$

where A is a symmetric matrix with explicit numerical values. Can it follow to write

$$\Sigma A^{-1} = I?$$

Solution No, it does not follow. Neither can it be written $A^{-1}\Sigma = I$. Beware. The correct expression is

$$A^{-1/2}\Sigma A^{-1/2} = I.$$

That $A^{-1/2}$ must be well-behaved is due to the fact that A not only has to be symmetric but also must be positive definite. It is intriguing that just by writing down an expression like $\Sigma = A$, it automatically already imposes by self-consistency that your A must obey some hidden conditions.

Homework 8.55

(a) Follow the text and solve the self-consistency condition of Σ_0 to verify Eq. (8.133).

(b) Then convince yourself that the solution (8.133) corresponds to a perturbed β-function (8.134).

(c) Given Eq. (8.133), find expressions of the other perturbed Courant–Snyder functions $\bar{\alpha}^*, \bar{\gamma}^*$. These expressions $\bar{\beta}^*, \bar{\alpha}^*, \bar{\gamma}^*$ refer to the location immediately ahead of the collision point.

8.6.3   Quadrupole mode instability — 1-D flat beam

Let us continue the analysis of two colliding flat beams. It is a simpler case because, without x-motion, it is a 1-D problem. The 12×12 maps mentioned earlier in Eqs. (8.131) and (8.132) reduce to 6×6.

Let us consider the $+$ beam first. Its total map for one superperiod is

$$T_{\text{tot}+} = T_0 T_{bb+}, \qquad T_{bb+} = \begin{bmatrix} 1 & 0 \\ -\frac{1}{f_-} & 1 \end{bmatrix},$$

where we want to express $\frac{1}{f_-}$ in terms of the elements of $\Sigma_- = \Sigma_0 + \Delta\Sigma_-$ and linearize it with respect to $\Delta\Sigma_-$. Here, Σ_0 and Σ_- are 2×2, referring only the y-elements. The result is, for a flat beam,

$$\frac{1}{f_-} = \frac{2Nr_0}{\sigma_{x0}\sigma_y} = \frac{2Nr_0}{\sigma_{x0}\sqrt{(\Sigma_-)_{11}}}$$

$$= \frac{1}{f_0}\frac{\sqrt{(\Sigma_0)_{11}}}{\sqrt{(\Sigma_-)_{11}}} \approx \frac{1}{f_0} - \frac{1}{2f_0}\frac{(\Delta\Sigma_-)_{11}}{(\Sigma_0)_{11}}.$$

The one-period total map for Σ_+ is therefore given by

$$
\begin{aligned}
(\Sigma_+)_{\text{after}} &= T_0 T_{bb+}\Sigma_+\tilde{T}_{bb+}\tilde{T}_0 \\
&\approx T_0 T_{bb0}\Sigma_+\tilde{T}_{bb0}\tilde{T}_0 \\
&\quad - \frac{1}{2f_0}\frac{(\Delta\Sigma_-)_{11}}{(\Sigma_0)_{11}} \times T_0\left(\begin{bmatrix}0 & 0 \\ 1 & 0\end{bmatrix}\Sigma_+\tilde{T}_{bb0} + T_{bb0}\Sigma_+\begin{bmatrix}0 & 1 \\ 0 & 0\end{bmatrix}\right)\tilde{T}_0,
\end{aligned}
$$

or, to 1st order in $\Delta\Sigma_\pm$,

$$
\begin{aligned}
(\Delta\Sigma_+)_{\text{after}} &\approx T_0 T_{bb0}(\Delta\Sigma_+)_{\text{before}}\tilde{T}_{bb0}\tilde{T}_0 \\
&\quad - \frac{1}{2f_0}\frac{(\Delta\Sigma_-)_{11}}{(\Sigma_0)_{11}} \times T_0\left(\begin{bmatrix}0 & 0 \\ 1 & 0\end{bmatrix}\Sigma_0\tilde{T}_{bb0} + T_{bb0}\Sigma_0\begin{bmatrix}0 & 1 \\ 0 & 0\end{bmatrix}\right)\tilde{T}_0.
\end{aligned}
$$

We can also work out the counterpart expression for $(\Delta\Sigma_-)_{\text{after}}$.

We know that Σ_0 obeys the steady-state condition (8.133). Substituting Eq. (8.133) into our results, and form the 6-D vector

$$\tilde{X} = (\Delta\Sigma_{+11}, \Delta\Sigma_{+12}, \Delta\Sigma_{+22}, \Delta\Sigma_{-11}, \Delta\Sigma_{-12}, \Delta\Sigma_{-22}),$$

we find the one-period map for X is

$$
X_{\text{after}} = M X_{\text{before}}, \qquad
M = \begin{bmatrix}
m_{11} & m_{12} & m_{13} & m_{14} & 0 & 0 \\
m_{21} & m_{22} & m_{23} & m_{24} & 0 & 0 \\
m_{31} & m_{32} & m_{33} & m_{34} & 0 & 0 \\
m_{14} & 0 & 0 & m_{11} & m_{12} & m_{13} \\
m_{24} & 0 & 0 & m_{21} & m_{22} & m_{23} \\
m_{34} & 0 & 0 & m_{31} & m_{32} & m_{33}
\end{bmatrix},
$$

$$m_{11} = \left(\cos\mu - \frac{\beta^*}{f_0}\sin\mu\right)^2,$$

$$m_{12} = 2\beta^*\sin\mu\left(\cos\mu - \frac{\beta^*}{f_0}\sin\mu\right),$$

$$m_{13} = \beta^{*2}\sin^2\mu,$$

$$m_{21} = -\frac{1}{f_0}\cos 2\mu - \frac{1}{2}\sin 2\mu\left(\frac{1}{\beta^*} - \frac{\beta^*}{f_0^2}\right),$$

$$m_{22} = \cos 2\mu - \frac{\beta^*}{f_0}\sin 2\mu,$$

$$m_{23} = \frac{\beta^*}{2}\sin 2\mu,$$

$$m_{31} = \left(\frac{\cos\mu}{f_0} + \frac{\sin\mu}{\beta^*}\right)^2,$$

$$m_{32} = -2\cos\mu\left(\frac{\cos\mu}{f_0} + \frac{\sin\mu}{\beta^*}\right),$$

$$
\begin{aligned}
m_{33} &= \cos^2 \mu \,, \\[2mm]
m_{14} &= \frac{\beta^*}{2 f_0^2} \sin \mu (-2 f_0 \cos \mu + \beta^* \sin \mu) \,, \\[2mm]
m_{24} &= \frac{1}{4 f_0^2} (-2 f_0 \cos 2\mu + \beta^* \sin 2\mu) \,, \\[2mm]
m_{34} &= \frac{1}{2 f_0^2 \beta^*} \cos \mu (\beta^* \cos \mu + 2 f_0 \sin \mu) \,.
\end{aligned}
\tag{8.135}
$$

It is straightforward to show that $\det M = 1$ (Homework 8.56).

Stability of the system is determined by the map M. For stability, all its eigenvalues must have absolute value of 1. The map M has six eigenvalues. Two of them are give by 1. The corresponding eigenvectors are (Homework 8.58)

$$
\begin{bmatrix}
-2\beta^{*2} \\
-\dfrac{3\beta^{*2}}{4 f_0} \\
-2 - \dfrac{3\beta^*}{2 f_0} \cot \mu \\
\beta^{*2} \\
0 \\
1
\end{bmatrix} , \qquad \text{and} \qquad
\begin{bmatrix}
\beta^{*2} \\
0 \\
1 \\
-2\beta^{*2} \\
-\dfrac{3\beta^{*2}}{4 f_0} \\
-2 - \dfrac{3\beta^*}{2 f_0} \cot \mu
\end{bmatrix} .
\tag{8.136}
$$

The meaning of these eigenvectors is that if the two beams are distorted in their distributions according to them, then these distortions will stay intact turn after turn as steady-state solutions. These distorted distributions are invariants of the motion.

The remaining four eigenvalues form two modes. Stability of these two modes determines the system stability. Eigenanalysis of the map M yields two instability stopbands due to these two modes, as illustrated in Fig. 8.50 (Homework 8.59). One set of stopbands corresponds to the integer resonances $\nu = m$; the other is of the half-integer resonances $\nu = \frac{m}{2}$, where $\nu = \frac{\mu}{\pi}$ is the total betatron tune of the collider ring. The integer resonance stopbands have the instability boundaries

$$
\xi_0 = \frac{10 \cos \pi\nu \pm \sqrt{98 + 2 \cos 2\pi\nu}}{24\pi \sin \pi\nu} \,.
\tag{8.137}
$$

The half-integer stopbands have the boundaries,

$$
\xi_0 = \frac{\cot \pi\nu}{3\pi} \,, \qquad \text{and} \qquad \xi_0 = \frac{\cot \pi\nu}{\pi} \,.
\tag{8.138}
$$

Figure 8.50 has also included the single-particle instability boundaries (8.139) of Homework 8.57 by dashed curves.

Homework 8.56

(a) Follow the text to derive Eq. (8.135).

(b) Show that $\det M = 1$. Is M symplectic?

Figure 8.50: Stability (ν, ξ_0) sawtooth diagram for the beam-beam quadrupole modes assuming equal flat beams with one bunch per beam, where $\xi_0 = \frac{N r_0 \beta^*}{2\pi \sigma_{x0} \sigma_{y0}}$ is the single-particle beam-beam parameter, ν is the total betatron tune for one full revolution. The shaded regions are unstable. Two sets of stopbands are shown per integer interval of ν, one set starts at $\nu = \frac{m}{2}$, the other starts at $\nu = m$. The dashed curves, plotted for reference, are the single-particle stability boundaries, given by Eq. (8.139) of Homework 8.57.

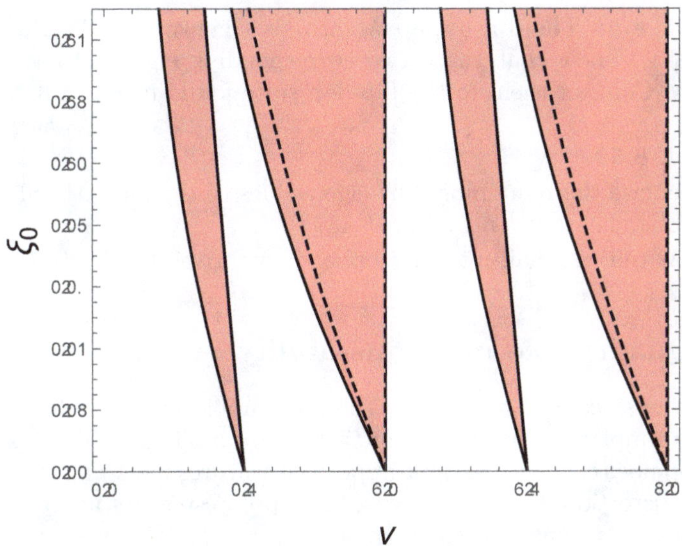

Solution (b) Symplecticity of a map makes sense only when dealing with the map on the canonical coordinates of a Hamiltonian system. Here the components of X do not even have a phase space to speak of. Nevertheless, $\det M$ is indeed equal to 1.

Homework 8.57 Consider the partial 3×3 map $M' = \begin{bmatrix} m_{11} & m_{12} & m_{13} \\ m_{21} & m_{22} & m_{23} \\ m_{31} & m_{32} & m_{33} \end{bmatrix}$ of the quadrupole mode dynamics.

   (a) Show that M' also has determinant equal to 1.

   (b) Find the eigenvalues of M'.

   (c) One of the eigenvalues of M' is equal to 1. Give the physical meaning of this particular mode.

   (d) The remaining two eigenvalues of M' constitute a mode. Find the stability condition for this mode. Give its physical meaning.

Solution

   (c) The eigenvector of this partial map with eigenvalue 1 is just Eq. (8.133).

Did you expect this?

(d) The stability condition is the familiar expression,

$$\xi_0 \;=\; \frac{\beta^*}{4\pi f_0} \;=\; \frac{\cot\frac{\mu}{2}}{2\pi} \quad \text{and} \quad -\frac{\tan\frac{\mu}{2}}{2\pi}\,. \tag{8.139}$$

This expression is shown as dashed curves in Fig. 8.50.

Homework 8.58 The 6×6 map M has six eigenvalues. Two of them have eigenvalues 1. Show that those two corresponding eigenvectors are given by Eq. (8.136). Connect them to the flip-flop state found in Sec. 8.3.4.

Homework 8.59

(a) Calculate the remaining four eigenvalues of the map M. Show that the integer stopbands are given by Eq. (8.137).

(b) Show that the half-integer stopbands are given by Eq. (8.138).

8.6.4 Quadrupole mode instability — 2-D x-y coupling

In the previous section, we studied the quadrupole instability for a flat beam case. It is a simpler case because there is no x-y coupling. As discussed, the flat-beam model does contain the most important features including the single-beam β-function distortion, the dynamic-β^*, the flip-flop steady state, and the integer and half-integer resonances $\nu_y = m$ and $\nu_y = \frac{m}{2}$. However, one noticeable feature of x-y coupling is missing, i.e. we have missed the resonance effects due to the $\nu_x \pm \nu_y = m$ resonances. This we will make up in this section. The price to pay is now we have to deal with 12×12 dynamics.

We now linearize the system of equations with respect to the infinitesimal elements in the matrices $\Delta\Sigma_{\pm}$ according to the algorithm Eqs. (8.128) to (8.132). Since each beam has six elements $\Delta\langle x^2\rangle, \Delta\langle xx'\rangle, \Delta\langle x'^2\rangle, \Delta\langle y^2\rangle, \Delta\langle yy'\rangle$, and $\Delta\langle y'^2\rangle$, the dynamics of the system is described by 12×12 transformation matrices. Eigenanalysis of the system then gives six modes. It is found[56] that four of the twelve eigenvalues are identically unity. They correspond to the two invariant beam distributions of the motion $\langle x^2\rangle\langle x'^2\rangle - \langle xx'\rangle^2$ and $\langle y^2\rangle\langle y'^2\rangle - \langle yy'\rangle^2$. The remaining four dynamic modes determine the dynamical stability of the system.

Equal tunes $\nu_x = \nu_y$ To simplify the analysis, let us assume a round beam with $\sigma_x = \sigma_y = \sigma_0, \xi_x = \xi_y = \xi_0$ and the unperturbed tunes $\nu_x = \nu_y = \nu_0$. With one bunch per beam and with two collision points, the unperturbed phase advance between collision points is $\mu_0 = \pi\nu_0$. The 12×12 system can be analytically solved. The eigenmode frequencies μ for the four dynamical modes

---

[56]Y. Kamiya and A. Chao, SLAC/AP-8 (1983); A.W. Chao, AIP Proc. 127, 201 (1983).

Figure 8.51: The sawtooth diagram shows the stability region in the (ν_0, ξ_0) space for the beam-beam quadrupole modes. Round beams are assumed with $\nu_x = \nu_y = \nu_0$ the total betatron tune for one full revolution, and $\xi_x = \xi_y = \xi_0$. The single-particle stability boundaries are shown as the dashed curves for comparison.

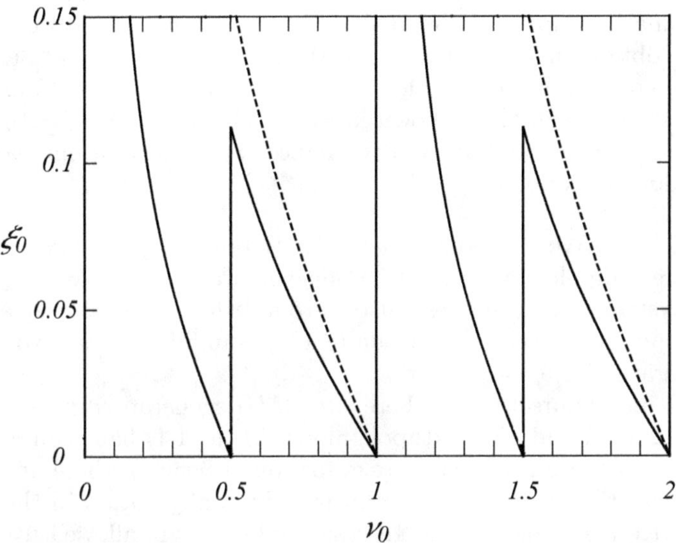

are given by

$$\cos \mu = \cos 2\bar{\mu} + \begin{bmatrix} 2 \\ 1 \\ -1 \\ -2 \end{bmatrix} 2\pi\xi_0 \sin \mu_0 \cos \bar{\mu}, \tag{8.140}$$

where $\bar{\mu}$ is defined by

$$\cos \bar{\mu} = \cos \mu_0 - 2\pi\xi_0 \sin \mu_0.$$

System stability requires all four modes to have real solution for the mode frequencies μ. Combining the stability conditions (8.140) of the four modes then establishes Fig. 8.51. The integer and half-integer resonances are apparent. Dashed curves are the single-particle instability for the incoherent weak-strong picture, shown for comparison. Analytical expressions of the sawtooth stability boundaries can be found in Homework 8.61.

Analytical results become cumbersome when there are more bunches per beam. A computer calculation was implemented to find the stable regions. Careful bookkeeping of the multiple bunches in both beams is required. The results are shown in Fig. 8.52. With M bunches per beam, there are $4M$ dynamic quadrupole modes, plus $2M$ static modes. Again, stopbands occur when the total tune is half-integral. As mentioned, the inclusion of coherent quadrupole modes introduces additional restrictions on the beam stability region, and the restriction increases as the bunch number is increased.

Coupling resonance $\nu_x + \nu_y = m$ The case of equal tunes $\nu_x = \nu_y$ does not allow us to study the sum coupling resonances. To do so, we have to relieve the equal-tune assumption. The analysis leading to Eq. (8.140) can in fact be extended to obtain analytical results for the cases $\nu_x \neq \nu_y, \xi_x \neq \xi_y$ and $\sigma_x \neq \sigma_y$ with the help of a computer to calculate the eigenvalues of 12×12 matrices (see footnote 56). One example is shown in Fig. 8.53 when $\xi_x = \xi_y = \xi_0$ and round beam $\sigma_x = \sigma_y$. The coupling sum resonances $\nu_x + \nu_y = m$ emerge clearly as bifurcated double stopbands.

Homework 8.60 We mentioned that with M bunches per beam, there will be $2M$ degenerate dipoles modes and $3M$ degenerate quadrupole modes in a 1-D flat beam system. These modes cluster around their natural frequencies of ν_y and $2\nu_y$, respectively. When condition (8.127) is fulfilled, these two clusters are cleanly separated from each other.

(a) Convince yourself that there are $2M$ degenerate dipole modes, $3M$ quadrupole modes, and $4M$ sextupole modes in the 1-D flat beam case.

(b) Show that in a 2-D system (e.g. for round beams), there are $4M$ dipole modes. In case the beams all maintain upright configuration in the x-y plane, there are $6M$ quadrupole modes. In case the beams are allowed to be tilted in their x-y phase space, there are $10M$ quadrupole modes.

(c) In each of the cases in (b), how many of those modes are static, and how many are dynamical?

Homework 8.61 For the equal-tune case, the text worked out the analytic expression (8.140) for the eigenmode frequencies of the four dynamical modes. Their combined stability diagram is provided in Fig. 8.51. The dashed curves in Fig. 8.51 are familiar, i.e.

$$\xi_0 = \frac{\cot \frac{\pi \nu_0}{2}}{2\pi}, \qquad \text{and} \qquad \xi_0 = -\frac{\tan \frac{\pi \nu_0}{2}}{2\pi}.$$

(a) Find the analytical expression of the sawtooth segment between $\nu_0 = 0$ and $\nu_0 = \frac{1}{2}$.

(b) Find the analytical expression of the sawtooth segment between $\nu_0 = \frac{1}{2}$ and $\nu_0 = 1$.

Solution

(a) $\xi_0 = \frac{\cot \pi \nu_0}{4\pi}$.

(b) $\xi_0 = \frac{6 \cos \pi \nu_0 \pm \sqrt{34 + 2 \cos 2\pi \nu_0}}{16\pi \sin \pi \nu_0}$.

Figure 8.52: The sawtooth quadrupole mode stability diagrams for 2, 3, and 4 bunches per beam (upper, middle, lower). Round beams are assumed with $\nu_x = \nu_y = \nu$ and $\xi_x = \xi_y = \xi$. Shaded regions are unstable. [Courtesy Yukihide Kamiya (2020).]

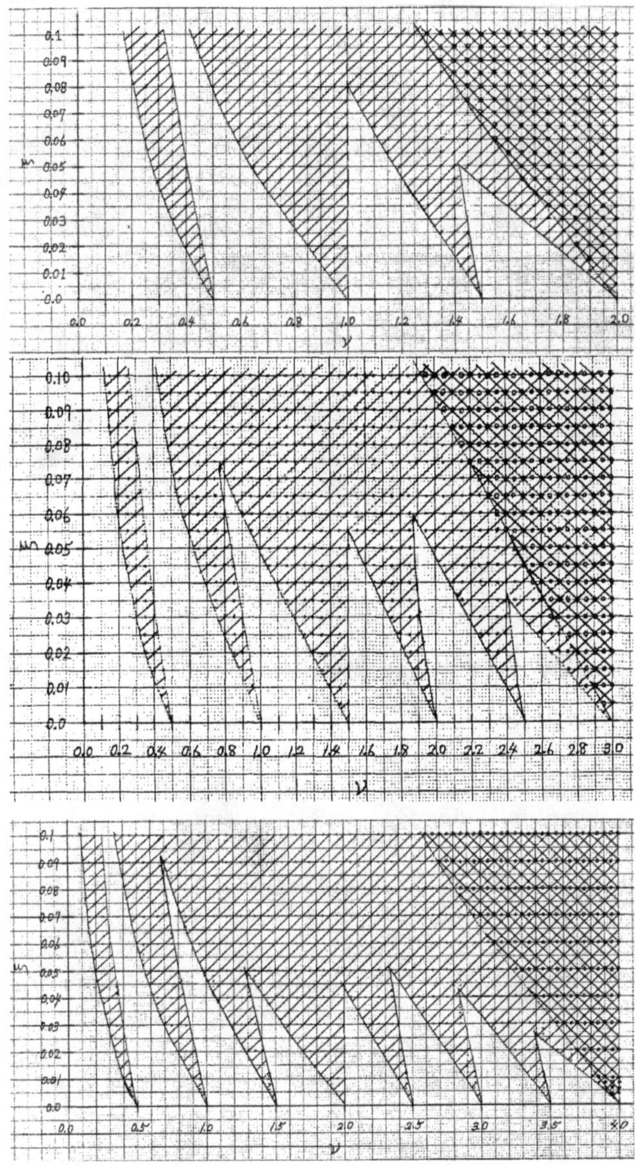

Figure 8.53: Stability diagram in the (ν_x, ν_y) space for the beam-beam coherent quadrupole modes. The case is for two round beams with $\sigma_x = \sigma_y$ and $\xi_x = \xi_y = \xi_0$. The left panel has $\xi_0 = 0.02$; the right panel has $\xi_0 = 0.06$. Shaded regions are unstable. In addition to the usual integer and half-integer resonances, double stopbands around the sum resonances $\nu_x + \nu_y = m$ also appear. [Courtesy Yukihide Kamiya (2020).]

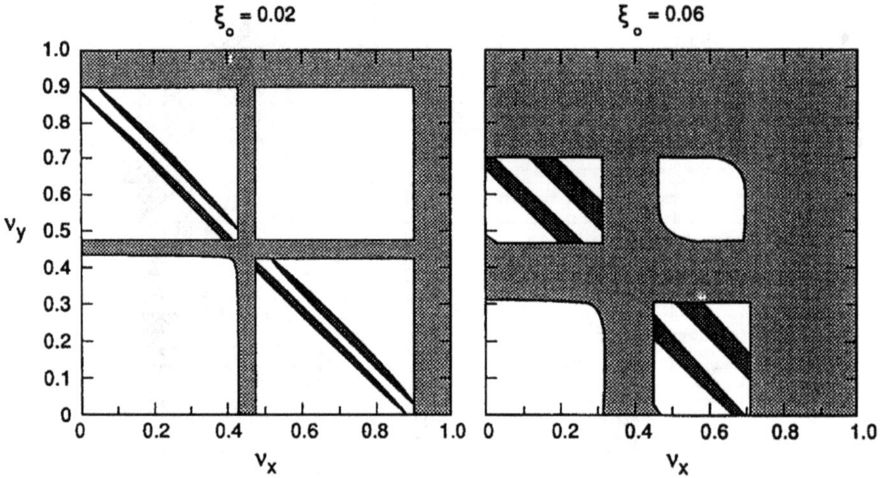

8.6.5   Quadrupole mode around dynamic-β^* and flip-flop states

The previous section addresses the quadrupole oscillations around the beam-beam perturbed equilibrium state Σ_0, which we identify by Eqs. (8.133) and (8.134). As discussed following Eq. (8.134), an equal-beam self-consistency condition then leads us to conclude that this Σ_0 is just the dynamic-β^* solution.

On the other hand, we also discussed in Sec. 8.3.4 and illustrated in Fig. 8.12 that there exists another steady-state solution when the two beams acquire different Σ_0's even when they have the same intensity and the same storage ring optics. Question remains what the strong-strong beams would choose to stay in when presented the choice between an equal-beam dynamic-β^* state or a flip-flop state. We shall now try to address this issue.

As mentioned before, linear beam-beam force is inconsistent with Gaussian distribution. A Gaussian beam carries an electromagnetic field that is linear in x and y only when x and y are much smaller than the beam sizes. Insisting on linear beam-beam force, strictly speaking, compromises the Maxwell equations, and the results at best apply only to particle motions very close to the beam cores. Extending the analysis to motion beyond the Gaussian beam core, however, must violate the approximation of linearized beam-beam force. To proceed, we seek the help of numerical computer simulations, and in addition, to simplify the picture by making some serious simplifications. Below we introduce two such efforts, each representing one simplifying model. We will briefly give their respective results and conclusions.

The first model In this model,[57] we make the simplifications, (i) the beam distributions are strictly Gaussian, (ii) the beam-beam force is that for Gaussian distributions, therefore nonlinear, but (iii) the beam-beam force affects only the second moments in such a way that the beam distribution remains always Gaussian. The beam-beam force is nonlinear, but effect of the nonlinear beam-beam force on higher moments are ignored. Obviously this model is still not self-consistent. Although step (ii) has rescued the Maxwell equations, step (iii) compromises the Hamiltonian dynamics and the Vlasov equation.

In this model, the infinitesimal quantities $\Delta\Sigma_{\pm}$ are not introduced. The problem is studied by numerical simulation of the $\Sigma_{\pm}$ matrices directly around the dynamic-β^* (the two beams assume the same size) and the flip-flop (the two beam sizes bifurcate) solutions to investigate the stability of $\Sigma_{\pm}$ around them. Figure 8.54 gives one of the results. It shows the steady-state solutions and their stability. We see from Fig. 8.54 that there is always a stable steady-state for the beams to stay in. Below $\xi \approx 0.05$, the beams stay in the equal-beam state, while they stay in the flip-flop state above $\xi \approx 0.05$.

The second model In this model,[58] the dynamics of the second moments is still given by Eqs. (8.129–8.130). In particular, the beam-beam force is still linearized and the focal length $f_{\pm}$ are still related to the components of $\Sigma_{\mp}$ in a nonlinear manner. The difference from the analytical model is that the infinitesimal quantities $\Delta\Sigma_{\pm}$ are not introduced and the problem is not linearized with respect to them. The problem is then studied by numerical simulation of the $\Sigma_{\pm}$ matrices. Figure 8.55 shows the equal-beam and the flip-flop solutions for the case of round beams. Solid and dotted sections indicate stability and instability of these steady-states, respectively.

Note that the equal-beam dynamic-β^* steady-state is unstable when the flip-flop state exists. In the region $\rho = 2\pi\xi$ less than approximately 0.25 (or equivalently $\xi = 0.04$), the equal-beam solution is stable against small perturbations. In the present model, $\rho = 0.25$ may be associated with the beam-beam limit observed in colliding beam storage rings. Beyond this point, the equal-beam solution is unstable except for a very small region around $\rho_0 = 0.5$. Between approximately $\rho = 1$ and $\rho = 2$, there are regions where the flip-flop state is stable, meaning the beams will choose to have one beam statically blown up while the other beam is pinched. In the rest of the regions, both the equal-beam and flip-flop steady-states are unstable.

---

[57]K. Hirata, Phys. Rev. Lett. 58, 25 (1987); 58, 1798(E) (1987); K. Hirata, Phys. Rev. D37, 1307 (1988).

[58]M.A. Furman, K.Y. Ng and A.W. Chao, SSC Laboratory report SSC-174 (1988); M.A. Furman, 3rd Advanced Beam Dynamics Workshop on Beam-beam Effects in Circular Colliders, 1989, Novosibirsk, p. 52; K.Y. Ng, SSC Laboratory report SSC-161 (1988).

Figure 8.54: Stability of the dynamic-β^* and the flip-slop steady-states are studied by a simulation that includes the nonlinear beam-beam force but restricts the beam response to strict Gaussian. The normalized steady-state beam size $\frac{\sigma^2}{\sigma_0^2}$ is plotted against the beam-beam parameter ξ. The solid and dashed curves mark the locations of the dynamic-β^* and the flip-flop solutions; solid curves mean stability around the solution, and dashed curve indicates instability around it. The two beams choose the dynamic-β^* state when $\xi \lesssim 0.05$ and the flip-flop state when $\xi \gtrsim 0.05$ according to this model. [Courtesy Kohji Hirata (2020).]

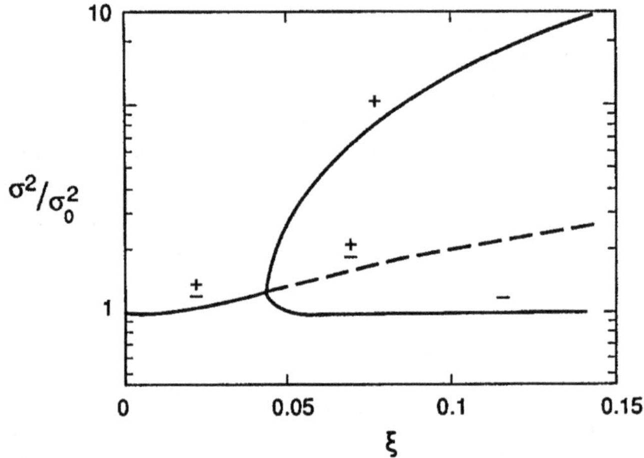

This model predicts the two beams cannot choose to stay stably in both the dynamic-β^* and the flip-flop states. This is not surprising. However, it also predicts regions where both steady-states are unstable. Simulations to explore these regions reveal more details of what is happening. First, the simulations reconfirms the steady-state as period-1 steady state when it is in the stable regions of Fig 8.55. In addition, states with higher periods are found. In a period-2 state, for example, the beams have two states (two sizes) to stay in, and the beams occupy these two states alternately from one interaction point to the next. Furthermore, there are also higher order steady states, e.g. period-3 and period-4 states. Still more, in the region approximately between $\rho = 0.25$ to 0.4, the motion of the second moments is apparently chaotic. Figure 8.55, quite differently from Fig. 8.54, thus contains underneath it a rather complex dynamics of the quadrupole modes.

The following table is a comparison between the two simplifying models.

Figure 8.55: Stability of the dynamic-β^* and the flip-slop steady-states are studied by a simulation that linearizes the beam-beam force and the $\Sigma_\pm$ are tracked directly. The horizontal axis labels the quantity $\rho = 2\pi\xi$. The solid portions of the curves indicate stability against small perturbations and represent the existence of period-1 steady states. Dotted portions indicate instability, chaos, or existence of states with higher order periods. [Courtesy Miguel Furman (2020).]

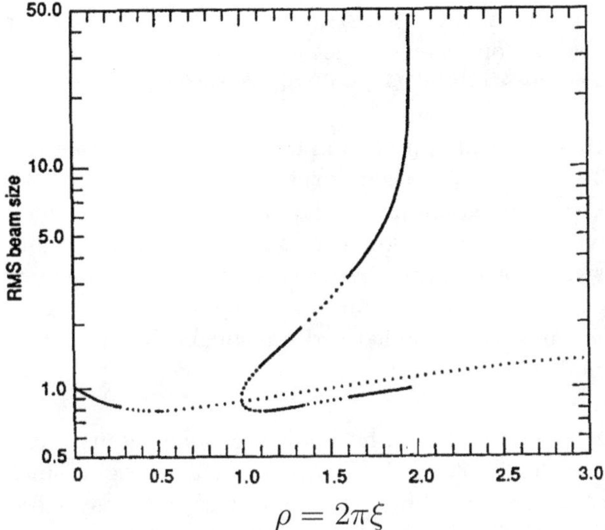

$$\rho = 2\pi\xi$$

	First model	Second model
Self-consistency		
Maxwell equations	yes	no
Hamiltonian dynamics	no	yes
Dynamic-β^*	yes	yes
Flip-flop	yes	yes
Chaos	no	yes
First instability encountered	flip-flop	chaos

It is not very useful to ask which of the two models is more correct because neither is, each having violated something fundamental. However, they do hopefully provide information from which one learns what to expect in an ultimate analysis.

8.7 Synchrobetatron mode

For the strong-strong picture, so far we have discussed the transverse rigid-dipole and transverse quadrupole mode instabilities. In this section we discuss the stability of a third possible coherent beam-beam mode, namely one that involves also the longitudinal dimension. We introduce this coherent synchro-betatron beam-beam effect by an example when the coupling is caused by a finite dispersion function at the interaction point, $D^* \neq 0$. Other synchrobe-tatron coupling effects, such as the hour-glass effect, can also produce similar beam-beam instabilities — both incoherent and coherent.

In Sec. 8.3.6, we had already studied the incoherent beam-beam synchrobe-tatron effect driven by a nonzero D^* and concluded that it was not a critical mechanism. After all, in the weak-strong picture, the beam-beam lens behaves no differently than a nonlinear magnet at a dispersive location. In this section, we shall complete the study of $D^* \neq 0$ considering now its coherent beam-beam effect.

We return to analytical approach in this section for the case when the beam-beam force is linearized. We assume the collider is operated above transition. As we will see, this time, the instability occurs near the difference resonances $\nu_x - \nu_s = m$ while the sum resonances $\nu_x + \nu_s = m$ are stable although also driven. This switch of stability between sum and difference synchrobetatron resonances above transition is similar to single-particle effects. If the beam-beam force is not linearized, higher order resonance stopbands will in principle be excited around $\nu_x - n\nu_s = m$.

Coupling mechanism   To see how coherent longitudinal motions can be ex-cited by the beam-beam interaction, consider first two rigid bunches with energy errors δ_+ and δ_-. The two bunches collide with a transverse separation that contains a term $D^*(\delta_+ - \delta_-)$ at the collision point. The separation gives rise to angular kicks that contain the contributions

$$\Delta x'_+ \;=\; -\Delta x'_- \;=\; -\frac{G}{f_x} D^*(\delta_+ - \delta_-) \,,$$

where f_x is the incoherent beam-beam focal length, G is the dipole kick form factor introduced in Sec. 8.5.2. We consider the dispersion D^* to be in the x-dimension and we consider only the x-motion of the two beams.

The kicks then contribute path length changes of the two bunches in their subsequent revolution by the amounts

$$\Delta z_+ \;=\; -\Delta z_- \;=\; -\frac{G}{f} D^{*2}(\delta_+ - \delta_-) \,.$$

If we consider the longitudinal motions of the two bunches, the beam-beam kicks therefore provide coupling mechanism among the longitudinal coordinates $z_+, \delta_+, z_-, \delta_-$ of the two beams. When $\frac{G}{f}D^{*2}$ becomes large enough, the coher-ent longitudinal motions can become unstable. Note the quadratic dependence on D^*.

The analysis here, however, has not included the betatron contributions, and has not properly treated possibility of synchrobetatron resonances. In fact, by dropping the betatron components, the above argument is not symplectic. A proper treatment invokes a symplectic matrix analysis that we make below.

Eigenanalysis Consider two beams, designated as $\pm$ beams, with one bunch each colliding at two interaction points in a storage ring collider. The bunches are considered rigid, executing center-of-charge motions. To avoid possible confusion, we shall not call them rigid-dipole motions as we did in Sec. 8.5.2 because here these center-of-charge motions are synchrobetatron coupled.

The state of the two-beam system is described by the vector

$$(x_+,\ x'_+,\ z_+,\ \delta_+,\ x_-,\ x'_-,\ z_-,\ \delta_-)\,.$$

We consider the equal-bunch case. The dynamics is described by 8×8 matrices. For the beam-beam kick, we have

$$T_{bb} = \begin{bmatrix} 1 & 0 & 0 & 0 & 0 & 0 & 0 & 0 \\ -\dfrac{G}{f_x} & 1 & 0 & 0 & \dfrac{G}{f_x} & 0 & 0 & 0 \\ 0 & 0 & 1 & 0 & 0 & 0 & 0 & 0 \\ 0 & 0 & 0 & 1 & 0 & 0 & 0 & 0 \\ 0 & 0 & 0 & 0 & 1 & 0 & 0 & 0 \\ \dfrac{G}{f_x} & 0 & 0 & 0 & -\dfrac{G}{f_x} & 1 & 0 & 0 \\ 0 & 0 & 0 & 0 & 0 & 0 & 1 & 0 \\ 0 & 0 & 0 & 0 & 0 & 0 & 0 & 1 \end{bmatrix}, \qquad \frac{G\beta^*}{4\pi f_x} = \xi\,.$$

To assure stability in the longitudinal dimension and to allow the study of synchrobetatron coupling resonances, we need to introduce an RF focusing cavity. The map for the cavity is

$$T_{\mathrm{cav}} = \begin{bmatrix} 1 & 0 & 0 & 0 & 0 & 0 & 0 & 0 \\ 0 & 1 & 0 & 0 & 0 & 0 & 0 & 0 \\ 0 & 0 & 1 & 0 & 0 & 0 & 0 & 0 \\ 0 & 0 & k_{rf} & 1 & 0 & 0 & 0 & 0 \\ 0 & 0 & 0 & 0 & 1 & 0 & 0 & 0 \\ 0 & 0 & 0 & 0 & 0 & 1 & 0 & 0 \\ 0 & 0 & 0 & 0 & 0 & 0 & 1 & 0 \\ 0 & 0 & 0 & 0 & 0 & 0 & k_{rf} & 1 \end{bmatrix}, \qquad k_{rf} = \frac{e\hat{V}_{rf}\omega_{rf}}{E_0 c}\,.$$

The transfer map from one collision point to the next (for half a revolution) is given by

$$T_0 = T_2 T_{\mathrm{cav}} T_1\,,$$

where T_1 is the map from the interaction point to the cavity, T_2 is the map from the cavity to the next interaction point,

$$T_1 = \begin{bmatrix} t_1 & 0 \\ 0 & t_1 \end{bmatrix}, \qquad T_2 = \begin{bmatrix} t_2 & 0 \\ 0 & t_2 \end{bmatrix},$$

with $t_{1,2}$ the 4×4 Courant–Snyder maps that involve the β-functions at the interaction point and the cavity, and involve the dispersion function D^*. Explicit expressions of $t_{1,2}$ are well-known and omitted here.

The total map for one superperiod (half a revolution) is

$$T_{\mathrm{tot}} = T_0 T_{bb} = T_2 T_{\mathrm{cav}} T_1 T_{bb}\,.$$

Stability of the system is given by the eigenanalysis of the map T_{tot} (see footnote 46). This involves eigenanalyzing an 8×8 system and is naturally rather involved. Fortunately this system is also highly degenerate. The reason is that the coherent beam-beam modes degenerate into one 0-mode when all bunches move transversely together in such a way that the beam-beam kicks do not have any effect, and one π-mode when bunches move transversely with opposite signs so that the beam-beam kicks have maximum effect. The 0-mode behaves the same as the weak-strong single-particle case.

The physics of the expected degeneracy leads to simplification of its mathematics. Mathematically, the degeneracy is evidenced by the fact that in the eigenvalue equation,

$$\text{Det}\,(T_{\text{tot}} - \lambda I) \; = \; 0\,,$$

the left-hand-side as an 8th order polynomial function of λ factorizes into a product of two 4th order polynomials. One of the factor polynomials is independent of ξ; it describes the eigenvalues of the 0-mode. The other factor polynomial depends on ξ and describes the π-mode. When $\xi = 0$, these two polynomials coincide.

An explicit demonstration of this analysis is given in Homework 8.62 for a slightly simplified hypothetical case when the RF cavity is located at the interaction point. It confirms a similar situation as in the weak-strong case, Eq. (8.44), i.e. the effects due to dispersive cavity and the dispersive beam-beam are mostly decoupled. The dispersive cavity is characterized by the parameter,

$$h \; = \; \frac{D^{*2}}{A\beta^*}\,,$$

of Eq. (8.45), where

$$A \; = \; \eta C - \frac{D^{*2}}{\beta^*}\sin\mu_x\,,$$

η is the phase slippage factor, C is the ring circumference, while the beam-beam effect is characterized by the beam-beam parameter ξ, both are small parameters. See also Homework 8.63.

Homework 8.62 As an explicit demonstration of the eigenanalysis of a linearly synchrobetatron coupled case, follow the text to analyze the case of two beams with one bunch each and $D^* \neq 0$, but simplify it by locating the RF cavity at the interaction point. Find the eigenmode frequencies analytically. Examine the four modes and identify two of them to be the 0-modes and the other two to be the π-modes.

Solution The maps T_{bb} and T_{cav} are given in the text. The map around the collider ring is

$$T_0 \; = \; \begin{bmatrix} t_0 & 0 \\ 0 & t_0 \end{bmatrix}, \quad t_0 \; = \; \begin{bmatrix} \cos\mu_x & \beta^*\sin\mu_x & 0 & D^*(1-\cos\mu_x) \\ -\frac{\sin\mu_x}{\beta^*} & \cos\mu_x & 0 & \frac{D^*}{\beta^*}\sin\mu_x \\ -\frac{D^*}{\beta^*} & D^*(\cos\mu_x - 1) & 1 & -A \\ 0 & 0 & 0 & 1 \end{bmatrix}.$$

The total map is $T_{\text{tot}} = T_0 T_{\text{cav}} T_{bb}$. Synchrotron phase advance is given by $\cos \mu_s = 1 - \frac{k_{rf}A}{2}$ with $\mu_s < \pi$. For small full-turn synchrotron tune $\nu_s = \frac{\mu_s}{\pi}$, we have $\nu_s \approx \frac{1}{\pi}\sqrt{k_{rf}A}$. For phase stability, we need $k_{rf}A > 0$.

As claimed in the text, the eigenvalue equation factorizes. The 0-mode eigenfrequencies μ satisfy the eigenvalue equation,

$$\cos^2 \mu - (\cos \mu_x + \cos \mu_s)\cos \mu$$
$$+ \left[\cos \mu_x \cos \mu_s + \frac{D^{*2}}{A\beta^*}\sin \mu_x (1 - \cos \mu_x)(1 - \cos \mu_s) \right] = 0.$$

This equation is independent of ξ; it describes the single-particle dynamics when the RF cavity is located at a dispersive location.

The π-modes have the eigenfrequencies satisfying

$$\cos^2 \mu - (\cos \mu_x + \cos \mu_s - 4\pi\xi \sin \mu_x)\cos \mu + \left[\cos \mu_x \cos \mu_s - 4\pi\xi \cos \mu_s \sin \mu_x \right.$$
$$\left. + \frac{8D^{*2}}{A\beta^*}\sin^2 \frac{\mu_s}{2}\sin^3 \frac{\mu_x}{2}\left(\cos \frac{\mu_x}{2} - 4\pi\xi \sin \frac{\mu_x}{2} \right) \right] = 0. \tag{8.141}$$

When $\xi = 0$, the two eigenfrequency equations coincide.

Equation (8.141) is to be compared with Eq. (8.44) for the single-particle weak-strong beam-beam result. As expected, Eq. (8.141) can be obtained from Eq. (8.44) by replacing ξ by $2\xi G$. For the π-mode, beam separation is effectively twice that of a weak-strong case, while the kick is to be reduced by the form factor G. The structure of synchrobetatron resonances can be referenced to Fig. 8.15.

Homework 8.63 This is a continuation of Homework 8.62. It is commented after Eq. (8.141) that the beam-beam effect, proportional to ξ, and the dispersive cavity effect, proportional to $h = \frac{D^{*2}}{A\beta^*}$, are mostly decoupled. The only potential term responsible for the synchrobetatron effect driven by dispersive beam-beam interaction is the term proportional to $h\xi$. Both h and ξ are small parameters. If we concentrate on the dispersive beam-beam driven effect and drop all other terms from Eq. (8.141), we have

$$(\cos \mu - \cos \mu_x)(\cos \mu - \cos \mu_s) - 32\pi h\xi \sin^2 \frac{\mu_s}{2}\sin^4 \frac{\mu_x}{2} = 0.$$

Show that this equation does provide synchrobetatron coupling. However, alas, show that it does not drive synchrobetatron instability stopbands!

8.8 Higher order mode

Let us return to the transverse beam-beam effects. We have now discussed the coherent dipole and coherent quadrupole instabilities driven by the beam-beam interaction. In doing so, we have implicitly assumed these modes do not couple

among themselves, and we have treated them separately and individually —
justified by the condition (8.127). To improve beyond the two models, one
has to deal with nonlinear beam-beam force, and to allow this beam-beam
force to distort the distribution from Gaussian. It is clear that a self-consistent
treatment is needed here. This can be provided by a Vlasov treatment that
takes into account of all distribution moments.[51,59] In other words, we have to
consider, in addition to the dipole and quadrupole modes, also sextupole and
octupole modes etc., and we need to treat them all together in one framework.
Also, ideally, a proper treatment will simultaneously be symplectic and satisfy
the Maxwell equations.

Take an octupole mode for example. It can be unstable due to the beam-
beam interaction according to the mechanism illustrated in Fig. 8.56. Let
Fig. 8.56(a) be the unperturbed beam distribution in the phase space (y, y')
for the two beams. To investigate the coherent octupole mode stability, perturb
one of the beams slightly in an octupolar manner as sketched in Fig. 8.56(b).
This perturbation will couple to the other beam through the beam-beam interaction. For example, one possible mode patterns could be such that the octupolar
modulations of the two beams interlace each other at the interaction point as
in Fig. 8.56(c). If the motion is stable, the coupled octupolar motion will stay
small, as Fig. 8.56(c). However, if the total betatron tune is sufficiently close
to a quarter-integer, i.e. when $\nu \approx \frac{m}{4}$, the small perturbations will grow exponentially as sketched in Fig. 8.56(d), leading to an octupole mode instability of
both beams.

We will see that the coherent beam-beam instability occurs when the betatron tune is close to a rational number just like in that of weak-strong case. The
stopband widths, however, are wider, and in case of multiple M bunches per
beam, there are M times more stopbands. The coherent beam-beam instability
imposes a much more stringent beam-beam limit than the weak-strong picture
predicts. This point has been emphasized earlier and will be raised again in this
section.

The difference among the weak-strong, the dynamic-β^*, and the coherent
strong-strong pictures was shown in Fig. 8.36 for the dipole mode. The analogous situation occurs for the octupole mode when $\nu \approx \frac{m}{4}$ and is illustrated in
Fig. 8.57. Note that we introduced the dynamic-β^* assuming linearized beam-
beam force, but the dynamic-β^* state mentioned in Fig. 8.57 has generalized it
to nonlinear beam-beam force.

8.8.1   Vlasov equation

The Vlasov equation (8.104–8.105) provides a tool for the general treatment of
higher order modes (assuming flat beams). Below, we will mention the analysis
as an illustration of the technique (see footnote 59). To proceed, let the beam

---

[59]H.S. Uhm and C.S. Liu, Phys. Rev. Lett. 43, 914 (1979); B. Zotter, CERN report LEP-
194 (1979); A.W. Chao and R.D. Ruth, Part. Accel. 16, 201 (1985); N.S. Dikansky and
D.V. Pestrikov, 3rd Advanced Beam Dynamics Workshop on Beam-beam Effects in Circular
Colliders, Novosibirsk, 76 (1989).

Figure 8.56: Schematic illustration of the coherent beam-beam octupole mode instability as observed at the collision point. (a) two beams unperturbed; (b) one beam with initial small perturbation; (c) two beams perturbed due to the mutual beam-beam coupling but perturbations remain small; (d) unstable case when small perturbations grow exponentially.

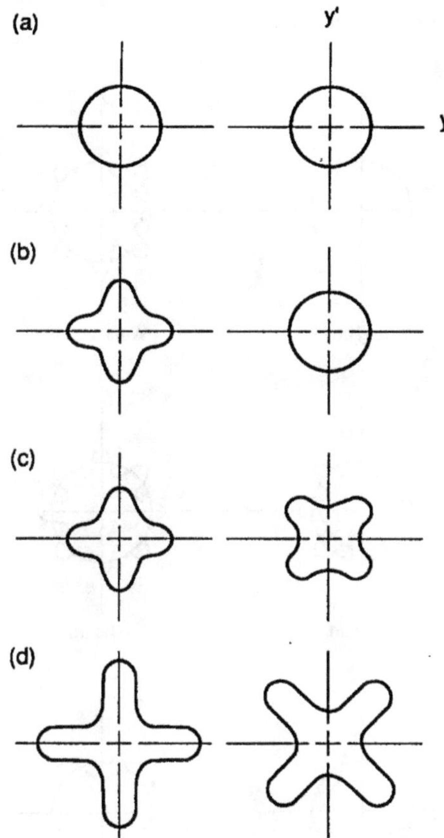

distributions be written as infinitesimal perturbations on top of a steady-state Ψ_0 as Eq. (8.103). The Vlasov equation is then linearize in $\Delta\Psi_{1,2}$. The perturbations $\Delta\Psi_{1,2}$ transform simply by a rigid rotation in phase space as the beams circulate around the storage ring and receive sudden changes at the collision points due to the beam-beam interaction. After linearization, the motion of $\Delta\Psi_{1,2}$ can be solved by a matrix technique to be described below.

We start by rewriting Eqs. (8.104–8.105),

$$\frac{\partial\Psi_1}{\partial s}+y'\frac{\partial\Psi_1}{\partial y}-\left[K(s)y+\frac{2\pi Nr_0}{L_x\gamma}\delta(s)\int_{-\infty}^{\infty}d\bar{y}H(y-\bar{y})\int_{-\infty}^{\infty}d\bar{y}'\Psi_2(\bar{y},\bar{y}',s)\right]\frac{\partial\Psi_1}{\partial y'}=0,$$

Figure 8.57: Beam behavior in phase space of two beams for (a) incoherent weak-strong case, (b) generalized nonlinear dynamic-β^* effect, (c) one of the coherent octupole modes. (b) and (c) are strong-strong cases. The solid and dashed curves indicate the unperturbed and the beam-beam perturbed distributions, respectively. The perturbed distribution moves with time in (c) but is static in (a) and (b). Note there are eight scalloping lobes in the static distortions but four such lobes in the dynamic distortions.

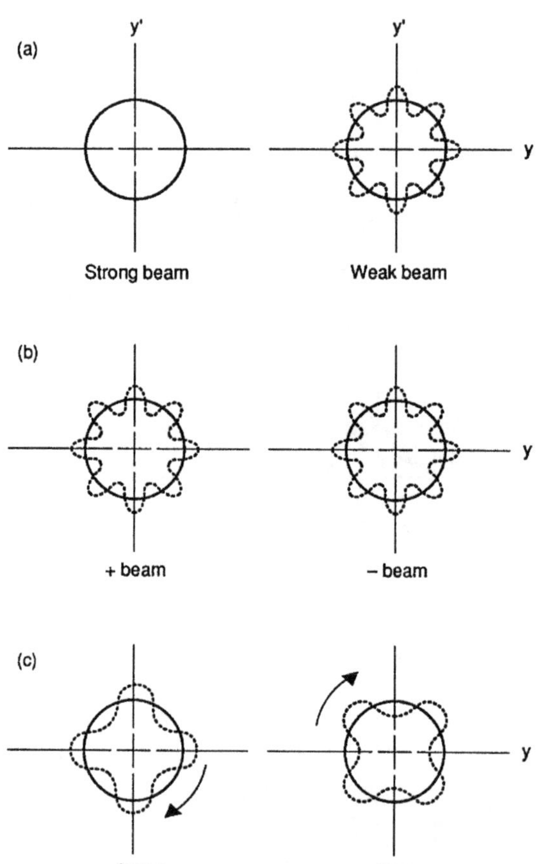

$$\frac{\partial \Psi_2}{\partial s} + y' \frac{\partial \Psi_2}{\partial y} - \left[K(s)y + \frac{2\pi N r_0}{L_x \gamma} \delta(s) \int_{-\infty}^{\infty} d\bar{y} H(y - \bar{y}) \int_{-\infty}^{\infty} d\bar{y}' \Psi_1(\bar{y}, \bar{y}', s) \right] \frac{\partial \Psi_2}{\partial y'} = 0,$$

where $K(s)$ is the betatron focusing function, $H(x)$ is the sign-function, and the distributions are normalized,

$$\int_{-\infty}^{\infty} dy \int_{-\infty}^{\infty} dy' \Psi_{1,2}(y, y', s) = 1. \tag{8.142}$$

We assume the two beams have opposite signs of charge but equal beam population N.

Steady state Ψ_0 The beam-beam perturbed equilibrium distribution Ψ_0 is determined by the self-consistency condition,

$$\frac{\partial \Psi_0}{\partial s} + y' \frac{\partial \Psi_0}{\partial y} - [K(s)y - F_0(y,s)]\frac{\partial \Psi_0}{\partial y'} = 0, \qquad (8.143)$$

where

$$F_0(y,s) = -\frac{2\pi N r_0}{L_x \gamma} \delta(s) \int_{-\infty}^{\infty} d\bar{y}\, H(y - \bar{y}) \int_{-\infty}^{\infty} d\bar{y}'\, \Psi_0(\bar{y}, \bar{y}', s) \qquad (8.144)$$

is the beam-beam force generated by the equilibrium distribution. Note that Ψ_0 is not considered s-independent. We consider Ψ_0 a steady-state in the sense that its dependence on time s is periodic, so that it repeats every revolution when observed at a given location in the collider.

Equation (8.143) is a nonlinear partial-differential integral equation, dependent on s, and is nonlinear in Ψ_0. Once a solution Ψ_0 is obtained, it can be referred to as a nonlinear generalized dynamic-β^* effect. A linearization in case of very weak beam-beam perturbation can be performed by decomposing the distortion into modes. One such decomposition was shown schematically as Fig. 8.57(b).

In what follows, however, admittedly with hesitation, we are not too interested in Ψ_0. We pay our attention to infinitesimal deviations from Ψ_0. Our goal is to study the time evolution of these infinitesimal deviations to see if they stay infinitesimal or grow exponentially with time when the beam-beam perturbation is imposed. For this purpose, in fact, we will later simply assume a simplified model for Ψ_0, a waterbag model. The step of actually calculating Ψ_0 is short circuited.

Let us make a comment before leaving the subject on Ψ_0. As mentioned, Ψ_0 can be found for small beam-beam strength ξ by decomposing it into a superposition of modes. The m-th mode will be sensitive if the tune is close to $\frac{n}{m}$. Figure 8.57(b), for example, shows the decomposition when $m = 4$. Note that Fig. 8.57(b) has 8 azimuthal modulations in the phase space for $m = 4$, although the dynamical modes have 4 modulations as shown in Fig. 8.57(c). Although short-circuited in our treatment, there is an understanding that Ψ_0 actually contains rich contents deserving exploration.

Linearized beam-beam force Even with this short circuiting of avoiding an explicit calculation of a beam-beam distorted steady state Ψ_0, it turns out, to readily yield a soluble case, we need to make one more approximation involving the beam-beam force F_0.

Consider for example a bi-Gaussian model,

$$\Psi_0(y,y') = \frac{1}{2\pi\sigma_y\sigma_{y'}} e^{-\frac{y^2}{2\sigma_y^2} - \frac{y'^2}{2\sigma_{y'}^2}}.$$

Here the model is needed only at the collision point, so we do not keep an s-dependence here. Substituting into Eq. (8.144) yields (Homework 8.65)

$$F_0(y) \;=\; -\frac{2\sqrt{2\pi}\,N r_0}{\gamma L_x}\,\delta(s)\int_0^{y/\sigma_y} dx\, e^{-x^2/2}\,. \tag{8.145}$$

Another model we might adopt is the waterbag model,

$$\Psi_0(y,y') \;=\; \begin{cases} \dfrac{\beta_y^*}{\pi \hat{y}^2}\,, & \text{if } (y^2 + \beta^{*2}y'^2) < \hat{y}^2, \\[2ex] 0, & \text{otherwise,} \end{cases} \tag{8.146}$$

where β_y^* is the β-function at the collision point, $\hat{y}$ is the maximum vertical beam size in the waterbag. In this case, we have, for $|y| < \hat{y}$ (Homework 8.66)

$$F_0(y) \;=\; -\frac{4N r_0}{\gamma L_x}\,\delta(s)\left(\frac{y}{\hat{y}}\sqrt{1-\frac{y^2}{\hat{y}^2}} + \sin^{-1}\frac{y}{\hat{y}}\right). \tag{8.147}$$

To facilitate a solution, we will make one more approximation with the steady-state force F_0 by linearizing it with respect to y. For small $|y|$, we let

$$F_0(y,s) \;\approx\; -\frac{y}{f_0}\,\delta(s)\,. \tag{8.148}$$

We then find

$$\frac{1}{f_0} \;=\; \begin{cases} \dfrac{2\sqrt{2\pi}N r_0}{\gamma L_x \sigma_y}\,, & \text{bi-Gaussian model, } |y| \ll \sigma_y, \\[2ex] \dfrac{8N r_0}{\gamma L_x \hat{y}}\,, & \text{waterbag model, } |y| \ll \hat{y}, \end{cases}$$

where f_0 is the focal length in the linearized model, L_x is the horizontal width of the flat beams. For a Gaussian horizontal distribution, $L_x = \sqrt{2\pi}\,\sigma_x$.

The force $F_0(y)$ represents the beam-beam induced steady-state force on particle motion. It is equivalent to introducing an additional steady-state potential well to the external focusing $K(y)$. The potential well itself does not participate in any dynamical effects. If we linearize $F_0(y)$ with respect to y, it means this steady-state beam-beam potential well has been approximated by a parabolic function. Taking the linear contribution into account is equivalent to taking into account of dynamic-β^* but dropping the nonlinear parts of the beam-beam potential. Since the potential well itself does not participate in the dynamics, and we have limited our interest to the beam-beam dynamical effects only, this approximation seems reasonable for our purpose, and will be what we assume in the rest of this section.

This approximation to linearize the steady-state beam-beam force, however, has a serious side effect. By ignoring the nonlinear terms in the potential well, even though these terms do not participate in the dynamics themselves, one consequence is that all particles move with exactly the same natural frequency. There is no detuning and no frequency spread like the one we emphasized in

Eq. (8.60). This frequency spread, even though not directly part of the dynamics, is in practice a very important stabilizing effect to instabilities through the mechanisms of detuning and Landau damping. The model we shall discuss therefore has the serious drawback that detuning and Landau damping effects, two most important beam-stabilizing mechanisms, have inadvertently been dropped.

Perturbations $\Delta\Psi_{1,2}$ The next step is to introduce $\Delta\Psi_{1,2}$ as infinitesimal perturbations to the beam distributions according to Eq. (8.103). Let us define

$$D_\pm(y, y', s) = \Delta\Psi_1(y, y', s) \pm \Delta\Psi_2(y, y', s) \,. \tag{8.149}$$

Substituting into the coupled Vlasov equations and linearizing with respect to $D_\pm$, we obtain

$$\frac{\partial D_\pm}{\partial s} + y'\frac{\partial D_\pm}{\partial y} - [K(s)y - F_0(y, s)]\frac{\partial D_\pm}{\partial y'}$$
$$\mp \frac{\partial \Psi_0}{\partial y'}\frac{2\pi N r_0}{L_x \gamma}\delta(s)\int_{-\infty}^{\infty} d\bar{y}H(y - \bar{y})\int_{-\infty}^{\infty} d\bar{y}' D_\pm(\bar{y}, \bar{y}', s) = 0\,, \tag{8.150}$$

where F_0 is the beam-beam force due to the steady-state Ψ_0, Eq. (8.144). The quantities $D_\pm$, unlike $\Delta\Psi_{1,2}$, are decoupled.

We then make the rather drastic approximation (8.148). As mentioned, this approximation is fair to the dynamics of the system, but in doing so has dropped the critical provision of detuning and Landau damping. Such a treatment is a familiar one in the treatment of the longitudinal microwave instability in storage rings.[60]

After making the approximation (8.148), the static part of the beam-beam force is included simply as a perturbation to the linear focusing structure of the storage ring lattice and can be absorbed into a perturbed Courant–Snyder analysis,

$$y = \sqrt{2\beta J}\cos\phi\,,$$
$$y' = -\sqrt{\frac{2J}{\beta}}\left(\sin\phi - \frac{\beta'}{2}\cos\phi\right)\,, \tag{8.151}$$

where the β-function β and the betatron phase ϕ have taken into account the dynamic-β^* effect due to the linearized steady-state beam-beam force F_0. The action-angle transformation (8.151) then transforms the phase space from (y, y') to (ϕ, J).

Assuming that the incoherent motion with the inclusion of the beam-beam focusing $F_0(y, s)$ is stable, and that we can indeed make the transformation (8.151), then the steady-state distribution Ψ_0 can only be a function of J — the perturbed J,

$$\Psi_0 = \Psi_0(J)\,.$$

---

[60]F. Sacherer, CERN report, CERN/SI-BR/72-5 (1972); A.W. Chao, Physics of Collective Beam Instabilities in High Energy Accelerators, Wiley (1993).

The Vlasov equation of motion after this transformation (8.151) reads (Homework 8.68)

$$\frac{\partial D_\pm}{\partial s} + \frac{1}{\beta}\frac{\partial D_\pm}{\partial \phi} \pm \Psi_0'(J)\sqrt{2\beta J}\sin\phi\,\frac{2\pi N r_0}{\gamma L_x}\delta(s)$$

$$\times \int_{-\infty}^{\infty} d\bar{y}\,H(y - \bar{y})\int_{-\infty}^{\infty} d\bar{y}'\,D_\pm(\bar{y},\bar{y}',s) \;=\; 0\,. \qquad (8.152)$$

Waterbag model　We need now to solve the partial-differential-integral equation (8.152). The problem becomes soluble if we assume a waterbag model of the steady-state distribution (8.146). Here in Eq. (8.146) we take β^* to be the β-function at the collision point including the dynamic-β^* distortion.

Let us rewrite the waterbag model (8.146) as

$$\Psi_0(J) \;=\; \begin{cases} \frac{1}{2\pi J_0}, & \text{if } J < J_0, \\ 0, & \text{otherwise,} \end{cases} \qquad J_0 = \frac{\hat{y}^2}{2\beta^*}\,.$$

The reason the waterbag model is most accessible to analysis is because

$$\Psi_0'(J) \;=\; -\frac{1}{2\pi J_0}\,\delta(J - J_0)\,,$$

which is proportional to the δ-function $\delta(J - J_0)$. This means all the dynamics occur only at amplitude $J = J_0$, i.e. at the edge of the waterbag. No dynamics will occur inside the beam core. In any case, physically, nothing can happen to the phase space distribution inside the uniform beam core due to the Liouville theorem. All dynamics has to happen only at the waterbag edge.

We now express the solution to Eq. (8.152) as a Fourier decomposition,

$$D_\pm(\phi, J, s) \;=\; \delta(J - J_0)\sum_{\ell=-\infty}^{\infty} g_\ell^\pm(s)\,e^{i\ell\phi}\,.$$

Our job is to solve for $g_\ell(s)$. For simplicity of notation, we shall drop the $\pm$ superscripts on $g_\ell^\pm$ in the following expressions. This should not cause confusion as $D_\pm$ are decoupled.

Substituting into Eq. (8.152) gives an infinite set of equations describing the coupled motion of the Fourier components $g_\ell(s)$ for all ℓ,

$$g_\ell' + \frac{i\ell}{\beta^*}g_\ell \mp \frac{N r_0}{\pi\gamma L_x}\sqrt{\frac{\beta^*}{2J_0}}\,\delta(s)\sum_{k=-\infty}^{\infty}\mathcal{M}_{\ell k}g_k \;=\; 0\,, \qquad (8.153)$$

where (Homework 8.70)

$$\mathcal{M}_{\ell k} \;=\; \int_0^{2\pi} d\phi\,e^{-i\ell\phi}\sin\phi\int_0^{2\pi} d\bar{\phi}\,H(\cos\phi - \cos\bar{\phi})\,e^{ik\bar{\phi}}$$

$$=\; \begin{cases} \dfrac{-32i\ell}{[(\ell+k)^2-1]\,[(\ell-k)^2-1]}, & \text{if } \ell + k = \text{even,} \\ 0, & \text{otherwise.} \end{cases} \qquad (8.154)$$

Here again, $H(x)$ is the sign-function.

The infinite matrix $\mathcal{M}$ has the following properties,

$$
\begin{aligned}
\mathcal{M} &= \text{purely imaginary}, \\
\mathcal{M}^2 &= 0, \\
\mathcal{M}_{\ell,-k} &= \mathcal{M}_{\ell k}, \\
\mathcal{M}_{-\ell,k} &= -\mathcal{M}_{\ell k}.
\end{aligned}
\tag{8.155}
$$

Between collisions, the different g_ℓ's are decoupled. A particular g_ℓ transforms by

$$
g_\ell(L^-) = g_\ell(0^+)\, e^{-i\ell\mu}, \tag{8.156}
$$

where L is the distance between collision points and

$$
\mu = \int_0^L \frac{ds}{\beta(s)}
$$

is the dynamic-β^*-perturbed betatron phase advance between collision points. The superscripts $\pm$ indicate immediately before or after the point referenced.

The beam-beam action is obtained by integrating Eq. (8.153) through $s = 0$, which yields

$$
g_\ell(0^+) - g_\ell(0^-) = \pm\frac{Nr_0}{\pi\gamma L_x}\sqrt{\frac{\beta^*}{2J_0}}\sum_{k=-\infty}^{\infty}\mathcal{M}_{\ell k}g_k(0^-), \tag{8.157}
$$

where β^* is the β-function at the collision point in the presence of the static linear beam-beam force. We have used $g_k(0^-)$ on the right-hand-side of Eq. (8.157). Using $g_k(0^+)$ would give the same result since $\mathcal{M}^2 = 0$.

Equations (8.156) and (8.157) can be combined to give a matrix transformation on the vector

$$
G = \begin{bmatrix} \vdots \\ g_2 \\ g_1 \\ g_0 \\ g_{-1} \\ g_{-2} \\ \vdots \end{bmatrix}
$$

from one collision point to the next. The transformation is given by

$$
T = R\left(I \pm \frac{Nr_0}{\pi\gamma L_x}\sqrt{\frac{\beta^*}{2J_0}}\mathcal{M}\right),
$$

where I is the unit matrix and R is a diagonal matrix describing the map in the collider arc,

$$
R = \begin{bmatrix}
\ddots & & & & & \\
& e^{-2i\mu} & & & & \\
& & e^{-i\mu} & & & \\
& & & 1 & & \\
& & & & e^{i\mu} & \\
& & & & & e^{2i\mu} \\
& & & & & & \ddots
\end{bmatrix} .
$$

Homework 8.64 An alert reader might question the applicability of the normalization condition (8.142). Use the coupled Vlasov equations to prove mathematically that once the normalizations (8.142) for the two beams are fulfilled at $s = 0$, they are assured to be fulfilled at all times s.

Homework 8.65
(a) Verify Eq. (8.145) for the bi-Gaussian model.
(b) Find its linearized expression and confirm that the linear coefficient relates to the beam-beam parameter with $\xi = \frac{\beta^*}{4\pi f_0} = \frac{N r_0 \beta^*}{\sqrt{2\pi} L_x \sigma_y}$.

Homework 8.66
(a) Show that the waterbag model (8.146) has been properly normalized by $\int dy \int dy' \Psi_0 = 1$. Show that its projected linear density is given by

$$
\rho_0(y) = \int_{-\infty}^{\infty} dy' \, \Psi_0(y, y') = \frac{2}{\pi \hat{y}} \sqrt{1 - \frac{y^2}{\hat{y}^2}} ,
$$

and show that the rms value of y is $\sigma_y = \frac{1}{2}\hat{y}$.
(b) Verify Eq. (8.147) for the waterbag model.
(c) Find its linearized expression and confirm that the linear coefficient relates to the beam-beam parameter with $\xi = \frac{\beta^*}{4\pi f_0} = \frac{2 N r_0 \beta^*}{\pi \gamma L_x \hat{y}} = \frac{2 N r_0}{\pi \gamma L_x} \sqrt{\frac{\beta^*}{2 J_0}}$.
(d) Calculate the beam-beam detuning of a waterbag model and verify Eq. (8.160) to be used in the next section.

Homework 8.67 The reader is expected not surprised by the introduction of $D_\pm(y, y', s)$ of Eq. (8.149) that leads to the decoupled Vlasov equations (8.150). Following the physical meaning of this redefinition, generalize the analysis to the case when the two beams have different intensities N_1 and N_2.

Homework 8.68 The action-angle transformation (8.151) is a Courant–Snyder procedure that is standard but worth going through. Derive Eq. (8.152). The β-function and the betatron phase ϕ are those having taken into account of the beam-beam force (8.148).

Homework 8.69 A fair question can be raised concerning the fact that the waterbag model we assumed would yield a beam-beam force given by Eq. (8.147), which conforms to a linear force (8.148) only when $|y|$ is small. So there is a certain self-inconsistency here.

(a) Show that there is indeed a distribution that yields exactly a linear beam-beam force, and it is given by

$$\Psi_0(J) = \begin{cases} \dfrac{\beta^*}{2\pi\hat{y}\sqrt{\hat{y}^2-2\beta^*J}}, & \text{if } J < \dfrac{\hat{y}^2}{2\beta^*}, \\ 0, & \text{otherwise.} \end{cases}$$

(b) Show that the beam-beam focal length is given by, without approximation,

$$\frac{1}{f_0} = \frac{2\pi N r_0}{\gamma L_x \hat{y}}.$$

Unfortunately if we adapt this model for $\Psi_0(J)$, we are a step closer to self-consistency but we are not able to solve the Vlasov equation (8.152) so readily.

Solution This case has $\rho_0(y) = \frac{1}{2\hat{y}}$, $|y| < \hat{y}$, and rms value $\sigma_y = \frac{1}{\sqrt{3}}\hat{y}$.

Homework 8.70

(a) Derive Eq. (8.154).
(b) Verify the properties (8.155) of the matrix $\mathcal{M}$.
(c) Show that

$$\det M = 0,$$
$$\det (I + \alpha M) = 1, \quad \text{for arbitrary } \alpha.$$

Solution A 7×7 visualization of $\mathcal{M}$ up to $|\ell| = 3, |k| = 3$ is like this,

$$i\begin{bmatrix} \frac{96}{35} & 0 & -\frac{32}{15} & 0 & -\frac{32}{15} & 0 & \frac{96}{35} \\ 0 & \frac{64}{15} & 0 & -\frac{64}{9} & 0 & \frac{64}{15} & 0 \\ -\frac{32}{45} & 0 & \frac{32}{3} & 0 & \frac{32}{3} & 0 & -\frac{32}{45} \\ 0 & 0 & 0 & 0 & 0 & 0 & 0 \\ \frac{32}{45} & 0 & -\frac{32}{3} & 0 & -\frac{32}{3} & 0 & \frac{32}{45} \\ 0 & -\frac{64}{15} & 0 & \frac{64}{9} & 0 & -\frac{64}{15} & 0 \\ -\frac{96}{35} & 0 & \frac{32}{15} & 0 & \frac{32}{15} & 0 & -\frac{96}{35} \end{bmatrix}.$$

8.8.2   Coherent beam-beam instability

In the absence of beam-beam collisions, the eigenvalues of the map T are

$$\lambda = e^{-i\ell\mu}, \qquad \ell = 0, \pm 1, \pm 2, \cdots,$$

corresponding to the motion of the ℓ-th Fourier component of the distributions. All eigenvalues have absolute value of unity provided that the beam-beam distorted potential well under the beam-beam force $F_0(y, s)$ is stable.

As the nonlinear perturbation is included, the eigenvalues are perturbed. A coherent instability occurs when any one of the eigenvalues acquires an absolute value larger than unity. Coherent beam-beam instability is most pronounced when the betatron phase advance between collision points is close to a rational number times π, i.e. near resonances

$$\nu = \frac{\mu}{\pi} \approx \frac{m}{\ell}, \tag{8.158}$$

where ν is the total tune around the collider ring with two collision points.

Close to the resonant condition (8.158), the Fourier components that perturb the beam motion most are g_ℓ and $g_{-\ell}$. Keeping only the ℓ-th and the $(-\ell)$-th elements ($\ell = 1, 2, 3, \cdots$ for dipole, quadrupole, sextupole $\cdots$ respectively) of T and defining the small distance between ν and $\frac{m}{\ell}$ to be

$$\Delta = \nu - \frac{m}{\ell},$$

the map T becomes

$$T = (-1)^m \begin{bmatrix} (1 \pm i\alpha)\, e^{-i\pi\ell\Delta} & \pm i\alpha\, e^{-i\pi\ell\Delta} \\ \mp i\alpha\, e^{i\pi\ell\Delta} & (1 \mp i\alpha)\, e^{i\pi\ell\Delta} \end{bmatrix},$$

with

$$\alpha = \frac{32\,\ell}{4\ell^2 - 1} \frac{Nr_0}{\pi\gamma L_x} \sqrt{\frac{\beta^*}{2J_0}}.$$

The eigenvalues of T are determined by the secular equation,

$$\lambda^2 - 2\lambda \left[\cos \pi\ell\Delta \pm \alpha \sin \pi\ell\Delta\right] + 1 = 0.$$

One of the eigenvalues has absolute value larger than unity, and therefore the beam-beam system is unstable, if

$$\left| \cos \pi\ell\Delta \pm \alpha \sin \pi\ell\Delta \right| > 1. \tag{8.159}$$

The $\pm$ signs here refer to the D_+ and D_- modes defined in Eq. (8.149).

Before we interpret Eq. (8.159), we need to calculate the dynamic tune ν in terms of the unperturbed tune in the absence of the beam-beam interaction. The dynamic tune is taken to include the detuning contribution from the force $F_0(y, s)$ of Eq. (8.147). Furthermore, it is to be evaluated at the edge of the waterbag, $J = J_0$ (instead of at $J = 0$), because that is where the dynamics takes place to 1st order in the beam-beam strength it is found from [see Homework 8.66(d)]

$$\cos \pi\nu = \cos \pi\nu_0 - \frac{32\, Nr_0}{3\pi\,\gamma L_x} \sqrt{\frac{\beta^*}{2J_0}} \sin \pi\nu_0. \tag{8.160}$$

Note that the tune shift evaluated at $J = J_0$ is a factor $\frac{8}{3\pi}$ times that of the tune shift at $J = 0$. A resonance occurs when the dynamic tune ν is close to a rational number and the parameter Δ is defined relative to ν. Be reminded again that ν is the total tune of the ring with two collision points. The tune between collisions is $\frac{\nu}{2}$.

Equation (8.159) determines the instability region near the resonance $\nu = \frac{m}{\ell}$ with a resonance stopband width approximately given by

$$\delta \nu_\ell \;=\; \frac{1}{\pi}\,\frac{32}{4\ell^2 - 1}\,\xi .\tag{8.161}$$

where we have defined the beam-beam strength parameter [Homework 8.66(c)]

$$\xi \;=\; \frac{2Nr_0}{\pi \gamma L_x}\sqrt{\frac{\beta^*}{2J_0}}\,.$$

The stopband centers around $\nu = \frac{m}{\ell}$. The $+$ mode is unstable on the side $\nu > \frac{m}{\ell}$ while the $-$ mode is unstable on the other side. Both modes need to be stable for the beam to be stable.

The higher-order resonance stopbands ($\ell > 1$) are affected by the tune shift (8.160) only in that their positions are shifted as the beam-beam strength is increased. On the other hand, for the lowest-order resonance $\ell = 1$, the instability condition (8.159) for the $+$ mode is always stable. The instability condition for the $+$ mode is eliminated while the instability strength for the $-$ mode is essentially doubled.

The results of this study are most conveniently shown by the sawtooth stability diagram, as we have done before, in terms of the two scaling parameters ν and ξ. The ℓ-th order resonance will then be apparent from the stopband that originates from the tune value of $\nu = \frac{m}{\ell}$ in these diagrams.

Figures 8.58(a)–(d) show the numerical results of the coherent beam-beam instability for the case of one bunch per beam (two collision points). As higher and higher order resonances are included, the stability diagram becomes increasingly fragmented.

In these numerical calculations, we have truncated the map T to the order of resonance considered. The results in general agree quite accurately with the approximate expression (8.159) except for some weak and detailed modifications resulting from the interference between adjacent resonances.

Multiple bunches per beam   The analysis can be extended to multiple bunches per beam (see footnote 59). The interaction matrix requires a careful bookkeeping of the bunches in the two beams taking turns to collide. The eigenmodes and their stability are evaluated by the eigenvalues of the total map T. One result is shown in Fig. 8.59 which shows the case when the modes are truncated to the 4th order while the number of bunches per beam is increased from 1 to 4. Case (a) for one bunch per beam reproduces that of Fig. 8.58(b). The higher-order resonance behavior can be rather complex when there are multiple bunches per beam.

Figure 8.58: Sawtooth stability diagrams for the case of one bunch per beam. Resonances up to order 2, 4, 6, and 8 are shown in (a), (b), (c), and (d), respectively.

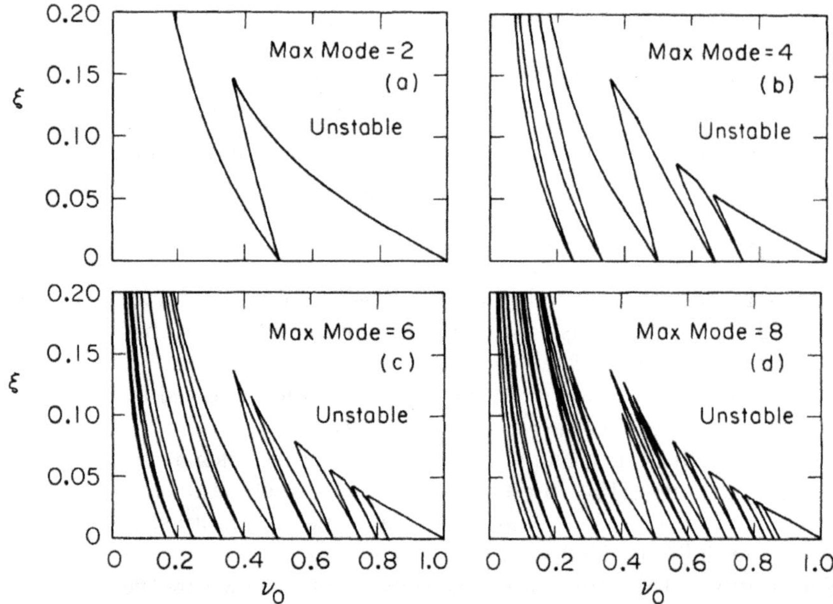

Recap

- As mentioned, an important drawback of linearized model is no frequency spread. The single-particle motion as well as motion of all modes have sharply defined frequencies. As a result, there is no provisions of detuning for single-particles and Landau damping for coherent modes. To extend the study, a more sophisticated analysis or a computer simulation in the strong-strong regime is required to look for the beam-beam coherent modes, and then look for their possible coherent effects when Landau damping is taken into account.

- Another disadvantage of the waterbag model is the absence of radial modes because all dynamics occurs at a single radial amplitude J_0 and all radial modes degenerate into $\delta(J - J_0)$.

- In spite of these limitations, the method used here to solve the Vlasov equation takes full account of the localized nature of the beam-beam kicks using a matrix mapping technique. It offers a simple description of the coherent beam-beam interaction and allows straightforward numerical calculations. In particular, all nonlinear resonances are treated simultaneously taking into account the coupling among the resonances.

Figure 8.59: Sawtooth stability diagrams for the cases with 1, 2, 3, and 4 bunches per beam in (a), (b), (c), and (d), respectively. In all cases, modes up to the 4th order are kept. Case (a) reproduces case (b) of Fig. 8.58.

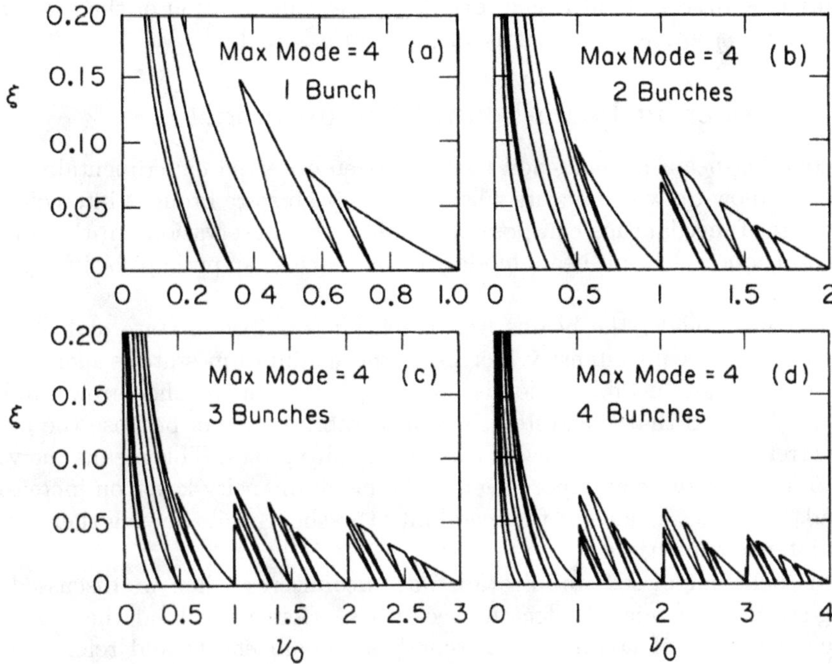

- Higher-order modes in a strong-strong picture can and have been studied by numerical simulations. This has been an extensive effort. More recent advances with the increasingly powerful computers have focused on particle-in-cell simulations allowing the study of detailed beam evolution under the beam-beam influence. One important advantage using simulations is the possibility to include the effects of Landau damping. It also allows the beam distribution to deviate from waterbag or Gaussian especially as the higher-order modes grow.[61] This effort is beyond the intended scope of the present chapter.

Homework 8.71 Follow the text, assuming only the ℓ-th mode is important and pay attention only to the two Fourier components g_ℓ and $g_{-\ell}$ to derive Eqs. (8.159) and (8.161).

---

[61]S. Myers, Proc. US-CERN School on Part. Accel., Sardinia, p. 205 (1985); K. Hirata and E. Keil, CERN report CERN/LEP-TH/89-57 (1989); S. Krishnagopal and R. Siemann, Phys. Rev. Lett. 67, 2461 (1991); J. Koga, Ph.D. thesis, Univ. Texas at Austin, 1990; Y. Cai, A.W. Chao, S.I. Tsenov, and T. Tajima, Phys. Rev. ST Accel. & Beams 4, 011001 (2001).

Homework 8.72 Let the tune be close to $\frac{1}{3}$. Consider a 4×4 interaction matrix for the variables $g_{\pm 3}$ and $g_{\pm 6}$. Use this model to study the effect of two interfering resonances.

Homework 8.73 Extend Homework 8.72 to a consideration of the case when ν is close to $\frac{1}{12}$ and a 4×4 matrix in terms of $g_{\pm 3}$ and $g_{\pm 4}$.

8.8.3   Coherent beam-beam blow-up model

One colliding-beam phenomenon that has been observed experimentally is the beam size blow-up which occurs when intense beams are brought into collision. A number of mechanisms can possibly explain this observation. In the present context of coherent beam-beam mode instability, this can perhaps be interpreted as follows.

As beams collide, the beams first reach an equilibrium state at low beam intensities. As beam intensity increases, the equilibrium state is increasingly distorted until an unstable region is entered. The beam size then grows rapidly, but only so much that the system becomes stabilized again because the beam-beam strength parameter ξ lowers when beam size grows. This means the value of ξ does not increase any more even if the beam intensity keeps on increasing. Instead, it reaches a value at the instability threshold while staying there as the beam intensity increases.

This behavior of self-stabilization was encountered when we discussed the strong-strong dynamic-β^* effect in Sec. 8.3.3. Here we extend the argument to include effects of nonlinear coherent beam-beam effects and refer to it as the coherent beam-beam blow-up model. For a given choice of tune away from resonances, the behavior is illustrated schematically in Fig. 8.60.

This coherent beam-beam blow-up model is analogous to the phenomenon of single-beam bunch-lengthening observed in electron storage rings. From the expression of ξ it can be deduced that in the blown-up regime, the beam size scales approximately linearly with the beam intensity N. This seems to agree qualitatively with beam-beam experiments in electron colliders.

In analogy to single-particle nonlinear dynamics, a Chirikov criterion may be applied by summing the stopband widths of all resonances. The total width is

$$\sum_{\ell=1}^{\infty} \ell \, \delta\nu_\ell ,$$

which diverges according to Eq. (8.161). This means that, for the waterbag model, no region in tune space is free of resonance effects. However, the waterbag model with sharp edge in beam distribution is expected to have overemphasized the effect of the high-order resonances.

Furthermore, it is important to keep in mind one special feature of the beam-beam interaction — a point we emphasized before. Unlike nonlinear fields due to magnet imperfections, the beam-beam effect is intrinsically stable in the sense that it perturbs only particles in the beam core and not particles with large

Figure 8.60: Behavior of beam-beam parameter, beam size and luminosity as functions of beam intensity for three cases. (a) unperturbed; (b) round beams; (c) flat beams. Dotted lines indicate a beam-beam threshold.

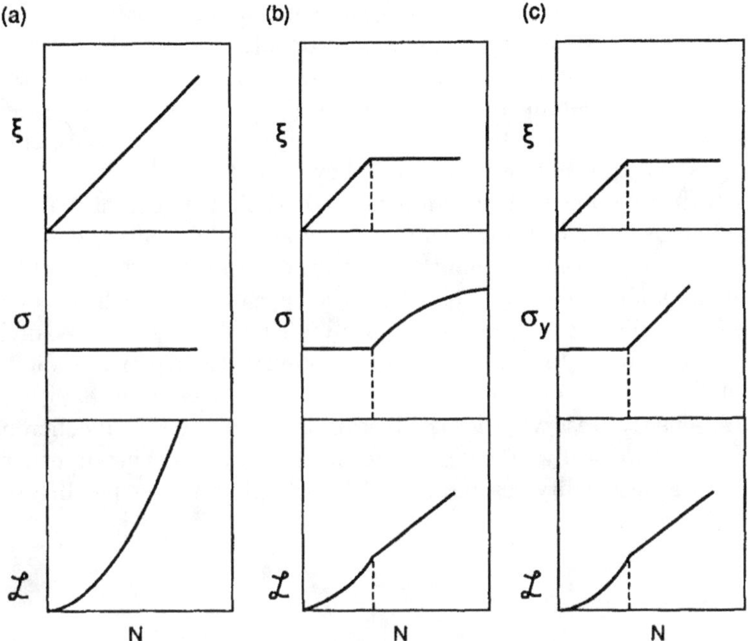

amplitudes. As a core particle becomes unstable, it necessarily grows out of the instability as its amplitude increases. This detuning effect plays an important role in determining the characteristics of the incoherent beam-beam effects and is expected to play a similar role in the coherent beam-beam effects.

Coherent or incoherent? A discussion is in order concerning a comment often been made, i.e. no coherent modes have been observed experimentally or in computer simulations other than the dipole modes — see Figs. 8.39 and 8.40, thus raising a question on whether the coherent modes actually play a role in beam-beam instability.

It should then be said that the observation of the dipole modes already is an evidence of the existence of the coherent beam-beam modes. Another strong indication of the failure to consider beam-beam effect as purely an incoherent beam-beam interaction has already been demonstrated by the DCI experiment as we discussed in Sec. 8.5.4.

We emphasized the fact that our analysis has ignored the key ingredient of Landau damping. Landau damping for the beam-beam interaction is particularly strong because the beam-beam tune spread ξ is large. Furthermore, the ℓ-th mode has a Landau damping rate ℓ times larger than ξ, so while the dipole mode $\ell = 1$ is clearly observable, the higher-order modes are damped too strongly to be detected, either experimentally or in numerical simulations. When Landau damping is included or when a coherent beam-beam blow-up is bubbling, all higher-order coherent structures are strongly Landau damped and smeared out quickly. In addition, in an experimental setting, the detector might be sensitive only to a center-of-charge motion and suppresses higher-order motions. As a result, coherent modes will not look too differently from incoherent motion. The fact that no clear appearance of coherent modes except for the dipole modes does not necessarily mean they do not play a role.

Still another evidence is shown when colliding two beams with multiple bunches. Let us take Fig. 8.38 as an example to illustrate the point. Landau damping is expected to blur the difference between the coherent and the incoherent instabilities in Fig. 8.38(a) for the case of one bunch per beam. But a notable difference occurs when there are multiple bunches per beam. In this case, the incoherent instability will have to observe the superperiodicity of the collider while coherent instability is insensitive to it, as Figs. 8.38(b), (c) (d) indicate. A sensitive test whether the instability is coherent or incoherent in origin, either experimentally or by simulation, is to examine the superperiodicity structure of the instability resonances with multiple bunches per beam.

Subject Index

CPSIA information can be obtained
at www.ICGtesting.com
Printed in the USA
JSHW040159030422
24286JS00001B/5